Vegetable Crops

— Volume 2 —

THE EDITORS

Professor T.K. Bose obtained his M.Sc. (Ag.) and Ph.D. degree from Calcutta University, West Bengal, India. Trained at the Royal Botanic Garden, Kew, UK. Former Secretary of the Agri-Horticultural Society of India; Project Coordinator, All India Coordinated Floriculture Improvement Project (ICAR); Professor and Head, Department of Horticulture; Dean, Faculty of Agriculture; Director of Research, Bidhan Chandra Krishi Viswavidyalaya (BCKV), West Bengal; Member of Scientific Advisory Committees of ICAR, CSIR (Govt. of India) and Government of West Bengal. Awarded (D.Sc.) Honoris Causa from OUAT, Bhubaneshwar and UBKV, Coochbehar, India. Guided around 45 Ph.D. students, published more than 250 research papers and authored 30 books on horticultural crops.

Prof. M.G. Som served as former Vice-chancellor, Dean, Faculty of Agriculture and Professor of Vegetable Science, BCKV, West Bengal. Obtained his M.Sc. and Ph.D. degrees from ICAR-IARI, New Delhi. Received fellowship in 1997 and Plaque in 1998 from the Indian Society of Vegetable Science (ISVS). Served as FAO expert in Bangladesh for increasing the vegetable production. Acted as a member of governing body of the ICAR, New Delhi; member of Advisory Research Council, ICAR-IIVR, Varanasi, and ICAR-CIARI, Port Blair, Andaman. Published more than 100 research papers, written 8 books, 2 monographs and few bulletins and several popular articles. Guided 14 Ph.D. students and more than 50 M.Sc (Ag.) students. In 2019, ISVS felicitated Prof. Som for his outstanding contribution in the field of vegetable research and education.

Prof. Arup Chattopadhyay obtained his M.Sc. and Ph.D. degrees from BCKV, West Bengal. He has 23 years of experience in Research, Teaching and Extension. Acting as the Officer-in-Charge, AICRP on Vegetable Crops, and served as Professor for more than 8 years. He has guided 15 M.Sc. and 9 Ph.D. students in Vegetable Science. Published more than 150 research papers, authored 05 books, 07 book chapters, and many popular articles, folders, leaflets. Developed nine trait specific varieties/hybrids of different vegetable crops. First recipient of 'Dwarikanath Young Scientist Gold Medal Award' for the best Ph.D. thesis in Vegetable Science; 'Fellows' of ISVS, Varanasi, and CWSS, West Bengal. BCKV has been felicitated 'Lt. Amit Singh Memorial Award' by the ICAR for the Best coordinating centre on Vegetable Crops in India for outstanding contribution in vegetable research under his able leadership.

Prof. Tapan Kumar Maity has acclaimed professional career of more than 35 years in the Department of Vegetable Science, BCKV, West Bengal. Acted as Nodal Officer of Quality Control Laboratory for Horticultural Produce, and PI of AINRP on Onion and Garlic (ICAR). Served as the Head of the Department of Vegetable Science and the Dean, Faculty of Horticulture. Published more than 100 original research papers, contributed 13 book chapters, edited 3 books. Supervised 12 Ph.D. and 32 M.Sc. students. Acted as Member of the Working Group of the Agriculture Commission, Government of West Bengal, Member of the Monitoring Team of AINRP on Onion and Garlic for different centers; acting as consultant of National Horticulture Board, Government of India.

Prof. Jahangir Kabir obtained his M.Sc.(Ag) and Ph.D.(Horticulture) degrees from BCKV) West Bengal. He has 33 years of varied experience in Teaching and Research in Vegetable Crops and Post Harvest Physiology and Technology of Horticultural Crops. Acted as former Head of the Department of Post Harvest Technology, , and served as Professor for more than 15 years. He has guided more than 30 M.Sc. and 10 Ph.D. students, authored more than 120 publications including books (edited), research papers, book chapters, review/technical and popular articles. Acted as reviewer of research papers, paper setter, and member of selection committee of Public Service Commission and several State and Central Universities.

Vegetable Crops

— Volume 2 —

— *Senior Editor* —

M.G. Som

Department of Vegetable Science
Faculty of Horticulture
Bidhan Chandra
Krishi Viswavidyalaya
West Bengal 741 252, India

— *Editor Emeritus* —

T.K. Bose

Faculty of Horticulture
Bidhan Chandra
Krishi Viswavidyalaya
West Bengal 741 252, India

— *Editors* —

A. Chattopadhyay • T.K. Maity • J. Kabir

Faculty of Horticulture
Bidhan Chandra Krishi Viswavidyalaya
West Bengal 741 252, India

2022

Daya Publishing House®
A Division of
Astral International Pvt. Ltd.
New Delhi – 110002

Fourth Revised & Illustrated edition 2022

Cataloging in Publication Data--DK

Courtesy: D.K. Agencies (P) Ltd. <docinfo@dkagencies.com>

Vegetable crops / senior editor, M.G. Som ; editor emeritus, T.K. Bose ; editors, A. Chattopadhyay, T.K. Maity, J. Kabir. -- Fourth revised & illustrated edition.

volumes cm

Includes bibliographical references and index.

ISBN 9789390435142 (HB)

1. Vegetables. 2. Vegetables--Pictorial works. I. Som, M. G., editor. II. Bose, T. K., editor. III. Chattopadhyay, A. (Arup), editor.

LCC SB320.9.V44 2021 | DDC 635 23

Published by : **Daya Publishing House®**
A Division of
Astral International Pvt. Ltd.
– ISO 9001:2015 Certified Company –
4736/23, Ansari Road, Darya Ganj
New Delhi-110 002
Ph. 011-43549197, 23278134
E-mail: info@astralint.com
Website: www.astralint.com

Digitally Printed at : **Replika Press Pvt. Ltd.**

From the Desk of Editor Emeritus

The undersigned had initiated the publications entitled 'Fruits: Tropical and Sub-tropical', 'Vegetable Crops' and 'Commercial Flowers' in one volume each in the 1980's in collaboration with co-editors, humbly expresses from the core of his heart that the updated revised editions of the above mentioned publications with colour illustrations in thirteen volumes which was planned in June, 2020, are being printed from October this year, a rare event in Horticultural Science has been possible by the blessing of Almighty God, with the cooperation of dedicated horticulturist as editors and contributors.

Kolkata

T.K. Bose
Editor Emeritus

From the Desk of Senior Editor

The book entitled "Vegetable Crops in India" was first published as a single volume in 1986, afterwards revised and published in 1993 and 2003 in three volumes by the name "Vegetable Crops". The previous three editions proved to be very useful to the students, teachers and researchers across the world. Concerted research in production and productivity of vegetable crops during last decades necessitates further revision of the book with coloured illustrations as the fourth edition in four volumes.

The second volume of this edition includes chapters on Cole crops, Okra and Tropical tuber crops. I am extremely happy to acknowledge the contributors for their dedicated efforts to make this volume very effective and meaningful. I would like to thank all the co-editors of this volume who tirelessly and sincerely participated in editing and proof reading of the manuscript. It is expected that this revised edition will act as a platform to enrich and strengthen the students and the researchers over their knowledge on vegetable crops.

I would like to specially appreciate Prof. (Dr.) T.K. Bose, who took active initiative and guided us to complete the fourth edition of this book. It would have been impossible for us to put this book into table without his encouragement and guidance.

Despite of our best efforts, the editors are conscious about the lapses that might have crept in the manuscript, however, we would like to emphasis on the fact that these lapses or errors are unintentional but circumstantial.

Kolkata

M.G. Som
Senior Editor

Preface to the Fourth Edition from Editors

The first edition of the book entitled 'Vegetable Crops in India' published in 1986 was greatly appreciated and proved valuable to the students, teachers, researchers and extension specialists in horticulture in general and vegetable science in particular, in different parts of the world. Then the book was revised twice in 1993 and 2003 with the title 'Vegetable Crops' in three volumes due to its increasing demand. All the previous editions, which contained updated scientific information of all commercially grown vegetable crops, were considered as valuable contribution on the subject. After about 17 years of the last revised publication in 2003, voluminous information pertaining to vegetable crop improvement, production, protection and post harvest management has been generated necessitating further revision in four volumes.

This particular edition has been enlarged by incorporation of up-to-date information on various aspects of vegetable crop production, protection and improvement particularly through biotechnological interventions with meaningful illustrations, and published in four volumes. Efforts have been made to cover as much information as possible. Second volume in this edition includes chapters on Cole crops, Okra and Tropical Tuber crops. Most of the contributors have taken care to review up-to-date research works and prepared the manuscript. The revision has been made by the editors and in few cases new contributors have been associated with some chapters.

We express our sincere thanks to all contributors and the Senior Editor Prof. M.G. Som and Editor Emeritus Prof. T.K. Bose who sincerely participated in revision, editing and proof-reading of the manuscript and we appreciate their commendable efforts to bring out this publication. We express our sincere gratitude to Dr. Praveen Kumar Maurya, Mr. Tridip Bhattacharjee, Mr. Soham Hazra, Mr. Subhrajyoti Chatterjee, Mr. Swadesh Banerjee and SK. Masudul Islam, Ph.D. scholars, Bidhan Chandra Krishi Viswavidyalaya, India for their help and cooperation for arranging illustrations, necessary typing and correction of citations.

Apart from own collection of photographs, editors have also taken some photographs which are in public domain from different websites as far as possible. Therefore we acknowledge with enormous thanks to all State Agricultural Universities and Research Institutes under Indian Council of Agricultural Research, New Delhi, India; Public and Private International Agricultural Institutes across the world for taking photographs from their websites to make this book a valuable knowledge for the readers across the world. We think images can be powerful teaching tools, as illustrations to in-class lectures, or for studying concepts outside of the classroom. We also think identification of vegetable crops, their package of practices, cultivars, utilization, diseases and pests with the help of suitable illustrations in a consolidated manner will make a dent for the students and teachers of vegetable science.

Our thanks are due to Shri Prateek Mittal, Director of Astral International (P) Ltd., New Delhi for his keen interest in bringing out this publication. Our special thanks go to Mr. Rajib Ray of Srijani Graphics, 24 Pgs (N), West Bengal for processing the manuscript.

Kolkata

A. Chattopadhyay

T.K. Maity

J. Kabir

Contents

List of Contributors

Aswath, C.

ICAR- Indian Institute of Horticultural Research, Hessarghata Lake Post, Bangalore 560089, Karnataka, India

Arora, S.K.

Department of Vegetable Crops, Dr. Y.S. Parmar University of Horticulture and Forestry, Naini, Solan 173 230, Himachal Pradesh, India

Bose, T.K.

Department of Horticulture, Bidhan Chandra Krishi Viswavidyalaya, Mohanpur, 741252, Nadia, West Bengal, India

Chatterjee, S.S.

Division of Vegetable Science, Indian Institute of Agricultural Research, Pusa 110012, New Delhi, India

Dey, Shyam Sundar

Division of Vegetable Science, ICAR-Indian Agricultural Research Institute (IARI), Pusa 110012, New Delhi, India

Hazra, S.

Department of Genetics and Plant Breeding, Faculty of Agriculture, Bidhan Chandra Krishi Viswavidyalaya, Mohanpur 741252, Nadia, West Bengal, India

Kabir, J.

Department of Post Harvest Technology, Faculty of Horticulture, Bidhan Chandra Krishi Viswavidyalaya, Mohanpur 741252, Nadia, West Bengal, India

Maharana, T.
Department of Horticulture, Orissa University of Agriculture and Technology, Bhubaneswar 751 002, Odisha, India

Mitra, S.
Department of Post Harvest Technology, Faculty of Horticulture, Bidhan Chandra Krishi Viswavidyalaya, Mohanpur 741252, Nadia, West Bengal, India

Naskar, S.K.
Regional Centre, Central Tuber Crops Research Institute, Bhubaneswar 751 001, Odisha, India

Nayar, T.V.R.
Central Tuber Crops Research Institute, Sreekariyam 695 017, Thiruvananthapuram, Kerala, India

Parthasarathy, V.A.
ICAR – Indian Institute of Spices Research, 32/482 C Narmada Nilayam, Bharathan Bazar, Chelavoor, Calicut 673571, Kerala, India

Roychoudhury, N.
Department of Horticulture, Bidhan Chandra Krishi Viswavidyalaya, Mohanpur, 741252, Nadia, West Bengal, India

Saha, Partha
Division of Vegetable Science, ICAR-Indian Agricultural Research Institute (IARI), Pusa 110012, New Delhi, India

Sen, H.
Department of Agronomy, Faculty of Agriculture, Bidhan Chandra Krishi Viswavidyalaya, Mohanpur, 741252, Nadia, West Bengal, India

Seth, T.
ICAR-Central Institute for Women in Agriculture, Bhubaneswar 751003, Odisha, India

Tarafdar, J.
Department of Plant Pathology, Faculty of Agriculture, Bidhan Chandra Krishi Viswavidyalaya, Mohanpur 741252, Nadia, West Bengal, India

Thakur, M.R.
Department of Vegetable Crops, Dr. Y.S. Parmar University of Horticulture and Forestry, Naini, Solan 173 230, Himachal Pradesh, India

Varma, S.P.
Regional centre, Central Tuber Crops Research Institute, Bhubaneswar 751 001, Odisha, India

COLE CROPS

S.S. Chatterjee, Shyam Sundar Dey, Partha Saha and J. Kabir

1.0 Introduction

The word 'cole' seems to have come from the abbreviation of the word 'caulis' meaning stem and it was variously spelt as Kale (English), Kohl (German), Kool (Dutch), Kai (Scandinavian), Chou (French), Col (Spanish), Cavolo (Italian) and Couve (Portuguese) and the word 'cole' was generally more accepted in the literature. It means a group of highly differentiated plants originated from a single wild *Brassica oleracea* var. *oleracea* (*sylvestris* L.), commonly known as wild cabbage or cliff cabbage or Colewort and according to an extension leaflet of Iowa State University, all cole crops are cultivated varieties of the species *Brassica oleracea* (Haynes *et al.*, 2009). According to the Royal Botanic Gardens, Kew all cole crops belong to the Brassicaceae family and contain over 330 genera and about 3,700 species. The largest genera are *Draba* (365 species), *Cardamine* (200 species), *Erysimum* (225 species), *Lepidium* (230 species), and *Alyssum* (195 species). The family contains the species mainly grown as vegetables such as *Brassica oleracea* (broccoli, cabbage, cauliflower, *etc.*), *Brassica rapa* (turnip, Chinese cabbage, *etc.*), *Raphanus sativus* (radish), *Armoracia rusticana* (horse radish). The extreme plasticity of the species has allowed differentiation, under human selection of this large number of forms due to the specialization of different plant organs that have given rise to various crops and uses. These include the leaves in the case of cabbage and leafy kale, the stem in kohlrabi and marrow stem kale, the inflorescences in broccoli and cauliflower, and the axillary (lateral) buds in the Brussels sprout. The variation in forms and colours is marked within each of these crop types (Maggioni *et al.*, 2010). Many centuries before the beginning of our era, some of these forms were introduced into cultivation, particularly for medicinal purpose. In both Greek and Roman writings, the merits of cole crops are highly praised.

The cole crops have spread all over Europe from the Mediterranean region, which is supposed to be the centre of origin. Curly kale, cabbage and kohlrabi were at first the best known cultivated types but due to the great diversity of the forms of *Brassica oleracea* in later centuries, different new crops were added to the old established ones. Cauliflower became widely known in the 18th century and Brussels sprouts in the 19th century. In the thirties of the 20th century, still another form, sprouting broccoli came into use, particularly in the USA. Cole crops are presently grown all over the world from the tropical to the Arctic where large kales may be seed growing in southern Greenland.

The ancient civilization of Egypt or Asia did not leave any information about cole crops, neither were these mentioned in the Bible. However, in Greek writings, there is mention of the coles and it was Theophrastus (371-286 B.C.), a pupil of Aristotle, who mentioned in his *Enquiry into Plants* about three types of cole. Cato (234-149 B.C.) is one of the Roman authors who described a cole type with large and smooth leaves and a thick stem, a type with curled leaves and a type having small plant and pungent taste-forms which were described by Theophrastus. In these classical writings, the cole crops are mainly recommended for medicinal purpose.

Cole crops are among the most widely grown vegetables in the temperate zones. But after the Second World War, they have spread to both tropical and subtropical areas fast. Among the cole crops in India, cauliflower and cabbage are the two most important crops grown while kohlrabi, broccoli and Brussels sprout are grown in certain areas, and yet to be as popular as the other two. Of the two, cauliflower is more important covering nearly 4.53 lakh hectares compared to 3.99 lakh hectares of cabbage which is more important in the world at large. There is a substantial increase (34 per cent) in the area of cauliflower and cabbage during the last decade in India. Some of the more important cole crops growing states in the country are Uttar Pradesh, Karnataka, Maharashtra, Bihar, West Bengal, Punjab and Haryana. These are grown almost round the year either in the hills or in the plains. Except potato and onion among vegetables, cole crops possibly occupy the largest area in India in winter.

2.0 Composition and Uses

2.1.0 Composition

Cole crops are known as the "Crown jewel of Nutrition" for their high vitamins, high fibre, low fat and low calorie properties. They are rich in vitamin C, beta carotene, fibre, antioxidants, and phytochemicals which help in preventing cancer. They are rich source of nutrients such as protein, carbohydrate, phosphorous and other minerals these crops are rich in glucosinolate, indoles and vitamin C, *etc.*

Cabbage is rich source of vitamin A (2000 IU) and vitamin C (124 mg/100g). It has fibre and calcium contents which reduces the risk of colon cancer. It protects against bowel cancer due to the presence of indole-3-carbinol. It also contains phosphorus which is helpful in utilization of calcium and assimilation of carbohaydrates and fats in human body. Cauliflower has high quality of proteins and peculiar in stability of vitamin C after cooking. It has potential chemopreventive agents including ascorbic acid, carotenoid, isothiocyanate, indoles and flavonoid. Broccoli and Kale is the most nutritious member of cole group. The anticarcinogenic effect of this crop is due to hydrolyzed glucosinolate (sulphoraphane) and indoles. It has role in reducing serum cholesterol and is highly rich in β carotene (pro-vitamin A) (125 mg/100 g fresh) and vitamin C (137 mg/100 g fresh) contents. In general, cole crops are used against such ailments as gout, diarrhoea, coeliac trouble, stomach trouble, deafness and headache. Cabbage juice is said to be a remedy against poisonous mushrooms and is also used as a gargle against hoarseness. The leaves are used to cover wounds and ulcers and are also recommended against a hangover. Cole products are eaten in the raw state as well as cooked. Investigations in different countries like Japan and the USA, suggested that there may be some especially protective properties against human bowel cancer in these vegetables (Ross, 1983). Glucosinolate present in all cole crops helps in the detoxification enzymes in liver and thereby help in boosting immunity. The isothiocyanates, dithiolthiones and sulforaphane are the bio-transformation products of glucosinolates that are involved in blocking enzymes which are responsible for tumourous growth in liver, lung, breast and gastrointestinal tracts (esophagus, stomach and colon).

The detail nutritional composition of cole crops in geven in Table 1.

Table 1: Nutritional Profiling of Cole Crops*

Name of the Crop	*Moisture (g)*	*Protein (g)*	*Energy (Kcal)*	*Calcium (mg)*	*Phos-phorus (mg)*	*Iron (mg)*	*β Carotene (mg)*	*Vitamin C (mg)*
Brussels sprouts	85.5	4.7	52	43	82	1.8	126	72
Cabbage	91.9	1.8	27	39	44	0.8	120	124
Cauliflower	90.8	2.6	4	33	57	1.2-3	30	56
Knol-Khol	92.7	1.1	21	20	35	1.5-4	21	85

*Gopalan *et al.* (2012)

2.2.0 Uses

In general, cole crops are used against ailments like gout, diarrhoea, coeliac trouble, stomach trouble, deafness and headache. Cabbage juice is said to be a remedy against poisonous mushrooms and is also used as gargle against hoarseness. The leaves are used to cover wounds and ulcers and are also recommended against a hangover. Cole products are eaten in the raw state as well as cooked. Investigations in different countries like Japan and USA, suggested that there may be some protective properties against human bowel cancer in these vegetables (Ross, 1983).

3.0 Origin and Evolution of *Brassica oleracea*

Various opinions about the antiquity of the members of the cabbage group are available. It appears that it was first collected for food at the beginning of Neolithic times before being cultivated. It is commonly believed that all cultivated forms of the cabbage group, *viz.*, cabbage, savoy cabbage, kales, collard, broccoli, Brussels sprouts, cauliflower and kohlrabi gradually originated from wild species through mutation, human selection and adaptation. A question that has stimulated the interest of scholars is related to the location of the initial introduction into cultivation of the wild plants and the subsequent domestication process (De Candolle, 1885; Song *et al.*, 1990; Gomez-Campo and Prakash, 1999). According to Maggioni (2010), the question is not merely academic, since understanding the history and evolution of crops can contribute to tracing existing wild and cultivated diversity, resulting in more efficient conservation and utilization. The general hypothesis is that ancient people were attracted by the fleshy leaves and tender shoots of wild *Brassica* species. These were introduced into home gardens or grown near the dwellings where less pungent and more luxuriant plants were selected, eventually giving rise to the domesticated leafy kale. This crop type is currently widespread in European gardens, often grown by elderly people, and several specific regional differences in the crop have developed (galega kale in Portugal, curly kale in Northern Europe, black kale in Tuscany, rashtan in Montenegro, and many other types).

The wild forms grow along the coast of the Mediterranean from Greece to England and are usually referred to as *B. Oleracea* var. *sylvestris* in literature. De Candolle (1883) suggested that cabbage was first domesticated somewhere in

Uses of Cabbage

Different recipes and value added products of Cabbage

Uses of Cauliflower

Different recipes and value added products of Cauliflower

Uses of Knolkhol

Different recipes of Knolkhol

Uses of Brussels Sprout

Different recipes of Brussels sprout

Uses of Broccoli

Different recipes of Broccoli

Uses of Kale

Kale Dishes **Kale Chips**

Different recipes and value added products of Kale

Uses of Chinese Cabbage

Different recipes and value added products of Chinese Cabbage

western Europe by Celts during the first millennium B.C. and was later introduced in the East from Europe. The invading Celts may also have found the cabbage already domesticated by the aboriginals, Ligurians and Iberians and adopted it from them. They also invented names for the cabbage or adopted those already used by the inhabitants. The different modern day European names for cabbage *e.g.*,'kopf kohl' (German), 'cabus' (French) and 'cabbage' (English) are properly derived from Celto-Slavic root cap or kap, the Celtic word for head. Hegi (1919) supposed var. *sylvestris* of Italy as probably the progenitor of various types. Schulz (1919) proposed that many of the crop types of this group were derived from the Mediterranean wild *oleracea* and suggested *B. cretica* as the probable ancestor of the cauliflower. Gates (1953) on the other hand, was of the opinion that white cabbage, Brussels sprouts and kale originated from *oleracea* in Western Europe, cauliflower and broccoli probably derived from a related species in the Eastern Mediterranean. He also mentioned that kohlrabi might have originated in the middle of the Mediterranean area. Broccoli possibly originated about 2500 years ago and the ancestral forms of modern cultivars were probably selected about the time of Christ (Schery, 1972). Hyams (1971) considered that Syria was the place where cauliflower was first noticed, selected and propagated. Similarly, Brussels sprouts were probably developed near Brussels in Belgium during the 14thCentury. A similar type of kale with finely dissected leaves and numerous buds was described in Gerarde's *Herbal* (1597) as 'parsley cabbage' (Henslow, 1908). These sprouts were reported to be served at the wedding feast of Alcande de Bred rode in 1481 (Hyams, 1971). The centre of origin of the cauliflower is considered by Siviero and Barbuzzi (1998) to be the Middle East and the Mediterranean Basin, with the first certain description of the crop in Syria coming from an Arab source in Spain in the 12th century. Subsequent spread of the crop is traced to west and central Europe, north Africa, India, China, the Americas and Australasia.

Helm (1963) proposed a scheme as to how the various taxa of the cabbage group developed. He suggested that the oldest type var. *ramosa* (thousand-head kale) which gradually gave rise to var. *gemmifera* (Brussels sprouts) probably in Belgium, were developed from wild cabbage var. *sylvestris*. From this developed demonstrosities such as var. *dalechampii*. The wild cabbage also developed into a stock var. *acephala* from which var. *costata* (Portuguese kale) was evolved in Portugal. Cabbage (var. *capitata*) and savoy cabbage (var. *sabauda*) are both derived from this Portuguese kale, the latter one in Italy. Similarly, the kales and collards (var. *acephala, sabellica, selensia and palmifolia*) were gradually developed from the same stock which also gave rise to the forage crop, marrow-stem kale (var. *medullosa*). Helm (1963) also suggested that Pliny's Pompeian cabbage could be an ancestor of this type. Kohlrabi (var. *gongylodes*) is derived from the same form or from marrow-stem kale. Sprouting broccoli (var. *italica*), initially developed from wild *sylvestris*, gave rise to cauliflower (var. *botrytis*). Both these types are supposed to have developed in the Eastern Mediterranean. Although the origin of these varieties is not fully understood, the conclusion which could be arrived at is the recognition of wild kale as the ancestral progenitor which in due course gave rise to different forms. This process might have been accomplished through macro and micro-mutations. Chromosome structural changes appear to have played a minimal part as the different forms are still very similar to the wild type in their genetic architecture (Harberd, 1972, 1975). At the same time, introgression from wild species such as *B. incana* from the West and South coasts of Italy, Sicily and Yugoslavia, *B. montana* from southern Europe, North-East Spain and southern Italy, *B. rupestris* and *B. insularis* from Corsica and Sardinia possibly helped in increasing the variability and adaptability. If and when a favourable mutant was detected it was selected, tested and propagated. The influence of human selection in maintaining the favourable mutants thus have played the major role in the development of the various forms.

Nieuwhof (1969) in his book on *Cole Crops* has opined that judging from the classical writings, a common southern European origin seems more likely for all the crop varieties. Besides, the linguistic relationship of words such as 'broccoli' (some times written as 'boroccoli' in old literature) and 'borecale' (kale) for types that differ morphologically, militatesagainsta polyphyletic origin.

According to Thompson (1976) although the wild cabbage is native to coasts of the Mediterranean and Western Europe, it is doubtful if many of the cultivated varieties of *B. oleracea* have evolved solely from the wild cabbage. Several other wild diploid relatives may have contributed to the development of the different varieties. Yarnell (1956) considered that *B. cretica, B. insularis* and *B. rupestris* may have contributed to this development. These species are inter-fertile with *B. oleracea* and, therefore, have been classified in one cytodeme by Harberd (1972). Some of them have characteristics quite unlike those of *B. oleracea*, for example, the short, fat siliqua of *B. macrocarpa*, corymbose inflorescence of *B. cretica, etc.* However, Thompson (1976) considered that the fixation of distinctive characters combined with the absence of mating barriers, may have resulted from geographical isolation of populations on the sea cliffs of Europe. According to him, the whole complex may possibly be classified as one species, *B. oleracea*. The wild related species are given in Figure 1.

The early evolution of the different cultivated types took place in the Mediterranean area and/or in Asia Minor. Thompson (1976), however, considered that the development of botanical varieties from wild species probably took place in several different areas of the Mediterranean, *viz.*; the Eastern Mediterranean was the evolutionary centre for the cauliflower (Helm, 1963). Kales used to be cultivated by the Greeks as early as 600 B.C. and several other types like heading cabbages and possibly kohlrabi were known to the Romans. Since these were highly self-incompatible and highly cross-fertilized, the maintenance of distinct types must have been difficult.

All *Brassica* spp. contain glucosinolates which in crushed leaves are broken down by the enzyme myrosinase to give bitter taste and goitrogenic substance. Hence the selection at early stage of domestication must have been for plants with less bitter taste than usual. Semi-quantitative estimations of these substances by Josefsson (1955) showed that the wild cabbage may contain four times the quantity present in the organs of cultivars eaten by man, for example cauliflower curd, cabbage heart and kohlrabi bulb. The Mediterranean wild species are annuals growing in the warm and moist winter. According to Thompson (1976) when

Figure 1: Plant Morphological Diversity of *Brassica* Wild Species (Branca and Cartea, 2011)

the early cultivars spread to North-Western Europe, biennial types with winter hardiness and considerable cold vernalization requirement for flowering were selected; this change probably resulted from outcrossing with the native biennial or perennial wild cabbages. He considered that less fibrous, thicker stems and more succulent storage organs were other early selection criteria.

The leafy unbranched kales and branching thin-stemmed kales were undoubtedly the earliest cultivated Brassicas. These types gave rise to forage kales which were later fed to livestock. Although cabbage 'tritianom' with an open head of 30 cm in diameter was described by Helm (1963) suggested that the hearting cabbage (var. *capitata*) was evolved in Germany. Both red and white cabbages were known about A.D. 1150 in Germany and in England by the 14th Century.Three types of savoy cabbage, varying from plants with loose rosettes of crinkled leaves to others with well-formed hearts but only slightly crinkled leaves were described in a German herbal in 1543. Savoy cabbage probably originated in Italy and spread to France and Germany in the 16th and 17th Centuries. Thompson (1976) proposed a scheme as to how the various cultivated forms of cabbage group developed which has been presented in Figure 2.

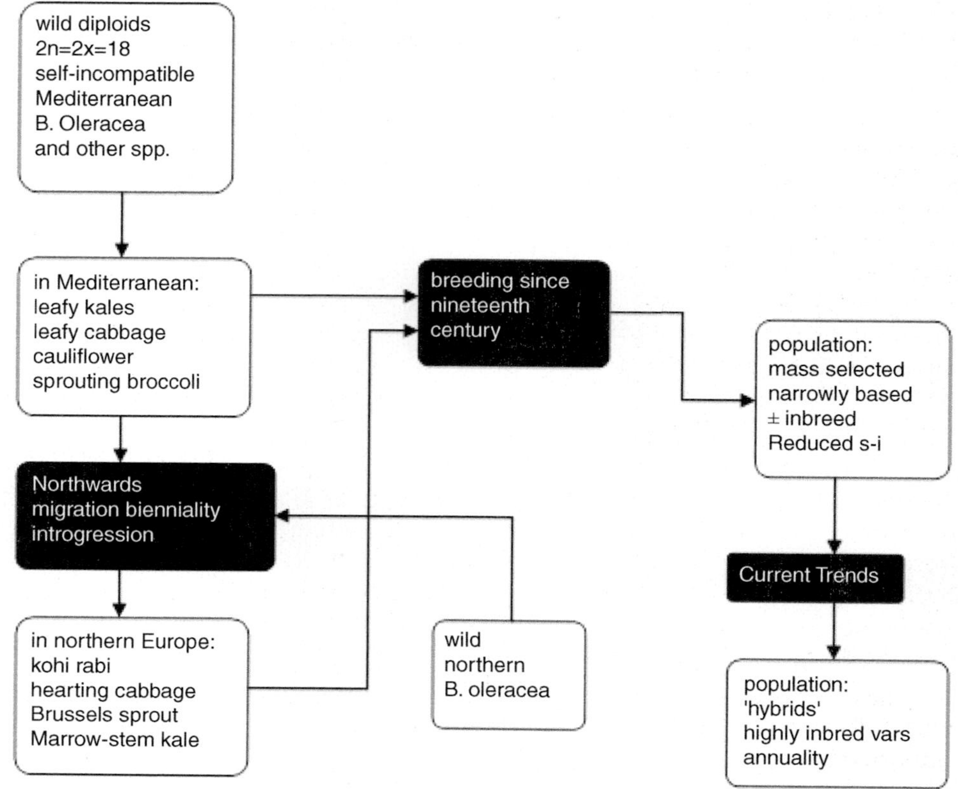

Figure 2: Evolution of Cultivated Forms of *Brassica oleracea* (Thompson, 1976)

Snogerup (1980) from the evidences of ancient literature, from the studies of the morphology, from very old cultivars still growing in field margins, waste places, building grounds and in similar localities, from crossing experiments of the wild and cultivated forms concluded that the headed cabbage have originated from West European *B. oleracea*. The savoy cabbage according to him, is the result of introgression from other coles. The branching bush kales has been interpreted as originated from *B. cretica* in Greece. They had chances of hybridization with other forms for more than 2000 years. Thompson (1976) stated that the marrow-stem kale is recorded with certainty only from the Vendee region of France in the early 19th Century and quickly spread to Germany and Denmark and was introduced in England about 1900. In New Zealand, it is still grown in its French name 'chou moellier'. According to Snogerup (1980), however, the marrow-stem kale or the stem kales have probably originated from *rupestris-incana* complex in the Adriatic area or Southern Italy. Among their many forms some indicate connection through hybridization to the cabbages and perhaps also to *B. cretica*.

Pliny described a Brassica in which the stem is thin and just above the root, it swells out in the region that bears the leaves, which are few and slender, a description which Pease (1927) considered had been referred to kohlrabi. This vegetable was described in the 16th Century herbals and Gerarde stated that it was grown in Italy, Spain and Germany (Helm, 1963).

Boswell (1949), however, thought that Pliny's description applied to the cauliflower and suggested that the kohlrabi developed in Northern Europe in the 15th Century. According to Helm (1963), the cauliflower in its present form was unknown before the early Middle Age. Three different cultivars of 'chou de syrie' were described in a work by an Arab botanist in Spain in the 12th century but it is not known if cauliflower was grown in Spain at the time. It was not mentioned by the earlier German herbalists and was only described in 1576 as *Brassica florida botrytis* and in 1583 as *Brassica cauliflora*. It was probably introduced into Italy by the Genoese from the Levant or Cyprus around 1490. It was grown for seed around the Gulf of Naples. At the beginning of the 17th century, it reached Germany, France and England.

Sprouting broccoli also came from Levant, Cyprus or Crete to Italy. It was not mentioned till 1660 and was referred to as sprout,'colli-flower' or 'Italian asparagus' in Miller's *Gardener's Dictionary* of 1724. As all flower buds open in the sprouting broccoli compared to prefloral fleshy meristem of cauliflower it is thought by many that sprouting broccoli may have preceded cauliflower in the evolution of species. It is, however, not known whether sprouting did not exist until the 17th century or whether it was not distinguished from cauliflower before that time.

These inflorescence kales, *i.e.*, cauliflower and sprouting broccoli are rather doubtful in origin. Schluz (1919) proposed an origin from *B. cretica* which is supported by evidence of crossings and the quick development of these forms. But at least cauliflower often has a leaf surface reminiscent of cabbage. According to Snogerup (1980) there is one possibility that these forms originated as abnormalities in crossings between early forms of branching shrub kale and other cultivars and followed by human selection.

Brussels sprouts in its present form appeared only in about 1750 in Belgium as a 'sport' but by 1800 had reached France and England and from 1820 was known as 'choude Bruxelles'. Its earlier history is uncertain. Records from 1231 and 1472 in Belgium to an 'en cent de sproq' may have been referred to *B. capitata polycephalos* illustrated by Dalechamp in 1587 (Helm, 1963).

Brassica alboglabra Bailey, a native of China, according to Snogerup (1980) originated from *B. cretica* sub-sp. *nivea*. This may have occurred through cultivation in ancient Greece, spreading of the plant to East Mediterranean trade centres and from there to further East. A search for cultivated material and traditions in the intermediate area, according to him is much needed. The 'karam sag' of Kashmir has been found to be a close associate of *B. alboglabra* (Snogerup, 1983).

Nuclear RFLPs were used to study the genetic relationship of Portuguese coles and other close related *Brassica* genotypes by Dias (1995). Preliminary results suggested *B. oleracea* domestication in Italy, Central-North Europe and Portugal. Kailan (*B. alboglabra*) evolved separately from the other *B. oleracea* cultivars in

Eastern Asia. Molecular data and fertility analysis of hybrids have recently shown that the North Atlantic wild types have a closer genetic relationship than the Mediterranean wild forms to all the currently cultivated forms of *B. oleracea* (Song *et al.*, 1990; von Bothmer *et al.*, 1995). Recent molecular and genetic studies provide evidence suggesting that the cauliflower curd arose in southern Italy from a heading Calabrese broccoli via an intermediate Sicilian crop type (Smith and King, 2000). However, the geographical origin of primitive domesticated leafy kale, from which all the other cultivated forms were probably derived according to our literary and linguistic findings, remains unclear.

Maggioni *et al.* (2010) synthesizes information from the linguistic, literary, and historical points of view that is compatible with domestication of *B. oleracea* in the ancient Greek-speaking area of the north-central and northeastern areas surrounding the Mediterranean. This area, during the 8th to 6th Century B.C.E., extended from Southern Italy to Greece and the west coasts of Turkey. There is a marked overlap of this area and the distribution area of the wild relatives of *Brassica oleracea* in the Mediterranean (Snogerup *et al.*, 1990). The absence of any trace of coles in ancient Near East documentary sources is therefore not completely surprising, but rather a confirmation that the origin of the cole crops is likely to be authentically North Mediterranean. The alternative hypothesis of a Northwest Atlantic origin, which mainly relies on the molecular findings of Song *et al.* (1990), is not supported by the study of Maggioni *et al.* (2010). According to them, the molecular data indicated that the *B. oleracea* populations growing in the wild along the North Atlantic coasts of Europe seem to be the closest to cultivated forms and, thus, they likely are the most closely related to the ancestors of cultivated *B. oleracea*. However, a correct interpretation of these data may not require that the domestication took place on the Atlantic coasts if we assume the possibility that the Atlantic wild *B. oleracea* are actually escapes from imported cultivated plants. This hypothesis has been already substantiated by Mitchell (1976), who observed that the wild derivatives of cultivated plants become morphologically indistinguishable from "native" plants within a few generations. On the basis of historical records of the wild populations of *B. oleracea* and their close proximity to villages and towns, he suggested that "cabbage" was brought to the British Isles by Romans or Saxons and that native population of *B. oleracea* might be regarded as introductions. The lack of correlation between genetic structure and geographic distribution of wild *B. oleracea* populations in England and Wales (Watson-Jones *et al.*, 2005) and the high degree of genetic homogeneity among the wild North Atlantic *B. oleracea*, contrasting with the substantial diversity detected within the Mediterranean wild brassicas (Allender *et al.*, 2007; Snogerup *et al.*, 1990), support Mitchell's hypothesis. Forthcoming studies focusing, inter alia, on the difference between populations that are truly wild versus those escaped from cultivation, will further test the hypothesis that the cultivated coles are of Mediterranean origin.

In any case, the *B. oleracea* gene pool is very rich in Sicily where *B. drepanensis*, *B. incana*, *B. macrocarpa*, *B. rupestris*, and *B. villosa* are widespread, and in some areas, they are strictly associated among themselves and with *B. oleracea* crops (Branca, 2008). Recent DNA analysis with molecular techniques support high similarity between Sicilian wild *Brassica* species (n = 9) and *B. oleracea* (Song *et al.*, 1990; Tatout *et al.*, 1999; Geraci *et al.*, 2001). Recently, more similarity was observed between *B. oleracea* and the Mediterranean wild Brassica species than the UK *B. oleracea* wild types (Allender *et al.*, 2007). Utilizing the simple sequence repeat (SSR) marker BoAP1, higher number of alleles were found in the wild Brassica species than in cabbage and cauliflower.

Detailed studies on molecular phylogenetics, diversification analysis, and historical biogeography (Arias *et al.*, 2014) established that the tribe Brassiceae originated in the intersection between the Arabian Peninsula and North Africa (Saharo-Sindian region) about 24 mya. The *nigra* lineage diverged from the *oleracea/rapa* lineage about 20 mya, with the former continuing to differentiate in southwestern parts of the Mediterranean region; the latter diversified in the Saharo-Sindian region and colonized the European Mediterranean through the east. The 'core *oleracea*' clade, containing *B. rapa* and the *B. oleracea*, originated in the northeastern Mediterranean ca. 6.5 mya, possibly at the time of the Messinian Salinity Crisis. At this time, the Mediterranean went through cycles of partial desiccation lasting several hundred thousand years, reducing its extension, with salt deposits accumulating and new drier and open environments being created. According to Arias *et al.* (2014), *B. oleracea* and its C-genome wild relatives diversified in the northeast Mediterranean region about 1.5 mya and then spread through the rest of Europe. Wild primitive forms of *B. oleracea* grow on cliffs of the Atlantic coast of Europe. Eventually, *B. oleracea* was brought to the east Mediterranean region where it became fully domesticated and gave rise to a wide range of cultivated forms. Variation within and between subspecies of *B. oleracea* and present-day cultivated cole crops evolved (Table 2) after a long time of natural/artificial hybridization, mutation, selection, and domestication. All forms are descended from a common kale like ancestor, the wild cabbage (*B. oleracea* L. var. *sylvestris* L.), which is still prevalent in western and southern Europe and North Africa.

Table 2: The Sequence of Evolution of Cultivated *B. oleracea* Crops*

Probable Sequence of Evolution	*Scientific Name (B. oleracea)*	*Common Name*	*Ancestor*[a]
1	var. *sylvestris* L.	Wild cabbage	—
2	var. *ramosa* DC.	Thousand-head kale, branching bush kale	1
3	var. *gemmifera* DC.	Brussels sprout	2
4	var. *dalechampii*		3
5	var. *costata* DC.	Portuguese tree kale, tronchuda kale	1
6	var. *medullosa* Thell.	Marrow-stem kale	1
7	Intermediate between 6 and 8		6
8	var. *gongylodes* L.	Kohlrabi	7
9	var. *sabauda* L.	Savoy cabbage	5
10	var. *capitata* L.	White cabbage	9

Probable Sequence of Evolution	Scientific Name (B. oleracea)	Common Name	Ancestor[a]
11	var. *capitata* L.	Red cabbage	10
12	var. *viridis* L., var. *sabellica* L., var. *palmifolia* DC.	Kale and collards	1
13	var. *italica* Plencks	Broccoli, calabrese	12
14	var. *botrytis* L.	Cauliflower (biennial)	13
15	var. *botrytis* L.	Cauliflower (annual)	14
16	var. *botrytis* L.	Cauliflower (Indian) or tropical cauliflower	15

a Ancestors represented by numerical letters (1 to 16) in order of evolution along with corresponding scientific name.

*Helm (1963), Prakash *et al.* (2011), Singh (2015), Singh *et al.* (2018)

4.0 Taxonomy, Classification and Gene Pool Concept

Brassica species belong to the Brassicaceae (formerly Cruciferae) family and the Brassica taxonomic studies started since 1700 by Tournefort and were continued by Linnaeus (1753), De Candolle (1821), Hooker (1862), Baillon (1871), Prantl (1891), Schulz (1919, 1936), Beilstein *et al.* (2006) and Branca and Cartea (2011). The genus *Brassica* is the most economically important genus within the Brassicaceae family and belongs to the subtribe Brassicinae, one of the nine subtribes of the Brassiceae tribe that shares with other 18 tribes a wide gene pool, which over time has been utilized directly or indirectly to improve several crops. Different species of the subtribes Raphaninae and Moricandiinae seem to be closely related to Brassicinae as confirmed by a long series of investigations on the chloroplast- DNA (cp-DNA) and restriction sites (Warwick and Black, 1991; Pradhan *et al.*, 1992; Warwick *et al.*, 1992; Warwick and Sauder, 2005). The cytogenetic relationship of the main species of the *Brassica* genus was depicted in the U triangle (U, 1935, Figure 3) in which *Brassica nigra* (L.) Koch (n = 8), *Brassica oleracea* L. (n = 9), and *Brassica rapa* L. (n= 10) represent the three diploid species in the vertices, and they developed by intercrossing to the three amphidiploid species, *Brassica carinata* A. Braun (n=17), *Brassica juncea* (L.) Czern. (n = 18), and *Brassica napus* L. (n = 19). The genome A was attributed to *B. rapa* L. (*B. campestris* in the past), the genome-B to *B. nigra* and genome-C to *B. oleracea*. The genome-A is carried by Chinese cabbage, sarson turnip, turnip greens, turnip, and turnip rape crops, which on a morphological basis are assigned, respectively, to the leafy, rapifera and oleifera types. The Chinese cabbage is economically important in Asia as salad; sarson turnip is a minor crop in Europe and in New Zealand where it is utilized for food purposes; turnip greens and turnip tops are highly used in Portugal and northern Spain for culinary uses (Padilla *et al.*, 2005); and turnip rape is widespread in the North America for oilseed production (McNaughton, 1995a). *B. campestris* has been renamed as *B. rapa* according to the International Code of Botanical Nomenclature. Oost *et al.* (1987) used the name since this variant has been largely adopted. The genome-B is possessed by black mustard, which is nowadays diffused in Europe as weed but was well known in the Middle Ages in Europe as condiment.

The genome-C is represented by *B. oleracea,* which diversified itself into several botanical varieties and related crops by domestication processes. Among the amphidiploid species, *B. carinata* is represented by Ethiopian mustard diffused in Abyssinian Plateau derived from the union of the BB and CC-genomes; *B. juncea* by Indian mustard cultivated in Asia derived from the union of AA and BB-genomes; and *B. napus* mainly by oilseed rape grown in Asia, Europe, and North America derived from the union of the AA and CC-genomes (McNaughton, 1995b).

The genetic resources available for the breeding of Brassica crops are regulated by the genetic boundaries of their primary, secondary, and tertiary gene pools (Harlan, 1975). *B. oleracea* represents the primary gene pool by itself, but several studies have been carried out to investigate the other gene pools and their potential utilization. The secondary gene pool was investigated by studies on the pachytene chromosome morphology, which permitted to identify the basic genomes of *Brassica* crops: AA (2n = 20) for *B. rapa,* BB (2n = 16) for *B. nigra,* and CC (2n= 18) for *B. oleracea.* Investigations on genomic libraries of *B. napus* and *B. oleracea* showed shared fragments among A, B, and C-genomes, suggesting their partial homology and the origin of the amphidiploid species *B. napus, B. carinata,* and *B. juncea* from the parental diploid ones (Hosaka *et al.,* 1990). The phylogenetic studies explain the evolution of *Brassica* and allied genera from a common ancestor with n = 6 through increase in the number of chromosomes and partial homology of A, B, and C-genomes (Prakash and Hinata., 1980; Song *et al.,* 1990). Finally, the tertiary gene pool includes species and genera related to *Brassica* crops in 36 cytodemes capable of genetic flux, such as *Diplotaxis, Enarthrocarpus, Eruca, Erucastrum, Hirschfeldia, Rhynchosinapis, Sinapis, Sinapodendron,* and *Trachystoma* genera (Harberd, 1976). These gene pools can confer favorable alleles and useful characteristics by using special methodologies. Tissue culture techniques, ovary and embryo rescue, and protoplast culture facilitated introgression of useful genes overcoming genetic boundaries.

Brassica oleracea L. is classified as follows:

Order: Brassicales (= Cruciales)

Family: Brassicaceae (= Cruciferae)

Tribe: Brassiceae

Subtribe: Brassicinae

Genus: *Brassica*

Section: *Brassica*

Species: *Brassica oleracea* L.

The family Cruciferae is characterized by 4 petals, standing opposite to each other in a square cross, 6 stamens of which 2 are short and a special kind of pod called siliqua.

Brassica genus has 100 species including coles and mustards, native to north temperate parts of the eastern hemisphere, while many of them produce important food materials, others and widely spread as weeds.

Cytological research of the various *Brassica* and their hybrids has provided the basis for a simple taxonomy of the genus in which three elementary monogenomic species and three amphidiploid digenomic species have been distinguished. These six species include the majority of the cultivated Brassicas.

	Species	Cultivated type	No. of chromosome	Constitution of Genome
I. Monogenomic Species				
	(i) *Brassica nigra* Koch.	black mustard	8	b
	(ii) *Brassica oleracea* L.	cole group	9	c
	(iii) *Brassica campestris* L.	turnip group	10	a
II. Digenomic Species				
	(iv) *B. carinata* Brown	Abyssinian	17	bc
	(v) *B. juncea* (L) Coss.	Mustard	18	ab
	(vi) *B. napus* L.	brown mustard	19	ac
		leaf mustard		
		swede group		

4.1.0 Nomenclature of *Brassica oleracea*

The cabbage group represents one of the highly polymorphic classes among *Brassica*, selected to give a wide range of crop types, *e.g.*, cabbage, cauliflower, broccoli, kales, kohlrabi, *etc.* Most of these forms are ancient and at present are usually referred to as varieties of *B. oleracea*. There is evidence that the Greeks and Romans were very familiar with these crops. Greek authors referred to the cole crop as κραμβη (krambe). Theophrastus (371-226 B.C.) in his *Enquiry into Plants* mentioned three types of cole crops: curly-leafed, smooth-leafed and a wild form. Cato (234-14 9 B.C.) in *De Agricultura* described three forms: one with large, smooth leaves and a thick stem (*B. levis*), one with curledl eaves (*B. crispa*) and one small plant with a pungent taste (*B. lenis*). Columellain his work *De Re Rustica* described 260 plants and mentioned *Brassica* several times, presumably referring to cole crops. Pliny, a contemporary of Columella, also described a large number of cole crops in his *Natural History*. For example, Pompei cabbage (Pompeianum) with a thickening between the leaves, Sabellian cabbage (Sabellica) with blistered leaves, Lacuturna cabbage (Lacuturres) with a head and abundant foliage, Tritian cabbage (Tritianom) with sprouts, Bruttian cabbage (Bruttiana) with very large leaves, a thin stalk and pungent taste, Cumae cabbage (Cumana) with leaves close to the soil and a spreading head and finally La Riccia cabbage (Aricinum), a sprouting form.

A reliable source of information regarding the cabbage crops is the various papers from the 8th Century A.D. For instance, *Capitualre de villis imperialibus*, where plants occurring in the garden of Charlemagne are described. 'Cauli' probably kale or cabbage and 'rava cauli' either swede or kohlrabi are mentioned. Among the several ancient Spanish and Arabic treatise, the best known are those of Ibn-al-Awam (12th Century) and Ibn-al-Baithar (3th Century), where these crops are mentioned. Both of

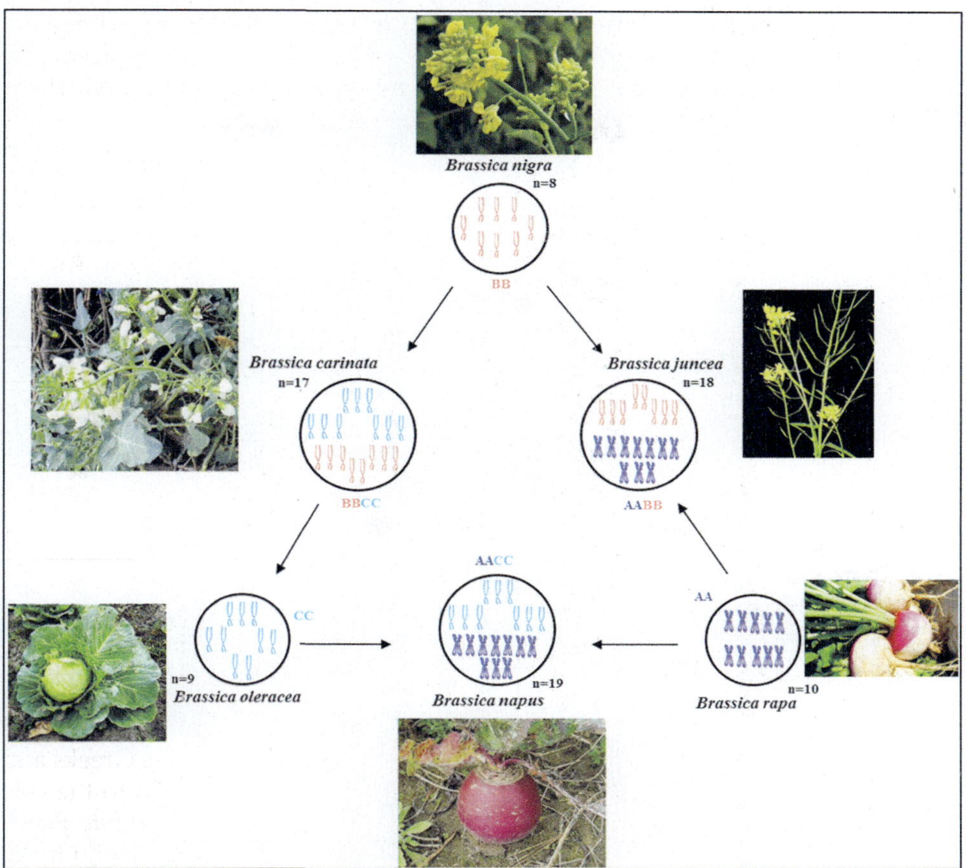

Figure 3: Triangle of U Diagram Illustrating the Genetic Relationships between Six Species of the Genus *Brassica*, such as Chinese Cabbage (AA), Cabbage (CC), Rapeseed (AACC), Indian Mustard (AABB), and other common vegetables, showing the genetic relationships between Chromosomes from each of the genomes A, B, and C are represented by different colors (Kumar *et al.*, 2015)

them were Moors living in Spain who obtained information from Greek and Roman writings. Ibn-al-Awamin his *Ki-tab-al-Falaha* described three kinds of cauliflower and referred to it as Syrian or Mosul cabbage. An additional name 'qarnabit', the word identical with the present-day Arabic word for cauliflower was also mentioned. The earliest attempt to illustrate *oleracea* was by Macer Floridus (1506) in *De Herbarum viribus* which suggests a wild plant and was called 'caulis' (Henslow, 1908). The cabbage group has been described extensively by various herbalists from the Middle Ages, particularly Dodoens (1953) and Gerarde (1597). Dodoens in *Stirpium Historia Pemptades* Sex, published in Dutch, described various forms of cultivated *oleracea*, among them a type was called 'flouriecoleivori'. The description the small stems grow together in the centre, thickest and fast throng together apparently refers to cauliflower (Giles, 1941). The name *B. cypria* indicates that it came from the land of Cyprus. Other types he identified with Pliny's names are as follows:

1. *B. sessilis or capitata* (corresponding with Pliny's tritianom)
2. *B. allobrogica or sabaudica* (Pliny's Lacuturres)
3. *B. cauliflores* (unknown to the ancients)
4. *B. patula* (with crispand rugose foliage)

Gerarde (1597) in *Herballor Geerall Historie of Plants* devoted a chapter to cabbages and gave 15 examples of 'coleworts' or forms of *oleracea*.

Since the 18th Century, several taxonomists have attempted to name different members of the cabbage tribe, resulting in a vast accumulation of synonyms. One of the early accounts of the tribe was given by Lamarck in *Encyclopedie Methiodique* (1784) where it was divided into six forms: (i) *arvensis*, very like the wild plant, (ii) *viridis*, the kales, (iii) *capitata*, the cabbage, (iv) *botrytis*, the cauliflower and broccolis, (v) *gongylodes*, the kohlrabi and (vi) *napobrassica*, the rutabagas. Linnaeus (1753) included all the forms in one species, *oleracea*. Apart from var. *sylvestris* which he regarded as the wild type, he recognized vars. *capitata* (cabbage), *botrytis* (cauliflower), *sabauda* (savoy cabbage) and *gongylodes* (kohlrabi). DeCandolle (1821) in addition, suggested vars. *Acephala* (kitchen kales), *bullata* (Brussels sprouts) and *caulorapa* (kohlrabi).

Hooker (1862) grouped together all the forms under five main varieties, *viz., acephala, gemmifera, capitata, botrytis* and *caulorapa*. However, the first major comprehensive taxonomic attempt was made by Bailey (1922, 1930). He included cultivated forms from China, Japan, Europeand North Americain this investigation. Various forms were classed into seven varieties with further subdivisions and included: *fruticosa* Metz. (branching bush kale), *acephala* DC. (kales or borecales and collards), *tronchuda* Bailey (tronchuda kale), *gemmifera* Zenker (Brussels sprouts), *capitata* L. (cabbage), *italica* Plenck (sprouting broccoli) and *botrytis* L. (cauliflower and broccolis).

Schulz (1919, 1936) classified the cultivated types in the following manner:

B. oleracea L. Kohl var. *ramosa* (DC.) Alefeld, Strauchkohl

var. *acephala* (DC.) Blattkohl. subvar. *viridis* (L.) Schulz

Sub var. *sabellica* (L.) Schulz, Grunkohl

Sub var. *laciniata* (L.) Schulz, Braunkohl

var. *bullata* DC.

Sub var. *sabauda* (L.) Schulz, Wirsingkohl, Welschkohl, Savoyer khol, Savoy, Sprouts

Sub var. *gemmifera* DC. Rosenkohl

Sub var. *palmifolia* DC. Palmkohl

var. *capitata* L. Kopikohl, Weiskohl, Rotkohl, Kraut, Kappus, Kunst, Cabbage var. *gongylodes* L. Kohlrabi, Oberrube, Ober-Kohlrabi

var. *botrytis* L. Blumenkohl

Sub var. *cymosa* Duchesne, Spargelkhol

Sub var. *cauliflora* (Gars.) DC. Blumenkohl,
Kasekohl, Cauliflower, Broccoli

Yarnell (1956) has classified the group into the following eight classes:

1. Kale, Collards, tree kale - var. *acephala* L.
2. Kitchen kale - var. *fimbriata* L.
3. Branching bush kale, thousand-headkale - var. *fruiticosa* L.
4. Cabbage - var. *capitata* L.
5. Cauliflower - var. *botrytis* L.
6. Brussels sprouts - var. *gemmifera* L.
7. Sprouting broccoli - var. *italica* L.
8. Knolkhol or kohlrabi - var. *gongylodes* L.

Helm (1963), however, grouped all the forms under 4 major convarieties, *viz.*, *oleracea*, *acephala*, *capitata* and *botrytis* which were further subdivided into varieties and of which 16 varieties were mentioned.

Nieuwhof (1969) in his book Cole Crops, recognized the following six cultivars:

B. oleracea L.

var. *capitata* L. f. *alba* DC.-white cabbage, *F. rubra* (L.) Thell. -red cabbage

var. *sabauda* L.-savoy cabbage

var. *botrytis* L.

subvar. *cauliflora* DC. -cauliflower

subvar. *cymosa* Lam.-sprouting broccoli

var. *gemmifera* DC.-Brussels sprouts

var. *gongylodes* L.-kohlrabi

var. *acephala* DC.

sub var. *laciniata* L.-curlykale

sub var. *plana* Peterm+smooth-leafedkale

sub var. *millecapitata* (Lev.) Thell+thousand-headkale

sub var. *palmifolia* DC.-treekale

sub var. *medullosa* Thell. -marrow-stemkale

There are many wild forms related to crop oleraceas which grow permanently along the coasts of Europe and the Mediterranean and bear a close resemblance, particularly to leafy kales. They have been awarded sometimes separate specific status. Onno (1933) reviewed a large number of wild forms from Italy and recognized two major species, the high growing *B. sylvestris* (L.) Mill and the smaller *B. cretica* Lam. Both of these species have the subspecies and varieties: *incana, robertiana, villosa* and *rupestris* under *sylvestris* and *macrocarpa, insularis* and *atlantica* under *cretica*. These various forms with the exception of *B. balearica* and *B. scopulorum* are easily crossed with *oleracea* crop types and regarded as belonging to the same *oleracea* cytodeme (Harberd, 1975).

Snogerup and Gustaffson (1982) have classified the different 9-chromosome species (n=9) and given the distribution as follows:

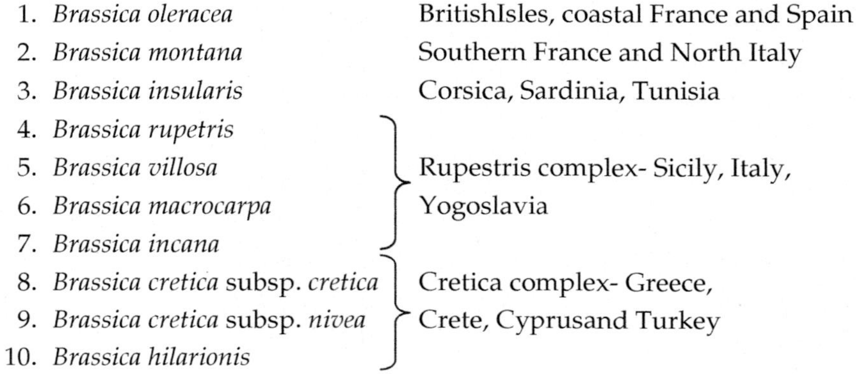

Species	Distribution
1. *Brassica oleracea*	BritishIsles, coastal France and Spain
2. *Brassica montana*	Southern France and North Italy
3. *Brassica insularis*	Corsica, Sardinia, Tunisia
4. *Brassica rupetris*	Rupestris complex- Sicily, Italy, Yogoslavia
5. *Brassica villosa*	
6. *Brassica macrocarpa*	
7. *Brassica incana*	
8. *Brassica cretica* subsp. *cretica*	Cretica complex- Greece, Crete, Cyprusand Turkey
9. *Brassica cretica* subsp. *nivea*	
10. *Brassica hilarionis*	

The relationship of these species also is being studied by them in regard to their possible role in the evolution of cole crops.

Thompson (1976) and Snogerup (1980) regarded the botanical variety sylvestris (wild cabbage) as *Brassica oleracea* var. *oleracea*. He also preferred the whole complex of wild relatives of *Brassica oleracea* (*B. cretica, B. insularis, B. rupestris* and many others) as only one species, *i.e., B. oleracea* in which morphological variability is not restricted to the cultivated types only.

Brassica alboglabra

It is a white flowering and very glaucous plant. The leaves are oval with slender petioles and have a pair of auricles at the base of the blade or on the petiole. Bailey (1922) found this species which is grown as a vegetable in China. Presently, it is known as Chinese kale.

The status of this species was, however, questioned by several authors like Sinskaia (1928), Burkill (1930), Sun (1946) and Yarnell (1956). Burkill (1930) regarded it as being only a form of *oleracea* while Sun (1946) treated it as *oleracea* var. *albiflora*. Herklots (1972) mentioned the existence of white, red and yellow flowered cultivars of *alboglabra*. Vavilov (1949) on the contrary, was in favour of assigning a separate status on the basis that *oleracea* originated in the Mediterranean area while *alboglabra* was of Chinese origin. Phelan and Vaughan (1976) supported a separate specific status on the basis of a study of volatile oils of the two species. Harberd (1972) preferred to keep this taxon in the same cytodeme as of *oleracea* as they cross easily with each other. Morinaga (1933) and Mizushima (1950) had used it as a source of C genome.

According to the Germplasm Resources Information Network nomenclature, there are several important cultivated *B. oleracea* groups (Table 3; Anonymous, 2015; Maggioni, 2015).

Table 3: Nomenclature of Cultivated *B. oleracea* L.*

Scientific name (B. oleracea var.)	*Common Name*	*Common Synonym*
botrytis L.	Cauliflower	
capitata L.	Red/white/Shetland cabbage	
sabauda L.	Savoy cabbage	
gemmifera DC.	Brussels sprout	
gongylodes L.	Kohlrabi	*B. caulorapa* (DC.) Pasq.
italica Plencks	Broccoli	
viridis L.	Kale, collard	var. *acephala* DC.
medullosa Thell.	Marrow-stem kale	
ramosa DC.	Thousand-head kale, branching bush kale	subsp. *fruticosa* Metzg.
sabellica L.	Curly kale	
palmifolia DC.	Palm kale, Jersey kale	
costata DC.	Portuguese tree kale, tronchuda kale	
alboglabra (L.H. Bailey) Musil	Chinese kale, kalian	*B. alboglabra* L.H. Bailey

*Maggioni (2015), Singh *et al.* (2018)

4.2.0 Cytology

Most of the research workers have suggested that the elementary species are secondarily balanced types. Studies regarding the secondary associations of bivalents, pachytene chromosome morphology and pairing behaviour in the haploids revealed the basic number of either 5 or 6. The first suggestion was by Catcheside (1934, 1937) who studied the secondary association in *napus* and *oleracea* and concluded that 6 is the basic number for the genus. He was of the opinion that the primitive haploid number in cruciferae is seven from which the fusion of two chromosomes could give rise to the basic number 6. Alam (1936) also on the basis of secondary association in 3 species supported Catcheside.

Richharia (1937) studied the problem of secondary association in *oleracea* and subsp. *chinensis, rapifera* and *pekinensis,* He recorded the maximum number of associations in *chinensis* and *rapifera* to be 6 and in the other two to be 5. On this basis, he was unable to come to any definite conclusion. Haga (1938), however, observed a frequent occurrence of 2 groups of two bivalents and four pairs in *nigra* and 3 groups of two bivalents and three pairs in *oleracea*. He proposed the basic number to be 6 and genomic formula ABCDEEFF for *nigra* and AABBCCDEF for *oleracea* and AABBCCDDEF for *campestris*. He considered that these three genomes were derived from a common origin.

Sikka (1940) classified the genome of 9 according to size and there are 1 very long, 4 long, 3 medium and 1 small chromosome. Studies on the secondary association in *oleracea* and other species of *Brassica* led Sikka to conclude that 5 is the basic number of the genus. He regarded *oleracea* as a modified amphidiploid from a cross between two primitive 5-chromosomc species with subsequent loss of one chromosome. The only attempt to describe the morphology of pachytene chromosomes in *Brassica* was made by Robbelen (1960) who studied the 3 elementary species. The chromosomes were classified on the absolute length as very short (upto 20 µm), short (20-25 µm), medium (25-30 µm), long (30-35 µm) and very long (40 µm and above). The oleracea genome consisted of 1 very long, 5 long, 2 medium and 1 short chromosome where the centromere of one chromosome was sub-telocentric, those of two were sub-metacentric and those of the rest were metacentric.

Robbelen's (1960) findings revealed that in all the three species 6 basic types can be recognized on the basis of characters like absolute length, symmetry of arms and the shape of the heterochromatic centromere region. In general, chromosome 1(A) was the satellite chromosome in all the three species. Chromosome 3 (C) in *nigra* also had a satellite. Nucleoli were attached to chromosome 1and 2 (both A) in *campestris*, to chromosome 1 and 4 (A and C) in *oleracea* and 1 and 3 (A and C) in *nigra*, He proposed genetic constitution of genomes to be AABCDDEFF for *campestris*, ABCDDEFF for *nigra* and ABBCCDEEF for *oleracea* which is a triple tetrasomic for chromosome types B, C and E. These results clearly established that these species are secondarily balanced polyploids with a basic chromosome number 6.

In 1970, Nwankiti reported 29 chromosome hybrids of *napus* x *campestris* and observed that at MI of meiosis, the number of scattered univalents was always 3, 5 or 9 suggesting that chromosomes of the C genome formed associations of 3 bivalents, 2 bivalents or no bivalent. The maximum pairing noticed was 3 bivalents and there were never less than 3 univalents in one cell. Based on such pairing, *i.e.*, 3 bivalents + 3 univalents, he suggested that in the course of evolution those of the original chromosomes of the C genome might have doubled and differentiated, thus interpreting that 6 was the basic genomic number of the C genome. Inter-strain differences in the karyotype of *B. oleracea* were studied by Mukherjee (1974). Sen (1957) while studying a claimed hybrid of broccoli and cauliflower observed a high degree of meiotic irregularity like univalent and multivalent formation which seems to be contradictory to all reports of *B. oleracea*. Datta and Deb (1974-75) studied seven different cauliflower cultivars and reported the cytotypes with 2n = 18, 2n = 20 and 2n = 36. From these studies they have concluded that both structural changes of chromosomes and polyploidy have played important role in the evolution of the horticultural cauliflower types.

In India, attempts were made to study the polyploidsin this species. Tandon (1951) induced tetraploidy in cauliflower by the treatment of 0.10-0.20 per cent aqueous colchicine solutionon 20 days' old seedlings for 24 hours. He found increased size of stomata, flower, pollen grain and seed. Pollen sterility was about 20 per cent and there was a considerable variability in the size of curds and time of their emergence. Seed number per siliqua was lower than the diploids. Tan *et al.* (2017) dissected the B genome from *B. nigra*, the triploid F_1 hybrid ($2n = 26$, CCB) obtained previously from the cross *B. oleracea* var. *alboglabra* ($2n = 18$, CC) x *B. nigra*. The monosomic alien addition lines (MAALs) with each of eight different B-genome chromosomes added to C genome ($2n = 19$, CC + 1B1–8), and seven different MAALs were established, except for the one with chromosome B2 which existed in one triple addition. Most of these MAALs were distinguishable morphologically from each other, as they expressed the characters from *B. nigra* differently and at variable extents. The alien chromosome remained unpaired as a univalent in 86.24 per cent pollen mother cells at diakinesis or metaphase I, and formed a trivalent with two C-genome chromosomes in 13.76 per cent cells.

4.3.0 Interspecific Hybridization

At the Horticultural Research Station, Hiratsuka, Japan, studies on the interspecific and intergeneric hybridization of *Brassica* relatives have been conducted since 1956 to obtain disease resistant Chinese cabbage cultivars. The most troublesome problem in this crop is the damage by bacterial soft rot, a soil-borne disease caused by *Erwinia aroideae* (Townsend) Hollander. The 9-chromosome *Brassica* cabbage group is highly resistant to this disease. Inter-specific hybridization between cabbage and Chinese cabbage has been done to breed soft rot resistant character in Chinese cabbage. The application of embryo culture technique gave unexpectedly good results in getting many hybrids and many of them had heading character in F_1 which had 19 diploid chromosomes. The F_1 hybrids were almost completely sterile in both self-pollination and cross-pollination with cabbage and Chinese cabbage. The F_1 hybrids were treated with colchicine and became artificial *B. napus* having 38 diploid chromosomes and most of them showed highly heading character. These showed high self-incompatibility and difficulty in further breeding programme. Further improvement and seed growing of this synthetic *B. napus* obtained in early experiments has been achieved in Perfectural Experiment Station and in seed companies and marketed under the name 'Hakuran' (Chinese cabbage-Hakusai and cabbage-Kau-ran) since 1966. It tastes like head lettuce with juicy, soft and Jess fibrous characters of leaves and is good for pickles and for fresh use. From the cultural point of view, 'Hakuran' is highly resistant to soft rot and other bacterial diseases and is very easy to grow because of their high resistance to drought and heat. Presently, it is cultivated in a limited area in Japan due to lack of ecotypes (Nishi, 1981).

Experimental results have suggested that the self-incompatibility of the parental strains may influence the production of F_1 embryos and their further growth and also that some parts of self-incompatible genes of both species are common to each. However, new breeding work is being done to improve the existing strains and also to breed new types of 'Hakuran'.

Inter-specific hybridization among the 6 *taxa* of *Brassica* revealed that hybrid seed set was highest when *B. campestris* subsp. *chinensis* or *pekinensis* were used as seed parent (Wojciechowski, 1985). Cultivar Osho was among the successful *B. pekinensis* seed parent for inter-specific hybridization with cabbage while Fujinase line D was most successful cabbage cultivar as female parent for hybridization with *B. pekinensis* (Sarashima and Matsuzawa, 1986). Von Bothmer *et al.* (1995) studied crossing experiments involving ten wild taxa and six major groups of cultivated *B. oleracea*, showed that the fertility of hybrids among all the cultivated forms and wild populations was high and that all the studied forms belong to the same biological species.

In India, interspecific hybridization was studied to transfer black rot resistance gene in cauliflower. Dey *et al.* (2015) attempted inter-specific hybridization for introgression of resistance genes from *B. carinata* into Indian cauliflower using three different embryo rescue approaches. Direct embryo culture was the most successful for development of F_1 and BC_1 population. All the F_1 hybrids were found to be highly resistant to *Xcc* race 1 and 4. Sharma *et al.* (2017) developed interspecific hybrid and between cauliflower "Pusa Sharad" and Ethiopian mustard "NPC-9" employing *in vitro* embryo rescue technique. Meiosis in the di-genomic (BCC) interspecific hybrid of *B. oleracea botrytis* group ($2n = 18$, CC) x *B. carinata* ($2n = 4x = 34$, BBCC) was highly disorganized and cytological analysis of pollen mother cells revealed chromosomes $2n = 26$ at metaphase-I. The F_1 inter-specific plants were found to be resistant to *Xcc* race 1.

5.0 Cabbage

Cabbage was introduced in our country much earlier than cauliflower. Though exact date of introduction is not available but it was reported to be grown during the Mughal period. It appears to have been introduced by the Portuguese, though the crop became popular during the British rule. The seeds of cabbage used to be imported from Europe and the attempts to produce seeds were undertaken during the Second World War when the supply was cut off. In spite of its long history of growing in this country, its popularity is somewhat less compared to cauliflower, though the position is reverse when the popularity is judged in the world as a whole.

5.1.0 Area and Production

The area and production of cabbage are given in Table 4 and Figure 4.

Table 4: Area, Production and Productivity of Cabbage of some Horticulturally Advanced Countries*

Country	*Area ('000 ha)*		*Production ('000 MT)*		*Productivity (kg/ha)*	
	2010	*2018*	*2010*	*2018*	*2010*	*2018*
China	910	967 F	30650	33189 F	336813	343458
India	301	402 F	7282	9035**	242309	224751
Russia	115	73 F	2733	345 F	236303	344601
South Korea	29	31	588	644 F	202623	207320
Ukraine	73	64 F	1523	1672 F	208345	262480
World	2312	2413	65911	69382	28520	28764

*FAO Stat. (2018)

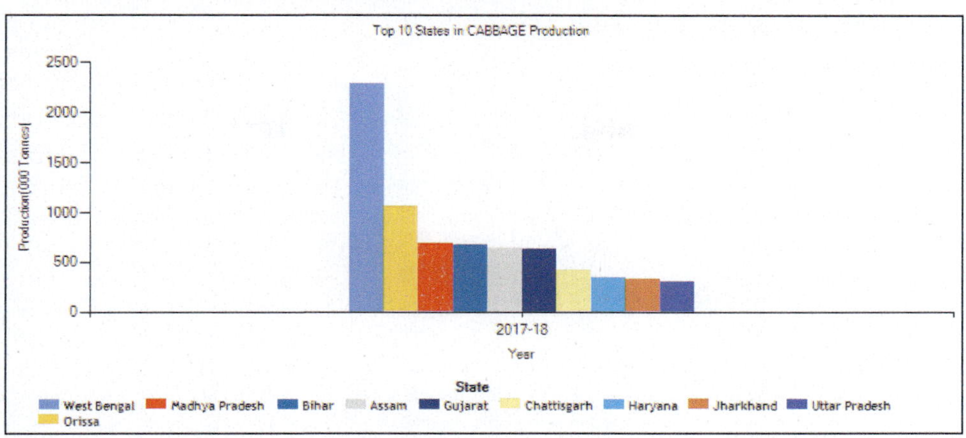

Figure 4: Top Ten Cabbage Producing States in India

The poor yield in the growing fields of India has been analyzed by Seshadri and Chatterjee (1983) and Swarup (1983). The former workers have pointed out that the low yield is due to poor quality of seeds as well as lack of suitable cultivars for the milder winter areas where the traditional cultivars perform poorly both in head formation and also in yield per unit area. Swarup (1983) brought out that the gap between the farm yield and experimental yield is due to improved cultivars (traditional) and efficiency of production technology. Further improvement is also possible by developing cultivars with stable production in different agro climatic regions of the country and also by developing disease resistant and F_1 hybrid cultivars.

5.2.0 Cultivars

In India, most of the cabbage cultivars that are grown belong to old European introductions. New cultivars were imported but research to develop cultivars for specific conditions or improvement over the old introductions is very limited. However, few cultivars/F_1 hybrids have been developed recently using indigenous research programme. In India, systematic research on development of cabbage cultivars is being taken up at ICAR-Indian Agricultural Research Institute, Regional Station, Katrain, Kullu, Himachal Pradesh.

The cabbage cultivars can be classified into three broad groups-i) white cabbage (*B. oleracea* var. *capitata* L. f. *alba* DC), ii) red cabbage (*B. oleracea* var. *capitata* L. f. *rubra* (L) Thell) and iii) savoy cabbage (*B. oleracea* L. var. *sabauda* L.). Of these, only white cabbage is of commercial importance in India. Very few red or savoy cabbages are grown in the country.

The white cabbages are available in three shapes, pointed, round and flat or drumhead. Among them round and flat headed cultivars are more popular but the recent trend is to grow more of round cultivars since the flat cultivars are often loose and voluminous.

5.2.1 Cultivars of White Cabbage

5.2.1.1 Early

The best known cultivars in this group are Golden Acre and Pride of India. Both are apparently early selection of the cultivar Copenhagen Market which was first introduced in Denmark in 1910. Of the two, Golden Acre is earlier by 5-7 days but lower in yield compared to Pride of India. These are characterized by small round; heads weighing 1.0 to 2.0 kg and can be cut within 60-70 days when transplanted in proper season. The outer leaves are few and may be slightly cup shaped. Under less favourable conditions, the head remains small and loose. Harvesting in good time is essential, otherwise the heads crack. Cultivars like Canadian Acre, Avant Premier, Primo, Ditmarsch, Wisconsin Golden Acre (yellows resistant) are almost similar. Late strains of Copenhagen Market are more leafy and come closer to Glory of Enkhuizen. A typical Copenhagen Market will be one to two weeks later than the Golden Acre or Pride of India. Its growing period under favourable condition varies from 75-85 days and the head weighs 1.5 kg to 3.0 kg. Yellows resistant cultivars released from Wisconsin are known as Copenhagen Resistant, Wisconsin Copenhagen, Marion Market *etc.*

Other early cultivars of limited distribution are Baby Head (syn. Early Vienna), Badger Market (yellow resistant), Early Flat Dutch, *etc.*, but most of these have a round or flat-round head, generally later in maturing than Golden Acre.

Qiangan1 is a new cabbage cultivar developed in China for spring cultivation with a tight and round head. The net head weight is 1.8 kg. It is tolerant to bolting when grown in spring, high-yielding and early maturing (Meng *et al.*, 1999).

5.2.1.2 Mid Season

A cultivar most widely grown in Glory of Enkhuizen (also known as Mid-season Market) which was first introduced in the Netherlands around 1900. The heads can be cut in 80-90 days after transplanting in favourable season. They are round and weigh 2.0 to 5.0 kg. Yellows resistant substitutes are Globe, Resistant Glory and Green Back. Badger Shipper, a cultivar resistant to yellows and some strains of club root was released in 1959 from Wisconsin. American cultivars, All Head Early (yellows resistant), All Head Select, Succession and All Seasons (yellows resistant Wisconsin All Season) belong to this group and of which the first name is earlier and the last two are somewhat later than Glory of Enkhuizen. In South America, an annual cultivar Sanare has been selected from Wiscons in All Seasons. In the Nilgiris, a cultivar known as September has become very popular which also seemed to belong to this group or slightly later than the Glory group. It produces a good crop in the Nilgiris and the head weighs from 3.0 to 5. 0 kg. This cultivar can be kept in the field after they are ready for harvest and does not show cracking.

5.2.1.3 Late

Most of the cultivars grown in the country are flat headed. Brunswick is one of the early introductions which form large, flat somewhat loose head. Comparatively better in tightness and yield are Success (flat-round) and Amager Short Stemmed (more or less round, also known as Danish Ballhead). Pusa Drumhead released by IARI long back is a very late selection from a F_1 hybrid of Japan, having some tolerance to 'back leg' disease. The heads are large, flat and somewhat loose. Sure Head (flat-round) and Late Flat Dutch are old established American cultivars. Most of these cultivars require long winter for a good crop, which is not available for majority of the cabbage-growing areas in the country.

In Norway, cv. Metino, a white cabbage cultivar showed particular promise, having a compact head, evenness of growth, good durability in the field and a high saleable yield (Dragland and Berentsen, 1991). In Belgium, Vanparys (1991) recommended cv. Predena for mid-early cropping; Bingo, Galaxy, Lennox and Strukton for mid-late to late cropping; Winston for late cropping and Zerlina for late to very late cropping.

5.2.1.4 Pointed Head White Cabbage

These cultivars have a special importance in Europe. For the very early crops, unlike in India, the pointed headed cultivars are mostly used. They bolt less rapidly than other cultivars and make head sooner. Besides, the pointed headed cultivars are used more often as raw due to their better palatability. Cultivars like Charleston Wakefield, Jersey Wakefield, *etc.*, belong to this group.

On the basis of trials during 1976-80, Bondarenko (1983) from the then USSR concluded that the highest percentage of marketable heads after mechanical harvesting was obtained from the interline hybrids TSKhA1789, TSKhA1781, TSKA1784 and TSKhA 1764 (93-100 per cent), followed by late Amager 611, Khar' Kovskaya Zimnyaya, Krasnodarskaya1 and Khidena F_1 and the early F_1 hybrids Druzhnyl and Rannil 3 (70-94 per cent). From Romania, Lunca and Andrei (1985) reported a high yielding cultivar De Buzau with a yield of 77.9 t/ha. Two new conical headed cabbage cultivars, Duncan and Bejo 1147 were found to withstand a winter with eleven degrees of frost satisfactorily (Wijk, 1985). Silva Junior and Miura (1986) reported that resistance to *Xanthomonas campestris* pv. *campestris* would make it possible to grow this crop in summer in the coastal low lands of Santa Catarina, Brazil, where at present it is solely a winter crop. In a trial in summer on a clay soil in Brazil with 31 cultivars they found that hybrid cultivars Mijabi, Superette, Shutoku and Kagero gave the highest yields. They also recorded that Kagero was 10 days earlier than others and Majabi was most resistant to *X. campestris*.

5.2.2 Cultivars for Storage

Generally, the cultivars for storage are not grown in India. However, in areas with high mountains, specially in Kinnaur in Himachal Pradesh and Ladakh in Kashmir, such cultivars may be better if storage is intended. Most of the cultivars have been derived from the Danish cultivars Amager Long Stemmed and Amager Medium Stemmed. Yellows resistant Improved Wisconsin Ballhead, Badger Ballhead, Hoollander, *etc.*, belong to this group. The heads of these cultivars are round to somewhat flat-round. These may show appreciable differences in keeping quality.

5.2.3 Cultivars of Red Cabbage

As has been pointed out, these are not popular in India. These cultivars are rich source of anthocyanin with high anti-oxidant property and are gaining popularity to add colour in vegetable salad. However, they can be used as source to extract anthocyanine and their use as functional food. The only cultivar available and sold is Red Acre. It has a distinct coat of wax and produces a head of 2.0-3.0 kg, taking about three months after transplanting. The shape is somewhat round.

In Belgium, best results with red cabbage cultivars were obtained with Rebusand Vorox (mid-early), Autoro, Pedro and Rodon (mid-late) and Raliant and Zorro (mid-late to late) (Vanparys,1996).

5.2.4 Cultivars of Savoy Cabbage

These are also not popular in India. All the cultivars have blistered leaf. This group unlike the red cabbage, has all the shapes like pointed, round and flat. The cultivar which is catalogued most commonly is Perfection (syn. Chieftain). This has fairly light green foliage and a large, rather smooth, flat-round head.

The following savoy cabbage cultivars gave the best result in Belgium: Hamasa, Saga, Tasmania, Denver, Polasa, Primavoy and Alaska (Vanparys, 1997). Further, Rooster and Spiessens (1999) reported some new savoy cabbage cultivars *viz.*, Salima, Premius and Kolosa which performed better as over wintering cultivation in Belgium.

5.2.5 Hybrid Cultivars

F_1 hybrid cultivars have been introduced which are more uniform than the common cultivars. In Japan, a whole range from tropical to temperate cultivars of white cabbages has been bred. The tropical F_1 hybrid cultivars have become very popular in tropical and subtropical areas of the world. These cultivars have made it possible to grow cabbages at a comparatively higher temperature (above 20°C). The most popular among these are K-K cross and K-Y cross. The hybrids have also been introduced by both the European and American breeders. Presently, it is estimated that nearly 100 per cent of the cabbage cultivars being marketed in Europe and the USA are F_1 hybrids (personal communication). Three hybrid cultivars of Indo-American Hybrid Seeds -Ganga, Kaveri and Yamuna were very popular in many parts of India in the last decade. With the advancement in development of cytoplasmic male sterile lines along with doubled haploid based inbreds, several F_1 hybrids have been developed by the private sectors predominantly cultivated in different parts of India. Development of F_1 hybrids in cabbage has been taken up systematically at ICAR-Indian Agricultural Research Institute, Regional Station, Katrain and one Self incompatible based and two CMS based F_1 hybrids have already been developed by this station.

During the last decade large number of F_1 hybrids has been developed in China, Korea, Taiwan and Japan. Some of these promising hybrids are described below:

The cv. Yanchun is an early spring cabbage bred from the self-incompatible lines D1-5-2-5-4 and H-1-6-1-1-10. It matures early (harvesting begins in mid-April), is high yielding, has a compact head and scarcity of premature bolting and is of good quality. F_1 hybrid 8398 is an early maturity good quality cultivar having high productivity. It can be harvested 50 days after transplanting, giving a yield of 49.5-57.0 t/ha (Fang *et al.*, 1996). Zhonggan 9 is an autumn cabbage hybrid cultivar. The heads weigh around 3 kg and are harvested about 85 days after transplanting. Heads are dark green, compact and round, 60-70 cm in diameter with 15-18 outer leaves. Storage quality is good. Yield is around 82.5-90 t/ha (Liu Yu Mei *et al.*, 1996). Chunbao is an early spring F_1hybrid; heads are pointed, resistant to bolting and of good quality. It produces 30-52.5 t/ha (Yin *et al.*, 1996).

The Bulgarian hybrid cultivar Karina form early uniform head with good resistance to bolting (Nankova and Alipieva, 1995).

In Brazil, Oliveira *et al.* (1999) evaluated cabbage hybrids under rainy condition and found that cv. Midore was the most suitable for cultivation giving a yield of 30.62 t/ha and a mean head weight of 1.314 kg, producing compact and round heads with high tolerance of black rot.

Parkash (2015) developed and evaluated large number of F_1 hybrids using 6 SI and one CMS lines. They have recorded highest heterosis for head weight was recorded in the hybrid, S-621 x GA-121 (37.9 per cent) followed by S-624 x GA-111 (34.1 per cent) and S-208 x GA111 (31.8 per cent) and it was concluded that the indigenously developed SI and CMS lines could be used widely in developing F_1 hybrids with higher hield and better quality. Kaur *et al.* (2018) evaluated 35 F_1 hybrids using SI lines and found that the hybrids SI-2008-09 x Glory1 (8.13 kg), II-12-4-10 x Glory1 (6.45 kg), IIICMS x Glory1 (6.28 kg), IICMS x E-1-3-1-2 (5.72 kg) and SI 2008-09 x GA(P) (5.41 kg) were superior to check varieties Varun (3.74 kg) and KGMR-1 (4.17 kg) for marketable head yield per plot.

In *B. oleracea* vegetables including cabbage, there is a consensus that CMS lines should be used widely in development of F_1 hybrids instead of the SI lines, because of poor stability and chances of sibs in SI based F_1 hybrids. Therefore, several SI based F_1 hybrids cultivated widely in the last decade were gradually replaced by the CMS based F_1 hybrids with better uniformity. Besides, completely homozygous DH lines have been developed and used widely in hybrid breeding of cabbage.

Popular Cultivars Developed by Indian Public Sector

Copenhagen Market

Introduced in 1909, it is one of the most common cultivars of cabbage. The round head fruit of this variety is large and weighs 2.5 -3 kilograms. The plants are ready to harvest in 75-80 days after their transplantation. This is still being grown but on a limited area. This variety has been replaced by 'Golden Acre' and 'Pride of India' because of their earliness and smaller head size.

Pride of India

It is an introduction, early in maturity and good in taste. The plant type of this variety is similar to 'Golden Acre' but it is about a week later in maturity. It has bigger sized heads, weighing about 1.5 to 2.0 kg per head. It is a high yielding variety of cabbage, with medium-large size fruits. The average yield is 20-29 tonnes per hectare, with a harvesting time of 75-85 days after transplanting.

Golden Acre

This is an earliest cultivar evolved by selection from the Copenhagen Market and recommended by IARI for cultivation in India. It takes about 60-65 days from transplanting to head formation. The heads are solid, short core and weighing about 1.0-1.5 kg. The heads should be harvested immediately after head formation otherwise, cracking may take place.

Jaersey Walkefield

It is an introduction with pointed head good in taste and preferred by some growers. It takes about 55-60 days for head formation. The outer leaves are few and slightly crinkled. Average weight is 1.0-1.5 kg.

Pusa Synthetic

This cultivar has been developed from IARI, Regional Station, Katrain, Kullu Valley, H.P. It is an early synthetic cultivar. The yield potential of this cultivar is 350-460 q/ha. The cultivar is slightly creamy with white cover and medium-size heads with an average yield of 35-46 tonnes per hectare.

Pusa Mukta

This is a new cultivar developed by Indian Agricultural Research Institute, Regional Station, Katrain by hybridization between EC 24855 x EC 10109. The light green foliage with wavy margins is a distinguishing character of the cultivar. The heads are compact, slightly flattish round with loose wrapper leaves at the top with an average head weight is 1.5-2.0 kg. It was ealier resistant to black rot, however, its resistance is not present now and it became moderately susceptible.

Pusa Ageti

This cabbage cultivar is an F_1 hybrid. The heads have grey-green leaves and are medium in size. It reaches maturity in 70-80 days from the date of transplantation, with an average weight of 600 grams to 1.5 kilograms.

Pusa Drumhead

It is a late-season cultivar. The heads are big, with each one weighing around 3-5 kilograms. Due to the larger size, the variety takes longer to reach maturity and requires a long winter to produce a good quality crop.

Pusa Cabbage-1

This the first F_1 hybrid developed by Indian public sector. This hybrid is developed using self-incompatible system. Heads round, solid, green in colour and coverd with outer leaves, possessed field resistance to black rot disease and high temperature. Maturity period 80-90 days after transplanting with an average yield of 35-40 tonnes/ha.

Public Sector Cultivars and Hybrids of Cabbage

Golden Acre

Copenhagen Market

Pusa Mukta

Pusa Ageti

Pusa Drumhead

Pusa Cabbage Hybrid-1

Pusa Cabbage Hybrid-81

Private Sector Hybrids of Cabbage

Mohor

Royal Challenger

Himrani

Sanjivani

Green Express

Green Boy

Pusa Cabbage Hybrid-81

This is the first CMS system based hybrid developed by Indian public sector. The hybrid has dark green, 12-14 non-wrapping waxy leaves, plant height is 22-25 cm, head is round in shape, very compact and covered with outer leaf. It matures in 60-65 days after transplanting and has very good field staying capacity (25-30 days) after head formation with average yield of 43.5 t/ha.

5.3.0 Soil and Climate

5.3.1 Soil

Cabbage can be grown on nearly all soils with a good water supply. Early crops are best grown on light soils, while late crops thrive better on heavier soils that are more retentive of moisture. On heavy soils, plants grown more slowly and the keeping quality is improved. Under high rainfall conditions, sandy or sandy loam soils are preferable because of improved drainage. The optimum pH for cabbages is about 6.0 to 6.5. Soils with a pH less than 6 should results in more occurance of club root and the soil should be limed if club root disease is prevelent.

Most cabbages are somewhat tolerant to salt. A measure of the salt content is the so-called C-index, which indicates the number of g of common salt per litre of soil moisture. For white cabbage and the savoys, the limit is C_6 while red cabbage is more sensitive with the normal limit at C_4. At C_8, the crop may show a reduction of 40 per cent yield. Plants grown in saline soils are more susceptible to diseases.

5.3.2 Climate

In the temperate zones, the cabbages are biennial. After sowing, the plants first develop vegetatively and then after exposure to low temperature in the winter, they pass into the generative phase.

As has been reported, the formation of flowers depends on the temperature and not on the day length. In Dutch experiments with adult plants of Brussels sprouts, flower initiation was observed after 4 weeks at 7°C and after five weeks at 4°C and 10°C whereas at 14°C and 23°C the plants remained vegetative. A longer period of cold treatment induces the plants to bolt sooner and flower more abundantly compared to exposure to shorter period. Further, intensity of flowering depends on the age of the plant and 3-5 month old plants are capable of transformation since they pass the juvenile phase. In the tropical and subtropical areas, the plants remain vegetative since they do not receive the required cold treatment.

In the early fifties, attempts were made to develop cabbage cultivars with low chilling requirement for transforming into generative phase by Chatterjee and Mukherjee (1962) at Kalimpong and a few strains were developed. Lines from cultivars, *e.g.*, English Ball, Eclipse Drumhead, Charleston Wakefield and a few others were developed as annual cabbage with low chilling requirement for generative phase. Most of these cultivars could be grown under moderate winter conditions.

Not much research work has been done on the effect of temperature on the growth and head formation of cabbages. Japanese investigations have shown that the optimum temperature for growth and heading is between 15° and 20°C. Above 25°C, the growth is arrested. The minimum temperature for growth of cabbages is just above 0°C. Winter cultivars of temperate zones can stand short spells of frost but severe frost damage may be noted when the temperature drops below -10°C and it continues for some time. Low temperatures are best withstood when the plants are half-grown and the temperature drops gradually.

Young plants stand a wider temperature range and they can be grown at higher temperature and planted later in a cool season. With the development of tropical cultivars in Japan, the range of temperature under which the cabbages can be grown has become much wider and these cultivars are capable of forming tight heads even at an average temperature of 25°C.

For high production the crop requires a cool, humid climate. The length of the total growing period varies between 90 (spring-sown) and 200 (autumn-sown) days, depending on climate, cultivar and planting date, but for good production the growing period is about 120 to 140 days. Most cultivar can withstand a short period of frost of -6°C, some down to -10°C. Long periods (30 to 60 days) of -5°C are harmful. The plants with leaves smaller than 3 cm will survive long periods of low temperature but when the leaves are 5 to 7 cm, the plant will initiate a seed stalk and this leads to a poor quality yield. Optimum growth occurs at a mean daily temperature of about 17°C with daily mean maximum of 24°C and minimum of 10°C. Mean relative humidity should be in the range of 60 to 90 percent. (http://www.fao.org/land-water/databases-and-software/crop-information/cabbage/en/).

In India, cabbage is grown in large areas of mild winter, especially in the eastern and western part. Most of the traditional European cultivars under such mild condition do not form tight heads, resulting in high percentage of loose and/or open heads. This is one of the reasons of poor average yield of cabbage in the farm land in India. But the situation may be improved significantly if the tropical cultivars of Japan are introduced and well adapted cultivars from this germplasm are developed for cultivation in the milder areas. Besides, these cultivars would extend the cultivation in areas not traditional for cabbage and also in season when cabbage is not generally grown in North India. The cultivars which have made such impact in the tropical and subtropical areas are K-K cross, K-Y cross and many others from Japan. In an initial testing programme of All India Vegetable Improvement Project, some of these cultivars have given very promising results. Attempts are being made at different centres like Kalimpong, Almora, *etc.*, to select more adaptable under warm Indian conditions. Three F_1 hybrids developed and recommended based on national trial are Pusa Cabbage-1, Pusa Cabbage-81 and KTCBH-822. All three F_1 hybrids are developed at IARI, Regional Station, Katrain and are suitable for cultivation during Rabi season throughout North Indian plain, However, they can be cultivated in summer seasons in high altitude areas of India.

In cabbage, differences in total and marketable yields and head traits among the planting dates were influenced by an interactive effect of air temperature and solar radiation as observed by Pranhos *et al.* (2016) at University of Florida. An average air temperature of 17.5°C combined with high solar radiation availability was ideal conditions for optimum cabbage head development. In this study, the September and October planting dates presented a more favorable average air temperature during the growing cycle, while the December planting date benefited most from an increase in solar radiation during head formation. The November planting date was the treatment most affected by adverse weather conditions.

5.4.0 Cultivation

The cultivation methods of cabbage are largely dependent on the climate in which they are grown. In India, cabbage is grown mainly in winter in the plains and also in the lower hills. In higher hills and in latitudes in the hills, it is grown in cool summer and rainy seasons. Though in mid-hills cabbage can be grown almost throughout the year, it has not become popular due to lack of suitable cultivars and also due to absence of suitable plant protection schedule, besides marketing difficulties. In addition, cabbage is also susceptible to heavy rains and hence research for development of cultivars for such conditions requires immediate attention.

5.4.1 Raising of Seedlings and Transplanting

5.4.1.1 Raising of Seedlings

Cabbage is grown from seed. In India, the seeds are generally sown in a seed bed and 4-6 week old seedlings are transplanted to the field. In the plains; the seed beds for early crops require cover to save small seedlings from rains, while in the high hills, it may be better to grow them under glass frames or polytunnels. However, the seedlings of mid season or late cultivars may be raised in the open.

The soil of the seed bed should be in good tilth, not liable to crusting, free from diseases and weeds. If necessary, the seed bed may be disinfected by soil drenching or sterilization. Manuring should be restricted to FYM (or cowdung manures) @ 6-8 tonnes/ha. If necessary, both phosphorus and potash may be added @ 50-75 kg and 150-200 kg/ha, respectively. Nitrogen should be added judiciously since over application will induce the seedlings to be weak and lanky.

For cultivation in one hectare, 200-500 g seed would be needed. The seed is sown in 200-300 sq m seed bed area at the rate of 1to 2 g per sq m. One gram seed would give rise to about 100 seedlings. Sowing should be done in rows and sowing too close may cause the seedlings to be lanky with poor health. Seeds should be sown in correct depth since deeper sowing will delay germination and will create other problems also. A depth of 1.5 to 2 cm would be optimum. A good moisture supply for rapid and optimum germination is essential. It may be a good idea to cover the beds which would prevent the washing away of seeds in case of rain, besides keeping the beds moist. The cover, if used, is removed just before the emergence of the seedlings.

Rapid germination avoids the chances of attack by soil organisms and a temperature of 15°-20°C would be quite optimum for germination. Attention should be paid for the control of pests and diseases after germination. The following seed treatment based schedule has been used widely for seed treatment (Table 5) and palleting of seeds for rsing healthy seedslings in nursery.

Table 5: Seed Treatment Schedule to Minimize Soil Borne Diseases

Fungicides	*Targeted Pests*
Thiram (2.50 g ai/kg)	Soil Borne Fungal Disease
Mefenoxam (0.075 g ai/kg), Fludioxonil (0.025 g ai/kg), Azoxystrobin (0.025 g ai/kg)	*Pythium/Phytophthora, Fusarium/Rhizoctonia*, Soil/Seed Borne Fungal Disease
Mefenoxam (0.075 g ai/kg), Fludioxonil (0.025 g ai/kg), Azoxystrobin (0.025 g ai/kg),	*Pythium/Phytophthora, Fusarium/Rhizoctonia,* Soil/Seed Borne Fungal Disease
Thiamethoxam (0.1 mg ai/seed)	Systemic Insecticide (aphids, cabbage maggots, certain seed and root maggots, flea beetles, white fly and wireworms)

In some areas of northern India, there is a tendency to sow seeds in situ to have an early crop. There are many limitations in such cultivation, *e.g.*, washing of seeds by rain, poor germination, uneven disstibution and growth.

Heavy rains pose difficulty in the preparation of land due to rains. However, in many foreign countries, the *in-situ* sowing has been standardized for all types of cole crops with the pelletted seeds and with the help of mechanical sowing. In India, *in situ* sown crop always suffers from many drawbacks. In addition to direct sowing *in situ* with pelletted seeds, a different system of sowing in thermocole plates with holes is also being done. In such cases, almost each seed is utilized for production. The method is now commercially followed in different countries in the West.

Mikhov *et al.* (1995) reported from Bulgaria that direct precision sowing of cv. Ditmarscher and simultaneously covering the rows with perforated plastic strips increased soil moisture in the top soil by upto 14.5 per cent and soil temperature by 0.5-1.6°C. This improved soil micro climate accelerated seed germination by 2 days and shortened the crop growing period by 21-23 days compared with using transplants.

Smith (1986) reported that cabbage seedlings grown in seedling trays produced equally good yields as those grown in seed beds provided that they were not held in the trays too long. In-tray time beyond 4 weeks resulted in progressively increases time to maturity from sown and lower yields.

Frantz *et al.* (1998) from USA compared seedling growth in different transplant production systems and concluded that both float-bed (FB) and ebb-and flood (EF) systems are suitable for cabbage transplant production. Float-bed (FB) is a simple hydroponic system used by tobacco industry for transplant production. Ebb-and-flood is a modified FB system with periodic draining of the bed to limit water availability and control plant growth.

To facilitate handling of transplants by machines, short, sturdy seedlings of plug grown vegetable crops are required. One major problem in commercial plug production is the control of the plant height when seedlings are grown in small plug cells, especially during summer. Due to legal restrictions of different chemicals, non-chemical strategies to control plant height include withholding water, light quality manipulation, shaking or brushing the shoot and temperature regulation in the greenhouse (Chen *et al.*, 1999). They noticed that irrigation with water at 5°C reduced stem length of cabbage seedlings 40 per cent, but use of water at 10 and 15°C did not (Figure 5).

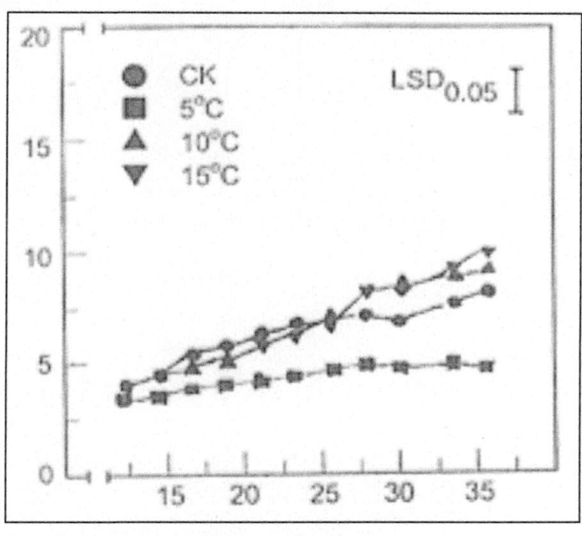

Days after sowing

Figure 5: Effect of Temperature of Irrigation Water on Stem Length of Cabbage Seedlings. Each point is the mean for 15seedlings. Least significant difference values are based on the pooled variance of all treatments (Chen *et al.*, 1999)

Sowing time in different parts of India varies. In the hills, specially in high hills, seeds are sown in May-June for summer/autumn crop. In the hilly areas which receive heavy rains, the summer or autumn crop is rather limited and sowing is undertaken in autumn to harvest them in late spring or early summer by over-wintering them.

In the plains of northern India, sowing *in situ* or in seed bed starts from early August and continues till November for the late cultivars. In eastern India, sowing is generally started from mid to late September. Sowing is delayed further for areas in the southern hills except in the western and southern peninsula where the crop can be grown almost round the year by selecting proper cultivars.

Han and Park (1985) studied the effect of sowing time on the heading and bolting of three over wintering cabbage cultivars grown to provide heads during April and May, a period of short supply. They obtained best results with Geumgye 201 sown during the last 10 days of September or the last ten days of October for harvesting April and May, respectively. Moel *et al.* (1986) reported that yield of root-ball transplanted plants was usually higher.

For assessing cultivars for the cropping of cabbage at Rumbeke, Belgium, seeds were sown between 12 and 18 April and the seedlings were transplanted between 6 May and 2 June (Vanparys, 1991).

In India, under Akola condition planting on 27 October resulted in a significant higher percentage of ratoon-forming plants, larger ratoon heads and higher yield (263.33 q/ha) than planting on the other dates. Danish Supreme was the most productive cultivar (173.3 q/ha) and Sri Ganesh Gol was the least productive (116.53 q/ha) (Futane *et al.*, 1995).

At Mizoram, an experiment was conducted with objective to examine the effects of different transplanting time on growth, yield and quality of various cabbage cultivars and hybrids in mild-tropical climate by Singh *et al.* (2010). All the parameters under study were significantly affected by both cultivars and transplanting dates. Maximum yield potential of various cultivars was realised by transplanting of cabbage seedlings on 1st December. The most compact head was observed for INDAM-1299 followed by Golden Acre, NS-160, KGMR-1 and lowest in Ryozeki.

5.4.1.2 Transplanting

In India, 4-6 weeks old seedlings are transplanted. Vaidya and Patel (1964-65, 1965) and Arora (1966) reported that the delay in transplanting of seedlings of cv. Golden Acre beyond and end of October resulted in the reduction in the number of marketable heads. This was mainly due to approaching warm weather in the western India or cold winter as in north India, which caused poor head formation in the cultivar and resulted in unmarketable heads. The same cultivar had been reported to perform better when transplanted in September in U.P. hills (Awasthi *et al.*, 1976) and in July in Kulu (Tewari *et al.*, 1977). The difference seemed to be due to altitude of the experimental area and the prevailing temperature.

Mann and Sandhu (1955a) did not find any difference in the performance when the seedlings were transplanted either on the ridge or on flat in the first year while in the second year, planting on ridges performed better. The difference in the result was probably due to either planting dates or to the intensity of rains. Arora (1966), however, suggested transplanting on one side of the ridge. For early crop, ridge planting may be beneficial, especially in areas where the operation is undertaken in rains but flat planting may be equally good if transplanting is done in dry period. Sowing, transplanting and choice of cultivar have to be done after very careful consideration of the area for growing in the country since the temperature plays the most important role on the performance. Most parts of the country have a comparatively short cooler period and, therefore, all cultivars belonging to early, mid-season or late may not do well. In such areas, cultivars maturing between 60-90 days of transplanting would do well, especially of the tropical types. Since these are not yet available, early cultivars like Golden Acre, Pride of India, Copenhagen Market, *etc.*, are likely to do well. In areas where the temperature fluctuation are very limited as is generally available in the southern and south-western peninsular region, tropical cultivars would be the ideal for repeated sowing and continued harvest. However, in North India and in the hills, especially in the northern hills, cultivars of all maturity may be grown by selecting suitable dates for sowing and raising the seedlings judiciously.

5.4.2 Spacing

Spacing is dependent on cultivars. The early cultivars may be spaced at 20-30 cm between the plants and 45-60 cm between the rows but the late cultivars require a wide spacing between the plants, mostly following 45-60 cm both ways. Trials on Drumhead cultivar under the All India Coordinated Vegetable Improvement Project at Orissa University of Agriculture and Technology, Bhubaneswar and Jawaharlal Nehru Krishi Viswavidyalaya, Jabalpur showed 30 cm x 75 cm to be the best. Similar results have been reported by Ghanti *et al.* (1982) from West Bengal on this cultivar.

Investigations on spacing of two white cabbage cultivars, Castello and Apex planted at 62.5 cm between rows and 60, 40, 27 or 20 cm within rows, equivalent to 27000, 40000, 60000 or 80000 plants/ha revealed that the proportion of desirable heads weighing 0.8-1.5 g increased with planting density, particularly above 60000 plants/ha (Titze and Alpers, 1987). Lawande *et al.* (1986) found that average yields rose with decreased spacing from 344.79 q/ha at 60 cm x 45 cm to 590.96 q/ha at 30 cm x 30 cm. However, closer spacing (30 cm x 30cm) appeared to favour higher incidence of cabbage looper resulting in higher yields of comparatively smaller unmarketable heads and prolong the duration of crop by about 28 days.

In Belgium, for late cropping a planting density of 25641/ha with plants 65 cm apart between rows 60 cm within rows was maintained by Vanparys (1991).

Bracy *et al.* (1993) in USA conducted trials to determine if cabbage cv. Headstart could be precision-sown to form a stand without subsequent thinning and to determine the optimum seed spacing for this. They found that cabbages directly sown at one seed per hill and at 30 cm spacing produced yields and head weights similar to or higher than those from cabbages sown 10 cm apart and thinned to 30 cm (the method currently used by some commercial operators).

A field experiment was carried out during rabi season 2010-11 to study the effect of different spacings on growth, yield and economics of cabbage by Prakash *et al.* (2014). The results revealed that significantly highest number of leaves (11.99), plant spread (56.43 cm) in North-South and (56.77 cm) in East–West, minimum days taken to first harvest after transplanting (92.25 days), maximum head diameter (14.23 cm) and average weight of head (0.80 kg) were recorded with plant spacing of 60 cm x 45 cm. Highest per cent of plant stands (77.02 per cent) was obtained with plant spacing of 45 cm x 45 cm. Maximum head yield per plot (25.00 kg) and per hectare (27777 kg) and fodder yield per ha (24194 kg) with maximum net realization of Rs. 171862 were obtained under plant spacing 30 cm x 30 cm.

A study to optimise the plant density in Red cabbage was conducted by Manasa *et al.* (2017) in the variety Red Jewel with 3 different spacing *i.e.* 45 x 45 cm, 60 x 45 cm and 60 x 60 cm. The experimental variables measured were vegetative characters (plant height, plant spread and number of leaves at 30, 60 DAT and at harvest along with number of days for head initiation and harvest) and yield (yield per plot and marketable yield). Plant spread, number of leaves, days for head initiation and days to harvest generally increased as the plant spacing increased, whereas, plant height, yield per plot and marketable yield were increased with a decrease in plant spacing.

At University of Florida, USA Pranhos *et al.* (2016) conducted an experiment in heading cabbage to determine the plant density under plastic mulch condition. It was observed that wider in-row spacing may be selected to decrease inter-plant competition to optimize the solar radiation absorption. The differences found for in-row plant spacing treatments were attributed to the effect of inter-plant competition. The 35 cm in-row spacing was the treatment that consistently produced greater marketable yields, and therefore may be selected to maximize productivity. However, if smaller heads become desirable in the market, narrower in-row spacing would be preferred.

5.4.3 Irrigation

Cole crops in general require a uniform level of moisture for steady stress-free growth, especially at the stage when the edible product is developing and enlarging. To meet crop moisture requirements, soils are frequently maintained near field capacity throughout growth. Moisture requirement varies with prevailing temperatures in addition to stage of plant development. The evapo-transpiration of some cole crops can be high. A rate of more than 4 mm per day has been reported for a cabbage crop.

Cabbage has an extensive, shallow root system. The majority of the roots are found in the top 0.4 to 0.5 m of the soil with a rapid decrease in root density with depth. Normally 100 percent of the water is extracted from this layer (D = 0.4-0.5 m). Under conditions when ETm = 5 to 6 mm/day, the rate of water uptake by the crop starts to reduce when the available soil water has been depleted by about 35 percent (p = 0.35). Furrow, sprinkler and trickle irrigation are used. However, the acreage under subsoil irrigation, which gives generally better results, is increasing. With subsoil irrigation, the depth of the water table is maintained between 0.3 and 0.7 m in fine sandy loam and between 0.7 and 1.1m in loam soils. Depending on climate, crop development and soil type, the frequency of irrigation varies between 3 and 12 days. If available water supply is limited, early irrigations should not be practised unless these can be continued until the end of the crop growing period. Water savings should preferably be made in the beginning of the crop growing period.

Water requirement of cabbage varies with climate, soil type, crop variety and management practices. On sandy soil frequent application of cabbage is highly productive in comparison to clay soil. This is because of downward and lateral movement of soil water is prevailed in sandy and clay soil, respectively. Thus, yield and water use efficiency of irrigated cabbage generally increased with increasing levels of management options. As a result, crop water requirement is inclusive to crop modeling strategies which are proactive to sustainable agriculture particularly in the view of climate change (Beshir *et al.*, 2017)

Cabbage has been classified as intermediately susceptible to water stress, with the head formation period being more sensitive than the period before (Adeniran *et al.*, 2010). The works of these authors were in line with which were reported that the

Seedling Growing, Transplanting and Field Establishment of Cabbage

Seedling Raising and Transplanting in Cabbage

Field Establishment of Cabbage

Furrow (Left) and Drip (Right) Irrigation in Cabbage Field

View of Cabbage Growing at Farmers' Field

most critical irrigation period for cabbage occurred during the last 3 to 4 weeks before harvest. Marketable yields of cabbage were similar when irrigations were applied at 100 kPa soil water tensions during growth and when cabbage was irrigated at 1000 kPa before head formation and 100 kPa after head formation. However, when cabbage was irrigated at 100 kPa before head formation and at 1000 kPa after head formation, marketable yields were similar to cabbage irrigated at 1000 kPa soil water tension during growth (Sharma, 1975). If available water supply is limited, early irrigations should not be practised unless these can be continued until the end of the crop growing period. Water savings should preferably be made in the beginning of the crop growing period. Water deficit during head formation period leads to shriveled and malformed heads. Different irrigation frequency differently influenced weight and number of cabbage at Chikwawa, Malawi on experimental site gross plot 5 m x 5 m and net plot 3 m x 3 m whereas 0.6 m spacing between rows and between plants with in the rows were used. The highest weight and number of marketable heads were observed when irrigation frequency is twice a week (Chafutsa *et al.*, 2007)

Arora (1966) suggested that the seedlings should be watered immediately after transplanting and thereafter irrigation at an interval of 15 days would be adequate. At the time of maturity watering should be avoided since this will cause splitting of heads. However, presently cultivars are available which would not split. Som *et al.* (1976) reported that irrigation at 0.25 atm and treatment with 140 kg N/ha gave the maximum yield of cabbage. Irrigation increased the yield of cabbage (cv. Prva Zetev) by 11 per cent compared with no irrigation (Cerne, 1991).

In an experiment in Germany, Hartmann and Zengerle (1987) observed that quick maturing (85 days) white cabbage cv. Grenitands low maturing (136 days) red cabbage cv. Autoro required 396 and 611 mm water and yielded marketable heads weighing 1.40 and 2.18 kg, respectively with soil moisture maintained at 75-100 per cent of capacity. While conducting experiment with cultivar Gloria, Fischer and Nel (1987) found that high soil moisture levels during vegetative growth improved leaf growth but not necessarily increased yields. Even at low moisture the crop gave satisfactory yields indicating considerable drought resistance.

According to Saha *et al.* (1998) total water use by cabbage for high yield, irrigation schedules ranged from 158.44 to 183.23 mm. Maximum efficiency of irrigation water in terms of marketable yield (602.35 kg/ha per mm) was achieved when irrigation led to 30 per cent DASM (depiction of available soil moisture). Comparative performance of cabbage (cv. Golden Acre) under different irrigation methods were studied by Srivastava and Chauhan (1999). The highest yield was obtained by micro sprinkler irrigation treatment (40.23 t/ha), followed by drip irrigation (38.97 t/ha), surface irrigation (33.76 t/ha) and microtube rrigation. Water use efficiency was highest for drip irrigation, followed by microtube irrigation, micro sprinkler irrigation and surface irrigation.

Under drip irrigation system, it was reported that the water requirement of the crop was computed on daily basis by using the following equation

$$V = Ep \times Kp \times Kc \times Sp \times Sr \times Wp$$

Whereas, V= Volume of water required (liter/day/plant); Ep = Pan evaporation as measured by Class-A pan evaporimeter (mm/day); Kc = Crop co-efficient (co-efficient depends on crop growth stage); Kp = Pan co-efficient; Sp = Plant to plant spacing (m); Sr = Row to row spacing (m); Wp = Fractional wetted area, which varies with different growth stage (0.3 to 1.0).

5.4.4 Manuring and Fertilization

The recommendations made by various workers with regard to manures and fertilizers per hectare are given below:

N (kg)	*P (kg)*	*K (kg)*	*Organic*	*Authors*
40-50	40	-	20 cartloads	Lal (1938)
150	-	-	-	Mann and Sandhu (1955b)
100	-	-	150 md cowdung	
150	90	-	15-20 tonnes FYM	Arora (1966)
120	-	-	-	Umale (1966)
120	120	-	-	Maharana and Naik (1967)
80	-	-	-	Sinhaand Bains (1967) for saline solis
150	90	40	-	Khokar and Arora (1968)
150	-	-	150 kg N (leaf compost)	Srivastava *et al.* (1970)
100	30	-	-	Agarwal *et al.* (1972) for loamy soil
100	30	30	-	Agarwal *et al.* (1972) for sandy loam soil
150	125	100	-	Thakur and Gill (1976)
150	80	75	-	For Bhubaneswar
180	50	50	-	For Jabalpur (under Coordinated Project)
200	-	-	-	Ghanti *et al.* (1982)
240	-	-	-	Kolota (1984) for Poland
252	-	-	-	Wiedenfeld (1986) for USA,
240	80	-	-	Nitrogen as slow release fertilizer
150	80	-	-	Lawande *et al.* (1986)
120	60	-	-	Prabhakar and Srinivas (1993) for high density planting
160	90	60	15-2- t/ha	Mohanty and Hossain (1998)
				Khan *et al.* (2002)

Umale (1966) observed that the application of P either alone or in combination with N did not give any significant increase in yield. Similar results have been reported by Ghanti *et al.* (1982). In Poland, Kolota (1984) concluded from his experiment with cv. Amager that the highest average commercial yield (65.7 t/ha) was obtained on irrigated plots where ammonium nitrate (240 kg N/ha) was applied in a split dose with one-half before planting and the other half as a top dressing. In California, Welch *et al.* (1985) noticed that application of nitrapyrin with ammonium sulphate at planting produced higher yields than ammonium sulphate alone, at planting or in split applications. Tissue levels of NO_3-N of cabbage were significantly higher in nitrapyrin treated plots than in those receiving comparable rates of N without nitrapyrin. Olenchenko *et al.* (1985) from the then USSR reported that on peaty podzolic light loamy soil the cv. Belorusskaya 455 produced high yields of optimum quality when N was applied at 120 kg/ha in addition to basal FYM at 120 kg and P_2O_5: K_2O at 120:120 kg/ha. Wiedenfeld (1986) from USA suggested application of N upto 252 kg/ha in the form of slow-release fertilizers. It increased yield, head weight and leaf N, in some instances over soluble N fertilizer. Pre-plant fertilizer application seemed advantageous on cabbage. Lawande *et al.* (1986) in their experiment with cabbage cv. Golden Acre, noticed that nitrogen 240 kg/ha and phosphorus 80 kg/ha produced higher yield of marketable heads under Pune conditions in India, but the response to added K_20 at 0-80 kg/ha was not significant.

Prabhakar and Srinivas (1993) recommended high planting density (50 cm x 30 cm spacing) and application of 150 kg N/ha and 80 kg P_2O_5/ha. Mohanty and Hossain (1998) obtained highest yield (242.80 q/ha) with cv. Pride of India by application of 50 q/ha FYM, 120 kg N and 60 kg P_2O_5/ha.

In a field trial in Poland on a sandy loam soil, and trickle rrigated with Delus F_1 white cabbage, Kaniszewski *et al.* (1999) found that reducing the N rate applied by fertigation from 200 to125 kg/ha had no deleterious effect on yield.

According to Arora (1966), there was no appreciable response to the application of K. In saline soils, Sinha and Bains (1967) found much higher concentration of salts on ridges than on furrows and recommended cultivation in furrows. They also observed that Pusa Drumhead was the most tolerant cultivar to saline conditions. Sabnis (1950) reported that the head formation was improved with the spray of 2 per cent common salt solution once or twice, especially in higher temperature conditions.

Ubugunov (1984) indicated the need for adequate applications of N, P and K supplemented by S, Zn, Cu and Co for obtaining high yield in cabbage. Alvares *et al.* (1985) from Brazil noted that of the four treatments with borax at 0, 10; 30 or 50 kg/ha, as furrow application on a yellow-red latosol soil, the two highest levels gave statistically higher yields than control.

A field experiment was carried out in Akure in the rainforest zone of Nigeria to determine the effect of poultry manure, wood ash and rice bran on the soil fertility improvement, growth and head yield of cabbage (*Brassica oleraceae* L). The three organic fertilizer treatments were applied each at 6 t/ha with a reference treatment NPK 15-15-15 applied at 300 kg/ha and a control treatment. The results showed significant increases ($P < 0.05$) in the growth and head yield parameters of cabbage crop under the different organic fertilizers compared to the control treatment. The application of poultry manure resulted in the highest values of cabbage plant height, stem girth, leaf number, leaf area, head weight, head length and head girth followed by NPK 15-15-15, wood ash and rice bran, respectively. Cabbage head weight, head girth, head length, plant height, stem girth, leaf number and leaf area increased by 17, 18, 8, 17, 19, 10 and 16 per cent, respectively, with application of poultry manure compared to NPK fertilizer. The poultry manure applied at 6 t/ha gave the best results in improving soil fertility, growth and head yield of cabbage, and this was because of its balanced nutrient composition and the least C/N ratio (Ibukunoluwa Moyin-Jesu, 2015).

Use of synthetic fertilizers can cause long-term problems for the environment. Developing a suitable strategy for agricultural application of alternative types of fertilizer is important. Commercial organic liquid fertilizers and a urea solution were utilized to investigate effects on growth and yield of Cabbage (*Brassica oleracea* var. *capitata* L.) by Manea (2017) in cultivars, Ramso and Copenhagen with the commercially available liquid fertilizers urea at 1 $g{\cdot}L^{-1}$; humic acid at 5 $mL{\cdot}L^{-1}$, Higro Amin-L 5 $mL{\cdot}L^{-1}$, and Maxinim at 5 $mL{\cdot}L^{-1}$. The interaction of 'Copenhagen' and Maxinim produced the highest percentage dry matter, widest heads and most marketable yield, longest stems, and highest percentage nitrogen. The interaction of 'Copenhagen' and Higro Amin-L produced the widest leaves and heaviest total and marketable head weights. The interaction of 'Ramos' and humic acid produced the fewest loose leaves. Use of 'Copenhagen' and Higro Amin-L fertilizer improves leaf width, total head weight, and marketable head weight. The interaction of 'Copenhagen' combined with Maxinim improves percentage dry matter, head diameter, and marketable yield. 'Ramos' combined with urea produced the fewest days to 50 per cent head initiation and days to 50 per cent marketable maturity. The best response in plant growth, increased head yield, and quality was with cv. Copenhagen fertilized with maxinim.

5.5.5 Weed Control

Mani *et al.* (1973) reported the usefulness of herbicides like treflan and desmetryne. Later, Gautam *et al.* (1976) and Gautam (1977) found that preplant application of either trifluralin (0.5 l/ha) or basalin (0.5 l/ha) or a pre-emergence application of desmetryne (0.25 l/ha) ensured excellent control of both dicot and monocot weed species with enhanced yield. Besides, the cost of chemical weed control is much less than manual weed control. Black polythene mulch proved effective in controlling weeds, conserving moisture, inducing better growth and higher yield of cabbage (Acharya, 1968). Kee (1985) from USA reported that pre-planting application of 0.25-0.375 lb oxyfluorfen + 0. 75 lb trifluralin/acre gave good weed control and crop safety. Marion *et al.* (1985) also found that oxyfluorfen alone, applied as pre-transplanting and in mixtures (sethoxydin) gave excellent broad spectrum weed control and increased cabbage yield. Kolesnikov and Ignatov (1984) from the then USSR reported that application of 1.5 kg treflan (trifluralin)/ ha before transplanting white cabbage seedlings and 0.25 kg simazine or 2 kg trazalox (nitrofen + simazine+ neburon)/ha, 7-10 days after transplanting followed by an irrigation with 2 days after their application, gave the most effective weed control, eliminated the need for hand weeding and resulted in yield higher than with 2 hand weedings. Dastgheib (1995) reported from New Zealand that highest crop yield was obtained with sequential application of aziprotryne + clethodim, or aziprotryne alone which resulted in more than a 2-fold increase in the fresh weight of cabbages compared with the untreated control. Ahuja and Sandhu (2003) recommended Pendimethalin at 0.56 kg a.i./ha and oxyfluorfen at 0.12 kg a.i./ha to decrease the weed population and dry weight accumulation by weeds at 45 DAT to get highest cabbage yield. DCPA (chlorthal-dimethyl) at 9 kg a.i./ha, napropamide at 1.1 kg a.i./ha, oxyfluorfen at 0.3 kg a.i./ha and oxyfluorfen at 0.2 kg a.i./ha + napropamide at 1.1 kg a.i./ha provided commercially-acceptable (>80 per cent) control of carpetweed (*Mullugo verticillata*), swineeress (*Coronopus didymus*), large crabgrass (*Digitaria sanguinalis*), red root pigweed (*Amaranthus retroflexus*) and common lamb squarters (*Chenopodium album*) under both conventional-tillage (100 per cent cultivated) and strip-tillage (25 per cent cultivated and 75 per cent residue) in North Carolina, USA (Hoyt *et al.*, 1996).

In the last decade the following herbicides were recommended for effective weed management in cabbage:

Herbicide	*Dose (kg/ha)*	*Mode of Application*	*References*
Pendimethalin	0.75-2.0	Pre-emergence	Kaur *et al.* (2015)
Alachlor	1.0	Pre-emergence	Nandal *et al.* (2005), Dhiman *et al.* (2005)
Oxadiazon	1.0	Pre-emergence	Nandal *et al.* (2005), Dhiman *et al.* (2005)
Oxyfluorfen	0.09-0.234	Pre plant incorporation	Nandanwar *et al.* (2006), Kaur *et al.* (2015), Kaur (2012)
Trifluralin	0.90	Pre-emergence	Kaur (2012)

5.5.6 Interculture

Post-planting care mainly concentrates on checking the weed growth and to keep proper moisture condition of the soil to promote sustained growth. Once the foliage has covered the soil, it is better to stop hoeing since it may damage the roots. Normally, the crop is kept free of weeds by 2-3 hand weedings and 1-2 hoeings. One day before transplanting of seedlings, the application of Pendimethalin (0.56 kg a.i./ha in 600-750 litres of water) followed by a hand weeding 40-60 days after transplanting effectively checks the weed population. If necessary, earthing up is done 30 days after transplanting. At the time of earthing up the plants are supported with soil to avoid toppling of the plant during head formation.

5.6.0 Harvesting and Yield

5.6.1 Harvesting

Majority of cabbage cultivars become ready for harvest at 90-120 days after planting. Cabbage crop should be harvested when the heads are firm and mature. Delaying harvest, even a few days beyond maturity can result in split heads and increased incidence of field disease. Harvesting immature heads, however, reduces yield, and the heads are too soft to resist handling damage. Immature heads also have a shorter shelf life than mature heads. The head is harvested by bending it to one side and cutting it with a knife. The stalk should be cut flat and as close to the head as possible, yet long enough to retain two to four wrapper leaves. Extra leaves act as cushions during handling and may be desired in certain markets. The head should not be removed by snapping or twisting it since this practice damages the head and results in inconsistent stalk length. Broken stalks are also more susceptible to decay. As the heads are not ready for harvest at the same time, therefore they are

View of Cabbage Growing at Experimental Field, BCKV, India

harvested in stages based on the maturity of the heads. Harvested produce should always be stored in shade before packing. Mechanical harvesting of cabbage heads is also practised which require uniform maturity of the heads besides longer stalk length for harvesting the produce without any damage or injury. In Japan, image processing system has been developed to determine the level of maturity of cabbage heads and forecast harvesting time (Hayashiand Sakaue, 1996).

5.6.2 Yield

The yield of cabbage varies between 20 and 50 tonnes per hectare. The variation in yield is due to the choice of cultivar, cultural practices and growing conditions.

In the early cultivars under comparatively higher temperature, the heads become somewhat loose at the initial stages, but become harder with the cooler season. As a result, the yield hardly exceeds 12 to 15 tonnes/ha in spite of closer spacing adopted by the farmers. In the mid season and late cultivars, the yield reaches between 20 and 30 t/ha, depending on the cultivar and compactness of head. The yields are optimum in the areas where moderate to longer winter conditions prevail, as in North India.

The average yield (tonnes/ha) of cabbage in different parts of the world is given below:

Country	*Yield (t/ha)*
China	25.22
Russia	18.91
India	18.26
Korea Republic	51.05
Japan	40.91
World	24.25

5.7.0 Storage

Cabbage can be stored for 159 days in regular atmosphere (RA) with a 39 per cent total mass loss and under control atmosphere (CA) (3-4 per cent CO_2 + 2-3 per cent O_2 + 93-95 per cent N_2) for 265 days with a 17 per cent total mass loss. The cabbage stored under CA showed better retention of green colour, fresh appearance and texture than under RA (Gariepy *et al.*, 1985). Berard (1985) reported from USA that CA storage usually reduced or eliminated grey speck disease and reduced the incidence and severity of vein streaking in most of the cases. The beneficial effect of CA persisted even after the removal of the stored produce. It considerably delayed degreening and eliminated abscission and loss of dormancy during the first 122 of 171 days of storage. Walker (1985) observed during 2 successive years, that cv. Mini cole could be stored for 2 or 3 months, using 9 different water vapour pressure deficits (14-88 Pa). These were applied in a flow through system regulating 3 different RH (98, 95 or 92) each at 3 temperatures (0.5, 4 or 8°C). Low vapour pressure deficits resulted in higher amounts of marketable heads at the 2 lower temperatures. Leaf yellowing and wilting are the common postharvest problems of cabbage in supply chain that are encountered at the collection or consolidation stage. Kramchote *et al.* (2012) studied the simulated conditions at the collection center and determined the effects of low temperature storage on cabbage quality. Storage at 4°C and 10°C effectively delayed leaf yellowing and maintained leaf chlorophyll content. Weight loss, respiration rate and ethylene production were also reduced at low temperatures and head firmness was maintained. Moreover, total soluble solids and ascorbic acid contents were higher in cabbages kept at 4 and 10°C than those stored at ambient condition (28°C). Cabbage can be stored successfully at 4°C for 18 days and 12 days at 10°C. At ambient, cabbage deteriorated rapidly and lasted for only 4 days. Tao *et al.* (1986) from China concluded that early cultivars (Bao Chun, Yuan Chun and Shuang Jin) are unsuitable for storage and recommended that organic fertilizer be used in preference to inorganic N fertilizer. The cultivar Jin Feng harvested in late June was considered suitable for long storage. Storage quality at room temperature was recorded highest with a high K: Mg application ratio (He Tian Xiu *et al.*, 1994). Development of black mid rib and necrotic spot on several cultivars was recorded by Berard *et al.* (1986), after 175 or 122 days in refrigerated (RS) and in CA storage. In comparison with RS, CA increased the incidence of black midrib and favoured the development of inner head symptoms in susceptible cultivars.

6.0 Cauliflower

Cauliflower was introduced in India in 1822 (Swarup and Chatterjee, 1972). The initial introduction was the Cornish types from England and later the other European types also were introduced. The typical Indian cauliflowers were the result of intercrossing between the Cornish and other European types. Thus, a new development took place in India resulting in tropical cauliflower resistant to hot weather and high rainfall. These types are normally grown in Indian plains from May-June to January which is followed by the only temperate cauliflower known as Snowball. All these cauliflowers grown in India are annuals, both Indian and Snowballs. Among the so-called winter vegetables in India, cauliflower is possibly the most popular. It is also very popular in Europe and in Australia. Since, 1930, in the USA the popularity of cauliflower is receding while its popularity has increased tremendously in the tropical and subtropical areas with the development of Indian cauliflowers.

6.1.0 Area and Production

World production of cauliflower including broccoli is 25.23 million tonnes (Anonymous, 2018). India is the second largest producer (8.20 million tonnes) of cauliflower after China but its productivity (19.25 t/ha) is very less as compared to other contries. Area and production of some leading producer of cauliflower over the years (including broccoli) is given in Table 6.

6.2.0 Morphology of the Curd of Cauliflower

Of the common terms applied for the edible part of cauliflower, *e.g.*, head, inflorescence, curd and flower cluster, 'curd' seems to be the most appropriate according to Sidki (1962). The term 'head' has been applied to the structurally

Table 6: Major Cauliflower and Broccoli Producing Countries in the World*

Country	*2002*			*2006*			*2012*			*2014*			*2016*		
	Prod. (Million Tonnes)	*Area (Million Ha)*	*Yield (T/Ha)*	*Prod. (Million Tonnes)*	*Area (Million Ha)*	*Yield (T/Ha)*	*Prod. (Million Tonnes)*	*Area (Million Ha)*	*Yield (T/Ha)*	*Prod. (Million Tonnes)*	*Area (Million Ha)*	*Yield (T/ Ha)*	*Prod. (Million Tonnes)*	*Area (Million Ha)*	*Yield (T/Ha)*
CHINA	6.59	0.30	21.69	7.78	0.38	20.30	9.08	0.45	20.26	9.37	0.48	19.58	10.26	0.52	19.73
INDIA	4.89	0.27	18.11	5.32	0.29	18.44	7.35	0.39	18.80	8.57	0.43	19.76	8.20	0.43	19.25
USA	0.28	0.02	17.00	0.32	0.02	19.84	0.30	0.01	20.78	1.22	0.07	18.72	1.32	0.07	19.40
SPAIN	0.49	0.03	19.46	0.43	0.03	16.91	0.51	0.03	17.14	0.60	0.03	17.98	0.61	0.03	18.35
MEXICO	0.27	0.02	14.17	0.30	0.02	13.86	0.40	0.03	14.84	0.50	0.03	15.41	0.58	0.04	16.40
ITALY	0.45	0.02	18.66	0.40	0.02	22.28	0.41	0.02	24.22	0.41	0.02	24.73	0.39	0.02	23.88
FRANCE	0.40	0.03	14.87	0.35	0.03	11.81	0.34	0.02	17.26	0.33	0.02	16.31	0.31	0.02	15.82
POLAND	0.18	0.01	18.14	0.25	0.01	17.40	0.31	0.02	19.81	0.32	0.01	22.29	0.31	0.01	22.34
PAKISTAN	0.20	0.01	18.19	0.21	0.01	18.45	0.23	0.01	17.20	0.22	0.01	17.39	0.22	0.01	17.46
EGYPT	0.11	0.00	24.66	0.13	0.01	24.90	0.17	0.01	28.61	0.16	0.01	29.93	0.17	0.01	30.30
OTHERS	1.86	0.13	14.31	2.31	0.23	10.04	2.48	0.26	9.54	2.56	0.27	9.36	2.81	0.16	17.54
WORLD	15.72	0.84	18.71	17.80	1.04	17.12	21.58	1.24	17.40	24.18	1.38	17.49	25.23	1.34	18.78

*Anonymous (2018)

distinct leafy portion of cabbage or lettuce. The other terms are inaccurate because they imply the presence of flowers. The curd of cauliflower is presently thought as the highly suppressed perfloral fleshy apical meristem.

The cauliflower curd has been described by De Candolle (1824), Masters (1869), Henslow (1908), Worsdell (1916) and Bailey (1930) as being composed of fasciated and hypertrophied flower stalks which produce only rudiments of abortive flowers. Masters (1869) and Worsdell (1916) attributed such abortion to excessive branching and hypertrophy of the flower stalks. Lund and Kjaerskou (1886) also described the curd as being excessively branched and thought this to be the result of suppression of the apex of main axis and the growth of the lateral branches whose apices were again suppressed. The repetition of this process of suppression was considered by them as the excessively branched curd. Aamlid (1952) disagreed with this view and according to him, it is the result of the absence of apical dominance of any of the apices of the inflorescence during the curd development phase.

Dark (1938) and Rao (1938) considered the curd as a hypertrophied inflorescence with normal flowers. Aamlid (1952) described the curd as an inflorescence which consisted of naked apices of peduncles surrounded by a whorl or two of indeterminate structures which develop either into apices of a new order of peduncles or to flowers.

From a detailed study of curd from morphological, anatomical and physiological points, Sidki (1962) concluded that a cauliflower curd consists of the repeatedly branched terminal portion of the main axis of the plant. The branching follows phyllotactic pattern characteristic of the whole plant and in this sense it is not excessive, contrary to earlier reports. The form of the curd is characterized by three main changes in the growth pattern of the plant: (i) leaf development is curtailed, so that only bracts are present throughout the curd, (ii) all lateral buds are elongated into shoots whose apices make up the convex surface of the curd and (iii) the internodes of the shoot system are shortened. Neither hypertrophy nor fasciation is involved. Thus, the curd consists of a shoot system with short internodes, branch apices and bracts. On this basis, the present understanding is that it is a prefloral fleshy apical meristem which invariably precedes floral initiation compared to the closely related sprouting broccoli (*B. oleracea* var. *italica*), whose heads are composed of flower buds. Wiebe (1975) has further expanded the idea that whereas broccoli forms inflorescence at the tips and in the nodes of the leaves, in cauliflower compact curds occur which first form fertile flower buds after the normal harvest stage, parallel to the elongation. A premature initiation of flower buds is characterized by riceyness in cauliflower.

Keiffer *et al.* (1998) decribed the curd development through electron microscopic view. According to them, at least five developmental stages describing shoot apical meristem activity can be observed: vegetative, curd induction, curd forming, curd thickening, and floral. During the vegetative stage the diameter of the shoot apical meristem increases in relation to the number of leaves initiated. It follows a quadratic progression and increases rapidly in size until curd induction when the rate of increase slows down. At curd induction cauliflower produce about 24 ± 25 true leaves and their shoot apical meristem has a diameter between 390 and 430 μm (Fig

6.). During curd induction, leaf production stopped while bracts were being initiated (Figure 6), and these bracts subtended developing branch primordia. During the curd-forming stage (Figure 6), branch primordia rapidly became more prominent than their subtending bracts. Eventually, by the end of the curd-thickening stage no more bracts are initiated. After a growth phase, each branch primordium become the apical meristem of a branch of Order II and start to produce branch primordia which in turn become apical meristems of branches of Order III and so on (Figure 6). During the thickening stage this iterative process induces a rapid increase in curd diameter. Virtually no stem internode elongation take place in curds and only a moderate stem thickening is observed despite the high number of inter nodes present in the headed phenotype.

Unlike other cole crops, most forms of Indian cauliflower do not require a specific vernalization for transformation to the reproductive stage. Winter cauliflowers and other biennial types require cold treatment for curding and flowering, whereas for summer cauliflowers cold treatment is only necessary for flowering. In the earliest group that is, high temperature curd size is smaller than that in later groups. The curd color of Indian cauliflower is yellowish to creamy white, and Snowball is bright white. Nevertheless, variants with orange, green,

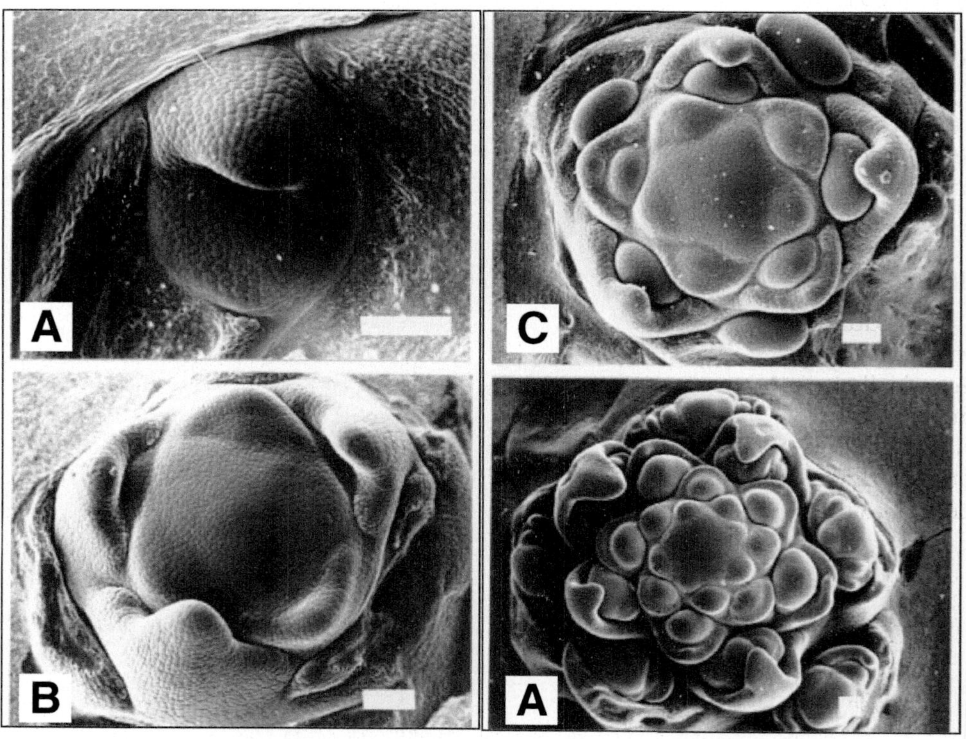

Figure 6: Scanning Electron Micrographs of Four Developmental Stages of Cauliflower Curd, A: Vegetative stage; B: Curd induction stage; C: Curd forming stage; D: Curd thickening stage; Bars: 100 lm (Kieffer *et al.,* 1996).

and purple-colored curds have been developed that are more nutritious than those with traditional white curds. An orange cauliflower mutant was first discovered in Bradford Marsh, Canada, in 1970. Through many years of work by Professor Emeritus Michael H. Dickson at Cornell University and other breeders, the orange cauliflower is now commercially available in India (Kalia *et al.*, 2020). The orange color results from a novel spontaneous mutation of a single gene designated as *'Or'* for orange gene (Crisp *et al.*, 1975). A purple cauliflower mutant displays vibrant violet hues resulting from a spontaneous mutation of a single semidominant gene designated *'Pr'* for purple gene.

6.3.0 Cultivars

The yield of cauliflower in Indian farm lands though slightly better than cabbage but still much lower than the other areas of the world as discussed in area and production. It is mainly due to the non standardized cultivars that are generally grown and also that a traditional approach of milder areas to grow only Snowball types which produce not only poor yield but also poorer quality. The yield in other tropical areas of the world has gone up due to the cultivation of F_1 hybrids introduced by the Japanese seed companies. These F_1 hybrids are not only better yielding but the quality is also superior. Incidentally, all these F_1 hybrids of tropical types are basically of the Indian materials.

In India, two separate groups of cauliflower grown are the Indian cauliflowers, which have mainly developed within India during nearly the last 165 years and the other annual temperate type, the Erfurts and Alpha strains, more commonly known as Snowball. Both are annuals and may be classified as tropical annuals and temperate annuals. The major differences between the two are in their adaptability in different temperature conditions, tropicals forming curds at a temperature around 20°C or above, while the temperate annuals requirea temperature range between 10° and 16°C. Further, the tropicals are resistant to high rainfall conditions, specially during vegetative phase.

In addition to these two annual types, there are other developments in this crop which took place independently in different regions of the world; they remained genetically isolated for a long period (except for the Italians or originals) and thus maintained their characteristic features. It would, therefore, be worthwhile to classify these broad groups so that a proper understanding and relationship of the present day cultivars are possible, Swarup and Chatterjee (1972) classified these groups as follows:

Cauliflower Types	*Country of Origin*	*Probable Period of First Cultivation*	*Characters*
Italians or Original	Mediterranean	16th Century	Plants short; leaves erect broad with rounded tips, bluish green; curds good not protected.
Cornish	England	Early 19th Century	Plants vigorous; long stalked; leaves loosely arranged, broadly wavy; curds flat, irregular, loose, not protected. Yellow, highly flavoured.
Northerns	England	19th Century	Leaves petiolate, broad, very wavy, serrated; curds good, well protected.

Cauliflower Types	*Country of Origin*	*Probable Period of First Cultivation*	*Characters*
Roscoff	France	19th Century	Plants short; leaves long erect, slightly wavy with pointed tip, mid-rib prominent, bluish green; curds white or creamy, hemispherical, well protected.
Angers	France	19th Century	Leaves very wavy, serrated, greyish green; curds solid, while, well protected.
Erfurt and Snowball	Germany and Netherlands	18th Century	Plants dwarf; leaves short, erect, glaucous green; curds solid, well protected.
Indian cauliflower	India	Late 19th Century	Plants short, long stalked; leaves loosely arranged, broadly wavy; curds flat, somewhat loose, yellow to creamy, not protected andhighly flavoured.

Crisp (1982) classified the groups of cauliflower according to their phylogeny in the following manner:

Group Name	*Characteristics*	*Common type*
Indian	Very diverse, include bothannuals and biennials and curds with peculiar conformations and colours	Jezi, Naples (Autumn Giant), Romanesco, Flora Blanca.
North-West European biennials	Derived with the last 300 years from Italian material	Old English, Walcheran, Roscoff, Angers, St. Malo
North European annuals	Developed in northern Europe for at least 400 years. Origin unknown, perhaps Italian or Eastern Mediterranean	Lecerf, Alpha, Mechelse, Erfurt, Danish.
Asian	Recombinants of European annuals and biennials, developed within 250 years, adapted to tropics	Four maturity groups are recognized by Swarup and Chatterjee (1972).
Australian	Recombinants of European annuals and biennials and perhaps Italian stock, developed during the last 200 years	Not yet been categorized.

The scheme of classification of Crisp (1982) requires further studies in respect to the origin of the North European annuals and the Australian types. It is, however, not very clear from the scheme, how Crisp (1982) suggested a separate origin of the annuals when both annuals and biennials are present in the Italians which according to him have given rise to the North-West European biennials. Besides, if the biennial Cornish could give rise to the typical tropical annuals in India, there may not have adequate justification for a separate origin of the North European annuals as claimed by Crisp (1982). Thompson (1976) suggested that as the early cabbage cultivars spread to North-Western Europe, biennial types have been selected with greater winter hardiness and a considerable cold vernalization of flowering. A similar adaptation may have taken place on biennial cauliflower also.

Besides, the Australian types are adequately categorized and different types from 3-4 month maturity to 8 month maturity are available, though typical winter hardy biennials are not present, since the environment does not permit such adaptation of cauliflowers by the growers. It appears that Australian development is a parallel development but independent of North-West European varietal influence though the base materials for both seem to be the Italians.

Recently, Dey *et al.* (2019) studied the relationship among different cultivated types of Indian cauliflowers which are adopted to higher temperature and humidity. The typical Indian cauliflowers have been used worldwide in developing cultivars with tolerance to higher temperature. They studied forty-five genotypes of cauliflowers including present day cultivars from different maturity groups of Indian cauliflowers and snowball types cultivated in India. To understand the relationship among the different cauliflowers a large number of genomic SSRs along with comprehensive morphological characterization was done. Two earliest groups of Indian cauliflowers (Group-I and II) were completely different from the snowball types cultivated in India. They are probably originated from the Cornish types. However, the widely cultivated genotypes have undergone several cycles of selection to acquire the desirable plant types. All genotypes of snowball types were clustered together along with genotypes from European origin. The last group of Indian cauliflowers share similarity with snowball types and other European genotypes both in molecular and agronomic level. The genotypes from Taiwan and China were more similar to the Indian cauliflower genotypes which establish the role of Indian cauliflowers in evolution of different tropical cauliflowers cultivated in different countries. This study establishes the frequent introgression of snowball types in the evolution of Indian cauliflowers with requirement of comparatively lower temperature. The role of the Indian cauliflowers with high temperature tolerance in the evolution of present-day tropical cauliflowers is also established.

6.3.1 Early Temperate Cultivars

Under each broader group, there are many cultivars. However, for early crops in North Europe extensive use is made of the early Erfurt strains. The curds of Erfurt are bright white, little knobbly but easily grow loose; sometimes much riceyness occurs. Under unfavourable conditions the plants soon button. In many regions, the Erfurt has been replaced by some what stronger Alpha, a type with wavy and more spreading leaves. As compared with Erfurt, the curds of Alpha are of better quality, do not grow loose quickly, are knobbly and less sensitive to riceyness but more susceptible to whiptail. With these early Alpha strains, it has become difficult to identify typical Alpha or Erfurt strains and most of the early strains of today belong to none of these two types. Strains like Aristo, Comet, Gloria, Lucullus (without anthocyanin), Bravo, Starla, *etc.*, are followed by Cambridge, King, Dominant, Danish Giant, Delfter Market, Idol, Igloo, Triumph, *etc.* Danish Giant is the typical representative of late, long-leafed Erfurt which is more commonly known as Snowball in the USA, India and other places. Mechlin is another cultivar of importance which has a wavy, dark green leaves with margins tending to curl inwards. The curd is markedly knobbly, very deep and of very good quality. These are resistant to spring frosts but are susceptible to buttoning.

6.3.2 Summer and Autumn in Temperate Regions

For summer and early autumn production, the French cultivar Lecerf is commonly used. Cultivars like All the Year Round, Diancourt, Omega, Reliance, Wiener, Le Normand, *etc.*, belong to this group. The plants of this group have dark green leaves with wavy margins. The curd is small, knobbly, fairly flat and soon grows loose. It is more reliable cropper and does not button so readily, though the qualityis rather lower. Some strains of Lecerf have a spherical curd but are more sensitive to unfavourable growing conditionsas they form less foliage.

Italian cultivars like Autumn Giantor Italian Giantare frequently used for late autumn crop. Representatives of this group are Allerhciligen and Tussensoort (Netherlands), Frankfurter Riesen (Germany), Autumn Mammoth, Invincible Cape, Majestic, Non Plus Ultra, Novo, Primus and Southern Cross. There are fairly large differences between these various types, like in earliness. The very late strains mature around December and form transitional group to the winter cultivars. Most of these make a sturdy plant with large, long erect leaves which are usually smooth. The curds are deep and knobbly. Flora Blanca, an introduction in the late forties or early fifties, is a rather early Dutch cultivar in this group with markedly undulating foliage. Gradually in recent years, a number of Flora Blanca strains have been introduced to cover the late summer and autumn. Cultivars like White Top, White Rock, Doc Elgon, Jura, Cervina, *etc.*, are taking up the production. These are either selection from Flora Blanca or some selections from cross breeding with Flora Blanca base. From Australia, number of late autumn cultivars have been introduced in recent years like Austral, Southern Pacific, Canberra, Melbourne Market, Sydney Market, *etc.* These Australian cultivars have a sturdy plant with large, long and erect wavy leaves. The curds are deep and knobbly.

Systematic and extensive cultivation of cauliflower occurred first in Italy where the 'Originals' or 'Italians' were developed. These 'Original' or 'Italian' types were taken to France, England, Germany, and The Netherlands where some important local types were developed such as the 'Northerns' in Yorkshire and Derbyshire; the 'Cornish' in Cornwall; the 'Angers' and 'Roscoff' in Brittany; and the 'Erfurt' or its allied 'Snowball' in Germany and The Netherlands. Broadly, on the basis of suitability of growing conditions, the European cauliflowers have been grouped into winter type; that is, winter cultivation ('Originals' or 'Italians', 'Cornish', 'Northerns', 'Roscoff', 'Angers') and summer type; that is, summer cultivation ('Erfurt' and 'Snowball'). The chief characteristics of important cauliflower types are broadly elaborated by Swarup and Chatterjee (1972).

6.3.3 Winter Temperate

The winter cultivars have clear-cut differences. Some have wavy, sometimes short foliage with strikingly thick veins while others have somewhat smoother leaves; sometimes these may have long petioles. However, the curds are well protected, especially at early stage, deep, firm with knobbles. The early types maturing in January, are not very hardy. The late cultivars are very hardy and can withstand temperatures of -10°C for some time. After the winter, they form new foliage and are harvested in May or June. However, mostly the medium-late maturing cultivars are grown which mature around February to April. These are fairly hardy, though a large part of the crop may be lost when severe frost occurs.

Both in France and in South-Western England, presently the Roscoff type developed in France are grown. These are not very winter hardy and consequently in the northern districts of England growers use the hardier strains of the old English cultivars, which have petiolate leaves with markedly wavy margins as in the case of Cambridge Hardy Late. These strains like those of the cultivar Angers which are grown in North-Western Europe, are rather sensitive to ring spot disease. In Netherlands, hardy strains of Angers are grown which are morphologically similar to the hardy English cultivars.

6.3.4 Tropical/Indian Cultivars

Most of the Indian cultivars are marketed by the name of the month they mature like *Kunwari* (earliest, in September-October), *Katki* (October-November), *Aghani* (November), *Poosi* (December), *Maghi* (January), *etc.* These cultivars are highly heterozygous in respect of all the characters whether vegetative or curd. Even for maturity, except the late cultivars, they are highly heterozygous. The earlier cultivars are short plant, bluish green leaves with waxy bloom and with very small to medium curds and tend to grow loose faster. The later cultivars are long-leafed sturdy plants with some what better quality curd but still not very deep and tend to grow loose fast. The early cultivars are also sensitive to button. These are highly heterozygous for maturity, plant growth, and curd traits.

The Indian or tropical cauliflowers grown in India are characteristically different from European types and are tolerant to high temperature and humid conditions, the earliest maturing types known and do not require vernalization for bolting. Swarup and Chatterjee (1972) hypothesized that Indian cauliflowers possessed most of the genes contributed by the 'Cornish' types and a few genes by 'Roscoff', 'Italian', and 'Northern' types. Certain favorable genes for tolerance to high temperature and rainfall, unknown in European types, have been successfully utilized in developing the improved cultivars Early Patna, Early Benaras, and Early Market in India; Pua Kea in Hawaii; Campinus in Brazil; Improved Japanese and D 96 in Israel; and Extra Early in Taiwan.

A few improved cultivars have been released from the Research Institutes. One of the early releases in fifties from IARI was **Pusa Katki** maturing in October-November, and having medium plant, bluish green and wavy leaves with medium curds. In the same season of maturity this cultivar has now been replaced by **Pusa Deepali**, an inbred with uniform, medium-long, erect foliage with rounded tip; curds are white and deep and more number of plants can be accommodated per unit area. Earlier than this cultivar is **Punjab Kunwari** from Punjab Agricultural University. This has somewhat bluish green leaves with waxy bloom and much wavy and maturing earlier than Pusa Deepali; this is, however, more sensitive to riceyness. **Improved Japanese** is a cultivar introduced from Israel, has long, erect narrow leaves with medium sized curds and maturingin November-December. Another introduced cultivar which gave good performance in December-January was **D-96** but due to its susceptibility to many diseases, was replaced now by **Pusa**

Synthetic from IARI. This has been synthesized from inbred lines with good general combining ability, having long wavy leaves with medium deep curds and maturing in December-January. In this maturity group, another somewhat earlier cultivar from Pantnagar, 114-S and two cultivars from Punjab, **Punjab Giant-26** and **Punjab Giant-35** are also being distributed. All these cultivars have medium to long leaves, wavy with medium to large curds. These cultivars are more sensitive to riceyness than Pusa Synthetic. In India, these cultivars are followed by Erfurt-Alpha group of cultivars. A cultivar known as **Dania Kalimpong** has proved very successful in the eastern region, more particularly in the eastern hills. The cultivar has very sturdy plant with wavy leaves and medium deep curd, but less sensitive to fluctuations of environmental conditions. Long-leafed Erfurt types, like **Snowball** are the other popular strains throughout the country. This is sensitive to temperature fluctuations and gives rise to pinkness or yellowing according to temperature and period of maturity. However, the curds are deep and bright white initially. In these Snowball types, **Pusa Snowball- I** has become popular, though it is susceptible to black rot. A new cultivar, **K-1** is rather tolerant to black rot. In the December maturity group, a new strain **6-1-2-1** from IARI has been recommended, which is resistant to black rot and curd and inflorescence blight.

Pusa Shubhra is an early maturing cultivar whichis field resistance to black rot and *Alternaria* blight. It produces medium-sized curds, weighing 700-800 g but are prone to looseness (Singh *et al.*, 1991).

Pusa Meghna is an extra early (Sept. maturity) cultivar developed by IARI and suitable for growing Delhi NCR region. The curds are white compact medium sized curd (350-400 g). The cultivar become matures in 95 days. Average yield is 125 q/ha.

Pusa Sharad is mid early (Nov.) cultivar developed by IARI suitable for growing Delhi NCR region. Foliage is bluish green, leaf with narrow apex and prominent mid-rib. Curds are semi-dome shaped white and very compact with an average weight of 900 g. Average yield is 24 t/ha.

Pusa Paushja is belonging to mid late maturity group (Dec.-Jan) developed by IARI and suitable for growing Punjab, U.P., Uttarakhand and Bihar. The curds are compact white. The variety becpme mature in 75 days. Average yield is 400 q/ha.

Pusa Shukti is a mid late cultivar (Dec.- Jan) developed by IARI and suitable for growing Delhi NCR region. The curds are compact white and maturity in 75 days. Average yield is 440 q/ha.

Recently a large number of cultivars have been released both from public and private sectors for commercial cultivation.

Pusa Ashwini is an early maturing cultivar. Its curds compact with retentive white colour. The average curd weight is 500-600 with a yield potential of 16.0-18.0 t/ha.

Pusa Kartiki is an early maturity group cauliflower cultivar. The curds are of retentive white colour and compact with net weight of 500-600 g and an average yield of 18 t/ha. It is suitable for early sown conditions beginning June under North Indian Plains and gets ready for harvesting during 2nd fortnight of October.

Pusa Betakesari-1 is the first ever indigenously bred bio-fortified beta carotene (800-1000 µg/100 g) rich cauliflower cultivar, an attempt to tackle beta carotene deficiency related malnutrition problem in India. Its curds are orange coloured, compact and very attractive with semi-self-blanching growth habit. It is suitable for September-January growing period. Average marketable curd weight is about 1.250 kg with an approximate marketable yield of 42.0-46.0 t/ha.

The cauliflower cultivars grown in India are classified into four groups, and the varieties are grouped on the basis of period of curd maturity (Chatterjee and Swarup, 1983; Singh and Devi, 2015). In north Indian plains, the temperature differs drastically from Punjab to Bihar and affects crop productivity if sowing or transplanting is not done at the correct time. Realizing the critical role of temperature for curding, the cultivars must be categorized on the basis of temperature requirement rather than the period or month of maturity. Hence, cultivars of Indian cauliflower should be reclassified as extra early (>30°C), early (25-30°C), mid (20–25°C), mid–late (15–20°C), and late or 'Snowball' (<15°C) to avoid ambiguity.

Table 7: Classification of Cauliflower Grown in India on the Basis of Maturity Period and Temperature Requirement and Recommended Varieties for Cultivation*

Maturity Group	*Period of Availability*	*Curd initiation Temperature*	*Cultivars*
IA (Early)	End of Aug. to Oct.	20-27	Punjab Kunwari, Early Kunwari, Pusa Early Synthetic, Pusa Kartik Sankar, Pusa Meghna, Kashi Kunwari, Pusa Ashwini, Pawas, Barsati, Riya
IA (Early)	Oct. - Nov.	20-25	Pusa Deepali, Pusa Katki, Pant Shubhra, Pant Gobhi-3, VRCF-86, Pusa Kartiki
II (Mid early)	Nov. to Dec.	16-20	Improved Japanese, Pant Gobhi-2, Pant Gobhi-4, Pusa Hybrid-2, Pusa Hybrid-3, Pusa Sharad, Punjab Giant-26, Pusa Agahani, Kashi Agahani, VRCF-50, VRCF-102
III (Mid late)	Dec. – Jan.	12-16	Pusa Synthetic, Pusa Shubhra, Pusa Himjyoti, Punjab Giant-35, Hissar-1, D-96, VRCF-22, VRCF-202, Ooty-1, Kala Patta, Pusa Paushja, Pusa Shukti
IV (Late)	Jan. – Mar.	10-16	Dania, Pusa Snowball 16, Pusa Snowball 1, Pusa Snowball 2, Pusa Snowball Cauliflower Hybrid-1, Pusa Snowball Kt-25, Snow Heart

*Chatterjee and Swarup (1983), Sureja (2014), Saha (2015), Singh *et al.* (2018)

In Belgium, the cultivars **Asterix, Aviso, Cadet, Fremont** and **Nautilusare** are suitable for early sowing from 1st to 3rd March (Vanparys, 1995b). The cultivars **Aviso, Fargo, Forrest, Marine** and **Nautilus** gave the best results when sowing was done on the third week of June for early autumn cultivation in Belgium (Vanparys, 1995c). Vanaerde *et al.* (1997), however, considered that in Belgium, for early autumn culture cv. RS915056 was the earliest and the best which was followed by Nautilus, Fremont, Aviso, Cadet and Marine.

In Belgium, cvs. **Trevi** and **Kosmos** had the best results when sown on 1st July and harvested between 13 October and 30 November (Rooster, 1998).

In USA, green cauliflower is a relatively new vegetable crop and experiments were initiated to investigate the yield potential of green cauliflower cv. **Alverda** during 1992 to 1993 and it was found that marketable yield was highest in the January planting (Csizinszky, 1996). In Belgium, three green cauliflower cultivars (Alverda, Compoverde and Kosmos) were grown in the open or under various coverings and it was noticed that in the open Compoverde was the earliest, followed by Alverda; under plastic fleece there was no difference in earliness, but Compoverde gave better quality and was hardier than Alverda. Kosmos was later than other cultivars (Benoit and Ceustermans, 1997).

6.3.5 Hybrids

In India, the heterozygous tropical cultivars distributed mainly by seed industry are basically poor in quality and yield. But the Japanese seed industry has introduced a number of F_1 hybrids in this group, improved both in quality and yield. These F_1 hybrids have been developed utilizing the strong self-incompatibility system generally available in this group. Some of these like, Snow Queen, Snow King, Snow Crown, White Contessa, Meigetsu, *etc.*, have given good performance in India. But none of these are available in the country in commercial quantities. Intensive work at the Indian Agricultural Research Institute has resulted in the isolation of large number of strong self-incompatible alleles and attempts are being made to develop F_1 hybrids in all the maturity groups. Some of them have been found to be highly promising. One F_1 hybrid from IARI has been reported to be resistant to multiple diseases like black rot, Alternaria disease and downy mildew, besides higher yield. This F_1 hybrid has not been tested in different regions. The F_1 hybrid for the early season would require intensive testing for their quality and yield before anything can be recommended.

Pusa Kartik Sankar (F_1 hybrid) is a hybrid belongs to early maturity (September) group suitable for Sub-humid Sutlej Ganga alluvial plains and Bengal Assam basin. The curd is compact, retentive white, medium in size, weighing about 475 g. It is resistant to downy mildew. The varity become matures in 95 days. Average yield is 149 q/ha. Pusa Snowball Cauliflower Hybrid 1 is the first indigenously developed CMS based hybrid. The curds are very compact, snow white in colour with self blanching habit. It is tolerant to black rot disease under field condition. Agverage curd weight is 1.5 kg with a yield of 50-55 t/ha.

An Australian company has released a few F_1 hybrids of Europe in the 3-5 months maturity range. Some of these F_1 hybrids are named as Flanca, Sebal, Gigo, Nedcha, *etc.* Most of these F_1 hybrids are the result of complex breeding programme involving Australian, European and Japanese materials.

Himani and Swathi, hybrid cultivars of Indo-American Hybrid Seeds are good tropical type cauliflower, and widely grown in tropical parts of India. Other private sector released varieties are Kimaya, Lucky, Monsoon Queen, Snow Fresh, Snow Heart, Snow Pearl, Suhasini, Tetris by Syngenta, Nuzi Snow White and Riya by Nuziveedu, Everest, Brahma, Damini, Mahi, Snow Peak by Hyveg, Aishwarya, Kanchan, Apsara, Kamini, Hot Shot by Doctors seeds, Anandi by Sungrow.

Important Cultivars of Cauliflower

Pusa Meghna

Pusa Kartik Sankar

Pusa Paushja

Pusa Deepali

Pusa Snowball K-1

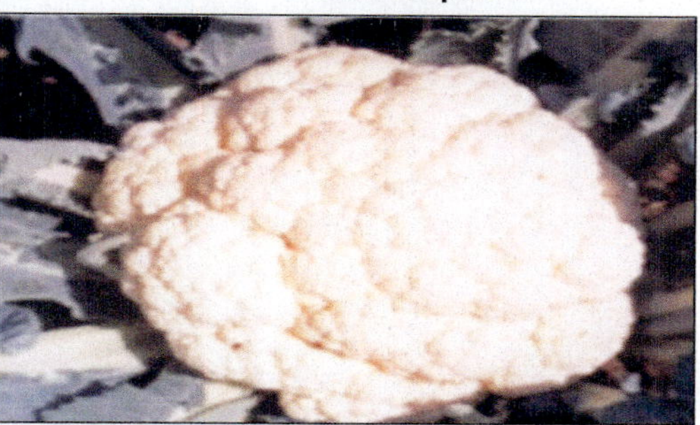

Pusa Sharad

Pusa Shukti

Pusa Snowball K-25

Important Cultivars and Public Sector Hybrids of Cauliflower

KTH-301 **Pusa Snowball-53**

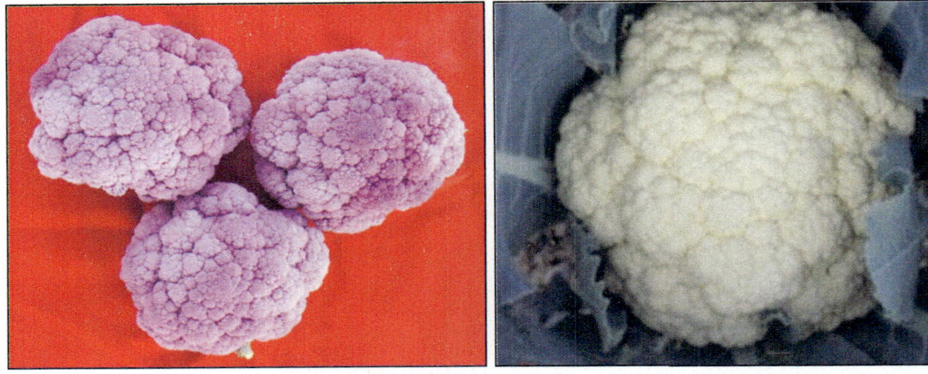

Pusa Purple Cauliflower-1 **Pusa Hybrid-2**

Pusa Hybrid-3 **Pusa Snowball Hybrid-2**

Private Sector Hybrids of Cauliflower

Himalini Pallavi

Himkiran Dawn-175

White Magic Sheetal

In China, Huang Cong Li *et al.* (1994) developed 2 elite F_1 hybrids of cauliflower from four highly stable self-incompatible lines. Hybrid 1 gave a mean yield of 45-52.5 t/ha and Hybrid 4 a mean yield of 52.5-60 t/ha. Both have good resistance to insect pests and black rot.

Recently, released/identified cultivars and F_1 hybrids from ICAR-Indian Agricultural Research Institite, New Delhi are given below.

Mid Maturity Group

Pusa Hybrid-301

The hybrid KTH-301 has been identified for commercial cultivation by XXXVII Group Meeting of AICRP (VC) held at TNAU, Coimbatore during 22-25th June, 2019. This is the first CMS based F_1 hybrid developed in the mid-late group of Indian cauliflower developed in Indian public sector. It was recommended for cultivation in the Zone I [Humid Western Himalayan Region *i.e.* Jammu and Kashmir (J and K), Himachal Pradesh and Uttarakhand] and Zone-VI Arid Western Plain: Rajasthan, Gujarat, Haryana and Delhi]. This hybrid is suitable for cultivation in the mid-season with harvesting of curd in the month of November-December. It has very compact white curd with average curd weight of 1.04 kg.

Pusa Hybrid-3

It is characterized by medium size open green foliage and compact white curd and suitable for sowing during first fortnight of September and transplanting during first fortnight of October month under north Indian plain conditions. It is suitable for sowing in July to August and harvesting in October to December months in mid-hill region. Optimum temperature for curd initiation is 12-16°C. Marketable curd weight is 1150-1250 g and marketable curd yield is 370-390 q/ha. This hybrid shows field resistance to downy mildew and moderate resistance to black rot diseases.

Snowball/Late Maturity Group

Pusa Snowball Hybrid-2 (KTH-DH-1)

It is suitable for growing during late season with harvesting from January end-March. It is the first doubled haploid (DH) based F_1 hybrid in any vegetable crop in India. It has unique advantage of extreme uniformity as one of the parents was completely homozygous DH line developed through isolated microspore culture. It has snow-white curd and retention capacity of curd colour. The curds are very compact with average weight of 1.41 kg.

Pusa Snowball K-53 (KT-DH-53)

It is an doubled haploid based open pollinated cultivar suitable for growing in late season with medium sized frame and suitable for dense planting (45 x 30 cm^2). It has been developed through isolated microspore culture of a male fertile hybrid, Lawyana Old x 13-85. It has the unique advantage of extreme uniformity. This line has distinct white compact curd of 0.72 kg and self-blanching habit. It could be a substitute to the F_1 hybrids to the small and marginal farmers who cannot afford the exorbitantly expensive F_1 hybrid seeds and in rural areas where improved seeds are not available.

Pusa Purple Cauliflower-1 (KTPCF-1)

First true purple cauliflower cultivar developed by Indian public sector. Curd is intense purple and the pigmentation is present deep inside the bracts below the curd. The average anthocyanin concentration 43.7 mg/100 g. Average marketable curd weight is 0.76 kg. It is suitable for cultivation during late season. It can play an important role in bio-fortification of vegetable brassica crops.

6.4.0 Soil and Climate

6.4.1 Soil

Cauliflower can be grown in all types of soil with good fertility and good water regime. However, for early crops in tropical and subtropial regions, a light to light-medium soil may be preferred so that the drainage is easier in the rainy season. The results may be disappointing at times if the growth is checked due to water stagnation or for any other reason, especially in the early crops. Therefore, in the cultivation of cauliflower, even steady growth is important since growth may retard the curd formation and it may also show growth of bracts. Under Indian conditions, especially in North India, the mid season and late crop will grow very well in medium, medium-heavy and heavy soils. However, it would still be needed to maintain an even growth. The optimum pH for cauliflower is like cabbage, *i.e.*, 6.0-6.5. The limit of salt tolerance is rather lower than cabbage and most of the cultivars lie between C_2 and C_4.

6.4.2 Climate

In principle, cauliflower now can be grown in a wide range of climate but during the transformation from vegetative to curding and curd development stages; the climatic factors play a very important role. The temperature in particular, plays an important part though some deviations may also be attributed to light conditions and also to water regime. The information on the influence of day length is however, lacking, but at higher latitude it has been noted that the longer day length promotes much faster growth of the plant.

Since temperature plays the most important role, it may be worthwhile to review its effect on vegetative, curding and generative phases. Cauliflower is one plant in the whole cole group which is physiologically different from others since an intermediate stage of curding between the vegetative and generative phase is absent in others where the transformation from vegetative to generative phase is direct. The curding precedes the floral initials in cauliflower and the plant may transform to curding from 5°C to nearly 28-30°C, depending on the cultivar. However, for this transformation a certain amount of vegetative growth would be necessary to pass over the juvenile phase of the plant. Sidki (1962, 1967) has reported that in case of cultivars like Snowball, four weeks growth and in case of cultivar maturing in February to early March (a somewhat winter type), 5-6 weeks growth would be sufficient for transformation from vegetative to curding phase.

For vegetative growth, it has been established that the young seedlings cease growth at temperature slightly above 0°C in case of early cultivars in temperate region. The optimum temperature for the growth of young plant is at first around 23°C and later drops to 17-20°C. The tropical cultivars show growth even at 32°C. However, at high temperature (than required for curding in specific cultivars), the plant remains vegetative and continues to form new foliage. Even when the plants become generative more leaves are usually formed at high than at low temperature. Thus it has been established that between 15° and 20°C the growth would be optimum for most of the temperate cultivars. However, this temperature would transform the tropical cultivars from vegetative to the curding stage. When these tropical cultivars are grown in lower temperature, not only that they button quickly after the passing of the juvenile phase but most of them show severe riceyness to typical green bud formation directly (Wiebe, 1975).

Chatterjee and Swarup (1983) have shown the optimum temperature requirement for curding for the recommended cultivars grown in India. Besides, they have classified the growth of cauliflower in three different stages, like vegetative phase, curd initiation phase and curd development phase. In the vegetative phase, the plants develop leaves and hence both water regime and fertility should be optimum along with the temperature, which should be a few degrees higher than the optimum needed for curd initiation. During the curd initiation phase, which may cover 10-15 days according to the cultivar and temperature, it is important that the temperature should not fluctuate too much so that the curd quality deteriorates. Similarly, during the curd development phase, *i.e.*, from initiation of curd till its harvest, wide fluctuation of the temperature would be detrimental for quality curd production. Wurr and Fellows (2000) reported that under UK conditions, the optimum temperature for curd induction of all maturity groups of cauliflower varied between 9°C and 14°C and the early summer crops took shorter period from planting to curd initiation whereas the winter cauliflower took the longest period. Ajithkumar and Savani (2007) reported that in Anand, Gujarat, for the completion of crop growth of cauliflower cultivar Snowball 16, the average values of accumulated growing degree days, heliothermal units and photothermal units were found to be 2173.5°C days, 20555°C day hours and 23308.4°C day hours, respectively. Rahman *et al.* (2007) observed that the cauliflower curds increased in size and shape with increase in mean growing temperatures after curd initiation up to an optimum and declined thereafter. The optimum temperature for curd growth was 21-22°C which was higher than that required for growth of vegetative components (19°C) except for stem length. They reported that the growth and development of cauliflower after curd initiation could be resolved into linear or curvilinear function of effective temperatures calculated with optimum temperature of 19-23°C. They also suggested that future warmer climates will be beneficial for winter cauliflower production rather than summer production. Suseela and Rangaswami (2011) reported that cauliflower gave better yield when maximum and average temperatures during vegetative and curd initiation stages inside the greenhouse were less than 30.5°C and 28.5°C, respectively during winter season and less than 33°C and 30°C during summer season. When maximum and average temperatures exceeded 33.5°C and 30.5°C, respectively, crop showed abnormal vegetative growth and resulted in low yield. Lin *et al.* (2019) studied curd initiation and transformation in tropical cauliflower cultivars under different temperature treatments in China. Two commercial cultivars, H-37 (early maturity) and H-80 (mid–late maturity), were used as the testing materials. 'H-37' reached the curd-initiation phase earlier than 'H-80' and showed superior growth during the curd's initial development phase under all temperature treatments. 'H-37' at a temperature of 18°C demonstrated the optimal transformation of apex development from the vegetative to reproductive stage. A temperature of 24°C promoted the apex development of 'H-37' at the curd initial development phase. A temperature 30°C accelerated the ending of juvenile stage and forward to curd initiation in 'H-80' and declined with temperature decreased. The results of this study also indicate that the number of leaves required to induce curd initiation is less than nine in tropical cauliflower at temperatures of 18 to 30°C. In conclusion, under nonvernalized high temperatures, different cultivars of tropical cauliflower can initiate curd development but with a different pattern from those cultivars grown in temperate zones.

The most well-known disorder of the cauliflower curd is riceyness, in which it acquires a somewhat velvety appearance owing to the development of small white flower buds at the curding stage. Also, the pedicels may show longitudinal growth. Although the influence of environment on the incidence of riceyness is not fully clear but more often its occurrence is related to fluctuation in temperature both higher and lower than the optimum required for curding of a particular cultivar. In addition, heavy nitrogenous dressing, rampant growth, high relative humidity, *etc.*, have also been reported to induce riceyness, though there are differences in the susceptibility among the cultivars.

6.5.0 Cultivation

The cultivation practices of cauliflower vary widely in the temperate and in the tropical conditions. Since they are more sensitive to temperature, the methods adopted are dependent on the local climate.

In Western Europe, all the different cultural methods are applied on a large scale. Winter cauliflowers are grown abundantly around the Mediterranean and in the coastal districts of France and southern England. Harvesting is done in winter and early spring. In Northern England and Southern parts of the Netherlands, half grown plants are overwintered in the field and harvesting is done from March to June. Further north winter cauliflower cannot be grown for obvious reasons of snow damage. However, in these areas early summer cultivars are grown both in glass and in the open. When grown under glass, the earliest crops are sown in September – October and the harvest coincides with the late winter crop. For the open cultivation, sowing is done generally in January-February under glass, planting out is done after winter and harvesting from May onwards. The summer and autumn crops are sown from February onwards under green house and from April it is sown in the open condition. In these northern regions, normally the harvest is over by September-October, according to location.

The conditions are different in the USA, especially in Salinas Valley of California. This valley is protected by nature and the climate is very suitable for the cultivation of only one type, *i.e.*, Snowball strains are grown almost nine months of the year, while for the production of cauliflower during February to April comparatively hardier cultivars are grown.

In the tropical region like India, cultivars have been developed for growing during different periods which are highly tolerant to rains and capable of producing curds at high temperature. Earliest of these cultivars are sown during the hottest period in North India and in rainy season in other areas. In Southern part of India, it is grown successfully. The early and mid season cultivars produce curds starting from August and continue till January while the long-leafed Erfurts commonly known as Snowball strains and also Alpha strains (adapted in Indian conditions) are harvested till March. Some of the lines perform very well during summer season which are sown during March and harvested during May-June which fetches premium prices in the market.

Earlier in India, cauliflower is used to be grown only during winter season because only European types were available during those days. With the development of tropical or Indian cauliflower it is now become possible to cauliflower throughout the year under North Indian condition. IARI, New Delhi has played pivotal role in developing specific cultivars for different temperature regimes, which can be grown at a particular time of the year. Based on temperature requirement for growth and curd development, cauliflower cultivars have been developed and by taking these cultivars, now it has become possible to grow cauliflower for almost all the year round. The following table describes about the

Maturity Group	*Sowing Time*	*Transplanting Time*	*Availability Period*
Early I(A)	Mid of May	Beginning of July	Mid September to Mid of October
Early I (B)	End of May to Mid of June	Mid of July	Mid October to Mid of November
Mid Early	End of July	Beginning of September	Mid November to Mid of December
Mid Late III	End of August	End of September	Mid December to Mid of January
Late	End of September to Mid of October	End of October to Mid of November	End of January to March

6.5.1 Raising of Seedlings and Transplanting

6.5.1.1 Raising of Seedlings

Cauliflower is grown from seed and the nursery bed preparation and raising of seedlings under Indian conditions depend mainly on the climatological factors. Chatterjee and Sharma (1970a) and Saha (2015) have reported the climatological conditions in North India for sowing of different cultivars covering the entire period. According to them, the conditions for early crop are not congenial since the period had high temperature, low humidity with dry hot winds. For such situation, nursery bed should be prepared in shade and seedling should be protected from high temperature by making Sarkanda roof. They have suggested that the beds should be made cool by maintaining moist beds and by covering them and sprinkling water during hotter period of the day. The nursery bed should be narrow (30-50 cm) and irrigation channel should be widened (75- 100 cm) for easy percolation of water. The period for midseason crop is, however, better from the point of temperature, but because of high rainfall and cloudy days, the beds require full protection from over-moist conditions of the seed beds and also to provide as much light as possible. In such situation, poly tunnel may be ideal for raising the plants. The nursery beds shoud be raised (15 cm height) and sloppy for easy drainage of water. Seedlings should be protected damping off disease because this disease is more problematic due to high moisture content. In case of mid-late or late crop, the period of sowing is rather more congenial and even may be ideal in North India, though it may still require some protection in other areas like Eastern India where monsoon continues till October. The seedling raising in any area in India and in other tropical countries will have to be followed in any of the above three methods.

The seed should be sown in well-drained seed beds of a deep and thoroughly cultivated fertile soil. About three weeks before sowing, the seed bed should be drenched with fungicides or may be sterilized with formalin, specially when sown during the rains. If sterilized with formalin, the bed should be aerated by turning over the soil with a clean fork or 'khurpi'. Care should be taken not to introduce any unsterilized soil and vegetable matter into the treated beds. The nursery soil may be solarised or formulation of *Trichoderma viride* may be mixed in cow dung manure @ 100 g per 5 kg before sowing of seed. Manuring should be restricted to FYM/cowdung manure @ 6-8 tonnes/ha before sterilization or before the drenching with fungicide. The surface of the bed is smoothed and firmed and the seed is sown thinly (about 15-20 seeds per 30 cm) in shallow drills, not more than 1.5 to 2 cm deep and about 7 cm apart. Seeds should be treated with captan or thiram @ 2.5-3 g/kg seeds. The seeds are sown in line at 7 cm apart in 1.5-20 cm deep. The seed may be covered after sowing by sieving some soil over the seed, making sure that the seed is not covered by more than 0.5 to 1 cm of soil. Mulching material should also be used to cover the seed. In areas where the weather is very dry during sowing, as in North India for early crops, the beds should be watered properly before sowing and mulching must be done with well-rotted manure, saw dust, dry grass or polythene to retain the moisture in the soil for quick and even germination of seeds. However, mulching of dry grass or polythene should be removed as soon as the seeds show sprouting. In case of mid season crop, the beds should be drenched with captan @ 2.5-3 g/liter of water or chlorothalonil (2 g) + carbendazim (1 g)/liter of water to avaoid damping off disease. Under normal conditions in tropical regions, the seeds will germinate in 3-4 days and rapid germination avoids the chances of attack by soil organisms.

For cultivation of cauliflower in one hectare, 200-500 g seed would be needed, though in the early crops, the requirement may be more, to the tune of 1 kg due to high mortality of seedlings. The seedlings should be watered frequently both before germination and when the plants are young so that the growth is not checked. When they are large enough to handle, the seedling may be thinned where the stand is dense, so that they will develop into sturdy plants. In North India, shedding will be needed during hotter part of the day, especially for the early crops while cover will be needed also for the seed bed in the rains. As pointed out earlier, it may be worth while to raise hardy seedlings in poly tunnel during the monsoon rains. In normal cases, water may be withheld 3-4 days before transplanting. However, before transplanting the beds are thoroughly soaked to facilitate the removal of seedlings from the seed bed without much injury to the roots.

6.5.1.2 Land Preparation

Soil should be prepared well and brought to a fine tilth before transplanting. While preparing, the manure and fertilizer should be applied as basal dose. Since drainage is a problem for early and early-mid season crops in North India and even for mid season crops in longer monsoon areas, beds should be prepared after the last cultivation and following fertilizer applications. It should be prepared in such a way that excessive water drains out rapidly but does not cause erosion. The size of the beds and the number of rows to be planted between the drainage furrows will vary according to the natural drainage of the land and the presence of fixed irrigation lines. The size of beds may vary from large one carrying 3-4 rows to single or double row beds of which the latter seems to be more workable.

6.5.1.3 Transplanting

Normally, a hardened seedling is likely to give better results and since the early crops are raised in the monsoon, 5-6 weeks old seedlings stand better chance of establishment.

In Terai region of West Bengal an experiment was undertaken by Chatterjee *et al.* (2012) to find out the most suitable time for planting of mid late season cauliflower. Highest curding percentage (75.67 per cent) was observed from 2nd planting date (30th November) whereas the highest net curd weight (620.67 g) was produced by the 15th November planting considering the yield and quality parameters it may be concluded that first planting date (15th November) was most suitable for cultivation of the cauliflower varieties *viz.* Girija, White Excell, Remi, Pushpa and White Marvel except Remi.

In Bangladesh, Ara *et al.* (2019) observed highest curd yield in 1st August planting. Rahman *et al.* (2016) found the highest plant height, number of leaves per plant, leaf length, leaf breadth at harvest, curd diameter, marketable yield per hectare were recorded when planting was done on 15th November.

6.5.2 Spacing

In general closure spacing are given for early and mid season crop and wider spacing for late season crop. The seedlings are planted at a specing of 45 x 45 cm for early crop and 60 x 45 cm for late crop. Chatterjee and Sharma (1970a) recommended 36-40 cm for early and 45 cm for other crops between the plants, should be followed in 60 cm beds/ridges or rows. Trials at Pantnagar under AICRP showed that planting of Pant Shubhra at 60 x 80 cm spacing with application of 150 kg N/ha was recommended. In Bhubaneswar and Kalyanpur, trial showed that a spacing of 40 x 60 cm was the best for Snowball.

Raising of Cauliflower Seedlings

Solarization of Seedbed

Preparation of Seedbed

Structure (Left) and Sowing of Seeds (Right) for Raising Seedlings of Early Cauliflower

Production of Cauliflower Seedlings in Nursery Bed

6.5.3 Irrigation

The seedlings just after transplanting should be irrigated. The early and mid-season crops require only protective irrigation depending upon the monsoon showers. However, moisture during both growth and curding phase should be adequate to maintainan even growth and proper development of curd. In Sikkim condition, Patle *et al.* (2019) found highest yield of about 23 ton/ha was under treatment with full volume of water through drip and mulch, whereas, the highest water use efficiency of about 248 kg/ha-mm was observed. Application of mulch showed significant increase in yield under furrow irrigation system. Use of mulching and gravity-fed drip irrigation system was technically viable option for the farmers of Sikkim, which can give them higher yields and returns under water stress periods as compared to the present practice of keeping their fields fallow.

6.5.4 Manuring and Fertilization

Since the early crops are grown under unfavourable conditions of rainy months with high temperatures and often fail, more attention should be paid for standardization of fertilizer requirement to these crops. Well rotten cow dung manure (25 t/ha) should be applied during land preparation. A fertilizer dose of NPK (120:80:40 kg/ha) should be applied before planting in which half dose of nitrogen through urea along with the full dose of phosphorous through single super phosphate and potassium through murate of potash. The remaining half dose of nitrogen should be applied in two split dose at 15 and 30 days of transplanting.

Chatterjee *et al.* (2012) reported that the application of inorganic fertilizers with organic manure and bio-fertilizers has significant effect on curd initiation and curd maturity in West Bengal.

In Uttar Pradesh a study conducted by Kumar *et al.* (2018) found that the treatment T_5 (4 per cent *Panchagavya* + 1:5 times dilution of Vermiwash) came out as the best combination for growth and yield characters of cauliflower for commercial production because organically produced vegetables fetch higher price in the market therefore, even if the yield is slightly low with organic inputs, it can be recommended for the commercial production of cauliflower on account of human health, soil health, environmental health and quality product instead of quantity.

Pawar and Barkule (2017) conducted an experiment to study the the effect of integrated nutrient management on growth and yield components in cv. Pusa snowball- 16. The growth parameters like height of plant, number of leaves per plant and diameter of stem at 40 and 60 DAT, respectively, the days required for curd initiation, days required for curd maturity and staying period of curd were best in treatment combination of 75 per cent RDF (120 N:80 P:40 K kg/ha) + FYM (10 t/ha) + *Azotobacter* (10 kg/ha) + *Azospirillum* (10 kg/ha). The yield parameters like weight of curd per plot and yield per hectare (180.04 q/ha) which was increased 44.10 per cent over RDF (Table 8).

Table 8: Effect of different INM Treatments on Curd Yield Parameters of Cauliflower*

Tr. No.	*Treatments*	*Weight of Curd (g)*	*Yield per Plot (kg)*	*Yield per Hectare (q)*	*Per cent Increase/ Decrease in Yield Over RDF*
T_1	100 per cent RDF (120:80:40 kg/ha.)	688.55	12.14	124.94	—
T_2	100 per cent RDF+ FYM + *Ab.* + *Ap.*	908.60	17.10	175.92	40.80
T_3	100 per cent RDF + *Ab.* + *Ap.*	888.60	16.54	162.13	29.77
T_4	100 per cent RDF + FYM + *Ab.*	904.15	15.50	159.48	27.65
T_5	100 per cent RDF+ FYM + *Ap.*	637.55	15.82	160.80	28.70
T_6	75 per cent RDF + FYM + *Ab.* + *Ap.*	933.30	17.50	180.04	44.10
T_7	75 per cent RDF + *Ab.* + *Ap.*	800.90	13.51	138.99	11.25
T_8	75 per cent RDF+ FYM + *Ab.*	877.65	16.65	171.32	37.12
T_9	75 per cent RDF+ FYM + *Ap.*	879.75	14.15	145.52	16.47
T_{10}	50 per cent RDF+ FYM + *Ab.* + *Ap.*	642.00	12.08	124.23	-0.57
T_{11}	50 per cent RDF+ FYM + *Ab.*	480.70	9.87	101.51	-15.55
T_{12}	50 per cent RDF + FYM + *Ap.*	569.25	10.44	107.43	-14.01
T_{13}	50 per cent RDF + *Ab.* + *Ap.*	528.60	9.57	98.41	-21.23
T_{14}	Control.	372.70	6.22	64.07	-48.72
	S.E. ±	58.70	1.37	5.31	—
	C.D. at 5 per cent	176.30	4.11	15.93	—

*Pawar and Barkule (2017)

6.5.5 Weed Control

Crop should be kept free from weeds either manually or using herbicides. Presence of weeds in the early stages reduces the yield due to poor growth of the plants. As soon as the weeds start appearing, shallow hoeing followed by earthing up should be done. Pre-planting application of 2- 2.5 litre Basalin (alachlor) or 3.3 litre Stomp (pendimethalin) per hectare controls dicot weeds for about 40 days. This crop has shallow root system, hence deep hoeing should be avoided. Application of plastic mulch helps in reducing weed population. Timely hoeings help to check the weed population.

At the Indian Agricultural Research Institute, New Delhi, it was found that basalin @ 2 kg a.i./ha before transplanting followed by1or 2 hand weeding gave better results for early crop.

6.5.6 Mulching

Application of plastic mulch helps in reducing weed population in cauliflower. In Hisar, Haryana condition black plastic mulch (50 micron) increased 31.9 per cent yield (Bhardwaj, 2016). The use of rice straw mulch doses of 5-10 t/ha can help maintaining water available for plants. The shelf life of curd was significantly more at mulched conditions under Assam condition (Savita *et al.*, 2017).

6.5.7 Interculture

After transplanting, the crop should be kept free from weeds by cultivation. If a surface crust forms on the soil, it must be broken up by cultivation, otherwise moisture penetration may be hindered. Cauliflower is a shallow rooted crop and hence it should never be cultivated to a depth greater than 5-6 cm close to the plant or serious damage will be done to the root system. During cultivation it would be necessary to incorporate side dressings of fertilizer into the soil and the plants at this time should be earthed up to provide stability to the plants. During monsoon, the roots may be exposed after every shower, especially in ridge planting and in such cases earthing up more than the usual number may also be necessary.

Cauliflower curds are exposed to the sun and they tend to discolour and lose the desired whiteness. Some cultivars have good wrapper leaves around the curd and are naturally well protected or self blanched type. Blanching is an essential operation to protect the curd from sunlight and yellowing, so that they may not look attractive in appearances. It is done by drawing and tieing of the upper leaves. If it is not possible, a cauliflower leaf is placed over the curd to protect it from the sun. This help in retaining milky white colour to the curd. Blanching should be done only when the curd is grown fully, and leaves should not be left tied over for more than 4-5 days. But these periods may be increased to one week in cold weather and reduces to 2-3 days during the hot weather.

6.5.8 Growth Substances

Among the various plant growth regulators, NAA and GA_3 are very popular and are used on a commercial scale in cauliflower. In an experiment conducted by Sahau *et al.* (2019) under Chattishgarh condition where they found that NAA @ 120 ppm was found to be highly effective for yield attributing characters as it recorded highest curd diameter and curd weight. Kaur and Mal (2018) found that GA_3@ 50 ppm gave maximum plant height (70.83 cm), minimum number of days taken to 50 per cent curd initiation, minimum number of days taken to 50 per cent marketable curd size and also increase yield and yield attributing characters such as curd diameter, individual curd weight, yield per plot and yield per hectare.

6.6.0 Disorders

6.6.1 Buttoning

The buttoning means formation of small button like curds. The causes for buttoning of cauliflower have been variously explained by many workers, *e.g.*, over-aged seedlings, poor nitrogen supply, wrong cultivars, *etc.* While many of these are partly correct, the general basis of buttoning may be explained that any check in the vegetative growth of the seedlings may induce buttoning. As has been explained earlier that the transformation from vegetative to curding in a cultivar of cauliflower is dependent on particular temperature, therefore, the check of vegetative growth followed by suitable temperature for transformation to curding may induce this malady. The check in growth may be caused by low nitrogen supply, root injury by insects or by some diseases (specially *Rhizoctonia* sp.), when early cultivars are planted late and transform easily due to lower temperature, over-aged

Open Field and Poly House Production of Cauliflower

Cauliflower Fields and Harvested Curd

seedlings after establishment do not get sufficient time to initiate growth before transformation, *etc.* Hot and dry weather is unfavourable for vegetative growth but favourable for inducing curd formation and inhibits further enlargement and produces button size curds. It may occur due to inadequate supply of N_2. When soil moisture is a limiting factor it checks the growth of the plant which, in turn, causes early formation of curd. Buttoning also occur when the seedlings are transplanted from poorly managed nursery soil. Buttoning can be overcome by growing specific cultivar suitable for the season. An adequate amount of N_2 and water should be applied. Transplanting should not be delayed. Cultural practices should be done in time and water logging and over-crowding should be avoided.

6.6.2 Riceyness

According to Wiebe (1975), a premature initiation of floral buds is characterized by riceyness in cauliflower and is considered to be of poor quality for marketing. Surface of the curd becomes loose and has velvety appearance due to elongation of pedicel. It is one of the most serious risks in cauliflower production, because riceyness influences on both quality and yield of curd.

According to the observations made at IARI, it has been found that such disorder may result from any temperature higher or lower than the optimum required for a particular cultivar. Apart from fluctuating and unfavourable temperature, heavy application of N and high humidity may cause riceyness. However, there are significant variations among the cultivars in their tolerance to such character. Such observation has been noted by European workers also and in the selection of cultivars, it has been brought out that some cultivars are particularly susceptible for such character.

Wiebe (1975) has also noted that the cauliflower cultivars show more bracteate curds at higher temperature than the optimum.

Ridge *et al.* (2015) reported that some pedicels start bolting accompanied with the differentiation of floral meristems only when the curd develops to the post-maturity stage, and then the plant begins to bloom. Therefore, the essence of riceyness is that the partial or total ability to differentiate floral meristems is restored in advance of the curd development stage, which is closely related to the regulatory network of plant flowering, especially the determinacy mechanism of floral meristems (Zhao *et al.*, 2020) (Figure 7) and they reported the riceyness as genetically controlled. Selection of proper cultivars for a particular time of cultivation, optimum application of nitrogenous fertilizer and planting of resistant and tolerant cultivars can reduce the problem of riceyness.

Blindness

Blind cauliflower plants are those lacking in terminal bud. The leaves are large, dark green, thick and leathery. The affected plants do not produce any curd and should be uprooted. Terminal buds may be damaged due to prevalence of low temperature reaching sub zero when the plants are still small, mechanical injury of the terminal bud, and injury caused by insect such as cutworm. Blindneess can be overcome by taking care full handling of the plants during transplant and good

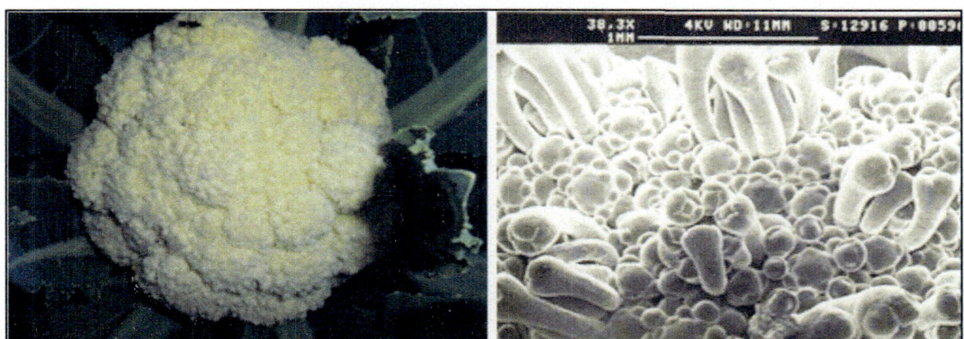

Figure 7: Naked Eye View of Riceyness in Cauliflower Curd (Kalia and Saha, 2014), Electron Microscopic View of Cauliflower Curd (Zhao *et al.*, 2020)

plant protection measure so that the terminal bud may not be injured. The exposure of plants to low temperature should be avoided.

Leafiness

This disorder is commonly seen by formation of small thin leaves from the curds, which reduces the quality of curds. Extremely small leaves appear in between curd segment may be due to heritable or non-heritable factors. Prevalence of high temperature during curding aggravates leafiness. Selection of proper cultivars helps to reduce it.

Pinking

It is the disorder of cauliflower where the curds show the pink tinge due to exposure of the curd to high light intensity. Under this condition, anthocyanin pigment develops which impart pink colour to the curds. This disorder is not so common. Covering the curd with leaf or cultivation of self blached cultivar may reduce the problem.

Leafiness or Bracting

This disorder of cauliflower is indicated by the appearance of green leafy bracts in-between the segment of curd. The main cause of leafiness is prevalence of high temperature than the optimum for a particular cultivar during curd development. Selection of right cultivar for the particular growing season can reduce the problem.

Multiple Curds

This disorder of cauliflower is marked by the appearance of a number of small button like cards (may be 4-5 in number) in the form of bunch. This disorder happens due to injury of terminal bud resulting in branching before curd initiation. Small curds are formed on each branch of the terminal bud. Careful handling of the plants and good plant protection measure so that the terminal bud may not be injured. Exposure of plants to low temperature should be avoided.

Black Speck

This disorder of cauliflower is marked by the development of black lesions with collapse of cells in interior of the curd of snowball (late) cauliflower. This disorder, though not very common, is caused when snowball (late) cultivars exposed to warm weather at curd development stage.

6.6.3 Nutritional Disorders

6.6.3.1 Boron

Browning

In cauliflower, boron deficiency has been reported very frequently. Till the curds start developing, external symptoms of boron deficiency is not very apparent in most cases. The first sign is the appearance of small water soaked areas in the centre of the curd. In later stages and in seriously affected plants, the stem becomes hollow with water soaked tissue surrounding the walls of the cavity. In more advanced stages, pinkish or rusty-brown areas develop on the surface of the curd and hence it is known as brown rot or red rot. Affected curds develop a bitter taste. This may be controlled by applying borax or sodium borate at the rate of 20 kg per hectare (Datta, 1963). On neutral or alkaline soil large quantitity of borax should be applied to make boron available as boron is more available in acidic soil.

In case of acute deficiency, spray of 0.25 to 0.50 per cent solution of borax at the rate of 1 to 2 kg per hectare would give a satisfactory control. In an experiment conducted at Ranchi, India with cv. Pushi revealed that application of 1 kg B/ha was sufficient for mid-season cauliflower, irrespective of the application method (broadcast and band placement in one and two instalments, and foliar spraying were tested) (Prasad *et al.*, 2000). Batabayal *et al.* (2015) reported from an experiment in cauliflower in Inceptisols of the Gangetic plains of India that boron application significantly increased cauliflower yield, plant B concentration and uptake of B. The critical plant B concentrations for deficiency, sufficiency and toxicity varied with the growth stages and the values being 26, 31 and 48 mg kg^{-1} at 50 days of growth and 17, 24 and 35.5 mg kg^{-1} at harvest, respectively. The study also recommends application of fertilizer B at the rate 0.9–4.5 kg ha^{-1} for optimum B nutrition. Ningawale *et al.* (2016) reported application of Borax @ 10 kg/ha were found best for cv Snowball 16 in Madhya pradesh.

Hollow Stem

This disorder is caused due to boron deficiency and or higher supply of nitrogen. Hollowness caused by boron deficiency may be indicated by water-soaked and discolor tissue and if the hollowness is caused due to excess nitrogen, the stem is perfectly clear white with no sign of disintegration. Hollow stem due to boron deficiency may be controlled by spraying of 0.25 to 0.50 per cent solution of borax. Hollow stem due to excessive growth can be controlled by reduncing nitrogen dose.

Disorders of Cauliflower

Boron Deficiency Causes Pinking Followed by Browning of Curd and Hollow Stem

Whiptail (Left) and Riceyness (Right) in Cauliflower

Blindness (Left) and Leafyness (Right) in Cauliflower

Buttoning (Right) in Cauliflower

Progressive Symptoms of N (Upper), P (Middle) and K (Lower) Deficiency in the Leaves of Cauliflower

Old (A), Intermediate (B), and Young (C) Leaves 15 Days after being Supplied with a Nutrient Solution without N.

Asymptomatic (A), Beginning of Yellowing (B), and Advanced Yellowing (C)

Chlorosis and Initial Necrosis of the Nervures (A), Advanced Stage of Necrosis of the Nervures (B), and Advanced Foliar Necrosis (C).

***Source*: Bianco *et al.* (2015)**

Progressive Symptoms of Ca and Mg Deficiency in the Leaves of Cauliflower

Ca Deficiency in the Leaves Culminating in Tip Burn, Supplied with a Nutrient Solution without Ca.

Tip Burn in Cauliflower

Mg Deficiency in the Leaves Supplied with a Butrient Solution without Ca

Source: Bianco *et al.* (2015)

6.6.3.2 Molybdenum

Whiptail

Cauliflower responds severely to the deficiency of molybdenum and the damage may be considerable. Young cauliflower plants in a shortage of this Element become chlorotic and may turn white, particularly along the leaf margins, they also become cupped and wither. Eventually, the leaf dies and the growing point also collapses. In older plants, the lamina of the newly formed leaves is irregular in shape, frequently consisting of only a large bare midrib and hence the common name 'whiptail' originated. According to Agarwal (1950), the nitrogen deficiency not only results in buttoning but it also develops the deficiency symptoms of molybdenum. The whiptail develops with high nitrate supply and low molybdenum (Agarwal and Hewitt, 1954). Since cauliflower has high nitrogen requirement, it may be useful to ensure an adequate supply of molybdenum to avoid whiptail. Deficiency of Mo causes whiptail syndrome especially in high acidic soil. High Manganese concentration in such soils hinders the uptake of Molybdenum which seldom occurs when pH is 5.5 or higher. Molybdenum should be applied in soil @ 1-2 kg/ha or spraying Na molybdate @ 3 g/m^2. Ningawale *et al.* (2016) reported application of Ammonium molybdate @ 2 kg/ha as soil application for the Snowball 16 cultivar.

6.7.0 Harvesting and Yield

6.7.1 Harvesting

When harvesting cauliflower, care should be taken to cut the curd immediately it reaches prime condition. If there is any doubt about the maturity, it is better to harvest early than late. If the harvests are late, flower stalk elongates and loose, leafy, ricey or fuzzy curds result. Such over-mature curds wilt rapidly after cutting. It is best to discard any such over-matured curds as they will spoil the appearance of the consignment. When harvesting, the curds are cut off the stalk with a large, sharp knife.

The large leaves are then trimmed away until only sufficient jacket leaves remain to protect the curd. When transporting loose, more jacket leaves should remain than when packing in crates. In crate, jacket leaves are trimmed leaving a fringe of leaves projecting 2-3 cm above the curds. When transported loose, the jacket leaves are only lightly trimmed until they reach the market.

6.7.2 Yield

In the early group of cauliflower, the yield varies depending upon the season of harvest and the cultivar. When the cauliflower curds mature at a higher temperature above 25°C, the curds are small, loose and yellow. In spite of closer spacing which is followed by the growers, the yield hardly exceeds 10 tonnes/ha. While in comparatively uniform cultivars, slightly later maturing curds between 20 and 25°C, the yield may vary between 12 and 15 tonnes/ha. With cooler season, the mid season cultivars can yield upto 20 tonnes/ha. The highest yield is, however, obtained from Snowball group because of its more compact curds and larger plant population per unit area, as comapared to mid season group and the yield may go upto 25 to 30 tonnes/ha. Incidentally, it may be pointed out that Snowball group is the only cauliflower which is grown almost under ideal conditions in North India. With the availability of hybrids in different group the yield has increased. In early group the yield may be 15-20 t/ha and in late group yield 50-60 t/ha.

Average yield (tonnes/ha) of cauliflower in different parts of the world are as follows:

India	17.24	Spain	20.29
China	25.95	U.S.A.	15.33
Italy	20.85	World	18.80

6.8.0 Storage

Postharvest losses in cauliflower at room temperature have been reported to be 10, 2.4, 5 and 15 per cent as decay, trimming, mechanical damage and weight loss, respectively (Dhall *et al.*, 2010). There is a need to store cauliflower at recommended storage conditions (0±1°C, 90–95 per cent RH) with appropriate packaging technique so that it can maintain its quality during storage till it reaches to consumer. In a study conducted by Dhall *et al.* (2010) shown that cauliflower curds individual packed in high density polyethylene bags (20 μm) with perforation (6 holes/bag) can be stored up to 21 days at 0±1°C and 90–95 per cent RH with maximum retention of white colour of curd, minimum spoilage, weight and firmness loss and good sensory quality attributes. Alibas and Kdksal (2015) reported tnat vacuum precooling was found to be most suitable method before cauliflower was submitted to cold storage and sent to market. Furthermore, the storage of cauliflower without precooling resulted in a significant decrease in quality parameters.

7.0 Knolkhol

The 'knol-khol' is a cool season vegetable cultivated for its edible knob (swollen portion above the ground). The knobs develop entirely above the ground. It is this portion which is used for vegetable, though young leaves are also used in some parts. It is one of the lesser known vegetables cultivated in limited area of India. It is also referred as Kholrabi, Navalkol and Ganth gobhi in different regions. The knob is harvested for consumption as raw for salad or cooked vegetable, though in some parts, young leaves are also used. The fleshy edible knob is an enlargement of stem, which develops entirely above ground and is used as a vegetable. The edible portion is globular to a slightly flattened stem. Kohlrabi has similar taste and texture as that of broccoli stem or cabbage, but milder and sweeter. The younger stems have crispy, pleasant taste, and rich flavor. Like other cole vegetables, Knol-khol is a good source of vitamin A and C, folic acid and dietary fibre. Moreover, it is an excellent source of vitamin B-complex such as niacin, pyridoxine, thiamin, pantothenic acid, *etc.* Besides that, it contains good levels of minerals *viz.* copper, calcium, potassium, manganese, iron and phosphorus especially in stem. It contains sulforaphane, which has anticancer properties. The cultivation of knolkhol or kohlrabi in India is not very popular except in Kashmir, West Bengal and some parts of the South India. Besides, it is hardly cultivated in large scale as is done in case of cauliflower or cabbage.

7.1.0 Cultivars

In India, only two cultivars are commonly cultivated. The most common cultivar is the White Vienna and the other Purple Vienna. White Vienna has many strains like early, late but the actual variation in these strains are more from the position of the whorls of leaves. Almost all the strains are globular to somewhat flat-round tubers. Purple Vienna is rather purplish blue in colour. All the cultivars available in the country are ready for harvesting within 50-60 days of transplanting. A brief description of the available cultivars is given below:

White Vienna

This is one of the popular cultivars marketed by different agencies. Plants are dwarf with light green leaves and stem. The knobs are globular, smooth and light green. Flesh creamy white, tender with delicate flavour. It takes 50-60 days after transplanting for formation of stem tuber. Average yield is 20-25 t/ha

Early White Vienna

An early strain isolated from White Vienna and it takes 50-55 days after transplanting for the development of knobs. Knobs are globular-round and green with fewer leaves. Flesh is creamy white, tender and crisp. Average yield is 20 t/ha.

Purple Vienna

Plants are medium in size with purple blue leaves and stem. Knobs are globular round, large and purple blue in colour. Flesh is light green. It takes 55-65 days from transplanting to knob formation. Average yield is 20-25 t/ha.

Early Purple Vienna

An early strain of purple Vienna ready for harvesting at 50-60 days. Foliage purple, knobs are globular round, large with purple skin and light green flesh. Average yield is 20-25 t/ha.

King of North

Plant height is 20-30 cm with broad and dark green foliage. Leaf sheath is large and well spread over the knob. Knobs are flattish round and dark green in colour. It takes 60-70 days after transplanting for development of knob.

Large Green

Plants are vigorous with dark green foliage. Knobs are flattish round and dark green in colour. It takes 60-70 days after transplanting for the development of the knob.

Pusa Virat

Developed by ICAR-IARI, Regional Station, Katrain. Plants are dwarf, semi-spreading with 13-14 dark green serrated leaves. It has bigger sized green knobs with 13-14 cm diameter and 12-13 cm length, weight 800 g. This cultivar is known for little or no fible at maturity. Harvesting may be started from 55-60 days after transplanting with average yield of 23.2 t/ha.

Palam Tender Knob

It is an **e**arly cultivar with light green knobs and gives average yield of 25-27 t/ha. It is reported to possess better shelf life as compared to White Vienna.

In Europe, a large number of cultivars are being marketed. The popular among them are Kossak, Rhein, Rowel, Korist, Wiesmoor Forcing White, Kolibri, Eder among others. Among the different European varieties, the cultivars developed by

Important Cultivars of Knol Khol

Early Vienna

Purple Vienna

Early White Vienna

Pusa Viraat

the Swiss breeder Roggli, are more popular. Among the early cultivars, Wiesmoor Forcing, White and Gaugels Forcing, White are resistant to bolting. Under favourable growing conditions, these two cultivars mature within 4 to 6 weeks of planting out and can be grown both in the glass and in the open in Europe. Roggli's Outdoor White is a few days later but is more suitable in less favourable conditions due to its high resistance to bolting and its tubers do not become fibrous soon. These Roggli cultivars are characterized by the horizontal position of the lower leaves. Santa (1985) from Hungary reported that the cultivar Szentesi Nyari Kek (Szentes Summer Blue) produced yields similar to or higher than those of the foreign summer cv. Becsi Kek (Vienna Blue). Cools (1985) assessed the cultivars Foran, Proloog and Express Forcer in glass house at several locations in Netherlands and found that Foran was the earliest, followed by Proloog, and their uniformity, tuber shape and leaf quality were better than those of Express Forcer. Foran was, however, the most susceptible to tuber splitting. In Belgium, cv. Express Forcer, Genta Rolpak, Korist, Proloog, Rhein and Rowel are recommended for fresh market (Vanparys, 1990). Vanparys (1996) further assessed 21 kohlrabi cultivars and found that Express Forcer, Bocal, Quickstar, Korist, Kolibri and Eder gave the highest yields. Another cultivar Kossakhas also been reported to give high yield (Vanparys, 1998). In Europe, there are many late cultivars maturing between 70 and 100 days after planting. These have larger tubers and are suitable for storing. One of the old cultivars in this group is Goliath, in which a white and a blue type occur.

7.2.0 Soil and climate

7.2.1 Soil

Knol khol can be grown on all types of soil, though it makes no high demands upon it. However, good condition and fertility favour growth in an uniform manner. Soil pH of 5.5 is ideal for its growth.

7.2.2 Climate

Though in temperate countries, early bolting is a problem, specially for the early cultivars but this does not seem to be of much significance under Indian conditions with the two cultivars that are generally grown. It is generally observed that the crop grows well for the tuber production at 10-20°C, though the quality is better under lower temperature. If temperature falls below 15°C for longer periods, early maturing cultivars develop flowering stalks without forming the knobs. At higher temperature, the knobs become fibrous and taste bitter. It has been reported, however, that the long oval tubers, which occasionally occur in cultivars with round or flat-round tubers, are probably vernalized tubers, whose development has been arrested. Habegger (1986) summarised that sensitivity to vernalization of cv. Express Forcer began after a short juvenile period. Vernalization of seed was not effective. High post-vernalization temperatures delayed flowering. The response to an interrupted chilling period was less than to a continuous chilling period with an identical amount of chilling. The devernalization effect of daily changing temperatures increased amplitudes of the day/night temperatures. Habegger and Weibe (1986) after extensive study concluded that cold stimulates kohlrabi to bolt.

7.3.0 Cultivation

7.3.1 Land Preparation

The land should be prepared to a fine tilth by disc ploughing followed by one or two harrowing. Well decomposed FYM @ 15-20 t/ha is incorporated at the time of land preparation to maintain sufficient organic matter in the soil. Application of organic manure or vermicompost improves plant growth, productivity and improves water holding capacity of field soil. The sterilization of soil by drenching, nursery beds with formalin @ 1:49, about 15-20 days before seed sowing is beneficial for preventing the attack of the fungal diseases. After drenching, seed beds should be covered with polythene for a week. Then beds are again dug and left open for 5-6 days to avoid injurious effect of formalin on seeds. The standard procedure for raising nursery should be followed. The beds should be covered with a proper mulching material before watering. Apply water with a water can over the grass mulch during initial stage *i.e.* 15-20 days of sowing, while during later stage watering should be done through furrows. The mulch should be removed as soon the emergence of seed sprouts. The beds should be provided with roof for shading against hot sunshine and rains.

7.3.2 Sowing Time, Seed Rate and Spacing

To raise nursery for one hectare 1.0- 1.5 kg seed is sufficient. As in other cole crops, knol khol is usually sown in a seed bed and the seedlings are transplanted. For early crops in the northern hills, sowing is done in March-April according to altitude with some protection at night and the seedlings are transplanted from April onwards. For summer and autumn crops in Indian hills with somewhat drier climate, sowing is done in open and 5-6 weeks old seedlings are transplanted. For the plains in India, sowing may be started in September in North India while for the milder winter regions, October is a better month. The precautionary measures for raising the seedlings should be taken as in other cole crops.

In European conditions, earliest crop in grown under glass, either in hot beds or in glass houses. The sowing is done in autumn if it is grown in glass house with heating arrangement; otherwise sowing is done in January-February and the seedlings are planted out in April for earliest crop in May. In Belgium highest yield of Kohlrabi cv. Kossak was obtained by sowing on 7 May and harvesting of 17 July and recommended for cultivation on light sandy loam soils (Vanparys, 1999).

Seed requirement for a hectare is at least 50 per cent more than cauliflower or cabbage since this crop is grown in much closer spacing. To raise nursery for one hectare 500-700 g seed is required. Generally, the seedlings are ready for transplanting 30-40 days after sowing. For harvest of the tender knobs, row to row spacing is maintained at 30 cm and plant-to-plant spacing at 10-20 cm by most of the farmers depending on the cultivars and season. The seedlings are also planted out at 25 x 25 cm, 25 x 30 cm, 25 x 40 cm or 30 x 45 cm according to the available intercultural facilities in different regions and countries. Closer spacing will decrease size of the knob but will increase the yield per hectare. If planted too close, some of the knobs may be harvested early to provide more space to the rest. Sande and

Seedling Raising, Transplanting and Field Growing of Knolkhol

Seedling Raising of Knolkhol

Transplanting of Knolkhol Seedlings

Field Growing of Knolkhol

Jeurissen (1986) from Netherlands noticed in hot house trials that average tuber diameter and weight at harvest declined as plant density increased. The highest returns were obtained with 22.5-23 plants/m^2. Kahn (1990) from U.S.A. observed that a population of 33 to 36 plants/m^2 of row (725,400 plants/ha) provide a balance between seed costs and yield. This can be achieved using sowing rates of 2.2 to 2.8 kg/ha.

Knol khol requires a steady growth and any check in the growth will induce the knobs to become rather fibrous, while too rapid growth may cause the knobs to crack, specially if the initial growth is slow. Closer planting also may reduce the growth but too close spacing will induce the tuber to be elongated, presumably due to lack of light. High temperature and nitrogen may also favour the production of elongated knobs, besides, check of growth due to vernalization. Research work on this crop is very limited.

7.3.3 Manures and Fertilizer

It is recommended to apply 20 tonnes well rotten FYM during field preparation. Besides, application of 90 kg N_2, 80 kg P_2O_5 and 80 kg K_2O per hectare is most common practice. The half dose of N and full doses of P and K should be applied before transplanting. The remaining half dose of N should be top dressed in two equal splits *viz.* after one month of transplanting and at the time of knob formation. For acid soils low in Ca and Mg, application of 'dolomite' as lime should be followed. Choudhuri and Som (1969) studied the response of knolkhol cv. Early White Vienna to nitrogen (0 to 120 kg/ha) and phosphorus (0 to 60 kg/ha) in sandy loam soil with low available nitrogen and phosphorus. Best yield was obtained with 100.7 kg nitrogen and 60 kg phosphorus/ha. Sande (1986) also reported that cultivar Foran, E295 and Express Forcer performed best when grown with solution culture with continuous circulation and knobs were largest when grown with the solution circulating for 15 minutes every 60 minutes. Knobs were of high quality, and the method of culture was considered promising. In India, Bairwa *et al.* (2017) reported a combined application of 40 kg sulphur ha-1 with 45 x 30 cm spacing best treatment for getting largest knob, whereas 40 kg sulphur/ha-1 with 30 x 30 cm spacing was most suitable for getting highest total yield.

7.3.4 Intercultural Operation

The intercultural operations are almost similar to that of other cole crops. Shallow frequent hoeing should be done in field by 'khurpi or hoe' to kill young weeds and provide soil mulch. Weeding should be started as soon as plants are set in the field. Four to five weeks after transplanting, plants should be slightly earthen up in the field. During inter-cultural operation care should be taken to keep the above grounds intact without causing any any mechanical injury to prevent formation of deformed knobs. For chemical weed control, Stomp 30 EC (Pendimethalin) 2.5 litre ha-1 can also be applied one day before transplanting of seedlings under moist soil conditions. In Belgium, Himme *et al.* (1984) recorded the best selective weed control in kohlrabi with1 kg Trifluralin incorporated presowing by 3 kg Propachlorpre-emergence.

Harvesting and Packaging of Knolkhol

Harvesting of Knolkhol

Packaging and Selling of Knolkhol in the Market

7.3.5 Water Management

Like other cole crops, knol khol also require a continuous supply of moisture for uniform growth and development. First irrigation is done immediately after transplanting and thereafter irrigation is done when needed, depending on soil and weather conditions. Subsequent irrigation can be given at an interval of 15 days. Heavy irrigation should be avoided however, there should be sufficient moisture in soil at the time of knob formation.

7.4.0 Harvesting and Yield

Knobs are harvested before they are fully grown, because old knobs eventually become fibrous. Generally, knobs of 5-8 cm diameter are favoured in the market with the foliage. Individual tuber may weigh 200-250 g while the yield per hectare may vary from 12 to 25 tonnes under Indian conditions. However, the cultivar Pusa Virat produce a larger knob at harsting stage with diameter of 10-12 cm and individual knob weight of 500-700 g and they remain tender without becoming fibrous at this stage.

8.0 Brussels Sprouts

The Brussels sprouts is a crop of rather recent development in the cole group. As the name implies, the crop was initially grown mainly in the area around Brussels in Belgium. It is grown for its green edible 'bud or sprouts' (small heads that resemble miniature cabbages). The stem of the crop makes normal longitudinal growth and, in some cultivars, the growth may reach a length of over one metre. The buds in the axils of the usually petiolate leaves develop into miniature heads, a few centimetres in diameter and it is those buds which are of commercial importance Brussels sprout is a hardy, slow-growing and long-season vegetable. The "sprouts" are produced in the leaf axils, starting at the base of stem and working upward. Brussels sprout are rich source of vitamin A and C, folic acid and dietary fibre. Moreover, it contains sulforaphane, which has anticancer properties. Likewise, other Brassicas, Brussels sprouts is also a rich source of indole-3-carbinol, which boosts DNA repair in cells and appears to block growth of cancer cells. The cool valleys in mountains have best potential for commercial production of Brussels sprouts followed by coastal areas, where long cool falls offer some opportunity. The major Brussels sprouts growing countries are China, Netherlands, Germany, United Kingdoms, United States, Mexico, Canada and India.

8.1.0 Cultivars

Brussels sprouts shows wide variation and the requirement of the growers of different regions are so specific that it has under gone extensive selection by the growers themselves and the resultant cultivars are not only important for a particular region but also adapted to soil and climate of that region itself and may fail when grown elsewhere. It has been reported that cultivars selected for a heavy soil may show rank growth and produce loose sprouts when grown in light soils and similarly cultivars of light soil requirement may remain short-stemmed on a heavy soil.

8.1.1 Dwarf Cultivars

These cultivars have short stem, mostly less than 50 cm in length. The two old cultivars, Herkules and Brussels Dwarf are no longer important now. The cultivar Improved Long Island is most widely grown cultivar. It is an early cropper and very suitable for areas with short growing season. In some areas, it is grown as a second crop after early potatoes. In these cultivars, the sprouts are of medium size and the distance between the sprouts is less making the picking rather difficult. Cultivars like Early Morn, Dwarf Improved, Friiher Zwerg, Kvik, *etc.*, also belong to this early type. Jade Cross, a Japanese F_1 hybrid cultivar has become popular for its short stemmed, early, high-yielding and single-harvest quality.

8.1.2 Tall Cultivars

In England, tall cultivars are preferred possibly for a longer season of growing. These cultivars are, however, classified in accordance with their maturity and the size of sprout. Strains of two open pollinated types, *e.g.*, Evesham and Bedfordshire in many other names were used to be grown. The former type is rather early, has fairly light green leaves; the latter is late in maturity and dark green in colour. Cambridge No. 1, 3 and 5 introduced long back from Horticultural Research Station at Cambridge, vary in their earliness in maturity but the sprouts were fairly light green. Two Irish cultivars, *e.g.* Irish Elegance and Sherrardian have become popular in recent years. Avoncross is the first F_1 hybrid cultivar released by National Vegetable Research Station at Wellesbourne, but it was withdrawn due to more percentage of sib-mated seeds. All these cultivars have large sprouts.

In continental Europe, cultivars with smaller sprouts (less than 4 cm in diameter) are grown. The early cultivars are Breda (in Netherlands), Weibulls Rapid (in Sweden) and Wilhelmsburg (in Germany), all have light green sprouts. None of these are hardy enough and should be harvested before the winter. Late and hardier cultivars are Red Vein, Hild's Ideal, De Rosny, Amager, Polarstjernen, *etc.* In the Netherlands, two F_1 hybrid cultivars known as Olaf and Thor have been bred at the Institute of Horticultural Plant Breeding at Wageningen. Both are medium-late and produce medium-sized sprouts.

In Poland, sprouts of cultivar Citadel contained high dry weight, protein, ascorbic acid and ash. It also possessed the best morphological features for freezing (Domagala and Kmiecik, 1985). In Netherlands, Rijbrock (1986) assessed and reported the following hybrids to be promising.

Early: Oliver, Dorema and Ottoline; moderately early: Roger, Asgard and Boxer; and late harvesting: Igor, Snowden and Gabion.

In Netherlands, early planting of cultivars Kundry, Lauris, Ajax, Philemon, Helemus, Stephen and Estate resulted in high yield (Vlaswinkel and Vanden, 1999).

Sharma and Gill (1986) reported that cv. Hilds Ideal is suitable for cultivation in northern India. It is 55-60 cm tall bearing sprouts, each weighing 6-7 g and yields 250-400 g/plant in 4 pickings at an interval of 10 days, the first bearing 115 days after transplanting. In Belgium, Vanparys (1986) listed cultivars which gave the best results at different months of harvest: for September harvesting, Oliver; for October,

Important Cultivars of Brussels Sprout

Hilds Ideal

Jade Cross

Dolmic, Jackeline, Nicoline and Roger; for November, Boxer, Cavlier, Fighter and Kundry; for November-December, Erwin, Igor, Lauris, Lunet, Rowena and Vitar; for late December, Fieldstar and Tardis, and for January, Fortress. Rijbrock (1988) observed that cv. Oliver was by far the earliest and Lauris was the most productive of the late cultivars.

Almost all the European seed companies are marketing F_1 hybrid cultivars of all the maturity period. Some of the important ones are Jade Cross from Japan, Doreman, Alkazar, Merilon, Poster, Rampart, Fortress, Predora, Rovoka, Sonora, Ladosa, *etc.* from the different European seed companies. These cover the entire season and the varying requirement of European growers.

The cultivar, Hilds Ideal was introduced in India and seed production are being done at ICAR Indian Agricultural Research Institute, Regional Station, Katrai, Kullu, HP and cultivated in parts of India for high end market. Besides, several other cultivars are marketed by the different private companies. Some of the popular cultivars available in India are Capitola, Confidant, Cryptus, Gigantus, Gladius, and Gustus.

Major cultivars grown in different parts of the world are:

Jade Cross E

The sprouts are larger in size, can be removed easily from stalk than with original strain. It is ready for harvest in about 90 days after transplanting.

Royal Marvel

The sprouts are compactly arranged. It is very productive and resistant to 'bottom rot' and 'tip burn'. The sprouts are ready to harvest after 85 days of transplanting.

Bubbles

It is hardy and able to tolerate warm weather. Moreover, it is resistant to rust and takes about 80-85 days to harvest.

Rubine

It is a novel cultivar. The sprouts give reddish appearance. It is a late maturing variety and takes about 105 days to harvest.

Oliver

The sprouts are very attractive in shape and can be removed easily from stalk. It takes 85 days to mature.

Prince Marvel

The sprouts are compact and sweet in taste. It is ready to harvest with in 90 days of transplanting.

8.2.0 Climate and Soil

The Brussels sprouts is a cool season crop and grown luxuriantly under cool humid climate. It grows best when exposed to an average daily temperature between 17 and 23 °C. However, it gives maximum yield of the edible sprouts at temperature range of 15–18°C. Superior quality sprouts are produced during cool or even lightly frosty weather. Brussels sprouts can be grown successfully on a wide range of soils; however, it performs best in well-drained loamy soil rich in organic matter. Sandy loams are preferred for early crops. Most ideal pH for its growth ranged from 6.0 to 6.8.

8.3.0 Cultivation

Brussels sprouts as in case of other coles are generally sown in a seed bed. For the F_1 hybrid cultivars, 300-400 g seeds are needed for a hectare. Sowing is done both before winter during August-September (as in England) and also during November to January. The early sowing is done in the open and over-wintered outside. However, very early sowing may show bolting in the spring and hence adjustment of sowing date in relation to a cultivar and area would be very much desirable. The November-January sowing is done in cold house or cold frame and the young seedlings are pricked off at a spacing of 5 x 5 cm. Most of the Brussels sprouts, however, is sown after the end of winter. This crop can be grown during off season (summer) in high altitudes areas of India like Leh, Srinagar, Lahauul and Spiti Valley, Chamba and some parts of Uttarakhand. Usually, 1g of seed will produce 100 seedlings. Growing of seedlings in the seed bed is generally followed in the same way as in other cole crops.

8.3.1 Land and Nursery Preparation

Generally, the land is prepared to a fine tilth by disc ploughing followed by one or two harrowing. It is recommended to incorporate well decomposed FYM @ 20 t/ha at the time of land preparation. Application of organic manure or vermicompost improves plant growth, productivity and improves water holding capacity of field soil.

The sterilization of soil by drenching, nursery beds with formalin @ 1:49, about 15-20 days before seed sowing is beneficial for preventing the attack of the soil borne diseases. After drenching, seed beds should be covered with polythene for atleast a week. Then beds are again dug and left open for 5-6 days to avoid injurious effect of formalin on seeds. The standard procedure for raising nursery should be followed.

The beds should be covered with a proper mulching material before watering. Apply water with a water can over the grass mulch during initial stage *i.e.*15-20 days of sowing, while during later stage watering should be done through furrows. The mulch should be removed as soon the emergence of seed sprouts. The beds should be provided with roof for shading against hot sunshine and rains.

8.3.2 Transplanting

The plants are set out on the flat or on ridges from February onwards for early crops which are harvested in July. The late crops are planted out in May or in early June. For cultivars with large sprouts, a wider spacing of about 90 x 90 cm is followed, while for smaller sprouts a spacing of 60 x 60 cm or 70 x 70 cm is followed. At wider

Field Growing of Brussels Sprout

Seedling Raising of Brussels Sprout

Bearing of Sprout in Brussels Sprout

spacing sprouts will be large but somewhat loose. When harvesting is done all at one time, a closer spacing would be more ideal, *e.g.*, 70 x 30 cm or 70 x 40 cm. In Netherlands, to obtain maximum yield of sprouts of size-class16-28 mm of cv. Ajax for the deep freeze industry, a plant density of 40000 plants/ha gives the best yield (Osinga, 1997). Bravo *et al.* (1986) from Chile reported that hybrid Jade Cross and the cultivars Isla and Emerald Ball transplanted on 15th February yielded 20.46, 14.89 and 17.09 t/ha, respectively, whereas the same hybrid and cultivars transplanted on 15th March yielded 11.41, 12.57 and 13.33 t/ha, respectively. No commercial yields were obtained from plants transplanted on 15th April. They also noticed that the cultivars, Valiant, Camelot, Merion and Lunet transplanted on 25th January yielded 11.8, 11.82, 12.85 and 11.84 t/ha, respectively. The cultivars transplanted on 17th February yielded 11.49, 9.85, 12.16 and 11.69 t/ha, respectively and those transplanted on 10th March yielded 4.06, 4.15, 3.54 and 4.13 t/ha, respectively. In Belgium, best results were obtained with cultivars Ajax, Lauris, Louis and Romulus by transplanting on 15th May and harvesting on 15th December at plant density of 70 x 45cm (Vanparys, 1998).

8.3.3 Manuring and Fertilization

Brussels sprouts are grown usually in the sameway as other cole crops. However, more intensive plant protection measures are needed. Since it is not grown commercially in India, information about its different requirements under Indian conditions are not available. However, it is well known that heavy dressing of nitrogen results in loose sprouts. Nevertheless, the amount usually applied from100-300 kg per hectare. Since even growth is required for firm sprouts, nitrogen is given in parts. Split doses are applied at the time of planting, a few weeks after planting, after first picking and also after the second picking. Application of nitrogen makes a sprout dark green which is normally preferred by the consumers. According to Welch *et al.* (1985) Brussels sprouts yield increased with N rate upto 90 kg/ha. Application of nitrapyrin had significantly higher yields than those with comparable N rates without nitrapyrin. However, the midrib NO_3-N levels were not a suitable guide to N needs of Brussels sprouts plants. In West European countries, 50-150 kg P and often more than 200 kg of K are applied for a hardier growth, sometimes top dressing of P and K is also suggested. Peck and McDonald (1986) noticed that cv. Jade Cross was less responsive to higher rates of concentrated superphosphate or KCl. Kolota and Biesiada (1990) obtained best yield with 200 kg N/ha. In a small pot trial Kolota and Dobromilska (1991) recorded 200 kg/ha N as optimum for high yields and best quality commercial sprouts for cv. Maczuga; higher N rates gave a greater number of loose heads. The crop reacted well to increasing P rates (30-300 mg/dm^3), which increased the yield and improved its quality. Babik *et al.* (1996) reported that plant height increased and number of senescent leaves decreased with higher N rates (levels of 100 to 600 kg/ha). Increased N application resulted in darker colour, higher chlorophyll concentration and undesired elongation of sprouts. Specific gravity, dry matter content and concentrations of total sugars and vitamin C (ascorbic acid) decreased with N levels > 200 kg/ha.

In Netherland, Booij (2000) treated cv. Kundry with N @ 0-300 kg/ha and analysed to establish the effect of N on the components of marketable yield.

Marketable yield (Y) was a function of radiation use efficiency (RUE), cumulative intercepted radiation (IPAR), partitioning of biomass (P), and the dry matter concentration in the buds (DMC), where Y= RUE x IPAR x P/DMC. RUE was not affected by N rate, while this only had a small effect on P and DMC. The effect of N availability on IPAR was the key factor in determining marketable yield. It had a strong effect on canopy development resulting in increasing interception of incoming radiation at higher N application rates due to an increasing 'leaf area duration (LAD) with increasing N availability.

8.3.4 Weed Control

In a trial in Scotland, Lawson (1984) observed that the competition between annual weeds and transplanted Brussels sprout reduced sprout yield by 13 per cent to 24 per cent; close-spaced crop plants suppressed weeds effectively but at the expense of yields. He emphasized the need for an early control of weeds. The herbicides recommended in cabbage are equally effective in Brussels sprout.

8.4.0 Harvesting, Grading and Yield

Brussels sprouts are picked for 3 to 5 times in common cultivars. The period of harvest can be manipulated to a certain extent by removing the growing points. With this topping treatment, the sprouts may be picked at one time after 4 to 8 weeks after topping. However, topping should be done after the lower sprouts have developed, especially when a single harvest is preferred. In normal crops also topping is done in some region but it is done only after the first or second pick and when the temperature is low.

Harvesting should commence when the lower sprouts have reached a certain size and firmness. In case of early dwarf cultivars, the harvest may be completed in two pickings. Late untopped cultivars may continue till the end of winter. In short-stemmed cultivars the sprouts are crowded on the stem and these become difficult to pick. However, in such cases once-over-harvest followed by indoor picking is possible.

Sprouts should satisfy the conditions for marketing: (i) should be solid and have their top leaves folded closely, (ii) should be dark or at least light green and not yellow, (iii) sprouts with small loose leaves at the base are considered poor in quality and (iv) similarly, open top is also considered poor in quality. Grading the sprouts is becoming more and more popular. Different sizes are used, *e.g.*, under 2 cm, 2-3 cm, 3-4 cm and over 4 cm in diameter. Loose sprouts are sometimes marketed as 'blowers'. The freezing industry requires a sprout less than 3 cm in diameter. The yield of a crop depends on various factors. Large sprouted cultivars yield on an average 10 tonnes per hectare, but normally spaced crop when picked only once, may yield between 3-5 tonnes/ha.

8.5.0 Storage

The sprouts retained good appearance for 4 weeks at 2.5°C whether stored in CA or in air. Storage in 0.5 per cent CO_2 occasionally induced a reddish tan discolouration of heart leaves and frequently an extremely bitter flavour in the non-green portion of the sprouts (Lipton and Mackey, 1987).

Harvesting and Packaging of Brussels Sprout

Harvesting of Brussels Sprout

Packaging of Brussels Sprout

9.0 Sprouting Broccoli

The name 'broccoli' refers to the young shoots which develop in spring on some species of the genus Brassica ('brocco' is Italian for a shoot). In Italy, these have been used as vegetable from early times but their economic importance became appreciable only since the thirties of last century when this vegetable became popular in the USA. A large part of the produce in the USA, goes to the freezing industry. It is becoming popular in other parts of the world only gradually.

Feher (1985) described the nutritional quality and suitability for export of cultivars Kayak, Corvett and Coaster. All had good organoleptic properties and high protein, vitamin C and mineral contents.

Sprouting broccoli is sometimes briefly called broccoli, though this name is also used for broccoli rape (the sprouts which develop on turnips) and for winter cauliflower (winter broccoli or heading broccoli). It would be desirable to confine the name 'broccoli' to the appropriate sprouting types of *Brassica oleracea*. In the USA, it is also known as 'Italian broccoli' pointing to its Italian origin while 'Italian Calabrese' was an initially widely used varietal name. Calabrese refers to the district of Calabria in Southern Italy.

Morphologically, sprouting broccoli resembles cauliflower. The plant forms a kind of head, consisting of green buds and thick fleshy flower stalks. The terminal head is rather loose, green in colour and the flower stalks are longer than cauliflower. The sprouts in the axils of leaves develop strongly, specially after the removal of the terminal head. Both terminal head and the sprouts with bud clusters are consumed as human food.

9.1.0 Cultivars

In broccoli, purple and green variants occur, though in the USA, it is only the green types that are grown. The purple types are fairly hardy and can be harvested during the winter. These are grown in limited scale in England.

The green sprouting broccolis are classified in accordance to their maturity, *e.g.*, early, medium and late cultivars. Though a very sharp distinction is not possible but the cultivars may differ in the size of plant, size of clusters and buds, firmness of the heads, colour of buds, and foliage. The demand is more for a firm head with small, dark green buds while the demand for the side shoots have decreased in recent years possibly due to harvesting difficulties.

Cultivars like De Cicco, Green Bud, Spartan Early are early cultivars and can be harvested in 6 to7 weeks after transplanting; however, these cultivars may not perform well under unfavourable conditions, specially at lower temperature. Cultivars like Waltham 29, Green Mountain, Coastal, Atlantic, *etc.*, are less sensitive to buttoning and can be grown for early or late crop. The medium strains like Green Sprouting Medium take about 100 days for maturity, while the late strains like Green Sprouting Late are biennials and may be harvested at the onset of winter or after its end.

Khattra *et al.* (1997) reported from Ludhiana, India that cvs. BI 80167 andBajaura-1 produced the highest yield/ha.

In New Zealand, the cultivars Premium Crop, PC72, Southern Comet, Mercedes, Skiff and Corvet produced heads better than standard Ford hook cultivars. They matured quicker, especially in summer and autumn harvests, and their percentage of saleable terminal heads are usually higher (Bussell and Dobson, 1985).

Benoit and Ceustermans (1986a) from Belgium found that in summer culture the cultivar Corvet performed best with an average crop weight of 342 g. In winter culture Southern Comet and Laser had the shortest interval between sowing and the first harvest and average crop weight was higher (261 g) with Southern Comet, followed by Skiff. Vanparys (1994) reported that the best performing broccoli cultivars for early culture in Belgium were Marathon, Samurai, Shogun and Skiff. For summer, cultivars Beaufort, Emperor and Shogun were, however, recommended by Vanparys (1995a) and cultivars Fiesta and Viking were considered best by Rooster and Vanparijs (1997). Vanparys (1999) assessed seven purple broccoli cultivars in Belgium and found that highest numbers of marketable spears (67 and 63 per cent, respectively) were produced by Violet Queen and RS. 84090.00 and recommended for cultivation on light sandy loam soils.

Solo head, a cultivar of broccoli evolved from the cross (Self-Blanche cauliflower x Early Fiji cabbage) x Spartan Early Broccoli, bears only a primary head, with no lateral shoot development prior to head maturity. It is resistant to *Xanthomonas campestris* (Honma, 1986). Shattuck *et al.* (1986) concluded that cv. Green Valiant was stable and better adapted to unfavourable environments.

In Denmark, McCall and Sorensen (1995) recommended Bacchus and Dandy as early cultivars. Another cultivar Marathon was notable for its high yield of heavy heads.

Guan Pei Cong (1995) from China described the plant growth and curd development characteristics of 4 broccoli cultivars, Li-Green, Jade-Crown, Tokyo-Green and Southern-Comel. The number of branch curds was 30-40, 40-45, 50 and 50-60 for Li-Green, Jade-Crown, Southern-Comeland Tokyo-Green, respectively.

An open pollinated broccoli cultivar Ramoso de Brasilia developed in Brazil is recommended for Brazilian winter cropping. It has a short growth cycle, high yields and produces a high number of lateral sprouts with soft and long stems. Inflorescences are bright green and of excellent market appearance and high cooking quality (Meloand Giordano, 1999).

In India, sprouting broccoli is hardly a commercial crop and hence the cultivars are also not properly classified. In North India, both early and mid season cultivars can be grown by sowing them in September and October and these will be ready for harvest from late November to early December and may continue till early February.

At present, a large number of F_1 hybrid cultivars are being marketed by the different seed companies of Japan, USA and Europe. Some of these F_1 hybrids are Southern Comet, Premium Crop, Clipper, Laser (extra early and early), Corsair,

Excalibur, Cruiser, Emerald Corona (mid season) and Late Corona, Stiff, Kayak, Green Surf (late), *etc.*

Some of the Important Cultivars Grown in Northern India

Pusa Broccoli KTS-1

Developed through selection at IARI, Regional Station Katrain (Kullu Valley). Compact light green heads, weighing around 250-400 g, plant height 40-50 cm and ready for harvesting in 85-95 days after transplanting.

Palam Samridhi (DPGB I)

Sprouting broccoli having compact green heads, free from yellow eyes and bracts, average head weight is 300-400 g, plant branched. It also bears sprouts in the axil of leaves which adds to the total yield, ready for harvest in 85-90 days after transplanting, average yield 150-200 q/ha.

Palam Haritika

Sprouting broccoli having dark green straight leaves, the main head formed deep inside the leaves. Late in maturity and gets 150 days for harvesting, free from yellow eyes, average yield is 200-250 q/ha.

Palam Kanchan

Yellow coloured heading type broccoli, leaves are large, green and straight with white midrib. The crop gets ready for harvesting in 140 days, average yield 250-300 q/ha.

Palam Vichitra

Purple coloured heading type broccoli, medium foliage, open green leaves having green tinge, heads are firm and solid. The crop gets ready for harvesting in 115-120 days, average yield 225-250 q/ha.

Punjab Broccoli 1

This cultivar is developed by PAU, Ludhiana, matures in 65-70days, bears large number of spears, foliage dark green with smooth leaf surface, heads and spears are dark green with bluish tinge.

9.2.0 Soil and Climate

9.2.1. Soil

The requirement of broccoli is not that specific as in cauliflower, though the soil should be sufficiently fertile for the rapid development of seedlings after transplanting. For the cultivation of broccoli, fertile soil with good moisture supply would be very much necessary. Broccoli can be grown in a wide variety of soils type. To get a better yield from broccoli crop sandy and silt loam soils are most preferred. The soil PH should be in between 5.5–6.5 pH. Before Planting broccoli Land is plowed 3-4 times then add compost or well rotten FYM 25-30 tons/hectare and mix thoroughly at the time of land preparation.

9.2.2 Climate

Since the cultivars are somewhat sensitive to temperature, warm weather will be disadvantageous as the bud clusters will grow loose quickly. Some cultivars of sprouting broccoli exhibit resistance to frost to a certain extent and that too dependent on the size of the plant. These are generally biennial, while the annual cultivars are somewhat sensitive to frost. In such cultivars even a light frost causes considerable damage to the buds. The other requirements are like that of cauliflower.

Benoit and Ceustermans (1986b) in a glass house study of winter broccoli maintained a minimum temperature of 6°C until flower head development started, after which a night/day temperature regime of 10°/15°C was maintained until the harvest.

Elson *et al.* (1992) suggested that high temperature inhibition of seed germination and seedling emergence is a potentially important fact or limiting direct-seeded broccoli stands in warm climates where the soil temperature often exceeds 20/32°C (day/night temperature) during summer.

9.3.0 Cultivation

In the USA, the crop is grown mainly for harvesting autumn and spring. The crop is sown in June-July and the seedlings are planted out from July to September. The cultivars are chosen on the basis of period and intensity of frosting.

9.3.1 Nutrient Management

Manure and fertilizer requirements in broccoli depend upon fertility status of the soil. So, a soil test is the most accurate guide to fertilizer requirements. Good management practices are essential, if optimum fertilizer responses are to be realized in the production of cole crops. Because of the influence of soil type, climatic conditions and other cultural practices, crop responses from fertilizer may not always be accurately predicted. Soil test results help determine the nutrients needed and the rate of application. Optimum fertilization is intended to produce top quality and yields in keeping with maximum returns. Apply 20 tonnes well rotten FYM during field preparation. Apart from application of manure, apply 100 kg N, 75 kg P_2O_5 and 50 kg K_2O per hectare. The half dose of N and full doses of P and K should be applied before transplanting. The remaining half dose of N should be top dressed in two equal splits *viz.* after one month of transplanting and at the of time head formation.

For acid soils low in Ca and Mg, application of 'dolomite' as lime should be followed.

The experiments carried out by Chung and Strickland (1986) in Tasmania, Australia revealed that sowing after February increased the length of growing period and reduced the spear yields for all 10 cultivars, compared with earlier sowings. It was also suggested that lower growing temperatures of later sowing might cause floral initiation at a younger physiological age and plants, therefore, develop heads before reaching full size and the spears are small and take longer to reach maturity. In a production schedule they further showed that yield in March would be about

Important Cultivars of Broccoli

Pusa Broccoli-KTS-1

Palam Kanchan

Palam Vichitra

Seedling Raising and Field Growing of Broccoli

Seedling Raising of Broccoli

Experimental Field of Broccoli, BCKV, India

14 t/ha while yield during May-August would be 5 tonnes/ha. In Australia, the effect of 2 planting dates (1 and 15 October) and accumulated heat units (AHU) on head quality of cv. Assiut 1 were investigated by Aboul-Nasr *et al.* (2000). Average head weight and total yield were greater with later planting, associated with 8.5 days longer growth before harvest. Head weight was positively correlated with AHU of the late vegetative stage. Chemical composition of the buds in cv. Assiut 1 was generally similar to one of its parents, cv. Parma. In Himachal Pradesh, India, the best time to plant broccoli was between 20 October and 3 November (Singh *et al.*, 1999). In a trial carried out during the winter season in Himachal Pradesh, India, by Sharma *et al.* (1995), cv. Green Head seedlings were transplanted on 16th August, 31st August, 15th September or 30th September at 45 x 45, 60 x 45 or 60 x 60 cm. Of the planting dates, 15th September resulted in the greatest head and seed yields (98.72q/ha and 17.94 g/plant, respectively). Among the spacing treatments, inflorescence yield was highest (104.97 q/ha) at 60 x 45 cm, while seed yield per plant was highest (20.51 g) at 60 x 60 cm. Under the climatic conditions of S.E. Poland, the effect of plant densities of 3.2, 4.0, 5.3 or 8.0 plants/m^2 was investigated by Kunicki *et al.* (1999) on the yield and quality of broccoli cultivars Kermit, Montilla and Skiff. Plant density had a significant effect on the marketable yield which ranged from 11.0 t/ha at 3.2 plants/m^2 to 16.6 t/ha at 8.0 plants/m^2. Head weight increased by 46 per cent as plant density decreased from 8.0 to 3.2 plants/m^2. Nitrate content was highest in Kermit and lowest in Skiff. In an experiment conducted in Italy, the effects of 3 planting dates (20 August, 24 September and 25 October), 4 topping regimes to stimulate growth of secondary stems (none, or topping at planting, 15 or 30 days later) and harvesting by hand (6-8 times) or mechanically (single harvest) on 4 broccoli cultivars were assessed (Dellacecca, 1996). Single mechanical harvest of inflorescences reduced yield by 49-60 per cent compared with hand harvesting. A combination of hand harvesting (primary inflorescences) and mechanical picking (secondary inflorescences) could reduce yield losses to about 23 per cent. Topping at planting and particularly in August resulted in the best and earliest yield of inflorescences with a relatively higher weight (about 140g), good firmness and small stem diameter. The hybrid XPH 4142 (Asgrow) averaged the highest yields (9.9 t/ha) and inflorescence weight (165 g) across the planting dates.

Broccoli is generally sown on seed bed like other cole crops. But more recently, the seeds are drilled directly to soil in the USA. The spacing in such sowings is maintained at 30 x 75 cm between the plants and rows, respectively. Crops grown in such close spacing yield more though main heads are smaller and these mature slightly later than when wider spacing is followed. Lancaster *et al.* (1985) suggested planting of single row, 15 cm between plants and double row, 30 cm between plants and rows of 5-week-old broccoli seedlings. Results of experiments conducted by Chung (1985) in Australia indicated that at a low density (2.8 plants/m^2) 85-89 per cent of the total yield can be obtained from 3 harvests. Yields could be increased by changing to a density of 8 plants/m^2 without significant cultural or marketing problems. Plant population of 45000/ha is suggested as being the optimum for cv. Corvet and Emperor (Brakeboer, 1990). Mullins and Straw (1990) from USA suggested that a spacing of 6 inches is desirable for fresh market production, where a small head size is required. The use of multiple plants transplant decreases mean head weight and diameter and reduces the incidence of hollow stem (Griffith and Carling, 1991). Increasing the plant density of cv. Premium from 40,000 to 125,000/ha increased total yield but reduced head diameter (Paschold, 1986). Broccoli plants removes a total of 15.3 g N, 2.1 g K, 12.0 g Ca, 1.1 g Mg, 18.9 g Mn, 10.8 g, Zn and 1.6 g Cu/m^2 for yield of 67g dry matter/m^2 (Shelp and Liu, 1992). According to Dufault and Waters (1985) increasing the N rate from 56 to 224 kg/ha at any population (24,000 to 72,000 plants/ha) linearly increased broccoli head weight and marketable yields, and decreased cull yields. Broccoli yields were highest at 72,000 plants/ha and 224 kg N/ha.

Gorbki and Armstrong (1985) reported from Ohio that closer spacings (40 cm to 20 cm) and higher plant densities increased yields/ha but the head size and incidence of hollow stem were decreased. Increasing N rate (0 to 224 kg/ha) increased yield, head size and incidence of hollow stem.

Everaarts *et al.* (1997) from Netherlands advised to apply a single dose of 260 kg N/ha (minus the amount of mineral N present in the top 0-60 cm layer of soil) by row application at the time of planting. A high yield crop takes up 200-250 kg N, 30 kg P and up to 350 kg K/ha. Everaarts and Willigen (1999), however, recommended the application of 270 kg N/ha for optimum yield minus the mineral N present in 0-60 cm soil layer through band placement at planting.

A close relationship existed between the texture and quality of the heads and N and K nutrition. Rational application of nitrogen and potash fertilizers regulated the plants, absorption of N, P and K promoted nitrate accumulation in leaves and spears, and increased spear yields (Yang *et al.*, 1995). The highest contents of ascorbic acid, sugars and nitrates were found in broccoli given 200, 400 and 600 kg N/ha, respectively (Elkner and Babik, 1999). In a field experiment in Ontario, Canada, both fresh and dry weights of shoot and inflorescence increased linearly with increasing nitrogen applied (70-130 kg/ha). Chloride application (0-450 kg/ha) did not significantly affect the final yield (Liuand Shelp, 1996).

Aftercare should be like that of cauliflower. Since a long stalk along with the terminal head is harvested, it is necessary that fertility of the soil and moisture should be maintained properly to avoid hollowness of the stalk. Irrigation favourably influenced the colour and chemical composition of spears. Experimental results indicated that water requirement of a broccoli cv. Clipper ranged from 322 to 426 mm (Rubino and Palma, 1986). Paschold *et al.* (2000) reported from Germany that total water requirements for broccoli were 230-322 mm (3.7-4.7 mm water/day), of which 90-218 mm was irrigation water.

In Poland, experiment was conducted in seedlings of cv. Corvet F_1 by Kunicki *et al.* (1996) with row covers treatments consisting of 0.04 mm thick perforated polyethylene plastic (100 holes of IO mm diameter/m^2) and polypropylene non-woven fabric Agryl P-17. Row covers promoted leaf area development and resulted in a harvest that was 3-4 days earlier than in controls. Both covers increased total and marketable yields of main heads. At the earlier planting date (first half of April) the yields from plastic film-covered plots were slightly higher than from Agryl plots. Plastic cover decreased the vitamin C (ascorbic acid) content.

9.3.1.1 *Effect of organic manures and inorganic fertilizers on vegetative growth*

Jigme *et al.* (2015) reported the highest value of all the vegetative parameters that were measured was found in the inorganic treatment (Table 9). The differences in the value were statistically significant over the rest of the treatments (P=0.05). Its mean plant weight was 1329 g. The control gave the least vegetative growth. However, its vegetative growth was statistically similar to treatment no. 1 (compost). Treatment no. 3 showed the highest vegetative growth among the organic treatments. However, the values were statistically similar to treatment no. 2. The treatment no. 2 in turn had all its vegetative growth values statistically similar to treatment no. 1 except for plant weight.

Table 9: Effect of Organic Manures and Inorganic Fertilizers on Vegetative Growth*

Treatment	*Stem Diameter (cm)*	*Leaf Number per Plant*	*Leaf Diameter (cm)*	*Plant Weight (g)*	*Plant Height (cm)*
Compost, Compost and CMT	3.60cd	20.73cd	22.38b	1062.18c	51.79cd
(100ml/week) Compost and CMT	3.70cb	22.13bc	22.48b	1145.67b	52.62cb
(200ml/week)	3.85b	23.18b	23.22a	1198.71b	54.50b
Inorganic fertilizers	4.03a	27.20a	23.76a	1329.23a	61.03a
Control	3.42d	19.36d	22.19b	1011.83c	49.98d

*Jigme *et al.* (2015)

9.3.2 Water Management

Broccoli needs sufficient moisture in the soil for uniform and continuous growth of plants. First irrigation should be given just after transplanting. First irrigation should be light to avoid the loss of freshly transplanted seedlings. Subsequent irrigation can be given at an interval of 7-8 days during summer and 10-15 days during winter depending upon soil type and weather. There should be sufficient moisture in soil at the time of head formation. The dry conditions adversely affect the quality and yield of shoots by being more fibrous.

9.4.0 Harvesting and Yield

Usually, only the central head is harvested. It is very important that broccoli is harvested at the correct time, *i.e.*, it is picked up before the buds open and when the bud clusters are compact. The heads are cut off with about 15 cm of the stem attached; after cutting, part of the foliage is removed from the harvested shoots. The head may be 15-25 cm in diameter and weighs 250 to 600 g. The yield per hectare averages about 5-10 tonnes. In the USA, pioneering attempts have been made to harvest broccoli shoots mechanically.

9.5.0 Storage

The yellowing of broccoli in storage is mediated by ethylene action and that

Harvesting and Packaging of Broccoli

Curd Bearing in Broccoli

Broccoli becomes Ready for Harvesting

Harvesting and Packaging of Broccoli

1-mcthyl cyclo propenc treatment inhibits yellowing and reduces respiration, even when broccoli is exposed to ethylene (Fan and Mattheis, 2000).

Tan *et al.* (1992) reported that after 6 weeks storage at 1°C, broccoli (cv. Green Belt) heads which were stored in unsealed polyethylene bags in waxed cartons without ice or in polystyrene boxes without the addition of ice (top ice) were still fresh with green compact heads and no sign of microbial infection. The use of a polystyrene box with top ice for broccoli is not necessary if the temperature during storage and transportation can be maintained below 2°C. Irrespective of packaging, all the broccoli head turned yellow and became unmarketable after transfer and storage for 2 days at 26°C, 70- 75 per cent RH, except for those stored in sealed polyethylene bags under MA conditions.

Barth *et al.* (1993) stored broccoli spears in polymeric film at 10°C for 96 h with low intensity fluorescent lighting and 80 ± 2 per cent RH. After 96 h storage, gas concentrations in the packages were 8 per cent CO_2 and 10 per cent O_2. Reduced ascorbic acid retention, moisture content, total chlorophyll and colour retention were greater in packaged broccoli compared to non-packaged.

Toivonen (1992) compared Chlorophyll A fluorescence (Fvar) with respiration and vitamin C content of broccoli during storage at 1°C. The amplitude of the Fvar maximum decreased in a similar manner to respiration and vitamin C content. The results demonstrated that Fvar can be used as a non-destructive indicator of early changes in tissue condition (*i.e.*, degree of freshness) of broccoli in storage.

The best O_2 and CO_2 levels for storing broccoli florets cv. Marathion were 0.5 per cent O_2 and 10 per cent CO_2 at 0 and 5°C, and 1 per cent O_2 and 10 per cent CO_2 at 10°C (Izumi *et al.*, 1996).

9.6 Physiological Disorders

Whiptail

The lamina of the newly-formed leaves become leathery, irregular and consisting of only mid-rib. It is caused due to molybdenum deficiency in plants.

Control: Soil application of molybdenum @ 1-1.5 kg ha-1 before transplanting reduces the occurrence of disorder. Foliar spray with 0.01 per cent Ammonium molybdate solution helps to check this disorder.

Browning Head

Firstly, there is appearance of water-soaked areas on bud clusters which later on turns pinkish or rusty-brown resulting in rotting. Browning head is a result of boron deficiency in plants.

Control: Soil application of borax or sodium borate @ 20 kg ha-1 prevents the disorder. Foliar spray of 0.25-0.5 per cent borax solution is very much for effective especially when the deficiency is acute.

Leafy Heads

There is presence of leaves within the head. This disorder is a result of high temperature coupled with lush growth due to excess water and nitrogen.

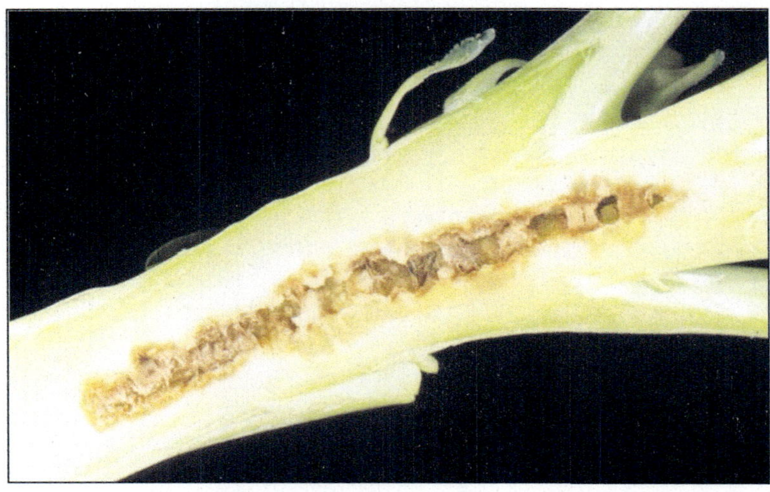

Boron Deficiency causes Hollow Stem in Broccoli

Control: Apply light frequent irrigation during high temperature conditions. Avoid indiscriminate use of nitrogen and it should be applied on the basis of soil test results.

Large and Coarser Buds

High temperature conditions and delayed harvesting resulted into excessively large or open buds renders it unfit for consumption.

Control: Apply light frequent irrigation during high temperature conditions. Harvesting should be done at proper stage *i.e.* when bud clusters are green and compact.

10.0 Kale

Amongst the kales, curly type is used for vegetable. It is closely related to wild cabbage. It is mainly a winter vegetable when other vegetables are scarce. Most forms of kale bear a rosette of leaves at the top of the stem, which is harvested and used as food and at times individual leaves are also picked up.

A large number of cultivars are available. Broadly, these can be divided into dwarf, medium-tall cultivars. The most important characters, for identification besides stem length, are the colour and curliness of the leaf and winter hardiness.

Kale is a very reliable cropper and can stand relatively unfavourable conditions. It grows very well in medium-heavy and fertile soils with good moisture supply, though it can also be grown successfully in lighter soils. Kale is more salt tolerant than cabbage and can also be grown at C_8. Kale can stand temperatures of 10 to 15°C. It even grows quite well in Southern Greenland.

Kales are normally sown in seed bed, though they are also sown direct. As it is harvested in autumn and winter, sowing is done from April to June in North-Western Europe. Planting distance is conditional on the time of planting, cultivar, soil and location. Dwarf cutlivars may be spaced at 70 cm and for late plantings, these are spaced more closely at 40-50 cm. Wide spacing increased mean fresh weight, branching and leaf production of kale plants whereas close spacing significantly increased total leaf yield per hectare but decreased leaf:petiole ratio per plant (Chweya, 1984). Weeds are controlled by hoeings as in other coles. Once the crop has covered the soil, further weeding is avoided.

In cultivars Blue Max and Vates maximum yields/ha were obtained with a spacing of 15cm within row (Dangler and Wood, 1993). Ne Smith (1998) suggested that population of 5-7 plants/m^2 for once-over harvest of whole plant bunch collards to optimize yield and crop quality, especially under favourable cropping conditions.

Recently one cultivar developed at ICAR-Indian Agricultural Research Institute named as **Pusa Kale-64.** The salient features of the cultivar are given below:

1. It has highly serrated, purplish green leaves, 40-50 cm in length and 15-20 cm in width, plant height is 50-60 cm.
2. Leaves are available throughout the winter season in multiple harvestings which can be consumed as cooked green vegetable like methi or mixed with other leafy vegetables, preparing soup and as a salad also.
3. It has higher contents of phenols, anthocyanin, ascorbic acid, chlorophylls, lycopene and total carotenoids than other cultivars of brassica vegetables

Dangler and Wood (1993) reported from USA that in autumn, the yield of cv. Blue Max and Vates were both highest (14.5 and 9.9 t/ha, respectively) with 112 kg N/ha.

Field study conducted in Alabama, USA with collards (Kale, *Brassica oleracea* var. *viridis*) cv. Vates and hybrid cv. Flash in autumn and spring revealed that maximum yield varied from by 222-250 lb/acre, depending on season and cultivar (Guertal and Santen, 1997).

Seedling Raising, Field Growing, Harvesting and Packaging of Kale

Seedling Raising in Kale

Kale Cultivar, Pusa Kale-64

Field Growing of Kale

Harvesting and Packaging of Kale

For fresh market, the wholetop rosette is frequently harvested while in tall cultivars the lower leaves, if adequately curled are also harvested. Yield of kale varies between 10 and 25 tonnes per hectare.

Chinese kale (*Brassica alboglabra*) is regarded as a separate species of Chinese origin. Present thinking, however, is that this should also be treated as *B. oleracea* due to its free inter-crossing and that it possibly has travelled from eastern part of the Mediterranean. Taxonomical studies carried out by Liu and Guan (1998) however, revealed that it is more reasonable to treat it as a separate species (*B. alboglabra*) than as a subspecies of *B. oleracea*. 'Karam sag' of Kashmir, used to be regarded as a kale, is also a very close relative of Chinese kale according to Snogerup (1983). In both these, the rosette of leaves at the top and the tender stems are edible. However, Chinese kale is a typical tropical or hot weather vegetable, while 'Karam sag' requires chilling for flowering. The variation in their flowering habit may be due to their adaptation in the two localities. The cultivation is more like that of other coles, though the Chinese kale should be grown at a comparatively higher temperature around 18-20°C, since there is chance of flowering very early. The flowers are white. Even at the bolting stage its edibility is not reduced.

11.0 Chinese Cabbage

Chinese cabbage, though not a member of the cole group, but it is regarded as a very closely allied crop due to similarity in breeding systems, cultivation practices, season of growing and other features besides being a member of Brassica vegetables. Thus, this crop is included in this chapter and while describing the different aspects of cultivation briefly, the origin, evolution and classification are narrated separately.

Heading Chinese Cabbage is one of the important vegetables in eastern Asia. In China, it is a widely grown vegetable and in northern parts it accounts for nearly twenty-five per cent of the total annual vegetable consumption. In Japan, it is also very popular, ranking third after radish and cabbage. In Korea, it is undoubtedly the most important vegetable both in terms of production, area and per capita consumption.

The crop is grown by small farmers due to its short duration. It is a valuable source of calcium, crude fibre and vitamin C in the human diet. In the tropics and subtropics, there has been a considerable increase in the area of production during the past decade or so due to the availability of new, tropically adapted cultivars though seeds are still imported in these areas.

11.1.0 Origin, Evolution and Classification of Caltivars

Heading Chinese cabbage is a member of the genus *Brassica* and is a native of eastern Asia. Its progenitor form B. *campestris*, is believed to have evolved in the Mediterranean area and later has been introduced into northern Europe where it was improved as an oilseed crop (Nishi, 1980). After introduction to China more than 2000 years ago, it differentiated into various sub-species (Lee, 1982). Both *B. campestris* sub sp. *rapa* (turnip) and *B. juncea* (mustard) were recorded in the 5th century B. C. in China. Until seventh century A.D. *B. campestris* sub sp. *chinensis* (non-heading Chinese cabbage) was cultivated only in South China, at least for 1600 years. The loose leaves heading Chinese cabbage (*B. campestris* subsp. *pekinensis*) was first illustrated in the 10th century A.D. and a clear illustration of the heading shape of Chinese cabbage with wrapping leaves was first recorded in 1330 (Jiang, 1981; Li, 1981). Much of the evolution of heading Chinese cabbage took place in China from about 600 years ago (Opena *et al.*, 1988). Li (1981) considered that Chinese cabbage has originated from hybridization of Pak-choi (non-heading Chinese cabbage) from South and turnip from North China when they were grown together in the city of Young-chou. To prove his point, Li (1981) had collected different forms of Pakchoi and turnip from south and north China, respectively and their artificial hybrids obtained by him showed typical morphological characteristics of Chinese cabbage (*B. campestris* ssp. *pekinensis*). This morphological evidence may be supported also by the cytological study by Riccharia (1937) who reported the chromosome karyotype of both ssp. *chinensis* and ssp. *pekinensis* which were designated as ABCDDEEFFF and that of ssp. *rapifera* as ABCDDEEEFF. The E and F chromosomes of ssp. *pekinensis* were morphologically alike to those two of ssp. *rapifera*. No doubt more detailed studies would be necessary for further proof of the hypothesis of possible hybrid origin of Chinese cabbage.

After studying the morphological, ecological and economical characteristics of many of local cultivars of China, Li (1981) has classified the Chinese cabbage broadly into four varieties:

11.1.1 Loose Leaved Caltivar (var. *dissoluta*)

Terminal bud under developed, forming no leaf bud; rosette leaves oblanceo late, spreading or erect and usually grown in spring and summer in all parts of China.

11.1.2 Semi-heading Caltivar (var. *infarcta*)

Terminal bud with its outer layer of leaves fairly well developed, forming head with a hollow centre; plants large and tall; rosette leaves oblanceo late and erect; usually grown in autumn and stored in winter in cold and arid areas where the growing season is too short for heading varieties.

11.1.3 Fluffy Topped Heading Caltivar (var. *laxa*)

Terminal bud well developed, forming rather a solid head with its leaf tips curling up wards and forming a fluffy top; plants small, rosette leaves obovate and spreading; usually grown in late summer and early autumn; keeping quality poor and cannot be over-wintered; sometimes grown in spring season.

11.1.4 Heading Caltivar (var. *cephalata*)

Terminal bud well developed, forming a solid head with leaf tips close to overlapping on the top; grown usually either as an early crop for autumn or a late crop for winter.

The *cephalata* has been further classified into three types depending on morphological and ecological characteristics (Li, 1981). These are:

11.1.4.1 Ovate Type (f. ovata)

Originated in Shandong peninsula, maritime climatic ecotype adaptable to temperate and fairly humid climates.

11.1.4.2 Flat Topped Type (f. depressa)

Originated in central part of Ho-nan province, continental climatic ecotype, adaptable to in land areas with fluctuating temperature, low relative humidity and with many sunny days.

11.1.4.3 Cylindrical Type (f. cylindrica)

Originated on the eastern coast of Ho-bei province with Pu-hai Gulf at its South and the inner Mongolian plateau at its North, thus making it adaptable to intermediate climate between maritime and continental; this is widely adapted to all parts of China.

Thus, according to Li (1981), the six basic forms of Chinese cabbage are *dissoluta, infarcta, laxa* and three forms, *e.g., ovata, depressa* and *cylindrica* or *cephalata* group (Figure 8).

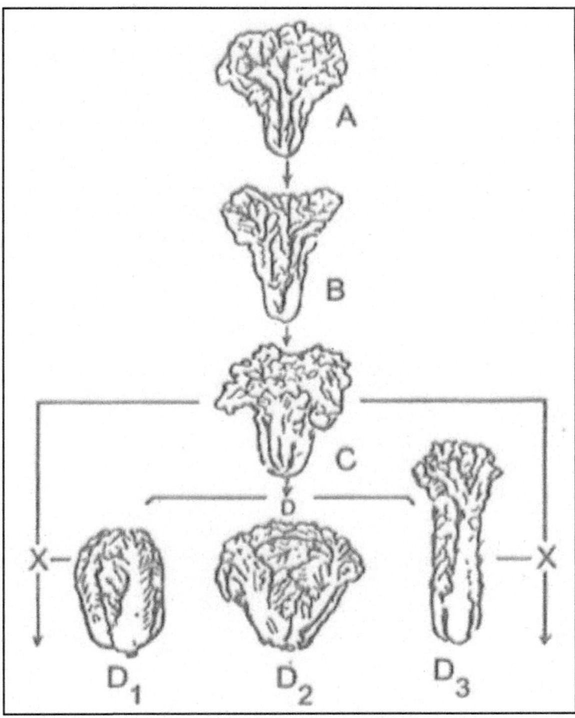

Figure 8: Evolution of Chinese Cabbage (*Brassica campestris* ssp. *pekinensis*). A. var. *dissoluta;* B. var. *infarcta*; C. var. *laxa*; D. var. *cephalata*; D_1. var. f. *ovata;* D_2. var. f. *depressa;* D_3. var. f. *cylindrica*. The 6 basic types of Chinese cabbage (Li, 1981).

Similarly, Cao (1986) recognized the following Pak-choi types - Pak-choi (var. *communis* or var. *erecta*), Tai Tsai (var. *rosularis* or var. *atrovirens*), Fen Tsai (var. *multiceps*) and Yu Tsai (var. *utilis* or var. *oleifera*).

The evolution of Chinese cabbage, as in other cultivated crops, is supposed to be the result of continuous improvement from the primitive forms through natural hybridization followed by human selection. As indicated earlier, the primary form var. *dissoluta* was first produced in a city between North and South China, possibly by hybridization of turnip and Pak-choi. In the South this type remained in its original form. The heading forms of the South were introduced from the North but not earlier than 19th century as the temperate autumn season in the South is neither long nor sunny enough for the plant to reach full growth and form a head as a storage organ. Comparatively, the climate in the North is more suitable for growing cool season crops since favourable climate lasts for 90-120 days. In Chinese cabbage excess photosynthetic product has been reported to be stored first in the petiole and hence the primary forms introduced in the North initially developed thick petioled cultivars as recorded in the literature of 12th century A.D. (Li, 1981). By further improvement of growing conditions through better nutrition and water supply, extra food was stored in the outer layer of leaves of the terminal bud. Thus, the semi-heading var. *irfarcta* appeared. Improved cultural practices and possible human selection for a solid head resulted in the development of var. *laxa*. This form was recorded in horticultural books in China in the 14th century AD. The form *laxa* produces quite solid heads but the tops remain loose and fluffy. Further human selection for a good solid head all through resulted the *cephalata* types. The first record of this type was found in a Chin dynastic geographic book entitled Shiun-Tian-Fu-Tse in the 17th century (Li, 1981). No doubt, a new form must be in existence in nature for many years before it is recorded in literature. The three forms of *cephalata* might have been influenced by the climate, human selection and their adaptation. The influence of human selection may also be seen that the seed of these types if raised without selection pressure, the mixture of all these different forms appear in the population more easily.

11.2.0 Cultivars

A systematic grouping of the original Chinese cabbage cultivars can be obtained only in Japan since scientific development of the cultivars were taken up early once the crop was introduced in mid-late Nineteenth century and became a very important vegetable crop. The early genotypes which were introduced from China were classified in the following groups (Watanabe, 1981).

11.2.1 Chiffu Group

Leaves thin and numerous; a strong savoic leaf form with many pubescences and blanching only at relatively low temperature; heads are cannon ball to round topped which is a distinct character of this group; leaf colour varying from dark green (in cooler areas) to yellow green (warmer areas).

11.2.2 Hotoren Group

Leaves thick and mostly light green and round topped head; these blanch easily and at relatively low temperature. These cultivars used to be considered of very high quality, but are susceptible to many common diseases. Due to its susceptibility these became uneconomic and disappeared from cultivation around 1925. However, the favourable genes remain in the progeny of hybrids with other groups.

11.2.3 Kaga Group

These cultivars are considered to have originated from a natural cross of Chiffu and Hotoren group and there is no specific characteristic associated with it. In general, Kaga group have thicker, somewhat larger, dark green leaves and round topped heads.

11.2.4 Aichi Group

These cultivars have thick light green leaves with less pubescence and a smooth surface. The head top is fairly large and round; these have good quality, low fibre content and blanch easily; generally, these are late bolters.

11.2.5 Kenshin Group

These are early maturing and require 40-60 days sowing to maturity. This is of subtropical origin and has moderate heat tolerance. They are unsuitable for traditional cold-season growing because their flower bud differentiation is extremely early.

Tropical group is of very recent origin and may have been developed from the Kenshin or some other allied group. AVRDC, Taiwan has done a pioneering work in developing a number of cultivars and F_1 hybrids which are capable of head formation even at a temperature above 20°C. Asveg I, an F_1 hybrid from AVRDC has shown promise in many tropical and subtropical regions. Tropical F_1 hybrids are now available from the private seed companies as well.

The present day Chinese cabbage cultivars are mainly F_1 hybrids within and between the groups and hence a proper classification may not be possible. The temperate cultivars may be grouped as spring-summer, summer, summer-autumn, autumn and autumn-winter cultivars and within each cropping season cultivars with varying maturity periods are available. Thus, once well-known open-pollinated cultivars, *e.g.*, Pe-Tsai (non-heading loose leaved), Michihili (heading type cylindrical) have been replaced or being replaced by the F_1 hybrid cultivars. Besides, in Europe this crop is now grown under cover (plastic tunnel or glass-house) early in spring and in open thereafter. Hence cultivars of Chinese cabbage and Pak-choi have been developed which are highly resistance to premature bolting covering almost the entire season of growing. Cultivars like Neon, Granado and similar other F_1 hybrids are becoming popular among Chinese cabbage while in Pak-choi, Hypro, Hylon and similar F_1 cultivars with resistance to premature bolting are being marketed. In Japan, F_1 hybrids with specific season of growing and wide adaptability are available. In Pak-choi, occurrence of broad white petiole is common which is the characteristic feature of Chinese cabbage (ssp. *pekinensis*).

In Belgium, cvs. Spring A-1 and Nagaoka are suitable for summer while Hong Kong for autumn cultivation (Benoit and Ceustermans, 1983). From Netherlands it has been reported that cvs. Kasumi, Nepos, Spring Sunand Nestor performed well in glasshouse (Bayense, 1986). In Indonesia, cvs. Waka and Nagarka have been recommended for cultivation (Sunarjono, 1980).

Quite a large numbers of Chinese cabbage cultivars have been developed in China. Some of these cultivars are mentioned below:

A high quality, cold resistant heading Chinese-cabbage cv. Huang Ya 14-1 has been reported by Xia, and Lian (1996). It produces stender and tasty leaves, the inner ones being bright yellow. Yields are in the region of 37.5-52.5 t/ha. Cv. Xiaoyeqing was bred by hybridizing two self-lines. It has characteristics such as good shape, short petiole and high disease resistance (Zhu *et al.*, 1996).

A new F_1 hybrid Zhengbai No. 4 has multiple disease resistance, and high and stable yields (106-147 t/ha) (Liu *et al.*, 1996). An early, and heat resistant, heading type F_1 cultivar Weibai1 was developed by crossing 2 self-incompatible lines Zhangpu 401 and Zhexiao 401 (Taili *et al.*, 1997). Hybrid Jinbai 56 was produced by crossing two self incompatible lines. It has good adaptability and can be planted in spring, summer and autumn. It is resistant to tipburn, virus, downy mildew and soft rot (Wang *et al.*, 1999).

Shilu 85 is a mid season maturing F_1 hybrid of heading Chinese cabbage obtained by crossing self incompatible lines 91-7-2 and 88-11-1-1. It has wide adaptability, disease resistance and high yield (Shou *et al.*, 1999).

In Korea, the autumn F_1 hybrid Hamhung 3 was bred by crossing 2 self incompatible lines to obtain hybrid vigour for plant type and compatibility. Yield is highest when sown in mid-July, transplanted in mid-August and fertilized with about 1050 kg N per ha (Han, 1995).

Chinese cabbage cultivars E53023, Yamiko, Kasumi and Nerva were most resistant both to club root and tip burn and they have been recommended for field cultivation in Poland (Borkowski, 2000).

11.3.0 Soil and Climate

Fertile soil with sufficient moisture supply and in good physical condition is required to raise a satisfactory crop of Chinese cabbage. Soil pH of 6.8 to 7.8 is suggested by Letard (1984).

As in cabbage, Chinese cabbage is a cold season crop in tropics and subtropics while in temperate climate it is grown as summer, autumn and spring crop (under protection).

In Chinese cabbage, yields decreased when air temperatures were either too high or too low, and highest yields would be obtained at a mean air temperature of 25°C during the growth period (Lin *et al.*, 1995).

11.4.0 Cultivation

As in cole crops, seeds are sown in raised seed bed and after about 4 weeks, seedlings are transplanted with a spacing of 30-40 cm with in and between the rows. Other cultural operations are same as those given for cabbage except that growing period of Chinese cabbage is less than cabbage or cauliflower.

The critical sowing date for obtaining marketable heads is mid-August in Japan (Seyama and Takai, 1982). In European countries, Chinese cabbage is grown throughout the year either in open or in plastic greenhouses or under perforated plastic protection. Lechl and Frenz (1977) in Germany suggested that in order to obtain good heads and high yields, seeds of Hong Kong F_1 and Nagaoka F_1 should are sown by 25 August for growing in greenhouse. Kiffner *et al.* (1983) also from Germany reported that when planting was carried out from early January to end of March in plastic greenhouses for harvesting at the end of March until May, yields were found to be increased with later sowing dates.

Coffey and Disney (1992) indicated potential for cultivation of Chinese cabbage in Tennessee, USA during spring and autumn season. Seedlings were raised in cellular trays under glass and transplanted in early to mid-April and mid-August for the spring and autumn crops, respectively; they were spaced 12 inches within the row, in rows 36 inches apart. In Taiwan, the cvs. Fengshan and Improvement, produced tallest plants and resulted in the highest leaf area index. Yields were also highest from planting in September, followed by October, August and July (Lin *et al.*, 1995).

Letard (1984) recommended a fertilizer dose of N_2 at 200-220 kg/ha, P_2O_5 at 100-150 kg/ha, K_2O at 200-250 kg/ha and MgO at 50-100 kg/ha for Chinese cabbage. Venter (1983) applied nitrogen @ 125 or 250 kg/ha in the form of calcium nitrate, ammonium sulphate, calcium cyanamide or Alzon and observed that higher N treatments produced higher yields and higher nitrate levels in the crop. Nitrate levels were lower with calcium cyanamide than with other N sources. Zheng *et al.* (1995) considered that Mo was advantageous for decreasing the accumulation of nitrate under high N supply.

In China, Liu *et al.* (1999) studied the effect of different ratios of NPK combination on yield and nitrate accumulation of Chinese cabbage. Best result was obtained with 360 kg N + 90 kg P and 180 kg K per ha having a plant density of 31500/ha. The nitrate accumulation was increased with the increase of the amount of N applied. Phosphate fertilizer had no significant effect on nitrate accumulation in plant; however, potassium fertilizer had a significant effect on nitrate content in plant. 30 and 50 days after planting were two key periods for fertilizer application on Chinese cabbage. Obreza and Vavrina (1993) reported from USA that yields (head weight/ha) and percentage marketable heads of cv. Tropical Quick increased with increasing N application rate (0 to 157 kg/ha). They recorded highest yield of 21.2 t/ha when at least part of the N was applied as NO_3.

Zheng *et al.* (1999) studied cultivating techniques of different spring Chinese cabbage cultivars in China. They found that application of calcium super phosphate at 750 and 1125 kg/ha increased yields by 33.19 and 51.17 per cent, respectively. Transplanting significantly increased yields of Zaoshu 6 compared to direct-sown crops, and use of soil surface mulch also significantly increased yields. The mean yields of crops transplanted into mulched beds was 21582 kg/ha compared to crops transplanted into non-mulched beds which yielded 12478 kg/ha. Overall best cultivars were Xiasheng, Zaoshu 6 and Chundajiang.

Field Growing and Protected Cultivation of Chinese Cabbage

Chinese Cabbage under Open Field Condition

Chinese Cabbage under Protected Cultivation

Letard (1984) advocated chlorthal at 7500 g/ha as pre-planting and propachlor at 4560 g/ha as post-planting for control of weeds in Chinese cabbage.

Felczynski (1995) observed in trials in Poland that Chinese cabbage cultivars grown with row covers of non-woven polypropylene advanced harvesting by 1 week and increased yield by 77 per cent compared with uncovered control. Under covers marketable yields ranged from 5.3 kg/m^2 in cv. Nepos to 8.2 kg/m^2 in cv. Nerva. Bolting and cabbage fly (*Delia radicum*) damage were less of a problem under covers.

11.5.0 Physiological Disorders

'Tip burn' is one of the most common physiological disorders of Chinese cabbage. It has been reported that calcium deficiency results in rotting of leaf margins. The deficiency results mainly from inhibited calcium absorption or translocation within the plant rather than calcium shortage in the soil. It is believed that high salt concentration (especially high NH_4-N) or high moisture tension-both of which inhibit calcium absorption from the soil induce calcium deficiency in the plant.

ShuFen (1996) described that during tip burn development, the photosynthetic rate decreased, and leaf sugar content and electrolyte leakage increased.

Disorder of Chinese Cabbage

1. Chinese cabbage plant without tip burn or rotting symptoms; 2. Tip burn symptoms; 3. Rotting leaf tissue; 4. bacteria visible in the rotting sector (Borkowski *et al.*, 2016)

Harvesting and Packaging of Chinese Cabbage

Ultrastructural studies showed that the plasmalemma became invaginated, small vesicles emerged and the organelles started to disintegrate. Finally, soluble pectin content increased markedly and that cell wall lost its integrity. Solid remains could be seen in the cell cavities and intercellular spaces with the appearances of the paper like symptoms. According to Maroto *et al.* (1996) tip burn is related to low Ca concentrations in young leaves and is precipitated by rapid growth under high temperatures and low R.H. Twice foliar applications of CCC at 1120 or 2240 ppm combined with $CaCl_2$ at 2 g/1 or 4 applications $CaCl_2$ alone reduced the appearance of tip burn in Spain. Wen *et al.* (1991) prevented tip burn in Chinese cabbage by spraying 5 times with 0.7 per cent calcium chloride + 50 ppm NAA.

Everaarts (2000) reported that the factors influencing the occurrence of internal tip-burn are disturbance of water balance in plants and high growth rate as well as early planting and late harvesting.

Boron deficiency symptoms which are also common in Chinese cabbage appear at the beginning of head formation. On the inner surface of the midrib's cracks appear and turn brown. If deficiency is severe, the growing point dies. Control measures are similar as in other brassicas.

Even excess of nitrogen may cause disorder. After head formation, many small black spots appear on the surface of midribs on the inner leaves. Heavy nitrogen application, especially as side dressing after heading formation appears to be the cause of this disorder.

11.6.0 Harvesting and Yield

Fully developed heads are cut from the stalk and loose outer leaves removed before marketing. Yield of Chinese cabbage varies from 25-50 tonnes/ha depending on the type, cultivar and growing season.

11.7.0 Storage

Shelf life of cvs. Osyo and Ozeki is 30 days at 5°C, 60 days at 2.5°C and 90-120 days at 0°C (Emura, 1978). Apeland (1985) reported that cvs. WR Green 60, Tip Top and Treasure Island can be stored for an average period of 90 days at different temperatures ranging from 0 to 10°C. The occurrence of physiological disorders like brown midribs is due to prolonged storage at 0°C. It has been reported from France that storage of Chinese cabbage from November to early January at 0.5 and 85 per cent R.H. is best (Letard, 1984).

Low O_2 (per cent) atmosphere at 0°C is effective to prolong the storage life of Chinese cabbage up to 5 months (Wang, 1983). It has been observed that cultivar Tip Top could be stored well at 5°C with 5 per cent CO_2 + 5 per cent O_2 (Apeland, 1985).

12.0 Diseases and Pests

The main emphasis of this chapter will be on the control of diseases and pests. Presently, the idea of controlling diseases and pests by a combination of cultural methods, fungicides and insecticides and by using resistant cultivars is gaining importance as the most effective measures. However, due to the introduction of newer chemicals, information on this subject gets out of date. Besides, residual toxicity of the chemicals is becoming a more severe concern of the consuming public, especially on vegetables and more particularly on cole crops, since some of them are consumed fresh. Therefore, the most sustainable approach is to develop resistant varieties or hybrids using identified resistance source and advanced breeding tools.

12.1.0 Diseases

Fungal Diseases

12.1.1 Black Leg

The disease is caused by dothideomycete fungal pathogens, *Leptosphaeria maculans* (Desm.) Ces. and de Not. (anamorph = *Plenodomus lingam*) and *Leptosphaeria biglobosa* Shoemaker and Brun (anamorph = *P. biglobosus*) (Fitt *et al.*, 2006; Mendes-Pereira *et al.*, 2003). The pathogen reproduces both sexually and asexually, can complete multiple disease cycles in a single growing season, and survives on infected crop stubbles as a saprophyte. Under favorable conditions, especially at moderate temperatures and high humidity during vegetative growth, the disease can cause up to 50 per cent yield loss. The fungus is carried by the seed and hence it may occur from the early stage. Stem of the affected plant when split vertically, shows severe black discolouration of sap stream. Whole root system decays from bottom upwards. Frequently, the affected plants fall over in the field.

As the main infection is through seeds, hot water treatment of seed at 50°C for 30 minutes is recommended. Seed can be treated with Captan @ 3 g/kg seed. Spraying infected plants with 0.2 per cent Captan is also effective to control the disease.

Pusa Drumhead, a cabbage cultivar has been reported to be tolerant under field condition.

12.1.2 Downy Mildew

It is caused by an obligate fungus *Hyaloperonospora parasitica* (Pers. Fr.). The disease is more severe in cool climates with frequent dew formation at night, which induces leaf and curd damage of cauliflower and makes the crop unmarketable. In India, this disease was first reported by Butler (1918). At present, it is prevalent in almost all states of the country wherever Brassica vegetables are grown. Under favorable conditions, the pathogen infects 50–60 per cent seed production of cole crops (Verma and Singh, 2018). It may attack young plants and also at the seed production stage as being commonly observed in Northern India in recent years, when high humidity prevails during seed production stage. The fungus when attacks the young seedlings, discolouration occurs and in severe cases the whole plant perishes. In adult plants, purplish leaf spots or yellow brown spots on the upper surface of the leaf appear, while fluffy downy fungus growth is found on the lower surface. The disease causes defoliation in adult plants in the field, thereby reducing the yield and quality and may become epidemic under favorable climatic conditions. During the bolting stage, the seed stalks show blackish patches and in severe cases the whole curd is spoiled.

To control the disease spraying of 0.3 per cent Mancozeb (Dithane M-45) has been recommended. In some cases, spraying may be necessary twice a week. Seed treatement in hot water at 50°C for 30 minutes is also recommended. Crop rotation with non *Brassica* vegetable crops and clean cultivation can reduce the infestation.

Recently from IARI large number of resistant lines has been identified through controlled inoculation. Singh *et al.* (2013) identified seven resistant genotypes *viz.* BR-2, CCm, 3-5-1-1, CCm-6, CCm-5, MGS-2-3, and cc-12. Saha *et al.* (2018) identified some resistant lines DMR-1-0-6, DMR-2-0-7, DMR-3-0-8, DMR-6-2-2-6-2. In Himachal Pradesh, Verma and Singh (2018) identified resistant lines DPCaY-8 and DPCaY-6 and also suggest the monogenic inheritance of resistance. Emphasis have been given to find out the gene (s) and inheritance which have been reported as single dominant genes in broccoli and cauliflower (Farnham *et al.*, 2002; Verma and Singh, 2018), a single recessive gene in cauliflower (Hoser-Krauze *et al.*, 1984), and multiple (additive) genes in cauliflower, broccoli and cabbage (Moss *et al.*, 1988; Hoser-Krauze *et al.*, 1995; Jensen *et al.*, 1999b). The genes for resitance have been idenetified at IARI as Ppa^{3} (Singh *et al.*, 2015) and Ppa^{207} (Saha, 2020).

12.1.3 Rhizoctonia

It is caused by *Rhizoctonia solani* (imperfect form) and may attack at different stages of growth. Young seedlings suffer the worst. Young plants show soft, water soaked spots on the skin just above the ground, the cotyledons wither and the plant eventually falls over and perishes. When infection occurs at a later stage of growth, the lower part shows discolouration over a length of several centimetres, becomes hard and woody, and thinner than usual as the cortical tissue dies and this phenomenon is known as wire stem. In older cabbage plants, the fungus causes a kind of hard rot, known as bottom rot. High temperature and high humidity favour the infection.

For effective control, both seed and plant bed should be disinfected and all affected plants should be discarded at the time of transplanting. Drenching of root zone with carbendazim 12 per cent + mancozeb 63 per cent WP @ 3 g/liter is also effective. In an experiment by Dabbas *et al.* (2009) showed that the seed treatment with carbendazim @ 2 g/kg seed, raising seedling in solarized beds, crop raising in green manure field + neem cake 25 kg/ha with soil treatment by *Trichoderma viride* @ 2 kg/ha gave the lowest disease intensity of disease with maximum curd yield.

12.1.4 Black Spot or Curd/Head and Inflorescence Blight

It is caused by species of the genus *Alternaria*. The most important species are *brassicicola* and *brassicae*. The disease occurs all over the world. The fungus is seed-borne. Young plants affected by the disease may resemble those resulting from *Rhizoctonia solani*. In older plants, leaves, petioles, and stems small, brown to black circular to slightly elongated spots appear. Sometimes, the spots join together. It causes damage to cabbage heads and cauliflower curds after maturity and during seed production stage. In cauliflower, at seed production stage, it has become quite a problem in all the areas. It is mainly caused by *A. brassicicola* and known as curd and inflorescence blight or brown rot. Seedling plants suffer badly and at later stage the siliquas turn yellow prematurely. Harvested seeds are contaminated and the germination capacity is reduced.

As a rule, spray of Captan @ 3 g/liter as against downy mildew should be followed. To reduce seed losses, affected seed crops are harvested prematurely followed by disinfection of seed before sowing. Some tolerance has also been recorded in few cultivars.

Sharma *et al.* (1972) reported that the resistance to the disease caused by *A. brassicicola* in cauliflower is controlled by a polygenic system and the resistance to the disease is dominant over susceptibility and in general, additive gene effects were found to be greater than dominance. The resistance was initially found in MGS-2-3, Pua Kea, 246-4 *etc.* The resistance has now been incorporated in selection1-6-1-4, 6-1-2-1 and 1-6-1-2 by back-cross method.

12.1.5 Yellows

It is caused by *Fusarium oxysporum* f. *conglutinans*. The foliage of affected plants turns yellow, dies and eventually drops off, the lowest being the first to show the symptoms. The vascular tissue becomes yellow to dark brown. Young seedlings may occasionally show signs of damping off. Since resistant cultivars are available, it is better to use them rather than to try other means of control. Addition of organic matter to the soil increased microbial activity due to addition of organic matter suppresses the fungal growth. Seed treatement with captan 2 g/kg seed is also effecitive.

Two types of resistance have been reported:

Type A- Single dominant gene controls the resistance, found first in cabbage cultivars All Head Early, Copenhagen Market, Glory of Enkhuizen (Walker, 1926, 1930, 1933; Walker and Wellman, 1928). It is stable at different soil temperatures. This apparently has been derived directly from *B. oleracea* var. *sylvestris*.

Type B- This is inherited as a quantitative character and is controlled by a number of genes. The resistance breaks down at or above 24°C. The cabbage cultivar Wisconsin Hollander has this type of resistance. Both A and B types are found in Wisconsin All Season. In Brussels sprouts and knolkhol, the disease has been reported, whereas Calabrese, De Cicco and Waltham 29 of broccoli and Early Purple cauliflower were found resistant by Natti and Atkin (1956). However, presently a large number of cultivars of cabbage are available which are resistant to the disease.

12.1.6 Club Root

It is caused by *Plasmodiophora brassicae*, a typical root parasite. As many as 9 races have been recognized. Once affected it causes irregular galls on the roots and those on the lateral roots take the shape of spindle. Leaves become yellowish and wilt on hot days. Club root is particularly prevalent on soils with a pH below 7.0, whereas it has been observed that the disease is often less serious on heavy soils and on soils containing little organic matter. Since the emergence, management of the disease has been a big challenge due to the obligate biotroph nature of the pathogen. The disease can be controlled by ploughing the field in summer months. Liming to correct the soil pH as the disease is particularly prevalent in acidic soil.

Treatment of seedbed with 1 per cent formaldehyde and soil drenching with 0.2 per cent captan considerably reduces the infestation.

Walker and Larson (1951) reported that susceptibility was dominant over resistance, though in F_2, 10-20 per cent resistant plants were recovered indicating a polygenic inheritance. Gallegly (1956) also supported polygenic theory. Standbeerg and Williams (1967) showed that resistance to race-6 and 7 was controlled by a single dominant gene. Vriesenga and Honma (1971) reported that resistance to race 1 and 3 involved two interacting genes with clubs on the main root dominant to dubs on lateral roots and to absence in roots. Dakouri *et al.* (2018) studied using diallel crossing methods or segregating populations and only one major resistance gene, *Rcr7*, has been identified, and it might be located on chromosome 7 (LG 7) in cultivars, Tekila and Kilaherb of cabbage. Mehraj *et al.* (2020) reported that about fifty QTLs have been identified in the studies using dierent populations wherein three QTLs in broccoli, five in cabbage, three in cabbage using the GBS technique. The identification of several club root loci indicates that club root resistance is controlled in a polygenic manner, confirming the complex genetic basis of the resistance, where a single resistance locus is not enough to confer sufficient resistance.

Somatic hybrids were produced between a Japanese radish with immunity to *Plasmodiophora brassicae* and susceptible cauliflower cultivar Nokaziwase by electrofusion of protoplasts from hypocotyls. Morphologically intermediate plants were immune to *P. brassicae* (Hgimori and Nagaoka, 1991).

12.1.7 Sclerotinia Drying or Watery Soft Rot

It is caused by *Sclerotinia sclerotiorum* and occurs all over the world. This disease is causing considerable damage in the seed crops of Snowball in Himachal Pradesh. Though earlier, it was not a problem in the growing areas of cole crops in the plains, but now a days it is causing huge loss in seed production of cauliflower.

In the seedling crop, the infection starts on the leaves, particularly on the base of petioles and passes into the adjoining part of the stem on which large, greyish-white, elliptical spots arise, sometimes involving the whole surface of stem. At the point of attack, the seed stalks can break, wither and eventually die. In cultivar Snowball, the losses may be even upto 50 per cent in some years. In seed crops, it is more commonly known as 'Sclerotinia stalk rot'. The symptoms are distinct from those of other diseases, in that this one involves the formation of white, often excessive tufts of mould and of hard, first white but later black sclerotia which are usually several milli metres in diameter. In cabbage, as the fungus advances to the bases of the outer leaves, the plant suddenly wilts. In cauliflower, watery soft rot symptoms are observed in the stem through the central portion of the curd and the plant may wilt. From distance the plant gives silvery appearance. The seedstalks break at the point of attack.

Generally, damp weather favours the occurrence of this disease. Infection is also affected through wounds. In dry soils, the sclerotia can remain viable for more than a year, though they are short-lived in wet soil and hence rotation with crops which are not affected by this disease would be desirable. The disease can be controlled by deep ploughing as the sclerotia of the fungus cannot survive below

15 cm. Flooding the field during summer as the sclerotia fail to germinate in free water and their population is thereby reduced. Application of sawdust of the soil @ 10 q/ha at least 15 days before planting reduces sclerotial population. Crop rotation with paddy, maize or Jowar for 3 years is found effective. Spraying of 64 per cent w/w mancozeb + 4 per cent w/w metalaxyl-M @ 0.2 per cent is very effective in controlling the disease.

12.1.8 Powdery Mildew

The disease is caused by *Erysiphe polygoni*, scarcely occurs in cole crops, though occasionally it has been known to cause limited damage. Initially, white tufts of mould arise on the upper surface of the leaves and later run together and the entire leaf becomes covered with greyish white mycelium. Siliqua also may show the symptoms.

Sulphur dust or spray with karathane has been reported to be very effective.

12.1.9 Pythium Disease

Many species of *Pythium* cause damping off of young seedlings. The best known species is *P. debaryanum*. The fungus hibernates in the soil and when seedlings are affected, they fall over and perish. The disease is promoted by high humidity and high temperature and the damage in the seed bed, in general, is considerable. Both soil sterilization and seed treatment are effective in reducing the attack.

Bacterial Diseases

12.1.10 Black Rot

Black rot caused by bacterium *Xanthomonas campestris* pv. *campestris* (Xcc) (Pammel) Dowson has been reported to be most devastating disease worldwide. It occurs everywhere in the world and particularly destructive in warm and wet climate. The disease has a wide geographical distribution and destructive to these vegetables causing yield (50-65 per cent) and quality loss due to cultivar susceptibility and environmental conditions. The disease spread through rain drops and the pathogen survive for long periods on crop debris, seeds, and weeds. The pathogen remains in the vascular system of the infected plants until favourable environment for bacterial growth take place. Since the first recognition of *Xcc* races, several studies analyzed black rot resistance in relation to race types. Initially, Vicente *et al.* (2001) identified six races (1–6), and later on, Fargier and Manceau (2007) added three additional races (7–9). Races 1 and 4 have been found to be the most virulent, widespread and accounting for more than 90 per cent of black rot disease around the world (Vicente *et al.*, 2001; Saha *et al.*, 2016).

The bacterium spreads through vascular tissue, clogg vessels. The infection of the foliage results in yellow 'V' shaped spots arising along the margin which extend in the direction of the midrib. These spots are associated with a typical black discolouration of the veins. The infection extends through the xylem to the stalk and the vascular bundles tum black. Marginal leaf chlorosis, necrosis and darkening of leaf veins and vascular tissue within the stem are main characteristic symptoms. In severe infestation, the whole leaf shows discolouration and eventually falls off.

In plants for seed production the disease may extend through the flower stalks to the siliquas. Plant sraised from infected seed may die prematurely. Infection by *black rot* may be followed by attack of soft rot becteria. The disease can be managed by growing resistant cauliflower varietie Pusa Shubhra and Pusa Snowball K-1 and cabbage cultivar Pusa Mukta. Seed treatment with hot water at 50-52°C for 30 minutes followed by a dip in 0.01 per cent Sterptocycline, for 30 minutes. Spraying of 0.01 per cent Streptocycline at transplanting, curd initiation and pod formation stages gives good control measure. Application of boron reduces the incidence as phenolic compounds that are accumulated in the leaf tissue due to application of boron check the disease. Bain (1951, 1955) established two cabbage cultivars resistant to black rot, *viz.*, Hugguenot and Early Fuji. From the crosses between the resistant and susceptible cultivars he concluded that the resistance was governed by one or more dominant factors. Jensen *et al.* (2005) found field resistance to black rot in hybrid cultivars T-689 F1, Gianty F1, No. 9690 F1, N 66 F1 and SWR-02 F1. Griesbach *et al.* (2003) reported strong resistance in cabbage inbred line AU 4518 and some degree of resistance in cv. Green Challenger, Utopia, Shinzan No. 2, Natsutae, Rotan and Tenacity against mixture of isolates containing races 1 and 4. Resistant source such as Pusa Snowball K-1 (late group cauliflower variety) was reported by Gill (1993) under temperate conditions. Sharma *et al.* (1995) reported cauliflower line EC-162587 line as highly resistant while RSK-1301 and MRS-1 as moderately resistant to black rot disease. Saha *et al.* (2014, 2016) identified black rot resistant line *viz.* BR-1, BR-161, BR-207, AL-15 agist race 1 in cauliflower.

Williams *et al.* (1972) studied the inheritance of resistance in Japanese Early Fuji and suggested that resistance is controlled by one major recessive gene and in heterozygous condition, influenced by two modifiers-one dominant and one recessive. The resistant genotype may be in any of the following:

ff	-	-
ff	Aa	BB
Ff	Aa	Bb

Tewari *et al.* (1979) studied the inheritance of resistance involving 3 resistant cultivars, *viz.*, Spitzkool, Greenland Hammerand Sci. 8 (24855 x 10109) and three susceptible cultivars Golden Acre, Pusa Drumhead and Desserto and concluded that the resistance was dominant and governed by polygenes. The gene effects showed that additive component was more pronounced.

Sharma *et al.* (1972) studied the inheritance of resistance to black rot in cauliflower involving 3 resistant lines, *viz.*, MGS-2-3, Pua Kea and Sel. 445 and 4 susceptible lines, *viz.*, Sel. 246, Sel.15, 12013 and 12012. In this study, the resistance to black rot was found to be dominant over susceptibility and governed by polygenes. The dominance component of variation was greater than the additive in almost all the crosses.

Sharma *et al.* (1995) reported *B. oleracea* var. *botrytis* line EC- 162587 line as highly resistant while RSK- 1301 and MRS-1 as moderately resistant to black rot. Kaur *et al.* (2009) reported single dominant gene in January King of cabbage. Vicente *et al.* (2002) reported resistance to race 3 in DH line derived from cv. Bohmerwaldkohl is controlled by single dominant gene. Saha *et al.* (2014) studied the inheritance in a cross of Pusa Himjyoti (susceptible) and BR-161 (resistant) and the result showed that resistance is dominant over susceptibility and the resistance is governed by single dominant gene *Xca1bo*. In aother study by Saha *et al.* (2016) in the cross of Pusa Sharad (susceptible) and BR-207 (resistant), the resistace is governed by single dominant gene. There have been several investigations which focused on improving the interspecific hybridization to transfer black rot resistant gene in cauliflower. Dey *et al.* (2015) reported introgression of black rot resistance genes into Indian cauliflower through inter-specific hybridization (*B. carinata* x *B. oleracea*) followed by embryo rescue. They utilized two *B. carinata* genotypes, NPC-9 and HCA-6, having very high resistance to *Xcc* race 1 and 4, for inter-specific hybridization to introgress resistance genes from *B. carinata* into Indian cauliflower employing three different embryo rescue approaches. They reported that all the F_1 hybrids and back cross were resistant to *Xcc* races 1 and 4. Recently, Sharma *et al.* (2018) developed interspecific hybrid and backcross progeny (B_1) between cauliflower "Pusa Sharad" and Ethiopian mustard "NPC-9" employing *in vitro* embryo rescue technique. Fertile giant pollen grain formation was observed frequently in interspecific F_1 hybrid and BC_1 plants. The F_1 inter-specific plants were found to be resistant to *Xcc* race 1.

12.1.11 Bacterial Soft Rot

It is caused by bacteria of which the most importantis *Erwinia carotovora*. It is a weak parasite, which only penetrates the plant through damaged tissue, *e.g.*, lesions caused by other pathogens. The infection is favoured by high humidity. Affected plants show a soft, slimy, bad smelling rot which under favourable conditions rapidly spreads throughout the entire plant. In stems, this organism causes 'stump rot'. Seedling plants, particularly cauliflower (Snowball) have been reported to be infected very badly, especially when damaged by hails. The whole plant in severe attack falls over.

It is one of the most destructive diseases of Chinese cabbage. Occurrence of soft-rot disease may involve essentially three processes: establishment of inoculum; its invasion in the stem of the host plant and its multiplication; inciting symptoms of soft rotting in the stem tissue. It has been found by Kikumoto (1981) that the bacteria invade the stem tissue of Chinese cabbage more or less easily, but the invasion will not necessarily cause the disease. He found a threshold value for the outbreak of the disease will vary with the specificity of a host-parasite system. He concluded that bacteriocin products are promising in controlling the disease. However, resistant cultivars would be the most effective means of control of this disease. In Japan, Hiratsuka No.1 has been reported to be highly resistant cultivar of Chinese cabbage to this disease. Putter (1999) recommended spray of $CaCl_2$ (20-40 kg/ha) every 5-6 days during cloudy weather in order to prevent bacterial soft rot (caused by *Pseudomonas fluorescens*) in broccoli.

The disease can be kept down by controlling other diseases and preventing damages. Avoid mechanical injury; removal of rotten leaves is the cultural method

Disorders and Diseases of Cabbage

Head Cracking (Left) and Tip Burn (Right) in Cabbage

Alternaria Leaf Spot (Left) and Yellowing (Right) in Cabbage

Black Rot of Cabbage

Damping Off (Left) and *Erwinia* Soft Rot (Right) of Cabbage

Diseases of Cabbage

Colletotrichum Blight of Cabbage

Powdery Mildew of Cabbage

A) Downy Mildew, B) Black Spot, C) Black Rot, D) Club Root in Cabbage

Diseases of Cauliflower

Sclerotinia (Left) and Black Leg (Right) Disease of Cauliflower

Alternaria Leaf Spot of Cauliflower

***Colletotrichum* Blight of Cauliflower**

Diseases of Broccoli and Brussels Sprout

Alternaria Leaf Spot of Broccoli

Downy Mildew of Broccoli

Sclerotinia Stem Rot (White Mold) (Left) and Ring Spot (Right) of Broccoli

Club Root in Broccoli Club Root (Left) and Black Rot (Right) of Brussels Sprout

of management. Spraying the crop with 0.01 per cent Streptocycline at weekly interval has been found to be effective.

12.2.0 Disease Control by Disinfection and Cultural Methods

12.2.1 Healthy Seed

Many pathogens are transmitted through seed in cole crops, *e.g.*, *Leptosphaeria maculans*, *Alternaria* sp., *Xanthomonas campestris* pv. *campestris*, *etc*. There is hardly any seed producing area in the world where conditions are completely unfavourable for these diseases. The conditions that can minimize the infection would be more or less rain free period during flowering and seed ripening.

The disinfection of seed before sowing would be a sound method to avoid effective pathogens to a great extent. In this method, 2-3 g of organomercuric compound per kg of seed may be useful for dry disinfection. Still better results may be obtained, especially in case of *Xanthomonas* or *Phoma* attack by immersing the seed for 30 minutes in water at 50°C but such treatment with seed of low germination capacity or old seed would reduce the germination considerably. In such cases, antibiotics may also be used.

After germination of seeds, the young seedlings may be attacked by *Rhizoctonia solani* and *Pythium sp*. both soil-borne fungi. In such cases, the treatment of seed bed by captan (@ 3 g/1 (drenching) or disinfection by 1 per cent formalin (drenching), followed by covering the bed with alkathene film for 4-5 days and then by aeration of soil would be very useful. Fungal culture of *Trichoderma viride* is mixed with cow dung manure (@ 1 kg per 5 kg) at least 15 days before sowing and then these are mixed with nursery soil. This treatment is very useful to control soil borne fungi. Gopalakrishnan and Artal (2013) reported that the combined treatment of hot water treatment of seeds at 50°C for 15 minutes followed by two applications of Streptocycline (100 ppm) + copper oxychloride (0.3 per cent) at fifteen days interval starting from fifteen days after transplanting recorded lowest disease incidence and significantly highest yield, as compared untreated control.

12.2.2 Seed Bed

The seed bed is preferably made on a well drained, healthy soil of good structure and the same spot should not be used for more than once in 3 years. The seed bed should be disinfected, the seeds be sown thinly and over-moist bed should be avoided.

12.2.3 Planting in the Field

Only healthy and stocky seedlings should be selected. Crop rotationis of utmost importance and the field should be used for Brassica once in 3 years. In case of club-root infection, a longer rotation is desirable besides the soil treatment with lime if the pH is low. Sanitary measures in case of implements, diseased parts of plants, *etc.*, would be advantageous in reducing the spread of disease.

Damping off in cauliflower caused by *Rhizoctinia solani* was appreciably reduced when population of the pathogen and other microflora in soil were manipulated by amending soil with powder of commercial cellulose, rice stubbles or water hyacinth biomass in combination with NH_4NO_3. Cellulose powder was most effective in reducing disease incidence. In general, when the C:N ratio in soil was increased, the fungal population decreased but actinomycetes and bacterial populations increased (Kunduand Nandi, 1985). Watery soft rot (*Sclerotinia sclerotiorum*) infestation in cauliflower could be reduced to 50 per cent by using pine needles or sunflower inflorescence residues as soil mulch. Crop rotation with rice proved highly effective in reducing the infection also (Sharma *et al.*, 1986).

Srivastava *et al.* (2018) found that treatment with all the isolates of *Trichoderma* spp. and *Pseudomonas fluorescence* resulted in significantly higher seed germination, less disease and higher dry biomass over the non treated control. Two neem based formulations also significantly suppressed the disease and were on par or superior in effectiveness as compared to the fungicides metalaxyl 8 per cent, mancozeb 64 per cent (Ridomil MZ 72) and captan 50 per cent WP (Captra 50 WP).

12.3.0 Pests

12.3.1 Caterpillars

There are many different caterpillars which attack cole crops, of which the best known is large white butterfly, *Pieris brassicae*. There are 2 other related species like *P. rapae* and *P. napi*; of these however, the large white butterfly is the most common in India. Caterpillars of the diamond back moth (*Plutella xylostella*), which occur all over the world, also cause heavy damage in certain areas like Ranchi in Bihar, Maharashtra and many other parts of India.

Among the other damaging caterpillars, special mention should be made of the cutworms (*Agrotis spp.*) which occur practically throughout the world. The large white butterfly is bigger than the small white butterfly which has a wing spread of 5 cm and the wings are yellowish white but with black markings along the top of the fore wings. The large ones lay its eggs in groups of several dozen on the leaf. The caterpillars are 3 cm long, green with black spots and yellow lines.

Diamond back moth derives its name from the white markings along the back of the fore wings which when folded form a diamond shaped pattern. The moths are grey or greyish brown, about 1cm long and have a wing span of 1 or 2 cm. The eggs are at first pale yellow but later greyish brown, 0.5 to 0.8 mm in diameter and deposited on the leaf either singly or in groups of at most 6. The caterpillars are 1cm long and light green or brownish green, when touched they make jumping motions.

Cut worms are greyish brown night moths, variously striped and mottled. Caterpillars are 3-4 cm long, grey or brown to almost black with various markings. The caterpillars of white butterflies are very voracious and cause much damage. They start feeding from the margin of the leaf and proceed to the centre; the whole leaf and the plant may be stripped resulting in poor yield and quality of the produce. The caterpillars of the diamond back moth at first tunnel inside the leaf tissue for some days and subsequently attack the lower surface of the leaf. When present in large number, they may also appear on the upper surface. The caterpillars of the cutworms cause damage by biting into the foliage of young plants and by cutting down the stems just above the ground level. Various species feed on leaves, young buds and fully grown heads of cabbage.

Pests of Cabbage

Diamond Back Moth (Left) and Cabbage Butterfly (Right)

Cabbage Aphids

Head Borer in Cabbage

Pests of Cauliflower

Tobacco Caterpillar in Cauliflower

Cabbage Webber (Left) and Aphids (Right) in Cauliflower

Diamond Back Moth in Cauliflower

Pests of Broccoli

Flea Beetle in Broccoli

Diamond Back Moth in Broccoli

Cabbage White (Cabbage Worm) Feeding in Broccoli

Pests of Brussels Sprout

Aphid Infestation in Brussels Sprouts

Cabbage Worm Feeding in Brussels Sprout

The diamond back moth remains a major pest of brassica crops worldwide because of non-availability of effective management strategies. DBM has been estimated globally to cost US$ 1 billion in direct losses and control costs. Chemical control of this pest remains difficult due to the rapid development of resistance to insecticides and to their effect on natural enemies. These problems are especially severe in South Asia and Africa where lack of knowledge, limited access to newer and safer insecticides, and a favourable climate result in DBM remaining a serious year-round pest which substantially increases the cost and uncertainty of crop production. Despite these problems, application of synthetic insecticides remains overwhelmingly the most common control strategy. Biologically-based efforts to control DBM in Africa and Asia have focused strongly on parasitoid introductions. However, despite the identification and deployment of promising parasitoids in many regions, these efforts have had limited impact, often because farmers continue early-season spraying of broad-spectrum insecticides that are lethal to parasitoids and thus exacerbate DBM outbreaks. A significant driver for this pattern of insecticide use is the presence of aphids and other pests whose appearance initiates inappropriate spraying. Despite often extensive training of producers in farmer field schools, many growers seem loath to discard calendar or prophylactic spraying of insecticides. The introduction of an IPM technology that could replace the use of broad-spectrum insecticides for DBM and other key Lepidoptera is crucial if the benefits of parasitoid introduction are to be fully realised. The deployment of DBM-resistant brassicas expressing proteins from *Bacillus thuringiensis* could help to break this cycle of insecticide misuse and crop loss, but their deployment should be part of an integrated pest management (IPM) package, which recognises the constraints of farmers while addressing the requirement to control other Lepidoptera, aphids and other secondary pests.

12.3.1.1 Chemical Control

The pests may be effectively controlled by different insecticides like parathion, malathion, diazinon, thiodan, *etc.* Spraying is to be done fortnightly or even weekly, the first spray being applied when eggs are deposited. However, the persistent insecticides should not be used after the formation of edible parts. Gupta *et al.* (1985) found spraying of 0.05 per cent chlorpyrifos most effective in controlling *P. brassicae* and *P. xylostella* on cauliflower seed crop. In a field trial in Japan, teflubenzuron controlled a strain of *P. xylostella* against which a mixture of pyrethroids and organophosphorus insecticides was not effective (Kohyama, 1986). In India, these insect growth regulators (teflubenzuron and flufenoxuron) were found to be highly effective against the cabbage pest *Plutella xylostella* which has developed resistance to a large number of insecticides (Peter and Sundarajan, 1991). Calderon and Hare (1986) reported that application of profenofosat 0.25-0.5 kg a.i./ha for controlling *P. xylostella* was most effective with an increased yield of 40-100 per cent compared to control in cabbage and Chinese kale. Considering the residual toxicity of 15 insecticides tested to control the larvae of *P. xylostella* in cabbage, Sinha and Rai (1984) suggested the safest dose for the application of deltamethrin and permethrin was 13.5 g a.i./ha. A field study in Florida revealed that chlorpyrifos, endosulfan, Mevinphos and *Bacillus thuringiensis* sub sp. *kurstaki* were more effective in controlling *P. xylostella* than cypermethrin, permethrin, methomyl and thiodicarb on cabbage (Leibee and Savage, 1992). Application of fipronil (at 25, 40 or 50 g a.i./ha) reduced the incidence of diamond back moth (*Plutella xylostella*) on cabbages by 67.8-84.8 per cent compared with an untreated control (Panda *et al.*, 1999).

Trials carried out in Punjab, India by Kandoria *et al.* (1999) during November to March 1995-96 revealed that planting one row of late season cauliflower (Snowball) with one row of main season tomato (cv. Punjab Kesri) significantly reduced the incidence of diamond back moth (*Plutella xylostella*) when the cauliflower was planted 30 days after the tomato.

Sharma *et al.* (2000) reported that bioasp and biolepat 2 kg/ha gave the highest larval mortality of diamond back moth in cauliflower which was statistically similar to that obtained with cartap hydrochloride 50 WP at 1.0 kg/ha. Applying *B. thuringiensis* also helped to conserve predatory coccinellids.

12.3.1.2 Biological Control

Earlier, ichneumon fly (*Angita cerophaga*) which oviposits on the caterpillars of the diamond back moth was reported to reduce the population. The use has been made of the spore forming bacterium *Bacillus thuringiensis* and of the virus *Bergoldia virulenta*. By spraying the bacterium at the rate of 10^{14} individuals per hectare it was possible to control the caterpillars of *Pieris* spp. and to a lesser extent of diamond back moth and cabbage looper. The feasibility of using *Trichogramma* spp. (mainly, *T. evanescens, T. confusum* or *T. dendrolimi*) to control *Pieris brassicae* and other lepidopteran pests was reported by Heiningen *et al.* (1985), Hassan and Rost (1985) and Atak and Atak (1986) while parasitoid *Diadegma euceropnaga* for the control of *Plutella xylostella* was suggested by Sastrodihardjo (1986) and Sastrosiswojo and Sastrodihardjo (1986). The importance of the bracnoid *Apanteles plutellae* in suppressing field populations of *P. xylostella* was evaluated using an insecticide check method in Malaysia and results showed that *A. plutellae* is an important parasite and can contribute substantially to the suppression of *P. xylostella* populations (Lim *et al.*, 1986). Contamination with virus is also lethal in its effect, but very slowly and hence appreciable damage may be caused before they die. Mateeva *et al.* (1997) controlled *Pieris brassicae* by plant extracts of *Delphinium elatum* and *Thermopsis lanceolata*.

Thuricide, a *Bacillus thuringiensis* formulation, is active against most of the lepidopteran pests in the larval stage. Efficacy of Thuricide is greatest when it is applied at hatching time and/or on 1st to 3rd larval instars during their intensive feeding period. Since Thuricide must be eaten by the insect to be effective, thorough leaf coverage is essential for best results. Thuricide HP (*B. thuringiensis* subsp. *kurstaki*) is used in most areas where *P. xylostella* is a serious pest of cruciferous plants (Brunner and Stevens, 1986; Nawrocka, 1986).

Chemical methods have failed to control diamond back moth pest and strategies have been developed to reduce dependence on insecticides. Talekar (1992) emphasised alternative control measures which included: identification of resistant cultivars; use of insect viruses and *Bacillus thuringiensis*; cultural practices; biological control with parasitoids such as *Diadegma semiclausum* and *Costesia plutellae*; and use of selective insecticides. Emamectin benzoate alone, *Bacillus thuringiensis* subsp. *Aizawai* alone, emamectin benzoate alternated with *Bacillus thuringiensis* subsp. *Aizawai* and mevinphos were shown to be effective at controlling diamond back moth population on cabbage in Florida (Leibee *et al.*, 1995).

12.3.2 Aphids

Aphids cause serious damage to cole crops all over the world. Among the different species, the most important is the cabbage aphid (*Brevicoryne brassicae*). In addition, green peach aphid (*Myzus persicae*) and turnip aphid (*Lipaphis erysimi*) also cause damage. Aphid infestation is not tolerated at the seedling stage and initial stage of cupping and <3000 aphid/plant are tolerable at precupping stage (Chen *et al.*, 2000). In late winter and spring, the seedling plants on which the eggs have been deposited, form sources of infection. Therefore, it is necessary to spray the insecticides on those parts and control measures should also be started whenever the aphids are observed in the fields. The lower surface of leaves and the heart need special attention and must be well covered with the spray liquid.

For the chemical control, use is made of various non-systemic insecticides like malathion, thiodan, diazinon, *etc.* and also with systemic insecticides like rogor, metacid, metasystox, *etc.* Spray of these chemicals would require a long period of safety for harvest of crops, though in seed crops it can be used safely but may affect the bees for pollination. Thiodan may be a comparatively safer chemical for use in seed crop also.

Khaire *et al.* (1985) obtained the best control (73.09 per cent) of aphid, *Brevicoryne brassicae* in cabbage cv. Golden Acre and the highest yield of marketable heads (392.35 q/ha) when chlorfenvinphos was applied at 0.15 per cent with no phytotoxicity of the chemical. In Poland, best control of the aphids was provided by the following insecticides: aztee 140 EC (triazamate), marshall 250 EC (carbosulfan), pirimor 50 DG (pirimicarb), danadum 400 EC (dimethoate) and talstar 100 EC (bifenthrin) (NarkiewiczJodko, 1995).

12.3.3 Cabbage Web Worm (*Hellula undalis*)

This is an important pest of Chinese cabbage. As the crop is being cultivated in recent years during hot rainy summer months due to introduction of tropical varieties, this insect has become a major pest of this crop. The insect lays eggs on the growing point or on older leaves and the larvae feed on the growing point or burrow into the stem of older seedlings. As a result, the infested plants are killed or give side shoots which do not form heads. On an active ingredient basis parathion was found to be most economical and effective insecticide. Intercropping with bitter gourd, watermelon, sweet potato and mung bean reduced cabbage web worm infestation markedly.

12.3.4 Snails and Slugs

These are found to damage the growing tips of the plants as also the surface or curd in cauliflower. These are more problematic when the crops are irrigated with sewerage water.

Control Measures

1. Trap them with the moisten gunny bag. Set up trap at 5-10 places during day as they like to hide under these empty moisten gunny bags and sprinkle slake lime or tobacco dust or salt (NaCl) or collect and kill them in 3 per cent brine solution and burry in the pit.
2. Apply EDTA Fe@1.5 -2 Kg/acre
3. Ferum Phosphate @ 1-2 kg/acre
4. Neonicotinoids like Thiomethoxam, Imidacloprid, Dontatsu *etc.* are known to disrupt their biorhythms after they feed or contact sprayed surface. If sprayed in the evening and are exposed to light during day and killed by predators
5. Use of beer trap
6. Use of ducks and Poultry as predator

12.3.5 Integrated Pest Management of Important Pests and Diseases

In cole crop IPM should be followed stage wise *i.e.* pre-sowing, seed sowing/ transplanting stage, sowing/planting, vegetative stage and head stage.

Management	*Activities*
	Pre-sowing
Nutrients	☆ Add well rotten farm yard manure (FYM) @ 8- 10 t/acre or vermicompost @ 5 t/acre treated with *Trichoderma* spp. and/ or *Pseudomonas* sp @ 2 kg/acre. Incorporate at the time of field preparation 1 week (vermicompost) or 2 to 3 weeks (FYM) before transplanting.
Weeds	☆ At the time of field preparation, adopt stale seed bed technique to minimize the weeds menace in field. ☆ Keep the nursery weed free by hand pulling of the weeds. ☆ Black plastic mulch prevents entry of light, which restricts germination of weed seeds and growth.
Resting stages of pests and nematodes	**Deep summer ploughing** ☆ **Soil solarization:** Cover the beds with polythene sheet of 45-gauge (0.45 mm) thickness for three weeks before sowing for soil solarization which will help in reducing the soil borne pests. ☆ Apply neem cake @ 100 kg/acre at the time of transplanting for reducing nematodes and borer damage. ☆ In nematode severe area apply carbofuran 3 per cent CG @ 20,000 g/acre

Management	*Activities*
DBM	**Cultural control:** ☆ Removal and destruction of plant remnants, stubbles, debris after harvest and ploughing the field. ☆ Trap crop: Sowing 2 rows of bold seeded mustard as a trap crop for every 25 rows of cabbage to attract moths to mustard. Plant the first row 12 days before transplanting and the second row 25 days after transplanting. ☆ Grow intercrops such as tomato, garlic, coriander and carrot in alternate rows with cabbage.
Seed Sowing/Transplanting stage	
Nutrients	☆ Before sowing, soil testing should be done to find out the soil fertility status. Nutrient should be provided as per soil test recommendations. ☆ For varieties apply 32 kg N/acre in three equal splits. The first one (33.3 per cent) at the time of transplanting as basal dose. ☆ For hybrids apply 48 kg N/acre in three equal splits. The first one (33.3 per cent) at the time of 30 transplanting as basal dose. ☆ For varieties apply entire quantity of P and K @ 60 and 40 kg/acre, respectively, at the time of sowing. ☆ For hybrids apply entire quantity of P and K @ 90 and 60 kg/acre, respectively, at the time of sowing. ☆ Micronutrient deficiency should be corrected by foliar application. ☆ Biofertilizers: For seed treatment with *Azotobacter* and phosphorous solubilizing bacteria (PSB) cultures @ 8-10 g/kg seed ☆ For seedling root dip treatment with *Azotobacter* and phosphorous solubilizing bacteria (PSB) cultures @ 250 g/acre seedlings
Weed management	☆ Keeps the nursery weed free by hand weeding. ☆ Avoid carrying weed seedlings along with cabbage seedlings.
Sowing/Planting	
Black rot	**Cultural control:** ☆ Crop sanitation ☆ Use of resistant varieties ☆ Crop rotation for 2-3 years with noncruciferous crops

Management	*Activities*
Damping off	**Cultural control:** ✰ Quality seed and a chemical or heat pasteurized planting medium should be used. ✰ Excessive watering and poorly drained areas of field should be avoided ✰ Use raised beds: more than 15cm height is better for water drainage or use pro trays for raising seedlings **Chemical control:** ✰ Soil drench with captan 75 per cent WP @ 1000 g in 400 l of water/acre. ✰ Treatment with captan 75 per cent WP @ 20-30 g/kg seed.
Alternaria leaf spot	**Cultural control:** ✰ Long rotations (3 years) without crucifer crops or cruciferous weeds such as wild mustard. ✰ Plant later plantings upwind of earlier plantings. ✰ Allow for good air circulation (*i.e.* wide spacings, rows parallel to prevailing winds, not close to hedgerows). **Chemical control:** ✰ Spray zineb 75 per cent WP @ 600-800 g in 300- 400 l of water/acre or mancozeb 75 per cent WP @ 600-800 g in 300 l of water/acre.
Vegetative stage	
Nutrients	**Top Dressing** ✰ Apply second dose (33.3 per cent) 30 days after transplanting and irrigate the crop immediately after fertilizer application. ✰ Micronutrient deficiency should be corrected by foliar spray of particular micronutrient.
Weeds	✰ Weeding and hoeing should be done once within 20-25 days after transplanting and second time 45 days after transplanting. Deep hoeing should be avoided. ✰ Mulching with black Low Density Polyethylene (LDPE) sheets of 30 micron thickness by burying both the ends into the soil to a depth of 10 cm will avoid weed growth.

Management	*Activities*
DBM	**Cultural control:** ☆ Installation of pheromone traps @ 4-5/acre for monitoring. **Biological control:** ☆ Release egg parasitoid, *T. chilonis/pretiosum* @ 20,000/acre 4-6 times at weekly interval. ☆ Release larval parasitoids, *Diadegma semiclausm* @ 1,00,000/acre (Hills – below 25–27°C) or *Cotesia plutellae* (plains) @ 20,000/acre from 20 days after planting ☆ Conserve other parasitoids such as *Brachymeria* spp., *Eriborus* spp. *etc.* ☆ Fungal pathogens, for example, *Paecilomyces* spp. and *Zoophthora radican* are effective. ☆ Foliar spray with 5 per cent NSKE or azadirachtin 0.03 per cent (300 ppm) neem oil based WSP @ 1000-2000 ml in 200-400 l of water/acre.
	Chemical control: ☆ Spray flubendiamide 20 per cent WG @ 15 g in 150 l of water/acre or lufenuron 5.4 per cent EC @ 240 g in 200 l of water/acre or spinosad 2.5 per cent SC @ 240–280 in 200 l of water/acre 32 or indoxacarb 15.8 per cent EC @ 106.4 ml in 200–400 l of water/acre or emamectin benzoate 5 per cent SG @ 60- 80 g in 200 l of water/acre or fipronil 5 per cent SC @ 320–400 ml in 200 l of water/acre or chlorantraniliprole @ 0.3 ml/l (last spray should be 15 days before harvesting).
Cabbage borer	**Cultural control:** ☆ Collect and destroy caterpillars mechanically in the early stages of attack. **Chemical control:** ☆ Malathion 50 EC @ 600 ml in 200-400 l of water/acre.
Cabbage leaf webber	**Cultural control:** ☆ Remove and destroy the webbed leaves with caterpillars within. ☆ Set up light traps @ 1/acre. **Biological control:** ☆ Conserve parasitoids such as *Cotesia crocidolomiae etc.*

Management	Activities
Cabbage Aphid	**Cultural Control:** ☆ Installation of yellow sticky traps, yellow water pan traps @ 12/acre to monitor alates (winged adult). **Biological control:** ☆ Conservation of parasitoids such as *Aphidius colemani* (adult and golden mummified aphid), *Diaeretiella rapae* (adult and golden mummified aphid), *Aphelinus* spp. (adult and golden mummified aphid) *etc.* ☆ Conserve predators such wasps, green lacewings, earwigs, ground beetles, rove beetles, spiders *etc.* **Chemical control:** ☆ Foliar spray with dimethoate 30 per cent EC @ 264 ml in 200-400 l of water/acre or fenvalerate 20 per cent EC @ 120-150 ml in 240- 300 l of water/acre or phosalone 35 per cent EC @ 571 ml in 200-400 l of water/acre or acetamiprid 20 per cent SP @ 300 ml in 200-240 l of water/acre. or imidacloprid 17.8 Sl @ 0.3 ml/l.
Tobacco caterpillar	**Cultural control:** ☆ Field sanitation and rouging. ☆ Repellant plants: Ocimum/Basil. ☆ Setting up light traps for adults @ 1/acre. ☆ Erecting of bird perches for encouraging predatory birds such as mynah, drongo *etc.* ☆ Use of ovipositional trap crops such as 33 castor @ 250 plants/acre and collection of larvae from flowers. ☆ Installing pheromone traps @ 4-5/acre for monitoring insect activity. **Biological control:** ☆ Spray NSKE 5 per cent against eggs and first instar larvae. ☆ Spray NPV @ 40LE/ac in combination with jaggery 1 kg, sandovit 100 ml or Robin Blue 50 g thrice at 10-15 days interval on observing the eggs or first instar larvae in the evening hours. ☆ Conserve parasitoids such as *Trichogramma chilonis* (egg), *Tetrastichus* spp. (egg), *Telenomus* spp. (egg), *Carcelia* spp. (larval-pupal), *Campoletis chlorideae* (larval) *etc.* ☆ Conserve predators such as lacewings (*Chrysoperla carnea*), coccinellids, king crow, dragonfly, spider, robber fly, reduviid bug, praying mantis, fire ants *etc.*

Management	*Activities*
	Chemical control: ☆ Spray trichlorfon 5 per cent GR @ 300 g/acre or thiodicarb 5 per cent GR @ 300g/acre or chlorfluazuron 5.4 per cent EC @ 600 ml in 200 l of water/acre.
Cabaage butterfly	**Cultural control:** ☆ Fine-mesh netting in nursery will stop butterflies from reaching the crop and lay eggs. Collect and destroy eggs or caterpillars mechanically by hand usually on the underside of the leaves. ☆ Intercropping cabbages with *Nasturtium* results in fewer eggs laid on cabbage by the butterflies. ☆ Manual collection and destruction of egg masses and early stage gregarious larvae. **Biological control:** ☆ Release of *Trichogramma* spp. ☆ Erection of bird perche ☆ Conserve parasitoids such as *Cotesia glomeratus* (larval), *Pteromalus puparum* (larval) *etc.* **Chemical control:** ☆ Same as DBM
Club root diseases	**Cultural control:** ☆ Use of disease free seedlings. ☆ A pH slightly above neutral (usually about pH 7.2) helps to minimize disease. ☆ Add hydrated lime to soil to increase pH to 7.2 34 (6 weeks before planting @ 1.5 t/ac). ☆ Avoidance of excess irrigation.
Downy mildew	**Cultural control:** ☆ Destruction of infected plant debris ☆ Avoid of thick sowing and excessive moist conditions ☆ Use a 3- year rotation without cruciferous crops ☆ Avoid overhead irrigation ☆ Allow for good air movement (*i.e.* wide spacing, rows parallel to prevailing winds, not close to hedgerows).
Powdery mildew	**Cultural control:** ☆ Destruction of infected plant debris. ☆ Maintain proper spacing.

Management	Activities
White rot	**Cultural control:** ✰ Sanitary measures and destruction of weeds ✰ Crop rotation with non-cruciferous crops
Black rot	✰ Same as in seedling stage
Alternaria leaf spot	✰ Same as in seedling stage
DBM and other lepidopteran insects	✰ Same as in vegetative stage
Head state	
Nutrients	The third dose (33.3 per cent) 50-60 days after transplanting and if they are long duration varieties third dose at 75-80 days after transplanting. Micronutrient deficiency should be corrected by foliar spray of particular micronutrient.

Since existing farmer's field practices are chemical intensive in nature, problems of pest resistance, resurgence and secondary pest outbreaks are real threat to profitable vegetable production. Hence, serious effort should be made towards sustainable pest management practices and IRAC guidelines should be followed as far as possible to avoid insecticide resistance development in multiple pests.

IRAC Guidelines for Insecticide Resistance Management

Follow label claim as far as possible

The principle is to provide clear IRM information using language in a format understandable to farmers. Moreover, IRAC member companies recommend stating on product labels the maximum number of applications. Implementation in countries depends on the local regulatory label guidelines.

Always use products at the recommended label rates and spray intervals with the appropriate application equipment

Insecticides used at rates higher or lower than recommended on the label can result in resistance and/or unwanted effects on non-target organisms and the environment. Ensure that all the spray equipment is well maintained and there are no blocked nozzles or filters since these results in incorrect rates. Target the most susceptible insect life stages whenever possible.

Rotation of insecticide, mode of action groups prevents rapid selection of resistant populations

Farmers can avoid prolonged selection for insecticide resistance by rotating and diversifying the insecticide modes of action used in a crop cycle. The recommended approach is to use products of the same MoA within a discrete period of time commonly called a "window". A window is defined by the approximate duration of an insect generation. For sucking pests this is generally simplified by using 15- or 30-day windows depending on the target species.

a) Avoid use of the same mode of action in consecutive windows.
b) If insecticides from several mode of action groups are available, then the use of multiple modes of action within a window is recommended provided that different modes of action are used in the following window.
c) Following a window of any mode of action group, rotate to a window of applications of effective insecticides with a different mode of action.
d) Multiple applications (generally less than 3) of the same MoA insecticide are acceptable if they are used to treat a single insect generation or are used within a window. Make sure that the residual activity of the multiple applications fits within the window.
e) For short cycle crops (*e.g.* lettuce), it may be necessary to consider the duration of the crop cycle as a window. In this case, it is recommended to alternate to different modes of action in the next crop cycle.
f) Avoid rotating products in different sub-groups of the same MoA except if there are no registered and effective alternatives.

Transplanted seedlings

It can be common for seedlings grown in plant nurseries to be treated with insecticides prior to transplant to the field. It is recommended that growers purchasing seedlings for transplanting enquire about insecticide use programs during the period of time the seedlings were held in the nursery, so that where possible they can avoid using sequential treatments of the same insecticide mode of action, upon transfer to the field. In addition, growers should be aware of the resistance status of pests in transplant production areas and to ensure those resistant insects are not re-distributed to local production fields.

Use Integrated Pest management (IPM) practices to protect crops from pest damage and reduce the risk of insecticide resistance

IPM discourages the development of pest populations by using diverse techniques that are economic, safe and environmentally sound. It does not exclusively rely on insecticides, hence in IPM systems selection pressure by specific modes of action is reduced and the risk of resistance minimized.

IPM strategies consist of basic components

a) Apply at established pest threshold levels like Economic Threshold Level (ETL) for insecticide applications. Monitor pest populations in the field to identify species, pest stages, population densities, and presence of natural enemies so that rational pest control decisions can be made.
b) Integrate effective control techniques including cultural, chemical, biological and plant biotechnology pest control measures, which minimize effects on non-target organisms:

- ☆ Use pest resistant or damage tolerant crop varieties.
- ☆ Practice sanitation and removal of infested postharvest crop residue and other pest insect host plants, including the consideration of weed management.
- ☆ Avoid year-round cultivation of susceptible crops to limit survival of treated pest populations.
- ☆ Consider deploying alarm or disruption pheromones where appropriate.

c) Conserve beneficial organisms, such a pollinators, insect predators and parasitoids.

Correct identification of the target pests can improve the effectiveness of the insecticide application

Although some sucking insect species can appear very similar at a quick glance, their susceptibility to different insecticide modes of action can be very different. It is therefore recommended to spend time to accurately identify the species of pests that require control and check the product label for guidance on the recommended product rate and timing. Insect species with similar appearance may also have a different insecticide resistance profile, which may affect product performance.

Consider the systemic properties of some soil and seed-applied products

The systemic properties of some active ingredients allow these products to be applied either directly to the soil, as a seed treatment or as foliar spray. Systemic activity may extend the residual efficacy and the length of the MoA spray window and needs to be considered when planning a program to minimize resistance development. If further insecticide treatment is required after either a seed treatment or a soil application (*e.g.* via drench, drip irrigation or granular in-furrow applications), it is recommended to use an effective foliar product with a different mode of action.

Ensure good coverage and retention of insecticide spray applications to optimize pest control

Sucking pest insects are often located in non-exposed parts of the plant, either on the underside of leaves, at the base of the plant or hidden in flowers. In order to achieve the optimal level of control of these difficult to reach pests, ensure that good spray coverage of the affected plant parts and optimal retention of the insecticide product are achieved by following label instructions on spray equipment, application volumes and the addition of adjuvants or additives.

Ensure that insecticide applications target the most appropriate target insect life stage

Some insecticide modes of action are active against specific life stages of the target insect pest (*e.g.* insect growth regulators such as inhibitors of chitin biosynthesis target juvenile insect stages such as nymphs/larvae) and therefore may either be ineffective or only provide partial control of other life stages if present in the target environment. Read the product label carefully and follow instructions to ensure products are applied optimally to control the target pest.

If in doubt about the timing of application of any product for best effects consult with the product manufacturer or local approved advisor. Also note that some forms of insecticide resistance are also life stage specific. For example, neonicotinoid resistance in tobacco whitefly (*Bemisia tabaci*) only affects pupae and adults. This can mean that although some populations of insects may develop resistance, the affected product may still provide some level of control against susceptible life stages.

Selection of the appropriate insecticide in order to maintain pest control and the preservation of non-target organisms

Different insecticide products have different selectivity profiles depending on their active ingredient and use pattern (*e.g.* foliar, soil, seed applied). The preservation of non-target organisms is beneficial in maintaining both a healthy environment and in maintaining control of pests through the preservation of predators and parasitoids. Where possible it is recommended to utilise target pest specific insecticides to minimise the impact on non-target organisms and limit the use of broader spectrum insecticides to incidences where multiple pest species may be present within the crop.

Using insecticide mixtures

IRAC has issued advice about the use of insecticide mixtures. For guidance refer to the IRAC-mixture-statement and IRAC leaflet-on-use-of-mixtures. As with applying single active ingredient products, insecticide mixture products should be used with careful consideration of the characteristics of the individual active substances, use pattern and pest complex targeted.

The use of insecticides of the same Mode of Action against different pests in the same crop

Multiple uses of different insecticides against more than one pest species in the same crop are feasible but should be considered within the framework of insecticide resistance management programs and developed at local level, taking into account changes in pest populations, overlapping of different species, the relative importance and the risk of resistance development. Good resistance management practices such as avoiding repeated applications of the same mode of action across multiple treatment windows due to application against multiple pest species are key to successful IRM implementation. Where two species appear simultaneously always use the higher recommended rate for the more difficult to control species.

Avoid using insecticides from the same Mode of Action where resistance is known

Continuous use of the same mode of action on a resistant population may escalate resistance levels and should be avoided, particularly if the product is used at higher than recommended rates. This recommendation is valid for solo and mixture products that contain the affected MoA.

Monitor problematic pest populations in order to detect first shifts in sensitivity

Baseline sensitivity data for representative field populations of pests should be established by industry experts before the products became widely used. Reexamining the insecticide sensitivity of pest populations at regular intervals can be used to detect changes in susceptibility. Monitoring methods for many of the major agricultural pests have been established by IRAC and can be found on the IRAC website www.iraconline.org/teams/methods/. Reporting of field failures to IRAC company representatives is also a good way to detect early shifts in pest sensitivity.

Where local information is known about cross-resistance between different MoA groups

Although in most situations rotation between different mode of action (MoA) insecticides will be useful, there have been some cases of metabolic cross resistance between molecules belonging to different groups. Therefore, it is recommended to consult local experts to find out the known status of resistance in your area. Avoidance of cross-resistance may help to build up a more effective rotation strategy.

Never use a product of questionable origin or composition

Products from unknown or non-approved sources may not have the advertised composition, in which case efficacy may be affected and IRM becomes impossible. Moreover, illegal products may pose risks for users and the environment.

13.0 Seed Production

Seed production is a specialized job and required technical knowledge of plant breeding, growing of seed crops and processing of seeds.

13.1.0 Floral Biology

Most of the cole crops, except some forms of sprouting broccoli and cauliflower require a specific cold treatment for the transformation of vegetative to the reproductive stage. In cabbage, kohlrabi or Brussels sprouts, it has been reported that a cold treatment of 4-7°C for about 6-8 weeks or more is required after it has passed its juvenile phase which varies according to the cultivar and crop. Cabbage passes its juvenile phase when its stem attains a thickness of pencil or when the plants are 10-12 weeks old, while in Brussels sprouts it takes 12-16 weeks to pass that stage. In recent years, the tropical cabbages have been reported which require only 10-13°C for 6-8 weeks for their transformation to reproductive phase.

Heading broccolis or winter cauliflowers also require cold treatment for its change to both curding and later to reproductive phase, while for the other cauliflowers such cold treatment is not necessary.

Different types of cauliflowers grown in India do not require any chilling although it is a general percention that snowball types have chilling requirement which is in correct. The seed production and breeding of snowball/late cauliflower are taken up in the higher hills not because of their chilling requirement for bolting but their inability to produce the fertile pollen grains at a temperature above 20°C.

The snowball cauliflower also bolts under North Indian plain but when they start flowering after bolting (mid of March onwards), the temperature goes above 30°C which causes non viable or sterile pollen grains. Therefore, there is no seed setting even after flowering under plain condition. Therefore, in contrast to the cabbage and other cole crops, there is no chilling requirement in any type of cauliflowers grown in India.

In cabbage, kale, Brussels sprouts, kohlrabi, *etc.*, the central stalk develops into reproductive shoots. If the heads are removed in cabbage, the axillary buds develop into reproductive shoots, while the cauliflower curd and sprouting broccoli heads give rise to multiple reproductive shoots.

The flowers are typical of cruciferous family having 4 petals, 6 stamens, of which 2 are short and 2 carpels with superior ovary. The typical structure of flowers is known as tetradynamous. The flowers are attached with the stalk with a short pedicel. The flower differentiates by the successive development of four sepals, six stamens, two carpels and four petals. The carpels form a superior ovary with a 'false' septum and two rows of campylotropous ovules. The nucellar tissue is largely displaced by the embryo sac and, when the buds open, the ovules mainly consist of the two integuments and the ripe embryo sac. The buds open under pressure from the rapidly growing petals. Opening starts in the afternoon, and usually the flowers become fully expanding during the following morning. The bright yellow petals grow to 10–25 mm long and 6–10 mm wide. The sepals are erect. Pollination of the flowers is by insects, particularly bees, which collect pollen and nectar. Nectar is secreted by four nectaries situated between bases of the short stamens and the ovary. The flowers are borne in racemes on the main stem and its axillary branches. The inflorescences may attain lengths of 1–2 m. The slender pedicels are 15–20 mm long. The anthesis of flowers depends on temperature and the flowers open mainly during hotter part of the day. Honey bees are the usual pollinating agents, though bumble bees and other may also be responsible for pollination. Myers and Fisher (1944) reported that as a pollinating agent wind can be practically disregarded. It has been found that stigma of *Brassica spp*. is receptive even 5 days before and 4 days after anthesis. The period from pollination to fertilization generally takes 24-28 hours, depending on temperature. The ideal temperature has been found to be 12-18°C but continuous foggy weather with lower temperature after both fertilization and development of seed. Higher day temperature is more harmful for fertilization since this causes pollen sterility. After fertilization, the endosperm develops rapidly, while embryo growth does not start for some days. The embryo is generally still small 2 weeks after pollination, and at this stage embryo rescue can be first attempted. The embryo fills most of the seed coat after 3–5 weeks, by which time the endosperm has been almost completely absorbed. Nutrient reserves for germination are stored in the cotyledons, which are folded together with the embryo radicle lying between them. The fruits of cole crops are glabrous siliquae (pods), 4–5 mm wide and 40–100 mm long, with two rows of seeds lying along the edges of the replus. One silique contains 10–30 seeds. Three to four weeks after the opening of a flower, the silique reaches its full length and diameter. When it is ripe, the two valves dehisce. Separation begins at the attached base and works towards the unattached distal end, leaving the seeds attached to the placentas. Physical force ultimately separates the seeds, usually by the pushing of the dehisced siliquae against other plant parts either by wind or in threshing operations. The pod maturity or harvest of pods may require 50-90 days from the date of flowering,

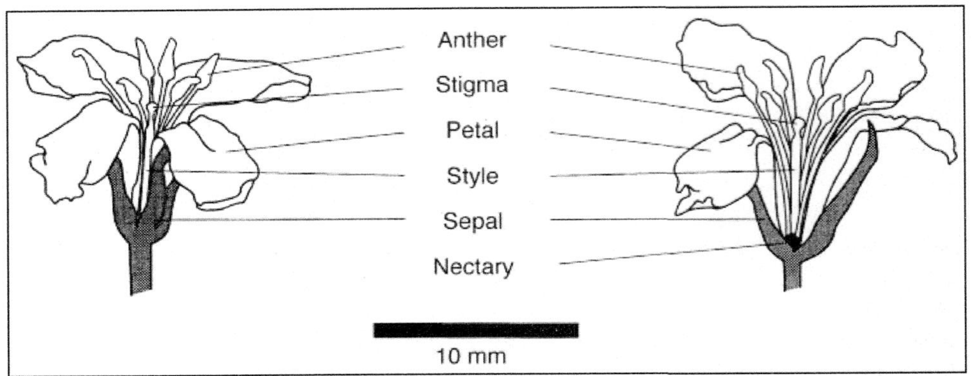

Flower Structure of *Brassica oleracea* (M. H. Dickson).

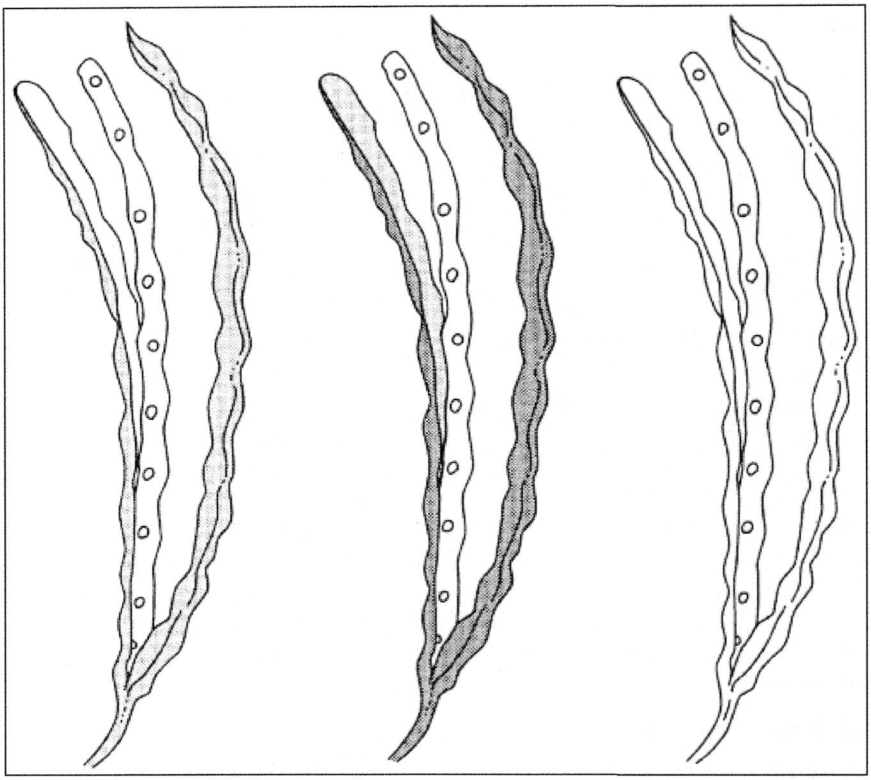

Generalised Structure of Dry Fruits of Brassicas (M. H. Dickson).

13.2.0 Isolation

Isolation is necessary in these crops for the maintenance of purity of cultivar. In *Brassica*, there are two distinct groups which do not inter-cross with each other, *Brassica oleracea* with all its botanical varieties and wild allies in one group and *B. rapa, B. juncea, B. chinensis etc.*, form another group. These intercrosses rather freely among the crops and cultivars within the same group and, therefore, isolation would be necessary when two crops of the same group are grown at the same location. Besides, within the same crop also, isolation would be essential for growing different cultivars. It would always be better to avoid more than one crop at a particular location.

Generally, at least 3000 m isolation is recommended for the production of breeder's seed, while at least 1500 m would be needed for certified seed. Though the isolation distances have been fixed arbitrarily on the basis of normal flight of bees, it is working well. The experimental evidence conducted in the Netherlands for 8 years did not provide more specific information in this respect (Nieuwhof, 1963). However, radish does not cross with the crops of these two groups of *Brassica* and thus one cultivar of each of the three groups (like *B. oleracea, B. rapa* and radish) may be grown side by side without any chance of intercrossing between them.

13.3.1 Cabbage

13.3.1.1 Climatic Requirements and Areas of Production

Most cabbages are biennial and do not bolt unless they exposed to required cold temperature. Besides, cool regions having fairly moderate rainfall (or fairly dry with irrigation arrangements), well distributed throughout the season in which the vegetative development takes place followed by the vernalization requirement and fairly dry and cool summers are favourable to cabbage for bolting. The day length has no effect on bolting or flower initiation. In India, cabbage seed production was mainly limited to Kashmir Valley. Presently, large scale seed production of cabbage is being done in Kullu, Lahaoul and Spiti and Kalpa valley of Himachal Pradesh. Locally adapted cultivars are also grown for seeds in Darjeeling hills and in the Nilgiris.

13.3.1.2 Methods of Production

There are 3 distinct methods of seed production, *e.g.*, head to seed method, seed to seed method and the stump method. Of these, the second method is mostly used for the production of market seed, while the first and the third methods provide a better scope for inspection and roguing. Hence for stock seed/breeder's seed, it is better to use the first and the third methods and the second method may be followed for both foundation and certified seed production by using high quality breeder's seed.

13.3.1.3 Head to Seed

The crop is grown in much the same way as that of market crop except that the period of transplanting the seedlings has to be adjusted in such a way, according to the location, that the full maturity of the head is reached prior to snowfall. If they mature early, the heads may begin to split before winter and in case the cropis not mature, proper inspection and roguing, *etc.*, may not be possible. However, early maturity may not pose any problem if the heads are stored in the cold storage which is not normally the case under Indian conditions except in Kalpa Valley. Since the heads are over wintered in the fields in Kashmir or Kulu Valley, adjustment of the planting time is of utmost importance.

***Roguing*:** Off-type plants are rogued out almost any time they are observed, but the general practice in this method is to rogue the plant at about the time, the head crop is mature. The off-type plants are removed and the field is inspected thoroughly for all the characters of the cultivar and those which do not conform to the accepted standard like the number of outer leaves, colour, shape, size, compactness, *etc.*, are discarded. The selected plants are lifted, the basal outer leaves are stripped off, exposing the head for more thorough examination and kept either for storage or for replanting.

***Storage*:** The necessary storage in most countries involves considerable risk, though in Indian condition such storage is unnecessary except in Kalpa Valley, where true-to-type heads after roguing are immediately stored in trenches for over wintering. A single layer of head is placed in a slant position in trenches 75 cm deep, 90 cm wide and 3 m long, with roots 5-7 cm deep in the soil. They are covered with wooden planks and about 15 cm of earth is spread over them. On both ends of the trenches small holes are provided for ventilation. The arrangement, according to report, keeps the heads in good condition during winter.

In cellar method of storage, the temperature and relative humidity recommended are 32°F (0°C) and 90-95 per cent, respectively. When cabbage is stored in cellars, it is desirable to place the entire plants with roots on wire mesh in tiers and as close as possible. During inspection, the decaying plants are removed. The soil is placed around the stem and roots after which the heads are covered by two boards in the form of an inverted 'V'. The boards in turn are covered with a layer of soil about 15 cm deep. If this becomes frozen, straw manure is put over it to a depth of 30-40 cm. This may be repeated, if necessary. The boards allow some ventilation at the top.

***Replanting*:** When the plants are kept in storage either in cellar or in trench, they are reset at the onset of spring when the soil is dry for work. The plants are spaced at 45 to 90 cm between the plants and 90 to 180 cm between the rows according to cultivar, location and method of cultivation. In Kashmir or Kulu, the plants are immediately reset in a manner that the whole stem below the head goes under ground with the head resting on the soil surface.

Dhesi and Padda (1964) suggested 60 x 60 cm for Pride of India and Early Drumhead and 60 x 75 cm for Late Drumhead cultivars. Saini and Rastogi (1976) reported that closer spacing of 30 x 30 cm gave significant higher seed yields. Hawthorn and Pollard (1954), however, advocated 90 cm between the rows, possibly for easiness of mechanized culture. After replanting, two cuts of 2.5-2.0 cm deep are given at right angles across each head so that the seed stalks may develop normally. Repeated incisions at later dates may be necessary.

13.3.1.4 Seed to Seed

In this method, the plants are allowed to seed in its place of transplanting. The most essential requirement of this method is the high quality foundation seed.

The time of transplanting is followed as in the head to seed method.

Roguing: Since cabbage is grown for above ground head, roguing is still possible, though not like the other method. Roguing is mainly done on the basis of outer leaves, their colour, shape, compactness, stem length, frame, *etc.*

Winter protection: After roguing in late autumn, a shallow furrow is ploughed on each side of the row and a large ridge of soil is thrown against the cabbage plants. Top is left exposed. The partially buried plants can withstand quite a low temperature and the plants are exposed in early spring.

13.3.1.5 Stump Method

In this method, the heads after their formation are beheaded keeping the outer whorl of leaves intact. The heads are marketed and the selected stumps are either kept in situ or transplanted during late autumn or early winter. The axillary buds of the outer leave develop and give rise to the flowering stalk. The method has several advantages like thorough inspection of the plant and head (even after harvesting) and early and increased yield of seeds; besides, extra income is possible from the sale of heads. The method has a disadvantage of decumbent flowering shoots which may require staking and that the inspection has to be made at the right time. The method is very useful for improvement of different characters by breeding (Millar, 1931).

13.3.1.6 Cultural Practices (for all the methods)

Cabbage is a heavy feeder and responds well to fertile soil and also to the application of fertilizers. Eguchi (1960) observed that side dressing with nitrogen at the bolting stage has an effective role in increasing the number of secondary branches, flowers and fruits. Thakur and Gill (1976) noted that a second dose of N, P and K at the rate of 250 kg/ha, 125 kg/ha and 100 kg/ha, respectively was optimum and economical for seed yield.

In regions, where the seed crop of cabbage is grown, the rainfall is usually well distributed and the crop may not require irrigation except of protective type if there is a long drought. However, the crop requires sufficient moisture, specially during flower, seed set and seed development stages. In drier regions, irrigation at close interval should be applied during spring and summer.

Boshnakov *et al.* (1986) found that treatment of CCC on established transplanted seedlings of cultivar Balkan improved the morphological and reproductive performance of plants and increased seed yields by 16.5-26.3 per cent. When plants grown from stecklings were sprayed with 1 per cent CCC with or without the addition of 0.25 per cent Wuxal (NPK + trace elements + IAA + vitamin B_1) at the start of anthesis, the seed yield was increased by 33.2 per cent.

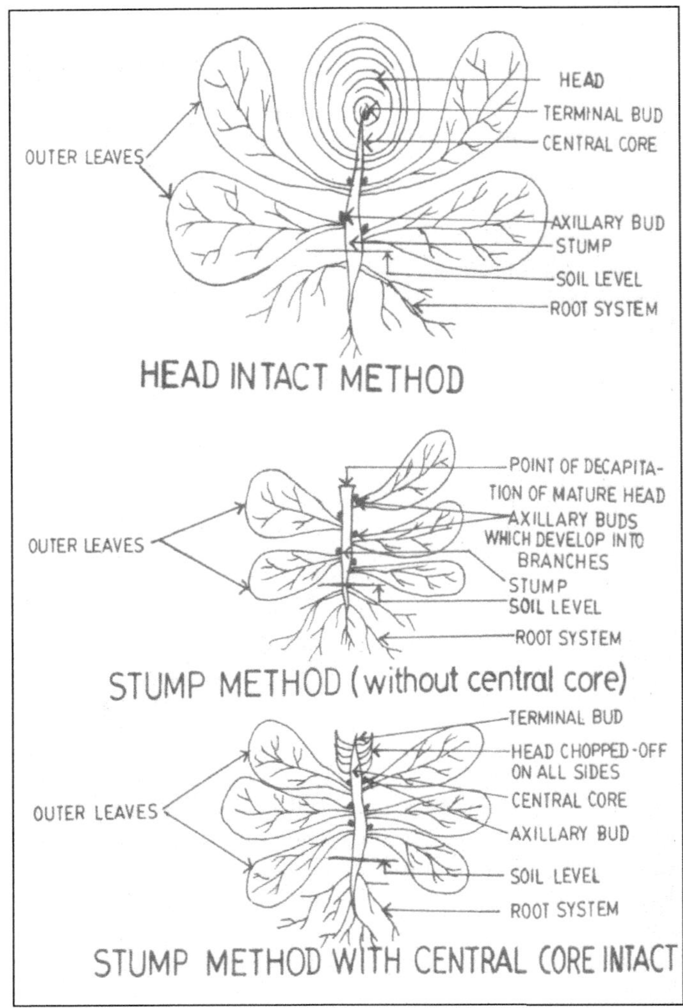

Different Methods of Seed Production Practised in Cabbage

13.3.1.7 Pre-mature Seed Stalk Formation

When the cabbage plants give rise to the seed stalk without forming the head, specially in commercial crop, it is known as premature bolting. Boswell (1929) while studying the over-wintered cabbage noted such phenomenon and observed that the primary cause of such behaviour is the fulfilment of chilling requirement of the plants after passing of juvenile phase followed by somewhat higher temperature. At this he concluded that adjustment of time of planting may reduce the changes of premature bolting. Millar (1929) also worked almost in the similar lines and concluded that the plant size and temperature are the two most important factors in inducing such premature bolting.

Different Stages of Green and Red Cabbage Seed Production

13.3.1.8 Harvesting

Cabbage seed pods are borne in a raceme and they ripen in the same sequence as the flowers open. As fully matured pod is likely to dehisce, the crop has to be harvested before the older pods become dry. Harvesting is generally started when a noticeable proportion of the pods have become yellow. After harvesting, curing should be in a shaded place and the harvested pods may be covered with a tarpauline. After curing for 4-5 days, when the pods show sign of drying and/or brownish appearance, they are dried in sun. It may take a few days or weeks to become completely dry depending on locality and climatic condition. However, dry weather will be an advantage. Threshing of cabbage pods is easy and may be done by a grain-threshing machine. After threshing, grading is done in specified sieves for separating the under developed grains.

13.1.1.9 Yield

At Katrain in Himachal Pradesh, the following seed yields were obtained:

(i) Golder Acre -500 - 600 kg/ha.

(ii) Drumhead -600 - 700 kg/ha.

However, it is not difficult to get a yield of 800-1000 kg/ha.

Orlowski (1985) recorded seed yield of cv. Ditmarska Najwczesniejsza of about 1166 kg/ha.

Two methods of cabbage seed production, *viz.*, leaving the plants *in situ* or replanting them, were compared by Verma *et al.* (1996) using cabbage cv. Golden Acre plants in which the heads destined for seed production were classified as small, medium or large. Seed yield parameters for each head size were greater for plants left *in situ* than replanted ones and also increased with head size; *in situ* plants with large heads yielded 54.76 g of seeds/plant.

13.3.2 Cauliflower

The seed production of cauliflower somewhat resembles with that of the cabbage, though the climatic requirements arc more exacting, especially temperature. It is much more subject of freezing injury than cabbage, yet the period during curd formation stage should be uniformly cool and somewhat humid to avoid the development of undesirable characters (riceyness/fuzziness, *etc.*).

In India, two different types of cauliflower, the early or Indian type and the Snowball or temperate summer type require different locations and also different climates. The areas of production of the former is in Bihar and U.P. while the later is mainly in Kulu, Saproon Valley in Himachal Pradesh and Kalimpong in Darjeeling hills. In recent years, some seed production is being taken up in the Nilgiri hills as well. The Haryana has also emerged as the leading state for undertaking seed production of early Indian cauliflowers.

Unlike cabbage, cauliflower varies from typical annual to typical biennial. The annual cultivars are dependent on varying temperature for curd initiation and development, while the biennials are fully dependent on vernalization for curding.

However, the snowball types grown in India do not require any vernalization for bolting but requires mild temperature at flowering for successful seed setting. The curd is an early stage of inflorescence development as its formation invariably precedes floral initiation. The inflorescence difference differs from that of cabbage is being more dwarf (except the earliest forms) and umbrella shaped. Such formation results from the absence of central main stem above the point where branching starts.

13.3.2.1 Methods of Production

Lancaster (1943) working on the Indian early types reported that the seeds produced *in situ* crop performed poorly compared to that for transplanted crops, *i.e.*, after the production of curd. But late, investigation by various research workers showed that the quality is dependent on the quality of stock seed rather than *in situ* or transplanted crop. The present practice of production is by seed to seed method only.

In case of cauliflower, unlike cabbage, the time of sowing and transplanting is just the same as the market crop, specially for the Indian types which are grown in the plains. In Muzaffarpur, India, Mishra (1989) recommended that sowing time of cv. Early Patna for the highest seed yield and good quality is 25 June. Chatterjee and Som (1990) obtained highest seed yield in cauliflower cv. KPS-1 from August sowing at Kalimpong, W.B., India and further concluded that seeds should be sown between August and October for seed production. However, the earliest crop suffers badly from the point of maintenance, size of the curd, infestation of diseases, especially *Alternaria, etc.*, and thus it results in poor yield of seed. In order to minimize the limitation, the seed growers raise the early cauliflowers slightly later than the market crop. But this practice, infact, has failed to serve the purpose of earliness, since continuous growing of a cultivar under cooler temperature is likely to induce the cultivar to be late. Possibly, it is one of the reasons that the earliest cultivars have the longest period of harvest as a market crop. At Kullu valley seed production of snowball cauliflower variety, Pusa Snowball K-1 is done by sowing in the first week of Septeber and transplanting before mid-October. Early results in injury of curds by snowfall as curing occurs by end of January in case of early sowing. It is advised that sowing and transplanting should be adjusted in such a way that curd formation and development occurs after snowfall in the month of January and February.

Two-year trials by Cuocolo and Duranti (1988) on cauliflower cv. Gigante di Napoliin the Sele river plain (Salerno), Italy, showed positive effects on N treatments (0, 100, 200 or 300 kg/ha) and plant density (1.50, 1.75, 2.00, 2.25 or 2.50 plants/m^2) on yields, but the intensity of the effects varied between years. The greatest increase in yield by 41 per cent was observed with the first 100 kg/ha N. Seed yield increased linearly with plant population, being more than 50 per cent higher (over 800 kg/ha) at 2.5 than at 1.5 plants/m^2. In cauliflower cv. Patna Main, application of 150 kg/ha N resulted in the highest 1000 seed weight and seed yield. For B, application of 10 kg/ha resulted in the highest 1000 seed weight and seed yield (Mishra, 1992). Seed yield of cv. Pusa Snowball-1 increased with application of upto 175 kg N/ha and up to 150 kg P_2O_2/ha (Jamwal *et al.*, 1995). Ved Prakash *et al.* (2000) in a field experiment

Different Stages of Seed Production of Cauliflower

at Almora, Uttarakhand in the winter seasons, Pusa Snowball-1 cauliflowers were spaced at 30, 45 or 60 cm apart in rows 60 cm apart and given 0, 50, 100, 150 or 200 kg N/ha. Seed yield increased with N rate and with closer spacing.

After the curd formation, roguing is done thoroughly for the various characters but mainly for the quality of the curd like size, colour, compactness and free from riceyness/fuzziness, *etc.* The plants are then left for bolting when roguing is done again for those which show bolting from the centre of the curd. This practice is not possible in Hazipur in Bihar, where central portion of the curd is scooped out and marketed. The plants come to flowering within 30-45 days, depending on the temperature and compactness of curd. If the day temperature is higher than 30°C, the inflorescence may show many deformities.

In case of the Snowball types including Dania, early October has been found to be best for transplanting the seed crop. In Saproon Valley or in Kulu, sufficient vegetative growth is completed before the cold temperature sets in and the curding starts only after the cold period is over in late February or early March. While at Kalimpong, there is no check of growth and the curding is completed by mid to late January. Chatterjee and Mukherjee (1962) suggested 'scooping and blanching' for highly compact cultivars like Snowball though the method is not so useful in Dania. Longitudinal half-cutting of the curd in very compact cultivars proved useful for easier bolting and early flowering, besides higher yield. However, in such operations, protection from infection of fungal or bacterial diseases should be provided by applying 0.1 per cent solution of Aureofung in at the cut portion, whether half or scooped followed by regular spraying of some copper fungicide.

The plants come to flower in April in Himachal Pradesh while in February to March at Kalimpong, West Bengal. As in cabbage, bees are the main pollinators. Prabucki *et al.* (1987) however, observed that beside bees, other insects' pollinators of the orders Hymenoptera, Diptera, Coleoptera and Lepidoptera affected seed yields of cauliflower. Harvesting is done in 2-3 shifts since all the pods do not mature at one time. Curing, drying and processing are followed as in the case of cabbage. The seed yield is however, less than that of cabbage. The Indian types may yield from 500 to 600 kg/ha while in Snowball it may vary between 200 and 300 kg/ha. Gurusamy (1999) reported that cauliflower cv. Giant could be safely harvested from 283 to 290 days after sowing without significant reduction in seed yield and quality.

13.3.3 Production of F_1 Hybrid Seeds in Cole Crops

In most of the developed countries, F_1 hybrid seeds of cole crops have become very popular. In almost all these F_1 hybrids, extensive use of self-incompatibility available in this group of crops has been made. Though there are many limitations in the use of this method but in countries like Japan, USA, Australia and in Europe, F_1 hybrid seed of cabbage, Brussels sprouts, sprouting broccoli and also of cauliflower have become very popular in the last decade and in some crops the F_1 hybrids marketed have already reached nearly 90 per cent. After the isolation of self-incompatible lines which have proved its merit in F_1 hybrid performance, the main constraint has been the maintenance and multiplication of these lines for the production of F_1 hybrid seed. An attempt to eliminate self-incompatibility temporarily has become rather successful, which has avoided the tedious bud pollination method in many cases. In this method, after flowering of a self-incompatible line, if the plant is placed in 2 to 5 per cent CO_2 gas within 2 to 6 hours of pollination, the self-incompatibility is temporarily eliminated. With this idea, a practical method has been devised and is in use by many seed producing companies of the world. The self-incompatible line placed in a greenhouse, which can tightly be closed, is pollinated with the help of bees when flowers are in bloom and then the greenhouse is treated with CO_2 gas; the fertilization takes place due to temporary elimination of self-incompatibility. In practice, the CO_2 gas treatment is done in the evening so that the gas can be diffused in the air outside through openings of greenhouse during the night and the condition of the greenhouse comes close to the natural one by the following morning. The treatment is performed several times during the full flowering period. However, it would be necessary that the concentration of the gas and the time of treatment have to be standardized for each self-incompatible line before large-scale multiplication is undertaken. It has been, however, observed that there are some lines whose self-incompatibility can not be eliminated by this method and hence has to be followed by bud-pollination method only.

When the parental lines are multiplied for production, the ratio of mixed planting will depend on the method which is to be followed for F_1 seed production. In case of single cross with both/all self-incompatible lines, 1:1 ratio may be suitable. Accordingly, the ratio may be changed to 2:1, 3:1 or even 4:1 depending on the difference in the lines and methods. In addition, the synchronization of flowering of the lines during production is another hurdle. Attempts by adjusting the time of transplanting of the different lines and also pinching the main stalk or picking off lower buds to synchronize flowering have been made with success. In F_1 hybrid seed production, use of bee hives for pollination invariably increases the seed yield.

However, because of poor stability of self-incompatible lines and occurance of 'sibs' in the hybrid population result in lesser use of this system in the recent times. At present cytoplasmic male sterility is the most preffered and widely used system in commecial hybrid seed production of cole crops.

Conner and Christey (1997) suggested a seed treatment method for eliminating non-hybrid plants while using atrazine resistance as a genetic marker for hybrid seed production in breeding programmes in broccoli. To improve pollen transfer, 2 atrazine-resistant CMS broccoli lines, 16A and 16D, were grown in a random mixture with the atrazine-sensitive, male fertile broccoli cv. Green Comet. Harvested seed consisted of hybrid seed, which is atrazine-resistant, and atrazine-sensitive seed resulting from these if pollination of the male parent Green Comet. Seed imbibition with 10g/l atrazine over night, followed by drying, prevented atrazine-sensitive broccoli plants from growing beyond the cotyledon stage after germination.

13.3.3.1 Schemes for Hybrid Seed Production using Self-incompatible Lines

Single cross- hybrids: In this case either one SI line or two SI lines are used as parent to develop F_1 hybrids. When the F_1 hybrid is produced using one SI line as female parent and aother open pollinated self-compatible line as pollen parent, the hybrids are called test cross derived hybrids. When both the female and pollen parent are SI lines, hybrid seeds can be harvested from both the parents. In general, only one SI parent is used in development of F_1 hybrid as development of two SI lines which are good combiners to give a superior F_1 hybrid is very difficult.

Triple crosss hybrid: The hybrid seeds are produced by crossing a single cross hybrid with another inbred which can be either self-compatible or incompatible. This is practised when the there is severe inbreeding depression in the course of developing the SI lines through selfing. The vigour of the SI line is restored through crossing with another SI line which should be closely related/derived lines to avaiod unacceptable segregation and less uniformity in F1 hybrid seeds.

Double cross hybrid: These types of hybrids are produced through crossing of two single cross hybrids which are self incompatible but compatible with earch other.

The uniformity is generally low in case of triple cross and double cross hybrids. Therefore, single cross F_1 hybrids are most popular in cole crops and triple cross and double cross hybrids are produced rarely.

13.3.4 General Scheme of Seed Production

The principle of managing the system of seed production of the different stages from breeder's seed to certified seed is shown in Figure 9.

In general, it is the fourth year when the grower is able to obtain the certified seed. However, the system can be modified, specially in the number of years on the basis of the efficiency of multiplication and in most vegetable seeds, the certified seed stage may be reached in 3 years except in case of bhindi, peas, cowpeas, French bean, *etc.* Most F_1 seed production, especially in single crosses, the hybrid seed may be produced in 2 years but for 3-way or doublecrosses more years may be needed.

14.0 Crop Improvement

In any improvement programme, the objectives for the breeding should be clear so that the correct methodology can be adopted for achieving the desired results. The major attention for any improvement programme is yield and disease and pest resistance. Whenever any attempt is made to improve the yield, the objective should cover the yield components either directly or indirectly. In cole crops, such components may include size of the edible portion and more important may be the compactness and their uniformity in many cases. In addition, along with yield a large number of other important characters may also require attention, *e.g.*, plant type, frame of plant, yield per unit area and per unit time (maturity) and also harvest index. Besides, resistance to common diseases and pests, the objectives may be on certain quality characters like taste, colour, flavour, *etc.* Even after attaining all these breeders may also see the potentiality of a cultivar for commercial seed yield, which should not be a limiting factor. If all these are taken together, the following characters are considered important in the improvement programme of cole crops:

Top cross hybrid
Single cross hybrid
o Self-incompatible inbred (female); S1S1
o Self-incompatible inbred (M/F); S1S1
• Open pollinated variety (Male); Various S alleles
• Self-incompatible inbred (M/F); S2S2

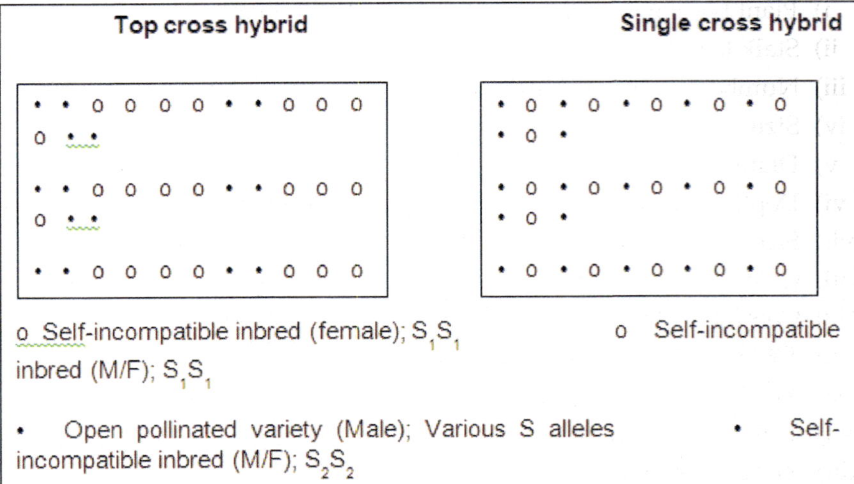

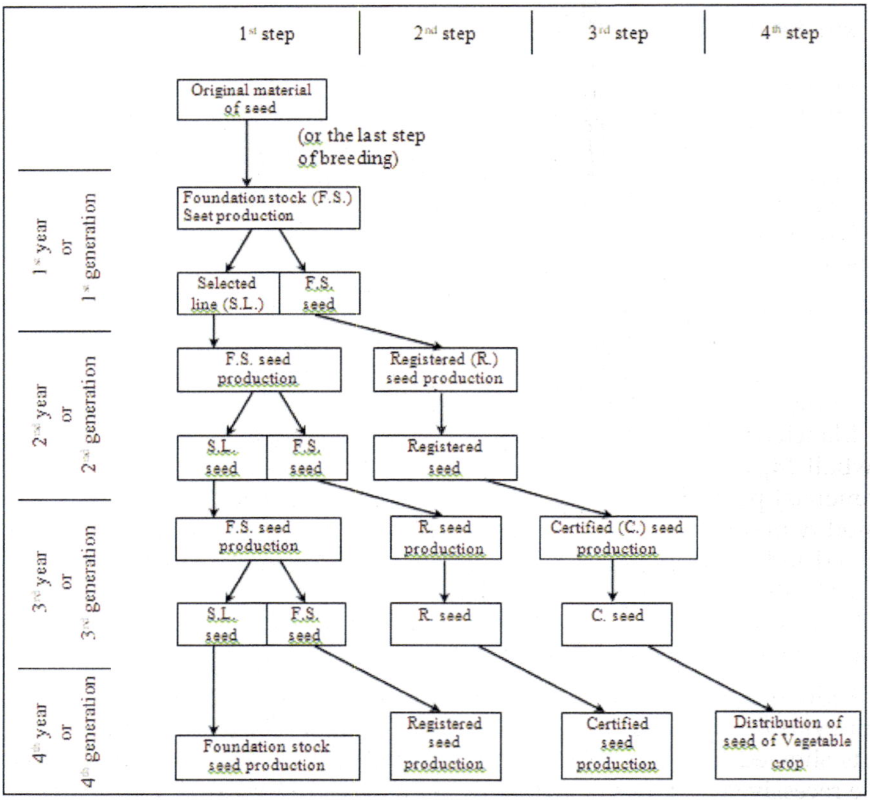

1st step
2nd step
3rd step
4th step
Original material of seed
(or the last step of breeding)
1st year or 1st generation
2nd year or 2nd generation
3rd year or 3rd generation
4th year or 4th generation
Foundation stock (F.S.) Seet production
Selected line (S.L.)
F.S. seed
F.S. seed production
Registered (R.) seed production
S.L. seed
F.S. seed
Registered seed
F.S. seed production
R. seed production
Certified (C.) seed production
S.L. seed
F.S. seed
R. seed
C. seed
Foundation stock seed production
Registered seed production
Certified seed production
Distribution of seed of Vegetable crop

Figure 9: General Scheme of Seed Production

i) Plant type including frame
ii) Stalk length
iii) Number of outer leaves/whorls
iv) Size of the leaves
v) Diameter of the edible portion
vi) Depth of the edible portion
vii) Size of the edible portion
viii) Colour of the edible portion
ix) Compactness of the edible portion
x) Taste/nutritive quality, *etc.*
xi) Yield of the edible portion
xii) Yield per unit area
xiii) Yield per unit time/maturity
xiv) Resistance to common diseases and pests
xv) Seed yielding capacity per unit area

Singh (1984) observed strongest positive correlation of yield with curd diameter in Early Kunwari and leaf weight in Snowball. Pandey and Naik (1985) found that the characters, day to curd initiation, curd weight, length and width of leaves and plant height affected yield of cauliflower, and in improvement, attention to these characters should be given. Radhakrishna and Korla (1995) reported that in cauliflower curd diameter was positively correlated with curd depth and correlated negatively with days to marketable maturity.

In recent years, attempts to incorporate favourable characters in cauliflower from both temperates and tropical have been made. Honma and Cash (1986) reported three varieties *viz.*, Supreme, Beta-White and One-up. While Supreme and Beta-White originated from Pua-kea (tropical) and Snowball (temperate summer) with Self- Blanche, the cultivar One-up was derived from a series of crosses involving Snowball M_1, Pua-kea, Avans Osena, February L, Early Fuji and Self-Blanche. Commercial possibilities of such varieties have to be seen. There is indication of possibility of developing tropical cabbage from a Sri Lankan line EC 16525 which flowered and set seed without chilling at Bangalore, India (Pandey and Naik, 1986b). Gill *et al.* (1987) isolated a pureline called Pusa Himjyoti from MG2-4 for July to October cultivation in hill area of Himachal Pradesh, India. The curd retains white colour even after exposure.

A glossy wax less mutation of cabbage cv. Yingchan, discovered in 1987, was used in various crosses to investigate the mode of inheritance by Xiang (1996). The results showed that the character was controlled by a simple mechanism with a single recessive gene. The possible use of this gene inbreeding new cultivars with improved quality and its use as a marker gene were considered.

14.1.0 Qualitative Characters

14.1.1 Nature of Gene Action for some Curd Characters in Cauliflower

Character	*Nature of Gene Action*
1. Curd weight	Dominance and epistasis (Swarup and Pal, 1966); epistasis (Kumaran, 1971); partial dominance and additive gene action (Deshpande, 1975); pronounced overdominance and epistasis (Singh *et al.*, 1975; Kale *et al.*, 1979); pronounced dominance and epistasis (Lal, 1977); additive and dominance gene action (Singh *et al.*, 1976; Chand, 1980; Yyoti and Vashistha, 1986).
2. Curd to plant ratio	Partial dominance (Kale *et al.*, 1979).
3. Curd depth	Pronounced additive gene action (Lal, 1977); additive and dominance gene action (Dadlani, 1977; Chand, 1980).
4. Curd diameter	Predominance of dominance gene action (Lal, 1977); additive and dominance gene action (Dadlani, 1977; Chand, 1980).
5. Curd size index	Epistasis (Kumaran, 1971); pronounced over dominance and epistasis (Singh *et al.*, 1975); partial dominance and additive gene action (Deshpande, 1975); dominance and epistasis (Lal, 1977); additive dominant gene action (Singh *et al.*, 1976); partial dominance (Kale *et al.*, 1979).
6. Curd angle	Pronounced additive gene action (Lal, 1977); additive and dominant gene action (Dadlani, 1977; Chand, 1980).
7. Curd compactness	Polygenic (Nieuwhof and Garretson, 1961); dominance and additive gene action (Lal, 1977); two genes Co. 1 and Co.2 (Singh *et al.*, 1978).
8. Maturity	
(i) Earliness	Partially dominant gene action (Watts, 1964); dominance and epistasis (Swamp and Pal, 1966); predominance of additive gene action (Kumaran, 1971; Singh *et al.*, 1975, 1976, Deshpande, 1975; Lal, 1977); additive gene action (Kale *et al.*, 1979); additive and dominant gene action (Dadlani, 1977; Chand, 1980).
(ii) Lateness	Recessive polygenes (Watts, 1963).
9. Riceyness	Polygenes, dominant to non-riceyness (Zhao *et al.*, 2020)
10. Purple curds	Incompletely dominant gene (Singh *et al.*, 2020)

Character	*Nature of Gene Action*
11. Curd architecture	Polygenic
12. Bracting	Monogenic dominant
13. Gross weight per plant	Additive gene action (Chatterjee *et al.*, 2018)
14. leaf number per plant	Additive gene action (Chatterjee *et al.*, 2018)
15. leaf size index	Additive gene action (Chatterjee *et al.*, 2018)
16. Stalk length	Non-additive gene action
17. Curd solidity	Dominance and epistatic effects (Chatterjee *et al.*, 2018)

Analysis on combining abilities of main economical characters in cauliflower indicated that heading stage, maturity, plant height, plant expansion, biological yield/plant and number of leaf blades were mainly controlled by non-additive genes while economic yield/plant, stem diameter, height and curd compactness were mainly controlled by additive genes (Li *et al.*, 1996).

14.1.2 Nature of Gene Action for some Head Characters in Cabbage

Character	*Nature of gene action*
1. Maturity 2. Net weight of head 3. No. of marketable head per unit area 4. Yield	Dominance and high epistasis (additive x additive) Dominance and high epistasis (additive x additive) Dominance and high epistasis (dominance x dominance) Dominance and high epistasis (additive additive) (Swamp *et al.*, 1963; Swamp and Sharma, 1965)
5. Head splitting	Additive-epistatic effects, major gene and polygenes (Su *et al.*, 2015)
6. Earliness of head formation	Additive and average dominant effects (Tanaka and Niikura, 2006)
7. Head shape	Quantitative (Zhang *et al.*, 2016)
8. Maturity	Quantitative
9. Frost hardiness	Quantitative

14.1.3 Nature of Gene Action in Broccoli

Trait	*Nature of Gene Action*
1. Bracting	Polygenic (Kop *et al.*, 2003)
2. Marketable yield per plant	High heritability, high genetic advance (Kalia and Shakuntla, 2002), additive gene action
3. Terminanl head weight	High heritability, high genetic advance (Kalia and Shakuntla, 2002), additive gene action

Trait	*Nature of Gene Action*
4. Head size index	High heritability, high genetic advance (Kalia and Shakuntla, 2002), additive gene action
5. Number of spears	High heritability, high genetic advance (Kalia and Shakuntla, 2002), additive gene action
6. Mean spears weight	High heritability, high genetic advance (Kalia and Shakuntla, 2002), additive gene action
7. Plant weight	High heritability, high genetic advance (Kalia and Shakuntla, 2002), additive gene action
8. Head compactness	Additive and non additive gene action
9. Days to marketable maturity	Non-additive gene action
10. Stem diameter	Non-additive gene action
11. Harvest index	Non-additive gene action

Analyses of genetic effects of some nutrient quality characters in non-heading Chinese cabbage by Ping and Chun (1997) revealed that dry weight, crude fibre, vitamin C, organic acid and soluble protein content were controlled by additive and partial dominant effects. Soluble sugar content showed positive heterosis. Narrow sense heritabilities of dry weight, crude fibre, vitamin C, organic acid, soluble protein and soluble sugar content were 90.1, 84.2, 82.9, 75.1, 67.3 and 27.9 per cent, respectively. The inheritance of high temperature tolerance in Chinese cabbage (*B. pekinensis*) indicated hat additive gene effects were predominant. Estimates of broad and narrow sense heritability were relatively high (Wu *et al.*, 1997).

14.2.0 Male Sterility and Cytoplasmic Differentiation

Male sterility system is widespread in angiosperms and it refers to the failure of plants to produce viable pollen or functional male gametophyte or microspores (Prakash *et al.,* 2009; Chen and Liu, 2014). It is an important genetic mechanism promoting allogamy and prevents autogamy. German botanist Kolreuter in 1763 reported first documentation of male sterility in flowering plants (Mayr, 1986) which later on recognized as the indispensable phenomenon for economic hybrid seed production. Male sterility trait has been documented in more than 610 species of flowering plants (Kaul, 1988, Chen and Liu, 2014) including rice, Brassica, Maize, cotton, capsicum, onion, carrot and others, arising mostly as spontaneous mutants. Kaul (1988) divided male sterility into genetic (spontaneous or induced) and nongenetic (induced) male sterility (Figure 10). On a phenotypic basis, genetic male sterility has been divided into sporogenous, structural and functional male sterility and non-genetic male sterility as chemical, physiological and ecological male sterility. Furthermore, on a genotypic basis, genetic male sterility has been categorized into genic (GMS), Cytoplasmic (CMS) and Cytoplasmic-genic male sterility (CGMS) (Kaul, 1988). Generally, there are three reported categories of male sterility phenomenon based on mode of inheritance *viz.* genic male sterility/nuclear male sterility (GMS/NMS), cytoplasmic male sterility (CMS) and cytoplasmic genic male sterility (CGMS) (Vedel *et al.*, 1994; Delourme and Budar, 1999; Chen and Liu, 2014; Rhee *et al.*, 2015). GMS is further subdivided into two classes, environment-sensitive genetic male sterility (EGMS) and environment-insensitive genetic male sterility. In addition, a novel type of dominant genetic male sterility has been developed through the genetic engineering approach referred to as transgenic male sterility. Male sterility that is not genetically inherited is also produced by chemicals known as male gametocides or chemical hybridizing agents (CHA) (Mc Rae, 1985; Kaul, 1988; Sharma and Sharma, 2005) or by altering atmospheric temperature, light intensity and field conditions (Nieuwhof, 1968). Both GMS and CMS types are prevalent in Brassica crops but, later is most common and utilized commercially for hybrid seed production in Brassica vegetables (Budar and Pelletier, 2001; Kucera *et al.*, 2006; Chase, 2007).

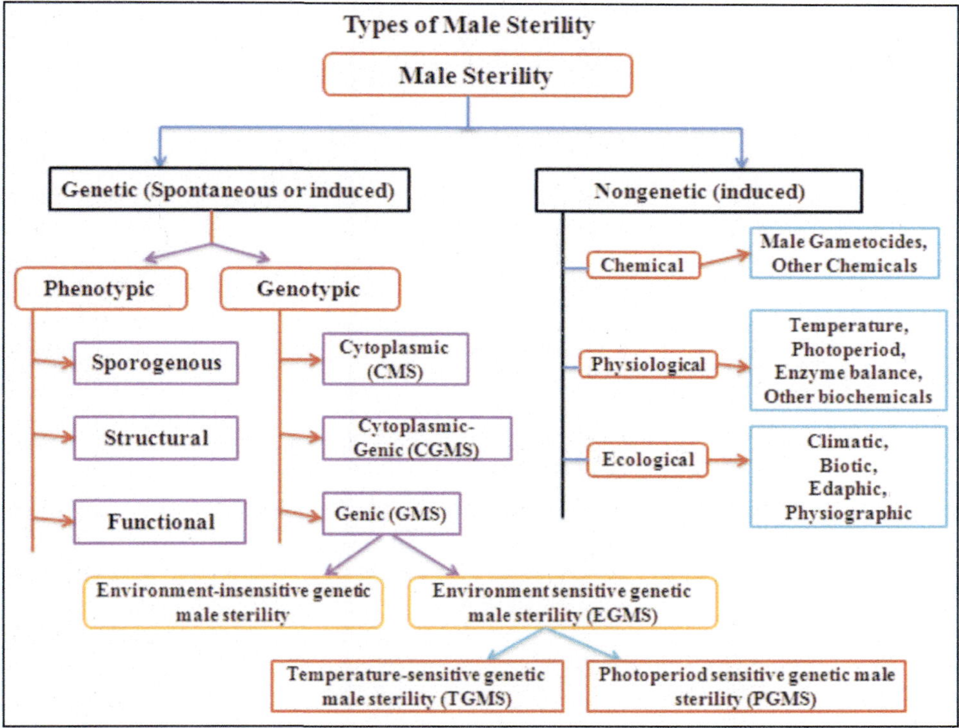

Figure 10: Classification of Male Sterility (Adapted and modified from Kaul, 1988)

14.2.1 Genetic Male Sterility

Genetic male sterility (GMS), governed by nuclear genes, ordinarily controlled by a single recessive gene, is of wide occurrence in angiosperms (Kaul, 1988). The most of sterile mutants have originated spontaneously having homozygous alleles (msms) and have been documented in about 175 plant species (Kaul, 1988) including Brassica vegetables. The pattern of inheritance of GMS in all cases is Mendelian

(Kaul, 1988; Colombo and Galmarini, 2017). Majority of the spontaneous or induced male sterile mutants represents monogenic recessive genes and occasionally, it is under the control of multiple recessive genes (Gottschalk and Kaul, 1974; Kaul, 1988). Among the Brassica species, the first report of genic male sterility (GMS) was documented in sprouting broccoli (*Brassica oleracea* var. *italica* L.) (Anstey and Moore, 1954) and during the past 60 years, other cases of GMS have been reported in most of the cultivated vegetable Brassicas. In broccoli the non-alleleic recessive genes have been designated as ms-1, ms-2, ms-4, ms-5 and ms-6 (Dickson, 1970). Out of these non-allelic recessive genes conditioning GMS in broccoli (*Brassica oleracea* var. *italica*), ms-6 gene, have been reported to be sensitive to low temperature (Dickson, 1970). Borchers (1971) suggested that temperature sensitive male sterile plants (ms-6 ms-6) homozygous for ms-6 gene could be enabled for development of hybrid seeds with few sibs at relatively high temperature. Dickson (1970) suggested that male sterile line could be maintained at low temperature through self-pollination.

Table 10: List of Genic Male Sterility in Cole Crops

(*ms-1*)	1. Sprouting broccoli	Anstey and Moore (1954); Cole (1957, 1959)
	2. Cabbage	Nishi and Hiraoka (1957); Kotatani and Yamada (1958); Rundfeldt (1960)
(*ms-4*)	3. Cauliflower	Borchers (1966)
	4. Brussels sprouts	Nieuwhof (1961)
	5. White cabbage	Nieuwhof (1961)
(*ms-2*)	6. Brussels sprouts	Johnson (1958)
(*ms-C*)	7. Cauliflower	Ahluwalia *et al.* (1977)

Chromosomal monogenic dominant male sterility in Chinese cabbage has been identified. It is expressed as much reduced stamens, found on backcrossing a hybrid from *B. rapa* sub sp. *pekinensis* (*B. pekinensis*), *B. rapa* subsp. *chinensis* to *pekinensis* parent (Meer, 1986, 1987).

EGMS

The plant reproductive development stages and sterility are extremely sensitive to environmental stresses such as high or low temperature, drought, cold salinity and other abiotic stresses (Nieuwhof, 1968; Giorno *et al.*, 2013; Zhang *et al.*, 2016). The type of male sterility that is sensitive to the environmental conditions and is the result of interaction of nuclear genes with environmental factors (Temperature, photoperiod and others) is termed as environment-sensitive genic male sterility (EGMS) (Virmani and Ilyas-Ahmed, 2001). At a particular range, duration, or concentration of an environmental factor, *mst* genes exhbit expression at a sensitive stage of plant development and induces male sterility. EGMS is categorized into three classes' *viz.* photoperiod sensitive genetic male sterility (PGMS), temperature sensitive genic male sterility (TGMS) and photo-thermo sensitive genetic male sterility (PTGMS). The EGMS plants during the sterile phase can be used as a female parent in the hybrid seed production programme, whereas, during the environmental conditions that induce fertility, these plants are maintained through selfing without the use of a maintainer line (B line) (Virmani and Ilyas-Ahmed, 2001). EGMS system is known as two-line system of heterosis breeding as only two lines are required for the maintenance and multiplication of male sterile plants and production of hybrid seed, in contrast to the three-line CMS system involving CMS (A), maintainer (B), and restorer (R) lines. TGMS expressed at specified critical temperature range is also regarded as an environmental-induced male abortion by exposure to mild temperature conditions (Storme and Geelen, 2014). Cell differentiation and male reproductive development in flowering plants is unique and highly coordinated biological process consisting of numerous distinct steps controlled by complex genetic regulatory network (Ma *et al.*, 2012; Xu *et al.*, 2014). If any of the pathways in this complex process is blocked, pollen grain production might be aborted, resulting in male sterility. EGMS system is of prime importance in plant science as it facilitates the exploitation of heterosis in crop plants (Ding *et al.*, 2012; Zhou *et al.*, 2014) and also the male sterility phenomenon is a model system to study the gametophyte development and complex genetic regulation network (Ma *et al.*, 2012). EGMS have been reported in wide range of crops species including vegetable crops and was first documented in rice from 'Nongken 58' a japonica cultivar. Among the vegetable crops the EGMS system was firstly reported in tomato and pepper and subsequently it was documented in Brassica vegetables.

Utilization of GMS in Hybrid Breeding

The cost effective and simplified hybrid development via natural genetic mechanisms of male sterility have encouraged the researchers to pay more attention to look for genetic resources of male sterility and their exploitation in heterosis breeding of Brassica crops. Both dominant (DGMS, DNMS) and recessive (RGMS) types of nuclear (genic) male sterility have been reported in Brassica crops and can be effectively utilized in hybrid breeding (Ji *et al.*, 2017). Borchers (1971) demonstrated the broccoli hybrid seed production utilizing homozygous recessive male sterile line "ms-6/ms-6". In the different planting ratios of male sterile to male fertile plants (1:1, 2:1 and 6:1), the range of hybrids in the progeny ranged from 91 to 100 per cent. Furthermore, Fang *et al.* (1997) reported the two step breeding programme (production, multiplication of male sterile line and subsequent F_1 hybrid seed development) with use of DGMS line in cabbage (Ms-cd1gene) for the production of hybrid seeds. The propagation of homozygous male sterile plants (Ms-cd1/Ms-cd1) was carried out with tissue culture approach and then via natural pollination of DGMS plants (Ms-cd1/Ms-cd1) with a sister inbred line, obtained the male sterile line. Subsequently, interplanting of male sterile line (as female parent) with male fertile inbred line (as male), having desirable features, in the ratio of 3:1 in panmictic plot was done and the F_1 hybrid seeds were collected from male sterile line (Fang *et al.*, 1997). In the recent years, in China, this two step method derived from DGMS line in cabbage has been utilized commercially in cabbage hybrid breeding (Liang *et al.*, 2017). On the other hand, RGMS also effective in escalating hybrid breeding in Brassica but no commercial utilization is reported yet owing to its limitation that it is not possible to derive 100 per cent male sterile plants. The homozygous recessive (RGMS) male sterile (msms) line is maintained by repeated backcrossing of it with heterozygous isogenic B line (also called as maintainer line) (Msms). Hence, in the hybrid seed production block, 50 per cent of male fertile segregants (Msms) in the male sterile line (female parent), should be identified and removed before pollen shedding (Figure 11). In some crops, the tight linkage of recessive male sterile gene 'ms' with recessive morphological markers, especially at seedling stage, is of prime significance in identification of male sterile and male fertile plants in female parent *e.g.* in broccoli (bright green hypocotyl). Hence, the time and labour needed for identification and removal of male fertile plants in female parent, in hybrid seed production block, is largely avoided. In cabbage, hybrids *e.g.* Zhonggan No 16, 17 and 18, have been bred in China utilizing DGMS (Ghani *et al.*, 2013).

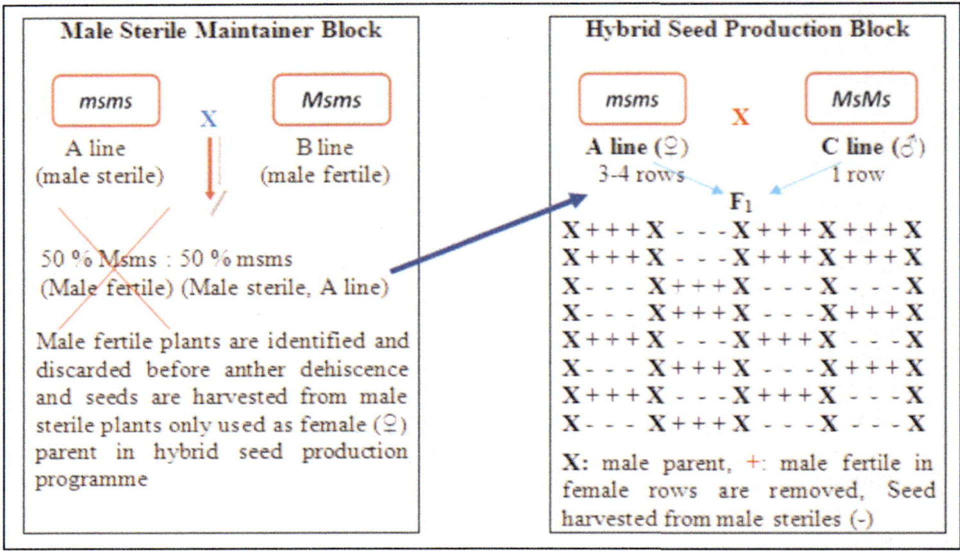

Figure 11: General Schematic Representation of Hybrid Seed Production using RGMS (Colombo and Galmarini, 2017)

14.2.2 Cytoplasmic Male Sterility

Raphanus sativus: Ogura (1968) first reported a new type of male sterility in an unidentified cultivar of Japanese radish. The sterility was induced by the interaction between the homozygous recessive nuclear genes (msms) and cytoplasm (S) Male sterile plants possessed the (S) *msms* gene constitution. Moreover, it was ascertained that all the cultivars of Japanese radish used, possessed a homozygous recessive gene, *msms* and no fertility restorer gene. The female organ showed normal fertility and an excellent productivity of seeds when they were pollinated by normal fertile pollens. Bonnet (1975) introduced this male-sterile radish and made a series of crosses with several lines derived from European cultivars of radish. Some sterility maintainers were observed and fertility restorer genes were found very frequently in the European materials. A good malesterile line was obtained by back-cross with European cultivars and F_1 seeds using this male-sterile line showed great vigour; the produce was homogenous, solid and tasty. However, the seed yield was poor in the F_1 and Bonnet (1977) suggested a 3-way cross for F_1 seeds for using this male sterility.

Bannerot *et al.* (1977) reported successful introduction of cabbage nucleus into the cytoplasm of the male sterile Japanese radish by means of repeated backcrosses. They observed that the level of 18 chromosomes was achieved after fourth backcross followed by *in vitro* culture of the immature embryos. All the *oleracea* genotypes that were introduced into this radish cytoplasm exhibited severe yellowing of the young leaves, when the young plants were grown at temperature below 12°C. It seemed that the nucleus of *oleracea* did not function properly in the radish cytoplasm. The degree of yellowing differed with recurrent male parents used. The most severe effect was observed with cauliflower while Brussels sprouts and savoy cabbage F_1 hybrids were greenish though yellowing remained. Bannerot *et al.* (1974) intended to replace the chloroplast set of radish origin by a set of cabbage chloroplast *via* protoplast fusion. So far nothing successful has come through. A newer source from *Brassica juncea* is also being tried in some centres, especially for Chinese cabbage. Cytoplasmic male sterility of cauliflower with *Raphanus sativus* cytoplasm was associated with chlorosis of seedlings and young leaves. Pelletier *et al.* (1995) gave information on a new source of cytoplasmic male sterility (CMS) through cybridization in *Brassica oleracea*; the use of cybrids in F_1 hybrid varieties in cabbage and molecular determinism of *Ogura* male sterility in *Brassica*. A cytoplasmic male sterile line of broccoli was developed by Lin *et al.* (1997) by crossing male-sterile cauliflower line MSA, used as a CMS with a fertile broccoli line followed by four generationsof backcrossing. TPS is a potential cytoplasmically male sterile line found in rape (*B. napus*). The TPS sterile cytoplasm has been transferred from rape to *B. campestris* sub sp. *chinensis* (*B. chinensis*) cv. Pakchoi by inter-specific crossing and backcrossing, which has resulted in the production of 3 Pakchoi CMS lines (Li *et al.*, 1998). A cytoplasmic male sterile Chinese cabbage line was derived from NY 8481. It showed stable sterility and good yields (Hu *et al.*, 1999).

Mitochondrial Genes Governing CMS in Brassica

The plant mitochondrial genome is highly complex and sensitive to frequent recombination, DNA rearrangements, incorporation of foreign DNA fragments, insertions/deletions and mutation (Chase, 2007; Bohra *et al.*, 2016). These characteristic features lead to generation of unusual chimeric open reading frames (orfs), which causes mitochondrial dysfunction and leads to cytoplasmic male sterility (Chase, 2007; Yamagishi and Bhat, 2014). About 14 mitochondrial genes determining CMS have been characterized as open reading frames (*orfs*) (Chase, 2007). Different orfs determining different origin of CMS have been characterized such as pol CMS contains a chimeric gene *'orf224'* which is co-transcribed with *atp6* gene, ogu CMS plants comprises *'orf138'* chimeric gene and co-transcribed with *atp8* gene, chimeric gene *orf222* determine nap CMS, similarly chimeric gene *orf288* co-transcribed with *atp6* gene determine *hau* CMS (Jing *et al.*, 2012; Yamagishi and Bhat, 2014). The association of a mitochondrial gene with CMS is inferred when the gene is expressed only in the male sterile plant but is suppressed in the fertility restored plant carrying a nuclear *Rf* gene.

Development of CMS Lines in B. oleracea

Systematic development of CMS lines in brassica vegetable like cabbage and cauliflower was started at ICAR-Indian Agricultural Research Institute, Regional Station, Katrain, Kullu, Himachal Pradesh, India since 1995. The *Ogura* based cytoplasm was introgressed into a large number of cabbage and cauliflower through back-cross introgression. Three CMS lines in cauliflower, *Ogu1A, Ogu2A* amd *Ogu3A* were developed by Dey *et al.* (2011) with good floral traits and seed setting capacity. These lines were evaluated and compared with their respective fertile maintainer (B) line for various agronomic traits like curd yield and size, floral like size of petals and style, shape of style and seed setting traits like number of seeds per pod and seed yield per plant. Nuclear-cytoplasmic interaction played an important role in determining various agronomic, floral and seed setting related traits. All the three CMS lines were similar with fertile maintainer lines for agronomic traits like days to curd maturity. Introgression of alien Ogura cytoplasm in the background of snowball cauliflower significantly reduced the petal size, filament length, style and stamen length. Number of pods per plant was significantly lower in all three CMS lines, when compared with their respective B lines. Number of seeds per pod and seed yield per plant in Ogu3A and its respective B line was at par but reduced significantly in Ogu1A and Ogu2A after introgression of Ogura cytoplasm. These Three CMS lines are being used widely in hybrid development in cauliflower. Two CMS lines Ogu3A (Snowball/late group) and Ogu33A (mid-late groups) were developed and registered with NBPGR by IARI, Regional Station, Katrai, Kullu. They are the first of CMS lines incole crops in India which are available in public domain and can be accesses by any breeders working on cauliflower improvement. Another set of 7 CMS lines were characterised by Dey *et al.* (2014) and found that they were suitable for use in hybrid breeding. However, introgression of alien cytoplasm resulted into reduced flower size in eaxh line because of nuclear-cytoplasmic interaction. Later, Dey *et al.* (2017) reported a large number (25) of cauliflower CMS lines developed after 9 generation of back-crossing in different maturity groups and characterised them for all important agronomic and floral traits. Besides, these lines were also grouped using a large number of SSR markers. These lines were also evaluated for combining ability to reveal their breeding potential. Two CMS lines, Ogu402-6A and Ogu76- 4A were distantly placed from rest of the CMS genotypes based on morphological characterization. Molecular analysis through SSR primers also revealed the genetic distance of the CMS lines Ogu402-6A from rest of the CMS pool. Three genotypes (Ogu402-6A, Ogu76-4A and Ogu119-2A) with early maturity were identified for their use in development of short duration hybrids. Similarly, 3 CMS lines Ogu118-2A, OguHL-3A and Ogu126- 1A were identified with yield potential of more than 60 t/ha. Based on 13 polymorphic SSR markers 25 CMS lines were grouped into 4 major clusters. Effects of improved cytoplasmic lines in agronomic and floral traits were studied in detail by Dey *et al.* (2013), and it was concluded the introgressed cytoplasm mainly alters the floral and reproductive traits and do not have any influence in performance for different agronomic traits of the CMS lines and F_1 hybrids developed using these CMS lines. In cabbage, one CMS line, CMS-1 was developed indigenously by Parkash *et al.* (2015) and was was found suitable

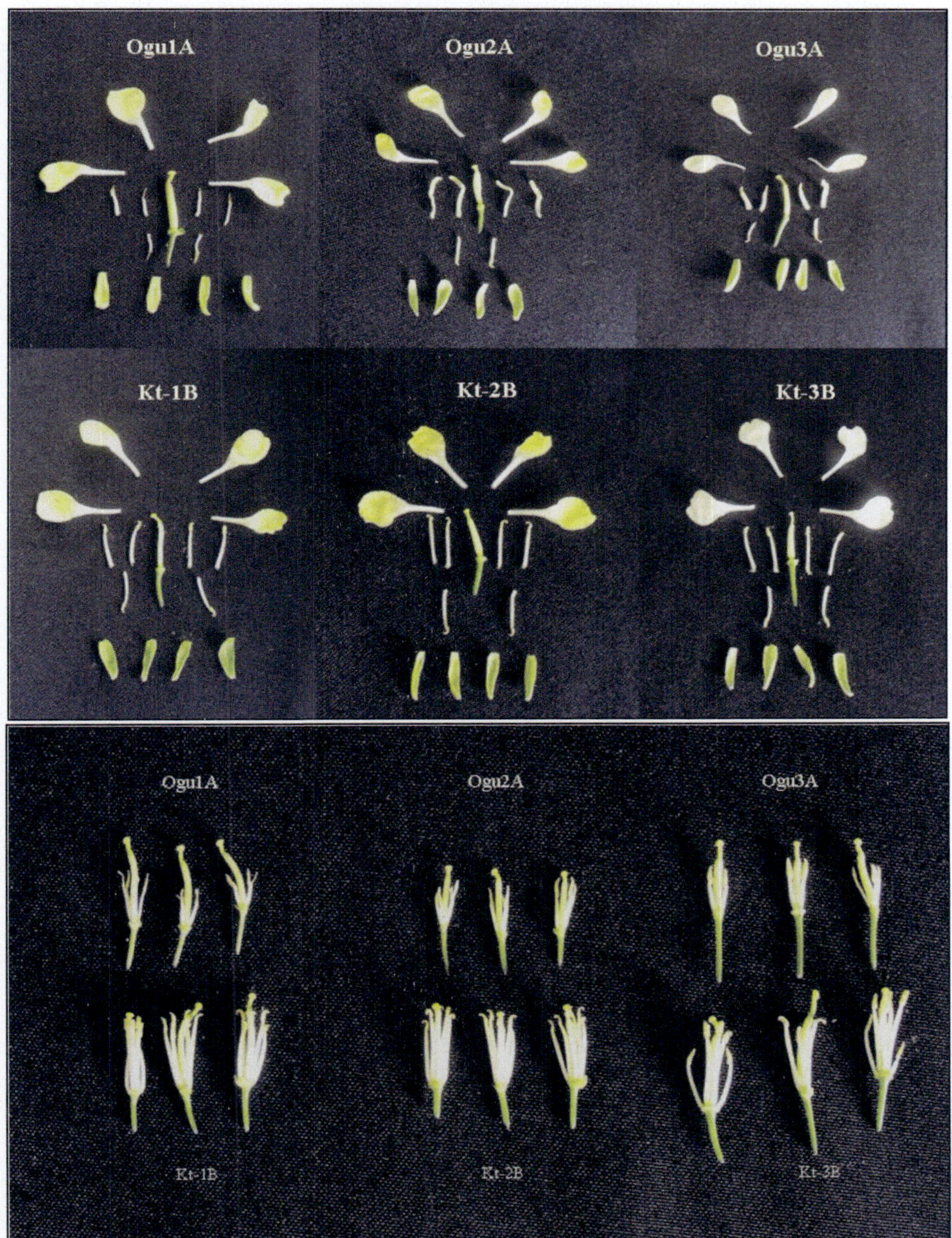

Floral Traits of 3 Cauliflower CMS Lines which are Used Widely in Hybrid Breeding in India (Dey *et al.,* 2011)

for use in commercial scle based on the performance of the F_1 hybrids developed using this line.

Development of F_1 Hybrids Using CMS Lines and Seed Production

Cytoplasmic male sterile ines are most preferred pollination control mechanisms in cole crops used in development of F_1 hybrids worldwide. Vegetative parts are the economically important edible portions in the cole crops and there is no need of successful fertilization for proper development of edible part unlike fruit vegetable. Therefore, restoration of male sterility is not necessary in the F_1 hybrids. In crops like *B. juncea* and *B. napus* which need successful development of seeds from the F_1 hybrids, restoration of male sterility is necessary in hybrid breeding and they need the restorer line with *Rf* gene used as pollen/male aprent in hybrid development. Howeber, in case of cole crops there is no need of any restorers with *Rf* gene to be used as male/parent. Any inbred which is good combiner with the male sterile female parent and gives superior hybrid with higher yield and quality is suitable as male parent.

Development of CMS lines have three major steps:

1. Identification of source of male sterile cytoplasm without any adverse effect when introgressed into the nuclear background of recurring parent.
2. Back-cross substitution of the nuclear genome of the souce line (non-recurreng parent) through 6-7 generations of back-crossing using the source parent only once as female parent.
3. Selection of the male sterile plants with higher resemblance to recurrent parent for different morphological traits and better floral traits and normal development of style and stigma.

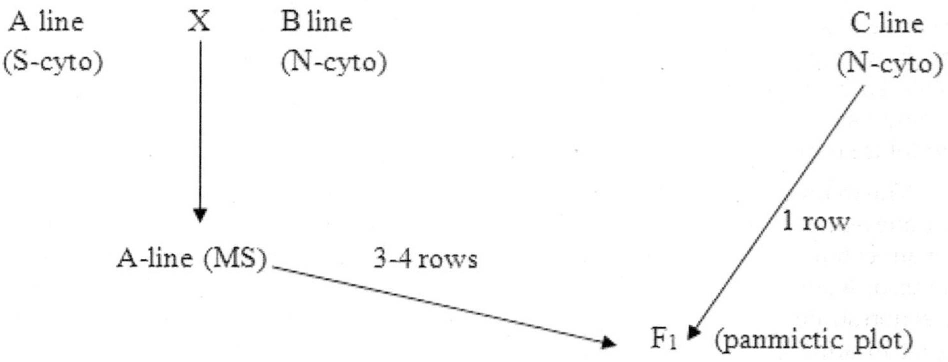

Hybrid Seed Production Using CMS System; Cyto-cytoplasm.

At the end of 6-8 generations of back-crossing the male sterile sterile lines developed will be the A line and the recurring male fertile parent will the maintainer of the male sterile line (B line). While introgressing the male sterile cytoplasm it is evident that introgression of alien cytoplasm results into reduced flower size. However, utmost care should be taken while selecting the plants after each generation of back-crossing. The back-progenies with normal flower structure and development of nectaries, secondly, cole crops are highly heterozygous with varied extent of self-incompatibility and expresses inbreeding depression while selfing or sibbing with limited number of plants. Therefore, development of strategy to select suitable number of plants used for back-crossing is very much essential. Unlike the other field crops, growing of large numner of plants in each generation of back-crossing is not possible. Moreover, while introgressing the male sterile cytoplasm into a large number of lines subsequently, it may not be possible to raise and pollinate a large number of plants.

Dey *et al.* (2018) developed a small population based breeding strategy for introgression of male sterile cytoplasm into a large number of genotypes (Figure 12). Three methods of back-crossings were adopted after 3 generations of single plant selection *viz.*, i. twelve plants both from recurrent and non-recurrent parents till BC9 (M1) ii. six plants both from recurrent and non-recurrent parents till BC9 (M2) and iii. single plant selection from recurrent and non-recurrent parent till BC9 (M3). Later on, two generations of bulking were practiced for their use in hybrid breeding. The agronomic and floral performance was tested at different stages. Inbreeding depression was calculated for various traits at different stage of development. Performance of the hybrids using the CMS lines was evaluated to find out the suitable method and ideal population structure. Among the different methods of backcrossing the M2 (6 plants from BC4 onwards) was found to be most effective in ensuring uniformity and complete recovery of recurrent genotype nuclear genome. However, in evaluating the performance of the hybrids M2 was at par with the M1. Inbreeding depression for various traits was highest at M3 followed by M2. However, the inbreeding depression was reversed when the plants generated through M2 undergone two generations of bulking. The plants generated through M3 were not able to recover even after bulking indicating the permanent loss in vigor because of several generations of inbreeding using single plant. Therefore, it was recommemded to adopt single plant selection for three generation of back-crossing followed by back-crossing using 6 plants in each generation. This model can be widely used in development of CMS lines in different cole crops. This innovative model is summerised as follows:

The main problem being faced is the lower quantity of hybrid seed on account of honey-bees preference for the pollen fertile C-line. This could be overcome to a certain extent with appropriate flower morphology and manipulating the ratio of plants of A and C lines, spacing and planting design. An unique honey bee mediated selection strategy was devised by Dey *et al.* (2019). Honey bees were used as selection agent in addition to manual selection to improve the flower structure and seed setting ability of the CMS lines. This study adopted an alternative selection strategy using honey bee (*Apis cerena indica*) as the selection agent in five selection cycles in three CMS lines of cauliflower (Ogu1A, Ogu2A and Ogu3A). Selected plants with the highest seed yield, determined through honey bee selection, were bulked to raise the plants for next generation. The reproductive structures of the CMS plants improved significantly after five cycles of honey bee-mediated selection. Seed yield/ plant, nectar quantity, size of nectary and flower size improved significantly in

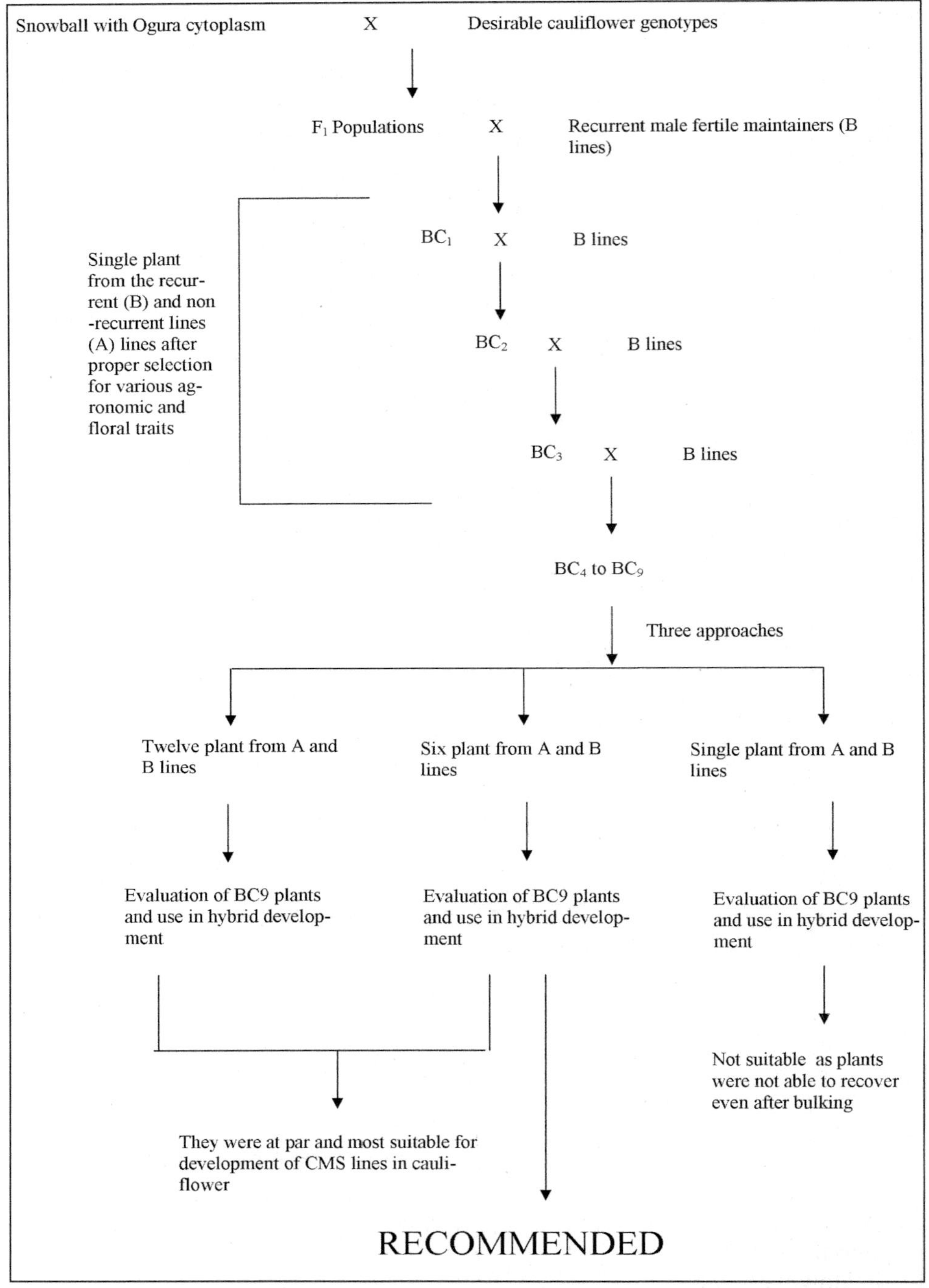

Figure 12: The Model Backcross Breeding Developed Based on the Experiments in the Last 12 Years for Introgression of Wild Male Sterile Cytoplasm into the Nuclear Background of *Brassica oleracea* (Dey *et al.*, 2018)

all three CMS lines at the end of the selection cycles. Quantity of nectar increased from 0.45 to 1.45 µl and from 0.37 to 1.49 µl per flower in CMS lines Ogu1A and Ogu3A, respectively, after five cycles of selection; similarly, quantity of seeds per plant improved from 13.48 to 20.71 g and from 12.77 to 17.57 g in the same CMS lines, respectively. The extent of improvement in nectar quantity and seed yield was comparatively lower in CMS line Ogu2A. The occurrence of different floral deformities was also reduced in all three CMS lines after five cycles of honey bee- mediated selection. CMS lines Ogu1A and Ogu3A responded better to honey bee-mediated selection than did CMS line Ogu2A. This novel strategy of honey bee-mediated selection can be used to develop CMS lines with higher seed yield in different crops where cytoplasmic male sterility is commercially used in hybrid seed production.

Once CMS lines with superior flower structure and, normal rectar production developed after complete nuclear substitution, they can be used successfully in development of F_1 hybrids. The following are the examples of successful use of CMS lines in development of F_1 hybrids:

Cauliflower

1. Pusa Snowball Hybrid-1: First CMS based F_1 hybrid in cole crops by Indian public sector
2. Pusa Snowball Hybrid-2: Developed using CMS lines as female parent and one doubled haploid line as male aprent
3. Pusa Hybrid-301: First CMS based F_1 hybrid in mid maturity group of cauliflower by Indian public sector

Cabbage

1. Pusa Cabbage Hybrid-81: First CMS based cabbage F_1 hybrid developed by Indian public sector
2. Pusa Cabbage Hybrid-822: It is a CMS based F_1 hybrid

14.3.0 Self-incompatibility

Incompatibility is the term used to describe the phenomenon of inability of plants having fuctional gametes to set seeds when either self-pollinated or crossed with some of their genetic relative. Most of the authors restricted the term self-incompatibility as the inability to set seeds after self-pollination. There are two types of incompatibility-(i) gametophytic and (ii) sporophytic. In the former system, the pollen pistil interaction is genetically governed by the haploid genome of each pollen grain and the diploid genome of the pistil tissue. In the latter system, the pollen-pistil interaction is determined by the genome of the somatic tissue (of the sporophyte) in which the pollen-grains are formed. In both systems, incompatibility is generally controlled by a series of alleles at a single locus S (Bateman, 1952; Lewis, 1956). However, the involvement of more than one loci has also been reported in some crops including the member of the family Cruciferae (Lundqvist, 1954; 1975; Lundqvist *et al.*, 1973; Larsen, 1974; Verma *et al.*, 1977; Lewis, 1977).

Bateman (1952, 1954, 1955) suggested that self-incompatibility system in all species of the family Cruciferae is of sporophytic type. Such a system was confirmed in kale by Thompson (1957), radish by Sampson (1957), broccoli by Sampson (1957) and Odland (1962), cabbage by Adamson (1965) and cauliflower by Hoser-Krauze (1979) and from India (Murugiah *et al.*, 1983). Thompson and Taylor (1966b) reported that ancestral *Brassica oleracea* was highly self-incompatible. On this basis, it is but natural that the cole crops would be self-incompatible.

The main genetic feature of the single locus multi-alleleic homomorphic sporophytic self-incompatibility may be summarized in the following (*cf.* Frankel and Gatun, 1977):

1. Incompatibility is controlled by 1 *S*-locus having several alleles.
2. The reaction of the pollen is determined by the genotype of the sporophytic tissue in which it was formed and thus it is controlled by 2 S-alleles.
3. Resulting from the above type of control, all the pollen grains of a plant have the same incompatibility reaction regardless of the S-allele received during meiosis.
4. The 2 S-alleles may react in dependently or they may interact by one being dominant over the other.
5. Active S-allele identity in both pollen and pistil leads to incompatibility.
6. The dominant/independence relationships of the S-alleles in the pollen and in the pistil may differ.

An individual plant has only l or 2 *S*-factors and so the number of *S*-factors or alleles in the line derived from it never exceed 2. Where only 1 *S*-factor occurs, all plants are homozygous for it and the line is self-incompatible and sib-incompatible. If the line in the progeny of a plant is heterozygous for the S-factor, the line is also heterozygous for the S-factor. The first generation of an S_xS_y plant consists of S_xS_x, S_xS_y and S_yS_y genotype (Nieuwhof, 1966).

The S-allele interactions after approach one of the two extremes (Thompson, 1957; Nieuwhof, 1966; Wallace, 1972, 1979b), (i) co-dominance, *i.e.*, simultaneous activity of both S-alleles of the heterozygote, (ii) dominance, *i.e.*, full activity of one S-allele of the heterozygote with complete inactivity of the other. The dominance is called partial if one allele has 60 to100 per cent activity and the other retains some activity. The S-allele interaction is called co-dominance when both alleles have near equal activities of about 75 to near 100 per cent (Wallace, 1979b).

Four classes of incompatibility in brassica were described as designated, as mode of inheritance for S-allele interaction typesI, II, III and IV (Thompson and Howard, 1959; Haruta, 1962; Wallace and Nasrallah, 1968; Mackay, 1977; De Nettancourt, 1977; Frankel and Galun, 1977; Wallace, 1979a and b; Hoser-Krauze, 1979). According to Wallace (1979b), the typesI, II, III and IV are derived from the two S-allele interactions (co-dominance and dominance) in factorial combination with the two sexual organs (pollen and stigma) as given below:

I	$S_a<S_b$	$S_a<S_b$
II	$S_a<S_b$	$S_a=S_b$
III	$S_a=S_b$	$S_a<S_b$
IV	$S_a=S_b$	$S_a=S_b$

Expected incompatible and compatible interpretation and reciprocal difference interpretations from reciprocal intercrosses among plants of the two homozygous and one heterozygous S-allele genotypes for sexual organ x S-allele interaction. Types I, II, III and IV are given below:

Sexual organ x S-allele Interaction	*S-allele Interaction*		*Female Genotype and Pheno-type*[b] *(♀)*	*Male Genotype and Phenotype*[b] *(♂)*		
	Stigma (♀)	*Pollen (♂)*		S_aS_a	$S_a<S_b$	S_bS_b
I	Dominance	Dominance	S_aS_a	Inc	Com	Com
			$S_a<S_b$	Com	Inc	Inc
			S_bS_b	Com	Inc	Inc
				S_aS_a	$S_a<S_b$	S_bS_b
II	Co-dominance	Dominance	S_aS_a	Inc	Com	Com
			$S_a=S_b$	Inc	Inc	Inc
			S_bS_b	Com	Inc	Inc
				S_aS_a	$S_a=S_b$	S_bS_b
III	Dominance	Co-dominance	S_aS_a	Inc	Inc	Com
			$S_a<S_a$	Com	Inc	Inc
			S_bS_b	Com	Inc	Inc
				S_aS_a	$S_a=S_b$	S_bS_b
IV	Co-dominance	Co-dominance	S_aS_a	Inc	Inc	Com
			$S_a=S_b$	Inc	Inc	Inc
			S_bS_b	Com	Inc	Inc

a. Reciprocal crosses with a reciprocal difference and indicated by ←→ and reciprocal crosses without a reciprocal difference by ←→

b. The *S*-allele phenotype of heterozygotesis indicated by the symbols < and=, where< specifies recessivevs. Dominant and = indicates co-dominance.

Com: Compatible; Inc: Incompatible.

It is possible to introduce active *S*-alleles from late varieties to early varieties in cauliflower by backcross method to make better use of heterozygosity (Kalloo, 1988). In Brussels sprouts weak S-alleles manifested lower self-incompatibility at high temperature (Visser, 1977), Murugiah *et al.* (1987) reported that self-incompatibility alleles of high rankings indominance series *SdSd* and *SvSv* lines retained their self-incompatibility throughout the flowering period, the mid-ranking *SmSm* lines gradually lost with age as did the low ranking *SaSa* lines though to a lesser extent. Nasrallah (1995) described the recent developments in the molecular and genetical analysis of the self-incompatibility system of *Brassica* and prospects for practical applications in light of the newly acquired information on the molecular basis of the S-locus activity. Distinct differences in the controls of amino acids were found between self-incompatible and self-compatible lines of cabbage by Jing *et al.* (1998). In self-incompatible lines threonine, tyrosine, glycine and alanine contents were higher than 0.223 per cent, 0.358 per cent, 1.593 per cent and 1.464 per cent, respectively, whereas in the self-compatible lines, they were less than 0.185 per cent, 0.164 per cent, 1.47 per cent and 1.006 per cent respectively.

Genes Governing Self-incompatibility Response in Brassica

The cell-to-cell communication of sporophytic self-incompatibility (SSI) system in *Brassicaceae* is governed by single polymorphic 'Mendelian S locus' (Bateman 1955). The molecular genetic analysis revealed that 'S-locus' is multigene, multiallelic highly complex locus, which spans many kilobase pairs, and encompasses various physically linked polymorphic genes that co-segregate with SI phenotype (Sehgal and Singh, 2018). Genomic analysis and physical mapping of S-locus in *Brassica* have identified three polymorphic physically linked loci *viz. SRK* (S-receptor kinase) (Stein *et al.*, 1991), *SCR/SP11* (S-locus cysteine rich protein/S-locus protein 11) (Suzuki *et al.*, 1999) and *SLG* (S-locus glycoprotein) (Sehgal and Singh, 2018). Besides these polymorphic linked genes, the other candidate genes playing role in downstream SI signaling pathway in *Brassicaceae* have been identified.

Molecular Mechanism of SI in Brassica

The SSI system in cole vegetables is controlled by single 'Mendelian S locus' with multiple alleles. Genomic analysis of S-locus in *Brassica* identified three physically linked loci *viz. SRK* (S-receptor kinase) (Stein *et al.*, 1991), *SCR/SP11* (S-locus cysteine rich protein/S-locus protein 11) (Suzuki *et al.*, 1999) and *SLG* (S-locus glycoprotein). *SRK* is the female determinant of self-incompatibility response in *Brassica*, a trans-membrane domain and an intracellular kinase domain (Stein *et al.*, 1991). The *S* domain of *SRK* is highly similar to *SLG*, the first S-locus gene to be identified. Another gene reported at S-locus by cloning and sequencing of S-locus region of *B. rapa*, encodes a protein expressed only in anthers, named as *SP11* (S-locus protein 11) or *SCR* (S-locus cysteine rich protein) and identified as male determinant of SI response in *Brassica* (Suzuki *et al.*, 1999). *SCR/SP11*, *SLG* and *SRK* are tightly linked at S-locus and alleles of these genes are inherited by progeny as one set, called S haplotypes. Recognition specificity is achieved by interaction of stigmatic protein *SRK* (female determinant) with its ligand, the male determinant *SP11/SCR* (Sehgal and Singh, 2018). The interaction of *SCR* and *SRK* from the same *S* haplotype induces autophosphorylation of *S*-receptor kinase in an *S*-haplotype specific manner. When there is self-pollination, pollen ligand (*SCR* protein) is bound to maternal gene *SRK*, which causes the activation of receptor and elicit a signal transduction cascade that results in the pollen rejection and inhibition of pollen tube penetration into the epidermal cell wall of stigma (Nasrallah and Nasrallah, 2014). Further, another protein *MLPK* (M-locus protein kinase) interact with *SRK*, and this complex induces signal transduction and result in self pollen rejection. In addition, the arm repeat containing protein 1 (*ARC1*), shows high specificity to *SRK*, and this *ARC1* is phosphorylated *by SRK-MLPK* complex, and phosphorylated *ARC1* degrades *Exo70A1*, thus inhibit the production of proteins for compatible reaction. Therefore, leads to self pollen rejection on selfing.

Constraints in Exploiting SI Lines

- ☆ Continuous inbreeding in many *Brassica* crops may lead to complete loss of inbred lines. Therefore, it is more desirable to use tissue culture techniques for propagation of self-incompatible inbred lines.
- ☆ Pseudo-incompatibility may lead to self seed instead of hybrid seed *e.g.* cauliflower.
- ☆ Maintenance of S-allele lines is costly (hybrid seed will be expensive).
- ☆ Effect of environmental factors on the level of self-incompatibility (high temperature and relative humidity may weaken S-allele action).
- ☆ Synchronization of flowering among the SI lines is difficult *e.g.* cauliflower.
- ☆ Difficulties to produce homozygous inbred lines.

Molecular Markers and S-haplotypes Identification

The genetic mechanism of SSI has been commercially utilized for hybrid seed production in *Brassicaceae*. Thus, the discrimination of S-alleles is useful not only for the study of self-incompatibility, but also for the commercial hybrid seed production. Traditionally, S-haplotypes have been identified by controlled pollination with S-allele tester lines, followed by fluorescence microscopic observation of pollen tube growth through stigma or by seed set data. The technique of counting seed set data is labor and time consuming, though, the fluorescent microscopy is relatively fast but it is essential to have the plants in flowering stage. In addition, the ideal plant selection having recessive traits, such as SI, is often difficult and thus a potential barrier in accelerating plant breeding. In this regard, the use of molecular marker technology for the S- haplotype identification is advantageous that it could be done on young plants and no need to have the S-allele plants in flowering phase. Hence, the development and identification of molecular markers linked with self-incompatible genes is of utmost value for geneticist and plant breeders to discriminate between class I and class II S-haplotypes. Different kinds of molecular markers have been reported for the identification of S-haplotypes in *Brassica* species by different group of researchers (Sehgal and Singh, 2018).

Development of SI Lines and Use in Hybrid Development

Six SI lines with strong incompatibility reaction were tested in Line x Tester design with six broad base testers by Parkash *et al.* (2015). Mean sum of squares of all the hybrids were significant for all the seven traits studied. Significant high GCA for earliness was recorded in the SI line, S-696 (–2.20). Significant high GCA for productivity was recorded in 3 SI lines, S-624, S-645 and S-208. Highest SCA effect for head size was recorded in S-208 x 83-6 followed by S-621 x GA-121. Highest heterosis for head weight was recorded in the hybrid, S-621 x GA-121 (37.9 per cent) followed by S-624 x GA-111 (34.1 per cent) and S-208 x GA 111 (31.8 per cent). Based on the CMS lines the following F_1 hybrids were developed in cole crops:

Cabbage

i. Pusa Cabbage-1: First F_1 hybrid in cabbage developed using SI line.

Cauliflower

i. Pusa Hybrid-2: This is a mid-maturity cauliflower F_1 hybrid.

14.4.0 Self-compatibility

In cabbage, self-compatibility has been reported by Kakizaki (1930) who found sympathetic *T*-alleles at a second locus which is epistatic to the *S*-alleles. *T*-alleles promoted the growth of pollen in the styles homozygous for the same *T*-alleles. Bateman (1954) found that self-compatibility was due to either alleles at the *S*-locus or recessive genes unrelated to S-allele system. Sampson (1957) showed that the phenomenon was not due to alleles at the *S*-locus. Nasrallah and Wallace (1968) reported that in cabbage, a self-compatible mutant from a self-incompatible inbred line homozygous for *S*-allele was determined by a dominant gene unrelated to the *S*-allele system. Similarly, Thompson and Taylor (1971) found that self-compatibility in the kale was determined by a single dominant gene which is independent of the *S*-locus and is expressed only in the absence of *S*-alleles high in dominance series. Pastuglia *et al.* (1997) reported that a functional *S*-locus anther gene is not required for the self-incompatibility response in *Brassica oleracea*.

14.5.0 Pseudo-compatibility

It refers to the phenomenon of plants with a functional incompatibility system but producing self-seed in amounts ranging from an occasional few to levels approaching true self-compatibility. The genetic causes of pseudo-compatibility are:

i) *S*-alleles that are weakly active may be associated with erratic seed set,
ii) Competitive interaction of complementary weakening may cause partial or complete self-compatibility,
iii) It may be the result of non-allelic gene affecting the normal *S*-gene activity,
iv) Modifier genes have also been found to affect the intensity and stability of the incompatibility,
v) It may also result from the genetic background in which the *S*-genes operate.

14.6.0 Break Down of Incompatibility

Besides the genetical factors, several environmental and physiological factors have been reported to affect the partial breakdown of the self-incompatibility. The two chief in trinsic factors are flower age and stage of flowering. Important extrinsic factors include high temperature, high humidity and high concentration of carbon dioxide. It has been reported by Palloix *et al.* (1985) that CO_2 concentrations of 4 to 6 per cent applied for 8, 16 or 24 ha t 100 per cent relative humidity proved to be the most effective treatment to reduce the self-incompatibility effect. It favours pollen adhesion, pollen germination, and increased the number of pollen tubes in the style.

14.7.0 Germplasm Pool

The importance of germplasm pool in breeding does not require any further emphasis since this will serve as a gene complex reservoir for all characters particularly for resistance to pests and diseases, hardiness, adaptability, *etc.* Besides, it forms the broad genetic base population from which to make further selections that are utilized in various methods for improvement of the crop.

14.8.0 Genetic Diversity

The genetic diversity has not been studied in details either in cabbage or in cauliflower. However, within the same species (botanical varieties) there is quite a good amount of variability. In cabbage, there are distinct groups depending on maturity, shape of head, colour of head, *etc.* Similarly, there are 5 different groups in cauliflower, such as Italian or Original, Northerns like Roscoff and Angiers from France, Erfurt and Alpha from Germany, Holland and Denmark, Cornish from England and lastly, Early, *i.e.*, heat and rainfall tolerant tropical Indian cauliflower. The genetic diversity within the same group is rather limited though among cauliflowers, Indian types show greater diversity than Erfurt-Alpha types (Swarup and Chatterjee, 1972). The hybrids between the groups show much superiority in earliness, yield and keeping quality. In view of the importance of genetic diversity, the germplasm pool becomes inevitable for a breeding programme.

14.9.0 Selection Procedures

Mass selection, a basic method for effecting improvement in cross-pollinated crops, has been found to be effective in cabbage-breeding for disease resistance (Walker, 1956) or ascorbic acid content. Useful results have also been reported in cauliflower. But it is not so useful in production breeding or in selection for polygenic characters. According to many workers it is a slow and not so effective method. Possibly, certain modified mass selection methods based on progeny testing would be better than mass selection. These may include mass pedigree method and family selection method as advocated by Nieuwhof (1959) in cabbage.

Inbreeding in general, induces depression in vigour though it is much less in cauliflower than in cabbage. Inbreeding helps to attain uniformity in head/curd shape and size, maturity and other characters, specially in heterozygous strains. The inbred lines developed in this way may be utilized for heterosis breeding, either for F_1 or for synthetic cultivars. However, this method has also been used for improving yield and quality, especially in cauliflower. Besides, they can be utilized for intervarietal crossing for improvement.

The inter-varietal crosses may be utilized for recombinations and fixing the favourable yield component genes by selection in F_2 and later generations. Those elite F_2 generations in which the decline in yield are least and have considerable amount of residual heterosis mainly due to the presence of pronounced additive variance, would favourably respond to selection. They can be used for producing composites as well. In cauliflower, there is evidence of transgressive segregations in favourable direction in F_2 generations of the intervarietal crosses for many economic characters. It may also be mentioned that useful results can be obtained by biparental progeny selections in the elite F_2 generations since these would be helpful in producing desirable recombinations and concentrating on additive variance.

Recurrent selection is practised in a single cultivar or a composite population to improve the yield by concentrating on additive variance. When yields of composites, synthetics or F_1 hybrids plateau off, recurrent selection can bring gains to put the genetic base higher. In maize such gains have been reported to be 15-20 per cent. Later, this new genetic base can be used in F_1 hybrids, synthetics or composites, for further improvementin yield.

The reciprocal recurrent selection proposed by Comstock *et al.* (1949) which is used in two source population utilizes both general and specific combining abilities.

The cultivar Ramosode Brasilia was developed through 6 cycles of selection between and within full-sib families derived from elite plants of a population collected in different parts of Brazil (Melo and Giordano, 1999).

The different selection procedures followed in cole crops are as follows:

14.9.1 Mass Selection

The mass-selection method is commonly used in the cross-pollinated crops for the improvement as well as maintenance of varieties. In this method the best individual plants are selected in the population and their seeds composited for raising the following generation. Since the mass selection is made exclusively on the phenotypic performance of the mother plant without any progeny test, the success of selection depends largely upon the heritability of the character under selection which is based upon the correlation between the genotype and the phenotype. For obtaining a closer relationship between the phenotype and the genotype in order to make the selection more effective, it is essential to grow the material under uniform conditions.

14.9.2 Line Breeding

When superior progeny lines are selected on the basis the progeny test and their seeds composited the procedure is known as line breeding progenies for testing can be developed by various methods like open-pollination, selfing, top cross, pair cross, diallel cross or polycross.

14.9.3 Family Breeding

In family breeding there are more elaborate testing of progenies not only in F_1 generation but also in later 2 or 3 generations and usually more than 1 cycle of selection. Family breeding procedure has been successfully practised in beet. This method can also be adopted in other root crops like radish, carrot as well as in cabbage and cauliflower.

14.9.4 Inbreeding

The loss in vigour following inbreeding is generally characteristic feature of cross-pollinated crops. However, the extent of inbreeding depression varies in different cross-pollinated crops. It is drastic in carrot while small extent in cabbage it is still less in onions. Hence, in carrot it is not advisable to practise inbreeding for more than 2 generations, about 2 or 3 generations in onion and 3 or 4 generations in cabbage. In cauliflower the loss in vigour due to inbreeding varies with the varieties, depending upon the proportion of self-compatible and self-incompatible plants occurring in the variety. The predominantly self-compatible varieties like the 'Early Summer Cauliflowers' of Europe do not show any loss in vigour following inbreeding while in winter cauliflower and heading broccoli which are self-incompatible the inbreeding depression is much pronounced. If the loss in vigour due to inbreeding becomes pronounced it can be restored by crossing occasionally. On the other hand, the cucurbits, though cross-pollinated in nature, do not show any loss of vigour due to inbreeding. In these crops inbreeding or individual plant selection can be practised as effectively as in a self-pollinated crop. However, in cucurbits selfing precedes selection unlike in other crops since in the former the first 1 or 2 fruits on the plant must be selfed; otherwise selfing in later stages often fails to set fruits.

The important uses of inbreeding in a cross-pollinated vegetable crop are:

a. To attain uniformity in plant characters,
b. To improve yield by individual plant selection as in cucurbits in which there is no inbreeding depression, or by selecting self-compatible plants in a predominantly self-incompatible variety, particularly in inbreeding.

Self-pollination followed by sib-mating between the best plants of the best lines was suggested by Camargo (1956) in Brazil as a useful method in cauliflower. Nieuwhof (1960) in Holland advocated methods of selection, namely inbreeding in autumn-sown crop of cauliflower followed by progeny evaluation in summer and vegetative propagation of the selected plants in winter followed by selfing and reselection from the progenies. Inbred lines are being utilized in commercial production in F_1 hybrid seeds in several cross pollinated crops like cabbage, Brussels sprout, broccoli, carrot and spinach.

14.9.5 Selfing and Massing

In some cross-pollinated crops, particularly those in which there is considerable loss of vigour due to inbreeding, improvement in a cultivar can be affected by selfing followed by massing. This method which has been suggested by Jones and Mann (1963) for onion can also be adopted in other root crops like radish, turnip, beet and carrot.

14.9.6 Recurrent Selection

The source population for recurrent selection which is heterozygous, may be an open-pollinated variety, single-cross or double-cross F_1 hybrid, intercrossed progenies of selected inbred lines, a synthetic variety or a composite. The method is effective in improving yield and other characters. Later this new genetic base can be used in hybrids, synthetics or composites for further improvement in yield. The resultant population of recurrent selection, which is a reservoir of superior germplasm, can be used in 3 different ways. Firstly, it may be utilized for obtaining homozygous inbred lines by selfing, for producing hybrid varieties. Secondly, it can be recombined with inbred lines, single crosses or other similar populations resulting from recurrent selection to produce F_1 hybrids. Thirdly it may be used as a source of foundation stock for a synthetic variety recurrent selection was found to be effective in selection for carotene concentration in carrot.

There are 2 different types of recurrent selection:

(a) **Simple recurrent selection:** In the simple recurrent selection procedure the best individual plants are selected and their selfed progenies intercrossed and further selections are made in the intercross progenies. The selfed progenies of the selected individuals are again intercrossed next year to complete the first cycle of selection. Usually more than one cycle of selection may be necessary to achieve the desired improvement in the character under selection. Since no test cross is made in the simple recurrent selection method, it is effective in the improvement of highly heritable characters which can be easily assessed by simple tests or on phenotypic evaluation. This method is, therefore, not effective in the improvement of yield.

(b) **Reciprocal recurrent selection:** The reciprocal recurrent selection method was proposed by Comstock, Robinson and Harvey (1949). The method is useful in estimating simultaneously for both general combining ability and specific combining ability in heterozygous and preferably genetically unrelated source populations, namely A and B. In the source population A, a number of individual plants with desired characters are selected, selfed and test crossed with several randomly chosen plants from the source population B.

14.9.7 Disruptive Selection

When a single population is maintained by choosing more than one class of individuals to provide the parents of each generation, such a selection is called disruptive selection (Mather, 1955). In its extreme form it involves choosing both extreme classes and discarding the intermediates. It is useful under certain situations, such as, when there is a tight linkage, correlated responses, latent variability in population which have failed previously to respond and also when it is desired to transfer polyploidy from wild type to cultivated plants. This may be useful in cole crops though not tested yet.

14.10.0 Synthetics and Composites

Promising inbreds or cultivars having good general combining ability can be combined to form synthetics or composites. On the basis of multi location testing of the F_1 and F_2 generations of the intervarietal crosses, the best F_2 generations can be selected for each zone. The elite F_2 generations which do not show much decline in yields largely due to the presence of additive gene effects, are selected for composites and tested further in advance generations before release. The elite base cultivars having wider genetic diversity and indicating good general combining ability in diallel or partial diallel crosses as judged by the performance of their F_1 hybrids, can be combined usually in combinations of 2 to 4 or more cultivars to form composites. These multiple varietal combinations may be tested further in advance generations.

14.11.0 Heterosis Breeding

Unlike synthetics and composites, the heterosis breeding utilizes mainly dominance variance. Both cabbage and cauliflower show pronounced heterosis in earliness, net weight of head or curd, yield per unit area, besides combining resistance to diseases, hardiness, uniformity, *etc.* In most of the advanced countries, the F_1 hybrids in cole crops are being preferred for uniformity of harvest and resistance to diseases. The yield may or may not be of much value since the level of yield other wise is also very high. This has become more popular, possibly due to comparatively easier method of F_1 seed production with the help of self-incompatibility and comparatively cheaper cost of seed production.

In cauliflower, Upadhyaya and Kumar (1983) found highest heterotic effect in cross between Early Kunwari x Punjab Giant 35 for yield, early maturity, curd weight, curd size and dry matter content.

Boos and Artemeva (1984) envisaged the significance of heterosis in breeding white cabbage for resistant to cracking and synchronous maturity for mechanical harvesting. There are reports of long storability of F_1 hybrid (Ceausescu *et al.*, 1987) and also hybrids of cabbage resistant to *Peronospora parasitica* (Alipieva and Dimitrov, 1984). On the basis of general combining ability, Pandey and Naik (1986a) recommended the following lines: IHR3, IHR4, IHR9 and IHR36, for their all round good characters. Cabbage hybrids cv. Karino, Hybrid 8398 and Chunbao has been developed by crossing self-incompatible lines (136 x 134; 8101-2-20-2 x 8180-1-2-1 and 441591 x 87123, respectively) (Nankova and Alipieva, 1995; Fang *et al.*, 1996 and Yin *et al.*, 1996). Fang *et al.* (1997) developed a male sterile line with dominant gene (*Ms*) in cabbage and described its utilization for hybrid seed production. The F_1 of Xia Qing Kou x Lu Da Xiao Gen was crossed and backcrossed with Tong Xiang Huang Ya Cai to give the new cold resistant Chinese cabbage cultivar Huang Ya 14-1 (Xia and Lian, 1996). According to Ping and Chun (1998) heterosis of non-heading Chinese cabbage was mainly evident in plant weight, leaf blade weight and petiole weight which were controlled by duplicate interaction genes.

At present CMS system is most commonly used in developing F_1 hybrids in cauliflower and cabbage. However, availability of strong S allele with good stability in crop like cabbage also offesrs good opportunity to develop F_1 hynrids.

Dey *et al.* (2011) studied the heterosis for yield and earliness using 3 CMS lines namely, Ogu1A, Ogu2A and Ogu3A as female parent. The parent Ogu2A exhibited highest gca effect for curd yield (4.51) and harvest index (1.97) while Ogu1A exhibited highest gca for earliness (-2.73). The parent, Ogu2A exhibited significant gca for curd length (0.39) while, none of the CMS lines showed significant gca for curd diameter and depth. Heterosis for curd yield was highest in the hybrid, Ogu2A x Kt-22 (63.5 per cent) followed by Ogu1A x WF (36.9 per cent) and Ogu1A x Kt-15 was the best hybrid for earliness followed by Ogu3A x Kt-22 with heterosis of -14.4 per cent and -11.7 per cent. However, the number of heterotic hybrids for yield and earliness was low indicating narrow genetic base of the snowball cauliflower. Heterosis for quality traits was also reported by Dey *et al.* (2014) using Five CMS lines and 8 male fertile lines in line x tester design to develop 40 hybrids. These hybrids along with parental lines were evaluated for different vitamins and anti-oxidant pigments to reveal extent of heterosis and genetic combining ability. The CMS line Ogu12A was good general combiner (gca effect) and Ogu16A was poor general combiner for most of the important traits under study. Most of the heterotic hybrid combinations were associated high specific combing ability (sca effect). However, gca effect was also important in developing quality heterotic hybrids. Highest number of heterotic hybrids in positive direction was recorded for ascorbic acid content followed by anthocyanin content. The accumulated average heterosis of the 40 hybrids was in positive direction for ascorbic acid, anthocyanin and lycopene concentration whereas it was in negative direction for carotenoids and chlorophyll pigments. Very high heterosis for ascorbic acid, anthocyanin and carotenoids in cauliflower indicated the scope for development of F_1 hybrids with higher concentration of these vitamins and antioxidant pigments. It is possible to develop heterotic hybrids for different vitamins and anti-oxidant plant pigments through selection of parental lines based on desirable genetic combing ability. Singh *et al.* (2018) studied the heterosis and combining ability for antioxidant compounds in cauliflower using CMS and doubled haploid lines. Twenty genetically diverse Ogura CMS lines of cauliflower and six DH male fertile inbred lines were crossed to develop 120 F_1 hybrids in line x tester mating design. The resulting 120 test cross progenies along with 26 parents and 4 standard checks were evaluated in 10 x 15 alpha lattice. The CMS lines Ogu33-1A, Ogu122-5A and Ogu119-1A were good general combiner and CMS line Ogu118-6A was poor general combiner for majority of traits. Highest number of heterotic hybrids with SCA effects in desired positive direction was recorded for ascorbic acid content and phenolic content followed by total carotenoid content. The F_1 hybrids with better combining ability and better per se performance could be useful in accumulation of favourable allele for higher concentration of important anti-oxidant compounds.

It is important to select suitable parents in developing heterotic hybrids for different yield related and quality traits. Both morphological diversity and molecular markers uniformly distributed throughout the genome can be employed for this purpose. Singh *et al.* (2019) reported the suitability of genomic- SSRs and EST-SSRs based genetic distance (GD) and agronomic trait based phenotypic distance (PD) for predicting heterosis in F_1 hybrids using CMS and DH based parents. 120 F_1 hybrids derived from 20 Ogura based CMS lines and 6 DH based testers were evaluated for 16 agronomic traits along with the 26 parental lines and 4 commercial standard checks. The genomic-SSRs and EST-SSRs based genetic structure analysis grouped the 26 parental lines into 4 distinct clusters. The CMS lines Ogu118-6A, Ogu33A, Ogu34-1A were good general combiner for developing early maturity hybrids. The SCA effects were significantly associated with heterosis suggesting non-additive gene effects for the heterotic response of hybrids. Less than unity value of σ^2A/D coupled with σ^2 gca/σ^2 sca indicated the predominance of non-additive gene action in the expression of studied traits. The correlation analysis of genetic distance with heterosis for commercial traits suggested that microsatellites based genetic distance estimates can be helpful in heterosis prediction to some extent. The accuracy of heterosis prediction can be enhanced with the application of large number of markers in each linkage group. Besides, re-sequencing of the parental lines through next generation sequencing can be employed for precise prediction of heterosis and selection of suitable parents.

14.12.0 Breeding for Disease Resistance

The development of insect and pest resistant varieties is a current priority in brassica breeding programs. The evaluation and screening of new sources of resistance to the most common pests and pathogens, the study of the genetic basis of resistance, and the identification of genes for resistance are present goals of breeding programs in cabbage and kale. Nevertheless, the genetic basis of multiple disease resistance is complicated by environmental interactions with pathogens and correlated gene expression of different resistance traits. Notable advances have been made in the use of molecular methods for locating the position of resistance genes on genetic maps of brassicas. Sources of resistance to most common diseases have been identified in cabbages and kales but have not been transferred into commercial varieties yet. The most important diseases for kales and cabbages, club root and black rot, are presently maintained below the threshold of economic damage with agronomic practices such as crop rotation and application of fungicides. The interest in releasing environmentally friendly varieties involving a minimum use of chemicals is related to the increasing present demand of crops resistant to several pests and diseases. Breeding for club root resistance, especially in *B. oleracea*, has been troubled by the complex of the plant-pathogen interaction and by inadequate characterization of the resistance nature. Different researchers have reported the gene introgression for resistance to black rot and club root in *Brassica oleracea* (Tonguc and Griffith, 2004; Taylor *et al.*, 2002).

A great effort has been made in the last decades to improve plant resistance against black rot and club root. To develop reliable host resistance, brassica breeders have been searching for resistant sources with high yield potential. Resistance against black rot has been identified in different genotypes of *B. oleracea*, including commercially available cultivars of cabbage and kales (Hansen and Earle, 1995; Taylor *et al.*, 2002; Tonguç and Griffiths, 2004). However, none of these sources have provided complete resistance to the disease and its quantitative genetic control complicates its use to produce resistant hybrid varieties. The cultivars carrying club root resistance have found limited applications because of their insufficient level of resistance or because of their insufficient quality. Resistance has come primarily from kale and rutabaga. High levels of resistance to club root within the kale group have been reported (Crisp *et al.*, 1989). A detailed review of inheritance of club root resistance in *B. oleracea* has been done by Voorrips *et al.* (2003). Resistance in *B. oleracea* usually involves several genes for control of each race. A group of major genes showing dominance effects in the resistance to race 6 of club root in cabbage was reported by Chiang and Crête (1976). Resistance to club root is quantitative under polygenic control and involves recessive genes. Resistance to *Alternaria* ssp. was not found in *B. oleracea* whereas high resistance was observed in the wild relatives of *Brassica* outside the tribe Brassicaceae. *Leptosphaeria maculans* is also a critical pathogen of horticultural *Brassica* species, although the search for sources of resistance on kale and cabbage cultivars has not been as extensive as that on rapeseed. Other fungal diseases, such as fusarium yellows, downy mildew, powdery mildew, and bottom rot, have been described in cabbage by Dickson and Wallace (1986). Brassica production is hampered by various diseases, especially club root and Fusarium wilt. Many types of R genes/QTLs have been identified in *Brassica* against the diseases and are being used for the improvement of resistance in cultivars. In case of club root disease, a total of 18 major CR loci have been identified in *B. rapa,* whereas only one major CR locus (Rcr7) and about 50 QTLs were detected in *B. oleracea.*

The role of glucosinolates on pest and disease resistance is another major aim of current breeding program. Many secondary metabolites found in plants have a role in defence against herbivores, pests and pathogens (Bennett and Wallsgrove, 1994). Extensive research work has been done on glucosinolates as plant defence compounds and their potential for the control of pests. Crucifer-specialized herbivores such as *P. xylostella, P. rapae,* and *B. brassicae* often use glucosinolates or their breakdown products as an attractive signal for the identification of suitable host plants. In addition, they are able to detoxify glucosinolates present in these crops. For instance, *P. xylostella* larvae feed exclusively on crucifers and produce a gut sulphatase that cleaves glucosinolates to form inactive desulfoglucosinolates (Ratzka *et al.,* 2002).

In breeding for disease resistance, the back-cross method of breeding is commonly used, especially in transferring one or two genes to an otherwise desirable cultivar. Two important advantages of the back-cross breeding over the pedigree method are (i) the smaller size of population required for selecting the desired genotypes and (ii) the feasibility of carrying out the back-cross programme under any environmental conditions. Generally, it takes about six generations of back-crossing to transfer a dominant or a recessive gene from non-recurrent parent provided the character under transfer can be distinguished before pollination. In polygenic system of control also this method is used. In this method, successive back-cross may not be needed unless the donor parent has many undesirable characters. The whole objective would be to develop families with the desirable characters and resistance and inter-mating between them to bring in more genes of resistance in the resultant cultivars. According to Stoner *et al.* (1986) resistance to thrips damage is inherited quantitatively with susceptibility dominant in cabbage. The cauliflower cultivar SN 445 carries a dominant gene for resistance to back rot (Jamwal and Sharma, 1986). In cabbage cultivar PI 436606, the back rot resistance is controlled by a single recessive gene while LAW 13 unrelated to PI 436606, appeared to carry 1or 2 modifier genes for seedling tolerance (Dickson and Hunter, 1987). In Brussels sprouts, significant additive and dominant variations for disease resistance to cauliflower mosaic virus and turnip mosaic virus (Pink *et al.,* l986b; Pink *et al.,* 1986a) are reported. Chinese cabbage cultivar Zhengabai No. 4 developed by using self-incompatible lines is multiple disease resistant (Hong *et al.,* 1996). Shanna *et al.* (1995) recorded that Early Winter Adam's White Head of cauliflower was highly resistant to *Sclerotinia* rot and downy mildew, whereas line EC162587 was highly resistant to *Sclerotinia* rot and black rot. Lines RSK1301 and MRS1 were moderately resistant to all 3 diseases. Chinese cabbage cultivar, Dwarf Resistant 4 was produced by crossing the genetically male sterile variety Aijiaohuang as the female parent with inbred line 93R-37. It is resistant to turnip mosaic poty virus and downy mildew and also resistant to *Alternaria* leaf spot as well (Chao *et al.*, 1998). In Chinese cabbage a hybrid (F_1) between a club root resistant line (930 WG) and a susceptible line (332 MS) showed a completely resistant reaction which indicated that club root resistance was governed by a dominant gene (Wook and Haing, 1997).

Inter-specific hybridization for transfer of genes from A and B genome of Brassicas into into C genome has been undertaken where the desirable source is not available. Dey *et al.* (2014) successfully attempted inter-specific hybridization between *B. caribata* and *B. oleracea* for introgression of black rot resistance genes from B. carinata line NPC-9 and cauliflower genotypes from different maturity groups. BC1 population were developed and progenies with resistance to race-1 (*Xcc1*) and Race-4 (*Xcc4*) were identified for further use in disease resistance breeding. Similarly, Sharma *et al.* (2017) successfully introgressed the *Xcc1* resistance from the B. carinata line, NPC-9 into cauliflower through direct ovule culture. They have also employed the linked SSRs to identify the resistant progenies in the back-cross population.

14.13.0 Phytochemical Value of *Brassicas*

The economically, nutritionally and scientifically important angiospermic plant family *Brassicaceae* comprises numerous species which are used as fodder, spice, oilseed and vegetable crops. The *Cruciferae* family constitutes about 4060 species under more than 372 genera. The genus *Brassica* belonging to this large plant family is the most diverse genus carrying many economically and nutritionally important vegetable crop species. In this context, the *Brassica oleracea* is the chief vegetable species and comprises different nutritionally important morphotypes (cabbage, cauliflower, kale, brussels sprout, broccoli and knol khol) known as 'cole vegetables. These *Brassica* vegetables are often entitled as "super foods" on account of their health-promoting properties due to the presence of potentially protective phytochemical and nutraceutical compounds. Ascribed to these functional food properties, the intake of cole vegetables is linked with the reduction of several types of cancer, chronic and cardiovascular ailments. Particularly, the antioxidant and anticancer properties of bioactive compounds present in *Brassicvaluea* vegetables has gained much attention, and has been investigated extensively over the last few years. Carotenoids (carotenes and xanthophylls) are lipid soluble antioxidants imparting yellow, orange and red pigments in fruits and vegetables. Some of the carotenoids are precursors of vitamin A (β-carotene, β-cryptoxanthin and γ-carotene). The anticancer activity of carotenoids is suggested to be due to different mechanisms such as reduction in cell proliferation, antioxidant activity on mutagenesis and genotoxicity, immunomodulatory efffects, enhancement of gap-junction cell communication, and increasing apoptosis of cancerous cells. The most abundant forms of carotenoids in *Brassica* vegetables are lutein and β-carotene. The highest content of lutein + zeaxanthin is recorded in kale (3.04-39.55 mg/100g), moderate in broccoli and brussels sprouts (0.78-3.50 mg/100g). In case of orange cauliflower, the single semi-dominant '*Or*' gene mutation confers β-carotene accumulation in the curd tissue of cauliflower. The orange colour in cauliflower exhibiting novel *'Or'* gene mutation is due to massive and highly ordered sheets doing carotenoid (β-carotene) sequestration in plastids, and classified as membranous chromoplasts. The advent of molecular markers has given the impetus to molecular breeding techniques and overcome the locks of conventional breeding. Although the conventional breeeding would be the important part of breeding techniques. Through the conventional breeding approach, a β-carotene rich orange cauliflower hybrid 'Jinyu 60' was developed involving CMS line as female parent and homozygous orange inbred 93-4 as donor parent. During the recent years, significant progress has been made in the molecular mapping and identifiation of carotenoid biosynthetic genes in plant species.

Anthocyanins are water soluble flavonoid antioxidants imparting red, blue, purple colour in many flowers, fruits and vegetables. They perfform the biological functions in plant defense mechanism against various biotic and abiotic stresses. Numerous regulatory genes including R2R3 MYB transcription factors (TF), basic helix-loop-helix (bHLH) transcription factors, and WD40 proteins have been cloned from different plant species, such as Arabidopsis, maize, petunia, *B. rapa,* cauliflower, red cabbage, purple chinese cabbage, purple kale and other *Brasssica species* and other species. In purple cauliflower a spontaneous mutant, single semi-dominant *Pr* gene encoding MYB transcription factor is responsible for giving tissue specific expression of anthocyanin accumulation in different plant parts. The major anthocyanin compound responsible for anthocyanin pigmentation in purple cauliflower is cyanidin, 3, 5-diglucoside. The R2R3MYB transcription factor, *BoPAP1* gene and transcriptional activation of *BoPAP2* and *BoTT8* accounting for upregulation of anthocyanin biosynthetic structural genes, is responsible for anthocyanin accumulation in purple kale and purple kohlrabi, respectively.

There is currently an increasing interest in the alteration of levels of specific glucosinolates in crop plants as certain glucosinolates have desirable properties in cancer prevention and crop protection. The study of inheritance of genes controlling glucosinolate biosynthesis will allow for manipulation of these genes and the development of lines of cabbage and kale with specific glucosinolate profiles. The biology and biochemistry of glucosinolates has been widely reviewed and significant progress has been made in understanding the genetics of glucosinolate biosynthesis. Recent studies have identified key genes responsible for structure regulation and glucosinolate content in different *Brassica* species. The cloning of some key genes provides the opportunity to engineer brassica crops with specific glucosinolate content by modifying their glucosinolate side-chains. This manipulation can have several possible applications. For example, down regulation of *BoGSL-ALK,* which controls side-chain desaturation and cosegregates with *BoGSL-OH,* responsible for side-chain hydroxylation, could produce brassica varieties lacking the antinutrient progoitrin and would simultaneously produce plants accumulating glucoraphanin as a source of anticarcinogens. Although considerable progress is being made in that field, presently, it is not possible to identify the genes responsible for the variations in glucosinolate content. Most of the genes controlling glucosinolate breakdown are still unknown. In the near future, the isolation of genes of glucosinolate metabolism could facilitate the development of plants with modified glucosinolate profiles. The glucosinolate metabolism will allow the study of the potential of glucosinolates to improve human nutrition and to improve the pest resistance.

Table 11: The List of the Genes Controlling Anthocyanin Concentration in different *Brassica* Species

Brassica Species	*Transcription Factor/Genes/Enzymes*
Brassica oleracea var. *botrytis*	*Pr* gene mutation encoding R2R3 MYB transcription factor
Red cabbage (*B. oleracea* var. *capitata*)	*BoMYB2* with *bHLH* transcription factor *BoTT8*
Purple heading chinese cabbage (*B. rapa* L. *ssp. pekinensis*)	BrPur
Ornamental kale (*B. oleracea* L. var. *acephala*)	BoPr
B. rapa L. ssp. *chinensis var. Purpurea*	*BrEGL3.1* and *BrEGL3.2* encoding *bHLH* transcription factor
Purple kale (*B. oleracea* var. *acephala* f. *tricolor*)	BoPAP 1
Brassica rapa	Anp
Brassica napus	BnaA.PL1
Brassica rapa var. *chinensis*	BrTT8
Brassica oleracea var. *gongylodes*	*BoPAP2* and *BoTT8*

Table 12: List of the Agronomically Important Traits, Associated Genes/ QTLs and their Inheritance in *Brassica olerace*

Crop	*Trait*	*Genes/QTLs*	*Inheritance*	*Reference (s)*
Cabbage	Head splitting resistance	Three major QTLs (*Hsr 3.2, 4.2, 9.2*)	additive-epistatic effects of three major gene pairs combined with those of polygenes	Su *et al.* (2015)
	Wax deposition in cabbage leaves	CER_3		Laila *et al.* (2017)
	Wax biosynthesis locus	*BoWax1*	Monogenic recessive for glossy green trait	Liu *et al.* (2018)
	Glossy trait	*Cgl1*	Single recessive	Liu *et al.* (2017)
	Waxless gene	*Bo-GL3*	Single dominant	Dong *et al.* (2019)
	Fusarium wilt resistace	*FOC1*	Single dominant	Lv *et al.* (2014)
	Black rot	*QTL-1,2,3*	Quantitative	Kifuji *et al.* (2012)
	GMS	*Ms-cd1*	Single dominant	Fang *et al.* (2004)
	GMS	*ms-3*	Single recessive	Han *et al.* (2017)
	Glossy green trait	*BoGL1*	Single dominant	Liu *et al.* (2017)
	Yellow green leaf	*ygl1*	Single recessive	Liu *et al.* (2016)
	Head shape	*Htd 7.1 and Hsi 7.2*	Quantitative	Zhang *et al.* (2016)
	Club root	*Rcr7*	Polygenic recessive	Dakouri *et al.* (2018)

Crop	Trait	Genes/QTLs	Inheritance	Reference (s)
	Club root	*QTL, pb-Bo(Anju)1, pb-Bo(GC)1*	Quantitative	Nagaoka *et al.* (2010)
Broccoli	Hollow stem	*QHS.C09-2*	Quantitative	Yu *et al.* (2019)
	Purple sepal	*a major locus, qPH.C01–2, two minor loci, qPH.C01–4 and qPH.C01–5*	Quantitative	Yu *et al.* (2019)
	GMS	*ms-1, ms-2, ms-4, ms-5, ms-6*	Monogenic recessive	Dickson (1970)
	Downy mildew resistance	*Pp523*	Single dominant	Farinho *et al.* (2004)
	Serrated leaf margin and leaf shape index	*BoLMI1*	Quantitative	Stansell *et al.* (2019)
	Flower color	*BoCCD4*	Quantitative	Stansell *et al.* (2019)
	Hooked sepal	*BoAP2*	Quantitative	Stansell *et al.* (2019)
	Head quality	*OQ_C03@57.0, OQ_C04@33.3, OQ_CC08@25.5, and OQ_C09@49.7* with *FLC homolog Bo9g173400 and Bo9g173370*	Quantitative	Stansell *et al.* (2019)
Kale	Pink leaf	*Pi*	Single dominant	Zhu *et al.* (2016)
	Purple leaf	*BoPr*	Single dominant	Liu *et al.* (2017)
	Red leaf	*Re*	Single dominant	Ren *et al.* (2015)
	Lobed leaf	*BoL1*	Single dominant	Ren *et al.* (2019)
Brussels Sprout	GMS	*ms-2*	Monogenic recessive	Johnson (1958)
Chinese kale	Yellow petal	*ckpc*	Single recessive	Xu *et al.* (2019)
Chinese cabbage	Purple heading	*BrPur*	Single dominant	Wu *et al.* (2016)
	Club root resistance	*Rcr2*	Single dominant	Huang *et al.* (2017)
Knol Khol	Purple trait	*BoMYB2*	Single dominant	Yan *et al.* (2019)
	Tuberous stem formation	*REnBo1, REnBo2, REnBo3, and REnBo4*	Quantitative	Hoque *et al.* (2017)
Cauliflower	Purple curd	*Pr*	Single semi-dominant gene	Singh *et al.* (2020)
	Large organ size and biomass	*CDAG1 (curd development* associated gene)	-	Li *et al.* (2017)
	Stalk length	(*qSL.C6–1, qSL.C6–2*	Quantitative	Zhao *et al.* (2020)
	Curd solidity	*qCS.C6–1 and qCS.C6–2*	Quantitative	Zhao *et al.* (2020)

Crop	Trait	Genes/QTLs	Inheritance	Reference (s)
	Reproductive development	*BoFLC2*	-	Ridge *et al.* (2015)
	Riceyness	*Bo4g024850*	Monogenic	Zhao *et al.* (2020)
	Orange color curd mutation	*Or*	Single dominant	Li *et al.* (2006)
	Arrest of curding, control leafy floral differentiation	*BoCAL-a* and *BoAP-1*	-	Labate *et al.* (2004)
	Black rot	*Xca1bo*	Monogenic dominant	Saha *et al.* (2014)

15.0 Biotechnology

15.1.0 Cauliflower

15.1.1 Tissue Culture

Researchon tissue culture of cauliflower has been going on since 1969 (Margara, 1969). The use of floral primordia for micropropagation, anther/microspore culture for production of haploids and callus culture for genetic transformation and improvement have been reported by several workers.

15.1.1.1 Micropropagation

Micropropagation is thought to be a useful technique for mass multiplication and virus elimination and floral meristems have been found to be ideal. The commonly used media are modified version of Murashiege and Skoog (MS) (Crisp and Walkey, 1974). For micropropagation various tissues have been used such as leaf veins, leaf lamina, excised petals, inflorescence stem, *etc.* But the simplicity and high rate of success with floral meristem has led to the report of investigations using the curd or floral explants (Grout, 1988).

Kumar *et al.* (1993) described a rapid single step method for complete plantlet regeneration. Curd explants cultured on MS medium supplemented with 1mg/l IAA at 3000 lux developed complete plantlets in 25 days.

Kieffer *et al.* (1996) removed the meristematic layer of curd from 7 cultivars and partially homogenized to disrupt the meristem clusters and then graded through precision sieves to produce several homogenous size classes. Over 4,00,000 explants of the optimal size class, 0.1-0.3 mm, were produced from a single curd. One to 3 shoots/explant were obtained following culture of 0.1-0.3 mm explants in agitated liquid MS medium supplemented with 2 mg kinetin and 1 mg IBA/l.

The *in-vitro* tissue culture offers a powerful aid for the accelerated production of parental lines of F_1 hybrids with strong SI background (Cristea *et al.*, 2009). It eliminates the need for the elaborate and expensive procedure for establishing and maintaining sib-incompatible nearly homozygous parental breeding lines. Various factors such as explants, hormone combination and concentration and cultural conditions affect the rapid in vitro propagation of self-incompatible line in headed cabbage (Zhang *et al.*, 2014). Shoot regeneration has been achieved from various tissues and organs including hypocotyls, cotyledons, roots, leaves, peduncle segments, callus and cell cultures, thin cell layers and protoplasts (Cardoza and Stewart, 2004). However, considerable variation has been observed by different groups, even when working with the same species or variety. With the aim to improve the breeding process of cabbage and to develop new cultivars in a shorter period of time, as well as to develop a basis for use of biotechnology methods, the *in-vitro* propagation of cabbage from different explants, was investigated. Micropropagation for large scale multiplication is also useful in crops like cabbage for maintenance of SI lines. This saves lot of time required for bud pollination of individual plants and when because of very strong 'S' allele, it become very difficult to maintain the SI lines even through bud pollination. Therefore, an *in-vitro* based protocol for mass multiplication of SI cabbage line, 'Selection-5' was optimised by Bhatia *et al.* (2013). Different types of explants, *viz.* apical bud, axillary bud and basal shoot sprout were tested for their *in-vitro* regeneration ability. The morphogenetic potential varied among the explants. Apical bud proved to be the most potent explant for initial culture establishment followed by axillary bud. Murashige and Skoog (1962) medium supplemented with 2 mg l-1 BA, 0.5 mg l-1 NAA and 0.1 mg l-1 GA_3 was optimum for culture establishment. The Maximum *in-vitro* shoot proliferation (5.89±0.38) was obtained on MS medium supplemented with 5 mg l-1 KIN + 0.1 mg l-1 NAA + 0.1 mg l-1 GA_3. The proliferation rate was significantly influenced by type and concentration of cytokinins with kinetin being more effective than BA. Half-strength MS medium supplemented with 1.0 mg l-1 IBA was most effective for rooting. The tissue cultured plants were successfully hardened and transferred to field condition with a survival rate of 76.67 per cent.

Micropropagation is also useful in multiplication of CMAS lines in brassica vegetablesas there is need to maintain them through crossing with the respective maintainer. As these crops are highly cross pollinated with significant heterozygosity, it becomes difficult to maintain the CMS lines in pure form. Bhatia *et al.* (2015) optimised protocol for multiplication of 3 cytoplasmic male sterile (CMS lines) of Indian cauliflower (*Brassica oleracea* var. *botrytis* L.; Ogu1A, Ogu2A and Ogu3A) for their use in hybrid development. They compared *in-vitro* regeneration ability using two types of explant, hypocotyls and curds *i.e.* pre-floral fleshy apical meristems. The highest rate of *in-vitro* establishment and multiplication index were recorded in Ogu3A. Ogu2A showed poor *in-vitro* establishment. The choice of explant had a significant impact on the morphogenetic potential of all three lines. Among the two types of explant, curds proved to be the more efficient explant for the initial establishment of cultures. The highest rate of *in-vitro* establishment and multiplication index were recorded for Ogu3A on Murashige and Skoog (MS) medium supplemented with 2.5 mg l -1 Kinetin, 0.2 mg l -1 α-naphthalene acetic acid and 0.2 mg l^{-1} gibberellic acid. The three CMS lines also showed significant variations in the numbers of roots induced per shoot. The maximum number of primary roots per shoot was recorded in Ogu3A on 0.5x MS medium supplemented with 1.0 mg l -1 indole-3-butyric acid. *In-vitro*-raised plants were hardened and used successfully for F_1 hybrid development. When F_1 hybrids developed using in vitro-maintained or conventionally maintained CMS lines were compared, it was observed that variations [standard deviation (SD)] in plant height, days to 50 per cent curd maturity (DCM), and net curd fresh weight (FW) were lower in those F_1 hybrids developed through in vitro-maintained CMS lines. The range of SD values as percentages of the mean value for DCM and net curd FW in F_1 hybrids developed from in vitro-raised CMS lines were 2.5 - 3.7 and 6.6 - 10.7, respectively. In contrast, the corresponding SD values ranged from 5.83 - 7.9 and from 13.6 - 19.9 for F_1 hybrids developed from conventionally maintained CMS lines. This *in-vitro* protocol, using curd explants, will be useful for the rapid multiplication of breeding lines of cauliflower without inbreeding depression, and for high quality hybrid seed production.

15.1.1.2 Embryo Culture

Embryo culture has been mostly used in Brassica for rescuing embryos of interspecific and intergenetic crosses. *Raphanus sativus* is resistant to pod shattering. In an attempt to transfer shattering resistance to *B. napus* (rape), Agnihotri *et al.* (1990) used *Raphanobrassica* [*R. sativus* cv. *Sweta* x *B. oleracea* (cauliflower) cv. Early Kunwari] as the male parent in crosses with *B. napus*. Plantlets were obtained by embryo rescue and were further multiplied *in-vitro* by micropropagation of nodal segments.

15.1.1.3 Development of Haploids and Double Haploids

This technology is extremely useful in cole crops as it is not easy to develop inbreds with complete homozygosity because of cross-pollinated in nature which favours high heterozygosity and inbreeding depression. Therefore, induction of haploids and their subsequent diploidization offers a single step approach to attain homozygosity at shortest possible time. Haploid production in cole crops is of significant importance as haploids can be used for the development of homozygous inbred lines. Earlier anther culture was used for induction of haploids and development of doubled haploids in cole crops. However, with the advancement in microspore culture, anther culture is rarekly practised because of presence of mixoploids, low regeneration rate and regeneration from the somatic tissues which make it difficult to identify the true DH lines.

Arnison and Keller (1990) conducted a survey on the anther culture response of different vegetable brassicas. They found that some microspore-derived embryos were recovered from the majority of cultivars tested, but rates of embryo formation were sometimes lower than those obtained under controlled environmental conditions.

Genotype, explant source, bud size and media affect the success of anther culture of cauliflower. Phippen and Ockendon (1990) tested eleven F_1 cultivars of cauliflower, representing a range of maturity types, for their responsiveness to anther culture. Embryoids were produced from each of the cultivars tested, mean yield ranging from 82.2 embryoids/100 anthers of cv. Dova to 0.6 embryoids of cv. Serrano. Variation between genotypes and between plants within a genotype was significant, both in terms of embryoid yield and percentage responsive anthers.

Ockendon (1988) studied the ploidy of anther culture-derived plants of 4 cultivars of autumn cauliflower. The cultured anthers produced 100 embryos per

100 anthers of cv. Nedcha, 9 embryos for a second cultivar, and almost zero for the 2 other cultivars and 986 embryos produced 418 plantlets. Duijs *et al.* (1992) found that microspore culture was applicable to a broad range of accessions belonging to six horticulturally important crop types of *B. oleracea*: broccoli, white cabbage, cauiflower, savoy cabbage, Brussels sprouts and curly kale. Of 64 accessions tested, 86 per cent were responsive. Large genotypic differences were found in the number of embryoids produced per flower bud, and in the frequency and mode of regeneration of plants from embryoids.

Microspore culture is successful in most crops' types of *Brassica oleracea* vegetables (Duijs, 1992). More than 400 regenerants of R_1 generation were derived in kohlrabi, cabbage and cauliflower by means of different modifications of microspore culture technique. Distinct genotype differences in embryogenic responsibility and regenerative ability of microspore embryos to whole plants were detected. Bhatia *et al.* (2016) optimized different factors like bud size, developmental stage of the microspores, culture density and heat shock treatments for efficient microspore embryogenesis in four different genotypes representing four different maturity groups of Indian cauliflower. Among different factors, the bud size and microspore developmental stage was found to be more critical for successful microspore embryogenesis. The microspore density of 6–8 x 10^4 cells per ml was found most suitable for efficient microspore embryogenesis. A post-culture incubation of microspores at 30.0 °C/32.5 °C for 24 h followed by maintenance at 25 °C was found optimum for induction of microspore embryos in different maturity group of Indian cauliflower. The flow cytometry analysis of 646 microspore derived plants revealed that more than 60 per cent plants were spontaneous diploids and less than 15 per cent were haploids. Cold pretreatment on flower buds have significant effect on microspore embryogenesis in cole crops (Klima, 2004). The addition of activated charcoal is reported to increase embryo yields in cole vegetables. In broccoli, embryo yield was significantly increased in most genotypes by incubating microspore cultures at 32.5 °C for 1 day as compared to standard incubation at 30 °C for 2 days (Dias, 2001). Whereas, in cabbage, the best yield in the temperature treatments were, when the microspores were cultured at 32 °C for two days. Recently, Bhatia *et al.* (2016) has optimized various factors affecting microspore embryogenesis in different maturity group of Indian cauliflower. Microspore culture is successfully used in development of doubled haploid in cabbage and cauliflower (Bhatia *et al.*, 2016; Bhatia *et al.*, 2017; Bhatia *et al.*, 2018). In all the studies it was apparent that genotypes and developmental stage (late uni-nucleate to early bi-nucleate) play pivotal role in successful generation of haploid embryos through isolated microspore culture. Besides, culture density (20000 to 40000 microspore/ml) and temperature pre-treatment (32.5°C for 24-48 hours) were also critical for high frequency regeneration of micvrospore derived embryos.

Important Factors Determining Successful Development of Haploids and Double Haploids

Explant Genotype

In cole crops androgenesis is successfully achieved by several workers, however. it is highly genotype specific.

Growth Condition of Donor Plant

The conditions under which donor plant are grown are an important consideration for successful haploid production. It is recommended to grow the donor in a controlled environment where the stress conditions could be minimized. Temperature is very crucial factor because sudden change in the temperature regime mainly after floral differentiation could greatly diminish the embryogenic response as stress alters the level of growth regulators, amino acids, carbohydrate or lipid in the plant. In autumn-heading cauliflower the best results were obtained during spring in a greenhouse where the temperature was maintained between 10 and 20°C. Overall winter and spring seemed more suitable than summer and early autumn for culture establishment.

Developmental Stage of the Pollen

The optimal bud development stage in cauliflower depends upon the genotype. In most of the cole crops the buds for microspore culture should contain microspore at mid to late uninucleate stage. In mid uninucleate stage microspores are round and developed a thick exine with the nucleus in the center. At late uninucleate stage, the nucleus migrates to the cell wall and exine become more yellow. In certain species, bud sizes can be correlated with the developmental stage of the pollen.

Pre-treatments

Pretreatment of the donor plants, floral organs, or anthers with growth regulators or environmental stress can influence embryogenesis. Abiotic stresses play an important role in androgenic induction. Low and high temperature shocks are applied as pretreatment or at early stage of induction in most protocols. Exposure of excised flower buds to a low temperature for some time, *e.g.*, 3-5°C for 2 days or at 7-8 °C for 12 days, prior to removal of anthers for culture may markedly enhance the recovery of haploid plants. Cold treatment of detached anthers is less effective than that of detached flowers. Low temperature kills weak or non viable anthers and microspores; it also arrests many of pollen grains in the first mitosis and blocks starch accumulation. Microspore and tapetal developments then become asynchronous, influencing the switch from gametophytic to sporophytic phase. Embryogenesis from eggplant anthers required essentially high temperature treatment combined with 2,4-D and kinetin (Matsubara *et al.*, 1992). In addition, treatments like centrifugation, irradiation with X-ray and gamma rays and reduced atmospheric pressure are reported to promote androgenesis.

Media Composition

The composition of culture media is an important factor influencing embryogenesis. For most species, a complete tissue culture medium is required. MS, LS (Linsmaer and Skoog) and some other tissue culture media are generally used. Media with dilute salt solutions, *e.g.* White's and Heller's media are ordinarily supplemented with coconut milk. The most often modified components are the source of organic nitrogen, carbohydrates and growth regulators. Sucrose plays a key role in induction of embryogenesis, while other medium constituents are needed for post induction development of embryo. High sucrose level may play an osmoregulatory role during induction, but it is not necessary or even detrimental during embryo development. For cole crops, the basal media used for washing is B5 (Gamborg *et al.*, 1968) and for culturing, NLN (Lichter, 1982). For microspore culture of *Brassica* species, sucrose concentration of culture medium is generally higher than the other species. This is usually 13 per cent or 17 per cent with media change to 10 per cent or 13 per cent sucrose. In anther cultures of paprika and eggplants, the inclusion of maltose in the induction medium improved the induction of microspore embryogenesis and raised the number of regenerant plants. Nitsch and Nitsch medium supplemented with glutamine, glutathione and serine is the most congenial for anther culture in Brassica. The use of ½ NLN-13 medium yielded greater number of embryos than the standard NLN-13 in broccoli. In general, organic nitrogen levels should be raised in the induction medium. The use of media with pH around 6 and addition of activated charcoal seems to promote embryo yield in cabbage.

Growth Regulators

Plant growth regulator (PGR) and their interaction with the plant's genotype and environmental factors play a crucial role in microspore embryogenesis (ME), controlling microspore-derived embryo differentiation and development as well as haploid/doubled haploid plant regeneration (Zur, 2015). Most aspects of microspore culture protocol have the capacity to cause stress to microspores, hence, less stressful treatments might be required to avoid deleterious effects. Polyamines are nitrogen containing compounds of low molecular weight found in plants and animals. The major forms: putrescine, spermidine and spermine are present in all parts of the cell including the nucleus and appear to participate in all cell processes. Polyamines (PA) and trehalose have been reported to act as compatible solutes or osmoprotectants by stabilizing proteins and biological membranes in stressed plants, (Redha and Suleman, 2013).

CENH3 Mediated Haploid Induction

Ravi and Chan (2010) discovered that haploids could be obtained in *Arabidopsis* through the manipulation of the centromere-specific histone 3 variant, CENH3. The CENH3, held promise of being translated to crop species, and has been successfully employed in maize (Kelliher *et al.*, 2016). Refinements of this technology have since been developed which indicate that non-transgenic modifications to CENH3 will also induce haploids. The complementation of a *cenh3* null by CENH3 from closely related plant species can result in plants that are fertile but haploid-inducing on crossing by CENH3 wt plants- suggesting that introgression of alien CENH3 may produce non-transgenic haploid inducers. Similarly, a remarkably wide variety of point mutations in CENH3, inducible by chemical agents, have recently been shown to result in haploid induction on crossing by wild-type CENH3 plants. These CENH3-variant plants grow normally, are fully fertile on self-pollination, and may be present in existing mutagenized collections. The alternative approaches to CENH3- mediated HI may also be employed. Non-GMO point mutants can be obtained by EMS mutagenesis, and may already be present in existing populations. The recent findings of complementation and HI by CENH3 from related species

(Maheshwari *et al.*, 2015) offers an alternative non-GMO approach for species in which EMS mutagenesis is inefficient.

Application of Haploids and Double Haploids in Cole Crops

Cultivar Development

In classical plant breeding program, a total of about eight inbreeding (selfing) generation is required to get an almost complete homozygous plant (99.2 per cent) with traits of interest. Through double haploid-based breeding, 100 per cent homozygosity for all the traits can be achieved in only one generation (Rahman and de Jimenez, 2016). One F_1 hybrid in cauliflower (KTH-DH-1) with proposed name of Pusa Snowball Hybrid-2 was developed at IARI, Regional Station, Katrai, Kullu using one DH line as male parent.

Haploid Mutant Production

The use of haploid system for mutant induction and selection is one of the most important applications of haploid technologies. Haploid tissue can facilitate the generation of genetic variation and its identification.

Some of the benefits of applying DH systems for induction and selection of mutants are:

1. Possibility to screen both recessive and dominant mutants in the the first generation after mutagenic treatment.
2. Immediate fixation of mutated genotypes, which saves time in production of pure mutant lines.
3. Increased selection efficiency of desired mutants due to the gametic versus zygotic segregation ratios (1:1 vs. 3:1, respectively) and the lack of chimerism.
4. Possibility of applying *in-vitro* selection methods at the haploid or double haploid level.

Mapping Quantitative Trait Loci

Most of the economic traits are controlled by genes with small, but cumulative, effects. Although the potential of DH populations in quantitative genetics has been understood for some time, it was the advent of molecular marker maps that provided the impetus for their use in identifying loci controlling quantitative traits.

Backcross Breeding

In backcross conversion, genes are introgressed from a donor cultivar or related species in to a recipient elite line through repeated backcrossing. A problem in this procedure is being able to identify the lines carrying the trait of interest at each generation. The problem is particularly acute if the trait of interest is recessive, as it will be present only in a heterozygous condition after each backcross. The development of molecular markers provides an easier method of selection based on the genotype (marker) rather than the phenotype. Combined with doubled haploidy it becomes more effective.

Bulked Segregant Analysis (BSA)

In bulked segregant analysis, a population is screened for a character of interest and the genotypes at the two extreme ends form two bulks. Then the two bulks are tested for the presence or absence of molecular markers. Since the bulks are supposed to contrast for alleles contributing positive and negative effects, any marker polymorphism between the two bulks indicates the linkage between the marker and character of interest. BSA is dependent on accurate phenotyping and the DH population has particular advantage in that they are true breeding and can be tested repeatedly. DH populations are commonly used in bulked segregant analysis, which is a popular method in marker assisted breeding.

Genetic Maps

Genetic maps are very important to understand the structure and organization of genomes from which evolution patterns and syntenic relationships between species can be deduced. Genetic maps also provide a framework for the mapping of genes of interest and estimating the magnitude of their effects and aid our understanding of genotype/phenotype associations. In genetic map construction a segregating population, such as F_2 or backcross or near isogenic lines or double haploid is required. DH populations have become standard resources in genetic mapping for species in which DHs are readily available. Doubled haploid populations are ideal for genetic mapping. It is possible to produce a genetic map within two years of the initial cross regardless of the species. Map construction is relatively easy using a DH population derived from a hybrid of two homozygous parents as the expected segregation ratio is simple, *i.e.* 1:1. Double haploid lines are used with high efficiency in genetic study for traits of interest. In conventional bi-parental cross based population, a trait controlled by a single dominant gene will segregate in the F_2 with a 1:2:1 ratio for homozygous dominant, heterozygous, and homozygous recessive, respectively. However, the same trait will segregate in the double haploid based population with a 1:1 ratio for homozygous dominant and homozygous recessive, respectively. In the case of multiple genes, double haploid technology could fix the genotypic recombinations for the homozygosity for many loci in a single generation. Thus, this method offers an advantage to use a smaller population for genetic study of quantitative traits.

Genomics

Although QTL analysis has generated a vast amount of information on gene locations and the magnitude of effects on many traits, the identification of the genes involved has remained elusive. This is due to poor resolution of QTL analysis. The solution for this problem would be production of recombinant chromosome substitution line, or stepped aligned recombinant inbred lines. Here, backcrossing is carried out until a desired level of recombination has occurred and genetic markers are used to detect desired recombinant chromosome substitution lines in the target region, which can be fixed by doubled haploidy.

Genetic Transformation

These days, in genetic transformation studies, haploid microspores are being used as targeted in transforming a gene prior to chromosome diplodization process that allow a stable fixation of homozygosity of the integrated gene. Haploid plants can also be used for gene transformation followed by the chromosome diplodization process. Drought tolerant gene (HVA1) was successfully transformed in haploid bread wheat followed by chromosome doubling to fix the homozygosity of the integrated gene

Limitations of Haploids and Double Haploids

1. Haploids can not be obtained in high frequency required for selection.
2. Selection cannot be imposed on the population in DH based breeding.
3. The cost benefit ratio in haploid breeding is often not favourable, thus discouraging the use of haploid breeding despite its obvious advantages.
4. Haploids will express recessive deleterious traits and deleterious mutations may arise during anther culture;
5. In haploids produced from anther culture, it is observed that some plants are aneuploids and some are mixed haploid-diploid types.
6. The over-usage of double haploidy may reduce genetic variation in breeding germplasm.

15.1.1.4 Protoplast Culture

Mesophyll protoplasts isolated from young and fully expanded leaves of seedlings from fertile and cytoplasmic male-sterile lines were cultured by Jourdan *et al.* (1985) employing an enzyme mixture consisting of cellulysin (0.5 per cent), driselase (0.01 per cent), and macerozyme along with 0.2 M mannitol, 0.8 M calcium chloride, 1mM KH_2PO_4, and 1mM MESat pH 5.6. Pelletier *et al.* (1983) cultured protoplasts at a plating efficiency of 1-2 per cent and obtained calli which regenerated to form plantlets from fused protoplasts.

15.1.1.5 Somatic Embryogenesis

Somatic embryogenesis has been induced from a range of vegetative explants such as leaf vein (Buiatti *et al.*, 1974), leaf lamina (Pareek and Chandra, 1978) and seedlings leaves (Jourdan *et al.*, 1985). Deane *et al.* (1997) derived embryogenic callus from *in-vitro*-grown leaf tissue of cauliflower cv. Dova.

15.1.2 Molecular Markers

Dias *et al.* (1991) used RFLP to study the genetic relationship of 31 coles (*Brassica oleracea*) representing different ecotypes of landraces grown in Portugal. Altogether 55 DNA probes, detecting 291 polymorphic restriction fragments were used. The results showed that the 48 accessions clustered into 5 groups: (i) a dispersed group with turnip and the wild brassicas, (ii) Chinese kales, (iii) broccolis and cauliflower, (iv) a dispersed group including Jersey kales, thousand head kale, cabbage and Savoy cabbage, and (v) a large and compact group containing all Portuguese tronchuda cabbage and kales. Demeke *et al.* (1992) studied the taxonomic differences in *Brassica* using RAPD markers. They could distinguish between cabbage and cauliflower

Protocol for Successful Development of Haploids and Double Haploids in Cauliflower through Isolated Microspore Culture (Bhatia *et al.,* 2016)
a. Growing microspores observed after 10 days of isolation,
b. The germinated microspore embryos 45 days after.

cultivars by RAPD's. King (1992) and Nienhuis *et al.* (1993) also studied genetic similaries in different groups of coles.

Figdore *et al.* (1993) reported identification of RFLP markers with loci affecting resistance to club root and morphological characters in *B. oleracea*. Gueguen (1995) identified RFLP markers of male sterility, which were used for marker-assisted selection for hybrid seed production in cauliflower.

15.1.3 Genetic Transformation

Eimert *et al.* (1992) described a simple procedure for gene transfer into cauliflower by macro injection. Seeds were injected with different *Agrobacterium tumefaciens* strains whose Ti-plasmid was having the kanamycin resistance (*npII*) gene. Integration and expression of *npII* gene was demonstrated.

He *et al.* (1994) reported transformation of kanamycin resistance (*kan*) gene into cauliflower through *A. rhizogenes* strain LBA9402 Bin 19. The transformed roots showed resistance upto 50 μg/ml kanamycin.

15.2.0 Cabbage

15.2.1 Tissue Culture

Establishment of an efficient regeneration protocol is a prerequisite step for transferring genes into plants. Plant tissue culture research in cabbage was reported by different scientists exploiting various explants such as mesophyll protoplast culture (Fu *et al.*, 1985; Chen *et al.*, 2004), microspore cultures (Rudolf *et al.*, 1999). Considering cabbage, many results pointed out that this crop regeneration and transformation efficiency greatly depended on the genotype (Gerszberg *et al.*, 2015). Thus, it is fully justified to conduct studies on regeneration potential of different cultivars of *B. oleracea* var. *capitata*. To develop the best regeneration protocol for cabbage, different factors such as the type of explant, its age, media supplemented with various plant growth regulators (PGRs) were examined. Recently, research was focused mainly on hypocotyl and cotyledon explants pointing to their huge potential for shoot organogenesis.

15.2.1.1 Micropropagation

The first reports on the micropropagation of cabbage were published by Primo-Millo and Harada (1975) and Bajaj and Nitsch (1975) using different explants. Leike (1981) induced multiple shoots from the axillary bud explants using MS medium supplemented with BA (1-3mg/I) and IBA or IAA (3 mg/1). Using this technique, mutiplica-tionrate in the ratio of 1:1 0,000 or more within 10 weeks could be obtained (Leikeand Wieczorrek, 1982).

15.2.1.2 Embryo Culture

Embryo culture to rescue embryos of inter-specific hybrids has been used by many workers working on *Brassica*. Inomata (1978b) excised ovaries from the inter-specific crosses of *B. campestris* x *B. oleracea* four days after pollination and cultured on White's medium containing case in hydrolysate. Morphological and cytological studies showed that plants raised from seeds and from excised embryos cultured *in-vitro* were hybrids. Hossain *et al.* (1989) obtained inter specific hybrids by embryo culture from all crosses when *B. campestris* was the female parent but only from 2 crosses in the reciprocal. Ayotte (1987) developed an embryo culture and regeneration protocol capable of producing *B. napus* x *B. oleracea* hybrids.

Takeshita *et al.* (1980) found ovule culture to be superior to embryo and ovary cultures in terms of the production rate of hybrids, when red cabbage was used as the female parent.

15.2.1.3 Anther Culture

Use of anther and microspore culture to produce doubled haploid has been found to be promising in cabbage. Kuo *et al.* (1983) cultured anthers of 14 breeding selections, from a cross of swede and cabbage and obtained three cabbage plants with *Plasmodiophora brassicae* resistance. There are many factors that determine the success of anther culture such as genotype, pre-treatment, anther stage, media composition, *etc.* Roulund *et al.* (1990) observed strong genotypic effects on embryoid formation fromcultured anthers.

Chang *et al.* (1996) cultured the anthers containing microspores between the tetrad and late uninucleate stages (flower bud about 2 to 4 mm in length) on MS medium supplemented with 2,4-D and BA for callus induction. Shoots regenerated through indirect organoenesis. Up to 98 per cent of the plantlets survived after transplanting using this method. Some haploid plants were detected by chromosome assay of root tips.

Microspore culture was reported to be applicable to a broad range of accessions belonging to six horticulturally important crop types of *B. oleracea* (Duijs *et al.*, 1992). Of 64 accessions tested by them, 86 per cent werer esponsive.

Rudolf *et al.* (1999) found that for the efficient application of haploid induction procedures in cabbage breeding, sufficient number of regenerants should be achieved in a broad spectrum of genotypes. However, the majority of genotypes are somewhat recalcitrant.

15.2.1.4 Protoplast Culture

Leaf mesophyll protoplasts have been successfully isolated and cultured (Bidney *et al.*, 1 983). Fu *et al.* (1985) isolated mesophyll protoplasts enzymatically from the first leaves of seedlings of the F_1 hybrid Baochun and cultured in modified liquid DPD medium. Intact plants were obtained when shoots were transferred to modified MS medium. Lillo and Shahin (1986) reported shoot formation from both protoplast-derived callus and root segments on a medium containing 4.5µM 2,4-D, 2.2µ MBA and 0.5µ M NAA.

The first fusion experiment in cabbage was reported by Schenck and Robbelen (1982) using protoplasts from sterile plants of interspecific hybrids of *B. oleracea* and *B. campestris* using PEG. From 1236 different combinations they regenerated 12 plants.

Efficient Plant Regeneration from Cotyledon and Hypocotyl Explants of Cabbage.

a. Aseptically germinated 7–9 days old seedlings of cabbage cv. Pride of India. b. Callus formation and shoot initiation from cotyledon explants on medium supplemented with (MS+ 0.33 mg/l TDZ + 79.9 mg/l Adenine). c. Shoot regeneration from cotyledon explants on medium supplemented with (MS+ 0.33 mg/l TDZ + 79.9 mg/l Adenine). d. Callus formation and shoot initiation from hypocotyl explants on medium supplemented with (MS+ 0.22 mg/l TDZ + 79.9 mg/l Adenine). e. Shoot regeneration from hypocotyl explants on medium supplemented with (MS+ 0.22 mg/l TDZ + 79.9 mg/l Adenine). f. Root regeneration from the in vitro developed shoot regeneration from explants on the medium supplemented with (MS + 0.10 mg/l NAA). g. Fully developed plantlets of cabbage taken out of medium (after proper washing) after 15 days of culturing showing well developed root system. Later on transferred to pots containing pre-sterilized cocopeat for acclimatization. h. Regenerated plantlets transferred to plastic pots containing mixture of pre-sterilized cocopeat covered with polythene bags for hardening. i. Plantlets transferred to potting mixture (containing sand + soil + FYM).

Gambhir *et al.* (2017).

Ryschka *et al.* (1999) crossed *Brassica oleracea* var. *capitata* cv. Korso with *Barbarea vulgaris, Mathiola incana* and *Raphanus sativus* genotypes carrying resistance genes against Plasmodiophora, Phoma and Alternaria. Protoplasts were fused and cultured on B5 medium containing 0.5 mg 2, 4-D, 0.2 mg NAA and 0.2 mg BAP/1. Several resistant somatic hybrids were identified.

Sigareva and Earle (1997) produced cold tolerant cytoplasmically male-sterile (CMS) cabbage by the fusion of leaf protoplasts from fertile cabbage and cold-tolerant Ogura CMS broccoli lines.

15.2.2 Molecular Markers

Cabbage germplasm can be characterized and conserved using molecular markers (Phippen *et al.*, 1997). For a quick and cost effective screening in order to quantify and partition the molecular variations among the accessions of cabbage, a RAPD assay was applied. Fourteen accessions of cabbage, which were phenotypically uniform and designated as cv. Golden Acre were tested. Total 110 fragments were generated and 80 of them were polymorphic.

McGrath *et al.* (1990) created a series of monosomic alien addition plants (2n=21). Chromosomes of *B. oleracea* (2n=18) were dissected for there synthesized amphidiploid *B. napus*, Hakuran (Cabbage x Chinese cabbage) by repeated back crossing to *B. campestris* (2n=20).

15.2.3 Genetic Transformation

Transformation studies in cabbage have been done with both *Agrobacterium tumefaciens* and *A. rhizogenes*. The main objective has been to develop transgenic cabbage plants carrying insecticidal crystal protein to protect against insect attack.

Agrobacterium-mediated transformation is widely employed to transfer a gene of interest (GOI) to crop species important from the economic point of view. However, most of them are recalcitrant to genetic modifications, since a lot of protocols are developed for model plants. Therefore, based on existing protocols, they should be properly refined for individual commercial plant species including cabbage. A review of the literature indicates the use of both *A. rhizogenes* and *A. tumefaciens* for the transformation of cabbage. The former pathway is rarely used for cabbage transformation. Considering *A. rhizogenes*-mediated transformation, it should be noted that transgenic roots of cabbage were successfully obtained, whereas difficulties appeared at the stage of plant regeneration. *A. tumefaciens*-mediated transformation is used more often. An effective genetic transformation of cabbage is strictly genotypically dependent; furthermore, some varieties are recalcitrant to transformation. So, it is clear that it is not possible to develop a universal cabbage transformation protocol. Therefore, it seems fully logical to establish a reliable transformation and regeneration procedure for specific cultivars. Most studies are dedicated to introducing genes conferring tolerance to insect or abiotic stresses. List of genes transferred in cabbage is mentioned below (Gerszberg, 2018).

Table 13: Gene Transfer List in Cabbage

Transformation Method	*Cultivar of Cabbage*	*Trait*	*Gene*
A. rhizogenes (A_4K, A_4H)	161	Resistance to kanamycin, hygromycin	*nptII, hph*
A. tumefaciens (AB1)	King Cole	Resistance to diamond back moth	*cry1a (c)*
A. tumefaciens (LBA4404)	Yingchun, Jingfeng	Insect tolerance to *Pieris rapae* L.	CpTI
A. tumefaciens (LBA4404)	Hercules, Brunswick, Cape spitz,	Resistance to kanamycin	*nptII*
A. tumefaciens (EHA105)	Scorpio, Testie	Resistance to diamond back moth (*P. xylostella*)	*cry1Ia3*
A. tumefaciens (LBA4404)	Uji	Enhanced tolerance to black rot disease by *Xanthomonas campestris* pv. *campestris*	*GO*
A. tumefaciens (LBA4404)	*B. oleracea var. capitata* cv. N/A	Induction of male sterile cabbage	*BcA9*
A. tumefaciens (GV2260)	*Golden Acre*	Enhanced salt tolerance	*beta*
A. tumefaciens (N/A)	*DTC 507*	Resistance to diamond back moth (*P. xylostella*)	*cry1 b*
A. tumefaciens (LBA4404)	*Xiaguang*	Increases submergence tolerance	*vhb*
Biolistic method	*Summer Summit, KY cross*	Resistance to diamond back moth	*cry1Ab*
A. tumefaciens (EHA105)	*CA21-3*	Resistance to diamond back moth	*cry1Ba3*
A. tumefaciens (N/A)	*Pride of India*	Study of the effect of antibiotic sensitivity on cabbage tissue	cry IAa

B. oleracea var. *capitata* as a crop plant is of great economic importance all over the world. Unfortunately, its yields are depleted by various environmental stresses (abiotic and biotic). However, great advance in the field of plant tissue culture, molecular biology as well as genetic engineering has offered new solutions for genetic improvement of significant vegetable crops including cabbage.

15.3.0 Brussels Sprouts

Like other vegetable brassicas, research on biotechnology of Brussels sprouts has been reasonably well reported. In the following, a brief account of the research is presented.

15.3.1 Tissue Culture

15.3.1.1 Micropropagation

Earliest report of micropropagation involving the meristem was reported by Clare and Collin (1974) and later by Dunwell and Davies (1975), Harada *et al.* (1985) and Williams and Pink (1987). The media used for tissue culture has been Murashige and Skoog (1962), Linsmaier and Skoog (1965) or B5 (Gamborg *et al.*, 1968). Skucinska *et al.* (1988) reported that genotype, explant size (using axial meristems), number of subcultures and composition of culture media all affected the success of *in-vitro* culture. Plantlets obtained from culture were rooted and transplanted into soil.

Generation of *Brassica oleracea* L. ('K-Y cross') Plants with Transgenic Plastids.

A. Plastid transformation of cabbage was achieved by biolistic bombardment of young sterile cabbage leaves. B. After bombardment leaf sample were cut into small pieces (3.3 mm^2). C. Plant regeneration into shoots from homoplasmic transplastomic callus tissue. D. Shoot regeneration 4 weeks after transfer of the callus to L1 media. E. Rooting of transplastomic cabbage shoots. F. Secondary selection of 200 mg/l spectinomycin resistant cabbage plants. G. Growth phenotypes of transplastomic plants over expressing aadA gene. Plants were transferred to soil and cultured to maturity in the green house (Liu *et al.,* 2007).

In order to hasten the breeding process, vernalization was induced in 90 per cent of plantlets by exposure to temperatures below10°C for several weeks. Borovec (1986) cultured explants from apical meristems of 5-6-days-old seedlings and from axillary buds of adult plants on MS mediums supplemented with BA at 0.2 mg/l to achieve multiple shoot formation with a view to utilizing clonal propagation in plant breeding. The same medium with 0.01mg/ladenine was used for rooting the shoots. Regenerants were subsequently planted in the field. No significant differences were found between the regenerants and the lines from which they were derived. Pawlowski (1990) cultured seedling tips (with cotyledons and hypocotyl segment) supplemented B_5 medium to induce shoot regeneration. Over 90 per cent of explants showed regeneration, except in Brussels sprouts (61 per cent). Shoots from the primary explants were propagated on medium containing benzyl adenine (especially at 2 mg/l) and the shoots obtained in culture were successfully rooted and planted in soil.

15.3.1.2 Anther Culture

Ockendon (1986) and Walkey and Pink (1988) reviewed the work on anther culture of Brussels sprouts. Vyvadilova *et al.* (1998b) studied some factors affecting the embryogenic processes in microspore cultures, namely donor plant genotype, donor plant growing conditions (growth chamber, glasshouse, field), developmental stage of the isolated microspore population (late uninucleate to early binucleate, predominantly binucleate) temperature (30, 32.5 and 35°C) and culture medium change. However, Ockendon (1984) opined that embryo yield were more consistent when the donor plants were raised under controlled conditions. Arnison and Keller (1990) also found rates of embryo formation were sometimes lower from field grown plants than those obtained under controlled environmental conditions. Bud size particularly ratio of petal to anther has been reported to influence the microsporogenesis (Ockendon, 1984; Leluand Bollon, 1985).

Temperature, particularly thermal shock, and the genotype have been found to be responsible for success of anther culture (Leluand Bollon, 1985, Duijs *et al.*, 1992). Biddington and Robinson (1990) observed that most somatic embryos were produced following 32 and 35°C, very few following 30°C and none following 38°C treatment. Experiments with cultivars Haland Gower suggested that high temperature was required for atleast 12h after anther removal, but beyond that time the optimum period varied from plant to plant. If the excised anthers were held at 25°C for 16 h or more with cv. Hal or 24 h or more with cv. Gower before being exposed to the high temperature treatment, embryogenesis tended to be reduced.

Variation between genotypes has been significance on the yield of embryos (Ockendon, 1984, 1986; Leluand Bollon, 1985; Duijs *et al.*, 1992; Vyvadilova *et al.*, 1998b).

Biddington *et al.* (1992) reported that gibberellic acid (GA_3) and fluridone, a non-specific ABA biosynthesis inhibitor, promoted embryo production in anther cultures of Brussels sprouts cv. Hal, but not in cv. Gower. The best GA_3 concentration was 10 μM compared with 1.0μM for fluridone. ABA and the gibberellin-biosynthesis inhibitor paclobutrazol inhibited embryo production in both cultivars.

The ethylene inhibitor silver nitrate ($AgNO_3$) is known to overcome the poor response of the Brussels sprouts cv. Hal to anther cuture. Biddington and Robinson (1991) found embryo production was inhibited by the inclusion of the ethylene precursor l-aminocyclopropane-1-carboxylic acid (ACC) or ethephon in the media. $AgNO_3$ and the ethylene biosynthesis inhibitor aminoethoxyvinylglycine (AVG) promoted embryogenesis but did not substitute for the high temperature treatment.

Werner *et al.* (1986) found from the evaluation of 1000 recombinant inbred lines (in the F_5 generation) produced by single seed descent from 4 source populations and133 produced by anther culture from 3 sources, that none of the lines gave a higher marketable yield. Werner *et al.* (1991) reported that inbred line performance in Brussels sprouts could be improved to levels comparable with those of commercial F_1 hybrids.

15.3.1.3 Protoplast Culture

Kik and Zaal (1993) reported for the first time on the successful regeneration of plants from mesophyll-derived protoplasts of 'rapid cycling' (60-days life cycle) *B. oleracea*. Comparative data are also presented on plant regeneration from mesophyll-derived protoplasts of 2 other varieties, namely vars. *botrytis* and *gemmifera*. It was found that a modified Pelletier protocol is highly beneficial for protoplast culture and plant regeneration from mesophyll-derived protoplasts. The planting efficiency of '*B. oleracea*' rapid cycling protoplasts was, in the best combination of isolation method, culture technique and culture media, 4.5 per cent ± 0.4 per cent and the plant regeneration frequency approximately 15 per cent. Plant regeneration was further improved by transferring the calluses from Pelletier's D medium to a callus growth medium (MS11) and subsequently to Glimelius' K3 regeneration medium.

15.3.2 Molecular Markers

Farnham (1996) evaluated a collection of collards (*Brassica oleracea, acephala* group) germplasm, including 13 cultivars or breeding lines and 5 landraces, using RAPD markers. He compared with representatives of kale (*acephala* group), cabbage (*capitata* group), broccoli (*italica* group), Brussels sprouts (*gemmifera* group) and cauliflowers (*botrytis* group) to assess genetic variation and relationships among collards and other crop entries to evaluate intra population variation of open-pollinated (OP) collard lines and to determine the potential of collard landraces to provide new *B. oleracea* genes.

Brace *et al.* (1994) identified S (self incompatibility)-alleles in 31 accessions using a method based on the amplification of S-sequences from genomic DNA, followed by digestion of the PCR products with selected restriction enzymes (PCR-FLP). The same *S*-allele was present in the homozygous state in a range of *B. oleracea* crop types. Results showed that, with minor exceptions, characteristic restriction patterns were obtained and therefore that it was possible to identify the *S*-allele.

Scutt *et al.* (1990) isolated a cDNA sequence homologous to the *Brassica* self-in-compatibility locus specific glycoprotein (SLSG) sequence from stigmas of *B. oleracea* (Brussels sprouts) plants homozygous for the S_5 allele, using a synthetic coligo nucleotide corresponding to a sequence which is highly conserved in other *S* alleles. Sequence analysis revealed an open reading frame of 1284 bases (427 amino acids) with a noncoding 3′ region of 178 bases and a poly (A) tail. A comparison of the S_5 and S_6 allelic forms of SLSG revealed 71 per cent homology at the DNA level and 64 per cent homology at the amino acid levels.

15.3.3 Genetic Transformation

Genetic transformation studies have been reported very little in Brussels sprouts and *Agrobacterium rhizogenes* has been used for transformation studies. Hosoki and Kogo (1994) introduced a *gus* gene [*uidA*] into Brussels sprouts using *A. rhizogenes* containing plasmid pBI121 that was transferred to *A. rhizogenes* IF013257 by the freeze-thaw method. When leaves of Brussels sprouts were inoculated with *Agrobacterium,* 34 hairy roots were produced showing GUS activity. From 7 root clones showing GUS activity, 14 shoots werer egenerated from calluses on MS medium supplemented with 1mg NAA and 5 mg zeatin/1. Christey *et al.* (1997) Co- cultivated leaf explants or petioles of intact cotyledons with *A. rhizogenes* strain A 4 harbouring various binary vectors. The T-DNA region of all binary vectors contained a *nptii* gene for kanamycin resistance, in addition to other genes. Hairy root lines grew prolifically on hormone-free medium containing kanamycin. Transgenic shoots were regenerated from all cultivars either spontaneously or after transfer of hairyroots to a hormone-containing medium. Southern analysis confirmed that the plants were transgenic. Plants from all brassica types were successfully transferred to greenhouse conditions. Christey *et al.* (1999) examined the field performance of 21 lines of transgenic vegetable brassicas from 9 cultivars representing five vegetable types and found that vegetable brassicas produced via *A. rhizogenes*-mediated transformation can produce healthy, vigorous plants.

15.4.0 Broccoli

Biotechnological research on broccoli is similar to that carried out in other brassica vegetables with similar objectives.

15.4.1 Tissue Culture

15.4.1.1 Micropropagation

Mix Wagner and Wang (1990) reported that IAA, benzyladenine and 2 I P enhanced regeneration of buds. Addition of various starch sources to the medium increased regeneration from 7.4 to 62.5 per cent and average number of shoots per explant from 1.9 to 4.1 per cent. Subjecting buds to 4°C raised shoot formation from 7.4 to 48.7 per cent. Well-rooted plants were transferred to soil with 60 to 80 per cent survival. Farnham and Nelson (1993) developed an *in-vitro* culture method for propagating unconditioned, field-grown broccoli from peduncle explants. The percentage of explant sregenerating shoots was significantly higher for cultivars grown in spring (17 to 100 per cent) than in the autumn (0 to 66 per cent).

15.4.1.2 Anther Culture

One of the earliest reports on anther culture was that of Quazi (1978). His study could not establish whether the regenerated plants were from sporophytic or gametophytic tissues. Various factors such as genotype, osmotic pressure, exogenous phytohormones, temperature, heat treatment, length of bud and factors related to the stage of pollen development on isolated microspore culture affect microsporogenesis. Embryoid formation by cultured broccoli anthers was best in the pH range 5.5 to 5.8 (Arnison *et al.*, 1 990a). Wide variation in plant to plant response was observed despite attempts to optimize the manipulation of physical parameters. Individual plants were identified which reliably formed many thousands of embryoids, whereas other plants failed to form embryoids under all tested conditions (Arnison *et al.*, 1990 b).

The optimum bud length for inducing embryogenesis was dependent on the plant genotype; for example, bc4 produced 37.08 embryoids per bud when microspores were obtained from buds 4.1-5 mm in length, when most microspores were at the late uninucleate and early binucleate phases (Zhang *et al.*, 1999). With the best conditions, 400- 700 embryoids per bud could be routinely obtained. Approximately 60 per cent of the embryoids germinated normally when subcultured onto a filter paper placed on top of agar medium. This frequency was 1.5- fold higher than for embryoids subcultured directly onto agar medium. More than 80 per cent of the regenerated plants were haploid (Takahata and Keller, 1991). Zhang *et al.* (1998) cultured 3.0-5.0 mm long buds of 13 genotypes of broccoli on modified liquid half strength NLN medium without growth regulators at 32°C for 24 h and subsequently at 25°C. Microspore-derived embryos were produced from 8 genotypes and 207 microspore-derived plants were produced from 5 genotypes. Balu broccoli produced the highest number (37.08) of embryos per bud.

Temperature has been found to influence embryogenesis (Keller and Armstrong, 1983; Orton and Browers, 1985). Embryo formation from microspores of broccoli is greatly enhanced by an initial incubation at elevated temperatures (*e.g.* 35°C) followed by continued incubation at 25°C. Fabijanski *et al.* (1991) observed that a 3 h high temperature treatment induced the formation of heat shock proteins in cultured anthers. Heat treatment at 32.5° C for 24 h was the most effective for inducing embryogenesis (Zhang *et al.*, 1999). Detailed studies of the effect of temperature on anther response showed the importance of 35°C treatments. An initial culture period at elevated temperature prior to maintenance at 25°C was required for induction of embryogenesis with incubation at 32.5°C for 1 day being the most effective (Takahata and Keller, 1991).

Silva-Dias (1999) studied the effect of the addition of a 0.1 ml drop of activated charcoal (AC) on microspore culture embryo genes is in nine morphotypes of *B. oleracea*. Embryo yields were significantly increased in all of the morphotypes by the addition of the 0.1 ml drop of AC to the microspore culture media.

Wang *et al.* (1999) conducted studies to characterize regenerants derived from microspore culture, to evaluate factors influencing these characteristics and to compare results with those from anther culture. Eight populations; four from each culture method, were generated simultaneously using the same four F_1 hybrids as donor parents. The ploidy level of all regenerants was determined by DNA flow cytometry the majority of them were diploid.

15.4.1.3 Protoplast Culture

Wang *et al.* (1992) isolated mesophyll protoplasts from the cultivar Futura. The callus developed adventitious roots after transfer to regeneration media. Shoots were obtained when the callus was cultured on MS medium containing 5 mg BA. Plantlets developed rapidly and could be transferred to the greenhouse after a short period of hardening.

15.4.2 Molecular Markers

Self-incompatibility in *B. oleracea* is controlled by a single genetic locus (the *S* locus) with nearly 50 different alleles. The S_2 allele is a pollen recessive self-incompatibility allele that exhibits weak DNA homology to the other previously sequenced *S* locus alleles (S_6, S_{13}, S_{14} and S_{22}, encoding the glycoproteins SLG6, SLG13, SLG22, respectively). Chen and Nasrallah (1990) isolated stigma cDNA clones with sequence homology to SLG13 cDNA from 2 S_2 homozygous strains of *B. oleracea* vars. *alboglabra* [*B. alboglabra*] (Chinese kale) and *italica* (broccoli). The S_2 cDNA sequences were 90 per cent homologous to each other, but only 70 per cent homologous to SLG 13 cDNA. The Chinese kale S_2 genetic background was used to demonstrate that the isolated alboglabra cDNA sequence is a transcript from a gene, designated SLG2A, that resides at the S locus and is proposed to be a putative determinant of S_2 allelic specificity. Sakamoto *et al.* (2000) evaluated several simple methods of DNA preparation from plant tissues for PCR-RFLP analyses of SLG and SRK alleles, which can be used for the identification of S (self-incompatibility) haplotypes of breeding lines in broccoli and cabbage (*Brassica oleracea*) and in purity tests of F_1 hybrid seeds. On the five methods tested, the NaI method was found to be the most suitable for the amplification of the *SLG* and *SRK* alleles. This method enables the use of a single seed as testing material. Using this method, we identified S haplotypes of 31 broccoli and 31 cabbage cultivars. Ninety-four percent of the cultivars of broccoli and 97 per cent of those of cabbage were single cross F_1 hybrids. Brace *et al.* (1994) identified *S* (self-incompatibility)-alleles in 31 accessions using a method based on the amplification of S-sequences from genomic DNA, followed by digestion of the PCR products with selected restriction enzymes (PCR-RFLP). The results showed that the standard method was not reliable for the identification of both of the *S*-alleles.

Location of the self-incompatibility locus in an RFLP and RAPD map of *B. oleracea* was reported by Camargo *et al.* (1997). They located a single locus controlling self-incompatibility (SI) on an RFLP and RAPD linkage map of *B. oleracea* by assaying for SI phenotype and segregation of an RFLP locus detected by the cloned SI gene *SLG6*. The map was developed using an F_2 population derived from a cabbage x broccoli cross and included 112 RFLP and 47 RAPD loci arranged into 9 main linkage groups covering 921 cM. Park *et al.* (2000) described a technique for the identification of the S-haplotype, which is involved in self-incompatibility in *Brassica*, based on PCR amplification of genomic DNA followed by restriction analysis. Polymorphism of SLG (S locus glycoprotein) and SRK (S locus receptor kinase) genes in broccoli and cabbage was analysed by PCR-RFLP using SLG and SRK specific primers.

Ramsay *et al.* (1996) described the methods for the construction of a substitution library of recombinant backcross lines in *B. oleracea* for the precision mapping of quantitative trait loci. The currently available methods for locating quantitative trait loci (QTLs) and measuring their effects in segregating populations lack precision unless individual QTLs have very high heritabilities.

Comparison of RAPD and RFLP genetic markers in determining genetic similarity among *B. oleracea* genotypes has been reported by Santos *et al.* (1994) and Thormann *et al.* (1994). RFLPs have been widely used to study the molecular taxonomy (Dias *et al.*, 1991), for phylogenetic relations (Song *et al.*, 1990), for genetic linkage (Slocum *et al.*, 1990). Later studies indicate the use of RAPD for above purposes (Massie *et al.*, 1999).

15.4.3 Genetic Transformation

Earle *et al.* (1996) presented a comprehensive review on the transformation of Brassica vegetables. Generally, the transformation is attempted for insect resistance derived from the *Bacillus thuringiensis* insecticidal protein transgene, prolonged shelf life and improved nutritive value (Coupe and Christey, 2000).

Generally, *Agrobacterium* is used for transformation work. However, some workers have used biolistic transformation also (Puddephat *et al.*, 1999). One of the earliest reports on the genetic transformation was that of Christey and Earle (1988). They described a procedure for regenerating peduncle explants. The procedures hold considerable potential for transformation of *B. oleracea* through co-cultivation with *A. tumefaciens*, despite yielding only 1 putative transformant with Green Comet upon testing. Besides, peduncle and hypocotyl explants, cotyledons and hypocotyl protoplasts have also been used for transformation (Li *et al.*, 1999). *A. rhizo genes* have also been used in transformation work on broccoli (Christey *et al.*, 1997; Henzi *et al.*, 2000).

Table 14: Genetic Transformation in Broccoli

Genes	*Traits improved*
SLG Gene	Flowering Control
Sadenosylmethionine hydrolase (*SAD*, EC-3.3.1.2)	Prevent postharvest losses and quality improvement
npt-II and *cryIA(c)*	Insect resistance
cryIA(c) and *npt-II*	Insect resistance
Bar gene	Herbicide resistance
Antisense pollen specific gene	Flowering control
cry1C	Insect resistance
npt and *ipt*	Prevent postharvest losses and quality improvement
ACC oxidase gene	Prevent postharvest losses and quality improvement
cry1A, *cry1C*, *hpt* and *npt-II*	Insect resistance
FCA and *CONSTANS*	Flowering control
Anti- gene *CYP86MF*	Flowering control

The transformation research has led to the development of potential herbicide resistant lines (Waterer *et al.*, 2000) and introduction of ACC oxidase genesfor better postharvest life (Henzi *et al.*, 1999). Mora and Earle (1999) introduced an endochitinase gene from *Trichoderma harzianum* into broccoli via *A. tumefaciens*-mediated transformation. The transgenic lines had endochitinase activity15-30 times higher than control lines indicating resistance to *Alternaria brassicicola* in only diploids.

15.4.3.1 Insect Resistance

The main objective of research on genetic transformation in broccoli is to obtain resistance to lepidopteran pests of crucifers such as diamond back moth (DBM, *Plutella xylostella*), cabbage looper (*Trichoplusia* sp.) and cabbage worm (*Pieris rapae*).

Metz *et al.* (1995) transformed flower stalk explants of five genotypes of broccoli with *A. tumefaciens* using the construct pMONI0517-lcontaining the *nptII* gene and the *bt* gene [*Cryla*(*c*) type] optimized for plant expression. Overall transformation efficiency was 6.4 per cent and 181 kanamycin-resistant plants were recovered. Of 162 kanamycin-resistant plants tested, 112 (69 per cent) caused 100 per cent mortality of Ist-instar larvae of a *Bt*-susceptible diamond back moth strain.

Cao *et al.* (1999) introduced *cryIC* gene into broccoli by *A. tumefaciens*-mediated transformation. Some 21 *CryIC* transgenic plants were regenerated from 400 hypocotyl and petiole explants. Variable amounts of stable steady-state *cryIC* mRNA accumulated in different transgenic plants. This study demonstrated that high production of *CryIC* protein can protect transgenic broccoli not only from susceptible or *CryIAR* DBM larvae but also from DBM selected for moderate levels of resistance to *CryIC*. *CryIC*-transgenic broccoli was also resistant to two other lepidopteran pests, cabbage looper and cabbage worm (*Pieris rapae*).

16.0 References

Aalbersberg, W. (1986) *Groenlen en Fruit*, **41:** 46-47.

Aamlid, K. (1952) *Ph.D. Dissert.*, Univ. Maryland, USA.

Abiko, K., Kubota, M., Yamada, K. and Hida, K. (1998) *Proc. of the Kansai Plant Protect. Soc.*, No.40, pp. 47-49.

Aboul-Nasr, M.H. and Ragab, W.S.M. (2000) *Assiut. J. Agric. Sci.*, **31:** 55-78.

Acharya, B.C. (1968) *Proc. Res. Found. Bihar Agri. Coll., Sabour*, **10:** 49.

Adamicki, F. and Elkner, K. (1985) *Biuletyn Warzywniczy, Instytut Warzywnictwa*, Poland, **28:** 197-224.

Adamson, R.M. (1965) *Canad. J Plant Sci.*, **45** : 493-497

Adeniran, K.A., Amodu, M.F., Amodu, M.O. and Adeniji, F.A. (2010) *AJAE*, **1:** 119-125.

Agarwal, M.L., Singh, C.N., Bhan, C. and Singh, N.N. (1972) 3rd *International Symp. Subtrop. Trop. Hort.*, p.142.

Agarwal, S.C. (1950) *Annual Report, Long Ashton Agri. and Hort. Sta.*, pp. 183-190.

Agarwal, S.C. and Hewitt, E.J. (1954) *J. Hort. Sci.*, **29:** 278-290.

Agnihotri, A., Shivanna, K.R., Raina, S.N., Lakshmikumaran, M., Prakash, S. and Jagannathan, V. (1990) *Plant Breed.*, **105:** 292-299.

Ahluwalia, K.S., Swarup, V. and Chatterjee, S.S. (1977) *Veg. Sci.*, **4:** 67-80.

Ahuja K.S. and Sandhu, S. (2003) *Ann. Biol.*, **19:** 27-30.

Ajithkumar, B. and Savani, M.B. (2007) *J. Agromet.*, **9:** 231-235.

Alam, Z. (1936) *Ann. Bot.*, **50:** 85-102.

Alipieva, M. and Dimitrov (1984) *Genetikai Selektsiya*, **17:** 95-99.

Allender, C.J., Allainguillaume, J., Lynn, J. and King, G.J. (2007) *Theor. Apppl. Genet.*, **114:** 609–618.

Alvares, M.C., Oliveira, S.A. De, Mattos, J.K. DeA. and Fo, M.V. DcM. (1985) *Horticultura Brasileira*, **3:** 18-21.

Anonymous (1998a) *FAO Production Yearbook*, **51:** 121-123.

Anonymous (1998b) *FAO Production Yearbook*, **51:** 126-127.

Anonymous (2018) Horticultural statistics at a glance. Horticulture Statistics Division, Department of Agriculture, Cooperation and Farmers Welfare, Ministry of Agriculture and Farmers Welfare, Govt. of India, New Delhi.

Anstey, T.H. and Moore, J.F. (1954) *J. Heredity*, **45:** 39-41.

Apeland, J. (1985) *Acta Horticulturae*, No. 157, 185-191.

Ara, N.K., Khalequzzaman, M.O., Kohinoor, H.K.M. and Ahamed, K.U. (2009) *J. Soil. Nature*, **3:** 16-19

Arias, T., Beilstein, M.A., Tang, M., McKain, M.R. and Pires, C. (2014) *Am. J. Bot.*, **101:** 86-91.

Arnison, P.G. and Keller, W.A. (1990) *Plant Breed.*, **104:** 125-133.

Arnison, P.G., Donaldson, P., Ho, L.C.C. and Keller, W.A. (1990a) *Plant Cell Tissue Organcue*, **20:** 147-155.

Arnison, P.G., Donaldson, P., Jackson, A., Semple, C. and Keller, W. (1990b) *Plant Cell Tissue Organ Cult.*, **20:** 217-222.

Arora, N., Yadav, N.R. and Chowdhury, J.B. (1996) *Cruciferae Newsletter*, **18:** 26-27.

Arora, N., Yadav, N.R., Yadav, R.C. and Choudhury, J.B. (1998) *Cruciferae Newsletter*, **20:** 31-32.

Arora, P.N. (1966) *Indian Hort.*, **10:** 44-48.

Arora, P.N., Jashiand, B.S. and Pandey, S.L. (1970) *Indian Hort.*, **15:**19-20.

Atak, U. and Atak, E.D. (1986) *Bitki Koruma Bulteni*, **24:** 173-199.

Awasthi, D.N., Joshi, S. and Ghildiyal, P.C. (1976) https://agris.fao.org/agris-search/search.do?recordID=US201303054810

Ayotte, R. (1987) *Diss. Abstr. Int. B. Sci. Eng.*, **48:** 17B.

Ayotte, R., Harneyu, P.M. and Machado, V.S. (1987) *Euphytica*, **36:** 615-624.

Babik, I., Rumpel, J. and Elkner, K. (1996) *Acta Hortic.*, **407:** 353-359.

Bai, Y.Y., Mao, H.Z., Cao, X.L., Tang, T., Wu, D., Chen, D.I., Li, W.G., Fu, W.J., You, C.B., Chen, Z.L. and Ding, Y. (1993) In: *Biotechnologyin Agriculture* (Eds.

You, C.B. and Chen, Z.I.), Proc. of the FirstAsia-Pacific Conf. on Agri, Biotech., Beijing, China, August 20-24,1992, pp.156-159.

Bailey, L.H. (1922) *Gentes Herbarum*, **1:** 53-108.

Bailey, L.H. (1930) *Gentes Herbarum*, **2:** 210-267.

Baillon, H.E. (1871) *Cruciferes. des plantes* (Paris) **3:** 188-195.

Bain, D.C. (1951) *U.S. Dept. Dis. Rept.*, **35:** 200-205.

Bain, D.C. (1955) *Phytopath.*, **45:** 35-37.

Bairwa, R.K., Singh, S.P., Mahawar, A.K., Das, K.K. (2017). *Int. J. Curr. Microbiol. Appl. Sci.*, **6**: 2438-2447.

Bajaj, Y.P.S. and Nitsch, J.P. (1975) *J. Exp. Bot.*, **26:** 883-890.

Balyan, D.S. and Singh, J. (1994) *Haryana Agril.Univ. Res. J.*, **24:** 88-92.

Bambal, A.S., Venn, R.M., Panchbhai, D.M., Mahorkar, V.K. and Khankhane, R.M. (1998) *Orissa J. Hort.*, **26:** 14-17.

Bannerot, H., Boulidard, L. and Cherpean, Y. (1977) *Cruciferae News Letter*, **2:** 16.

Bannerot, H., Boulidard, L., Cauderon, Y. and Temp, J. (1974) Eucarpia Meeting Cruciferae, Dundee, Scotland, pp. 52-54.

Barth, M.M., Kerbel, E.L., Perry, A.K. and Schmidt (1993) *J. Food Sci.*, **58:** 140-143.

Bastelaere, H. Van., Geenew, J., Pelsmaeker, M. De and Verbruggen, J. (1985) *Boer en de Tuinder*, **91:** 12-13.

Batabyal K., Sarkar, D. and Mandal, B. (2015) *J. Plant Nutr.*, **38:** 1822-1835.

Batal, K.M., Granberry, D.M. and Mullinix, B.G. Jr. (1997) *HortScience*, **32:** 75-78.

Bateman, A.J. (1952) *Heredity*, **6:** 285-310.

Bateman, A.J. (1954) *Heredity*, **8:** 305-332.

Bateman, A.J. (1955) *Heredity*, **9:** 53-68.

Bathkal, B.C. (1960) *Nagpur Agri. Coll. Mag*, **34:** 1-9.

Bathkal, B.C. and Lal, S.B. (1961) *Nagpur Agri. Coll. Mag.*, **35:** 1-11.

Bayense, B. (1986) *Groenten en Fruit*, **41**: 38-39.

Beilstein, M.A., Al-Shehbaz, I.A. and Kellog, E.A. (2006) *Amer. J. Bot.*, **93:** 607-619.

Bennett, R.N. and Wallsgrove. R.M. (1994) *New Phytologist*, **127**: 617-633.

Benoit, F. and Ceustermans, N. (1983) *Boer en de Tuinder*, **89:** 29.

Benoit, F. and Ceustermans, N. (1997) *Proeftuinnieuws*, **7:** 38-40.

Benoit, F. and Cuestermans, N. (1985) *Boer en de Tuinder*, **91**:21.

Benoit, F. and Cuestermans, N. (1986a) *Boer en de tuinder*, **92:** 20.

Benoit, F. and Cuestermans, N. (1986b) *Boer en de tuinder*, **92:** 25.

Benson, L. (1957) Plant Classification, Raytheon Education Company, Boston, Massachusetts, USA.

Berard, L.S. (1985) *4th Nat. Cont. Atmos. Res. Conf.*, Releigh, North Carolina State University, USA, pp.150-159.

Berard, S.L., Vigier, B. and Dubuc-debreux, M.A. (1986) *Phytoprotection*, **67**: 67-73.

Beshir, S. (2017) *J. Geosci. Environ. Prot.*, **5:** 59-69.

Bhan, K.C. (1996) *Indian Hort.*, **10:** 68-69.

Bhardwaj, R.L. (2013) *Agric. Rev.*, **34:** 188-197.

Bhatia, R., Dey, S.S., Parkash, C., Sharma, K., Sood, S., and Kumar, R. (2018) *Sci. Hortic.*, **233:** 178-187.

Bhatia, R., Dey, S.S., Sharma, P.C. and Kumar, R. (2015) *J. Hortic. Sci. Biotechnol.*, **90:** 171-179.

Bhatia, R., Dey, S.S., Sood, S., Sharma, K., Parkash, C. and Kumar, R. (2017) *Sci. Hortic.*, **216:** 83–92.

Bhatia, R., Dey, S.S., Sood, S., Sharma, K., Sharma, V.K., Parkash, C. and Kumar, R. (2016) *Euphytica*, **212:** 439–454.

Bianco, M.S., Filho, A.B.C., Carvalho, L.V.D. (2015) *PLoS ONE*, **10**(4): e0123500.

Biddington, N.L. and Robinson, H.T. (1990) *Plant Cell Tissue Organ Cult.*, **22:** 48-54.

Biddington, N.L. and Robinson, H.T. (1991) *Plant Cell Tissue Organ Cult.*, **25:** 169-177.

Biddington, N.L., Sutherland, R.A. and Robinson, H.T. (1992) *Plant Growth Reg.*, **11:** 81-84.

Bidney, D.L., Shepard, J.E and Kaleikau, E. (1983) *Protoplasma*, **117:** 89-92.

Biever, K.D., Hostetter, D.L. and Kern, J.R. (1994) *American Entomologist*, **40:** 103-108.

Bigelli, G. and Paradisi, U. (1985) *Informatore Agrario*, **41:** 31-34.

Bohra, A., Jha, U.C., Adhimoolam, P., Bisht, D. and Singh, N.P. (2016) *Plant Cell Rep.*, **35:** 967-993.

Bondarenko, L.D. (1983) *Kartofel i Ovoshchi*, No. 8, pp. 27-28.

Bonnet, A. (1975) *Ann. Amelier Planter*, **25:** 381-397.

Bonnet, A. (1977) *Cruciferae News Letter*, **2:** 5.

Booij, R. (1989) *Sci. Hort.*, **38:** 23-32.

Booij, R. (1990) *Neth. J. Agri. Sci.*, **38:** 641-651.

Booij, R. (2000) *Gartenbauwissenschaft*, **65:** 30-34.

Boos, G.V. and Artemeva, A.M. (1984) *Selektsiya ovoschhnykh Kul'tur*, pp. 33-37.

Borchers, E.A. (1966) *Proc. Amer. Soc. Hort. Sci.*, **88:** 406-410.

Borchers, E.A. (1971) *J. Amer. Soc. Hort. Sci.*, **96:** 542-543.

Borkowski, J. (2000) *Ochrona Roslin*, **44:** 45-46.

Borkowski, J., Dyki, B., Oskiera, M., Machlañska, A. and Felczyñska, A. (2016) *J. Hortic. Res.*, **24:** 47-56.

Boshnakov, P., Kerin, V. and Sivek. P. (1986) *Rasteniev dni Nauki*, **23:** 74-78.

Boswell, V.R. (1929) *Md. Agri. Exp. Sta. Bull.*, **313:** 69-145.

Boswell, V.R. (1949) *Nat. Geog.*, **96:** 145-217.

Bothmer, R., von, Gustafsson, M., and Snogerup, S. (1995) *Genet. Res. Crop Evol.*, **2:** 165–178.

Bottcher, H. (1986) *Archiv fur Gartenbau*, **34:** 409-422.

Bottcher, H. (1987) *Archiv fur Gartenbau*, **35:** 33-45.

Brace, J., Ryder, C.D. and Ockendon, D.J. (1994) *Euphytica*, **80:** 229-234.

Bracy, R.P., Parish, P.L., Bergeron, P.E., Moser, E.B. and Costantin, R.J. (1993) *HortScience*, **28:** 179-181.

Brakeboer, T. (1990) *Groenten enfruit*, **45:** 56-57.

Branca, F. (2008) In: *Vegetables* I. *Handbook of Plant Breeding*, (Eds. J. Prohens, F. Nuez), vol 1. Springer, New York, NY.

Branca, F. and Cartea, E. (2011) In: Kole C. (eds.) *Oilseeds*, 17-36.

Bravo, M.A., Alegre, M. and Campos, P. (1986) *Ciencia e Investigacion Agraria*, **13:** 3-8.

Brown, H.D. and Hutchison, C.S. (1949) *Vegetable Science*, J.B. Lippincot Company, NewYork, pp. 21-30.

Brunner, E. and Stevens, R.F.E. (1986) In: *Diamond-back Moth Management*, Proc. First Int. Work-shop, Tainan, Taiwan, 11-15 March, 1985, pp. 213-217.

Budar, F. and Pelletier, G. (2001) *Comptes Rendus de l'Académie des Sciences-Series III-Sciences de la Vie*, **324**: 543-550.

Buiatti, M., Baroncelli, S., Benniei, A., Pagliai, M. and Tesi, R. (1974) *Z. Pflanzenzhucht*, **72:** 269-274.

Burkill, I.H. (1930) *Gadi Bull.*, **5:** 99-117.

Bussel, W.T. and Dobson, B.G. (1985) *New Zealand Commercial Grower*, **40:** 44-45.

Butler, E.J. (1918) *Spink and Co., Calcutta* p. 547.

Calderon, J.I. and Hare, C.J. (1986) In: *Diamond-back Moth Management*, Proc. First Int. Workshop, Tainan, Taiwan, 11-15 March, pp. 267-269.

Camargo, L. S. (1956). *Bragantia*, **14**: 315–328

Camargo, L.E.A., Savides, L., Jung, G., Nienhuis, J. and Osborn, T.C. (1997) *J. Hered.*, **88:** 57-60.

Campbell, R.N., Greathead, A.S., Myers, D.F. and Boer, G.I. De (1985) *Phytopath.*, **75:** 665.

Cao, J., Tang, J.D., Strizhov, N., Shelton, A.M. and Earle, E.D. (1999) *Mol. Breed.*, **5:** 131-144.

Cao, S.C. (1986) *Cruciferae Newsletter*, No. 11, pp. 8-11.

Cardoza, V. and Stewart, N.C. (2004) *In Vitro Cell Dev. Biol. Plant*, **40:** 542-551.

Catcheside, D.G. (1934) *Ann. Bot.*, **48:** 601-633.

Catcheside, D.G. (1937) *Cytologia, Fujii Jub.* Vol. I, pp. 366-378.

Ceausescu, M.F., Tudor, T.A., Cosmin, S. and Amariutei, A. (1987) *Productia Vegetala, Horticultura*, **36:** 34-39.

Cerne, M. (1991) *Bioloski Vestnik*, **39:** 95-98.

Chafutsa, W., Fandika, I.R., Kadympakeni, D.M., Chiipanthenga, H.M. and Mafunga, G. (2007) Irrigation Drainage Team, Kasinthula Agricultural Experiment Station, Chikwawa.

Chand, J. (1980) *Ph.D. Thesis*, Indian Agricultural Research Institute, New Delhi.

Chang, Y.M., Liou, P.C. and Hsiao, C.H. (1996) *J. Agri. Res. China*, **4:** 35-46.

Chao, S.C., Hou, X.L., Zhang, S.N., Zhu, Y.L. and Liu, K.J. (1998) *J. Nanjing Agri. Univ.*, **21:** 24-29.

Chase, C.D. (2007) *Trends Genet.*, **23:** 81-90.

Chatterjee, R. and Som, M.G. (1990) *Veg. Sci.*, **17:** 66-69.

Chatterjee, R., Jana, J.C. and Paul, P.K. (2012). *Indian J. Agric. Sci.*, **82:** 323-327.

Chatterjee, S., Aralikatti, O., Sharma, S., Mukherjee, D., Patil S., *et al.*, (2018). *Int. J. Curr. Microbiol. Appl. Sci.*, **7**: 82-92.

Chatterjee, S.S. (1963) *Indian J. Hort.*, **19:** 80-81.

Chatterjee, S.S. and Mukherjee, S.K. (1962) *Indian J Agric. Sci.*, **32:** 201 -207.

Chatterjee, S.S. and Sharma, R.R. (1970a) *Indian Fm. Digest*, **3:** 37-40.

Chatterjee, S.S. and Sharma, R.R. (1970b) *Indian Hort.*, **15:** 14-16.

Chatterjee, S.S. and Swamp, V. (1983) *Seeds and Farms*, **9:** 15-16.

Chen, C.H. and Nasrallah, J.B. (1990) *Mol. Gen. Genet.*, **222:** 241-248.

Chen, J.J., Sun, Y.W. and Sheen, T.F. (1999) *HortScience*, **34:** 852-854.

Chen, L.P., Zhang, M.F., Xiao, Q.B., Wu, J.G., Hirata, Y. (2004). *Plant Cell Tiss. Org. Cult.*, **77**: 133–138.

Chen, L.T. and Liu, Y.G. (2014) *Annu. Rev. Plant Biol.*, **65:** 579-606.

Chen, Y.N., Chen, C. and Ma, J. (2000) *Plant Protection*, **26:** 10- 13.

Chiang, M. S. and R. Crête, (1976) *Can. J. Plant Sci.*, **56**: 865–868.

Chonnkar, V.S. and Sharma, S.B. (1966) *Indian J. Hort.*, **23:** 81 -87.

Choudhuri, B. and Som, M.G. (1969) *Hort. Sci.* (Calcutta), **2:** 26-29.

Christey, M.C. and Earle, E.D. (1988) *Cruciferae Newslett.*, No. 13, pp. 94-95.

Christey, M.C., Braun, R.H. and Reader, J.K. (1999) *SABRAO J. Breed. Genet.*, **31:** 93-108.

Christey, M.C., Sinclair, B.K., Braun, R.H. and Wyke, L. (1997) *Plant Cell Rep.*,**16:** 587-593.

Chung, B. (1985) *Aust. J Expt. Agric.*, **25:** 959-962.

Chung, B. and Strickland, H.L. (1986) *Australian J. Expt. Agric.*, **26:** 497-500.

Chweya, J.A. (1984) *Acta Hortic.*, **163**: 295-301.

Clare, M.V. and Collin, H.A. (1974) *Hort. Res.*, **13:** 111-118.

Coffey, D.L. and Disney, G.W. (1992) *Tennessee Farm Home Sci.*, **164:** 17-22.

Cole, K. (1957) *Proc. Genet. Soc. Can.*, **2:** 44.

Cole, K. (1959) *Cand. J. Genet. Cytol.*, **1:** 203-207.

Colombo, N., Galmarini, C.R. (2017) *Palnt Breed.*, **136:** 287-299.

Comstock, R.E., Robinson, H.F. and Harvey, P.H. (1949) *Agron. J.*, **41:** 360-367.

Conner, A.J. and Christey, M.C. (1997) *Ann. Bot.*, **80:** 561-564.

Cools, M.H. (1985) *Groenten en Fruit*, **40:** 45-46.

Corcuff, R., Arul, J., Hamza, F., Castaigne, F. and Makhlouf, J. (1996) *Postharvest Biol. Tech.*, **7:** 219-229.

Coupe, S. and Christey, M. (2000) *Commercial Grower*, **55:** 28-30.

Crisp, P. (1982) *Euphytica*, **31:** 725-734.

Crisp, P. (1984) *J. Agri. Sci.*, **102:** 405-413.

Crisp, P. and Walkey, D.G. (1974) *Euphytica*, **23:** 305-313.

Crisp, P., Crute, I.R., Sutherland, R.A. Angell, S.M., Baloor, K., Burgess, H. and Gordon, P.L. (1989) *Euphytica*, **42:** 215-226.

Crisp, P., Walkey, D.G.A., Bellman, E. and Roberts, E. (1975) *Euphytica*, **24:** 173–176.

Cristea, T., Ambarus, S., Stoian, L., Falticeanu, M. and Calin, M. (2009) *Acta Hortic.*, **812:** 343-348.

Crockett, P.A., Bhalla, P.L., Lee, C.K. and Singh, M.B. (2000) *Genome*, **43:** 317-332.

Csizinszky, A.A. (1996) *HortScience*, **31:** 930-933.

Cuocolo, L. and Duranti, A. (1988) *Rivista di Agronomia*, **22:** 203-207.

Dabbas, M.R., Singh, D.P. and Yadav, J.R. (2009) *Int. J. Plant Protec.*, **2:** 128-130.

Dadlani, N.K. (1977) *Ph.D. Thesis*, Indian Agricultural Research Institute, New Delhi.

Dakouri, A., Zhang, X., Peng, G., Falk, K.C., Gossen, B.D., Strelkov, S.E. and Yu, F. (2018) *Sci. Rep.*, **8:** 17657.

Dangler, J.M. and Wood, C.W. (1993) *HortScience*, **28:** 701-703.

Dark, S.O.S. (1938) *Ann. Bot., NS.*, **2:** 751-752.

Dastgheib (1995) In: *Proc. of the forty eight New Zealand Plant Protect. Conf*, Angus Inn, Hastings, New Zealand, August 8-10, pp. 331-332.

Datta, A.K. (1963) *Extn.*, **3:** 11-12.

Datta, P.C. and Deb, A. (1974-75) *Genet. Iber.*, **26-27:** 16-18.

De Candolle, A. (1821) *Mem. Mus. Hist. Nat., Paris*, **7:** 169-252.

De Candolle, A. (1824) *Trans. Hort. Soc.*, London, **5:** 1-43.

De Candolle, A. (1883) *l'Origine des Plantes cultivees*, Paris.

De Candolle, A. (1885) In: *Origin of Cultivated Plants*, The New York Public Library, p. 468.

De Nettancourt, D. (1977) *Self-incompatibility in Angiosperm. Springer, Berlin-Heidelberg*, New York.

Deane, C.R., Fuller, M.P. and Dix, P.J. (1997) *Cruciferae Newslett.*, **19:** 43-44.

Deleseny, M., McGrath, J.M., This, P., Chevre, A.M. and Quiros, C.F. (1990) *Genome.*, **33:** 733-744.

Dellacecca, V. (1996) *Acta Hortic.*, **407:** 347-351.

Delourme, R. and Budar, F. (1999) Amsterdam, Elsevier Science, pp. 185-216.

Delpierre, N. and Doccon-Gibod, J. (1992) *Plant Cell Reports.*, **11:** 351-354.

Demeke, T., Adams, P.R. and Chibbar, R. (1992) *Theor. Appl. Genet.*, **84:** 990-994.

Deshpande, A.A. (1975) *M.Sc. Thesis, Indian Agricultural Research Institute*, New Delhi, India.

Dey, S.S., Bhatia, R., Parkash, C., Sharma, S., Dabral, M., Mishra, V., Bhardwaj, I., Sharma, K., Sharma, V.K. and Kumar, R. (2017) *Plant Breed.*, **136:** 400-409.

Dey, S.S., Bhatia, R., Sharma, K., Bhardwaj, I., Pandey, C.D., Parkash, C. and Kumar, R. (2019) *Euphytica*, **215:** https://doi.org/10.1007/s10681-019-2346-0.

Dey, S.S., Bhatia, R., Sharma, S.R., Sharma, K., Parkash, C. and Kumar, R. (2018) *Sci. Hortic.*, **232:** 231-239.

Dey, S.S., Sharma, K., Bhatia and Dey, R. (2015) *Euphytica*, **204:** 149-162.

Dey, S.S., Sharma, S.R., Bhatia, R., Kumar, P.R. and Parkash, C. (2011) *Indian J. Genet.*, **71:** 37-42.

Dey, S.S., Sharma, S.R., Bhatia, R., Parkash, C. and Barwal, R.N. (2011) *Euphytica*, **182:** 187-197.

Dey, S.S., Sharma, S.R., Bhatia, R., Parkash, C. and Sureja, A.K. (2013) *Sci. Hortic.*, **157:** 45-51

Dey, S.S., Singh, N., Bhatia, R., Parkash, C. and Chandel, C. (2014) *Euphytica*, **195:** 169-181.

Dhall, R. K., Sharma, S. R. and Mahajan, B.V.C. (2010) J. *Food Sci. Technol.*, **47:** 132-135.

Dhesi, N.S. and Padda, D.S. (1964) *Punjab Hort. J.*, **4:** 137-141.

Dhesi, N.S., Padda, D.S., Kumar, J.C. and Malik, B.S. (1964) *Punjab Hort. J.*, **4:** 176-179.

Dhiman, N.K., Nandal, T.R. and Rajender, S. (2005) *Crop Res.*, **30:** 73-76

Dias (2001) it is da Silva Dias, J.C. (2001) *Euphytica*, **119:** 389–394.

Dias, J.S. (1995) *Genet. Resour. Crop Evol.*, **42:** 360-369.

Dias, J.S., Lima, M.B., Song, K.M., Monteiro, A.A., Williams, P.H. and Osborn, T.C. (1991) *Euphytica*, **58:** 221-229.

Dickson, M.D. and Hunter, J.E. (1987) *HortScience*, **22:** 108-109.

Dickson, M.H. (1970) *J. Amer. Soc. Hortic. Sci.*, **95:** 13-14.

Dickson, M.H. and Wallace, D.H. (1986) In: Breeding vegetable crops (Ed. M.J. Bassett), pp. 395-432.

Ding, J., Lu, Q., Ouyang, Y., Mao, H., Zhang, P., Yao, J., Xu, C., Li, X., Xiao, J. and Zhang, Q. (2012) *Proc. Natl. Acad. Sci., USA*, **109:** 2654-2659.

Ding, L.C., Hu, C.Y., Yeh, K.W. and Wang, P.J. (1998) *Plant Cell Rep.*, **17:** 845-860.

Djatnika, I. (1991) *Cruciferae Newsletter*, No.14-15, P. 142.

Dodoens, R. (1953) *Cruydeboeck, Plantijn*, Antwerpen.

Domagala, F. and Kmiecik, W. (1985) *Acta Agraria et Silvestria, Agraria*, **24:** 119-129.

Dong, X., Ji, J. and Yang, L. (2019) *Mol. Genet. Genomics*, **294:** 1231-1239

Dore, C. (1986) *Agronomie*, **6:** 797-801.

Dragland, S. and Berentsen, E. (1991) *FAGINFO*, No.16, P. 12.

Dufault, R.J. and Waters, L.Jr. (1985) *HortScience*, **20:** 127-128.

Duijs, J.C., Voorrips, RE., Visser, D.L. and Clusters, J.B.M. (1992) *Euphytica*, **60:** 45-55.

Dunwell, J.M. and Davies, D.R. (1975) *Grower*, pp. 105-106.

Dutta, J.P. (1999) *J. Agric. Sci. Soc. N.E. India*, **12**:38-42.

Earle, E.D., Metz, T.D., Roush, R.T., Shelton, A.M., Dias, J.S., Crute, I. and Monteiro, A.A. (1996) *Acta Hortic.*, **407:** 161-168.

Eguchi, T. (1960) *Proc. Amer. Soc. Hort. Sci.*, **76:** 425-435.

Eimert, K., Schrodes, C. and Siegemund, F. (1992) *J. Plant Physiol.*, **140:** 37-40.

Elkner, K. and Babik (1999) *Veg. Crops Res. Bull.*, **50:** 93-105.

Elson, M.K., Morse, R.D., Wolf, D.D. and Vaughan, D.H. (1992) *Hort. Techol.*, **2:** 417-419.

Emura, K. (1978) *Bull. Saitama Hort. Exp. Sta.*, No. 7, pp. 41-47.

Everaarts, A.P. (1999) *PAV-Bulletin Vollegrondsgroenteteelt*, December, pp. 2-3.

Everaarts, A.P. (2000) *PAV Bulletin Vollegrondsgroenteteelt*, No. 3, pp. 24-26.

Everaarts, A.P. and Moel, C.P. De (1995) *Verslag-Proefstation voor de Akkerbouw en de Groenteteelt in de Vollegrond*, No. 198, P. 60.

Everaarts, AP., Moel, C.P. Dc, Willigen, P. De (1997) *PAV Bulletin Vollegrondsgroenteteelt*, February, 16-17.

Everaarts, AP. and Willigen, P. De (1999) *Neth. J. Agric. Sci.*, **47:** 123-133.

Fabijanski, S.F, Altosaar, I. and Arnison, P.G. (1991) *Plant Cell Tissue Cult.*, **26:** 203-212.

Fang, Z.Y., Liu, Y.M., Yang, L.M., Wang, X.W., Zhuang, M., Zhang, Y.Y. and Sun, P.T. (2004) *Sci. Agric. Sin.*, **37:** 717-723.

Fang, Z.Y., Sun, P.T., Liu, Y.M., Yang, L.M., Hou, A.F. and Wang, X.W. (1996) *China Vegetables*, No. I, 5-8.

Fang, Z.Y., Sun, PT., Liu, Y.M., Yang, L.M., Wang, X.W., Hou, A.F. and Bian, X.S. (1997) *Euphytica*, **97:** 265-268.

FanXue, Tong and Mattheis, J.P. (2000) *HortScience*, **35:** 885-887.

FAO Stat (2018) http://www.fao.org/faostat/en/#home

Fargier, E. and Manceau C. (2007) *Plant Pathol.*, **56:** 805-818.

Farinhó, M., Coelho, P. and Carlier, J. (2004) *Theor. Appl. Genet.*, **109:** 1392-1398.

Farnham, M.W. (1996) *J. Amer. Soc. Hort. Sci.*, **121:** 374-379.

Farnham, M.W. and Nelson, B.V. (1993) *HortScience*, **28:** 655-656.

Farnham, M.W., Wang, M. and Thomas, C.E. (2002) *Euphytica*, **128:** 405–407.

Feher, K.P. (1985) *Kerteszeti Egyetem Kozlemenyei*, **47:** 183-188.

Felezynski, K. (1995) In: *Materialy ogolnopolskiej Konfereniji naukowej Nauka Praktyee Ogrodniczej Zokazji XXV - lecia Wydzialu Ogrodniczego Akademii; Rolniczej w Lublinie*, pp. 625-628

Figdore, S.S., Ferrire, M.E., Slocum, M.K. and Williams, P.H. (1993) *Euphytica*, **69:** 33-34.

Fischer, I-I.H. and Nel, P.C. (1987) *Appl. Plant Sci.*, **1:** 28-33.

Fitt, B.D.L., Brun, H., Barbetti, M.J. and Rimmer, S.R. (2006) *Eur. J. Plant Pathol.*, **114:** 3-15.

Frankel, R. and Galun, E. (1977) *Theor. Appl. Genet.*, p. 2.

Frantz, J.M., Welbaum, G.E., Shen ZhengXing and Morse, R. (1998) *HortScience*, **33:** 976-979.

Fu, Y.Y., Jia, S.R. and Lin, Y. (1985) *Theor. Appl. Genet.*, **71:** 495-499.

Futane, D.N., Kulwal, L.V and Tayade, G.S. (1995) *PKV Res. J.*, **19:** 135-139.

Gallegley, M.E. (1956) *Phytopath.*, **46** : 467.

Gambhir, G., Kumar, Pankaj and Srivastava, D.K. (2017) *Biotechnology Reports*, **15:** 107-113.

Gamborg, O.L., Miller, R.A. and Ojima, K. (1968) *Exp. Cell. Res.*, **50:** 1-158.

Gariepy, Y., Raghavan, G.S.V., Plasse, R., Theriault, R. and Phan, C.T. (1985) *Acta Horticulturae*, **157:** 193-201.

Gates, R.R. (1953) *J. Genet.*, **51:** 363-372.

Gautam, K.C. (1977) *Veg. Sci.*, **4**: 49-52.

Gautam, K.C., Mani, V.S., Das, Bhagwan and Khokhar, N.S. (1976) *Indian Fmg.*, **26:** 11-13.

Geraci, A., Divaret, I., Raimondo, F.M. and Chèvre, A.M. (2001) *Plant Breed.*, **120:** 193-196.

Gerarde, J. (1597) *The Herballor Generall Historie of Plants*, London.

Gerszberg, A., Hnatuszko-Konka, K. and Kowalczyk, T. (2015) *In Vitro Cell Dev. Biol. Plant.* **51**:80–87.

Ghani, M.A., Zhang, L., Li, J., Liu, B. and Chen, L. (2013) *Biotechnology of Crucifers*, pp. 209-220.

Ghanti, P., Sounda, Gopeshwar, Jana, P.K. and Som, M.G. (1982) *Veg. Sci.*, **9:** 1-4.

Giles, W.F. (1941) *J. Royal Hort. Soc.*, **69:** 265-277.

Gill, H.S. (1993) In: Advances in Horticulture-Vegetable Crops, (Eds. K.L. Chadha and G. Kalloo), Vol. 5, Malhotra Publishing House, New Delhi, pp. 287-303.

Gill, H.S., Sharma, S.R. and Kapoor, K.S. (1987) *Indian Hort.*, **32:** 13-16.

Gill, H.S., thakur, P.C. and Bhullar, B.S. (1975) *Indian J. Hort.*, **32:** 94-97.

Giorno, F., Wolters-Arts, M., Mariana, C. and Rieu, I. (2013) *Plants*, **2:** 489–506.

Gómez-Campo, C. and Prakash, S. (1999) *Elsevier*, pp. 33-52.

Gopalakrishnan, C. and Artal, R.B. (2013) *Pest Mgt. Hortic. Syst.*, **19:** 220-224.

Gopalan, C., Ramasastri, B. and Balasubramanian, S. (2012) National Institute of Nutrition, ICMR, India.

Gorbki, S.F. and Armstrong, D.M. (1985) *Res. Cir., Ohio Agricultural Research and Development Centre, No.* 288, pp. 16-18.

Gorecka, K. and Krzyzanowska, D. (1997) *Acta Hortic.*, **447:** 335-337.

Gottschalk, W. and Kaul, M.L.H. (1974) *Nucleus*, **17:** 133-166.

Griesbach, E., Loptien, H., and Miersch, U. (2003) *J. Plant Dis. Protec.*, **110:** 461-475.

Griffith, M. and Carling, D.E. (1991) *Canadian J. Plant Sci.*, **71:** 579-585.

Gronborg, H. (1985) *Gartner Tidende*, **101:** 391.

Grout, B.W.W. (1988) In: *Biotech. in Agri. Forest.*, Vol. 6, (Ed. Bajaj, Y.P.S.), Springer-Verlag, Berlin, Germany, pp. 211-225.

Guan, PeiCong, Chen, YuDi and Yang, X. (1995) *Acta Hortic.*, **402:** 194-199.

Gueguen, M. (1995) *Biofutur.*, **148:** 43-45.

Guertal, E.A. and Santen, E. Van (1997) *J. Prod. Agric.*, **10:** 438-441.

Gupta, O.P., Singh, R.P. and Singh, R. (1967) *Labdev J. Sci. Technol.*, **5:** 157.

Gupta, P.R., Verma, A.K. and Mishra, P.C. (1985) *Veg. Sci.*, **12:** 49-54.

Gurusamy, C. (1999) *Seed Sci. Tech.*, **27:** 929-936.

Habegger, R. (1986) *Gartenbauwissenchaft*, **51:** 184-189.

Habegger, R. and Wiebe, H.J. (1986) *Gemuse*, **22** : 434-435.

Haga, T. (1938) *Japanese J. Genet.*, **8:** 227-283.

Han, F., Zhang, X. and Liu, X. (2017) *Genes*, **8:** 159.

Han, L.G. (1995) *Acta of Academy of Agricultural Science of the Democratic Peoples Republic*, Korea, pp. 63-68.

Han, W.T. and Park, Y.B. (1985) *J. Korean Soc. Hort. Sci.*, **26:** 231-238.

Hansen, M. and Javornik, B. (1994) *Proc. Internatl. Colloq. on Colloq. on Impact of Plant Biotech. on Agric., Rogla*, Slovenia, December, 5-7.

Harada, T., Inaba, A., Yakuwa, T. and Tamura, T. (1985) *HortScience*, **20:** 678-680.

Harberd, D.J. (1972) *Bot. J. Linn Soc.*, **65:** 1-28.

Harberd, D.J. (1975) *The Biology and Chemistry of Cruciferae*, London, pp. 47-68.

Harberd, D.J. (1976) In: *The biology and chemistry of the Cruciferae.* (Eds. J.G. Vaughan, A.J.MacLeod, M.G. Jones), *Academic, London*, pp. 47-68.

Harlan, J.R. (1975) Crops and man. American Society of Agronomny, Crop Science Society of America, Madison, WI.

Hartmann, H.D. and Zengerie, K.H. (1987) *Gemuse*, **23:** 58-61.

Haruta, T. (1962) *Res. Bull. Takii Plant Breed. Sta.*, **2:** 1-169.

Hasan, S.A. and Rost, W.M. (1985) *Mededdingen Van de Faculteit Landbouw-wetens chappen, Rijksuni versiteit genet.*, **50:** 389-398.

Hawthorn, L.R. and Pollard, L.H. (1954) *Vegetable and Flower Seed Production*, Blackiston Co. New York.

Hayashi, S. and Sakaue, O. (1996) *J. Soc. High Tech. Agric.*, **8:** 20-27.

Haynes, C., Everhart, E., Jauron, R., Nelson, D. and Lenahan, J. (2009) Cole Crops. Iowa State University Extension.http://www.extension.iastate.edu/Publications/PM1896.pdf.

He, TianXiu, He, ChenHui and Hardter, R. (1994) *Potash Review, Subject 9, J111 Suite*, No. 2, 1-5.

Hegi, G. (1919) *Illustrierte Flora Von Mittel-Europe*, Lehmanns Verlag, Munchen.

Heiningen, T.G. Van, Park, G.A., Hassan, S.A. and Lanteren, J.C. Van (1985) *Mededelingen Van de faculteit Landbouwwetenschappen, Rijksuniversiteit Genet.*, **50:** 379-388.

Helm, J. (1963) *Kulturpflanze*, **11:** 92-210.

Henslow, G. (1908) *J. Royal Hort. Soc.*, **34:** 15-23.

Henzi, M.X., Christey, M.C. and McNeil, D.L. (2000) *Plant Cell Rep.*, **19:** 994-999.

Henzi, M.X., McNeil, D.L., Christey, M.C. and Lill, R.E. (1999) *Australian J. Plant Physiol.*, **26:** 179-183.

Herklots, G.A.C. (1972) Cited by Prakash and Hinnata, *Opera Botanica*, **55:** 1-57.

Hgimori, M. and Nagaoka, M. (1991) *Physiologia Plantarum*, **82:** A 25.

Himme, M. Van, Stryckers, J. and Bulcke, R. (1984) *Mededelingen van het Centrum voor Onkruidonderzoek van de Rijksuniversiteit Gent.*, No. 40, pp. 90-97.

Hong, L.W., Wei, Z.X. and Lin, S.B. (1996) *China Vegetables*, **1**: 9-11.

Honma, S. (1986) *Cruciferae News Letter*, No. 11, P. 43.

Honma, S. and Cash, J.N. (1986) *HortScience*, **21:** 1076.

Hooker, J.D. (1862) In: Genera plantarum, (Eds. G. Bentham, J.D.Hooker), vol 1. Lovell Reed, London, pp. 57-102.

Hoque, M., Shea, D.J. and Asada, M. (2017) *Mol. Breed*, **37:** 109.

Horne, F.R. (1952) *Sci. Hort.*, **11:** 128-131.

Hosaka, K., Kianian, S.F., McGrath, J.M. and Quiros, C.F. (1990) *Genome.*, **33:** 131–142.

Hosen-Krauze, J. (1987) *Genetica Polonica*, **28:** 101-108.

Hoser-Krauze, J. (1979) *Genetica Polonica*, **20:** 341-367.

Hoser-Krauze, J., Lakowska-Ryk, E. and Antosik, J. (1984) *Cruciferae Newsletter*, **9:** 92–93.

Hoser-Krauze, J., Lakowska-Ryk, E. and Antosik, J. (1995) *J. Appl. Genet.*, **36**(1): 27–33.

Hosoki, T. and Kigo, T. (1994) *J. Jap. Soc. Hort. Sci.*, **63:** 589-592.

Hossain, M.M., Inden, H. and Asahira, T. (1989) *Mem. Coll. Agric., Kyoto Univ.*, **135:** 21– 30

Howe, T.K. and Waters (1983) *Proc. Fl. State Hort. Soc.*, **96:** 60-63.

Hoy, C.W. and Glenister, C.S. (1991) *Entomophaga*, **36:** 561-573.

Hoyt, G.D., Bonanno, A.R. and Parker, G.C. (1996) *Weed Tech.*, **10:** 50-54.

Hu, Q.Z., Wei, S. L., Song, M. X. and Hu, T.H. (1999) *Zhejiang Nongye Kexue*, **6:** 285-287.

Huang, Z., Peng, G., Liu, X., Deora, A., Falk, K.C., Gossen, B.D., McDonald, M.R. and Yu, F. (2017) *Front. Plant Sci.*, **8:** 1448.

Huang, Cong Li, Xu, RuiDe, He, DaJing and Xu, WeiDong (1994) *China Vegetables*, **6:** 27-29.

Humpherson-Jones, F.M., Dixon, G.R., Craig, M.A. and Ann, D.M. (1992) In: *Brighton Crop Protect. Con, Pests and Diseases*, Vol. 3, Farnhan, U.K.

Hyams, E. (1971) *Plants in the Service of Man, London*, pp. 33-81.

Ibukunoluwa, M.J.E. (2015) *Int. J. Recycl. Org. Waste Agric.*, **4:** 291–298.

Ilknur, Alibas and Nezihe, Kdksal (2015) *Food Sci. Technol.*, **35:** 45-50.

Inomata, N. (1978a) *Japanese J. Genet.*, **53:** 161-173.

Izumi, H., Watada, A.E. and Douglas, W. (1996) *J. Amer. Soc. Hort. Sci.*, **121:** 127-131.

Jaggi, R.C. and Dexit, S.P. (1995) *J. Indian Soc. Soil Sci.*, **43:** 229-231.

Jamwal, R.S. and Sharma, P.P. (1986) *Euphytica*, **35:** 941-943.

Jamwal, R.S., Thakur, D.R. and Jagmohan Kumar (1995) *Himachal J. Agric. Res.*, **21:** J 8-41.

Jensen, B.D., Massomo, S.M.S. and Swai, I.S. (2005) *Eur. J. Plant Pathol.* **113:** 297–308.

Jensen, B.D., Vaerbak, S., Munk, L. and Anderson, S.B. (1999) *Plant Breed.*, **118**: 549–554.

Ji, J.L., Yang, L.M., Fang, Z.Y., Zhuang, M., Zhang, Y.Y., Lv, H.H., Liu, Y.M. and Li, Z.S. (2017) *Theor. Appl. Genet.*, **130:** 1441-1451.

Jiang, M.C. (1981) *Chinese Cabbage Cultivation*260 p. Agr. Publishing Co., Beijing.

Jigme, J. N., Sutigulabud, P., Inthasan, J. and Sakhonwasee, S. (2015) *J. Org. Syst.* **10**: 9-14.

Jing, L.B., Min, S., Qiong, L.C. and Jia, W.X. (1998) *Acta Botanica Sinica*, **40:** 1028-1034.

Johnson, A.G. (1958) *Euphytica*, **10:** 351-356.

Joko, M. (2019) *Agric. Fin. Rev.*, **79:** 85-106.

Jones, H.A., and Mann, I.K. (1963) Onions and Their Allies. Leonard Hill Publishers, New York.

Josefsson, E. (1955) *Phytochem.* **6**: 1617-1627.

Joshi, R.P., Singh, R.D. and Adhikari, K.S. (1992) *Prog. Hort.*, **21:** 235-238.

Jossefson, A. (1955) *Hereditas*, **41:** 285-287.

Jourdan, P.S., Earle, E.D. and Mutschler, M.A. (1990) *Plant Cell Tissue Organ Cult.*, **21:** 227-236.

Jourdan, P.S., Earle, E.X. and Mutschler, M.A. (1985) *Cruciferae Newslett.*, **10:** 94-95.

Kahn, B.A. (1990) *HortScience*, **25:** 179-180.

Kakizaki, Y. (1930) *Jap. J. Bot.*, **5:** 133-208.

Kale, P.N., Chatterjee, S.S. and Swarup, V. (1979) *Veg. Sci.*, **6:** 11-20.

Kalia, P. and Saha, P. (2014) *Personal communication.*

Kalia, P. and Shakuntla. (2002) *Indian J. Hortic.*, **59**: 67-70.

Kalia, P., Aminedi, R., Golz, J., Russell, D., Choudhary, P., Rawat, S., Bansal, S., Tenazas, F., Singh, A., Panwar, A., Kalia, V., Behera, T. and Bhatacharya, R. (2020) *Acta Hortic.*, **1282**:237-246.

Kalloo, G. (1988) In: *Vegetable Breeding*, CRC Press Inc. Boca Raton, Florida, P. 15.

Kandoria, J.L., Singh, Gurdeep and Singh, Labh (1999) *Insect Environ.*, **5:** 137-138.

Kaniszewski, S. and Rumpel, J. (1998) *J. Veg. Crop Prod.*, **4:** 67-75.

Kaniszewski, S., Rumpel, J. and Dysko, J. (1999) *Veg. Crops Res. Bull.*, **50:** 21-30.

Kaul, M.L.H. (1988) Monographs on Theor Appl Genet 10. Springer-Verlag, Berlin.

Kaur, M. (2012) *M.Sc. Thesis*, Punjab Agricultural University, Ludhiana, India.

Kaur, M., Chadha, C., Kumar, N., Sehgal, N. and Kanwar, S. (2018) *J. Hill Agric.*, **9:** 35-38.

Kaur, N., Bhullar, M.S. and Gill, G. (2015) *Crop Prot.*, **74:** 18-23.

Kaur, P. and Mal, D. (2018) *J. Pharm. Phytochem.*, **7:** 2805-2807.

Kaur, R., Shivani, Saxena, B., Kanwar, H.S., Dohroo, N.P., Majeed, S. and Sharma, D.R. (2009) *Fruit Veg. Cereal Sci. Biotechnol.*, **3**: 12–15.

Kee, W.E. (1985) In: *Proc. 39th Annual Meet. of Northeastern Weed Sci. Soc.*, pp. 142-146.

Keller, W.A. and Armstrong, K.C. (1983) *Euphytica*, **32:** 151-159.

Kelliher, T., Starr, D., Wang, W., McCuiston, J., Zhong, H., Nuccio, M.L. and Martin, B. (2016) *Front. Plant Sci.*, **7:** 414.

Khaire, VA., Salunkhe, G.N., Lawande, K.E., Patil, J.D. and Kolhe, D.S. (1985) *South Indian Hort.*, **33:** 339-340.

Khattra, A.S., Singh, G., Thakur, J.C. and Singh, G. (1997) *Punjab Vegetable Grower*, **32**:15-19.

Khokhar, K.S. and Arora, P.N. (1968) *Indian Hort.*, **12:** 7-8.

Kieffer, M., Fuller, M.P. and Jelling, A.J. (1996) *Acta Hortic.*, **407:** 361-367.

Kiffner, W., Lanckow, J. and Mobius, G. (1983) *Gartenbau*, **30:** 43-45.

Kifuji, K., Hanzawa, H., Terasawa, Y., Ashutosh and Nishio, T. (2013) *Euphytica*, **190:** 289-295.

Kik, C. and Zaal, M.A.C.M. (1993) *Plant Cell Tissue Organ Cult.*, **35:** 107-114.

Kikumoto, T. (1981) *Chinese cabbage* (Eds. T.S. Talekar and T.D.P. Griggs), Proc.First Intl. Symp.AV RDC, Taiwan, pp.113-128.

Kim, B.S. (1986) *Korean J. Plant Path.*, **2:** 96-10 l.

King, G.J. (1992) *Phytoparasitica*, **20:** 93-97.

Klasse, H.J. (1996) *Acta Hortic.*, **407**: 403-409.

Klima, M., Vyvadilová, M. and Kučera, V. (2004) *Hort. Sci.* (PRAGUE), **31**(4):119–123.

Kohyama, T. (1986) In: *Diamond-back Moth Management*, Proc. First Int. Workshop, Tainan, Taiwan, March11-15, pp. 267-269.

Kolesnikov, V.A. and Ignatov, V.A. (1984) *Khimiya v Sel'skom Khozyaistve*, No.7, pp. 41-43.

Kolota, E. (1984) *Biuletyn Warzywniczy*, **27:** 203-235.

Kolota, E. and Biesiada, A. (1990) *Biuletyn Warzywniczy*, No. 36, pp. 107-127.

Kolota, E. and Dobromilska, R. (1991) *Roczniki Nauk Rolniczych. Seria A., Produkcja Roslinna*, **109:** 95-107.

Kop, E.P., Teakle, G.R. McClenaghan, E.R., Lynn, J.R. and King, G.J. (2003) *Pl. Sci.*,**164**: 803-808

Kototani, M. and Yamada, K. (1958) *Jappanese J. Breed.*, **8:** 66.

Kramchote, S., Srilaong, V., Wongs-Aree, C. and Kanlayanarat, S. (2012) *Int. Food Res. J.*, **19:** 759-763.

Kucera, V., Chytilova, V., Vyvadilova, M. and Klima, M. (2006) *Hort. Sci.* (Prague), **33:** 148–152.

Kumar, A., Kumar, V.A. and Kumar, J. (1993) *Plant Sci. Limer.*, **90:** 175-178.

Kumar, M., Choi, J.Y., Kumari, N., Pareek, A. and Kim S.R. (2015) *Front. Plant Sci.*, **6:** 688.

Kumar, S., Trivedi, H., Sah, R., Verma, A.K. and Yadav, A. (2018) *J. Pharma. Phytochem.*, pp. 769-772.

Kumar, V.A., Kumar, A. and Kumar, J. (1995) *Curr. Sci.*, **69:** 543-545.

Kumaran, N. Mohan (1971) *Ph.D. Thesis*, Indian Agricultural Research Institute, New Delhi, India.

Kundu, P.K. and Nandi, B. (1985) *Plant and Soil*, **83:** 357-362.

Kunicki, E., Capecka, E., Siwek, P. and Kalisz, A. (1999) *Folia Hortic.*, **11:** 69-79.

Kunicki, E., Cebula, S., Libik, A. and Siwek, P. (1996) *Acta Hortic.*, **407:** 337-383.

Kuo, C.G., Delafield, S., Chong, C. and Chiang, M.S. (1983) *Cruciferae Newslett.*, **8:** 56-57.

Labate, J.A., Robertson, L.D, and Baldo, A.M. (2006) *J. Amer. Soc. Hortic. Sci.*, **131**: 667-673.

Laila, R., Robin, A.H.K., Yang, K., Park, J.I., Suh, M.C., Kim, J. and Nou, I.S. (2017) *Front. Plant Sci.*, **7**: 1972.

Lal, D. (1938) *Nagpur Agri. Coll. Mag.*, **12:** 198.

Lal, M. (1977) *Ph.D. Thesis*, Indian Agricultural Research Institute, New Delhi.

Lancaster, D.M., Johnson, C.E. and Young, W.A. (1985) *Louisiana Agriculture*, **29:** 15-16.

Lancaster, P. (1943) *Indian.J. Hort.*, **1:** 129-131.

Larsen, K. (1974) *Incompatibility News Letter*, **4:** 20-22.

Lawande, K.E., Bhore, D.P., Kale, P.N. and Patil, J.D. (1986) *J. Maharashtra Agri. Univ.*, Pune, **11:** 192-196.

Lawson, I.I.M. (1984) In: *Comptes rendus du 7 eme colloque international sur l'ecologie, la biologie et la systematique des mauvaises herbes*, pp. 315-322.

Lech, P.and Frenz, F.W. (1977) In: *Jahresbericht, Fachhochschule Weihenstephan Munick*, 33-35.

Lee, S.H. (1982) *Sci. Hort.*, **17:** 201-209.

Leibee, G.L. and Savage, K.E. (1992) *Florida Entomologist*, **75:** 585-559.

Leibee, G.L., Jansson, R.K., Nuessly, G.S. and Taylor, J.L. (1995) *Florida Entomologist*, **78:** 82-96.

Leike, H. (1981) *Tag. Ber. Akad. Landwirtsch Wiss DDR*, **191:** 19-27.

Leike, H. and Wieczorrek, G. (1982) *Arch Zuchtungsforsch*, **12:** 375-383.

Lelu, M.A., Bollon, H. (1985) *C.R. Acad.Sci. Paris*, **300:** 71-76.

Leskovar, D.J. and Boales, A.K. (1996) *HortScience*, **31:** 405-409.

Letard, M. (1984) *Chinese cabbage*. Infos-CTIFL No.5, pp.7-11.

Lewis, D. (1956) *Brookhaven Sym. Biol.*, **9:** 89-100.

Lewis, D. (1977) *Proc. Royal Soc. Lond. J.J.*, **196:** 161-170.

Li, C.W. (1981) *Chinese cabbage* (Eds. T.S. Talekarand and T.D.P.Griggs), *Proc. First Intl. Syrup.* AVRDC, Taiwan, pp. 3-10.

Li, G.L., Zhong, Z.X. and Li, X. (1999) *Acta Agric. Shanghai*, **15:** 28-32.

Li, L., Lu, S., Cosman, K.M., Earle, E.D., Garvin, D.F., and O'Neill, J. (2006) *Phytochemistry*, **67:** 1177–1184

Li, S.L., Zhou, X.R., Li, Q.F. and Zhou, Z.J. (1998) *Acta Hortic. Shanghai*, **14:** 9-12.

Li, S.W., Sunnr Zhang, B.Z. and Chen, X.L. (1996) *Acta Agric. Boreali-Sinica*, **11:** 104-108.

Li, X. and Zhong, Z.X. (1996) *Acta Agric. Shanghai*, **12:** 23-27.

Liang Z., Chen K., Li T., Zhang Y., Wang Y., Zhao Q., *et al.*(2017) *Nat. Commun.*, **8**:14261.

Lichang, Gong, Zhou, Chang and Yang, Hong Yun (1994) *Acta Botanica Sinica*, **36:** 905-910.

Lichter, R (1982) *Z. Pflanzenphysiol.*, **105**: 427–434.

Lillo, C. and Shahin, E.A. (1986) *HortScience*, p. 21.

Lim, G.S., Sivapragasam, A. and Ruwaida, M. (1986) In: *Diamond-back Moth Management, Proc. First Int.Workshop*, Tainan, Taiwan, March 11-15, pp.185-194.

Lin, B.Y., Wei, W.L. and Gao, S. (1997) *J. Acad. Agric. Sci.*, **12:** 24-26.

Lin, C.Y. Chen, K.S. Chen, H.P. and Lee, H.I. (2019) *HortScience*, **54:** 1351-1356.

Lin, M.H., Hsu, Y.P. and Chang, S.K. (1995) *Chinese J. Agromet.*, **2:** 121-128.

Linnaeus, C. (1753) Species Plantarum II: 561. Stockholm.

Linsmaier, E.M. and Skoog, F. (1965) *Physiol. Plant*, **18:** 100-127.

Lipton, W.J. and Mackey, B.E. (1987) *J. Amer. Soc. Hort. Sci.*, **112:** 491-496.

Liu, C., Lin, C., Chen, J.J. and Tseng, M. (2007) *Plant Cell Reports*, **26:** 1733-1744.

Liu, D., Dong, X., Liu, Z., Tang, J., Zhuang, M., Zhang, Y., Lv, H., Liu, Y., Li, Z., Fang, Z., and Yang, L. (2018) *Front. Plant Sci.*, **9**: 309.

Liu, H.T. and Guan, P.C. (1998) *J. South China Agri. Univ.*, **19:** 82-86.

Liu, I.J., Song, M., Li, C.Q. and Wang, X.J. (1998) *Acta Botanica Sinica*, **40:** 1028-1034.

Liu, L. and Shelp, B.L. (1996) *Canadian J. Plant Sci.*, **76:** 367-377.

Liu, W.H., Zhang, X.W. and Song, B.L. (1996) *China Vegetables*, **1:** 9-11.

Liu, X., Gao, B., and Han, F. (2017) *BMC Genomics*, **18:** 230.

Liu, X., Yang, C. and Han, F. (2016) *Mol. Breed.*, **36:** 82.

Liu, Y.J., Cao, Y.P., Xia, J. and Wu, J.F. (1999) *Soils and Fertilizers* (*Beijing*), **4:** 26-29.

Liu, Y.M., Fang, Z.Y. and Sun, P.T. (1996) *China Vegetables*, **4:** 4-6.

Liu, Z., Fang, Z., Zhuang, M., Zhang, Y., Lv, H., Liu, Y., Li, Z., Sun, P., Tang, J., Liu, D., Zhang, Z. and Yang, L. (2017) *Front. Plant Sci.*, **8:** 239.

Lunca, L. and Andrei, P. (1985) *Productia Vegetala Horticultura*, **34:** 15-18.

Lund, S. and Kjaerskou, H. (1886) *Bot. Tydskr.*, **15:** 1-151.

Lundqvist, A. (1954) *Hereditas*, **40:** 278-294.

Lundqvist, A. (1975) *Proc. Royal Soc.Lond. B.*, **188:** 235-245.

Lundqvist, A., Osterbye, U., Larsen, K. and Lindehaursen (1973) *Hereditas*, **74:** 161-168.

Lv, H., Fang, Z. and Yang, L. (2014) *BMC Genomics*, **15:** 1094.

Ma, X., Feng, B. and Ma, H. (2012) *BMC Plant Biol.*, **12:** 23.

Macer, F. (1506) In: V.B. de Vitalibus, P. 96.

Mackay, C.R. (1977) *Heredity*, **38**: 201-208.

Maggioni, L., Von Bothmer, R., Poulsen, G. and Branca, F. (2010) *Econ. Bot.*, **64**: 109-123.

Maggioni, L. (2015) *Acta Universitatis Agriculturae Sueciae*, pp. 1652-1680.

Maharana, D. and Naik, C. (1967) *Indian J. Hort.*, **24:** 517-522.

Maheshwari, S., Tan, E.H., West, A., Franklin, F.C.H., Comai, L. and Chan, S.W.L. (2015) *PLoS Genet.*, **11**:e1004970.

Manasa, S., L. Mukunda, L., Syed, S., and Rajasekharam, T. (2017) *Int. J. Curr. Microbiol. App. Sci.*, **6**: 3143-3147.

Manea, A.I. (2017) *Int. J. Veg. Sci.* **23**: 567-574.

Mani, V.S., Khokhar, K.S. and Das, B. (1973) *Proc. 3rd Weed Control Seminar*, Hissar, p. 57.

Mann, K.S. and Sandhu, B.S. (1955a) *Indian J. Hort.*, **12:** 188-195.

Mann, K.S. and Sandhu, B.S. (1955b) *Indian J. Hort.*, **12:** 196-200.

Manton, I. (1932) *Ann. Bot.*, **46:** 509-556.

Margara, J. (1969) *Ann. Physiol. Veg.*, **11:** 95-112.

Marion, G.R., Ilnicki, R.D., Kupatt, C. and Vitolo, D.B. (1985) *In Proc. 3 91h Annual Meet.of Northeastern Weed Sci. Soc.*, pp.147-150.

Maroto, J.V., Lopez-Galarza, S., Pascual, B., Alagarda, J., SanBautista, A., Bardisi, A. and Rubio, M.C. (1996) *Acta Hortic.*, **407:** 333-338.

Massie, I.H., Astley, D., King, C.U. and Ingrouile, M. (2000) In: *Proc. III Internatl. Symp, onthe Taxonomy of Cultivated Plants*, (Eds. Andrews, S., Leslieand, A.C. and Alexander, C.), pp. 443-445.

Masters, M.T. (1869) *Vegetable Teratology*, R. Hardwicke, London, P. 534.

Mateeva, A., Andreev, R. and Vantscheva, V. (1997) *Mededelinge - Faculteit Landbowvkundige en Toegepaste Biologische Wetenschappen, Universiteit Genet*, **62:** 7-511.

Mather, K. (1955) *Evolution*, **9:** 52-61.

Mayr, E. (1986) *Osiris*, **2:** 135-176.

McRae, D.H. (1985) *Plant Breed. Rev.*, **3:** 169-191.

McCall, D. and Sorensen, L. (1995) *SP Rapport-Statens Planteavlsforsog*, No. 25, P. 16.

McGrath, J.M., Quiros, C.F., Harads, J.J. and Landry, B.S. (1990) *Mol. Gen. Genet.*, **223:** 198-204.

Mckay, A.G., Floyd, R.M. and Boyd, C.J. (1992) *Aus. J. Expt. Agric.*, **32:** 127-129.

McNaughton, I.H. (1995a). In: *Evolution of Crop Plants* (Eds. J. Smartt and N.W. Simmonds), Longman, London, pp. 62-68

McNaughton, I. H. (1995b) In: *Evolution of Crops Plants*, 2nd edn (Eds. J. Smartt and N.W. Simmonds), Longman, London, pp. 68–75.

Meer, Q.P.V. (1986) *Cruciferae NewsLetter*, **11:** 44-45.

Meer, Q.P.V. (1987) *Euphytica*, **36:** 927-931.

Mehraj, H., Akter, A., Miyaji, N., Miyazaki, J., Shea, D.J., Fujimoto, R. and Asadud Doullah, Md. (2020) *Plants*, **726,** doi:10.3390/plants9060726

Melo, P.E. De and Giordano, L. De B. (1999) *Horticultura Brassileira*, **17**: 172-173.

Mendes-Pereira, E., Balesdent, M.H., Brun, H., and Rouxel, T. (2003) *Mycol. Res.*, **107**: 1287– 304.

Meng, P.H., Wu, K.Y. and Li, Q.T. (1999) *Acta Hortic. Sinica*, **26:** 414.

Metz, T.D., Dixit, R. and Earle, E.D. (1995) *Plant Cell Rep.*, **15:** 287-292.

Metz, T.D., Roush, R.T., Tang, J.D., Shelton, A.M. and Earle, E.D. (1995) *Mol. Breed.*, **1:** 309-317.

Mikhov, M., Antonova, G. and V'Lchev, S. (1995) *Rasteniev dni Nauki*, **32:** 52-55.

Millar, J.C. (1929) *Agri. Exp. Sta. Bull.*, **137:** 15.

Millar, J.C. (1931) *Proc. Amer. Soc. Hort. Sci*, **28:** 341-342.

Mishra, H.P. (1989) *Indian J. Hort*, **46:** 230-233.

Mishra, H.P. (1992) *Indian J. Hort.*, **49:** 83-86.

Mishra, T.B., Singh, I.J., Ohekar and Singh, S.M. (1959) *B.V.J. Agri. Sci. Res.*, **1:** 17-24.

Mitchell, N.D. (1976) *Watsonia*, **11:** 97-103.

Mix Wagner, G. and Wang, H.M. (1990) *Landbauforschung Volkenrode*, **40:** 251-256.

Mizushima, U. (1950) *Tohuku J. Agri. Res.*, **1:** 1-14.

Moel, C.P.Dc, Booy, R. and Schroen, G. (1986) *Groenten en Fruit*, **41:** 64-65.

Mohanty, B.K. and Hossain, M.M. (1998) *Orissa J. Hort.*, **26:** 106-108.

Mora, A. and Earle, E.D. (1999) *Cruciferae Newslett.*, **21:** 57-58.

Morinaga, T. (1933) *Japanese J. Bot.*, **6**: 467-475.

Moss, N.A., Crute, I.R., Lucas, J.A. and Gordon, P.L. (1988) *Cruciferae Newsletter*, **13:** 114-116.

Mukherjee, P. (l 974) *Curr. Sci.*, **43:** 592-594.

Mullins, C.A. and Straw, R.A. (1990) *Tennessee Farm Home Sci.*, **156:** 11-15.

Murashige, T. and Skoog, F. (1962) *Physiol. Plant*, **15:** 473-497.

Murugiah, S., Vidyasagar, Chatterjee, S.S. and Swamp, V (1983) *15th Int. Cong. of Genet.* (India), *Abstr. No.* 446.

Murugiah, S., Vidyasagar, Chatterjee, S.S. and Swamp, V. (1987) *Indian J. Agric. Sci.*, **57:** 845-847.

Myers, C.H. and Fisher, W.J. (1944) *Mem. Cornell Agri. Exp. Sta.*, **259:** 29.

Nagaoka, T., Doullah, M.A.U., Matsumoto, S., Kawasaki, S., Ishikawa, T., Okazaki, K., *et al.* (2010) *Theor. Appl. Genet.*, **120:** 1335–1346.

Nandal, T.R., Dhiman, N.K. and Sharma, R. (2005) *Indian J. Weed Sci.*, **37:** 240-243.

Nandanwar, A., Gonge, V.S., Warade, A.D., Mohariya, A. and Jagdale, Y.L. (2006) *Int. J. Agric. Sci.*, **2:** 93-94.

Nankova, T. and Alipieva, M. (1995) *Rasteniev dni Nauki*, **32:** 237-238.

Narkiewicz Jodko, J. (1995) *Mededelingen -Faculteit Landbouwkundige en Toegepaste Biologische Wetenschappen, Universiteit Gent*, **60:** 941-943.

Nasrallah, J.B. (1995) *Acta Hortic.*, **392:** 61-67.

Nasrallah, J.B. and Nasrallah, M.E. (2014) *Biochem. Soc. Trans.*, **42:** 313-319.

Nasrallah, M.E. and Wallace, D.H. (1968) *Euphytica*, **7:** 170-178.

Natti, J.J. (1956) *Pl. Dis. Reptr.*, **40:** 591-595.

Natti, J.J. and Atkin, J.D. (1956) *Fm. Res.*, **22:** 6.

Natti, J.J. and Atkin, J.D. (1960) *Phytopath.*, **50:** 241.

Nawrocka, B. (1986) *Biuletyn Warzywniczy*, **26:** 159-169.

Ne Smith, D.S. (1998) *HortScience*, **33:** 36-38.

Nienhius, J., Slocum, M.K., Devos, D.A. and Moren, R. (1993) *J. Amer. Soc. Hort. Sci.*, **118:** 298-303.

Nieuwhof, M. (1959) *Zaadbelangen*, **13:** 88-90.

Nieuwhof, M. (1961) *Meded. Dir. Tuinb.*, **24:** 371-376.

Nieuwhof, M. (1963) *Euphytica*, **12:** 17-26.

Nieuwhof, M. (1966) *Acta. Agricultural Scandinavica Suppl.*, **16:** 69.

Nieuwhof, M. (1968) *Euphytica*, **17:** 265-273.

Nieuwhof, M. (1969) *Cole Crops - Botany, Cultivation and Utilization*, World Crops, Book, Leonard Hill, London. Nieuwhof, M. and Garretson, F. (1961) *Euphytica*, **10:** 301-306.

Ningawale, D.K., Singh, R., Bose, U.S., Gurjar, P.S., Sharma, A. and Gautam, U.S. (2016) *Indian J. Agric. Sci.*, **86**: 825-829.

Nishi, S. (1980) *Brassica Crops and wild allies* (Eds. Tsunoda, S., Hinata, K. and Gomez-Campo, C.P.), *Jpn. Sci. Soc. Press. Tokyo*, pp. 133-150.

Nishi, S. (1981) *Proc. 1st. Int. Symp.* AVRDC, Shanhua, Taiwan.

Nishi, S. and Hiraoka, T. (1957) *Bull. Natn. Inst. Agri. Sci.*, **E6:** 1-4.

Obreza, T.A. and Vavrina, C.S. (1993) *Communications in Soil Science and Plant Analysis*, **24:** 1465-1479.

Ockendon, D.J. (1984) *Ann. Appl. Biol.*, **105:** 285-291.

Ockendon, D.J. (1985) *Ann. Appl. Biol.*, **107:** 1-104.

Ockendon, D.J. (1986) In: Genetic *manipulationin breeding* (Eds. Horn, Jensen, J., Odenback, Shieder, O.), DeGruyter, Berlin, NewYork, pp. 265-272.

Ockendon, D.J. (1988) *Ann. Appl. Biol.*, **113:** 319-325.

Odland, M.L. (1962) *Proc. Amer. Soc. Hort. Sci.*, **80:** 387-400.

Ogura, H. (1968) *Memo. Fac. Agric. Kagoshima Univ.*, **6:** 39-78.

Oliveira, A.P. De, Costa, J.S. and Costa, C.C. (1999) *Horticultura Brasileira*, **17:** 164-166.

Onno, M. (1933) *Ost. Bot. Z.*, **82:** 309-334.

Oost, H., Brandenburg, W.A., Reuling, G.T.M. and Jarvis, C.E. (1987) *Taxon,* **36:** 625-634.

Opena, R.T., Kuo, C.G. and Yoon, J.Y. (1988) *Tech. Bull.*No.17, AVRDC, Taiwan.

Orlowski, M. (1985) *Tiuletyn Warzywniczy,* **27:** 47-49.

Orton, T.J. and Browers, M.A. (1985) *Theor. Appl. Genet.,* **69:** 637-643.

Osinga, K.J. (1997) *PAV Bulletin Vollegrondsgroenteteelt,* February 3-5.

Padilla, G., Cartea, M.E., Rodrý´guez, V.M. and Orda´s, A. (2005) *Euphytica,* **145:** 171-180.

Paksoy, M., Ellialtioglu, S., Sari, N. and Abak, K. (1995) *Turkish J. Bot.,* **19:** 281-286.

Palloix, A., Herve, Y., Knox, R.B. and Dumas, C. (1985) *Theor. Appl. Genet.,* **70:** 28-33.

Panda, S.K., Behera, U.K. and Nayak, S.K. (1999) *Insect Environ.,* **5:** 104-105.

Pandey, S.C. and Naik, G. (1985) *Cruciferae News Letter,* No.10, pp. 68-70.

Pandey, S.C. and Naik, G. (1986a) *Cruciferae News Letter,* No.11, pp. 36-37.

Pandey, S.C. and Naik, G. (1986b) *Cruciferae News Letter,* No.11, pp. 56-57.

Paranhos, L.G., Barrett, C.E., Zotarelli, L., Darnell, R., Migliaccio, K. and Borisova, T. (2016) *Sci. Hortic.,* **202:** 49–56.

Pareek, L.K. and Chandra, N. (1978) *Plant Sci. Lett.,* **11:** 311-316.

Park, J.I., Yang, S.H., Nou, I.S., Park, Y.S. and Lee, S.S. (2000) *J. Korean Soc. Hort. Sci.,* **41:** 21-26.

Parkash, C., Dey, S.S., Bhatia, R. and Dhiman, M.R. (2015) *Indian J. Hortic.,* **72:** 212-217.

Paschold, P.I. (1986) *Gartenbau,* **33:**113-114.

Paschold, P.J., Kleber, J. and Zengerle, K.H. (2000) *Gemiise (Miinchen),* **26:**11-12.

Pastuglia, M., Ruffio-Chable, V., Delorme, V., Gaudi, Y., Dumas, C. and Cock, J.M. (1997) *Plant Cell,* **9**: 2065-2076.

Patel, R.M. and Jyotishi, R.P. (1969) *Madras Agric. J.,* **56:** 181-184.

Pawar, R. and Barkule, S. (2017) *J. Appl. Nat. Sci.,* **9:** 520-525.

Pawlowski, W. (1990) *Folia Hortic.,* **2:** 53-61.

Pease, M.S. (1927) *J. Genet.,* **17:** 253-267.

Peck, N.H. and MacDonald, G.E. (1986) *J. Amer. Soc. Hort. Sci.,* **111:** 195-201.

Pelletier, G., Ferault, M., Lancelin, D., Boulidard, L., Dore, C., Bonhomme, S., Grelon, M. and Budar, F. (1995) *Acta Hortic.,* **392:** 11-17.

Pelletier, G., Primard, D., Vedel, F., Chetrit, P., Remy, R., Rouselle and Renaud, M. (1983) *Mol. Gen. Genet.,* **191:** 151-152.

Peter, C. and Sundararajan, R. (1991) *Indian J. Plant Protect.,* **19:** 161-166.

Petrova, S. and Antonova, G. (1996) *Cruciferae Newsletter,* **18:** 31.

Phelan, J.R. and Vanghan, J.G. (1976) *Biochem. Syst. Ecol.,* **4:** 173-178.

Phippen, C. and Ockendon, D.J. (1990) *Theor. Appl. Genet.,* **79:** 33-38.

Phippen, W.I., Kresovich, S., Candelas, F.G. and McFerson, J.R. (1997) *Theor. Appl. Genet.,* **94:** 227-234.

Ping, Z.G. and Chun, C.S. (1997) *Acta Hortic. Sinica*, **24:** 43-47.

Ping, Z.G. and Chun, C.S. (1998) *J. Nanjing Agri. Univ.*, **21:** 31-35.

Pink, D.A.C., Carter, P.J. and Walkey, D.G.A. (1986a) *Annual Rep. 1985, Nat. Veg. Res. Sta.*, Wellesbourne, U.K.

Pink, D.A.C., Sutherland, R.A. and Walkey, D.G.A. (l 986b) *Ann. Appl. Biol.*, **109:** 199-208.

Politanskaya, V.V. (1985) In: *Sovershenstvovanie Tekhnologicheskogo Vyrashchivaniya Ovosluchnykh Kul'tur V Othrytomi ZashchischennomGrunte Severo-ZapadnoiZony*, pp. 39-42.Porter, I.J., Merriman, P.R. and Keane, P.J. (1991) *Aus. J Expt. Agric.*, **31:** 943-851.

Prabhakar, D.S. and Srinivas, K. (1993) *Prog. Hort.*, **22:** 112-116.

Prabucki, J., Chuda-Mickiewicz, B. and Woyke, H.W. (1987) *Biuletyn Warzywniczy*, **30:** 71-91.

Pradhan, A. K., Prakash, S., Mukhopadhyay, A. and Pental, D. (1992) *Theor. Appl. Genet.*, **85**: 331-340.

Prakash, S. and Hinata, K. (1980) *Oper Bot.*, **55:** 1-57.

Prakash, S., Bhat, S.R. and Fu, T.D. (2009) In: Biology and Breeding of Crucifers (Ed. S.K. Gupta),CRC Press, United States, pp.113-135.

Prakash, S., Wu, X.M. and Bhat, S.R. (2011) *Plant Breed. Rev.*, **35:** 19-84.

Prakash, S.C., Thakor, C.M., Morwal, B.R., and Jat, M.L. (2014) *Hort. Flora Res. Spect.*, **3:** 197-198.

Prantl, K. (1891) Cruciferae. In: Nat€urlichen Pflanzenfamilien (Eds. A. Engler, K.D. Prantl), Wilhelm Englmann, Leipzig, pp. 145-208.

Prasad, K.K., Singh, K.P., Chowdhary, B.M. and Singh, S. (2000) *J. Res. Birsa Agril. Univ.*, **12:** 117-118.

Primo-Millo, W. and Harada, H. (1975) *C.R. Acad. Sci., Paris Ser D.*, **280:** 2847.

Puddephat, L.J., Thompson, N., Robinson, H.T., Sandhu, P. and Henderson, J. (1999) *J. Hort. Sci. Biotechnol.*, **74**: 714-720.

Putter, I.I.De (1999) *PAV-Bulletin Vollegrondsgroenteteelts*, October, 15-17.

Quazi, H.M. (1978) *Ann. Bot.*, **42:** 473-475.

Radhakrishna, V. and Korla, B.N. (1995) *Hort. J.*, **8**: 61-65.

Rahman, F. (1965) *Indian Agric.*, **9**: 59-71.

Rahman, H.U., Hadley, P. and Pearson, S. (2007) *Plant Growth Regul.*, **52:** 61-72.

Rahman, M., Michalak, de., and Jiménez, M. (2016) *J. Plant Sci Mol. Breed.*, **5**, doi. 10.7243/ 2050-2389-5-1.

Rahman, M.A., Imran, M., Ikrum, M., Rahman, M.H. and Rabbani, M.G. (2016) *Environ. Sci. Nat. Resour.*, **9:** 143-150.

Ramsay, L.D., Jennings, D.E., Bohuon, E.J.R., Arthur, A.E., Lydiate, D.J., Kearsey, M.J. and Marshall, D.F. (1996) *Genome*, **39:** 558-567.

Randhawa, K.S. and Bail, A.S. (1976) *Indian J. Hort.*, **33**: 83-91.

Rao, L.N. (1938) *Curr. Sci.*, **7**: 237-238.

Ratzka, A., Vogel, H., Olds, T.M., and Kroymann, J. (2002) *Proc. Natl. Acad. Sci. USA*. **99**: 11223-112.

Ravi, M. and Chan, S.W.L. (2010) *Nature*, **464:** 615-8.

Razaullah, K., Sher, A., Salimullah, K., Fayaz, A., Meer, Z. and Bashir, A.K. (2002) *Asian J. Plant Sci.*, **1:** 548-549.

Redha, A. and Suleman, P. (2013) *Plant Cell Tiss. Org. Cult.*, **105**: 345-353.

Ren, J., Liu, Z. and Du, J. (2019) *Mol Breed.*, **39:** 40.

Ren, J., Liu, Z., Niu, R. and Feng, H. (2015) *Plant Breed.*, **134:** 494-500.

Reycke, L. De and Vanaerde, H. (1997) *Proeftuinnieuws*, **7**: 16-17.

Rhee, S.J., Seo, M., Jang, Y.J., Cho, S. and Lee, G.P. (2015) *BMC Genomics*, **16:** 914.

Richharia, R.H. (1937) *J. Genet.*, **34**: 19-44.

Richharia, R.H. (1937) *J. Genet.*, **34**: 45-55.

Ridge, S., Brown, P.H., Hecht, V., Driessen, R.G. and Weller, J. L. (2015) *J. Exp. Bot.*, **66**: 125-135.

Rijbroek, P.V. (1986) *Groenten en Fruit*, **41** : 49-51.

Rijbroek, P.V. (1988) *Groenten en Fruit*, **43** : 48-49.

Robak, J. (1994) *Acta Hortic.*, **371**: 223-226.

Robbelen, G. (1960) *Chromosoma*, **11**: 205-228.

Rooster, L. De (1998) *Proeftuinnieuws*, **8**: 24.

Rooster, L. De and Spiessens, K. (1999) *Proeftuinnieuws*, **9**: 31.

Rooster, L. De and Vanparijs, L. (1997) *Proeftuinnieuws*, **7**: 52.

Ross, Walter (1983) *Reader's Digest*, October, pp. 111-119.

Roulund, N., Hansted, L., Andersen, S.I. and Farestveit, B. (1990) *Euphytica*, **49:** 237-242.

Roy, H.K. (1981) *Veg. Sci.*, **8:** 75-78.

Roy, R.S. and Choudhury, B. (1950) *Indian J. Hort.*, **7:** 1-6.

Rubino, R. and Palma, E.De. (1986) *Irrigazione e Drenaggio*, **36**: 5-8.

Rudolf, K., Bohanec, B. and Hansen, N. (1999) *Plant Breed.*, **118:** 237-241.

Rundfeldt, H. (1960) *Z. Pflanzenz*, **44**: 30-62.

Ryschaka, U., Schumann, G., Klocke, E., Schoize, P. and Kramer, R. (1999) In: *Plant Biotechnology and in vitro Biotechnology in the 21st Century*, Kluweracademic Publishers, pp. 205-208.

Sabnis, V.R. (1950) *Poona Agri. Coll. Mag.*, **12**: 7.

Sagave, B.N. and Badhe, N.V. (1976) *Nagpur Agri. Coll. Mag.*, **49:** 11-16.

Saha, P. (2014) *Personal communication.*

Saha, P. (2015) Krishisewa. https://www.krishisewa.com/articles/production-technology/631-growing-cauliflower-for-higher-nutrition.html.

Saha, P. (2015) *Personal communication.*

Saha, P. (2016) *Personal communication.*

Saha, P. (2020) *Personal communication.*

Saha, P., Kalia, P., Sharma, M. and Singh, D. (2016) *Euphytica*, **207**: 35–48.

Saha, P., Kalia, P., Sonah, H. and Sharma, T.R. (2014) *Plant Breed.*, **133**: 268-274.

Saha, U.K., Hasnat, M.A., Haider, J., Saha, R.R. and Kawai, S. (1998) *Japanese J. Trop. Agric.*, **42**: 71-77.

Sahu, T.K., Dixit, A., Sharma, P.K. and Siwna, Y. (2019) *Int. J. Chem. Stud.*, **7**: 01-03.

Saimbhi, M.S., Singh, K. and Padda, D. S. (1969) *Punjab Hort. J.*, **9**: 198-203.

Saini, S.S. and Rastogi, K.B. (1976) *Haryana J. Hort. Sci.*, **5**: 233-235.

Sakamoto, A., Valverde, R., Alia, Chen, T.H.H. and Murata, N. (2000) *Plant J.*, **22**: 449– 453.

Sampson. D.R. (1957) *Genetics*, **42:** 253-263.

Sande, J. VanDe (1986) *Groenten en Fruit*, **41**: 28-29.

Sande, J. VanDe and Jeurissen, J. (1986) *Groenten en Fruit*, **41:** 36-37.

Santa, M. (1985) *Kerteszet es Syoleszet*, **34**: 4.

Santos, J.B. Dos., Nienhuis, J., Skroch, P.,Tivang, J. and Slocum, M.K. (1994) *Theor. Appl. Genet.*, **87:** 909-915.

Sarashima, M. and Matsuzawa, Y. (1986) *Bull. Cell. Agri.*, Utsunomiya Univ., **13**: 1-10.

Sastrodihardjo, S. (1986) In: *Diamond-back Moth Management, Proc. Frist Int. Work*shop, Tainan, Taiwan,11-15 March,1985, pp. 35-41.

Sastrosiswojo, S. (1996) In: *Biological control introductions- opportunities forimproved cropproduction. Proceeding of an International Symposium Brighton*, U.K., 18 November, Farnham, U.K.

Sastrosiswojo, S. and Sastrodihardjo, S. (1986) In: *Diamond-back Moth Management. Proc. First Int.Workshop*, Tainan, Taiwan, 11-15 March, 1985, pp.185-194.

Savita, B., Luchon, S., Bonti, G. and Sonbeer, C. (2017) *Int. J. Agric. Sci.*, **9:** 4226-4228.

Schenck, H. R. and Röbbelen, G. (1982) *Z. Pflanzenzüchtg.*, **89**: 278–288.

Schery, R.W. (1972) *Plantsfor Man*, pp. 508-509.

Schultz, T., Gabrielson, R.L. and Olson, S. (1986) *Plant Disease*, **70**: 1027-1030.

Schulz, O.E. (1919) Cruciferae–Brassicaceae. Part I: Brassicinae and Raphaninae. In: Das Pflanzenreich (Ed. A. Engler), vol 68–70. Wilhelm Engelmann, Leipzig, pp. 1-290.

Schulz, O.E. (1936) Cruciferae. In: (Eds. A. Engler, P. Prantl), Die Nat€urlichen Pflanzenfamilien, Wilhelm Engelmann, Leipzig, pp. 227–658.

Scutt, C.P., Gates, P.J., Gatehouse, J.A., Boulter, D. and Croy, R.R.D. (1990) *Mol. Gen. Genet.*, **220:** 409-413.

Sehgal, N. and Singh, S. (2018) *3 Biotech*, **8:** 347.

Sen, S. (1957) *Bull. Bot. Soc. Bengal*, **9**: 17.

Seshadri, V.S. and Chatterjee, S.S. (1983) *30th Year Commemoration Issue, South Indian Hort.*, pp. 52-58.

Seyama, N. and Takai, T. (1982) *Bull. Veg. Ornament. Crops Res. Stat.*, No. 4, pp. 27-46.

Shah, A., Srivastava, K.K., Roy, A.J. and Bora, S.S. (1985) *Progress. Hort.*, **17**: 72-74.

Shanna, D.K., Chaudhary, D.R. and Narayan, R. (1995) *South Indian Hort.*, **43**: 59-61.

Shanna, R.P. and Prashar, K.S. (1982) *Veg. Sci.*, **9**: 75-81.

Shanna, S.L., Kaith, D.S. and Khan, M.A. (1986) *Indian Phytopath.*, **36**: 601-603.

Sharma, B.B., Kalia, P., Singh, D. and Sharma, T.R. (2017) *Front. Plant Sci.*, **8:** 1255.

Sharma, B.R., Swarup, V. and Chatterjee, S.S. (1972) *Canadian J. Gent. Cytol.*, **14:** 363-370.

Sharma, D. (1975) MSc. Thesis, Submitted to Faculty of Agriculture University of Udaipur, Rajasthan, India.

Sharma, J.N. and Singh, M. (1963) *Indian J. Agron.*, **7:** 285-289.

Sharma, S.R. and Gill, H.S. (1986) *Indian Hort.*, **31**: 17-18.

Sharma, S.R., Kapoor, K.S and Gill, H.S. (1995) *Indian J. Agric. Sci.*, **65**: 916-918.

Sharma, S.S., Kalra, V.K. and Mrig, K.K. (2000) *Ann. Agri. Bio. Res.*, **5:** 67-70.

Sharma, Y., Sharma, S.N. (2005) *Agric. Rev.*, **26:** 114-123.

Shattuck, V.I., Shelp, B.I., Laughton, A. and Baker, R. (1986) *Canadian J. Plant Sci.*, **66**: 683-688.

Shelp, B.J. and Liu, L. (1992) *Plant and Soil*, **140:** 151-155.

Shiau, F.L., Chung, W.C., Huang, J.W. and Huang, H.C. (1999) *Canadian J. Plant Path.*, **21**: 368-374.

Shou, Z.M., Hong, S.J. and Qing, Z.S. (1999) *China Vegetables*, **4**: 22-23.

ShuFen, W., Yi, Z. and Yan, S.Z. (1996) *Acta Hortic. Sinica*, **23**: 37-44.

Shukla, K.N., Singh, P.K. and Singh, K.K. (2001) Plasticulture Development Centre, GBPUAT, Pantnagar.

Sidki, Sadik (1962) *Amer. J. Bot.*, **49**: 290-297.

Sidki, Sadik (1967) *Proc. Amer. Soc. Hort. Sci.*, **90:** 252-259.

Sigareva, M.A. and Earle, E.D. (1997) *Theor. Appl. Genet.*, **94:** 213-220.

Sikka, S.M. (1940) *J. Genet.*, **40:** 441-509.

Silva Dias, J.C. (1999) *Euphytica*, **108**: 65-69.

SilvaJunior, A.A. and Miura, L. (1986) *Pesquisa em Andamento*, EMPASC No. 54, P. 4.

Singh, A.K., Khan, A.R. and Singh, A. (1999) *Crop Res. (Hissar)*, **17**: 104-106.

Singh, A.K., Singh, J. and Saini, S.S. (1978) *Hort. Res.*, **18**: 1-6.

Singh, B.K. (2015) TM No. 66, Indian Council ofAgricultural Research–Indian Institute of Vegetable Research, Varanasi, Uttar Pradesh, pp. 101-115.

Singh, B.K. and Devi, J. (2015) IIVR Training Manual No. 59, Indian Institute of Vegetable Research, Varanasi, pp 102–119.

Singh, B.K., Pathak, K., Sarma, A. and Thapa, M. (2010) *Indian J. Hill Farm.*, **23**: 1-5.

Singh, B.K., Singh, B. and Singh, P.M. (2018a) *Int. J. Veg. Sci.*, **24:** 58-84.

Singh, D.P., Swarup, V. and Chatterjee, S.S. (1975) *Veg. Sci.*, **2**: 1-7.

Singh, D.P., Swarup, V. and Chatterjee, S.S. (1976) *Veg. Sci.*, **4**: 47-53.

Singh, Gurmail and Thakur, J.C. (1997) *Punjab Veg. Growers*, **32:** 15-19.

Singh, I.J. and Singh, A. (1970) *J. Agri. Sci. Res.*,**12:** 93-99.

Singh, K.P. and Rajput, C.B.S. (1976) *Expt. Agr.*, **12:** 195-199.

Singh, R., Chatterjee, S.S., Swarup, V., Trivedi, B.M. and Sen, B. (1991) *Indian Hort.*, **36**: 33-34.

Singh, R., Jain, P.C. and Singh, S.M. (1960) *Indian J. Hort.*, **17**: 69-73.

Singh, R.D., Tewari, S.N., Singh, B.B. and Seth, J.N. (1976) *Progessive Hort.*, **8:** 29-34.

Singh, R.P. (1984) *South Indian Hort.*, **32**: 236-238.

Singh, R.V. and Naik, L.B. (1993) *Progress. Hort.*, **22:** 160-163.

Singh, S., Bhatia, R., Kumar, R., Sharma, K., Dash, S. and Dey, S.S. (2018b) *Euphytica*, **214:** 207.

Singh, S., Dey, S.S., Bhatia, R., Kumar, R., Sharma, K. and Behera, T.K. (2019) *PLoS one*, **14**: e0210772.

Singh, S., Kalia, P., Meena, R.K., Mangal, M., Islam, S., Saha, S. and Tomar, B.S. (2020) *Front. Plant Sci.*, https://doi.org/10.3389/fpls.2019.01766

Singh, S., Sharma, S.R., Kalia, P., Deshmukh, R., Katara, J., Sharma, P. and Sharma, T.R. (2015) *Indian J. Biotechnol.*, **14:** 455-460.

Singh, S., Sharma, S.R., Kalia, P., Deshmukh, R., Kumar, V., Sharma, P. and Sharma, T.R. (2012). *J. Hortic. Sci. Biotechnol.*, **87**: 137-143.

Singh, S.B. and Mishra, R.S. (1975) *Progess. Hort.*, **7:** 65-71.

Singh, S.D. (1978) *Agron. J.*, **70**: 951-964.

Singh, S.N. and Singh, S.M. (1963) *Indian J. Hort.*, **20**: 104-110.

Singh, V.P., Sanyal, N.B. and Ram Kumar (1978) *Indian Hort.*, **35:** 138-141.

Sinha, A.K. and Rai, B.K. (1984) *Indian J. Entom.*, **46**: 385-388.

Sinha, U.N. and Bains, S.S. (1967) *Indian J. Hort.*, pp. 652-658.

Sinskaia, E.N. (1928) *Bull. Appl. Bot. Genet. Pl. Breed.*, **19:** 1-48.

Siviero, P. and Barbuzzi, G. (1998) *Informatore Agrario Supplemento*, **54**: 5-6.

Skucinka, B., Miszke, W., Kruczkowska, H. and Pawlowska, H. (1988) *Acta Agraria et Silvestria, Agraria*, **27**: 101-118.

Skucinska, B., Miszke, W., Kruczkowska, H. and Pawlowska, H. (1988) https://agris.fao.org/agris-search/search.do?recordID=PL9000294

Slocum, M.K., Figdore, S.S., Kennard, W.C., Suzuki, J.Y. and Osborn, T.C. (1990) *Theor. Appl. Genet.*, **80:** 57-64.

Smith, L.B. and King, G.J. (2000) *Mol. Breed.*, **6:** 603-613.

Snogerup, S. (1980) *Brassica Crops and Wild Allies* (Eds. S.Tsunoda, K. Himda and C. Gomez-Campo), Japan Scientific Societies Press, Tokyo.

Snogerup, S. (1983) *Personal communication.*

Snogerup, S. and Gustafsson, M. (1982) *Personal communication.*

Snogerup, S., Gustafsson, M. and Bothmer, R. von (1990) *Willdenowia.*, **19:** 271-365.

Som, M.G., Jana, P.K. and Pal, S.S. (1976) *Veg. Sci.*, **3**: 131-137.

Som, M.G.and Jana, P.K. (1975) *Veg. Sci.*, **2:** 35-36.

Song, K., Osborn, T.C. and Williams, P.H. (1990) *Theor. Appl. Genet.*, **79**: 497-506.

Srivastava, D.C. Verma, A.N., Mishra, H.R. and Sharma, R.K. (1970) *Pl. Sci.*, **2**: 132-134.

Srivastava, J.N., Singh, A.K., Kumar, M., Kumai, A. and Kumar, R. (2018) *J. Pharmacog. Phytochem.*, **SP1:** 3000-3003

Srivastava, P. and Chauhan, H.S. (1999) *J. Appl. Hort.* (Lucknow), **1:** 137-138.

Standberg, J.O. and Williams, P.H. (1967) *Phytopath.*, **57**: 330.

Stansell, Z., Farnham, M., Bjorkman, T. (2019) *Front. Plant Sci.*, https://doi.org/10.3389/fpls.2019.01104

Stein, J.C., Howlett, B., Boyes, D.C., Nasrallah, M.E. and Nasrallah, J.B. (1991) *Proc. Nat. Acad. Sci. USA*, **88:** 8816-8820.

Stipic, M. and Campion, B. (1997) *Plant Breed.*, **116:** 153-157.

Stoner, K.A., Dickson, M.H. and Shelton, A.M. (1986) *Cruciferae News Letter*, No.11, pp.102.

Storme, N.D. and Geelen, D. (2014) *Plant Cell Environ.*, **37:** 1-18.

Su, Y., Liu, Y., Li, Z., Fang, Z., Yang, L. and Zhuang, M. (2015) *PLoS One*, **10**: e0138073.

Sun, U.G. (1946) *Bull. Torrey Bot. Club.*, **73:** 244-281.

Sunarjono, H. (1980) *Indonesia Agric. Res. and Dev. J.*, **2:** 92-94.

Sureja, A.K. (2014) In: *Training manual on teaching of post-graduate courses in Horticulture* (*Vegetable Science*). (Eds. A.D. Munshi, A.K. Sureja, S.K. Singh and K.V. Prasad) Post Graduate School, IARI, New Delhi, India.

Suseela, P. and Rangaswami, M.V. (2011) *Karnataka J. Agric. Sci.*, **24**: 706-708.

Suzuki, G., Kai, N., Hirose, T., Fukui, K., Nishio, T., Takayama, S., Isogai, A., Watanabe, M. and Hinata, K. (1999) *Genetics*, **153:** 391-400.

Swamp, V. and Pal, A. B. (1966) *Indian J. Genet.*, **26**: 269-281.

Swaroop, V. and Sharma, B. R. (1965) *Indian J. Genet.*, **25**: 56 – 64.

Swarup, V. (1983) *Contribution in the* 7th *Workshop of All India coordinated Veg. Improvement Project*, held at Hissar, May 1983.

Swarup, V. and Chatterjee, S.S. (1972) *Econ. Bot.*, **26:** 381-393.

Swarup, V. and Pal, A. B. (1966) *Indian J. Genet.*, **26** : 269-281

Swarup, V., Gill, H.S. and Singh, D. (1963) *Indian J. Genet.*, **23** : 90-100.

Taili, H., LeRong, F. and Hua, Y.C. (1997) *China Vegetables*, **2**: 30-31.

Takahata, Y. and Keller, W.A. (1991) *Plant Sci., Limerick*, **74**: 235-242.

Takeshita, M., Kato, M. and Tokumasu, S. (1980) *Japanese J. Genet.*, **55**: 373-387.

Talekar, N.S. (1992) In: *Integrated Pest Management in the Asia - Pacific region* (Eds. P.A.C. Ooi, G.S. Lim, T.H. Ho, P.L. Manalo and J.K. Waage), Wallingford, U.K., CAB International, pp. 37-49.

Tan, C., Cui, C., Xiang, Y., Ge, X. and Zaiyun, L. (2017) *Theor. Appl. Genet.*, **130:** 2491.

Tan, S.C., Berston, J. and Hynes, Y. (1992) *New Zealand J. Crop Hort. Sci.*, **20**: 167-172.

Tanaka, N. and Niikura, S. (2006) *Breed. Sci.* **53**: 325-333.

Tanaka, S., Kochi, S.I., Kunita, H., Ito, S.I. and Kameya-Iwaki, M. (1999) *European J. Plant Path.*, **105:** 577-584.

Tandon, S.L. (1951) *Sci. Cult.*, **16**: 483.

Tao, X.Q., Zhao, H., Cheng, F. and Hu, H. (1986) *Acta Hortic., Sinica,* **13:** 42-48.

Tatout, C., Warwick, S., Lenoir, A. and Deragon, J.M. (1999) *Mol. Biol. Evol.*, **16**: 1614–1621.

Taylor, J.D., J. Conway, S.J. Roberts, D. Astley and Vicente J.G. (2002) *Phytopathology,* **92**: 105–111.

Tesi, R., Nencini, A., Botrini, L. and Graifenberg, A. (1996) *Rivista di Agronomia,* **30:** 174-180.

Tewari, R.N., Chatterjee, S.S. and Swarup, V. (1 979) *Veg. Sci.*, **6**: 27-36.

Tewari, R.N., Gill, H.S. and Singh, R. (1977) *Progess. Hort.*, **8:** 35-38.

Thakur, P.C. and Gill, H.S. (1976) *Indian J. Hort.*, **33**: 262-265.

Thomas, D.S., Turner, D.W. and Tan, S.C. (1992) *New Zealand J. Crop Hortic. Sci.*, **20**: 147 -151.

Thompson, K.F and Taylor, J.P. (1966) *Heredity*, **21**: 637-648.

Thompson, K.F. (1957) *J. Genet.*, **55**: 45-60.

Thompson, K.F. (1976) *Evolutional Crop Plants* (Ed. N.W. Simmonds), Longman, London and New York, pp. 49-52.

Thompson, K.F. and Howard, I.I.W. (1959) *J. Genet.*, **56**:325-340.

Thompson, K.F. and Taylor, J.P. (1971) *Heredity*, **27:** 459-471.

Thormann, C.E., Ferreira, M.E., Camargo, L.E.A., Tivang, J.G. and Osborn, T.C. (1994) *Theor. Appl. Genet.*, **88**: 973-980.

Titze, W. and Alpers, S. (1987) *Gemuse*, **23:** 54-56.

Toivonen, P.M.A. (1992) *HortScience*, **27:** 1014-1015.

Tonguc, M. and Griffiths, P.D. (2004) *Euphytica*, **137**: 193-201.

U.N. (1935) *Japanese J. Genet.*, **9:** 159-160.

Ubugunov, L.L. (1984) *Soviet Agric. Sci.*, No.9, pp. 72-75.

Umale, S.B. (1966) *Diamond Jubilee Special Research Number, NagpurAgril. Coll. Mag.*, I.

Upadhyaya, D. and Kumar, J.C. (1983) *J. Res., Punjab Agric. Univ.*, **20:** 448-453.

Vaidya, S.J. and Patel, A.V. (1964-65) *Nagpur Agri. Coll. Mag.*, **38**: 30-43.

Vaidya, S.J. and Patel, A.V. (1965) *Indian J. Agron.*, **10**: 420-422.

Vanaerde, H., Reycke, L. De and Janssens, M. (1997) *Proeftuinnieuws*, **7**: 34-35.

Vanparys, L. (1986) *Boer ende Tuinder*, **92**: 11-12.

Vanparys, L. (1990) *Mededeling - ProvinciaalOnderzoek-en Voorlichtingscentrumvoor Land-en Tuinbouw Beitem-Roeselare*, No.309, P. 4.

Vanparys, L. (1991) *Mededeling - Provinciaal Onderzoek-en Voorlichtingscentrum voor Land - en Tuinbouw, Beitem-Roeselare*, No. 323, P. 4.

Vanparys, L. (1994) *Mededeling - Provinciaal Onderzoek-en Voorlichtingscentrum voor Land-en Tuinbouw, Beitem- Roeselare,* No. 354, P. 4.

Vanparys, L. (1995a) *Mededeling - Provinciaal Onderzoek-en Voorlichtingscentrum voor Land-en Tuinbouw, Beitem- Roeselare,* No.356, P. 4.

Vanparys, L. (1995b) *Mededeling - Provinciaal Onderzoek-en Voorlichtingscentrum voor Land-en Tuinbouw, Beitem-Roeselare,* No. 360, P. 4.

Vanparys, L. (1995c) *Mededeling - Provinciaal Onderzoek-en Voorlichtingscentrum voor Land-en Tuinbouw, Beitem-Roeselare,* No. 362, P. 4.

Vanparys, L. (1996) *Mededeling - Provinciaal Onderzoek-en Voorlichtingscentrum voor Land-en Tuinbouw, Beitem-Roeselare,* No.367, P. 4.

Vanparys, L. (1997) *Mededeling - Provinciaal Onderzoek-en Voorlichtingscentrum voor Land-en Tuinbouw, Beitem - Roeselare,* No. 379, P. 4.

Vanparys, L. (1998) *Proeftuinnieuws,* **8**: 28-29.

Vanparys, L. (1999) *Proeftuinnieuws,* **9:** 38-39.

Vavilov, N.I. (1949) *Chronica Bot.,* **13:** 1-364.

Ved Prakash, Pandey, A.K., Singh, R.D. and Mani, V.P. (2000) *Crop Res.,* **19:** 54-58.

Vedel, F., Pla, M., Vitart, V., Gutierres, S., Chetrit, P. and De Paepe, R. (1994) *Plant Physiol. BioChem.,* **32:** 601-618.

Venter, F. (1983) *Gartenbauwissenchaft,* **48**: 9-12.

Verma, A. and Singh, Y. (2018) *Crop Protect.,* **106:** 132-138.

Verma, S.C., Malik, R. and Dhir, I. (1977) *Proc. Royal Soc. Lond,* **196:** 131-159.

Verma, T.S., Joshi, S. and Thakur, P.C. (1996) *Recent Hort.,* **3:** 147-150.

Vicente, J.G., Conway, J., Roberts, S.J. and Taylor, J.D. (2001) *Phytopathol.,* **91:** 492-499.

Vicente, J.G., Taylor, J.D., Sharpe, A.G., Parkin, I.A.P., Lydiate, D.J. and King, G.J. (2002) *Phytopathol.,* **92**: 1134-1141.

Virmani, S.S. and Ilyas-Ahmed, M. (2001) *Adv. Agron.,* **72:** 139-195.

Visser, D.L. (1977) *Euphytica,* **26:** 273.

Vlaswinkel, M. and Berg, W. Van Den (1999) *PAV-Bulletin Vollegrondsgroenteteelt,* No. June, 2-4.

Vlaswinkel, M.E.T. (1999) *PAV Bulletin Vollegrondsgroenteteelt,* October, 30-31.

Voorrips R. E., Jongerius M. C. and Kanne H. J. (2003) In: *Biotechnology in Agriculture and Forestry,* (Eds. T. Nagata and S. Tabata), vol 52. Springer, Berlin, Heidelberg.

Vriesenga, J.D. and Honma, S. (1971) *HortScience,* **6**: 395-396.

Vyvadilova, M., Klima, M. and Kucera, V. (1998b) *Zahradmictvi.,* **25**: 137-144.

Vyvadilova, M., Kucera, V. and Tomaskova, D. (1998a) *Zahradnictvi,* **25**: 9-14.

Walker (1985) *Acta Hortic.,* **157**: 256.

Walker, J.C. (1926) *Phytopath.,* **16**: 87.

Walker, J.C. (1930) *J. Agric. Res.,* **40:** 721-745.

Walker, J.C. (1933) *J. Agric. Res.,* **46:** 539-549.

Walker, J.C. (1956) *Diseases of Veg. Crops,* McGrow-Hil1 and Co., New York.

Walker, J.C. and Larson, R.H. (1951) *Phytopath.*, **41**: 37.

Walker, J.C. and Larson, R.H. (1960) *Bull. Wis. Agri. Exp. Sta.*, **547**: 12-16.

Walker, J.C. and Wellman, F.L. (1928) *J. Agri. Res.*, **37:** 233-241.

Walker, J.C. and Williams, P.H. (1965) *U.S. Dept. Agri. Dis. Reptr.*, **49:** 198-201.

Walkey, D.G.A., Neely, I.I.A. and Crisp, P. (1980) *Scientia Hortic.*, **12**: 99-107.

Wallace, D.H. (1972) *Incompatibility News Letter*, **1:** 25-26.258

Wallace, D.H. (1979a) *Theor: Appl. Genet.*, **54**: 193-201.

Wallace, D.H. (1979b) *Theor. Appl. Gent.*, **54:** 249-265.

Wallace, D.H. and Nasrallah, M.E. (1968) *Cornell Univ. Agri. Exp. Memoir*, **406**: 23.

Wang, C.Y. (1983) *J. Amer. Soc. Hort. Sci*, **108:** 125-129.

Wang, H.M., Wager, G.M. and Yang, Y.L. (1992) *Acta Agric. Boreadi Sinica*, **7:** 61-72.

Wang, M., Farnham, M.W. and Nannes, J.S.P. (1999) *Plant Breed.*, **118**: 249-252.

Wang, Y.L., Song, L.J., Zhang, B., Wen, F.Y. and Liu, X.H. (1999) *Acta Hortic. Sinica*, **26**: 66.

Warwick, S. I. and Sauder, C.A. (2005) *Can. J. Bot.*, **83**: 467-483.

Warwick, S.I. and Black, L.D. (1991) *Theor. Appl. Genet.*, **82**: 81-92.

Warwick, S.I., Black, L.D., and Aguinagalde, I. (1992) *Theor. Appl. Genet.*, **83**: 839–850

Watanabe, E. (1981) *Chinese cabbage-Proc. First Intl. Symp.* (Eds. Talekar, T.S. and Griggs T.D.), AVRDC, Taiwan.

Waterer, D., Lee, S., Scoles, G. and Keller, W. (2000) *HortScience*, **35**: 930-932.

Watson-Jones, S., Maxted, N. and Ford-Lloyd, B. (2005) *Crop Wild Relative*, **3:** 20–24.

Watts, L.E. (1963) *Euphytica*, **12**: 323-340.

Watts, L.E. (1964) *J. Hort. Sci.*, **39:** 84-91.

Wei, Z.M., Huang, J.Q., Xu, S.P. and Xue, H.W. (1998) *Acta Agriculture, Shanghai*, **14:** 11-l8.

Welch, N.C., Tyler, K.B. and Birie, D. (1985) *HortScience*, **20**: 1110-1112.

Wen, E.Y., Sun, D.J., Ju, P.H., Su, Y.M. and An, Z.X. (1991) *Acta Hortic. Sinica*, **18:** 148-152.

Werner, C.P., Smith, B.M. and Kearsey, M.J. (1986) In: *Biometrics in plant breeding*, pp. 125-139.

Werner, C.P., Smith, B.M., Godwin, R.M. and Kearsey, M.J. (1991) *Ann. Appl. Biol.*, **119**: 331-337.

Whitewell, J.D. and Senior, D. (1985) *Proc. British Crop Protection conference*, Weeds, **3**: 833-840.

Wiebe, H.J. (1975) *Sci. Hort.*, **3**: 95-101.

Wiedenfeld, R.P. (1986) *HortScience*, **21**: 236-238.

Wijk, C. Van (1985) *Groenten en Fruit*, **41**: 39.

Wijk, C.A.P. Van and Neuvel, J.J. (1999) *PAV-Bulletin Vollegrondsgroenteteelt*, Dec., 23-24.

Williams, P.I., Strub, T. and Sutton, J.C. (1972) *Phytopath.*, **62**: 247-252.

Wojciechowski, A. (1985) *Genetica Polonica*, **26**: 423-436.

Wook, K.D. and Haing, O.J. (1997) *Korean J. Plant Path.*, **13**: 95-99.

Worsdell, W.C. (1916) In: *Principles of Plant Teratology*, Vol. 2, Dalhan and Co. Ltd., London.

Wu, G.S., Wang, Y.J., Jiang, Y.W. and Zhang, L.R. (1997) *Acta Hortic., Sinica*, **24:** 141-144.

Wu, J., Zhao, J., Jing, Q., Meiling, R. and Zang, Y. (2016) *Hortic. Plant J.*, **2:** 351-356.

Wurr, D.C.E. and Fellows, J.R. (2000) *Acta Hortic.*, **539:** 69-74.

Xia, S.M. and Lian, W.S. (1996) *China Vegetables*, **3**: 14-16.

Xiang, C.L., Hong, Z.X. and We, W.Y. (1996) *Hereditas (Beying)*, **18**: 26-28.

Xu, J., Ding, Z., Vizcay-Barrena, G. and Shi, J. (2014) *Plant Cell*, **26:** 1544-1556.

Xu, X., Luo, W. and Guo, J. (2019) *Mol Breed.*, **39:** 96.

Yamagishi, H. and Bhat, S.R. (2014) *Breed Sci.*, **64**: 38-47.

Yan, C., An, G. and Zhu, T. (2019) *Theor. Appl. Genet.*, **132:** 895-906.

Yang, Xiang and Guan, PciCong (1995) *Guangdong Agricultural Sciences*, No. 6, pp. 21-23.

Yang, C.H. and Hsieh, W.H. (1985) *Plant Protec. Bull.*, **27:** 225-231.

Yang, Q., Chauvin, J.E. and Herve, Y. (1992) *Plant Cell Tissue Organ Cult.*, **28**: 289-296.

Yarnell, S.H. (1956) *Bot. Rev.*, **22:** 81-166.

Yin, Y.H., Zhu, Z.Y. and Zhu, L. (1996) *China Vegetables*, **4**: 7-8.

Yu, H., Wang, J. and Sheng, X. (2019) *BMC Plant Biol.*, **19**: 228.

Yu, H., Wang, J., Zhao, Z., Sheng, X., Shen, Y., Branca, F. and Gu, H. (2019) *Front. Plant Sci.*, **10:** 45.

Yyoti, S. and Vashistha, R.N. (1986) *Haryana J. Hort. Sci.*, **15**: 263-266.

Zhang Feng Ian and Takahata, Y. (1999) *Breed. Sci.*, **49**: 161-166.

Zhang, D.S., Cao, M.Q. and Qin, Z.W. (1998) *Acta Hortic., Sinica*, **25:** 201-202.

Zhang, D.S., Cao, M.Q. and Qin, Z.W. (1999) *Acta Agric. Boreali Sinica*, **14:** 68-72.

Zhang, J., Liu, Z., Liu, X., Dong, J., Pang, H. and Yu, C. (2016) *Plant Breed.*, **135:** 191-199.

Zhang, Y.Y., Zhuang, M. and Sun, S. (2014) *China Vegetables*, **1**: 87-88.

Zhao, Z., Sheng, X., Yu, H., Wang, J., Shen, Y. and Gu, H. (2020) *Int. J. Mol. Sci.*, **21:** 1999.

Zheng, J.H., Ye, L.Y. and Rao, L.B. (1999) *China Vegetables*, 4: 12-15.

Zhou, H., Zhou, M. and Yang, Y. (2014) *Nat. Commun.*, **5:** 4884.

Zhu, P., Cheng, M. and Feng, X. (2016) *Euphytica*, **207:** 377-385.

Zhu, Y.Y., Zhang, S.Q., Lin, C., Yao, W.Y. and Shen, F.Y. (1996) *Acta Agric. Shanghai*, **12**: 36-40.

OKRA

T. Seth, S. Hazra, M.R. Thakur, S. K. Arora and J. Kabir

1.0 Introduction

Okra (*Abelmoschus esculentus* L. Moench) is an important annual vegetable crop of the family Malvaceae, grown from seed in tropical and subtropical parts of the world. The word *Abelmoschus* possibly originated from the Arabian word "*abul-l-mosk*" meaning "father of musk, source of musk," referring to the musky smell of the seeds of the genus (Yadav *et al.*, 2014). Its tender green fruits are used as a vegetable and are generally marketed in the fresh state, but sometimes in canned or dehydrated form. It is grown commercially in India, Western Africa, Bangladesh, Afghanistan, Pakistan, Burma, Japan, Malaysia, Brazil, Ghana, Ethiopia and the Southern United States. Globally, okra occupies an area of 2.02 million ha, having production of 9.87 million metric tons (MT) with an average yield of 4.66 MT/ha. India ranks first in the world having a production of 6.13 million MT (62 per cent of the total world production) from over 0.51 million ha area with 11.92 MT/ha productivity (FAOSTAT, 2018).

2.0 Composition and Uses

2.1.0 Composition

Okra is rich in vitamins, calcium, potassium and other mineral matters. However, the data in Table 1 will show its nutritive value.

Table 1: Composition of Okra (per 100 g edible portion)*

Moisture	89.6 g	Iron	1.5 mg
Carbohydrates	8.2 g	Sodium	6.9 mg
Protein	2.1 g	Potassium	103 mg
Fat	0.2 g	Copper	0.19 mg
Fibre	1.7 g	Sulphur	30 mg
Minerals	0.7 g	Vitamin A	88 IU
Calcium	84 mg	Thiamine	0.07 mg
Magnesium	43 mg	Riboflavin	0.80 mg
Oxalic acid	8 mg	Nicotinic acid	0.60 mg
Phosphorus	90 mg	Vitamin C	47 mg

*Aykroyd (1963), Gopalan *et al.* (1989), Gemede *et al.* (2015)

2.2.0 Uses

Okra fruits can be cooked in a variety of ways. It can be fried in butter or butter oil and cooked with necessary ingredients. They can be boiled and served as salad or cut into pieces and served with soup. The roots and stems of okra are used for clearing the cane juice from which *gur* or brown sugar is prepared (Chauhan, 1972). In some places, the plants are soaked in water and the resulting solution is used as a clarifier in the manufacture of jaggery. Its ripe seeds are roasted, ground and used as a caffeine-free substitute for coffee in Turkey (Mehta, 1959; Calisir and Yildiz, 2005). Matured fruits and stems containing crude fibre are used in the paper industry.

Okra is said to be very useful against genito-urinary disorders, spermatorrhoea and chronic dysentery (Nadkarni, 1972).

The mucilage of okra is a polysaccharide associated with proteins and minerals (Woolfe *et al.*, 1977; Mishra *et al.*, 2017) and is added to different recipes like stews and sauces to increase the consistency of the food (Gemede *et al.*, 2015). It is considered as an important component for balanced food due to its dietary fibres and amino-acid composition which is rich in lysine and tryptophan (Hughes, 2009). It contains various ethno-pharmacological and medicinal properties against cancer, high-cholesterol and *Diabetes mellitus* (Jenkins *et al.*, 2005; Sabitha *et al.*, 2011). It is helpful in curing ulcers and haemorrhoids (Adams, 1975). It has high antioxidant activity in terms of total phenolics and total flavonoids content (Gemede *et al.*, 2015). It is full of both folate and vitamin C which helps in the growth and development of the fetus during pregnancy (Zaharuddin *et al.*, 2014). In Asian medicine the mucilage is used as a plasma replacement or blood volume expander (Akinyele and Temikotan, 2007) and as a protective food additive against irritating and inflammatory gastric diseases (Lengsfeld *et al.*, 2004). Dry seeds contain 13-22 per cent edible oil and refined oil may be used as a substitute of common edible oil particularly cotton seed oil. Oil from seeds is also used in perfume industry (Hazra *et al.*, 2011).

3.0 Origin and Taxonomy

3.1.0 Origin

The cultivated okra is of old world origin and the genus *Abelmoschus* is accepted to be of Asiatic origin, though opinions differ for the origin of *A. esculentus* as India (Masters, 1875), Ethiopia (De Candolle, 1883; Vavilov, 1951), Abyssinia (Darlington and Wylie, 1955), West Africa (Chevalier, 1940; Murdock, 1959), Tropical Asia (Grubben, 1977) and Hindustani Centre of origin chiefly India, Pakistan, Burma (Zeven and Zukovsky, 1975). According to Vavilov, it was probably domesticated in the Ethiopian region but according to Murdock it was in West Africa (Joshi *et al.*, 1974). There is a late 19th century record of its occurrence in the wild state in White Nile of Sudan (Singh *et al.*, 1975). No dates can be given for the migration of cultivated okra from Africa to the Mediterranean region and to Asia. It was apparently introduced to tropical and subtropical America during the early post-Columbian period.

The accepted view that okra is of African origin should not, however, be taken as established. The cytotaxonomy of *Abelmoschus* is yet so confused that an Asian origin of whole or part of the cultigen does not seem to be impossible. It is borne out of the fact that Indian okra bears genomes (T^1, Y) that show a varying degree of homology with the genomes of two Indian wild species (T, F) and that African-Indian hybrids show some sterility. The crop could perhaps yet turn out to be polyphyletic (Joshi and Hardas, 1976).

Uses of Okra

Okra Fries

Okra Soup

Okra Stew

Baked Chicken with Okra

Frozen Okra

Pickled Okra

Vaccum Fried Okra Chips

Okra Powder

3.2.0 Taxonomy

Okra is known by many local names in different parts of the world. For example, it is called lady's finger in England, gumbo in the United States of America, guino-gombo in Spanish, guibeiro in Portuguese, bamiah in Arabic and bhindi in India. Again in the Indian subcontinent it is given different names in different regional languages (Chauhan, 1972). The botanical name generally associated with it is *Abelmoschus esculentus* L., but earlier it was designated as *Hibiscus esculentus* (L.) Moench under the section *Abelmoschus* of *Hibiscus*, established by Linnaeus in 1937. The generic name, *Abelmoschus*, proposed by Medikus, was upheld by Hochreutiner (Joshi *et al.*, 1974) on the basis of the constant distinguishing features of caducous calyx. Kundu and Biswas (1973) and Terrel and Winters (1974) distinguished the genus *Abelmoschus* from *Hibiscus* by the characteristics of the spathulate five teeth calyx that connate to the corolla and caduceus after flowering. The calyx, corolla and staminal column are fused together at the base and fall off as one piece after anthesis (Hochreutiner, 1924; Bates, 1968; Ige and Eludire, 2014), whereas in case of *Hibiscus* these floral parts are persistent (Sharma, 1993). Index Kewensis lists over 30 species of *Abelmoschus* in the old world, 4 in the new world and 4 in Australia. In his revision of South-East Asian Malvaceae, Van Borssum Waalkes (1966) had a more conservative point of view, retaining only six species. These are *Abelmoschus moschatus* Medikus, *A. manihot* (L.) Medikus, *A. esculentus* (L.) Moench., *A. ficulneus* (L.) Wt and Art. ex Wt., *A. crinitus* Wall. and *A. angulossus* Wall. ex Wt. and Art. The former 3 species consist of wild and cultivated forms while the latter 3 species consist of wild forms only. The former 3 species are also endemic in the West Africa; *A. esculentus* and *A. caillei* are also reported to be cultivated there (Hamon and Hamon, 1991).

During the international workshop on okra genetic resources at New Delhi (1990), a new classification was adopted on the basis of cytogenetical evidences, comprising nine species (IBPGR, 1991). These are *Abelmoschus moschatus* Medikus, *A. manihot* (L.) Medikus, *A. tetraphyllus* (Roxb. ex Hornem.) R. Graham var. *tetraphyllus*, *A. esculentus* (L.) Moench, *A. tuberculatus* Pal and Singh, *A. ficulneus* (L.) Wt and Art. ex Wt., *A. crinitus* Wall., *A. angulossus* Wall. ex Wt. and Art and *A. caillei* (A. Chev.) Stevels. In this workshop it was also concluded that 'Guinean' type okra and *A. manihot* var. *caillei* (A. Chev.) were the same and named as *A. caillei* (A. Chev.) Stevels whereas, 'Soudanian' type of okra was named as *A. esculentus*. The species, *A. caillei* was identified wrongly earlier as *A. manihot* subsp. *manihot* by Flora of Tropical West Africa, Kew Botanic Gardens.

Whereas Van Borssum Waalkes (1966) considered *A. tuberculatus* a wild form of *A. esculentus* and cytogenetical studies justified its recognition as a distinct species (Pal *et al.*, 1952). The studies also suggested that *A. manihot* in a wide sense is actually a polyspecies complex and that it has many forms.

4.0 Botany

Okra is an erect, herbaceous annual, 1-2 metre tall. Stem green or with reddish tinge. Leaves alternate, broadly cordate, palmately 3-7 lobed, hirsute, serrate. Flowers solitary, 5 to 8 cm in diameter (Dhankhar and Singh, 2009), axillary with

Wild Species of Okra

Abelmoschus caillei

Abelmoschus manihot

Abelmoschus moschatus

Abelmoschus angulosus (left) and *Abelmoschus tuberculatus* (right)

Immature and Mature Fruits of different *Abelmoschus* spp.

(a) ***A. angulosus*** var. ***angulosus***; (b) ***A. angulosus*** var. ***grandiflorus***; (c) ***A. angulosus*** var. ***purpureous***; (d) ***A. angulosus*** var. ***tetraphyllus***; (e) ***A. palianus***; (f) ***Abelmoschus*** sp. nov.; (g) ***A. manihot***; (h) ***A. ficulneus***; (i) ***A. caillei***; (j) ***A. esculentus***; (k) ***A. crinitus***; (l) ***A. moschatus***; (m) ***A. enbeepeegearense***; (n) ***A. rugosus***; (o) ***A. rhodopetalus***; (p) ***A. tuberculatus*** (Yadav *et al.*, 2014).

about 2 cm long peduncle; epicalyx up to 10, narrow hairy bracteoles which fall before the fruit reaches maturity; calyx split longitudinally as flower opens; petals 5, yellow with crimson spot on claw, 5-7 cm long; staminal column united to the base of petals with numerous stamens; ovary superior, stigma 5-9 deep red. Fruit a capsule, light green, or sometimes red in colour pyramidal-oblong, beaked, longitudinally furrowed, 10-30 cm long, dehiscing longitudinally when ripe. Seeds green to dark brown, black, greyish or greenish in colour, rounded with a hard coat and average weight is 1.6 to 8.1 g/100 seeds (Martin and Rhodes, 1983). Seeds reach maturity at 40 days after fertilization (Dhankhar and Singh, 2009).

The greatest increase in fruit weight, length and diameter occurs during 4th to 6th day after pollination (Sistrunk *et al.*, 1960; Iremiren *et al.*, 1991, Dhankhar and Singh, 2013). Generally, the fibre formation in the fruit starts from the 5th to 6th day of fruit formation and a sudden increase in fibre content from the 9th day is observed (Nath, 1976). Therefore, the picking should be done every second day (Moekchantuk and Kumar, 2004).

4.1.0 Floral Biology

A flower bud appears in the axil of each leaf above 6th to 8th leaf depending upon the cultivar and is hermaphrodite. The crown of the stem at this time bears 3-4 underdeveloped flowers but later on during the period of profuse flowering of the plant there may be as many as 10 underdeveloped flowers on a single crown. As the stem elongates, the lower most flower buds open into flowers. There may be a period of 2, 3 or more days between the times of development of each flower but never does more than one flower appear on a single stem. The buds measuring 2.82 x 1.55 cm open into flowers (Purewal and Randhawa, 1947). Flower bud initiation generally takes place 22-26 days after sowing (Sulikeri and Rao, 1972) and a flower bud takes another 22-26 days from initiation to full bloom (Sharma, 1993). The first flower openes 41-48 days after sowing and continues for 40-60 days (Tripathi *et al.*, 2011; Ige and Eludire, 2014). The time of anthesis varies with the cultivar, temperature and humidity and it ranges from 8 to 10 a.m. (Purewal and Randhawa, 1947; Srivastava, 1964; Ramu, 1976; Sharma, 1993). However, Hamon and Koechlin (1991a) reported that anthesis takes place at the end of the night and the flower opens at dawn. According to Hao *et al.* (1999), flowering lasts for 18.00-20.00 h, generally commencing at 18.00-19.00 h, with flowers fully opened at 8.00-9.00 h the following day and closed by 14.00-15.00 h. The dehiscence of anthers is transverse and occurs 15 to 20 minutes after anthesis. The dehiscence is complete within 5-10 minutes (Purewal and Randhawa, 1947). The pollen fertility is maximum during the period between an hour before and an hour after opening of the flower (Srivastava, 1964). It takes 2-6 hours for fertilization after pollination. The flowers remain open for a short time and they wither late in the afternoon. The stigma is receptive only after opening of the flower (90-100 per cent) (Purewal and Randhawa, 1947), hence pollination at bud stage is not successful. However, stigma receptivity was also observed the day before flowering (50-70 per cent) and the day after flowering (1-15 per cent) (Tripathi *et al.*, 2011; Ige and Eludire, 2014). Reports also suggest that the stigma is receptive before anthesis (protogyny) and self-pollination can occur before anthesis (Hamon and Koechlin, 1991b). Studies on the influence of ovule position on pollen tube entry and seed set revealed that ovule positions 7 to 14 received the maximum number of pollen tubes, *i.e.*, about 90 per cent (Chandra and Bhatnagar, 1974). The pollen can remain viable in storage at 50 per cent RH for 55 days (Dubey and Singh, 1968b; Dhankhar and Singh, 2009).

The style is surrounded by a staminal column and the number of anthers on individual flower ranges from 62-100 (Hamon and Koechlin, 1991a; Rithichai *et al.*, 2004). An average of seven pollen grains is required for fertilizing one ovule (Ter Avenesian, 1978). The pollen of 3-11 anthers having 280-1120 pollen grains could be enough to fertilize 40-160 ovules per okra flower (Hamon and Koechlin, 1991a). The pollens come in contact with the stigma through lengthening of the staminal column thus the flowers of okra are self-fertile. The pollen grain is large with many pores and every pore is a potential tube source therefore, many tubes can develop from one pollen grain (Purewal and Randhawa 1947; Dhankhar and Singh, 2009).

Experimentally it has been found that there is no significant difference in fruit set under open-pollinated, self-pollinated (by bagging alone) and self-pollinated (hand pollination of bagged flowers), indicating that it is potentially a self-pollinated crop (Purewal and Randhawa, 1947). The inbreeding depression well pronounced in cross-pollinated crops has not been reported in this crop (Duranti, 1964-65). Though essentially self-pollinated, because of its showy corolla, the possibility of cross-pollination by insects cannot be ruled out. Okra falls under the category of often cross-pollinated vegetable crop (Dhankhar and Mishra, 2005). Consequently, cross-pollination to the extent of 4.0-19.0 per cent (Purewal and Randhawa, 1947; Choudhury and Anothai-Choomsai, 1970; Shalaby, 1972) with maximum of 42.2-63.0 per cent (Mitidieri and Vencovsky, 1974; Chandra and Bhatnagar, 1975; Martin, 1983; Aken'Ova and Fatokun, 1984; Hamon and Koechlin, 1991a) has been reported. Okra flower produces nectar which attracts various insect such as butterflies, houseflies, honeybees and ants around the flower. Pollen grains are sticky and heavy, so can not be readily transferred by wind, therefore insects are responsible for the cross pollination in okra plant (Hamon and Koechlin, 1991b; Ige and Eludire, 2014). The extent of cross-pollination in a particular place will depend upon the cultivar, competitive flora, insect population and season, *etc.* Emasculation and pollination events are easier due to large flower and monoadelphous stamens (Kumar *et al.*, 2013).

5.0 Cytology and Genetics

5.1.0 Cytology

The chromosome numbers (2n) of *A. esculentus*, a widely grown species throughout the world have variously been reported to be 66, 92, 108, 118, 120, 122, 124, 126, 130, 132, 134 and 144 by various authors (Siemonsma, 1982a). The majority of investigators, however, agree that the species has 2n = 130 chromosomes. The chromosome numbers of other species reported are 60, 66 and 68 for *A. manihot*, 72 for *A. moschatus*, 72 and 79 for *A. ficulneus*, 56 for *A. angulosus*, 58 for *A. tuberculatus*, 138 for *A. pungens*, and 130 and 138 for *A. tetraphyllus*. The latter two species have been considered as forms of *A. manihot* (Siemonsma, 1982a). Singh and Bhatnagar (1975) observed 2n = 194 chromosomes in an okra from Ghana (Africa) the highest number reported so far in the genus. This is probably the same as Guinean (forest), a non-conventional type having chromosome numbers varying from 185-196 as reported by Siemonsma (1982b). The Ghana type has been identified by the Central National Herbarium, Howrah as *A. manihot* (L.) Medikus (Arumugam *et al.*, 1975) and as *A. manihot* (L.) subsp. *manihot* by the Royal Botanical Gardens, Kew, England (Thakur, 1976).

Allopolyploid nature of *A. esculentus* has been reported by Joshi and Hardas (1956). They obtained fertile plant from a colchicine treated sterile F_1 hybrid between *A. esculentus* (n = 65) and *A. tuberculatus* (n = 29). They further observed that most frequent associations in the sterile F_1 hybrid between both these species of $29_{11}+ 36_{1}$, suggesting that the haploid complement of *A. esculentus* contains one genome with 29 chromosomes and another with 36, the former being homologous with one haploid set in *A. tuberculatus*. Hence they postulated that *A. esculentus* arose through hybridization of one species with n = 29 and another with n = 36 followed by doubling of the chromosome numbers. The second parental species of *A. esculentus* with n = 36 is still unknown. The 36 chromosomes (genome Y) of *A. esculentus* showed greater (but still incomplete) homology with 36 chromosomes of *A. ficulneus* as compared to those of *A. moschatus*, the two Indian species (Joshi and Hardas, 1956; Joshi *et al.*, 1974). These studies confirmed that cultivated okra is an amphidiploid between *A. tuberculatus* and *A. ficulneus* (29 T + 36 Y) (Figure 1). Charrier (1984) also opined that *A. esculentus* is a polyploid species originated through amphidiploidy. In his research, a large number of interspecific crosses between potential parental species such as between *A. manihot* (2n = 68) and *A. tuberculatus* (2n = 58) were executed but the resulting plants were sterile. In the meiosis in hybrids of the former crosses, only univalents and rarely 1-2 bivalents

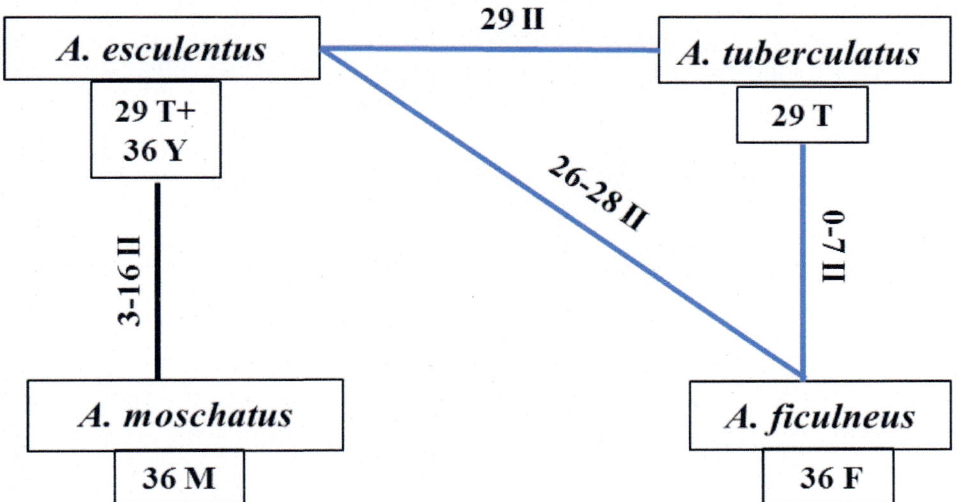

Figure 1. Bivalent Formation Indicating Chromosomal Homologies Observed in Pollen Mother Cells of F_1 Hybrids Involving *A. esculentus* (2n = 130), *A. moschatus* (2n = 72), *A. ficulneus* (2n = 72) and *A. tuberculatus* (2n = 58) (Joshi *et al.*, 1974)

and in the hybrid of the latter cross, 0-7 bivalents were observed (Pal *et al.*, 1952; Kuwada, 1974; Joshi and Hardas, 1956; Joshi *et al.*, 1974). The genomes of these species seem, therefore, to be little related. This is also borne out of the fact that an artificial amphidiploid of *A. ficulneus* x *A. tuberculatus* (2n = 130) realized by Joshi *et al.* (1974) was genetically unbalanced and completely sterile.

On the basis of crossing behaviour Siemonsma (1982a,b) hypothesized that the Guinean type (*A. caillei*) which is gradually replacing the ordinary okra (Soudanian -*A. esculentus*) in the humid parts of West Africa, is a natural amphidiploid of *A. esculentus* (2n = 130-140) and *A. manihot* (2n = 60-68). Amphidiploidy was also induced in the cross *A. esculentus* (2n = 130) x *A. manihot* ssp. *manihot* (2n = 194) by colchicine treatment. The amphidiploid had 88.12 per cent fertility as against only 7.07 per cent in the F_1 and the chromosome configuration at metaphase was more or less regular forming 162_{11} (Jambhale and Nerkar, 1981). This gives credence to Siemonsma's hypothesis of Guinean being a natural amphidiploid.

A semi-wild form of okra, Thamaravenda occurs widely in Kerala, India. The chromosome number was confirmed as 2n = 184 and it formed 92 bivalents at metaphase I. Morphological observations and cytological data reveals that this semi-wild form of okra belongs to *Abelmoschus caillei*, which has a long period of domestication in West Africa (Chacko *et al.*, 1996).

5.2.0 Genetics

Monogenic control of pigmentation of calyx, corolla and fruit colour in okra has been reported by Kolhe and D'Cruz (1966). The cut leaves are dominant to lobed leaves and the white fruit colour is dominant to green fruit colour and both are monogenically controlled characters. The fruit shape (angular vs round) is digenically inherited and it involves epistasis. The heterozygotes for plant height and spininess are intermediate indicating polygenic control (Jasim, 1967). The inheritance of fruit hairiness and leaf lobing characters are monogenic with incomplete dominance (Kalia and Padda, 1962; Nath and Dutta, 1970). The male sterility in okra is controlled by a single recessive gene (Dutta, 1980).

The susceptibility to yellow vein mosaic (YVM) disease has been reported to be controlled by two complementary dominant genes, Yv_1 and Yv_2 in intervarietal crosses (Singh *et al.*, 1962) but reverse has been found true in interspecific crosses involving *Abelmoschus esculentus* and *A. manihot* ssp. *manihot* in which case two complementary dominant genes are said to condition the resistance (Thakur, 1976; Sharma and Dhillon, 1983; Sharma and Sharma, 1984; Pullaiah *et al.*, 1998; Dhankhar *et al.*, 2005). Contrary to the latter, a single dominant gene control of resistance to YVM has also been reported (Jambhale and Nerkar, 1981; Dutta, 1984; Arora *et al.*, 2008, Boonsirichai *et al.*, 2009; Bharathkumar *et al.*, 2019). In some studies resistance found to be controlled by two dominant genes (Sharma and Dhillon, 1983; Adiger, 2015). Sadashiva (1988) reported that YVMV resistance in okra is imparted only when at least one of the genes is in homozygous dominant condition. However, resistance to YVMV disease was conditioned by two duplicate dominant genes in resistant x resistant cross have also been reported (Pullaiah *et al.*, 1998; Seth *et al.*, 2017).

Ali *et al.* (2000) hypothesized from the inheritance pattern that the tolerance of yellow vein mosaic virus of cultivar IPSA Okra-1 is quantitative, with possibly two major factors, and dependent on gene dosage with incomplete dominant gene action. They also observed that tolerance in IPSA Okra-1 is genetic and not due to escape. Whereas, another report by Arora *et al.* (2008) depicted that resistance to YVMV is due to the presence of major gene(s) along with minor genes.

Whereas, Vashisht *et al.* (2001) based on 9 generations derived from crosses involving resistant (Parbhani Kranti) and susceptible cultivars (Punjab 8, Punjab Padmini, Pusa Makhmali and Pusa Sawani), reported that additive gene effects were more significant than dominance gene effects.

The resistance to okra enation leaf curl virus disease has been reported to be controlled by single dominant gene in tolerant x tolerant cross whereas, two genes with duplicate recessive epistasis are said to condition the resistance in tolerant x susceptible crosses (Devi *et al.*, 2019a).

Powdery mildew resistance is controlled by a single incompletely dominant gene (Nerkar, 1990).

Resistance to cotton jassid (*Amrasca bigutulla bigutulla*) is polygenic but controlled by dominant genes. Both additive and dominant gene effects are significant but additive gene effects and dominant x dominant type of interactions appears relatively more important (Sharma and Gill, 1984).

Very few studies on the genetic of quantitative characters in okra have been made. Studies on the estimates of genetic parameters in three components of yield, *viz.*, plant height, days to flowering and number of fruits per plant revealed that both additive and non-additive type of gene actions were operating, dominant gene action being more pronounced. One to three groups of dominant genes were found to control days to flowering and number of fruits per plant, while four to five groups of dominant genes controlled plant height. All the three characters showed over-dominance (Kulkarni, 1976). Further analysis of the gene effects showed that duplicate and complementary types of epistasis were more pronounced (Kulkarni *et al.*, 1973 and 1978). Additive genetic variance was important for yellow vein mosaic virus incidence while dominance variance was predominant for yield and individual fruit weight (Veeraragavathatham and Irulappan, 1990).

Effect of additive gene action for days to first flowering (Arora *et al.*, 2008; Kishor *et al.*, 2013) and days to 50 per cent flowering (Sood and Kalia, 2001; Khanpara *et al.*, 2009; Seth *et al.*, 2016a) were reported. Node at first flowering had a preponderance of both additive and non-additive genetic effect (Jindal *et al.*, 2009; Seth *et al.*, 2016a), whereas, the response of non-additive gene actions for internodal length, plant height, number of branches per plant, fruit length, number of fruits per plant and fruit yield per plant has been observed (Dhankhar and Dhankhar, 2001, Das *et al.*, 2013; Adiger *et al.*, 2013; Seth *et al.*, 2016a). Presence of both additive and non-additive gene interactions and complementary and duplicate type of epistasis were conditioned for days to first appearance and percent disease index (PDI) of YVMV disease (Seth *et al.*, 2017). Whereas, non-additive gene action and duplicate epistasis were the most predominant types of gene effects controlling days to first appearance of ELCV disease and PDI of ELCV traits (Devi *et al.*, 2019a).

6.0 Cultivars

The recommended cultivars of okra grown in different parts of India are briefly described below:

Pusa Makhmali

This is a 5-ridged cultivar developed through pure line selection from a collection from West Bengal and released from Indian Agricultural Research Institute, New Delhi, India in 1955 (IBPGR, 1991). It is highly susceptible to yellow vein mosaic virus and produces light green fruits. This is considered good for yellow vein virus free period, *i.e.*, in spring-summer in Northern India and rainy season in Southern India.

Punjab No. 13

It was released by Punjab Agriculture Department in India. This cultivar produces light green and 5-ridged fruits of medium length. It is susceptible to mosaic and hence only suitable for spring-summer season.

Pusa Sawani

It was developed through intervarietal hybridization in 1960 from a cross between IC 1542 (symptomless carrier for yellow vein mosaic from West Bengal) and Pusa Makhmali and released from Indian Agricultural Research Institute, New Delhi. The fruits are dark green, smooth with 5 ridges and about 10-12 cm long at the marketable stage. It takes about 50 days from sowing to marketable maturity. The cultivar is characterized by the presence of purple patch at the base of the yellow petal on both the sides and a slight purple pigmentation on the stem, petiole and base of the leaf. The stem and leaves are moderately hairy. The top leaves are deeply lobed (3 to 5 lobes) and the first fruit is at 8th node. It is suitable for both spring-summer and rainy season crops. This cultivar when released was tolerant to yellow vein mosaic virus but its tolerance has since broken down.

Pusa A-4

It was developed by Indian Agricultural Research Institute, New Delhi, India in 1994 for cultivation in Delhi, NCR and throughout India. Fruits are dark green, 12-15 cm long, first picking after 45 DAS. It shows resistant to yellow vain mosaic virus and tolerant to aphids and jassids. Yield potential is 14 t/ha.

IHR 20-31

It produces long fruits with thick flesh and lush green skin colour (Dutta and Singh, 1970). Average number of fruits per plant varies from 20-25 under Bangalore conditions. Fruits are 5-ridged.

Red Bhindi (Selection AE 106)

The fruits are red, long and slender, fleshy and have fewer seeds than Pusa Sawani. It produces as much as Pusa Sawani in southern plains (Anon., 1967).

Sel-2

It is developed through intervarietal hybridization involving multiple cross (Pusa Sawani x Best-1 x Pusa Sawani x IC 7194) and released for cultivation from National Bureau of Plant Genetic Resources in 1985. It is fairly resistant to yellow vein mosaic virus and has long attractive fruits (Rana *et al.*, 1990). Yield potential is 11-13 t/ha.

Punjab Padmini

It has been evolved as a result of interspecific hybridization between *Abelmoschus esculentus* and *A. manihot* ssp. *manihot* in 1982 at Punjab Agricultural University, Ludhiana. Plants are tall with mild purple tinge on the stem and petiole, leaves are dark green, deeply lobed and hairy, fruits are quick growing, dark green, hairy, 5-ridged and remain tender for a longer period. It attains marketable maturity in 55-60 days. It shows field tolerance to yellow vein mosaic virus as low intensity of virus and symptoms appear on the new shoot growth quite late in the season. It is also suitable for both spring and rainy season. Yield potential is 10-12 t/ha.

Parbhani Kranti

This YVMV resistant cultivar is developed at Marathwada Agricultural University, Parbhani in 1985 through pedigree selection from interspecific cross between *A. esculentus* cv. Pusa Sawani x *A. manihot* ssp. *manihot*, an African species. Leaves are deeply lobed, fruits dark green, smooth, tender, 5-ridged. Fruits are dark green, smooth, tender and slender.

Punjab 7

It is developed through backcross method from a cross between *A. esculentus* cv. Pusa Sawani x *A. manihot* ssp. *manihot* cv. Ghana at Punjab Agricultural University, Ludhiana in 1986. Fruits are medium long, green, tender, 5-ridged, top of the fruit is blunt and slightly furrowed. It shows resistant to YVMV. It is resistant to yellow vein mosaic virus, jassids and boll worm. The plants are medium tall with splashes of purple pigmentation present on the stem. Leaves are deeply lobed up to the base of the petiole and the margins are less serrated. The basal portion of the petiole is deeply pigmented. It is suitable for both spring-summer and rainy season cultivation.

Punjab 8 (EMS 8)

It is an induced mutant of Pusa Sawani using EMS (1 per cent) and developed at Punjab Agricultural University, Ludhiana in 1995. Plants are medium tall with splashes of purple pigmentation present on the stem. Leaves are deeply lobed and less serrated. Leaves, stem and petiole are less hairy. Fruits are thin, long, 5-ridged, dark green, highly resistant to YVMV and tolerant to fruit borer. It is suitable for both spring-summer and rainy season cultivation.

Punjab Suhawani

It was released by Punjab Agricultural University, Ludhiana in 2017. Plants are medium tall, leaves deeply lobed, dark green with serrated margins. Fruits are medium long, dark green, tender and 5-ridged. It has tolerance to YVMV.

Perkins Long Green

It was introduced from USA and recommended by IARI, New Delhi in 1976. Plants are tall, leaves large, with rough surface. Fruits long tapered, bright green, spineless mostly eight edged. It is suitable for cultivation in slightly cooler climate in lower hilly areas of north India. Yield is about 9-15 t/ha.

CO 1

It was released by TNAU, Coimbatore, India in 1978 as pureline selection from Red Wonder collected from Hyderabad. Fruits are long, slender, 5-ridged, glossy, smooth and scarlet red but colour disappears on cooking. The flesh content is 72 per cent with a mean seed content of 28 per cent, crude fibre of 14.06 per cent, starch and protein content of 14.80 per cent and 18.08 per cent, respectively. It produces yield of 12 t/ha.

MDU 1

It was developed at TNAU, Coimbatore in 1978 as an induced mutant from Pusa Sawani through gamma rays. It can be cultivated throughout the year under irrigated conditions and is adaptable to southern districts of Tamil Nadu. Plants are dwarf with shorter internodes, stem is green with light purple pigmentation. Fruits are light green and about 20 cm long with less crude fibre content (12.3 per cent). The crop duration is 90 days.

Arka Anamika

It was developed by interspecific hybridization of (*Abelmoschus esculentus* x *A. manihot* ssp. *tetraphyllus* var. *tetraphyllus*) followed by back cross method and released by IIHR, Bangalore in 1995. Fruits are free from spines having 5-6 ridges, lush green, tender, long and having delicate aroma. Good keeping and cooking qualities. It shows field tolerant to YVMV with yield potential of 20 t/ha.

Arka Abhay

This cultivar was developed by interspecific hybridization of (*Abelmoschus esculentus* x *A. manihot* ssp. *tetraphyllus* var. *tetraphyllus*) followed by back cross method and released by IIHR, Bangalore in 1995. Fruits are free from spines, lush green, tender and long. Fruits have delicate aroma, good keeping and cooking qualities. It is tolerant to YVMV under field condition and gives yield of 18 t/ha.

Varsha Uphar

This cultivar was developed through pedigree selection (Lam Selection 1 x Parbhani Kranti) and released by CCS HAU, Hisar. Plants are medium tall with short internodes, leaves deeply lobed and dark green colour. Fruits are long, dark green, 5-ridged and appear on 4th node. It has tolerance to YVMV and suitable for rainy season cultivation. Average yield is 10 t/ha.

Hisar Unnat

It was developed from the cross Sel 2-2 x Parbhani Kranti at CCS HAU, Hisar. Plants are medium tall with 2-3 branches. Fruits are long, green, 5-ridged and appear on 4th to 5th node. It is resistant to YVMV and suitable for sowing in rainy season. It gives yield of 10-12 t/ha.

Hisar Naveen

This cultivar was released by CCS HAU, Hisar in 2004. Plants are medium tall (110-120 cm) with short internodes, leaves deeply lobed and green colour. Fruits are long, dark green, 5-ridged and appear on 4th node. It has tolerance to YVMV and suitable for rainy season cultivation. Average yield is 10 t/ha.

Utkal Gaurav

This cultivar was developed from the cross Pusa Sawani x Goudiapada Local and released by OUAT, Bhubaneswar in 1994. Plants are medium tall (100-125 cm). Fruits are long, green, 5-ridged. It is resistant to YVMV and moderately tolerant to fruit borers. It is suitable for sowing in spring-summer and rainy season. It gives yield of 8-10 t/ha.

Susthira (AE-286-1)

It is a high yielding perennial type cultivar released from Kerala Agricultural University in 2006. Fruits are 5-ridged, 22 cm long and green in colour. It responds well to pruning. It is resistant to YVMV and suitable for rainy season. It gives average yield of 18-19 t/ha.

Aruna

It is a red coloured cultivar released by KAU. It is a high yielding cultivar with long red fruits. It is resistant to YVMV. Average yield is 15.8 t/ha.

Salkeerthi

This cultivar was released by KAU. Fruits are light green, 24-30 cm long, 5-ridged. It produces average yield of 16.2 t/ha.

Kiran

This cultivar was released by KAU. Fruits are light green and long fruits with average yield of 11.21 t/ha.

Gujarat Okra-2 (GO-2)

This cultivar was released by Gujarat Agricultural University, Anand in 2001. Plant height is 120 to 125 cm. Fruits are long (16-17 cm), green, tender and attractive. This is suitable for cultivation during rainy and spring summer season. It shows tolerance against YVMV and produce yield of 17 t/ha.

Gujarat Okra-3 (GO-3)

This cultivar was developed through hybridization followed by pedigree selection from cross between HRB 55 x Arka Anamika and released by Junagadh Agricultural University, Junagadh in 2006. Plant height is 130-135 cm. Fruits are smooth, green, tender, 13-14 cm long and ready to harvest within 53-55 DAS. This is suitable for rainy season cultivation. It shows tolerance against YVMV disease and jassids and fruit borer.

Kashi Mohini (VRO-3)

This is a 5-ridged cultivar released from Indian Institute of Vegetable Research (IIVR), Varanasi, India in 2001. Plants are tall (110-140 cm), flowers at 4-5 node during summer and 5-7 nodes during rainy season after 39-41 DAS, fruits are 11.3-12.6 cm long at marketable stage and gives yield of 13-15 t/ha. It is suitable for both spring-summer and rainy season cultivation. It shows tolerance to high temperature during summer season and shows field tolerance to yellow vein mosaic virus (YVMV) and okra leaf curl virus (OLCV).

Kashi Mangali (VRO-4)

It was developed through pureline selection and released by IIVR, Varanasi, India in 2001. Plants are medium tall (120-125 cm), fruits are light green with five ridges. It takes about 45-50 days from sowing to marketable maturity and produces yield of 13-15 t/ha. It shows resistant to YVMV and OLCV under field conditions and can withstand high temperature during summer season. It is suitable for both spring-summer and rainy season cultivation.

Kashi Vibhuti (VRO-5)

It was developed through pedigree selection (NIC 9303 x IC111547) x IIVR 20 and released from IIVR, Varanasi, India in 2006. This cultivar has dwarf growth habit (60-70 cm), bushy type with short internodal length. A plant bears 20-25 dark green fruits with 11-14 cm length at marketable stages. It is suitable for both spring-summer and rainy season cultivation and yield varies from 17-18 t/ha. It is resistant to YVMV and OLCV.

Kashi Pragati (VRO-6)

This cultivar was developed through pedigree selection (NIC 9303 x PK 20) and released from IIVR, Varanasi, India in 2006. Plants are tall (120-140 cm) with short internodes. Early flowering, first flower appears after 36-38 DAS. Fruits are dark green in colour, 5-riged with 11-13 cm in length. Average number of fruits per plant varies from 20-25 per plants and yield 16-18 t/ha. This cultivar shows resistant to YVMV and OLCV.

Kashi Satdhari (IIVR-10)

It was developed through pedigree selection (PK x IC111542) x IIVR-20 and released from IIVR, Varanasi, India in 2006. Fruits are 7-ridged, 13-15 cm long, dark green in colour. It is resistant to YVMV and OLCV. It gives yield of 11-14 t/ha and recommended for cultivation in Uttar Pradesh, Bihar and Jharkhand.

Kashi Lila (IIVR-11)

This cultivar was developed through pedigree selection (NIC 9305 x HRB 9-2) x IIVR-50 and released from IIVR, Varanasi, India in 2006. Plants are of medium tall (110-130 cm), fruits are of 5-ridges, dark green and 10-13 cm long. It is suitable for both spring-summer and rainy season cultivation. This cultivar is resistant to YVMV and OLCV and gives yield of 16-18 t/ha.

Promising Cultivars of Okra

Kashi Lalima (VRO-157)

Kashi Chaman (VRO-109)

Kashi Vardaan (VRO-25)

Kashi Satdhari

Arka Abhay

Pusa Sawani

Arka Anamika

Parbhani Kranti

Kashi Shristi (VROH-12)

Arka Nikitha

Kashi Kranti (VRO-22)

This cultivar was developed through pedigree selection (VRO-6 x 67-10-02) and released from IIVR, Varanasi, India in 2011. It is an early, medium tall variety have resistant to YVMV and OLCV under field condition. The fruits are dark green and average yield is 12-14 t/ha. It is suitable for both spring-summer and rainy season cultivation in Uttar Pradesh, Bihar, Jharkhand.

Kashi Vardaan (VRO-25)

It was released from IIVR, Varanasi, India in 2015. Plants are medium tall (120-125 cm) with short internodes and 2 to 3 branches, flowering starts on 30-35 DAS. Fruits are 5-ridged, dark green, 10-11 cm long, easily cooked, good for both table and export purposes. Resistant to YVMV and okra enation leaf curl virus (OELCV) under field condition and giving yield of 14-15 t/ha. It is suitable for both spring-summer and rainy seasons cultivation in Uttar Pradesh, Bihar, Jharkhand and Punjab.

Kashi Lalima (VROR-157)

It is a red coloured cultivar developed by IIVR, Varanasi, India in 2019 for cultivation in Uttar Pradesh. Plants are medium tall and short internodes. Fruits are reddish purple in colour and rich in anthocyanin and phenolics. Suitable for both spring-summer and rainy season cultivation. It is tolerant to YVMV and OLCV and gives yield of 14-15 t/ha.

Kashi Chaman (VRO-109)

It is developed by IIVR, Varanasi, India in 2019. Plants are medium tall (120-125 cm), flowering starts in 39-41 DAS and fruiting period lasts from 45-100 days. Fruits are dark green with fruit length of 11-14 cm. Plants are resistant to YVMV and OLECV under field conditions. Suitable for both spring-summer and rainy season cultivation. It is recommended for cultivation in Uttar Pradesh. Yield potential is 15-16 t/ha.

There are other cultivars like **Lucknow Dwarf, Long Green, Velvet Green, Smooth Velvet, Long White, Red Wonder** and **Samalkota** in seed catalogues of various seed companies. The cultivars **Dwarf Prolific** is recommended in Trinidad and **White Velvet** in Puerto Rico.

Hybrids

CO 2

It is a F_1 hybrid between AE 180 x Pusa Sawani released by TNAU, Coimbatore in 1987. Fruit surface is less hairy with a better consumer's appeal and market preference. Fruits are very long (22-25 cm) and contain 12.6 per cent crude fibre, 15.2 per cent starch, 19.2 per cent protein, 2.84 per cent fat and 50.5 mg of ascorbic acid in 100 g of pulp. It has a yield potential of 15-16 t/ha in 90 days.

CO 3

It is a F_1 hybrid between Parbhani Kranti x MDU 1 released by TNAU, Coimbatore in 1991. Fruits are dark green, medium in size. It is suitable for cultivation during rainy as well as summer season. It produces yield of about 16-18 t/ha and shows moderately resistant to YVMV.

COBh H 1

This hybrid is a cross between Varsha Uphar selection x PA 4 (T) and released by TNAU, Coimbatore in 2007. The fruits are dark green, slender, less fibrous and sparsely pubescent. It can be cultivated during May-June, November-December and February-March. Total crop duration is 120 days. It has resistance to YVMV and gives yield of about 22.1 t/ha under normal conditions.

Shitla Uphar (DVR-1)

It was developed through heterosis breeding by IIVR, Varanasi, India in 2001. Plants are medium tall (110-130 cm), flowering starts at 38-40 days after sowing at 4-5th nodes. Fruits are green, 11-13 cm long at marketable stage and yields 15-17 t/ha. It is suitable for both spring-summer and rainy season cultivation in Punjab, U.P., Bihar, M.P. It is resistant to YVMV and OLCV.

Shitla Jyoti (DVR-2)

This hybrid was developed through heterosis breeding and released by IIVR, Varanasi, India in 2001. Plants are medium tall (110-150 cm), flowering starts on 30-40 days after sowing at 4-5th nodes. Fruit are green, 12-14 cm long at marketable stage and yields 18-20 t/ha. This hybrid is suitable for both spring-summer and rainy season cultivation and resistant to YVMV and OLCV.

Kashi Bhairav (DVR-3)

It was developed through heterosis breeding by IIVR, Varanasi, India in 2006. Plants are medium tall with 2-3 branches, fruits are 5-ridged, dark green with 11-14 cm length at marketable stage. It is suitable for both spring-summer and rainy season cultivation in all the okra growing areas of the country and gives average yield of 19-22 t/ha. This is resistant to YVMV and OLCV under field conditions.

Kashi Mahima (DVR-4)

This hybrid was developed through heterosis breeding and released by IIVR, Varanasi, India. Plants are tall (130-170 cm), flowering starts at 36-40 DAS at 4-5th nodes, fruits dark green with 11-13 cm of length at marketable stage and yields 20-23 t/ha. This has shown field resistance against YVMV and OLCV.

Kashi Shristi (VROH-12)

This hybrid was developed through heterosis breeding (Kashi Kranti x VRO-102) and released by IIVR, Varanasi, India in 2019. Plants are medium tall (125-135 cm) with short internodes and narrow angled 2-3 branch. Fruits are 5-ridged, dark green in colour and produce yield of 18-19 t/ha. This hybrid is suitable for both summer and rainy season cultivation. It is tolerant to YVMV and recommended for cultivation in Uttar Pradesh.

HBH-142

This is a F_1 hybrid between the cross HRB 105-2-2 x BB1 and released by CCS HAU, Hisar in 2006. Plants are tall (125 cm) with short internodes, leaves are deeply lobed and green in colour. Flowering starts at 3-4 nodes, fruits are smooth, green, and 5-ridged. It has resistance to YVMV and shows tolerance to fruit borer and leaf hopper. It is suitable for rainy season cultivation and produces yield of about 13.5-15 t/ha.

Gujarat Okra Hybrid-2 (JNDOH-02-2)

This F_1 hybrid was developed through heterosis breeding between KS-404 x HRB-108-2 and released by Junagadh Agricultural University, Junagadh in 2008. Plants are tall (158 cm), fruit are dark green, 11 cm long at marketable stage. This hybrid is suitable for rainy season cultivation and tolerant to YVMV, jassids and fruit borer.

Arka Nikitha

It is a F_1 hybrid between GMS-4 x IIIHR-299-14-11-585 developed through genetic male sterile line and released from Indian Institute of Horticultural Research, Bangalore, India in 2017. Early flowering and first female flower appears after 39 DAS on 9th node from the base of the vine and takes 43 days for first picking of fruits. Fruits are dark green, medium, smooth and tender. It has excellent cooking quality and is nutritionally rich in antioxidant activity, high mucilage content (1.08 per cent FW) and high edible fiber content (8.85 per cent DW). It is rich in minerals like potassium (3.7 per cent), calcium (997 mg/100 g), magnesium and iodine (33.31µg/kg). It gives yield of 21-24 t/ha in 125-130 days duration.

Hybrid okra marketed in India by Indo-American Hybrid seeds are **Varsha, Vijaya and Vishal.**

Some cultivars developed and assessed by other countries are given below:

Jade

It is an early maturing, semi-dwarf cultivar of USA and produces dark green, 47.7 fruits per plant on average (Scott *et al.*, 1989).

BARI Dherosh-1

It was developed by Bangladesh Agriculture Research Institute (BARI), Gazipur, Bangladesh in 1996. Flower initiation starts at 45 days after sowing. Plants are 2.0-2.5 m tall and bear 25-30 fruit/plant. The identical characteristic of this variety is presence of brown hairy coating in mature and dry seed. Yield is about 14-17 t/ha.

Cvs. **Ela** and **Beatrice**, now widely grown have been evolved from Research Institute of Vegetables and Floriculture in Vidra, Romania (Ionescu and Ionescu, 1995).

Damarany and Farag (1994) assessed some okra cultivars and reported that **Blondy** was the earliest flowering cultivar and produced the most fruits per plant and greatest total yield.

From Chile, Pertierra and Melin (1998) reported that cv. **NN Claudia** had a yield of 32.9 t/ha in 2 months of harvest, while cv. **Dwarf Green Long Pod** produced only 6.7 t/ha. Harvesting began 69 and 71 days after sowing for **NN Claudia** and **Dwarf Green Long Pod**, respectively.

7.0 Soil and Climate

7.1.0 Soil

Sandy to clay soils, so long as those are well manured, supplied with enough organic matter and with good drainage is fit for okra cultivation. However, loose, friable and well-manured loam soils are the best. For best yield, soil pH should range between 6.0 and 6.8.

7.2.0 Climate

Being a warm season crop it is susceptible to frost. The sowing times and number of crops in a year are determined by the temperature conditions. It thrives best during warm, moist season, although it grows fairly well in the hottest summer. The seeds do not germinate below 17°C and optimum temperature for seed germination is 29°C. When day temperature exceeds 42°C, flowers drop (Chauhan, 1972).

8.0 Cultivation

8.1.0 Land Preparation

The land should be well prepared with 2-3 ploughings. Well rotten farmyard manure should be incorporated at the time of soil preparation. The total amount of phosphorus, potash and half of nitrogen should be added at the time of final preparation of soil.

8.2.0 Sowing Season

In South India, where the winter is mild, it is possible to grow okra all the year round (Singh and Bhagchandani, 1967). In the northern plains, usually two crops, *i.e.*, spring-summer and rainy season crops are raised. The spring-summer crop is sown from the beginning of February to the end of March. It can, however, be sown a little earlier in Assam, West Bengal, Orissa and Bihar. The rainy season crop is sown during May-July. In the hills, the crop is sown in May-June (Anon., 1978, 1983).

Leonardi (1995) studied the adaptability of okra during autumn, winter and spring season grown in a cool greenhouse (0.15 mm thick polyethylene as cladding) and in outdoors. He found that low light intensities affected plant growth and development both in outdoors and under protection, with both winter crops and the spring outdoor crop failing to produce flowers. Autumn and spring greenhouse crops yielded 72.6 and 65.8 g/plant and the autumn outdoor crop only 2.1 g/plant.

Okra sown during 2nd week of June minimizes the incidence of both jassids (*Amrasca biguttula biguttula*) and shoot and fruit borer (*Earius sp.*) (Hazra *et al.*, 2011).

In Bangladesh, sowing on 15th February produces the highest green fruit yield of 9.11 t/ha in cv. Annie Oakley (Dash *et al.*, 2013). Whereas, in Pakistan, 30th March sown crop produces maximum fresh fruit yield (16.24 t/ha) and seed yield (16.01 q/ha) in cv. Sabz Pari (Shahid *et al.*, 2015).

8.3.0 Seed Rate and Seed Treatment

It requires about 18-22 kg seed/ha during summer season, whereas 8-10 kg seed/ha is required for rainy season crop (Nath, 1976). During summer season the vegetative growth is relatively less and hence the seed rate is more than rainy season crop. One ounce of seed contains about 500 seeds with an average germination of 50-60 per cent (Sundaraj *et al.*, 1965). Soaking of seeds for 24 hours in water before sowing will enhance germination. Dhumal *et al.* (1993) observed that seed treatment with 25 ppm GA_3 plus application of fertilizer at 70 per cent of the recommended rate, gave a significant increase in fruit yield (16.48 t/ha) compared with full recommended rate of fertilizer alone (9.64 t/ha).

Soaking of seeds in 45°C water for 1.5 hours before sowing results in rapid field emergence (Onwueme, 1975).

Sharma *et al.* (2014) reported that hydro-priming of okra seeds for 12 hours is beneficial from farmer's point of view. However, hydro-priming for 12 hours and solid matrix (SM) priming with calcium aluminium silicate (1:0.4:1; seed:SM:water) for 24 hours significantly increased the seed germination percentage, seedling vigour, mean germination time and fruit yield up to 55 per cent in comparison to control in northern India and similar environments in other parts of the world in cv. Hisar Unnat.

Osmo-priming of seeds with 5 per cent polyethylene glycol for 24 hours enhancing germination percentage (94.39 per cent) and better seedling growth (28.96 cm) in cv. Punjab 8 (Kaur *et al.*, 2015).

Soil application of vermicompost @ 5 t/ha 10 days before sowing increase the germination percentage (100.00 per cent), emergence speed index (15.17 plants/day) and mean emergence time (6.40 days) (Sarma and Gogoi, 2015).

8.4.0 Method of Sowing and Spacing

Various practices of sowing like broadcasting, line sowing and dibbling methods are followed in various parts of the country (Singh and Bhagchandani, 1969). The seed should be sown on ridges in February-March on the side facing the sun and on flat beds in May-July. The vegetative growth during summer is relatively less, hence a distance of 30 x 15 cm is recommended for spring-summer crop (Singh and Bhagchandani, 1967). For May-July sowing, a distance of 45-60 cm between rows and 25-30 cm between plants is considered best (Anon., 1983). Under Coimbatore conditions, Kamalanathan *et al.* (1970) recommended a distance of 60 x 20 cm for getting better yield. They also recommended its sowing on ridges. Gupta (1990) observed that closest plant spacing of 50 x 10 cm gave the lowest mean fruit weight (11.6 g) and the highest yield/ha (150.8 q) in cv. Pusa Sawani. Raghav (1996) reported that green fruit yield of cv. Pusa Sawani was highest with sowing on 1st March (57.32 q/ha) and at widest spacing of 45 x 30 cm (47.67 q/ha).

Seed Sowing and Germination in Okra

Okra Seed

Line Sowing of Okra Seed

Seed Germination of Okra

In field trials of Ludhiana and Jalandhar, India with cv. Punjab 7, Saimbhi *et al.* (1997) recorded that the fruit yield increased as plant spacing decreased in both spring and rainy season. The highest fruit yields in spring (45 x 15 cm) and rainy (30 x 15 cm and 45 x 15 cm) seasons were significantly higher than those at the recommended row and plant spacings (45 x 25 cm and 60 x 30 cm in spring and rainy seasons, respectively). The 45 x 15 cm spacing was considered most suitable for okra cultivation in both spring and rainy seasons, as a closer row spacing of 30 cm in the rainy season would make fruit picking difficult (Saimbhi *et al.*, 1997).

The effect of two dates (1st and 15th April, 1996) with or without a transparent poly- ethylene film mulch (1.2 m wide and 0.05 mm thick) was evaluated in a field trial in Sicily, Italy by Incalcaterra *et al.* (2000). In mulched plot, plant emergence after 10 days from sowing (100 per cent), fruit production (35 t/ha) and number of fruits per plant (220) were higher compared to those observed in unmulched plots (25 per cent, 24 t/ha and 178 fruits per plant, respectively). Differences in plant emergence were more pronounced for first sowing date than for the second. Sowing on 1st April resulted in higher plant height, yields and number of fruits per plant.

Whitehead and Singh (2000) suggested from their study that okra plants reached maximum CO_2 exchange rate and leaf area index earlier and produced higher fruit yield per unit area when spaced closed together in the row.

Olasantan (2001) reported from Nigeria that when rainfall is limiting, okra could be intercropped with cassava using population density up to 35000 plants/ha to allow the vegetable to develop and to maximize fruit yield under the relatively dry conditions. The optimum target suggested is 50000-60000 plants/ha during the wet periods.

In the field trial under Bangladesh condition with cv. BARI Dherosh-1, Talukder *et al.* (2003) observed that sowing on 15th April with close spacing (60 cm x 30 cm) and picking of fruits at 2 and 3 days interval produce higher fruit yield of 24.94 and 23.79 t/ha, respectively. Similarly, in Barbados, cv. Clemson Spineless gave maximum green fruit yield (23.99 t/ha) at 2 days picking interval with closer spacing of 30 cm x 45 cm in April sown crop (Maurya *et al.*, 2013).

Ram *et al.* (2013) studied that the cv. VRO-6 when sown at a distance of 30 cm x 45 cm produces maximum plant height, number of leaves per plant and number of branches per plant. Whereas, maximum fruit yield of 10.80 t/ha was obtained at closer spacing of 30 cm x 15 cm in eastern U.P during spring season.

Wider spacing of 60 cm x 45 cm and sowing during 2nd week of June the cv. Arka Anamika produces tallest plant (105.63 cm), maximum number of fruits per plant (16.73), highest seed yield (15.57 q/ha) and highest value of vigour index-1 (1348.93) in Madhya Pradesh (Singh *et al.*, 2018).

8.5.0 Manuring and Fertilization

The importance of organic manures and fertilizer in okra production had been reported by Yawalkar *et al.* (1962), who recommended 8 tonnes of farmyard manure, 45 kg N, 22 kg each of P and K per hectare for the production of 2725 kg green fruits.

Agronomic Practices of Okra

Manuring and Fertilization in Okra

Plastic Mulching in Okra

Surface Irrigation (left), and Field view (right)

Hazra *et al.* (2011) reported that open pollinated varieties required 125 kg N, 75 kg P_2O_5 and 60 kg K_2O per hectare whereas, hybrid required requires 200 kg N, 100 kg P_2O_5 and 100 kg K_2O per hectare.

Nitrogen application markedly increased fruit and shoot dry weights whereas phosphorus increased them only moderately. Leaf and primary branch production and plant height were also enhanced by nitrogen fertilization up to 100 kg per ha but were not influenced by phosphorus application (0 to 26 kg/ha). Treatment with N increased the concentration of N, P and K in fruits and N and Mg in leaves while P and K concentrations in leaves were depressed (Majanbu *et al.*, 1986). Ahmed and Tullock-Reid (1968) studied the response of okra to nitrogen, phosphorus, potassium and magnesium fertilization at Trinidad on loam soil and best yields were obtained with 112 kg N, 168 kg P, 280 kg K and 112 kg Mg per hectare. Under Coimbatore conditions, Kamalanathan *et al.* (1970) recommended 40 kg N and 30 kg K per hectare along with 25 tonnes of farmyard manure at the time of soil preparation. In Rajasthan, at Udaipur, Verma *et al.* (1970) recommended the application of 90 kg N and 80 kg P and 10 tonnes of farmyard manure per hectare during rainy season for improving the qualitative and quantitative yield of okra. Shivananda and Iyengar (1990) found that placement at 10 or 15 cm depth below the seed resulted in maximum utilization of applied phosphorus fertilizer.

Ahmed *et al.* (1999a) recorded maximum plant height (1.85 m), number of fruits per plant (24.59) and the highest fruit yield per hectare (16.9 tonnes) from plots with 120 kg N + 90 kg P + 60 kg K/ha. According to Bhardwaj *et al.* (2000) 50 per cent recommended NPK + 50 per cent neem cake (0.72 t/ha) in okra gave higher yield which is statistically at par with recommended doses of chemical fertilizers.

At Kanpur (UP), Singh and Singh (1965a, b), compared the various sources of nitrogen and reported that a dose of 68 kg N per hectare in the form of urea improved the yield. Similarly, Verma *et al.* (1974) conducted experiments at Bichpuri, Agra, on sandy loam soil with slightly alkaline reaction and suggested application of 150 kg urea per hectare for higher yield during summer season and for obtaining best results, one-fourth dose should be applied through soil and remaining three-fourth as foliar spray. In an investigation at the Punjab Agricultural University, Ludhiana, Singh *et al.* (1967) observed 113 kg N per hectare to be an economic dose. It was also recommended that 62.5 kg N per hectare is sufficient on soils of good fertility for both spring and rainy season crop (Anon., 1983). Saimbhi and Padda (1967-68) reported that under Hissar conditions, an increase in the levels of nitrogen application

from 67 kg to 134 kg per hectare increased the yield of okra only when phosphorus level was kept at 28 kg per hectare. In Bangalore, the best yield of fruits was produced with 60 kg N in soil and 20 kg N/ha as foliar spray. Similarly, 80 kg P in soil and 20 kg as foliar spray gave the best yield (Dhuria and Shukla, 1972). In Himachal Pradesh, NPK at the rate of 60, 50 and 30 kg per hectare, respectively, were recommended for getting the best yield (Anon., 1978).

Dwivedi *et al.* (1996) observed that seed yield of cv. Pusa Sawani were highest at 120 kg N and 100 kg P/ha. Pandey and Dubey (1997) reported from an experiment conducted at Port Blair, India that N uptake of cv. Pusa Sawani was highest with an application of 120 kg N and 80 kg P_2O_5/ha planted at a spacing of 30 cm. P uptake was highest at 120 kg N + 120 kg P_2O_5/ha with intra-row spacing of 30 cm while K uptake was highest with an application of 60 kg N + 80 kg P_2O_5/ha at 30 cm intra-row spacing. Somkumar *et al.* (1997) found that fruit yield per plant and yield per ha increased with an increase in nitrogen dose. Parbhani Kranti produced the highest fruit yield (117.11 per plant) and yield per ha (77.70) at 75 kg N/ha. The application of 90 kg N/ha increased plant height by 14.03 per cent, advanced flowering by 4.08 days and increased fruit yield by 67.20 per cent as compared to control (Singh *et al.*, 1998). Increase in nitrogen rates is associated with an improvement in the crude protein content (Sharma and Prasad, 1973), more crude fibre content (Mani and Ramanathan, 1981), and total carbohydrates (Singh, 1979) in fruits.

Seed inoculation with *Azospirillum* increased green fruit yield by 195 per cent as compared to untreated control and by 54.5 per cent compared to the 50 per cent NPK treatment when applied alone (Mishra and Patjoshi, 1995). Ganeshe *et al.* (1998) reported that N at 40 kg/ha in combination with soil-applied *Azospirillum* gave a yield of 56.33 q/ha that was statistically equivalent to that obtained with 80 kg N/ha (56.78 q/ha) and also gave similar fruit nutritional quality. N fertilizer application however decreased the total bacterial soil population.

Selvi *et al.* (1997) observed that foliar (inceptisol) or soil (alfisol) application of micronutrients, with or without composted coir pity (CCP) and biofertilizers, increased the dry matter yield at all stages of growth (vegetative, flowering and harvest) compared with other treatments.

Ushakumari *et al.* (1999) studied the efficiency of vermicompost on growth and yield of okra cv. Anamika and observed that vermicompost applied as an organic source (at 12 t/ha) + the full recommended dose of inorganic fertilizers (50 kg N, 8 kg P_2O_5 and 25 kg K_2O/ha) resulted in the highest yield (5663 kg/ha). Yields were also high following treatment with vermicompost (12 t/ha) + three-quarters of the recommended NPK dose (4680 kg/ha) and treatment with an NPK dose of 25: 8: 25 kg/ha followed by 2 weeks later with vermicompost treatment *in situ* by releasing *Eudrilus eugeniae* earthworms at a rate of 50 worms per sq. m. (4183 kg/ha).

The cv. Parbhani Kranti under Maharashtra condition, produce maximum fruit yield when fertilized with 50 per cent recommended dose of NPK (100:50:50 kg/ha) + 50 per cent N through neem cake (Bharadiya *et al.*, 2007).

In Nigeria, Moyin-Jesu (2007) conducted an experiment to study the effect of plant residues on soil fertility, fruit nutrients, root growth and fruit weight of okra in cv. NHAe 47. They recommend that spent grain, wood ash and cocoa husk @ 6 t/ha as fertilizer helped in improving soil chemical composition, root length and composition of fruit nutrients as compared to sawdust and rice bran. Spent grain, wood ash and cocoa husk due to low C/N ratio showed beneficial effect in reducing soil bulk density and immobilization and detoxification of the toxic cations (*e.g.*, Fe, Mn and Al).

In Bangladesh, Firoz (2009) studied the combined effects of nitrogen and phosphorus during the rainy season under hill slope condition and observed that

N @ 100 kg/ha and P @ 120 kg/ha in combination gave maximum yield (19.22 t/ha) and highest B:C ratio (4.08).

At Pakistan, the cv. Subz Pari produced early flowering (40.2 days), highest plant height (98.4 cm), number of fruits/plant (15.0) and fruit yield (12.9 t/ha) under soil application of chemical fertilizer (nitrogen @ 150 kg/ha and phosphorus @ 75 kg/ha) along with foliar application of super-dawn liquid fertilizer in four equal splits at 30, 45, 60 and 75 days after plant emergence (Abbasi *et al.*, 2010). Maximum uptake of NPK was achieved with application of NPK @ 150:75:75 kg/ha and spacing of 60 cm in cv. Sabz Pari at Tandojam, Pakistan (Solangi *et al.*, 2015).

Highest seed yield (18.88 q/ha) was obtained from the closest spacing (45x20 cm) under application of higher dose of nitrogen fertilizer (150 kg N/ha). Least leaching loss of nitrogen in closest spacing might help in availability of nutrients in adequate amount to increase growth and yield attributing characters (Parmar *et al.*, 2016).

Ghosh *et al.* (2018) reported that maximum fruit weight (11.57 g) and fruit yield (20.75 t/ha) were recorded with an application of 50 per cent vermicompost + 50 per cent recommended dose of NPK.

Integrated nutrient management using organic, inorganic fertilizers and biofertilizers helps in improving the fruit yield and other yield components. Bairwa *et al.* (2009) studied the effectiveness of integrated nutrient management on growth, yield and quality of okra cv. Arka Anamika and found that neem cake at 6 q/ha + vermicompost 10 q/ha + *Azotobacter* + PSB + 60 per cent recommended dose of NPK through inorganic fertilizers resulted in the highest plant height, number of fruits/plant (18.36), fruit weight (17.65 g), fruit yield (13.51 t/ha), protein content of fruits (1.86 g/100 g), total chlorophyll content and NPK content of leaves and higher B:C ratio (3.19). The green fruit yield was increased by 29.30 per cent as compared to recommended dose of NPK (60:30:30 kg/ha) when applied alone. At Bhubaneswar (Odisha), Mal *et al.* (2013) evaluated the performance of okra F_1 hybrid (Mahyco-10) over integrated use of diazotrophs, vermicompost and chemical fertilizers and obtained maximum fruit yield (22.83 t/ha) in treatment with FYM @ 10 t/ha + 100 per cent recommended dose of NPK (200:100:100 kg/ha) + vermicompost @ 5 t/ha + biofertilizers (*Azotobactor, Azospirillum* and PSB (1:1:1) @ 6 kg/ha. However, Phonglosa *et al.* (2015) reported that integrated nutrient management through application of 75 per cent recommended dose of NP (120:60 kg/ha) + total K (50 kg/ha) + *Azospirillum*, PSB, VAM (7.5 kg/ha) + FYM (10 t/ha) restored better soil nutrient availability and higher fruit nutrient accumulation in okra F_1 hybrid (OH-152).

Nagegowda *et al.* (2020) studied the efficiency of combined effect of fertigation and mulching against fertilizer applied in soil in cv. Arka Anamika. They found that NPK fertigation @ 150:75:150 kg/ha in 16 equal splits at 7 days interval with black polythene mulching resulted in significantly higher seed yield (14.05 q/ha) and higher NPK uptake in various plant parts than the fertilizer applied through soil. Maximum fertilizer use efficiency (31 per cent over control) and water use efficiency (3.51 kg ha-mm^{-1}) were also observed with integrated application of fertigation and mulch.

8.6.0 Irrigation

Very little information on water requirements of okra is available under Indian conditions. Seed should be sown when the soil is moist. Enough soil moisture should be maintained to get a good germination stand. The crop should be irrigated at an interval of 4-5 days in summer and whenever required in rainy season. Moisture stress at both flowering and fruit formation stages resulted in a reduction of more than 70 per cent fruit yields (Mbagwu and Adesipe, 1987). At Coimbatore, Sivanappan *et al.* (1974) recommended drip system of irrigation in place of conventional furrow system for economy in water utilization to the extent of 84.7 per cent without any loss of yield. At IIHR, Bangalore, Gupta (1990) observed that irrigation at 20 mm cumulative pan evaporation resulted in the highest mean yield.

In USA, irrigation equal to 60 per cent pan evaporation produced the highest yield (Singh, 1987). In Nigeria, seed yield of cultivar NHae 47-4 was best with 15 mm water/week along with dry grass mulch (Aiyelaagbe and Ogbonnaya, 1996).

Rathod and Patel (1996) observed highest plant height, number of branches/plant, number of leaves/plant, number of fruits/plant and fruit yield with the irrigation treatment of IW/CPE ratio 1:1. Tiwari *et al.* (1998) indicated that 100 per cent irrigation requirement met through drip irrigation along with black plastic mulch gave the highest yield (14.51 t/ha) with 72 per cent increase in yield as compared to furrow irrigation.

Sharma *et al.* (2016) reported that cv. Punjab-8 produced maximum number of fruits/plant (18.67), fruit length (10.83 cm), fruit weight (9.7 g) and fruit yield (15.89 t/ha) when irrigated with 80 per cent drip irrigation of cumulative pan evaporation and fertigation with 80 per cent of recommended dose of N fertilizer.

8.7.0 Weed Control

To keep weeds under control, three to four hoeings are required. The first hoeing may be given when the seedlings are two weeks old and subsequent hoeings be given at fortnightly interval. According to Iremiren (1988) four times weed removal at 3, 6, 9 and 12 weeks enhanced crop growth and fruit yield to a greater extent.

In okra, shorter the initial weed-free period poorer the plant growth and fruit yield. A weed-free period of up to 7 weeks from sowing resulted in yields comparable with those obtained in a weed free situation throughout the period of crop growth. In unweeded plots fruit yield was reduced to 63.9-71.7 per cent (Gogoi *et al.*, 1996).

Application of alachlor @ 1.5 kg/ha immediately after sowing was the cheapest and equally effective as manual weeding (Leela, 1989). Adejonwo *et al.* (1991) reported that metolachlor + prometryn at 2.0 + 1.0 kg/ha followed by supplementary hoeing gave excellent reduction of weeds and highest yield. Pre-emergence application of basalin (fluchloralin) @ 2 kg/ha, N @ 50 kg/ha and polyethylene mulching increased uptake of N, P and in plants and also improved quality of fruits with high ascorbic acid content and low crude fibre content (Vethamoni and Balkrishnan, 1990).

Efficiency of weed control was greatest with pre-emergence application of pendimethalin (0.75 kg/ha) as compared to other herbicides (Saimbhi *et al.*, 1994). Gupta *et al.* (1999) reported that application of alachlor @ 1.5 kg/ha + one hand weeding produced almost similar green fruit yield as that of fluchloralin @ 1.0 kg/ha + one hand weeding and fluchloralin @ 1.0 kg/ha alone.

Pre-emergence application of pendimethalin (1 kg/ha) followed by hand weeding at 30 DAS and post-emergence application of quizalofop-ethyl (75 g/ha) followed by hand weeding at 30 DAS significantly reduced the weed population (Sharma and Patel, 2011). Jalendhar *et al.* (2016) observed that pre-emergence application of oxyfluorfen @ 0.15 kg a.i./ha + one hand weeding at 30 DAS effectively controlled the weed population and influenced the crop growth characters like plant height, leaf area index and total fruit yield (13.27 t/ha).

Oyewole and Obaweda (2020) observed that weeding at every 2 weeks interval gave the highest number of fruits per plant (20.80), fruit length (6.70 cm), fruit diameter (5.75 cm), fruit weight (98.04 g), fruit yield (20.8 t/ha) and significantly reduced the number of pests (red spider mites, aphids, vegetable leaf miner and white fly).

8.8.0 Mulching

In general, mulching (with sawdust, polyethylene film or straw) increased production in okra and polyethylene mulch gave the best performance (Batra *et al.*, 1985).

Saikia *et al.* (1997) reported that black, low density polyethylene film (LDPE) mulch promoted the growth of okra and resulted in yields of 223.7 q/ha, as compared to 31.1 q/ha for unmulched controls. Black LDPE also had the highest cost-benefit ratio of 1:3.1, gave the maximum weed control efficiency (96.5 per cent) and maintained higher moisture and temperature regimes in the soil. Transparent film was only better than black film when combined with herbicides.

Patel *et al.* (1999) reported that among different mulches, black plastic mulch enhanced maximum green fruit yield in cv. GOH-1 with 12.04 per cent increase over the control.

In Sri Lanka, okra grown on temperature regulated poly tunnels (34°C) with straw mulch produce 37 per cent higher fruit yield in Red Okra (indigenous variety) (Godawatte and De Silva, 2016) and 27 per cent higher in cv. Haritha (Gunawardhana and De Silva, 2011). The high temperature stress hastens early harvest of fruits at 5 to 6 days after full bloom without significant yield reduction.

Gonzaga *et al.* (2017) reported that among different mulches, black plastic mulch produced the highest fruit yield (26.87 t/ha) followed by rice straw mulch (22.95 t/ha) under marginal upland conditions of Philippines. However, use of silver plastic mulch enhanced plant height (134.9 cm), leaf number (26.33), leaf length (40.98 cm), stem girth (2.49 cm) and fruit yield (8.10 t/ha) of okra in Nepal (Jha *et al.*, 2018).

In Nigeria, use of legume mulching during the wet season (early March sowing) and grass mulching in the dry season (early October sowing) maximized the fruit yield in cv. NHAe-47-4. Due to low C:N ratio, legume mulches (*Pueraria phaseoloides* and *Mucuna pruriens*) decomposed faster in wet season as compared to grass mulches (*Pennisetum purpureum* and *Panicum maximum*) with high C:N ratio. But use of mulch during the dry season had a much greater economic benefit compared with wet season (Adekiya *et al.*, 2017).

8.9.0 Growth Substances

The highest average fruit set and yield were obtained with cycocel at 100 ppm as seed soaking for 24 hours + foliar treatment at 20 and 40 days after sowing (Arora *et al.*, 1990).

According to Rai *et al.* (1998) fibre quality of okra was generally improved by foliar application of 10 ppm NAA. Rathod and Patel (1996) noticed that application of cycocel at 750 ppm produced the lowest plant height and the highest number of branches, leaves and fruits/plant. Treating seeds with GA_3 @ 45 ppm increased plant height by 8.97 per cent, advanced flowering by 3.33 days and increased fruit yield by 30 per cent (Singh *et al.*, 1998).

Foliar application of cultar (100 ppm) in cv. T13 gave the highest yield/ha of 14.32 t/ha as compared to 11.87 t/ha with cultar at 200 ppm and 11.97 t/ha in case of control (Asghar *et al.*, 1997).

Gasti *et al.* (1997) reported that application of 175 ppm mepiquat chloride at 45 days after sowing resulted in the highest yield of 13.60 t/ha. The increase in yield was 44 per cent over that of the unsprayed control as a result of retarded vegetative growth, which in turn enhanced the performance of yield-related characters. Application of NAA 50 mg/l along with black plastic mulch recorded highest green fruit yield in cv. GOH-1 (Patel *et al.*, 1999).

Kokare *et al.* (2006) reported that foliar spray of NAA @ 200 ppm produced highest fruit yield (13.88 t/ha) in cv. Parbhani Kranti. The cv. Utkal Gaurav gave maximum fruit yield (8.92 t/ha) when seeds were treated with NAA @ 50 ppm (Sanodiya *et al.*, 2017a). However, it produced fruit yield of 9.66 t/ha when 50 ppm NAA was applied as foliar spray at 40 days after sowing (Sanodiya *et al.*, 2017b).

Combined application of GA_3 and NAA @ 200+100 ppm applied at 2-leaf stage increased the plant height and number of leaves per plant. Whereas, the number of fruits per plant, fruit length, fruit fresh and dry weight, seed yield, germination percentage and 1000-seed weight was maximum in plants receiving foliar spray of both GA_3 and NAA @ 200+200 ppm (Shahid *et al.*, 2013).

Rani *et al.* (2013) reported that spraying of 50 ppm GA_3 at 20 and 40 days after sowing results in early 50 per cent flowering (36.6 days), maximum number of fruits per plant (22.6) and fruit yield (19.9 t/ha) in cv. Arka Anamika. They also observed that foliar application of micronutrient (0.4 per cent $FeSO_4$) produced maximum fruit yield of 12.98 t/ha.

Foliar application of GA_3@ 100 ppm + soil application of borax @15 kg/ha gave high fruit yield (14.85 t/ha) and seed yield (77.48 g/plant) with higher B:C ratio (3.14) in cv. Kashi Kranti (Kumari, 2016).

Chormule and Patel (2018) reported that seed treatment with GA_3 @ 150 ppm before sowing for 8 hours had produced the maximum seed yield per plant and good quality seed after harvest in terms of 100 seed weight, seed germination percentage, seedling length, seedling dry weight and seedling vigour index in cv. GJO 3.

Pateliya *et al.* (2008) reported that the foliar application of CCC @ 300 ppm twice at 25 and 50 days after sowing recorded early flowering and significant response on all yield characters with higher B:C ratio (3.53) in cv. GO-2 during the summer season. Whereas, higher concentration of CCC @ 600 ppm at 30 days after sowing resulted in early flowering, fruit production, fruit yield (16.15 t/ha) and fruit quality (vitamin A and crude fibre per cent) in cv. Varsha Uphar (Kumar *et al.*, 2018).

The cv. BARI Dherosh-1of Bangladesh, produced 19.62 t/ha fruit yield when sprayed with 100 ppm GA_3 + soil application of poultry manure @ 11.5 t/ha (Chowdhury *et al.*, 2014.) whereas, the same cultivar gave fruit yield of 16.25 t/ha when NAA @ 100 ppm was applied as foliar application at 25 days after sowing + soil application of 70 kg P_2O_5 per ha (Afrin, 2016).

9.0 Harvesting, Yield and Storage

9.1.0 Harvesting

The tender young fruits, 7-10 cm long should be harvested every alternate day. Frequent picking promotes fruit development and hence yield is increased. The total number of fruits per plant increased with the increase in number of pickings to a maximum of 11.58 fruits per plant with 6 pickings in cv. Kiran (Isaac *et al.*, 1998). The time of harvesting for the production of maximum yield of edible fruits is 7-8 days after pollination for cv. Pusa Sawani (Kolhe and Chovan, 1967). Singh *et al.* (1990) however indicated that fruits should be harvested 6-9 days after fruit set for maximum crop yield and nutrient values. Iremiren *et al.* (1991) noticed that fruits harvested more than 7 days after fruit set were of poorer quality mainly due to an increase in crude fibre and reduction in the moisture, crude protein and ash contents of older fruits. According to Hazarika *et al.* (1997) the fruits of okra (cv. Pusa Sawani) are ready for harvest from the 6th day of flowering onwards, but fruits harvested on or after the 8th day of flowering became fibrous with prominent seeds and were not suitable for consumption.

Piloo and Kabir (2011) reported that up to 5th day of fruit set fruits became excellent to good in terms of sensory quality (sensory score 1 to 2), after that fruits deteriorated in quality and became more fibrous and unsuitable for consumption.

Barnwal *et al.* (2017) reported that harvesting of fruits between 4-8 days after flowering was ideal in terms of optimum fresh weight, length, diameter, free from thick pericarp, maximum ascorbic acid content, phenol content and thus making it consumable in cv. Kashi Pragati and Kashi Kranti. Whereas, optimum seed yield and seed quality (germination percentage and vigour) were obtained when fruits were harvested at 38 days after flowering.

Harvesting of okra is laborious process and the hair like bristles on the fruit causes irritation to the fingers therefore, to minimize this problem sometimes gloves are used.

Harvesting and Packaging of Okra

Harvesting of Fruits

Packaging of Okra

Selling of Okra in the Market

9.2.0 Yield

On an average, the yield of okra varies from 65 to 75 quintals of green fruits per hectare during spring-summer and 115-125 quintals per hectre during the rainy season.

9.3.0 Storage

Okra fruits are highly perishable due to high moisture content and high respiratory activity (>120 mg CO_2/kg/hr) (Burzo, 1980) and are susceptible to rapid deterioration soon after harvest. Therefore, proper packaging and storage of fruits is necessary if not consumed immediately.

Okra can be stored for 7-10 days at a temperature of 8.7-10°C and relative humidity at 90-95 per cent (Rice *et al.*, 1990). Baxter and Waters (1990) observed that fruits stored in a CA of 5 per cent O_2 and 10 per cent CO_2 at 11 ± 1°C had greater retention of sugars, soluble proteins, amino acids and citric acid, malic and ascorbic acid than air-stored fruits.

According to Ratnapala *et al.* (1997) longest storage life was exhibited by fruits stored in polyethylene bags at 11 ± 2°C; immature and mature fruits maintained an acceptable appearance for 5 and 7 days, respectively. In addition, this storage condition reduced weight loss and crude fibre content when compared with the other storage conditions tested.

Finger *et al.* (2008) observed that wrapping of fruits with PVC film immediately after harvest and storing them at 5 or 10°C reduces the symptom of chilling injury (pitting in fruits) in Brazilian okra cv. Amarelinho.

According to Ngure *et al.* (2008) dipping of fruits in hot water (50°C for 1 min) followed by storage at room temperature (15-20°C) increase the shelf life of fruits for 21 days. This treatment enhanced visual appearance and reduced weight loss, electrolyte leakage, off odour and decay of fruits in cv. Pusa Sawani.

LDPE packaging bags (15 x 15 cm) extended the marketable life of fruits with lowest weight loss up to 9 days at room temperature (28 ± 2°C) and more than 9 days under refrigeration (15 ± 2°C) (Babarinde and Fabunmi, 2009). However, fruits can be stored at low temperature (8 ± 1°C) up to 12 days in polypropylene packets (100 gauge; 20.5 cm x 25.0 cm) with 0 per cent perforation (Piloo and Kabir, 2015).

Chukunda and Nwonuala (2013) however indicated that fruits were stored for an optimum period of 9-10 days when packed in cartons lined with bitter leaf (*Vernonia amygdalina*) and stored at room temperature (20°C and 78 per cent relative humidity). The carton storage method also helped in transportation of fruits from the outlying villages to the market with minimum percentage loss of moisture.

10.0 Diseases and Pests

10.1.0 Diseases

Yellow vein mosaic virus and *Cercospora* diseases are often severe in Northern India, while powdery mildew is a serious disease in the southern part of the country. Yellow vein mosaic virus disease is almost a limiting factor for successful cultivation of okra in India. However, the important diseases affecting this crop are discussed below.

Fungal Diseases

10.1.1 Damping Off

The disease is mostly caused by *Pythium* sp., *Rhizoctonia* sp. and *Fusarium* sp. and it results in both pre- and post-emergence death of seedlings. The treatment of seed with 3 g captan or thiram per kg of seed before sowing is recommended for its control.

10.1.2 Powdery Mildew (*Erysiphe cichoracearum*)

White minute patches first appear on the upper surface of older leaves and then gradually spread to younger ones. As the disease progress, white greyish powdery coating is seen on the surface of severely affected leaves. The leaves turn yellow and finally drop. The disease causes yield losses of about 17 to 86.6 per cent (Sridhar and Sinha, 1989). Application of 0.2 per cent wettable sulphur at 1 or 2 week interval will control the disease.

Three sprays of 0.05 per cent tridemorph or 0.2 per cent sulfur after the appearance of disease symptoms can be recommended for powdery mildew control in seed crop (Singh *et al.*, 1998).

Two sprays of 0.1 per cent Bavistin followed by two sprays of 0.1 per cent karathane or tridemorph effectively control the disease (Hazra *et al.*, 2011). However, lowest disease incidence (18.51 per cent) and disease severity (9.86 PDI) with maximum disease control (84.09 per cent) were obtained with four spray of 0.1 per cent karathane (Dinocap 46 per cent EC) (Khalikar *et al.*, 2011).

Application of talc-based biological fungicide Bio-Dewcon (*Ampelomyces quisqualis* 1.15 per cent WP) @ 3.0 kg/ha and wettable sulphur 0.4 per cent at 30 and 45 days after sowing were found to be equally effective in controlling powdery mildew disease in okra (Gopalakrishnan and Valluvaparidasan, 2009). Early sowing (1st July) of okra reported least disease index under Dharwad, Karnataka condition (Sataraddi *et al.*, 2009).

Derbalah *et al.* (2011) reported that four sprays with 100 per cent culture filtrate of *Epicoccum nigrum* and combination of 75 per cent culture filtrate of *E. nigrum* + 25 per cent recommended dose of penconazole gave higher efficiency against powdery mildew disease in Egypt.

Three sprays of sterol biosynthesis inhibiting (triazole group) fungicides like difenoconazole 25 per cent EC, hexaconazole 5 per cent EC and propiconazole 10 per cent EC were effective against powdery mildew disease (Siddappa *et al.*, 2014; Dahivelkar *et al.*, 2017; Sharma *et al.*, 2017; Waychal *et al.*, 2018).

10.1.3 Cercospora Leaf Spot (*Cercospora* sp.)

The disease is mostly caused by two species *Cercospora abelmoschi* and *C. malayensis*. Sooty black angular spots are the first sign of infection by *C. abelmoschi* whereas, small brown irregular spots surrounded by red or pink colour are typical symptom for leaf spot caused by *C. malayensis*. Generally, symptoms first appear on the older leaves and later newer lesions appear on the upper leaves. As the disease advances, the leaf spots enlarge and cover the entire leaf, which then turns necrotic and often rolls as it dries, but remains attached to the stem. Rainfall and high humidity favour infection, disease development and sporulation of the pathogens on the okra leaves. It causes complete defoliation in seed crop with a yield loss of up to 43.6 per cent (Pandey and Pandey, 2003).

Foliar spray of tebuconazole (0.1 per cent) three times at 10 days interval recorded minimum disease incidence (0.17 per cent) and maximum fruit yield 14.51 t/ha (Kumar, 2015). However, foliar spray of *Pseudomonas fluorescens* @ 10 g/kg was effective to minimise the disease intensity (19.67 per cent) (Bochalya *et al.*, 2017).

Naik and Jayalakshmi (2017) observed lowest disease severity of 11.79 and 16.31 with two sprays of propiconazole 25 EC @ 0.1 per cent and carbendazim 12 per cent + mancozeb 63 per cent WP @ 0.2 per cent, respectively at 8 days interval.

Viral Diseases

10.1.4 Yellow Vein Mosaic Virus

The disease was first reported in Bombay, India by Kulakarni (1924) and the viral nature of the disease was established by Uppal *et al.* (1940) and the disease was named as "yellow vein mosaic".

Sastry and Singh (1974) reported that yellow vein mosaic is the most destructive viral disease in okra, infects at all the stages of the crop and severely reduces growth and yield. This viral disease occurs throughout India wherever it is grown. This disease was described by Capoor and Varma (1950). Investigations carried out to find out the effect of yellow vein mosaic virus infection on the growth and yield of okra at different stages of plant growth revealed that the growth of the plants is very much retarded and the plants produced comparatively fewer leaves and fruits when the infection occurred within 35 days following germination.

The disease is now wide-spread in subtropical regions during rainy season from June to September and in tropical region during spring-summer season from February to June (Capoor and Verma, 1950; Sastry and Singh, 1974).

Varma (1952) studied the relationship of the yellow vein mosaic virus and its vector, the white fly (*Bemisia tabaci*). Though a single insect was able to transmit the virus, the minimum number of flies required to produce 100 per cent infection was about 10. The white flies could secure the virus from diseased plants after feeding for one hour and the viruliferous insects could transmit the virus to healthy plants after feeding on them for 30 minutes. The white flies which had fed on diseased plants for 12-24 hours remained infective throughout their lives. The female white flies have been found to be slightly better than males with regard to the transmission of the virus. A late attack of the disease reduces the yield of fruit by over 25 per cent but when the attack is at early stage, the crop is often a total loss. YVMV disease infects the crop at all growth stages (Verma, 1952) and 100 per cent plant infection in the field is quite usual with yield loss ranging between 50 and 94 per cent (Sastry and Singh, 1974; Hamer and Thompson, 1957). Plants infected in the early stages

Fungal Diseases of Okra

Fusarium Wilt of Okra

Powdery Mildew of Okra

Cercospora Leaf Spot of Okra

remain stunted and leads to yield losses of 12 per cent -19.5 per cent (Stephan *et al.*, 2008). The disease caused 50-100 per cent loss in yield if the plants infected within 20 days after germination (Sanwal *et al.*, 2014a). Plants infected 50 and 65 days after germination suffer a loss of 84 per cent and 49 per cent, respectively (Sastry and Singh, 1974). Nath and Saikia (1992) reported the maximum (94.42 per cent) and minimum (32.65 per cent) yield losses when plants were infected at 35 and 63 days after sowing, respectively.

The first visible symptom is the clearing of small veins, which usually starts at various points near the leaf margins in about 15-20 days after inoculation of plants. The vein clearing within 24 hours of its appearance distinguishes itself as vein-chlorosis which rapidly extends to most of the leaf. If the infection takes place at the later stage, the earlier formed leaves on the main stem remain green, whereas the top leaves and the flowering part and also the secondary branches show clear symptoms. On such plants although fruits are formed, they are uniformly yellow at the picking stage and hence unmarketable.

At present no definite control measures against yellow vein mosaic are known. Protection of the crop from white flies and other insects by spraying with suitable insecticides may reduce the disease incidence. Sastry and Singh (1974) have stated that spraying at the initial stage of the crop, just after germination, is most important. If the crop is not sprayed within 20 days after germination disease incidence remains high.

Pun *et al.* (1999) found that spraying of 10 per cent leaf extract of *Prospos chilensis* and *Bougainvillea spectabilis* on the primary leaves of seven days old okra seedlings @ 2.5 ml/plant reduced the YVMV infection by 83.3 per cent and 81.7 per cent, respectively.

In pot experiment in glasshouse, Pun *et al.* (2000) observed that yellow vein mosaic virus inhibitory chemicals, barium chloride was the most effective (76.67 per cent reduction in infection in comparison with the control) followed by acetyl salicylic acid @ 200 ppm and salicylic acid @ 500 ppm, which gave reductions of 75.0 and 73.3 per cent, respectively. Neem oil (3 per cent) and neem kernel extract (5 per cent) reduced virus infection by 88.3 and 86.7 per cent, respectively.

Destruction of weed hosts, whenever possible, should also be given importance. No cultivar of okra is immune to the disease. In the year 1962, Pusa Sawani cultivar was evolved as tolerant to this disease, but gradually it has become completely susceptible. Immunity to the virus has been reported in several wild species of *Abelmoschus* and *Hibiscus* (Nariani and Seth, 1958) and these can be used for breeding disease resistant cultivars. However crossibility among the wild types and cultivated *Abelmoschus* poses problem of transferring resistance from the wild types to the cultivated ones. Mathew *et al.* (1993) recorded lowest incidence of yellow vein mosaic gemini virus in cv. Sel 4 (2.31 per cent) and Arka Anamika (3.02 per cent).

Atiri (1990) suggested that expensive control measures against leaf curl virus disease or its vector is unnecessary for lines in which the symptoms appear only after the commencement of fruiting because it does not significantly reduce fruit yield. This trait may be incorporated into commercial cultivars. Since both mosaic

Viral Disease of Okra

Okra Plants Infected with YVMV Disease

Progress of YVMV Disease on Leaves of Okra

YVMV Disease Symptom on Plants and Fruits of Okra

Viral Disease of Okra

Enation Leaf Curl Virus Disease of Okra

Different Stages of Enation Leaf Curl Virus Infection

and leaf curl viruses cause more damage in mixed infections, it may be preferable to breed for combined resistance, particularly if the genes for resistance are linked (Atiri and Ibidapo, 1989).

Ansar *et al.* (2014) reported that seed treatment with imidacloprid @ 0.5 g/l and sowing of two rows of maize border along with spraying of imidacloprid @ 0.5 ml/l + neem oil @ 2.0 ml/l until fruit formation showed least incidence (15.47 per cent) of disease.

10.1.5 Okra Enation Leaf Curl Virus

Okra enation leaf curl virus disease (OELCV/OELCD/OELCuD) was first reported from Karnataka (Bangalore), south India in the early 1980s (Singh and Dutta, 1986; Singh, 1996). The disease caused by monopartite single-stranded DNA begomovirus and is neither sap nor seed transmitted, transmission occurs through the whitefly insect vector (Venkataravanappa *et al.*, 2014). This disease can cause significant yield losses ranging from 30-100 per cent (Singh, 1996; Chakraborty *et al.*, 1997) depending upon the age of the plant at the time of infection. It is an emerging problem for okra cultivation and going to be future menace of okra cultivation (Yadav *et al.*, 2018). The disease initially causes small pinhead enations on the under surface of the leaves, followed by a warty and rough texture of leaves. The characteristic symptoms are vein-thickening, upward curling of leaves and decrease in leaf surface area. Infected plants show twisting of the stem, petioles and lateral branches with thick and leathery leaves. In severely infected plants, newly emerged leaves also exhibit bold enations, fruit setting is drastically reduced and seeds may be aborted. The disease can be managed by controlling the whitefly population.

10.2.0 Pests

There are few insect pests such as jassid, shoot and fruit borer, white fly, aphid and red spider mite which are important in okra. Of these, jassids and shoot and fruit borer cause severe damage to the crop.

10.2.1 Jassid (*Amrasca biguttula biguttula*)

Both nymphs and adults of this wedge shaped pale green leaf hopper cause damage to the crop by sucking sap and injecting some toxic material, with the result that the leaves turn pale and curl upwards. In case of severe infestation, the leaves have a burnt look and fall down. Ultimately, the yield is very much reduced up to 54-66 per cent (Satpathy *et al.*, 2004).

Seed treatment with imidacloprid 70 WS or thiamethoxam 25 WG @ 3 g/kg of seed was found to be effective against sucking pests of okra up to 25-30 days after sowing (Rai *et al.*, 2014).

Dhanalaxmi and Mallapur (2008) observed that among the different newer molecules tested, imidacloprid 200 SL @ 0.5 ml/l and acetamiprid 20 SP @ 0.2 g/l were the most effective against aphids and jassids in okra. In another study, Patil *et al.* (2014) reported that thiamethoxam 25 WG @ 0.006 per cent and 0.008 per cent were effective against jassid population in okra with three sprays given at fortnightly interval.

In search of newer eco-friendly safer molecules, Gosalwad and Wadnerkar (2009) reported that fipronil @ 11.25 g a.i./ha favoured maximum population of the predatory lady bird beetle and *Chrysoperla* spp., followed by acetamiprid @ 15 g a.i./ha when sprayed against the sucking pests of okra.

Kodandaram *et al.* (2017) suggested a chemical module comprising seed treatment with imidacloprid 70 per cent WS @ 3 g/kg of seeds + sprayings of thiamethoxam 25 per cent WG @ 0.5 g/l at 25 days after sowing (DAS) + spraying of indoxacarb 14.5 per cent at 35 DAS + spraying of dimethoate 30 per cent EC @ 2 ml/l at 45 DAS + emamectin benzoate 5 per cent SG @ 0.5 g/l at 55 DAS + spraying of cypermethrin 25 per cent EC @ 1 ml/l at 65 DAS for effective reduction of jassid population (74.26 per cent reduction over control).

10.2.2 Shoot and Fruit Borer (*Earias vittella* and *E. insulana*)

At the larval stage, they bore into the growing shoots, flower buds, flowers and fruits of okra, either killing the plants or causing severe shedding of fruit. The affected fruits become distorted and unfit for human consumption. Its infestation starts from the beginning of the crop growth and damage due to the borer varies from 21.33 to 43.99 per cent in shoots and 21 to 51.3 per cent in case of fruits (Singh *et al.*, 2007a). Hollyhock (*Alcea rosea* Linn.) which often acts as off-season (November to February) alternate host plant by providing food and shelter to bollworms (Halder *et al.*, 2016a) should be discarded from major okra growing regions. Similarly, ratooned cotton also act as an alternate source of this borer pests and may be uprooted. The infested fruits should be removed regularly and buried deep into the soil.

Hirekurubar and Ambekar (2008) assessed the bio-efficacy of novel insecticides against *Earias* spp. infesting okra. They found out that indoxacarb was the most effective insecticide in reducing the okra shoot and fruit borer damage by registering 5.14 and 4.16 per cent damage on number and weight basis, respectively.

In another study, spraying of cypermethrin 25 per cent EC @ 0.5-0.75 ml/l or emamectin benzoate 5 per cent SG @ 0.35-0.5 g/l or lambda-cyhalothrin 5 per cent EC @ 0.75-1.0 ml/l of water was found to be effective against okra shoot and fruit borer (Rai *et al.*, 2014).

Birah *et al.* (2012) reported that integrated module including seed treatment with imidacloprid @ 5 g/kg seed one day before sowing + sowing of maize at the borders as barrier crop + weekly clipping of infested shoots and fruits + erection of pheromone trap @ 100 traps/ha + foliar spray of NSKE @ 30 ml/l, spinosad 45 SC @ 0.5ml/l and karanj oil @ 30 ml/l at 45, 60 and 75 days after sowing, respectively significantly reduces the incidence of shoot borer (4.23 per cent), fruit borer (5.64 per cent), jassids (3.32 jassids/leaf) and whitefly population in okra. Recently, Kodandaram *et al.* (2017) also reported that integrated module comprising sprayings of chlorantraniliprole 18.5 per cent SC @ 0.15 ml/l at 25 days after sowing (DAS) + NSKE 4 per cent at 35 DAS + emamectin benzoate 5 per cent SG @ 0.5 g/l at 45 DAS + *Bacillus thuringiensis* (*Bt*) @ 2 ml/l at 55 DAS and nimbecidine @ 5 ml/l at 65 DAS and their need based rotation was most effective in reducing the fruit borer damage (71.74 per cent) and yellow vein mosaic disease (17.75 per cent) with significant increase in the yield (17.77 t/ha) over control.

Acaro-insecticide bifenthrin 8 SC @ 100 to 140 g a.i./ha effectively controlled shoot and fruit borer without any adverse effect on population of beneficial insects like spider, predatory mite and *Braconid* parasitoid in okra crop ecosystem (Roy *et al.*, 2017).

10.2.3 White Fly (*Bemisia tabaci*)

It generally does not cause significant damage by direct feeding but serves as vector and spreads yellow vein mosaic and enation leaf curl disease in okra. If this pest is not controlled in time, the diseases spread and cause extensive damage and severely reduce the yield (Anon., 1983).

Praveen and Dhandapani (2001) observed that application of the predator *Chrysoperla carnea* (25,000 larvae/ha/release) plus econeem 0.3 per cent (0.5 l/ha) thrice at 15 days interval starting from 45 days after sowing could effectively checked the population of sucking pests (*Amrasca biguttula biguttula, Bemisia tabaci, Aphis gossypii*) as well as fruit boring insects, *Helicoverpa armigera* and *Earias vitella*.

Venkataravanappa *et al.* (2011) reported that the seed treatment with imidacloprid FS (5 g/kg) and 4 sprayings of imidacloprid SL (0.5 ml/l) in combination with neem oil 0.03 EC (2.0 ml/l) at 15 days interval starting from two weeks after germination significantly reduced the whitefly population. Similar results were reported by Dhar and Bhattacharya (2015) that single application of imidacloprid 17.8 per cent SL followed by two applications of spinosad 45 per cent SC gave maximum reduction in infestation of whitefly in okra. This treatment was also effective against reduction of disease incidence of yellow vein mosaic and enation leaf curl virus disease in okra due to effective management of whitefly acting as vector.

Rai *et al.* (2014) documented that inoculate release of *Chrysoperla zastrowi sillemi* @ 50,000 first instar grubs/ha at weekly intervals were effective in controlling sucking pests complex of okra as well as eggs and early instars larvae of fruit borers. They also reported that seed treatment with imidacloprid 70 WS or thiamethoxam 70 WS @ 3 g/kg of seeds was effective in reducing the infestation of whiteflies up to 25-30 DAS.

In a recent study, spinosad 45 SC was found to be the best treatment against whiteflies in okra followed by acetamiprid 20 SP and sequential spray of neem oil @ 2 per cent and buprofezin 25 SC. The lowest okra YVMV incidence was reported in thiamethoxam 70 WS (16.56 per cent) followed by imidacloprid 70 WS (16.90 per cent) and spinosad 45 SC (19.43 per cent). Maximum marketable fruit yield of 12.2 t/ha was recorded in thiamethoxam 70 WS followed by imidacloprid 70 WS (11.5 t/ha) (Manju *et al.*, 2018).

10.2.4 Red Spider Mite (*Tetranychus cinnabarinus* and *Tetranychus urticae*)

The nymphs and the adults attack leaves and suck cell-sap. The affected leaves turn yellow, dry up and finally wither away giving an unhealthy appearance and ultimately result in defoliation. Dicofol alone and in combination with neem pesticide (azadirachtin 0.03 per cent) resulted in 100 per cent mortality of spotted spider mite (*Tetranychus urticae*) after 14 days (Singh and Singh, 1999).

Spiromesifen 22.9 per cent SC @ 0.8-1.0 ml/l or fenpropathrin 30 per cent EC @ 0.3 ml/l or dicofol 18.5 per cent EC @ 2.5 ml/l of water was found to be effective against red spider mite infesting okra (Rai *et al.*, 2014). Similar report was also presented by Shukla (2018) that two sprays of fenpyroximate 5 EC @ 25 g a.i./ha significantly controled red spider mite.

Roy *et al.* (2017) reported that foliar spray of bifenthrin 8 SC @ 100 to 140 g a.i./ ha was found to be effective against red spider mite population in okra. Rajeshwari *et al.* (2018) reported that organic salt 30 WS @ 5 ml/l effectively reduced the mite population (2.59 mites/cm^2 leaf) on okra.

10.2.5 Nematodes

10.2.5.1 Root-knot Nematode *(Meloidogyne incognita)*

Root-knot nematodes are one among major group of plant parasitic nematodes and have become major threat to okra cultivation. They are obligate, sedentary endoparasite and cause annually an average of 19.5 per cent crop loss and an amount of 2480.86 million rupees of annual monetary loss in okra crop (Kumar *et al.*, 2020). Root-knot nematodes typically detected first in localized areas exhibit patchy growth within infected field. Affected plants usually express formation of knots or galls on roots and above ground symptoms mostly like nutrient deficiency or drought stress symptoms such as chlorosis, yellowing of leaves, wilting and stunted growth.

Jain and Gupta (1997) reported 73.3 per cent reduction in the *M. javanica* population in the summer ploughing treatment and 78.5 per cent reduction in ploughing, followed by covering with transparent polythene sheet for 2 weeks treatment. Fallowing without ploughing recorded a reduction of 60.7 per cent in the nematode population. Summer soil solarisation treatment together with the use of carbosulfan-treated okra seed at 3 per cent a.i. (w/w) resulted in increased yield/plot and lowered root-knot index.

In a field experiment conducted by Siddiqui and Alam (1997), oil-seed cakes of castor, neem, groundnut, mustard, rocket lettuce and aldicarb and carbofuran were found to be highly effective against plant parasitic nematodes inhabiting the rhizosphere of okra cv. Pusa Sawani. The nematicides were found slightly more effective than the oil-seed cakes and compost manure failed to reduce the population build-up of nematodes.

In pot experiment, Wani *et al.* (1999) observed that soil amendments with urea coated with Nimin (a neem – *Azadirachta indica* based product rich in triterpenoids) was the most effective treatment for the control of root-knot nematode.

At Nigeria, Tanimola and Akarekor (2014) revealed that poultry manure @ 100 kg/ha after three weeks of sowing suppressed the incidence of *Meloidogyne incognita* by reducing the gall index by 65.2 per cent and nematode population by 86.7 per cent.

Singh *et al.* (2014) reported that the combination of two bioagents such as egg parasitic fungus (fungal bio-agent), *Purpureocillium lilacinum* (*Paecilomyces lilacinus*) and plant growth promoting rhizobacteria, *Pseudomonas fluorescens* applied as seed

Insect-Pests and Root-Knot Nematode of Okra

Whitefly-Vector of YVMV Disease

Jassids Infestation

Okra Shoot and Fruit Borer Infestation and Root-knot Nematode

treatment (10 g/kg seed) as well as soil application (10 kg/ha) with 1.5 t/ha FYM significantly reduced 82.0 per cent of root galls, 82.4 per cent of egg masses/plant and 77.0 per cent of final nematode population of *M. incognita* in okra crop under field condition. However, aqueous extracts of *Datura stramonium* (20 per cent w/v) at 1:5 concentrations recorded maximum juvenile mortality (97 per cent). Soil application of *Trichoderma viride* @ 20 g/kg and *Azadirachta indica* leaves @ 30 g/kg of soil reduced the nematode population and increases the plant growth parameters (Sidhu *et al.*, 2017).

Gowda *et al.* (2017) reported that seed treatment with *Pseudomonas fluorescens* 1 per cent W.P. (2×10^8 cfu) @ 20 g/kg seed and soil application of FYM (5 t/ha) enriched with 2.5 kg of each *Paecilomyces lilacinus* (2×10^6 cfu/g) + *Pseudomonas fluorescens* (2×10^8 cfu/g) have been found to be highly effective in reducing the root knot nematode disease which ultimately results in increasing in marketable yield of okra. They also found that seed dressing with carbosulfan 25 DS @ 3 per cent a.i. (w/w) managed root knot nematode incidence in okra.

Application of *Bacillus thuringiensis* isolates (B.t-64) reduced gall formation up to 70 per cent and 29 per cent under greenhouse and field conditions, respectively. Population of root knot nematodes and eggs/egg mass in the rhizosphere were decreased up to 61 per cent and 56 per cent, respectively under field condition in Pakistan (Abbasi *et al.*, 2019).

11.0 Seed Production

No special agronomic practices are needed to be followed for the crop raised for seed production. It is an often cross-pollinated crop and an isolation distance of 200 metres between cultivars is recommended for maintaining the purity of seed (Anon., 1971). The field isolation distance of 500 meters in the case of foundation seed and 250 meters in the case of certified seed is maintained for the production of genetically pure seeds (Trivedi and Gunasekaran, 2013). The growth habit, leaf and fruit shape, the extent of hairiness and pigmentation are useful to locate off-type plants and rouging should be done before flowering. It has been observed that growers raise a dual purpose crop and no difference in seed yield has been found if no picking is done or only one or two pickings are made.

Soaking seeds of okra cv. Pusa Sawani in GA_3 at 150 ppm or NAA at 20 ppm has been reported to give high seed yields (20.4 and 19.4 q/ha, respectively) (Lal and Lal, 1997-98).

The seed crop should be sown by 20th June before the onset of monsoon with a spacing of 60 x 45 cm in Northern India (Grewal *et al.*, 1973). Singh and Pandita (1981) found 120 kg N per hectare alone as optimum dose for getting higher seed yield under Hissar (Haryana) conditions. However, Kumar (1983) found 90 kg N and 30 kg P per hectare for the highest seed yield at Punjab Agricultural University, Ludhiana. Chattopadhyay and Sahana (2000) noticed that most of the seed quality and seed yield parameters improved significantly with increasing rates of N and P, the optimum N and P rates, being 100 and 60 kg/ha, respectively. Germination percentage and 100-seed weight were not significantly affected by N or P. Sarnaik *et al.* (1986) reported highest seed yield (10.2-17.7 q/ha) in Pusa Sawani at spacing 60

Seed Production of Okra

Mature Pods of Okra Ready for Harvesting of Seeds

x 20 cm. Khan and Jaiswal (1988) noticed that Pusa Sawani gave highest seed yield (8.33-9.02 q/ha) when spaced at 30 x 15 cm, N applied at 150 kg/ha and picked twice. Pandey and Manocha (1989) obtained optimum seed production from plants spaced at 45 x 15 cm, receiving N at 136 kg/ha and irrigated at 46 mm CPE (cumulative pan evaporation). High seed yield (7.9 q/ha) in cv. Parbhani Kranti was obtained by applying 100 kg N and 30 kg P_2O_5/ha (Lenka *et al.*, 1989).

In okra, retaining 12 fruits per plant along with apical pinching treatment resulted in high (1.24 t/ha) seed yield with medium seed quality. Further the seeds from 6 fruits retained per plant were superior in terms of quality but the yield was low (0.88 t/ha) (Reddy *et al.*, 1998). Singh *et al.* (1999) reported that treatments of 150 ppm GA_3 and 20 ppm NAA applied by seed soaking and foliar application of 20000 ppm urea increased seed yield by 64.4, 55.9 and 42.8 per cent, respectively.

According to Singh and Singh (1999), seed yield of cv. Parbhani Kranti was higher than that from cv. Pusa Sawani. Seed yield increased with increasing N fertilizer rate and reached a maximum at 120 kg N/ha. Seed quality characteristics were not affected significantly and seed yields were higher at the closer spacing.

Prabhakar *et al.* (1985) noticed best seed quality and yields were produced by fruits on the 3rd and 4th node. Kanwar and Saimbhi (1987) reported that although the number of seeds/fruit reached a maximum after 28 days of anthesis, the number of matured seeds, seed weight/fruit and germination percentage were highest in seeds from fruits harvested 35 days after anthesis. The ripe fruits loose their green colour and become hard. To avoid shattering of seeds, it should be picked periodically before the fruit bursts from the top. The fruits should be kept for a week for seed drying. These are then threshed, cleaned and stored at a cool and dry place. The seeds remain viable for two years under ordinary storage conditions. An average seed yield of 12-15 quintals can be expected from a hectare. Seed yield to the extent of 27 quintals per hectare has been reported by Singh *et al.* (1964) in Kulu Valley, Himachal Pradesh, India.

Under northern dry zone of Karnataka, the optimum period for harvesting of okra seeds is between 35-40 days after anthesis and slow drying method (drying of seeds within the fruits) was the most useful practice in cv. Arka Anamika (Sajjan and Jamadar, 2003)

Chattopadhyay *et al.* (2011) studied the effects of sowing time on seed yield and quality of okra in cv. Parbhani Kranti. They observed that mean seed yield was highest when seeds were sown on 16th February (6.84 q/ha) followed by 1st March (6.18 q/ha). Seeds sown on either 16th February or 1st March produced the best quality seed (85.30 and 80.0 per cent germination; 45.0 and 44.0 g test weight; 11.79 and 11.64 vigour index, respectively). However, for seed crop, sowing on 16th February is more beneficial in terms of income per rupee investment (2.40) followed by sowing on 1st March (1.90) under West Bengal condition.

In Madhya Pradesh, the cv. Arka Anamika when sown on 2nd week of June produced quality seeds in terms of test weight (64.31 g), seed germination percentage (75.73), vigour index-1 (1345.66) and seed yield (18.61 q/ha) as compared to other sowing dates *i.e.*, 4th week of June, 2nd week of July and 4th week of July (Singh *et al.*, 2018).

In Pakistan, the plant spacing of 40 cm resulted in the maximum number of seeds/fruit (67.28), seed weight/fruit (4.61 g), 100 seed weight (6.87 g) and seed yield (44.2 q/ha) during spring season at a spacing of 60 cm (Khan and Rab, 2019).

12.0 Crop Improvement

Since okra is an autogamous but often cross-pollinated crop, the breeding methods suitable for improving the self-pollinated crops can usefully be employed in this crop also. The most important methods are plant introduction, inbreeding or pure line selection and hybridization-pedigree method, interspecific crosses and back-cross technique.

A cultivar from Ghana (Africa) belonging to the species *Abelmoschus manihot* ssp. *manihot* has, however, been introduced by the National Bureau of Plant Genetic Resources, New Delhi (Singh and Bhatnagar, 1975). Since the species carries resistance to yellow vein mosaic disease, it has extensively been used to breed cultivars resistant to the disease. The cultivar Perkins Long Green was introduced from USA and recommended by IARI, New Delhi in 1976.

Inbreeding or pureline selection is mainly used for maintaining nucleus seed or stock seed or breeders seed of recognized cultivars. The cultivar Pusa Makhmali has, however, been bred out through this method from the local material from West Bengal (Singh and Sikka, 1955). Other high yielding cultivars *viz.*, Co-1 and Kashi Mangali were also developed using the single plant selection approach *viz.* pureline selection method.

The method of hybridization, the size of population to be grown in each generation and the characters to be selected has thoroughly been described by Swarup (1977). Yield of okra is said to have a positive and significant correlation with plant height, number of fruits/plant and fruit length (Swamy Rao and Ramu, 1975; Mahajan and Sharma, 1979; Arora, 1980; Yadav, 1986; Sogalad, 2008; Kumar *et al.*, 2012; Kumar *et al.*, 2019; Walling *et al.*, 2020). High heritability of these yield components as reported by (Mahajan and Sharma, 1979; Arora, 1980; Prakash and Pitchaimuthu, 2010; Nwangburuka *et al.*, 2012; Das *et al.*, 2012; Duggi *et al.*, 2013; Seth *et al.*, 2016b; Yadav *et al.*, 2018a; Singh *et al.*, 2020) makes it possible to select these characters from the very early generation onwards for breeding high yielding genotypes. Number of fruits per plant and length of pods gave moderately high genotypic coefficient of variation and high heritability (Yadav, 1986). Although edible fruit weight was the best index of yield, but as it cannot be assessed visually in the field, edible fruit width may be a better criterion for selection in view of its close association with fruit weight (Ariyo *et al.*, 1987).

Simon *et al.* (2013) reported that seed size showed high positive direct effect and positive genotypic association with seed yield, indicating that seed size had strong influence on seed yield.

Panda and Singh (1997) estimated high genotypic and phenotypic coefficients of variation for number of branches, number of fruits and total yield per plant.

Further, high heritability coupled with high genetic advance was observed for plant height, number of fruits and total yield per plant, indicating that these traits are suitable for improvement through selection as they are governed by additive genes.

Temam *et al.* (2020) also reported higher genotypic and phenotypic coefficients of variation coupled with high heritability and genetic advance for number of primary branches, plant height, internode length, peduncle length, number of tender fruits per plant, mucilage content, fruit length, average fruit weight, fruit yield per plant and seed yield per fruit. However, moderate heritability and genetic advance was observed for number of fruit ridge, hundred seed weight and percentage of dry matter content.

The pedigree method of selection after hybridization has been used for development of many cultivars through intervarietal hybridization.

12.1.0 Hybrid Vigour

Hybrid vigour has also been exploited in this crop and as much as 60 per cent increased yield was recorded by Joshi *et al.* (1958). Emasculation and pollination by hand may not be an economic proposition due to less number of seeds per fruit. Use of male sterility in this case will be very effective for commercial hybrid seed production. Induction of male sterility by use of chemical (Dubey and Singh, 1968a; Deepak *et al.*, 2007; Naveen *et al.*, 2017) and by irradiation (Dutta, 1971; Bandyopadhyay *et al.*, 1986; Thombre and Deshmukh, 2006; Boonsirichai *et al.*, 2009; Phadvibulya *et al.*, 2009) has indicated the possibility to exploit commercial hybrid seed production.

In okra, the male sterility is controlled by a single recessive gene (ms1), it is expressed when the gene is present in homozygous (ms1ms1) condition and the gene was found to be stable and not much influenced by the environmental factors (Dutta, 1980). Pitchaimuthu *et al.* (2012) reported that anthesis was normal in genic male sterile (GMS) plants but anther dehiscence was partial and microsporogenesis was normal up to the tetrad stage. Under open field, alternate planting of 2 rows of male sterile plants and 1 row of fertile plants (2:1 planting ratio) can accommodate about 4166 male sterile plants/ha and produce 52.56 per cent higher seed yield (5.66 q/ha of hybrid seed) as compared to fertile control. For production of 1 kg of hybrid seed using male sterile line, it required 3 hours to hand pollinate 274 flowers, whereas, by hand emasculation and hand pollination it took over 9 hours, around 70 per cent saving in time and manual labour can be achieved (Dutta, 1980).

Agarrado and Rasco (1986) observed the potentiality of heterosis over better parent (heterobeltiosis) for yield, and other characters like fruit weight, length and diameter in okra. The crosses Pusa Sawani x Clemson Spineless and Pusa Sawani x IC 8911 gave superior performance for fruit yield and most of the yield component characters except fruit length (Vijay and Manohar, 1986a). They (1986b) measured heterobeltiosis for fruit yield of 64.93 per cent and 66.81 per cent in cross Pusa Sawani x Clemson Spineless and Pusa Sawani x IC 8911, respectively. Arora (1993) reported that Pusa Sawani, Vaishali Vadhu and Foam Barelley were the best general combiners for most characters. The crosses Pusa Sawani x Vaishali Vadhu,

Emasculation and Pollination Technique in Okra

Vaishali Vadhu x Foam Barelley and Pusa Sawani x Foam Barelley had the highest SCA values for most characters. Both additive and non-additive genetic variance components were important for controlling inheritance of all characters.

Vaishali Vadhu x Local Akola was identified as an outstanding hybrid for fruits/plant, yield/plant and seven other characters which could be exploited commercially for higher gain (Wankhade *et al.*, 1996). Ahmed *et al.* (1999b) reported that the hybrids SB-3 x Punjab-7, SB-5 x Pusa Sawani, Perkins Long Green x SB-5 and Perkins Long Green x Parbhani Kranti could be considered as potential hybrids for commercial exploitation.

Several researches were conducted to study the hybrid vigour of different hybrid combinations over better parent and standard check for yield and its components. The details are listed below.

Table 2: Studies Showing Promising Heterobeltiosis

F_1 Hybrid Combinations with Heterobeltiosis (Per cent) for Fruit Yield	*Improved Characters*	*Source*
Arka Anamika x IC 112475 (39.6 per cent)	Fruit yield/plant	Shoba and Maria-ppan (2007)
VRO-6 x Parbhani Kranti (55.67 per cent), Daftari-1 x Arka Abhay (54.31 per cent), Daftari-1 x Parbhani Kranti (53.35 per cent)	Fruit weight, fruit yield/plant	Mehta *et al.* (2007)
IC-90273 x IC-90044 (56.48 per cent), IC-90273 x IC-111478 (53.98 per cent)	Number of fruits/plant, number of seeds/fruit, fruit yield/plant	Weerasekara *et al.* (2008)
Pusa A4 x Punjab Padmini (43.23 per cent)	Fruit yield, number of branches/plant, number of fruits/plant	Kumar (2011)
IC29119-B x IC99716 (33.67 per cent)	Number of marketable fruits/plant, marketable yield/plant	Reddy *et al.* (2012)
Arka Anamika x Azad Bhindi-2 (63.82 per cent), KS-442 x Azad Bhindi-2 (43.84 per cent), KS-450 x KS-419 (38.35 per cent)	Fruit yield and other desired characters	Singh *et al.* (2013)
Arka Anamika x IC-331217 (72.94 per cent), Arka Abhay x IC-331217 (69.58 per cent)	Fruit yield and its contributing characters	Jagan *et al.* (2013)
VNR Green x Shagun (36.27), Barsha Laxmi x Parbhani Kranti (29.69 per cent)	Plant height, number of leaves/plant, nodes at first flowering, number of fruits/plant, fruit diameter, fruit yield/plant	Das *et al.* (2013)
BCO-1 x Arka Anamika (44.88 per cent), VRO-6 x 11/RES-6 (44.11 per cent), VNR Green x 10/RES-4 (30.79 per cent)	Fruit yield/plant	Seth (2014)
FB10 x Parbhani Kranti (32.54 per cent), BO2 x Azad Krishna (31.81 per cent), VRO5 x Azad Krishna (27.41 per cent)	Fruit yield and other desired characters	Singh *et al.* (2016)
JOL-09-8 x Pusa Sawani (57.29 per cent), JOL-09-7 x Pusa Sawani (41.72 per cent), JOL-10-17 x GJO-3 (41.72 per cent)	Days to 50 per cent flowering, plant height, number of fruits/plant, fruit length, fruit weight, fruit diameter, fruit yield/plant, number of seeds/plant and 100-seed weight	Patel and Patel (2016)

F_1 Hybrid Combinations with Heterobeltiosis (Per cent) for Fruit Yield	Improved Characters	Source
IC-282280 x EC-329380 (82.42 per cent)	Fruit yield and number of fruits/ plant	Kumar *et al.* (2017a)
Balady Long x Balady Short (38.59 per cent), Balady Short x Ladyfingers (29.07 per cent)	Number of fruits/plant, weight of fruits/plot, number of seeds/fruit	Shamloula and Moghazy (2017)
285-1-5-1-16-23 x VRO-178 (45.30 per cent), 285-1-3-1-15-2-1 x VRO-178 (41.12 per cent)	Fruit yield/plant	Yadav *et al.* (2018b)
IC7952 x Hissar Unnat (72.50 per cent), IC1543 x Arka Anamika (62.30 per cent), IC1543 x Punjab Padmini (57.32 per cent)	Number of fruits/plant, single fruit weight and fruit yield/plant	Joshi *et al.* (2019)
EC359637 x IC014026 (50.13 per cent)	Fruit yield/plant	Chowdhury and Kumar (2019)
2014/OKYV RES-5 x 2014/OKYV RES-10 (45.89 per cent), 2014/OKYV RES-5 x BCO-1 (37.48 per cent)	Plant height, node at first flowering and fruit yield/plant	Devi *et al.* (2019b)

Table 3: Studies Showing Promising Standard Heterosis

F_1 Hybrid Combinations with Standard Heterosis (Per cent) for Fruit Yield	Improved Characters	Source
IC 169340 x IC 112475 (49.3 per cent)	Plant height, number of nodes/ plant, number of fruits/plant, single fruit weight, protein content and yield/plant	Shoba and Mariappan (2007)
MDU 1 x Hissar Unnat (65.23 per cent)	Fruit yield, and most of the yield component characters except number of branches per plant and fruit length	Kumar and Sreeparvathy (2010)
IC282248 x IC45732	Days to first flowering and fruiting nodes	Reddy *et al.* (2012)
IC-282280 x EC-329380 (35.66 per cent)	Fruit yield and number of fruits/ plant	Kumar *et al.* (2017a)
EC359637 x IC014026 (76.78 per cent)	Fruit yield/plant	Chowdhury and Kumar (2019)
IC1543 x Arka Anamika, TCR2086 x Hissar Unnat, TCR2086 x Parbhani Kranti, EC306741-A6 x Hissar Unnat	Days to first flowering	Joshi *et al.* (2019)
IIHR-875 x IIHR-478 (168.55 per cent)	Fruit yield and number of fruits/ plant	Kerure and Pitchaimuthu (2019)

12.2.0 Resistant Breeding

12.2.1 Yellow Vein Mosaic

Among wild species, *A. tuberculatus* carries symptomless type of resistance, while *A. pungens* carries true resistance to the disease (Nariani and Seth, 1958). Because of prevalence of high sterility in hybrids between *A. esculentus* and these species, transfer of resistance to the cultivated type has not been possible (Pal *et al.*, 1952). However, an accession of okra (EC 31830) under the name of Asuntem Koko from Ghana introduced by the Division of Plant Introduction, IARI, New Delhi and identified as *A. manihot* ssp. *manihot* has been reported to be immune or highly resistant to the disease. The species carries symptomless type of resistance (Sandhu *et al.*, 1974; Thakur, 1976). Two more accessions of *Hibiscus manihot* one each from Africa and Japan has also been found to be highly resistant to the disease (Arumugam *et al.*, 1975). The species *A. manihot* has widely been used in crosses with *A. esculentus* for the transfer of resistance genes. In most cases only one way cross, *i.e.*, *A. esculentus* as a female parent was possible. The F_1 was invariably sterile but it was partially compatible when back-crossed to *A. esculentus* (Thakur, 1976; Sharma and Dhillon, 1983). The cultivar Punjab Padmini, a tolerant type, is a derivative of such a cross, bred and released at the Punjab Agricultural University, Ludhiana, India.

Varietal resistance to yellow vein mosaic in *A. esculentus* is rare. A partial resistance of symptomless carrier type reported in a multi-ridged Bengal cultivar, IC 1542 was used in resistance breeding and as a result cultivar Pusa Sawani which became popular throughout the country was bred out (Singh *et al.*, 1962). Resistance in Pusa Sawani has since been broken down. Parbhani Kranti (PB No. 57) is another cultivar reported to be resistant to yellow vein mosaic virus. In India, yellow vein mosaic virus resistant cultivars has been developed through interspecific hybridization followed by back crossing like Punjab Padmini and Punjab 7 (using *A. manihot* ssp. *manihot*) and Arka Anamika and Arka Abhay (using wild taxa *A. manihot* ssp. *tetraphyllus* var. *tetraphyllus*). But the characteristics of YVMV resistance of these cultivars were broken down due to appearance of new viral strains, recombination in viral strains and emergence of new biotypes of whitefly (Sanwal *et al.*, 2014a).

Singh and Singh (1986) found that cultivar Kb S 312 was the most resistant to yellow vein mosaic virus, with 3.3 per cent infection in rainy season. Screening for yellow vein mosaic virus by Dhankar *et al.* (1989) revealed that cultivars IC 9273, Baunia, 3(I) and IC 23592 were in the resistant category.

Wild species like *A. manihot, A. tuberculatus, A. tetraphyllus, A. caillei, A. crinitus, A. angulosus, A. pungens, A. vitifolius* and *A. panduraeformis* have been used in breeding programs for the transfer of resistance genes to the cultivated *A. esculentus* (Dhankar and Mishra, 2005; Singh *et al.*, 2007b). But the botanical distance and sterility problems act as a pre-zygotic barrier during the passage of pollen tube to the ovary resulting in few viable seed setting in F_1 hybrid. However, Joseph *et al.* (2013) reported that colchicine treatment was successful for fertility restoration in the interspecific crosses of *A. esculentus* with *A. tuberculatus* and *A. tetraphyllus*. They also reported that *A. caillei* can be used as a potential bridging parent to transfer desirable genes in okra breeding programmes.

Due to the emergence of different novel viral strains (Venkataravanappa *et al.*, 2012) in different parts of the world, at present no cultivated okra variety or hybrid had shown absolute resistance. The resistance to begomoviruses is found unstable with frequent breakdowns. The high recombination rate of begomoviruses, emergence of the polyphagous 'B' biotype of *B. tabaci* with its increased host range of more than 600 plant species and moderate to strong resistance of whiteflies to various insecticides are the main hindrance in YVMV resistant breeding programme (Chowda-Reddy *et al.*, 2012).

About 27 begomoviruses were reported to infect okra of which, yellow vein mosaic disease (YVMD) and okra enation leaf curl disease (OELCD) most severely affect its production in terms of yield and fruit quality (Swanson and Harrison, 1993; Sanwal *et al.*, 2014a). Bhendi yellow vein mosaic virus (BYVMV) belongs to the genus Begomovirus and family Geminiviridae. This disease is caused by a complex consisting of the monopartite begomovirus BYVMV encapsulated within a typical geminated particle (Jose and Usha, 2003) and a small satellite DNA with additional molecules known as betasatellites and/or alpasatellites (Briddon and Stanley, 2006). The begomoviruses native to the New World have only bipartite genomes (having DNA-A and DNA-B components) whereas, of Old World have both bipartite and monopartite (have DNA-A homolog and lacks DNA-B) genomes (Brown *et al.*, 2012). Alpha satellites (DNA-A) are self-replicating (autonomous) circular ssDNA molecules, encodes factors required for the viral encapsidation, replication, suppression of host defence and play no role in symptom induction (Briddon *et al.*, 2004). Whereas, betasatellites (DNA-B) associated with the monopartite viruses, encodes factors essential for viral systemic movement, host range determination and induce typical disease symptoms in their original hosts (Briddon *et al.*, 2003; Rojas *et al.*, 2005; Venkataravanappa *et al.*, 2012).

In Bangladesh, Salehuzzaman (1985) tested 301 accessions collected from 29 countries against the YVMV disease in six sowings. The study reported that accessions number 18 from Ivory Coast, 5 from Ghana and 1 from Liberia remained unaffected in every sowing. Arora *et al.* (1992) evaluated 157 advanced germplasm and 7 cultivars of okra for two years and they observed that Punjab Padmini and EMS-8 remained free from the YVMV. Bora *et al.* (1992) reported that genotypes Pb-7, GOH-6, GOH-4, AROH-1, Pb-57 and 74-11 showed resistance to YVMD. Sharma *et al.* (1993) evaluated okra varieties for four years and reported that Punjab Padmini and Punjab-7 were found to be high yielding and resistant to yellow vein mosaic virus. Another study comprising 51 okra hybrids and their 20 parents were screened for YVMD and only one parent (Prabhani Kranti) and four hybrids were found to be highly resistant, while P 7 showed moderate resistance to YVMD (Dhankhar *et al.*, 1996).

In Haryana, during rainy season, open pollinated okra varieties (Okra No. 6, LORM-1, VRO-3 and P-7) were found free from this disease, whereas VRO-4 showed mild reaction and among hybrids, DVR-1 and DVR-2 expressed resistance against the disease (Batra and Singh, 2000). Rashid *et al.* (2002) evaluated 12 okra germplasm under field conditions for two years and identified two resistant lines OK-292 and OK-285.

Abdul *et al.* (2004) conducted an experiment in the Regional station of National Bureau of Plant Genetic Resources, Thrissur, Kerala for screening of 941 indigenous and exotic okra germplasm for fruit yield, quality and field resistance to YVMV. They observed three accessions (IC 218887, IC 69286 and EC 305619) showed resistance and 43 lines expressed moderate resistance to YVMD.

In a screening experiment in Pakistan, the cultivar Surkh Bhindi was found to be highly resistant and two cultivars Subz Pari and Safal showed were graded as moderately resistant (Ali *et al.*, 2005).

Deshmukh *et al.* (2011) screened 35 lines/varieties under field conditions for eight seasons and lines NOL-285 expressed high resistance to YVMD, while NOL-303 and NOL-145 showed moderate reaction. Venkataravanappa *et al.* (2013) screened 29 genotypes of okra (popular variety/hybrid and advanced breeding lines) under both artificial and natural conditions. None of the genotypes showed immunity to the disease. However, the genotypes Nun 1145 and Nun 1144 showed moderate resistance reactions under both glass house and field conditions.

Singh *et al.* (2007b) screened 347 lines of *A. esculentus* and 119 lines of different wild species and observed that the wild taxa of *A. tetraphyllus* and *A. caillei* showed resistance to YVMV. Prabu *et al.* (2007) screened wild and cultivated lines in four seasons and reported wild species *A. angulosus* completely free from YVMV symptoms. *A. tetraphyllus* lines (1, 2, 3 and 4), *A. moschatus* lines (1, 2, 3, 4 and 5), *A. caillei*-2 and *A. manihot* ssp. *tetraphyllus* were found to be highly resistant, while the wild lines *A. tetraphyllus*-5, *A. manihot* (L.) Medikus and *A. manihot* ssp. *manihot* were found to be resistant. At National Bureau of Plant Genetic Resources, New Delhi under field epiphytotic conditions, some accessions of *A. angulosus* var. *grandiflorus, A. crinitus, A. ficulneus, A. tetraphyllus* var. *pungens* and *A. enbeepeegeearense* did not express any YVMV symptom (Joseph *et al.*, 2013).

Sanwal *et al.* (2014b) screened 219 accessions of cultivated and wild species. Among the cultivated lines VRO 109, VRO 104, VROB 178, 307 10-1 and No. 315 were completely free from YVMD. The wild species accessions IC 582757 (*A. enbeepeegeerense*), NIC5952 (*A. moschatus*), Jpn/N-2176 (*A. manihot*), IC-90340 (*A. tuberculatus*) and IIVR-Tube-1 (*A. tuberculatus*) expressed resistant to YVMD.

Screening of twenty four species of *A. esculentus* along with one each species of *A. manihot* and *A. caillei* for yellow vein mosaic virus during rainy season was done by Seth *et al.* (2016b). They found that BCO-1, 11/RES-6, VNR Green, 12/RES-2, *A. caillei, A. manihot* showed disease resistance under field condition.

The biochemical parameters also impart resistance to YVMV disease in okra. Resistant cultivar contains higher amount phenols, orthodihydroxy phenol, peroxidase, polyphenol oxidase, phenylalanine ammonialyase (PAL), flavonols, ascorbic acid and total chlorophyll content than susceptible cultivars (Arumugam and Muthukrishnan, 1977; Ahmad *et al.*, 1994; Mahajan *et al.*, 2004; Prakasha, 2009; Seth *et al.*, 2017). The leaves of susceptible cultivars contained higher amount of soluble sugar than the resistant cultivar (Bhagat and Yadav, 1997). The wild parents resistant to okra YVMV had maximum phenolics, peroxidase, polyphenol oxidase activity and seed soluble protein content and lower reducing, non-reducing, total sugars and total nitrogen content compared to cultivated okra (Sharma *et al.*, 1981; Prabu and Warade, 2009).

Mutation breeding is also a viable alternative for resistance breeding of YVMV disease. Phadvibulya *et al.* (2009) reported seeds of two okra varieties, Annie and

Okura, irradiated with Gamma rays at doses of 400 and 600Gy. One M_4 plant of Okura (B-21) irradiated at 400Gy was found to be highly resistant, but none of Annie.

12.2.2 Enation Leaf Curl Virus

Singh *et al.* (2007b) reported that *A. esculentus* lines VRO-5, VRO-6, KS-410, VRO-4, 111-8-9 (Red), 131-8-1 (Red) and IIVR-10 were free from infection. Among wild taxa, infection was absent in *A. pungens, A. manihot, A. crinitus, A. ficulneus, A. moschatus* and *A. angulosus.* Nine accessions of *A. tetraphyllus* and eight accessions of *A. tuberculatus* expressed resistance (less than 10 per cent infection) against okra enation leaf curl virus disease infection.

Under West Bengal condition, among 35 okra genotypes, only 3 genotypes (285-1-3A-1-17-1, VRO-178 and 285-1-3-1-15-2-1) expressed appreciable ELCV disease tolerance during rainy season (Yadav *et al.*, 2018a).

Devi *et al.* (2018) conducted an experiment for screening of 560 germplasm of okra against enation leaf curl virus disease over two consecutive years and observed only 20 genotypes were infected with ELCV disease. Genotypes 10/OKYVRES-5 and 10/OKYVRES-10 showed high tolerance and 10/OKYVRES-11 and IC112449 expressed moderate tolerance.

Jamir *et al.* (2020) screened 565 genotypes during rainy season and found that two genotypes IC111551 and IC433616 were moderately susceptible against ELCV disease up to 60 days after sowing with disease severity (15.87 PDI).

12.2.3 Fungal Diseases

Resistance to *Corticium solani* is reported in cultivars and lines Red Ghana, Sel. 7-1, B.H. 27, IC 12029 and IC 17252, resistance to *Erysiphe cichoracearum* in cultivar Nigeria and resistance to *Cercospora abelmoschi* in cvs. Sel. 7-1, Round Selection, Nigeria, EC32598, IC 10238, IC 8248 and IC 1542 (Sohi and Sokhi, 1974).

Wild taxa namely *A. moschatus, A. manihot* and *A. pungens* showed resistance (Prabhu *et al.*, 1971) and *A. manihot* and *A. tetraphyllus* were immune to *Erysiphe cichoracearum* (Joi and Shende, 1979). Singh *et al.* (2007b) observed complete resistance to powdery mildew in *A. tetraphyllus, A. moschatus, A. crinitus* and moderate resistance in *A. angulosus, A. pungens* and *A. manihot*. However, under Rahuri, Maharashtra conditions, the wild lines *A. caillei*-2 and *A. moschatus*-4 were found to be immune to powdery mildew (Prabu, 2005). Vidhate *et al.* (2014) evaluated 20 germplasm lines of cultivated okra during rainy season in Parbhani, Maharashtra. Eight lines (PBN OK 4, PBN OK 5, PBN OK 6, PBN OK 8, PBN OK 9, AOH 23, Surya and NOH 100) showed moderate resistance to powdery mildew disease.

Resistance to cercospora leaf spot was also reported in wild species of *A. crinitus, A. moschatus, A. pungens, A. manihot* and *A. angulosus* (Singh *et al.*, 2007b).

12.2.4 Jassid

Resistance to cotton jassid (*Amrasca bigutulla bigutulla*) has been reported in cultivars and lines like Pusa Sawani, Crimson Smooth Long, IC 7194, IC 8899 of *A. esculentus* and in Ghana of *A. manihot* ssp. *manihot* (Sandhu *et al.*, 1974) but no commercial resistant type has been reported so far.

Hooda *et al.* (1999) recommended backcross method to improve leafhopper resistance in commercially grown cultivar of okra.

Nerkar (1990) reported that due to the presence of laminal hairiness in the wild species *A. moschatus*, it was found to be highly resistant to jassids. The wild relatives *A. moschatus*, *A. ficulneus* and *A. crinitus* species expressed resistance to jassid infestation (Singh *et al.*, 2007b). At Rahuri, Maharashtra, the wild species *A. moschatus* lines 1, 2, 3, 4 and 5 recorded least jassid (nymph) population per leaf (Prabu *et al.*, 2009).

Halder *et al.* (2016) studied the physical and biochemical basis of resistance against *Amrasca bigutulla bigutulla*. They identified one tolerant genotype VROB-181 which had relatively higher number of trichomes on leaf lamina (11.85/cm^2), mid-rib (9.17 cm) and vein (9.95 cm), mid-rib thickness (1.58 mm), leaf length (22.43 cm) and total phenol content in green leaves (75.04 mg/100 g).

12.2.5 White Fly

Prabu *et al.* (2009) reported that wild taxa *A. angulosus* and *A. moschatus* lines 1, 2, 3, 4 and 5 recorded minimum mean whitefly (adult) population per leaf. These wild species could be utilized in the breeding programme to develop YVMV resistant okra variety.

12.2.6 Shoot and Fruit Borer

Singh *et al.* (2007b) screened 119 lines of wild okra and observed that 16 accessions of *A. tuberculatus* recorded less than 10 per cent fruit damage. It may be due to the presence of tubercles on the surface of the fruit wall and toughness of the fruit wall does not permit the neonate larvae to feed and survive. Whereas, Prabu *et al.* (2009) reported that *A. tuberculatus* lines 1, 2 and 3 were found to be immune to fruit borer infestation while *A. tetraphyllus* lines 2, 3, 4 and 5 and *A. manihot* ssp. *tetraphyllus* were found to be completely free from shoot borer infestation.

Halder *et al.* (2015) screened 10 genotypes to study the relationship between physico-morphic and biochemical characters against *Earias vittella*. They identified one tolerant genotype SB 6 which had dense trichomes (51.8/cm^2), wider fruit angle (26.4^o), thick stem (1.28 cm), fewer leaves (20/plant), fewer fruits (13.67/plant) and branches (1.4/plant), higher total phenol content (34.78 mg/100 g) and anthocyanin content in fresh fruits (0.047 OD value).

12.2.7 Red Spider Mite

Singh *et al.* (2007b) reported that three accessions of wild species *A. caillei* (EC-305656, EC-305664, EC-305696) did not harbour any mite population, exhibiting complete resistance.

12.2.8 Root-knot Nematodes

Meloidogyne javanica and *M. incognita* are two important species of nematodes which infest okra and reduce its yield considerably. Unfortunately, no source of resistance against both the species is so far available. Source of resistance to *M. hapla* and *M. arenaria* has been reported in cultivar Clemson Spineless of *A. esculentus*

(Sasser and Kirby, 1979). The species *M. arenaria* has not been reported in India. The species *M. hapla* has, however, been found in the hills of North India. Cultivars 118/82-74, 92/82-2 and IC 52314 are reported to be resistant towards *M. incognita* (Darekar and Ranade, 1990).

Out of 56 okra germplasms/cultivars, five lines *viz.*, IC-111472, IC-128372, IC-169482, IC-411698 and IC-433686 exhibited moderately resistant reaction with minimum gall index in Odisha, India (Mohanta and Mohanty, 2012).

Mukhtar *et al.* (2014) evaluated 12 okra cultivars against root-knot nematodes in Pakistan and identified that five cultivars *viz.*, Sanam, Dikshah, Arka Anamika, Ikra-1 and Ikra-2 as moderately resistant.

Screening of 7 okra cultivars for resistance against root knot nematode in Karnataka, India was done by Kedarnath *et al.* (2017). Two cultivars Arka Abhay and Anoop were identified as moderately resistant to *M. incognita* with root-knot index of 3.0 both at 60 days after sowing and at harvest.

13.0 Biotechnology

Work on biotechnology of okra is very less. Even work pertaining to micropropaga-tion is poorly reported. Mangat and Roy (1986) cultured hypocotyl, cotyledonary node and leaf segment explants were excised from aseptically grown seedlings. Callus formation and root differentiation occurred in MS medium containing NAA or IAA. There was a greater proliferation of roots on medium supplemented with NAA. The addition of 2,4-D to the growth medium suppressed root formation. No shoot bud or shoot development was observed at any of the auxin levels tested. Shoots were produced on cotyledon and cotyledonary node explants cultured in a medium supplemented with BA and NAA. These shoots developed roots on the same medium. The plantlets, on transfer to soil, grew normally. In a subsequent study, they (Roy and Mangat, 1989) found that the presence of kinetin or zeatin in the induction medium stimulated callus formation on hypocotyl explants. The cotyledonary axil explants failed to yield callus when cultured on media containing kinetin or zeatin or the combination of either of these two cytokinins with IAA or NAA. Rapid callus induction resulted from all explants cultured on MS medium supplemented with BA (1.0 mg/l). The hypocotyl-derived callus remained non-organogenic when subcultured on fresh BA-supplemented MS medium, whereas cotyledonary-axil-derived callus produced a number of buds. The transfer of the organogenic calluses to BA-enriched MS media containing various concentrations of silver nitrate (0-40 mg/l) stimulated multiple shoot development within 2-3 weeks. The presence of silver nitrate at 10 or 20 mg/l significantly increased the number of calluses producing multiple shoots when myo-inositol and casein hydrolysate were omitted from the growth medium. Roots developed quickly when regenerated shoots were subcultured on MS medium containing BA (1.0 mg/l) and NAA (1.0 mg/l). The plantlets, on transfer to soil, grew normally.

In case of *Abelmoschus esculentus*, micropropagation using single-node explants from seedlings as the primary explants was reported by Akogo *et al.* (1996). They cultured explants on solidified MS medium supplemented with 30 g sucrose/l at pH 4.7. Single-node explants were taken from the shoots, which developed and were subcultured at 4-weekly intervals. Root, leaf and node number, and plantlet height were assessed at weekly intervals. Initially, both species grew and developed well.

Attempts were made to transfer the resistant gene from wild species to the cultivated okra but the interspecific crosses showed post-zygotic incompatibility and crossed seeds were shrivelled and non-viable. To overcome these problem Rajamony *et al.* (2006) studied the *in-vitro* embryo rescue of interspecific crosses (*A. esculentus* x *A. moschatus*) for transferring virus resistance in okra. They observed that 12-15 days old embryos of the interspecific crosses cultured on MS medium supplemented with BA 0.5 mg/l + CW 150 ml/l yielded transplantable hybrids. Multiple shoots were produced when cultured on MS medium supplemented with BA 1 mg/l + CW 150 ml/l. Rooting of individual shoots was achieved in MS medium supplemented with IAA 1 mg/l. Embryo derived plants of interspecific crosses recorded 30-36 per cent survival under *ex-vitro* condition. However, embryos of reciprocal crosses (*A. moschatus* x *A. esculentus*) turned brown on 11th day of pollination and could not be cultured *in-vitro*.

Ganesan *et al.* (2007) studied plant regeneration protocol of okra through somatic embryogenesis from suspension cultures. Embryogenic callus from hypocotyl explants were obtained on liquid medium fortified with MS salts, Gamborg (B5) vitamins, 30 g/dm^3 sucrose, 2.0 mg/dm^3 2,4-D and 1.0 mg/dm^3 kinetin. The best conversion of embrya into plantlets (67.3 per cent) was achieved after 4 weeks of culture on media with half strength MS salts, B5 vitamins, benzylaminopurine (BAP) (0.2 mg/dm^3) and GA_3 (0.2 mg/dm^3). The regenerated plantlets showed 100 per cent survival rate in the greenhouse conditions after proper acclimatization.

Similarly, callus induced organogenesis in okra using leaf disc and hypocotyl explants was reported by Anisuzzaman *et al.* (2008). They reported that hypocotyl was better source of explant for callogenesis (95 per cent) in MS medium augmented with NAA (2.0 mg/l) + thidiazuron (TDZ) (0.5 mg/l). The highest shoot regeneration (65 per cent) and highest average number of shoot per callus mass was obtained with 2.0 mg/l BAP + 0.1 mg/l IBA. Whereas, MS medium with NAA (1.5 mg/l) resulted in highest root formation. About 80 per cent of regenerated plantlets were survived and showed normal growth and development under *ex-vitro* condition. Rizwan *et al.* (2020) also reported that hypocotyl cultured in MS media supplemented with 1.0 mg/l BAP and 1.5 mg/l 2,4-D gave the maximum callus induction. The best shoot induction was achieved in the MS media containing 40 mM total nitrogen supplemented with 1.5 mg/l KIN and 0.5 mg/l IBA. Regenerated shoots rooted better on half strength MS media supplemented with 2.0 mg/l IBA and 100 mg/l activated charcoal. Okra due to its recalcitrant nature creates problem in regeneration through *in-vitro*. This problem can be overcomed by using MS media with 40 mM total nitrogen which favours friable embryogenic callus (FEC) formation without whitish compact non-embryogenic callus growth with high shoot induction frequency.

Anisuzzaman *et al.* (2010) standardized a protocol for *in-vitro* propagation of disease free plantlets using meristem culture. Meristems were isolated from shoot tips of 3 week old *in-vitro* grown seedlings. The isolated meristems (0.3-0.5 mm in

Micropropagation of Okra through Hypocotyl Explants

Shoot Differentiation from Callus through Organogenesis in Okra
(A) initiation of callus from hypocotyl explant, (B) organogenic calli showing green points, (C) development of plantlets from callus and (D) regenerated plantlets with developed roots (Anisuzzaman *et al.,* 2008).

Micropropagation of Okra through Hypocotyl and Cotyledon Explants

***In-vitro* produced seedlings of okra (a) Callus induction in hypocotyl and cotyledon explants; (b) Explants showing whitish compact callus on full nitrogen composition (60 mM); (c) Greenish friable organogenic callus formation on modified MS (40 mM) medium; (d) Shoot induction from hypocotyl derived calli at 40 mM of nitrogen level; (e) Individual shoot elongation containing GA_3; (f) Rooting of individual elongated shoots; (g) Acclimatization of rooted plantlets in plastic cups under growth chamber conditions and (h) Bar = 0.5 cm (Rizwan *et al.,* 2020).**

Micropropagation of Okra through Meristem Culture

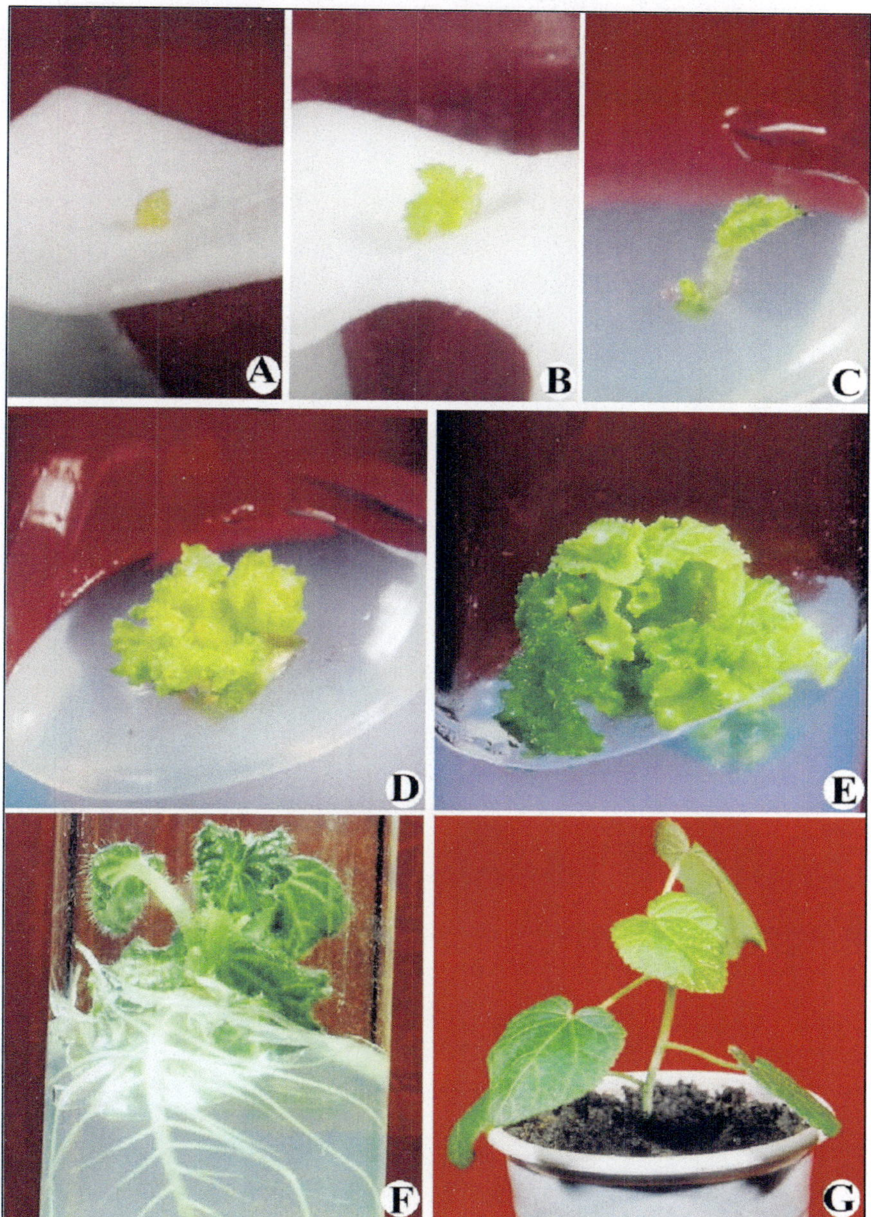

Different Stages of Meristem Culture in *Abelmoschus esculentus*.

(A) Growth of isolated meristem (five days old) on filter paper bridge in liquid medium; (B) Developed (21 days old) meristem in liquid medium; (C and D) Shoot initiation from developed meristem in semi-solid medium; (E) Shoot multiplication from nodal segment of mericlones; (F) Rooting in meristem-derived plantlet and (G) Acclimatization of plantlet in soil (Anisuzzaman *et al.,* 2010).

Micropropagation through Apical Shoot Culture

Apical shoot inoculation on MS medium containing IBA + NAA

Regeneration of apical shoots on MS medium containing IBA + NAA

Rooting on MS medium containing IAA+ Charcoal 1.0 g/L

Elongation of regenerated shoots on MS medium containing Kinetin

Callus Formation and Root Differentiation in Okra (Dhande *et al.,* 2012).

size) were established rapidly in MS liquid medium supplemented with 1.0 mg/l of BAP. Rapid shoot multiplication of mericlone and maximum number of shoots per node (8.86) were achieved from node cutting cultured in semisolid MS medium containing 1.0 mg/l BAP + 0.5 mg/l GA_3. Effective root development from the developed plantlets was successful in 1.0 mg/l IBA. More than 75 per cent of the micropropagated plantlets were successfully acclimatised in soil up to maturity and was found to be healthy.

Regeneration of okra through shoot tip explants of 9-10 days old germinating seedlings was reported by Dhande *et al.* (2012). Combination of 1.0 mg/l IBA and 0.5 mg/l NAA in MS medium were found to be most effective for apical shoot regeneration (98.5 per cent). Best shoot elongation (5.60 cm) was observed in MS medium supplemented with 0.5 mg/l kinetin. Highest frequency of rooting (99.0 per cent) was observed in media containing 1.0 mg/l IAA and 1.0 g/l activated charcoal.

Kabir *et al.* (2016) cultured cotyledonary nodes and hypocotyl explants from aseptically grown seedlings for organogenesis in okra. They found cotyledonary node explants showed better shooting response (100 per cent) with callus than hypocotyls on various concentrations of thidiazuron (TDZ). The induction of high number of multiple shoots per explant (6.8) with maximum shooting percentage (64.0 per cent) was observed in 8.4 days in MS medium containing 0.044 µM TDZ. The maximum root induction (83.3 per cent) in 9.7 days was recorded in 2.46 µM IBA. Shorter time (about 3 months) was taken from culture initiation to plant establishment in the greenhouse.

Rizwan *et al.* (2018) observed that BAP at 2.0 mg/l was found to be most effective for shoot bud initiation (75 per cent) from cotyledonary nodal explants. MS medium containing BAP (1.5 mg/l) + KIN (1.5 mg/l) + $AgNO_3$ (3.0 mg/l) resulted in massive shoot multiplication (85 per cent) and maximum number of shoots (6.75 shoots/explants, 3.90 cm in length). Best result of individual shoot elongation (80 per cent) and shoot length (5.60 cm) were achieved in MS medium supplemented with 1.5 mg/l BAP + 1.5 mg/l GA_3. Elongated shoots when rooted on half strength MS medium + 1.0 mg/l IBA + 200 mg/l activated charcoal produced highest root (83 per cent) and maximum roots per shoot (5.20) with 4.8 cm length.

Tahir *et al.* (2019) studied the effects of plant growth regulators on the *in-vitro* propagation of okra using seed explants. The MS media supplemented with 0.15 mg/l NAA in single treatment had the best performance on the days to germination, vigour, number of leaves, germination percentage, seedling height and root length as compared to different treatments combination of BAP and NAA.

Mallela *et al.* (2009) examined the potentiality of plant regeneration from different explants (hypocotyls, cotyledonary nodal meristems, cotyledons and shoot tips). They reported a combination of different concentrations of cytokinins (1.0 mg/l NAA + 1.0 mg/l BAP + 0.04 mg/l TDZ) in the regeneration media resulted in higher frequency of organogenic callus induction, shoot regeneration and root induction. They also studied the genetic transformation by infecting the explants with *Agrobacterium tumefaciens* strain EHA105 carrying the plasmid pBI121 with a selectable marker gene for *npt*II (neomycin phosphotransferase). Transformed cells were sub-cultured on kanamycin (50 mg/l) and cefatoxime (300 mg/l). Proliferation of callus was achieved with complete suppression of *Agrobacterium*. About 50-60 per cent calli showed GUS (β-glucuronidase) expression, confirming transformation. Hypocotyls showed highest number of GUS+ calli (6.6) as compared to cotyledons. The regenerated plants were successfully hardened in earthen pots after adequate acclimatization. Narendran *et al.* (2013) standardized a tissue culture protocol for okra using zygotic embryo explants and studied the *Agrobacterium*-mediated transformation protocol for developing transgenic *Bt* okra having resistance to *Earias vittella* (shoot and fruit borer). Isolated plumule of embryos were wounded and inoculated in *Agrobacterium* suspension (EHA105 carrying *cry1Ac* and *npt*II genes) for 15 min. The plumules were transferred on MSB2 medium (MS salts, B5 vitamins, Zeatin 2 mg/l, agar 0.8 per cent, sucrose 3 per cent, kanamycin 50 mg/l, cefotaxime 500 mg/l) for getting multiple shoot buds. The multiple shoot buds were elongated and rooted on MSB1 medium (MS salts, B5 vitamins, agar 0.8 per cent, sucrose 3 per cent, kanamycin 50 mg/l, cefotaxime 500 mg/l). Through this transformation protocol, phenotypically normal and fertile transgenic okra plants were produced in short period (4-5 months). In-plants *Agrobacterium* mediated genetic transformation was also studied by Menon *et al.* (2018). Leaf and hypocotyl from *in-vitro* grown seedlings were used as explant. Result showed that hypocotyl was more callogenic (83.3) as compared to leaf and MS medium containing NAA (3.0 mg/l) + BAP (0.5 mg/l) was the best medium for callus induction.

Few reports on application of DNA markers in the improvement and characterization of *Abelmoschus* have been made. Martinello *et al.* (2001) studied the genetic diversity of 39 *Abelmoschus* accessions consisting 15 *A. esculentus*, 13 *A. caillei*, 7 *A. moschatus*, 2 *A. manihot* and one each of *A. ficulneus* and *A. tetraphyllus*, at the DNA level with the random amplified polymorphic DNA (RAPD) marker and at the phenotypic level with stable and highly heritable morphological characters. Thirty-one random decamer primers were used to amplify DNA by the polymerase chain reaction (PCR) and 103 RAPD fragments were generated. Thirteen quantitative and 14 qualitative characteristics were measured on the genotypes in field experiments. They found that the quantitative descriptors and the molecular approach have high potential in the okra germplasm morpho-agronomic characterization.

Engels *et al.* (1991) suggested a conservation strategy to meet the needs of the gene-pool using a balanced application of technologies such as *in situ*, *in vitro*, seed, pollen and DNA conservation for *Abelmoschus* ssp.

The utilization of molecular biotechnology in okra is limited due to the availability of few molecular markers, the absence of molecular genetic map, chromosome linkage-groups and other molecular tools (Mishra *et al.*, 2017) and complex polyploidy nature of the okra genome (Sastry and Zitter, 2014). High mucilage content in all plant parts and presence of secondary metabolites (polysaccharides, alkaloids, flavonoids, phenols and terpenes) creates difficulties in the extraction of good quality genomic DNA (Seth *et al.*, 2018). These also creates problem during pipetting and amplification of genomic DNA by inhibiting Taq polymerase action (Bayer *et al.*, 1999). However, reports on isolation of good quality DNA from different plant parts like dark grown seedlings (Kochko and Hamon,

1990; Seth *et al.*, 2018), fresh leaves (Khanuja, 1999; Singh and Kumar, 2012; Ahmed *et al.*, 2013; Narendran *et al.*, 2013; Adiger and Sridevi, 2014; Seth *et al.*, 2018) and dried leaves (Meena *et al.*, 2014) have been made.

Various DNA based marker system like RAPD (Bisht *et al.*, 1995; Martinello *et al.*, 2001 and 2003; Aladele *et al.*, 2008; Nwangburuka *et al.*, 2011; Prakash *et al.*, 2011; Kaur *et al.*, 2013; Haq *et al.*, 2013), SSR markers (Sawadogo *et al.*, 2009; Schaeitner *et al.*, 2013; Fougat *et al.*, 2015; Kumar *et al.*, 2017b; Ravishankar *et al.*, 2018; Ouedraogo *et al.*, 2018), SRAP (Gulsen *et al.*, 2007), AFLP (Akash *et al.*, 2013; Kyriakopoulou *et al.*, 2014; Salameh, 2014; Younis *et al.*, 2015; Massucato *et al.*, 2020), ISSR (Yuan *et al.*, 2014; Younis *et al.*, 2015; Ahiakpa *et al.*, 2017; Surendran and Udayan, 2020), SDS-PAGE (Osawaru *et al.*, 2014) and inter-primer binding site (iPBS)-retrotransposon (Yildiz *et al.*, 2015) were used for genetic diversity study in cultivated *A. esculentus* and its wild species all over the world.

Patil *et al.* (2018) studied the YVMV resistance in okra genotypes using gene-specific markers. Transcriptomic study using RNA sequencing of combined leaf and pod was reported by Schaeitner *et al.* (2013).

14.0 References

Abbasi, M.W., Khan, M.Q., Zaki, M.J., Shaukat, S.S., Rauf, A., Ahmed, N., Azeem, M. and Tariq, M. (2019) *Acta Phytopathol. Entomol. Hung.*, **54**: 173-186.

Abbasi, F.F., Baloch, M.A., Zia-ul-hassan, Wagan, K.H., Shah, A.N. and Rajpar, I. (2010) *Pak. J. Agril. Engg. Vet. Sci.*, **26**: 11-18.

Abdul, N.M., Joseph, J.K. and Karuppaiyan, R. (2004) *Indian J. Plant Genet. Resour.*, **17**: 241-244.

Adams, C.F. (1975) *Agric Handbook No. 456*, U.S. Department of Agriculture, Washington, D.C., p. 29.

Adejonwo, K.O., Ahmed, M.K., Lagoke, S.T.O. and Karikari, S.K. (1991) *Trop. Pest Manag.*, **37**: 91-95.

Adekiya, A.O., Agbede, T.M., Aboyeji, C.M. and Dunsin, O. (2017) *Sci. Hortic.*, **217**: 209-216.

Adiger, S. (2015) *Ph.D. Thesis*, University of Agricultural Sciences, Dharwad, India.

Adiger, S. and Sridevi, O. (2014) *Trends Biosci.*, **7**: 2306-2309.

Adiger, S., Shanthakumar, G. and Salimath, P.M. (2013) *Karnataka J. Agric. Sci.*, **26**: 6-9.

Afrin, R. (2016) *M.Sc. Thesis*, Sher-e-Bangla Agricultural University, Dhaka, Bangladesh.

Agarrado, R.E. and Rasco, E.T. (1968) *Philippine J. Crop Sci.*, **11** (Suppl 1) Abst. 3B-6a.

Ahiakpa, J.K., Magdy, M., Werner, O., Amoatey, H.M., Yeboah, M.A., Appiah, A.S., Quartey, E.K. and Ros, R.M. (2017) *Ann. Agric. Sci.*, **62**: 131-138.

Ahmed, F., Ashtiaq, M., Muhammad, S., Ahmad, F., Ashtiaq, M. and Muhammad, S. (1999a) *Sarhad J. Agric.*, **15**: 405-407.

Ahmed, N. and Tullock-Reid, L.T. (1968) *Agron. J.*, **60**: 353-356.

Ahmed, N., Hakim, M.A. and Gandroo, M.Y. (1999b) *Indian J. Hort.*, **56**: 247-251.

Ahmed, N., Nawaz, S., Iqbal, A., Mubin, M., Butt, A., Lightfoot, D.A. and Maekawa, M. (2013) *Biosci. Met.*, **4**(4): 19-22.

Ahmed, N., Thakur, M.R., Bajaj, K.L. and Cheema, S.S. (1994) *Plant Dis. Res.*, **9**: 20-25.

Aiyelaagbe, I.O.O. and Ogbonnaya, F.C. (1996) *Res. Bull.*, Nat. Hort. Res. Inst., Ibadan, No. 18, p. 13.

Akash, M.W., Shiyab, S.M. and Saleh, M.I. (2013) *Life Sci. J.*, **10**: 2771-2779.

Aken'Ova, M.E. and Fatokun, C.A. (1984) *Beitrage zur tropischen Landwritschaft und Veterinarmeidzin*, **22**: 149-153.

Akinyele, B.O. and Temikotan, T. (2007) *Int. J. Agric. Res.*, **2**: 165-169.

Akogo, Y., Aidam, A., Odah, K. and Quashie, A.M. (1996) *Cahiers Agril.*, **5**: 109-111.

Aladele, S.E., Ariyo, O.J. and Pena, R.D.L. (2008) *Afr. J. Biotechnol.*, **7**: 1426-1431.

Ali, M., Hossain, M.Z. and Sarker, N.C. (2000) *Euphytica*, **111**: 205-209.

Ali, S., Khan, M.A., Habib, A., Rasheed, S. and Iftikhar, Y. (2005) *Int. J. Agri. Biol.*, **7**(1): 145-147.

Anisuzzaman, M., Jarin, S., Naher, K., Akhtar, M.M., Alam, M.J., Khalekuzzaman, M., Alam, I. and Alam, M.F. (2008) *Asian J. Plant Sci.*, **7**: 677-681.

Anisuzzaman, M., Kabir, A.H., Sarker, K.K., Jarin, S.and Alam, M.F. (2010) *Arch. Phytopathology Plant Protect.*, **43**: 460-466.

Anonymous (1967) *Madras Agric. J.*, **54**: 549.

Anonymous (1971) *Indian Minimum Seed Certification Standard*, Central Seed Committee, Min. Food and Agric., Comm. Dev. Co-op., New Delhi.

Anonymous (1978) *Package of Practices for Kharif Crops of Himachal Pradesh*, Himachal Pradesh Krishi Viswavidyalaya, Palampur.

Anonymous (1983) *Package of Practices for Kharif Crops of Punjab*, Punjab Agric. Univ., Ludhiana.

Ansar, M., Saha, T., Sarkhel, S. and Bhagat, A.P. (2014) *Trends Biosci.*, **7**: 4157-4160.

Ariyo, O.J., Aken'ova, M.E. and Fatokun, C.A. (1987) *Euphytica*, **36**: 677-686.

Arora, D., Jindal, S.K. and Singh, K. (2008) *SABRAO J. Breed. Genet.*, **40:** 93-103.

Arora, S.K. (1980) *Ph.D. Thesis*, Himachal Pradesh Krishi Viswavidyalaya, Solan, India.

Arora, S.K. (1993) *Punjab Hort. J.*, **23**: 116-122.

Arora, S.K., Dhanju, K.C. and Sharma, B.R. (1992) *Plant Dis. Res.*, **7**: 221-225.

Arora, S.K., Dhankar, B.S. and Sharma, N.K. (1990) *Res. Dev. Rept.*, **7**: 123-129.

Arumugam, R., Chelliah, S. and Muthukrishnan, C.R. (1975) *Madras Agric. J.*, **62**: 310-312.

Arumugam. R. and Muthukrishnan, G.R. (1977) *Madras Agric. J.*, **64**: 839-840.

Asghar, S., Hussain, S.A., Ali, N., Asghar, S. and Ali, N. (1997) *Sarhad J. Agric.*, **13**: 449-453.

Atiri, G.I. (1990) *Sci. Hort.*, **45**: 49-53.

Atiri, G.I. and Ibidapo, B. (1989) *J. Agric. Sci.* (UK), **112**: 413-418.

Aykroyd, W.R. (1963) *ICMR Special Rept. Series*, No. 42.

Babarinde, G.O. and Fabunmi, O.A. (2009) *Agricultura Tropica Et Subtropica*, **42**: 151-156.

Bairwa, H.L., Shukla, A.K., Mahawer, L.N., Kaushik, R.A. Shukla, K.B. and Ameta, K.D. (2009) *Indian J. Hort.*, **66**: 310-314.

Bandyopadhyay, B.B., Banerjee, J. and Mukherjee, T. (1986) *Phytobreedon*, **2**: 49-56.

Barnwal, A.K., Pal, A.K., Tiwari, A., Pal S. and Singh, A.K. (2017) *Int. J. Agric. Environ. Biotechnol.*, **10**: 695-701.

Bates, D.M. (1968) *Baileya*, **16**: 99-112.

Batra, B.R., Malik, Y.S. and Pandita, M.L. (1985) *Agric. Sci. Digest.*, **5**: 8-9.

Batra, V.K. and Singh, J. (2000) *Veg. Sci.*, **27**: 192-193.

Baxter, L. and Waters, L.L.R. (1990) *J. Amer. Soc. Hort. Sci.*, **115**: 452-454.

Bayer, C., Fay, M.F. and DeBruijn, A.Y. (1999) *Bot. J. Linnean Soc.*, **129:** 267-303.

Bhagat, A.P. and Yadav, B.P. (1997) *J. Myco. Plant Pathol.*, **27**(1)**:** 94-95.

Bharadiya, P.S., Kalalbandi, B.M. and Shinde, V.N. (2007) *Asian J. Hort.*, **2**: 199-201.

Bharathkumar, M.V., Dhankhar, S.K., Dahiya, M.S. and Srikanth, M. (2019) *Indian J. Agric. Sci.*, **89**: 640-645.

Bhardwaj, M.L., Raj, H., Koul, B.L. and Raj, H. (2000) *Indian J. Agric. Sci.*, **70**: 653-656.

Birah, A., Srivastava, R.C., Kumar, K., Singh, P.K. and Bhagat, S. (2012) *Indian J. Agric. Sci.*, **82**: 470-472.

Bisht, I.S., Mahajan, R.K. and Rana, R.S. (1995) *Ann. Appl. Biol.*, **126**: 539-550.

Biswas, S., Das, K. and Ghosh, S.K. (2009) *J. Entomol. Res.*, **33**(4): 331-333.

Bochalya, S.C., Lal, A.A., Simon, S. and Meena, N.K. (2017) *The Allahabad Fmr.*, **73**: 61-63.

Boonsirichai, K., Phadvibulya, V., Adthalungrong, A., Srithongchai, W., Puripunyavanich, V. and Chookaew, S. (2009) *Induced Plant Mutations in the Genomics Era* (Q.Y. Shu ed.), Food and Agriculture Organization of the United Nations, Rome, pp. 352-354.

Bora, G.C., Saikia, A.K. and Shadeque, A. (1992) *Indian J. Virol.*, **8**: 55-57.

Briddon, R.W. and Stanley, J. (2006) *Virology*, **344**: 198-210.

Briddon, R.W., Bull, S.E., Amin, I., Idris, A.M., Mansoor, S., Bedford, I.D., Dhawan, P., Rishi, N., Siwatch, S.S., Abdel-Salam, A.M., Brown, J.K., Zafar, Y. and Markham, P. G. (2003) *Virology*, **312**: 106-121.

Briddon, R.W., Bull, S.E., Amin, I., Mansoor, S., Bedford, I.D., Rishi, N., Siwatch, S.S., Zafar, M.Y., Abdel-Salam, A.M. and Markham, P.G. (2004) *Virology*, **324**: 462-474.

Brown, J.K., Fauquet, C.M., Briddon, R.W., Zerbini,M.,Moriones, E., and Navas-Castillo, J. (2012) *Virus Taxonomy* (A.M.Q. King, M.J. Adams, E.B. Carstens, and E.J. Lefkowitz eds.), Elsevier Academic Press, London, pp. 351-373.

Burzo, I. (1980) *Acta Hortic.*, **116**: 61-64.

Calisir, S. and Yildiz, M.U. (2005) *J. Food Eng.*, **68**: 73-78.

Capoor, S.P. and Varma, P.M. (1950) *Indian J. Agric. Sci.*, **20**: 217-230.

Chacko, R.S., Babu, K.V.S. and Rajan, S. (1996) *J. Trop. agric.*, **34**: 138-139.

Chakraborty, S., Pandey, P.K. and Singh, B. (1997) *Veg. Sci.*, **24**: 52-54.

Chandra, S. and Bhatangar, S.P. (1975) *Acta Bot. Indica*, **3**: 104-113.

Chandra, S. and Bhatnagar, S.P. (1974) *Naturwissenschaften*, **61**: 688.

Charrier, A. (1984) *Genetic Resources of Abehnoschus (Okra)*, IBPGR, Rome, Italy. p. 61.

Chattopadhyay, A. and Sahana, B.C. (2000) *Res. Crop*, **1**: 176-180.

Chattopadhyay, A., Dutta, S. and Chatterjee, S. (2011) *Afr. J. Biotechnol.*, **10**: 5461-5467.

Chauhan, D.V.S. (1972) *Vegetable Production in India* (3rd edn.), Ram Prasad and Sons, Agra.

Chormule, S.R. and Patel, J.B. (2018) *J. Pharmacogn. Phytochem.*, **7**: 3041-3048.

Choudhury, B. and Anothai-Choomsai, M.L. (1970) *Indian J. Agric. Sci.*, **40**: 405-812.

Chowda-Reddy, R.V., Kirankumar, M., Seal, S.E., Muniyappa, V., Valand, G.B., Govindappa, M.R. and Colvin, J. (2012) *J. Integr. Agric.*, **11**: 235-248.

Chowdhury, M.S., Hasan, Z., Kabir, K., Shah Jahan, M. and Kabir, M.H. (2014) *The Agric.*, **12**: 56-63.

Chowdhury, S. and Kumar, S. (2019) *Int. J. Chem. Stud.*, **7**: 853-857.

Chukunda, F.A. and Nwonuala, A.I. (2013) *Int. J. Adv. Biol. Res.*, **3**: 322-324.

Dahivelkar, P., Atre, G.E., Gawande, P.V. and Mate, G.D. (2017) *Int. J. Curr. Microbiol. Appl. Sci.*, **6**: 3189-3193.

Damarany, A.M. and Farag, I.A. (1994) *Assiut J. Agric. Sci.*, **25**: 57-70.

Darekar, K.S. and Ranade, M.S. (1990) *Int. Nematol. Net Work Newsletter*, **7**: 6-7.

Darlington, C.D. and Wylie, A.P. (1955) *Chromosome Atlas of flowering Plants*, George Allen and Unwin, London, p. 519.

Das, S., Chattopadhyay, A., Chattopadhyay, S.B. and Hazra, P. (2013) *Int. J. Veg. Sci.*, **19**: 58-77.

Das, S., Chattopadhyay, A., Chattopadhyay, S.B., Dutta, S. and Hazra, P. (2012) *Afr. J. Biotechnol.*, **11**: 16132-16141.

Dash, P.K., Das, S., Rabbanil, M.G. and Mondal, M.F. (2013) *Int. J. Biosci.*, **3**: 123-131.

Deepak, K.D., Deshpande, V.K., Vyakarnahal, B.S., Ravi Kumar, R.I., Uppar, D.S. and Hosamani, R.M. (2007) *Karnataka J. Agric. Sci.*, **21**: 202-205.

Derbalah, A.S., Kot, G.A.El. and Hamza, A.M. (2011) *Arch. Phytopathol. Plant Prot.*, **44**: 2012-2023.

Deshmukh, N.D., Jadhav, B.P., Halakude, I.S. and Rajput, J.C. (2011) *Veg. Sci.*, **38**: 79-81.

Devi, A.P., Banerjee, S., Bhattacharjee, T., Maurya, P.K., Mandal, A.K., Dutta, S. and Chattopadhyay, A. (2019) *Int. J. Veg. Sci.*, **26**: 150-162.

Devi, A.P., Banerjee, S., Maurya, P.K., Bhattacharjee, T., Jamir, I., Chatterjee, S., Mandal, A.K., Dutta, S. and Chattopadhyay, A. (2018) *Agri. Res. and Tech.*, **18**: 556072.

Devi, A.P., Bhattacharjee, T., Banerjee, S., Maurya, P.K., Chatterjee, S. and Chattopadhyay, A. (2019b) *Int. J. Veg. Sci.*, **26**: 163-189.

Dhanalakshmi, D.N. and Mallapur, C.P. (2008) *Ann. Plant Prot. Sci.*, **16**: 29-32.

Dhande, G.A., Patil, V.M., Raut, R.V., Rajput, J.C. and Ingle, A.G. (2012) *Afr. J. Biotechnol.*, **11**: 15226-15230.

Dhankar, B.S., Chauhan, M.S. and Nand Kishore (1989) *Indian J. Virology*, **5**: 94-98.

Dhankhar, B.S. and Dhankhar, S.K. (2001) *Haryana J. Hort. Sci.*, **30:** 230-233.

Dhankhar, B.S. and Mishra, J.P. (2005) *J. New Seeds.*, **6**: 195-209.

Dhankhar, B.S. and Singh, R. (2009) *Okra Handbook: Global Production, Processing, and Crop Improvement*, New York, HNB Publishing.

Dhankhar, S. and Singh, S. (2013) *J. Agrometeorol.*, **15**: 39-42.

Dhankhar, S.K, Dhankhar, B.S. and Saharan, B.S. (1996) *Ann. Biol.*, **12**: 90-92.

Dhankhar, S.K., Dhankhar, B.S. and Yadava, R.K. (2005) *Indian J. Agric. Sci.*, **75**: 87-89.

Dhar, T and Bhattacharya, S. (2015) *Int. J. Bio-Resource Environ. Agric. Sci.*, **1**: 126-131.

Dhuria, H.S. and Shukla, V. (1972) *Proc. 3rd Int. Symp. Subtrop. and Trop. Hort.*, Bangalore, p. 152.

Dubey, R.S. and Singh, J.P. (1968a) *Indian J. Agric. Sci.*, **38**: 108-114.

Dubey, R.S. and Singh, S.P. (1968b) *Indian J. Hort.*, **25**: 210-213.

Duggi, S.L., Magadum, S., Srinivasraghavan, A., Kishor, D.S. and Oommen, S.K. (2013) *Int. J. Agric. Environ. Biotech.*, **6**: 45-50.

Dumal, B.S., Belorkar, P.V., Patil, B.N., Golliwan, V.J. and Gaikwad, Y.S. (1993) *J. Soil and Crop*, **3**: 27-29.

Duranti, A. (1964-65) *Ann. Fac. Sci. Agri. Portici*, **30**: 393-406.

Dutta, O.P. (1971) *Proc. 3rd Int. Symp. Subtrop. and Trop. Hort.*, Bangalore, India, pp. 141-156.

Dutta, O.P. (1980) *Ph.D. Thesis*, University of Agricultural Sciences, Bangalore, India.

Dutta, O.P. (1984) *Annual Report*, IIHR, Bangalore, India.

Dwivedi, Y.C., Sengupta, S.K., Naidu, A.K. and Choubey, P.C. (1996) *JNKVV Res. J.*, **27**: 53-55.

Engels, J.M.M., Chandel, K.P.S. and Withers, L.A. (1991) *Rep. Internatl. workshops on okra genetic resources*, National Bureau for Plant Genetic Resources, New Delhi, India, 8-12 October, 1990-91, pp. 119-124.

FAOSTAT (2018) Available online at: http://faostat.fao.org.

Finger, F.L., Della-Justina, M.E., Casali, V.W.D. and Puiatti, M. (2008) *Sci. Agric. (Piracicaba, Braz.)*, **65**: 360-364.

Firoz, Z.A. (2009) *Bangladesh J. Agric. Res.*, **34**: 713-722.

Fougat, R.S., Purohit, A.R., Kumar, S., Parekh, M.J. and Kumar, M. (2015) *Indian J. Agric. Sci.*, **85:** 1223-1228.

Ganesan, M., Chandrasekar, R., Kumari, B.D.R. and Jayabalan, N. (2007) *Biol. Plant.*, **51**(3): 414-420.

Ganeshe, R.K., Pandey, R.P. and Rawat, A.K. (1998) *Hort. J.*, **11**: 53-58.

Gasti, V.D., Madalageri, B.B., Ryagi, Y.H. and Dharmatti, P.R. (1997) *Adv. Agric. Res. India*, **8**: 21-25.

Gemede, H.F., Ratta, N., Haki, G.D., Woldegiorgis, A.Z. and Beyene, F. (2015) *J. Food Process. Technol.*, **6**: 458.

Ghosh, C., Biswas, P., Mahato, S., Rana, D.K. and Mahato, B. (2018) *SATSA Mukhapatra - Annual Tech. Issue,* **22**: 96-102.

Godawatte, V.N.A. and De Silva, C.S. (2016) *OUSL J.*, **10**: 41-57.

Gogoi, S., Hatibarua, P., Mazumder, A. and Shadeque, A. (1996) *J. Agric. Sci. Society of N.E. India*, **9**: 115-118.

Gonzaga, Z.C., Obeda, W.L., Gorme, A.L.G., Rom, J.C., Jr. Abrantes, O.F. and Capuno, O.B. (2017) *Ann. Trop. Res.*, **39**: 44-54.

Gopalakrishnan, C. and Valluvaparidasan, V. (2009) *J. Biol. Control*, **23**: 325–327.

Gopalan, C., Rama Sastri, B.V. and Balasubramanian S. (1989) *Nutritive Value of Indian Foods*, National Institute of Nutrition (NIN), Publishing House, ICMR, Hyderabad, India.

Gosalwad, S.S. and Wadnerkar, D.W. (2009) *J. Plant Prot. Environ.*, **6**: 11-16.

Gowda, M.T., Rai, A.B. and Singh, B. (2017) *IIVR Technical Bulletin No. 76*, IIVR, Varanasi, pp. 32.

Grewal, B.S., Nandpuri, K.S. and Kumar, J.C. (1973) *Punjab Hort. J.*, **13**: 247-252.

Gulsen, O., Karagul, S. and Abak, K. (2007) *Biologia*, **62:** 41-45.

Gunawardhana, M.D.M. and De Silva, C.S. (2011) *Trop. Agric. Res.*, **23**: 77-83.

Gupta, A. (1990) *Veg. Sci.*, **17**: 16-19.

Gupta, C.R., Singh, V.K. and Dwivedi, R.K. (1999) *Orissa J. Hort.*, **27**: 35-42.

Halder, J., Kushwaha, D., Ranjan, P., Rai, A.B. and Gupta, S. (2016a) *Pest Manag. Hortic. Ecosyst.*, **22**: 96-98.

Halder, J., Sanwal, S.K., Deb, D., Rai, A.B. and Singh, B. (2016b) *Indian J. Agric. Sci.*, **86**: 481-484.

Halder, J., Sanwal, S.K., Rai, A.K., Rai, A.B. Singh, B. and Singh, B.K. (2015) *Indian J. Agric. Sci*, **85**: 278-282.

Hamer, C. and Thompson, (1957) *Vegetable Crops*. McGraw Hill Co., Inc. N. X. Toronto, London.

Hamon, S. and Hamon, P. (1991) *Euphytica*, **58**: 101-111.

Hamon, S. and Koechlin, J. (1991a) *Euphytica*, **53**: 41-48.

Hamon, S. and Koechlin, J. (1991b) *Euphytica*, **53**: 49-55.

Hao, C.X., Fu, C.Y., Min, L.M., Chen, X.H., Cheng, Y.F. and Liao, M.M. (1999) *China Vegetables*, No. 4, pp. 33-34.

Haq, I.U., Khan, A.A. and Azmat, M.A. (2013) *Pak. J. Agric. Sci.*, **50**: 655-662.

Hazarika, D.N., Talukdar, M. and Mohan, N.K. (1997) *Indian J. Hill Fmg.*, **10**: 96-97.

Hazra, P., Chattopadhyay, A., Karmakar, K. and Dutta, S. (2011) *Modern Technology in Vegetable Production*, New India Publishing Agency, New Delhi, India.

Hirekurubar, R.B. and Ambekar, J.S. (2008) *Crop Res. (Hisar)*, **36**: 302-307.

Hochreutiner, B.P.G. (1924) *Candollea*, **2**: 79-90.

Hooda, V.S., Dhankhar, B.S., Dahiya, B.S., Singh, R. and Singh, R. (1999) *Indian J. Hort.*, **56**: 73-76.

Hughes, J. (2009) *Acta Hortic.*, **806**: 39-47.

IBPGR (1991) *International Crop Network Series 5.* Report on International Workshop on Okra Genetic resources held at the National bureau for Plant Genetic Resources, New Delhi, India.

Ige, O.E. and Eludire, M.O. (2014) *American Int. J. Biol.*, **2**: 1-9.

Incalcaterra, G., Vetrano, F., Stoffella, P.J., Cantlifee, D.J. and Damato, G. (2000) *Acta Hortic.*, **553**: 329-336.

Ionescu, C. and Ionescu, A. (1995) *Anale Institutul de Cercetari-pentru Legumicultura si Floricultura, Vidra*, **13**: 75-82.

Iqbal, J., Sagheer, M., Hasan, M. and Nadeem M. (2012) *Pak. J. Agric. Sci.*, **49**: 179-184.

Iremiren, G.O. (1988) *Exp. Agric.*, **24**: 247-252.

Iremiren, G.O., Osara, A.W. and Okiy, D.A. (1991) *Exp. Agric.*, **27**: 33-37.

Isaac, S.R., Pushpakumari, R. and Saraswathy, R. (1998) *J. Trop. Agric.*, **36**: 50-52.

Jagan, K., Reddy, K.R., Sujatha, M., Sravanthi, V. and Reddy, S.M. (2013) *IOSR J. Pharm. Biol. Sci.*, **7**: 69-70.

Jain, R.K. and Gupta, D.C. (1997) *Indian J. Nematol.*, **26**: 121-122.

Jalendhar, G., Reddy, C.S.K., Srinivas, A.and Rao, M.A. (2016) *Int. J. Sci. Nature*, **7**: 165-167.

Jambhale, N.D. and Nerkar, Y.S. (1981) *Theor. Appl. Genet.*, **50**: 313-316.

Jamir, I., Mandal, A.K., Devi, A.P., Bhattacharjee, T., Maurya, P.K., Dutta, S., Chattopadhyay, A., Pramanik, K. and Banik, S. (2020) *Indian Phytopathol.*, DOI: 10.1007/s42360-019-00183-0

Jassim, A. (1967) *Diss. Abst.*, **28**: 3-13.

Jenkins, D.J.A., Kendall, C.W.C., Marchie, A., Faulkner, D.A., Wong, J.M.W., de Souza, R., Emam, A., Parker, T.L., Vidgen, E., Trautwein, E.A., Lapsley, K.G., Josse, R.G., Leiter, L.A., Singer, W. and Connelly, P.W. (2005) *Am. J. Clin. Nutr.*, **81**: 380-387.

Jha, R.K., Neupane, R.B., Khatiwada, A., Pandit, S. and Dahal, B.R. (2018) *J. Agric. Nat. Resour.*, **1**: 168-178.

Jindal, S.K., Arora, D. and Ghai, T.R. (2009) *Crop Improv.*, **36**: 59-66.

Joi, M.B. and Shende, R.V. (1979) *Veg. Sci.*, **6**: 126-129.

Jose, J. and Usha, R. (2003). *Virology*, **205**: 310-317.

Joseph, J.K., Nissar, V.A.M., Latha, M., Shrikant, S., Patil, P., Malik, S.K., Negi, K.S., Keisham, M., Rao, S.R., Yadav, S.R. and Bhat, K.V. (2013) *Curr. Hortic.*, **1**: 35-46.

Joshi, A.B. and Hardas, M.W. (1956) *Nature*, **178**: 1190.

Joshi, A.B. and Hardas, M.W. (1976) *Evolution of Crop Plants* (N.W. Simmonds ed.), Longmans, London, pp. 194-195.

Joshi, A.B., Gadwal, V.R. and Hardas, M.W. (1974) *Evolutionary Studies in World Crops, Diversity and Change in the Indian Subcontinent* (J.B. Hutchinson ed.), Cambridge, pp. 99-105.

Joshi, B.S., Singh, H.B. and Gupta, P.S. (1958) *Indian J. Genet.*, **18**: 57-68.

Joshi, J.L., Rajan, R.E.B. and Kumar, C.P.S. (2019) *Plant Arch.*, **19**: 539-543.

Kabir, M.R., Ahmed, S. and Akhond, M.A.Y. (2016) *Bangladesh J. Agril. Res.*, **41**: 521-528.

Kalia, H.R. and Padda, D.S. (1962) *Indian J. Genet.*, **22**: 248-251.

Kamalanathan, S., Sundarrajan, S. and Thambjraj, S. (1970) *Madras Agric. J.*, **57**: 10-17.

Kanwar, J.S. and Saimbhi, M.S. (1987) *Punjab Hort. J.*, **27**: 234-238.

Kaur, H., Chawla, N., Bassia, G. and Pathak, M. (2015) *Int. J. Curr. Sci.*, **15:** 51-58.

Kaur, K., Pathak, M., Kaur, S., Pathak, D. and Chawla, N. (2013) *Afr. J. Biotechnol.*, **12**: 3160-3170.

Kedarnath, Ravichandra, N.G., Preethi, D.M., Reddy, B.M.R., Pavithra, B.S. and Pavithra, R.S. (2017) *Int. J. Curr. Microbiol. App. Sci.*, **6**: 3420-3426.

Kerure, P. and Pitchaimuthu, M. (2019) *Electron. J. Plant Breed.*, **10**: 248-255.

Khalikar, P.V., Jagtap, G.P. and Sontakke, P.L. (2011) *Indian Phytopathol.*, **64**: 286-290.

Khan, A.R. and Jaiswal, R.C. (1988) *Veg. Sci.*, **15**: 8-14.

Khan, M.A. and Rab, A. (2019) *Sarhad J. Agric.*, **35**: 751-756.

Khanpara, M.D., Jivani, L.L., Vachhani J.H., Shekhat H.G. and Mehta, D.R. (2009) *Int. J. Agric. Sci.*, **5**: 554-557.

Khanuja, S.P., Shasany, A.K. and Kumar, S. (1999) *Plant Mol. Biol. Rep.*, **17**: 1-7.

Khulbe, D., Sasmal, A., Panda, S. and Sarangi, P.K. (2018) *J. Plant Prot. Environ.*, **15**: 47-51.

Kishor, D.S., Duggi, S., Arya, K. and Magadum, S. (2013) *Bioinfolet*, **10**: 490-494.

Kochko, A.D. and Hamon, S. (1990) *Plant Mol. Biol. Rep.*, **8**: 3-7.

Kodandaram, M.H., Halder, J., Singh, N. and Rai, A.B. (2017) *Pestic. Res. J.*, **29**: 153-157.

Kokare, R.T., Bhalerao, R.K., Prabu, T., Chavan, S.K., Bansode, A.B. and Kachare, G.S. (2006) *Agric. Sci. Digest*, **26**: 178-181.

Kolhe, A.K. and Chovan, V.M. (1967) *Indian J. Agric. Sci.*, **37**: 115-166.

Kolhe, A.K. and D'Cruz, R. (1966) *Indian J. Genet.*, **26**: 112-117.

Kulkarni, C.S. (1924) *Poona Agric. Coll. Mag.*, p. 16.

Kulkarni, R.S. (1976) *Mysore J. Agric. Sci.*, **10**: 332-333.

Kulkarni, R.S., Rao, T.S. and Virupakshappa, K. (1978) *Indian J. Genet. Plant Breed.*, **38**: 160-162.

Kulkarni, R.S., Swamy, Rao, T. and Virupakshappa, K. (1973) *Res. J.*, Kerala, **14**: 13-20.

Kumar S., Parekh M.J., Fougat R.S., Patel, S.K., Patel, C.B., Kumar M. and Patel, B.R. (2017b) *J. Plant Biochem. Biotechnol.*, **26**: 172-178.

Kumar, A., Baranwal, D.K., Aparna, J. and Srivastava, K. (2013) *Madras Agric. J.*, **100**: 30-35.

Kumar, A., Kumar, M., Sharma, V.R., Singh, M.K., Singh, B. and Chand, P. (2019) *Prog. Agric.*, **19**: 140-145.

Kumar, Narinder (1983) *M. Sc. Thesis*, Punjab Agric. Univ., Ludhiana.

Kumar, P., Haldankar, P.M. and Haldavanekar, P.C. (2018) *The Pharma Innov. J.*, **7**: 180-184.

Kumar, P., Singh, K.V., Singh, B., Kumar, S. and Singh, O. (2012) *Prog. Agric.*, **12**: 354-359.

Kumar, P.S. and Sreeparvathy, S. (2010) *Electron. J. Plant Breed.*, **1**: 1431-1433.

Kumar, S., Dabbas, M.R. and Tiwari, P. (2015) *Int. J. Plant Protec.*, **8**: 384-388.

Kumar, S., Singh, A.K., Yadav H. and Verma, A. (2017a) *J. Appl. Nat. Sci.*, **9**: 774-779.

Kumar, S.N. (2011) *Plant Arch.*, **11**: 683-685.

Kumar, V., Khan, M.R. and Walia, R.K. (2020) *Natl. Acad. Sci. Lett.* DOI. 10.1007/s40009-020-00895.

Kumari, R. (2016) *M.Sc. Thesis*, Bihar Agricultural University, Sabour, India.

Kundu, B.C. and Biswas, C. (1973) *Proc. Indian Sci. Congr.*, **60**: 295-298.

Kuwada, H. (1974) *Japanese J. Breed.*, **24**: 207-210.

Kyriakopoulou, O.G., Arens, P., Pelgrom, K.T.B., Karapanos, I., Bebeli, P. and Passam, H.C. (2014) *Sci. Hortic.*, **171**: 58-70.

Lal, G. and Lal, G. (1997-98) *Recent Hortic.*, **4**: 135-137.

Leela, D. (1989) *Indian J. Agron.*, **34**: 147-149.

Lengsfeld, C., Titgemeyer, F., Faller, G. and Hensel, A. (2004) *J. Agric. Food Chem.*, **52**: 1495-1503.

Lenka, P.C., Das, D.K. and Mishra, H.N. (1989) *Orissa J. Agric. Res.*, **2**: 125-127.

Leonardi, C. (1995) *Colture Protette*, **24**: 79-83.

Mahajan, M., Ghai, T.R. and Chawla, N. (2004) *J. Res. Punjab Agric. Univ.*, **41**(3): 345-351.

Mahajan, Y.P. and Sharma, B.R. (1979) *Sci. Hort.*, **10**: 135-139.

Majanbu, I.S., Ogunlela, V.B. and Ahmed, M.K. (1986) *Fert. Res.*, **8**: 297-306.

Mal, B., Mahapatra, P., Mohanty, S. and Mishra, H.N. (2013) *J. Crop Weed*, **9**: 109-112.

Mallela, R.R., Vutukuri, P.R. and Kanuri, M. (2009) *Fruit, Veg. Cereal Sci. Biotech.*, **3**: 1-6.

Mangat, B.S. and Roy, M.K. (1986) *Plant Sci.* (*Irish Republic*), **47**: 57-61.

Mani, S. and Ramanathan, K.M. (1981) *South Indian Hort.*, **29**: 100-113.

Manju, K.P., Lakshmi, K.V., Babu, S.B. and Anitha, K. (2018) *Int. J. Curr. Microbiol. Appl. Sci.*, **6**: 1676-1681.

Martin, F. and Rhodes, A. (1983) *Plant Foods Hum. Nutr.* (*Formerly Qualitas Plantarum*), **33**: 41-49.

Martin, F.W. (1983) *J. Agric. Univ. Puerto Rico*, **67**: 50-52.

Martinello, G.E., Leal, N.R., Amaral Junior, A.T. and Pereira, M.G. (2001) *Acta Hortic.*, **546**: 101-104.

Martinello, G.E., Leal, N.R., Amaral, J.A.T., Pereira, M.G. and Daher, R.F. (2003) *Hortic. Bras.*, **21**: 20-25.

Massucato, L.R., Nakamura, K.K., Ruas, P.M., Zeffa, D.M., da Silva D.J.H. and Gonçalves, L.S.A. (2020) *Acta Sci. Agron.*, **42**: e43426.

Mathew, S.K., Vahab, M.A., Devadas, V.S. and Cherian, A. (1993) *J. Trop. Agric.*, **31**: 215-218.

Maurya, R.P., Bailey, J.A. and Chandler, J.St.A. (2013) *American J. Agric. For.*, **1**: 48-54.

Mbagwu, J.S.C. and Adesipe, F.A. (1987) *Scientia Hort.*, **31**: 35-43.

Meena, R.K., Chhatterjee, T. and Thakur, S. (2014) *Indian J. Appl. Res.*, **4**: 57-59.

Mehta, N., Asati, B.S. and Mamidwar, S.R. (2007) *Bangladesh J. Agric. Res.*, **32**: 421-432.

Mehta, Y.R. (1959) *Vegetable Growing in Uttar Pradesh,* Bureau of Agric. Inf., U.P., Lucknow.

Menon, R., Sarao, N.K. and Pathak, M. (2018) *App. Biol. Res.*, **20**: 221-228.

Mishra, G.P., Singh, B., Seth, T., Singh, A.K., Halder, J., Krishnan, N., Tiwari, S.K. and Singh, P.M. (2017) *Front. Plant Sci.*, **8** (360).

Mishra, M. and Patjoshi, A.K. (1995) *Environment and Ecology*, **13**: 732-733.

Mitidieri, J. and Vencovsky, R. (1974) *Revista de Agriculture*, Brazil, **49**: 3-6.

Moekchantuk, T. and Kumar, P. (2004) *Export okra production in Thailand.* Inter-country programme for vegetable IPM in South and SE Asia phase II Food and Agriculture Organization of the United Nations, Bangkok, Thailand, p. 56.

Mohanta, S. and Mohanty, K.C. (2012) *J. Plant Prot. Environ.*, **9**: 66-68.

Moyin-Jesu E.I. (2007) *Bioresour. Technol.*, **98**: 2057-2064.

Mukhtar, T., Hussain, M.A., Kayani, M.Z. and Aslam, M.N. (2014) *Crop Prot.*, **56**: 25-30.

Nadkarni, K.M. (1927) *Indian Meteria Medica*, Nadkarni and Co., Bombay.

Nagegowda, N.S., Hebbar S. S. and Shilpashree, V.M. (2020) *Curr. Agric. Res. J.*, **8**: 25-29.

Naik, G.R. and Jayalakshmi, K. (2017) *Int. J. Chem. Stu.*, **5**: 1210-1212.

Narendran, M., Deole, S.G., Harkude, S., Shirale, D., Nanote, A., Bihani, P., Parimi, S., Char, B.R. and Zehr, U.B. (2013) *Plant Cell Rep.*, **32**: 1191-1198.

Nariani, T.K. and Seth, M.L. (1958) *Indian Phytopath.*, **11**: 137-143.

Nath, P. (1976) *Vegetables for the Tropical Region*, ICAR, New Delhi.

Nath, P. and Dutta, O.P. (1970) *Canad. J. Genet. Cytol.*, **12**: 589-593.

Nath, P. and Saikia, A.K. (1992) *New Agric.*, **3**: 199-202.

Naveen, N., Pallavi, H.M., Satish, D., Cholin, S.S., Ganiger, V.M. and Babu, A.G. (2017) *J. Exp. Biol. Agric. Sci.*, **5**: 623-630.

Nerkar, Y.S. (1990) *International Crop Network Series 5.* IBPGR Report 1991, p. 111.

Ngure, J.W., Aguyoh, J.N. and Gaoquiong, L. (2008) *J. App. Biosci.*, **6**: 173-179.

Nwangburuka, C.C., Denton, O.A., Kehinde, O.B., Ojo, D.K. and Popoola, A.R. (2012) *Spanish J. Agric. Res.*, **10**: 123-129.

Nwangburuka, C.C., Kehinde, O.B., Ojo, D.K., Popoola, A.R., Oduwaye, O., Denton, O.A. and Adekoya, M. (2011) *Acta SATECH*, **4**: 1-14.

Olasantan, F.O. (2001) *J. Agric. Sci.*, **136**: 207-214.

Onwueme, I.C. (1975) *J. Agric. Sci., Camb.*, **85**: 233-239.

Osawaru, M.E., Ogwu, M.C. and Omologbe, J. (2014) *Egypt. Acad. J. Biol. Sci.*, **5**: 55-65.

Ouedraogo, M.H., Sawadogo, N., Batieno, T.B.J., Zida, W.P.M.S.F., Bougma, A.L., Barro, A., Kiebre Z. and Sawadogo M. (2018) *Afr. J. Biotechnol.*, **17**: 126-132.

Oyewole, C.I. and Obaweda, O.D. (2020) *GSC Biol. Pharm. Sci.*, **11**: 106-112.

Pal, B.P., Singh, H.B. and Swarup, V. (1952) *Bot. Gaz.*, **113**: 455-464.

Panda, P.K. and Singh, K.P. (1997) *Madras Agric. J.*, **84**: 136-138.

Pandey, K.K. and Pandey, P.K. (2003) *Veg. Sci.*, **30**: 128-134.

Pandey, U.C. and Manocha, V.P. (1989) *Agric. Res. Digest (Karnal)*, **9**: 91-93.

Pandey, V.B. and Dubey, R.P. (1997) *Agric. Sci. Digest*, Karnal, **17**: 114-116.

Papal, S.A. and Bharpoda, T.M. (2009) *Karnataka J. Agric. Sci.*, **22**: 707-709.

Parmar, P.N., Bhanvadia, A.S., Chaudhary, M.M. and Patel, A.P. (2016) *J. Pure Appl. Microbiol.*, **10**: 485-488.

Patel, B.G. and Patel, A.I. (2016) *Ann. Agric. Environ. Sci.*, **1**: 15-20.

Patel, S.R., Shah, R.R. and Raman, S. (1999) *Gujrat Agric. Univ. Res. J.*, **24**: 74-78.

Pateliya, C.K., Parmar, B.R. and Tandel, Y.N. (2008) *The Asian J. Hort.*, **3**: 317-318.

Patil, S., Bagde, E., Tatte, R., Mane, S.S., Naik, R.M., Gawande, P.P. and Jadhav, P.V. (2018) *Int. J. Chem. Stud.*, **6**: 2809-2812.

Patil, S.R., Lande, G.K., Awasthi N.S. and Barkhade, U.P. (2014) *The Bioscan*, **9**: 1597-1600.

Pertierra, L.R. and Melin, M.P. (1998) *Agro-Ciencia*, **14**: 23-28.

Phadvibulya, V., Boonsirichai, K., Adthalungrong, A. and Srithongchai, W. (2009) In: *Induced Plant Mutations in the Genomics Era* (Ed.Q.Y. Shu), Food and Agriculture Organization of the United Nations, Rome, pp. 349-351.

Phonglosa, A., Bhattacharyya, K., Ray, K., Mandal, J., Pari, A., Banerjee, H. and Chattopadhyay, A. (2015) *Sci. Hortic.*, **187**: 1-9.

Piloo, N. and Kabir, J. (2011) *J. Environ. Res. Dev.*, **5**: 615-622.

Piloo, N. and Kabir, J. (2015) *The Asian J. Hort.*, **10**: 134-138.

Pitchaimuthu, M., Dutta, O.P. and Swamy, K.R.M. (2012) *J. Hort. Sci.*, **7**(2): 199-202.

Prabhakar, R.S., Hegde, D.M. and Srinivas, K. and Doijode, S.D. (1985) *South Indian Hort.*, **33**: 115-117.

Prabhu, A.S., Pathak, K.D. and Singh, R.P. (1971) *Indian J. Hort.*, **2**: 310-312.

Prabu, T. (2005) *Ph.D. Thesis*, Mahatma Phule Krishi Vidyapeeth, Rahuri, Maharashtra, India.

Prabu, T. and Warade, S.D. (2009) *Veg. Sci.*, **36**: 283-287.

Prabu, T., Warade, S.D. and Ghante, P.H. (2007) *Veg. Sci.*, **34**: 119-122.

Prabu, T., Warade, S.D., Saidi, M. and Baheti, H.S. (2009) *Indian J. Plant Protec.*, **37**: 87-91.

Prakash, K. and Pitchaimuthu, M. (2010) *Electron. J. Plant Breed.,* **1**: 1426-1430.

Prakash, K., Pitchaimuthu, M. and Ravishankar, K.V. (2011) *Electron. J. Plant Breed.,* **2**: 80-86.

Prakasha, T.L. (2009) *M.Sc. Thesis,* University of Agricultural Sciences, Dharwad, India.

Praveen, P.M. and Dhandapani (2001) *J. Veg. Crop Prodn.,* **7**: 3-11.

Pullaiah, N., Reddy, T.B., Moses, G.J., Reddy, B.M. and Reddy, D.R. (1998) *Indian J. Genet. Plant Breed.,* **58**: 349-352.

Pun, K.B., Doraiswamy, S., Jeyarajan, R. and Doraiswamy, S. (2000) *Indian Phytopath.,* **53**: 95-96.

Pun, K.B., Sabitha, D. and Jeyaran, R. (1999) *Indian Phytopathol.,* **52**: 221-223.

Purewal, S.S. and Randhawa, G.S. (1947) *Indian J. Agric. Sci.,* **17**: 129-136.

Raghav, M. (1996) *Recent Horticulture,* **3**: 99-101.

Rahman, M.M., Uddin, M.M. and Shahjahan, M. (2013) *Int. Res. J. App. Life Sci.,* **2**: 1-9.

Rai, A.B., Loganathan, M., Halder, J., Venkataravanappa, V. and Naik, P.S. (2014) *Technical Bulletin No. 53, ICAR-IIVR,* Varanasi, Uttar Pradesh, p. 104.

Rai, S., Hossain, M. and Hossain, F. (1998) *Hort. J.,* **11**: 77-82.

Rajamony, L., Chandran, M. and Rajmohan, K. (2006) *Acta Hortic.,* **725**: 235-238.

Rajeshwari, G., Bhanu, K.R.M., Chakravarthy, A.K., Sridhar, V. and Kumar, S.M. (2018) *J. Entomol. Zool. Stud.,* **6**: 1030-1033.

Ram, H., Khan, M.M. and Singh, P.K. (2013) *The Asian J. Hort.,* **8**: 507-511.

Ramu, P.M. (1976) *Mysore J. Agric. Sci.,* **10**: 146.

Rana, R.S. Thomas, T.A. and Arora, R.K. (1990) *International Crop Network Series 5.* IBPGR Report 1991, p. 43.

Rani, M.U., Jyothi K.U. and Kumar M.K. (2013) *Bioinfolet,* **10**: 1079-1081.

Rashid, M.H., Yasmin, L., Kibria, M.G., Mollik, A.K.M.S.R. and Monowar-Hossain, S.M. (2002) *Plant Pathol. J.,* **1**: 61-62.

Rathod, R.R. and Patel, C.L. (1996) *Gujrat Agril. Univ. Res. J.,* **22**: 109-111.

Ratnapala, D.G.S., Peiris, B.C.N. and Ariyaratne, A.R. (1997) *Trop. Agric. Res.,* **6**: 1-16.

Ravishankar, K.V., Muthaiah, G., Pitchaimuthu, M. and Gundale, S.K. (2018) *J. Genetics,* **97**: e39-e47.

Reddy, D.M.V., Chandrashekara, B.P. and Chandrashekara, R. (1998) *Seed Res.,* **25**: 41-44.

Reddy, M.T., Haribabu, K., Ganesh, M. and Begum, H. (2012) *Chil. J. Agric. Res.,* **72**: 316-325.

Rice, R.P., Rice, P.W. and Tindall, H.D. (1990) *Fruit and Vegetable Production in warm climates,* MacMillan Publishers Ltd., London and Basingstoke.

Rithichai, P., Fujime, Y., Terabayashi, S., Okuda, N and Date, S. (2004) *J. Japan Soc. Hort. Sci.*, **73**: 421-428.

Rizwan, H.M., Irshad, M., He, B., Liu, S., Lu, X., Sun, Y. and Qiu, D. (2020) *S. Afr. J. Bot.*, **130**: 300-307.

Rizwan, H.M., Irshad, M., He, B.Z., Liu, S., Debnath, B., Mitra, S., Li, M., Lu, X.C., Sun, Y.T. and Qiu, D.L. (2018) *Appl. Ecol. Environ. Res.*, **16**: 3421-3435.

Rojas, M.R., Hagen, C., Lucas, W.J. and Gilbertson, R.L. (2005) *Annu. Rev. Phytopathol.*, **43**: 361-394.

Roy, K., Rathod, A. and Devi M.S. (2017) *J. App. Nat. Sci.*, **9**: 344-350.

Roy, M.K. and Mangat, B.S. (1989) *Plant Sci. (Irish Republic)*, **60**: 77-81.

Sabitha, V., Ramachandran, S., Naveen, K. R., and Panneerselvam, K. (2011) *J. Pharm. Bioallied Sci.*, **3**: 397-402.

Sadashiva A.T. (1988) *Ph.D. Thesis*, University of Agricultural Sciences, Dharwad, India.

Saikia, S., Saikia, A. and Shadeque, A. and Gogoi, S. (1997) *Annals of Biology*, Ludhiana, **13**: 253-257.

Saimbhi, M.S. and Padda, D.S. (1967-68) *Ann. Prog. Rept. Dept. Hort.*, Punjab Agric. Univ., Ludhiana, pp. 189-190.

Saimbhi, M.S., Daljit, S., Sandhu, K.S., Kooner, K.S., Dhillon, N.P.S. and Singh, D. (1997) *Agric. Sci. Digest*, Karnal, **17**: 40-42.

Saimbhi, M.S., Sandhu, K.S., Singh, D., Kooner, K.S., Dhillon, N.P.S. (1994) *J. Res.*, Punjab Agril. Univ., **31**: 38-41.

Sajjan, A. and Jamadar, M.M. (2003) *Karnataka J. Agric. Sci.*, **16**: 461-464.

Salameh, N.M. (2014) *American-Eurasian J. Agric. Environ. Sci.*, **14**: 155-160.

Salehuzzaman, M. (1985) *Bangladesh J. Agric.*, **10**: 1-8.

Sandhu, G.S., Sharma, B.R., Singh, B. and Bhalla, J.S. (1974) *Crop Improv.*, **1**: 77-81.

Sanodiya, K., Pandey, G., Kacholli, P. and Dubey, A.K. (2017b) *Int. J. Curr. Microbiol. App. Sci.*, **6**: 3551-3556.

Sanodiya, K., Pandey, G., Saklesh, Rajput, P.S. and Verma, A.K. (2017a) *Int. J. Chem. Stud.*, **5**: 2301-2304.

Sanwal, S.K., Singh, M., Singh, B. and Naik, P.S. (2014a) *Curr. Sci.*, **106**: 1470-1471.

Sanwal, S.K., Venkataravanappa, V. and Chauhan, N.S. (2014b) *Int. Conf. on Horticulture for Nutritional Livelihood and Environmental Security in Hills: Opportunity and Challenges*, Kalimpong, Darjeeling, India.

Sarma, B. and Gogoi, N. (2015) *Cogent Food Agric.*, **1**: 1030906.

Sarnaik, D.A., Baghel, B.S. and Singh, K. (1986) *Journal of Farming Systems*, **2**: 13-15.

Sasser, J.M. and Kirby, M.F. (1979) *Crop Cultivars Resistant to Root-knot Nematodes Meloidogyne species*, Raligh, North Carolina, USA.

Sastry, K.S., and Zitter, T.A. (2014) *Plant Virus and Viroid Diseases in the Tropics, Vol. 2, Epidemiology and Management*, Springer Netherlands, pp. 149-480.

Sastry, K.S.M. and Singh, S.J. (1974) *Indian Phytopath.*, **27**: 294-297.

Sataraddi, A.R., Shivanna, E., Ningraddi, N.P., Jamadar, M.M., Kulkarni, S. and Prabhu, V. (2009) *5th Int. Conf. on Plant Pathology in the Globalized Era*, Indian Phytopathological Society, New Delhi, India, Souvenir Abstract Volume, 210.

Satpathy, S., Rai, S., Nirmal, D. and Singh, A.P. (2004) *Indian J. Plant Protec.*, **32**: 22-25.

Sawadogo, M., Ouedraogo, J.T., Balma, D., Ouedraogo, M., Gowda, B.S., Botanga, C. and Timko, M.P. (2009) *Afr. J. Biotechnol.*, **8**: 2476-2482.

Schafleitner, R., Kumar, S., Lin, C.Y., Hegde, S.G. and Ebert, A. (2013) *Gene*, **517**: 27-36.

Scott, S.J., McFerran, J. and Goode, M.J. (1989) *Arkanasas Farm Res.*, **38**: 8.

Selvi, D., Rani, P. and Perumal, R. (1997) *Madras Agric. J.*, **84**: 625-626.

Seth, T. (2014) *Ph.D. Thesis*, Bidhan Chandra Krishi Viswavidyalaya, West Bengal, India.

Seth, T., Chattopadhyay, A., Chatterjee, S., Dutta, S. and Singh, B. (2016b) *J. Agric. Sci. Tech.*, **18:** 751-762.

Seth, T., Chattopadhyay, A., Dutta, S., Hazra, P. and Singh, B. (2016a) *Vegetos*, **29**: 3.

Seth, T., Chattopadhyay, A., Dutta, S., Hazra, P. and Singh, B. (2017) *Euphytica*, **213**: 30.

Seth. T., Mishra, G.P., Singh, B., Kashyap, S., Mishra, S.K., Tiwari, S.K. and Singh, P.M. (2018) *Veg. Sci.*, **45**: 1-6.

Shahid, M., Rehman, A., Malik, A.A., Khan, M.S. and Zakaria. (2015) *J. Biol. Agric. Healthc.*, **5**: 172-176.

Shahid, M.R., Amjad, M., Ziaf, K., Jahangir, M.M., Ahmad, S., Iqbal, Q. and Nawaz, A. (2013) *Pak. J. Agric. Sci.*, **50**: 387-392.

Shalaby, G.I. (1972) *Asiut. J. Agric. Sci.*, **3**: 381-386.

Shamloula, G.M. and Moghazy, A.M. (2017) *Proc. IX Int. Symp. on In Vitro Culture and Horticultural Breeding* (A.A. Abul-Soad *et al.*, eds) *Acta Hortic.*, 1187.

Sharma, A.D., Rathore, S.V.S., Srinivasana, K. and Tyagi, R.K. (2014) *Sci. Hortic.*, **165**: 75-81.

Sharma, B.K. and Gill, B.S. (1984) *Euphytica*, **33**: 215-220.

Sharma, B.R. (1993) In: *Genetic Improvement of Vegetable Crops* (Eds.) G. Kalloo and B.O. Bergh), Pergamon press.

Sharma, B.R. and Dhillon, T.S. (1983) *Genet. Agr.*, **37**: 267-276.

Sharma, B.R. and Sharma, D.P. (1984) *Indian J. Agric. Sci.*, **54**: 917-920.

Sharma, B.R., Arora, S.K., Dhanju, K.C. and Ghai, T.R. (1993) *Indian J. Virol.*, **9**: 139-142.

Sharma, B.R., Kumar, V. and Bayay, K.L. (1981) *Genet. Agrar.*, **35**: 121-130.

Sharma, J.K., Solanki, V.A., Sharma, K.K., Deshmukh, N. and Patel, H.P. (2017) *Int. J. Chem. Stud.*, **5**: 1424-1430.

Sharma, J.P. and Prasad, R. (1973) *Gartenbenwissinzchaft*, **38**: 163-169.

Sharma, P., Kaushal, A., Singh A. and Garg, S. (2016) *Int. J. Eng. Res. App.*, **6**: 85-91.

Sharma, R.P. and Bhati, K.K. (2008) *Pestology*, **32**: 47-49.

Sharma, S. and Patel, B.D. (2011) *Indian J. Weed Sci.*, **43**: 226-227.

Shivananda, T.N. and Iyengar, B.R.V. (1990) *J. Nuclear Agric. Biol.*, **19**: 148-1151.

Shoba, K. and Mariappan, S. (2007) *Acta Hortic.*, **752**: 437-439.

Shukla, A. (2018) *J. Entomol. Zool. Stud.*, **6**: 43-46.

Siddappa, B., Amaresh, Y.S., Naik, M.K., Sunkad, G., Sreenivas, A.G., Hussain, A. and Asawathanarayan, D.S. (2014) *J. Plant Dis. Sci.*, **9**: 185-189.

Siddiqui, M.A. and alam, M.M. (1997) *Proc. Indian Nat. Sci. Academy, Part B., Reviews and Tracts Biological Sciences*, **63**: 545-550.

Sidhu, H.S., Kumar, V. and Madhu, M.R. (2017) *Int. J. Pure App. Biosci.*, **5**: 569-574.

Siemonsma, J.S. (1982a) *Euphytica*, **31**: 241-252.

Siemonsma, J.S. (1982b) *The Cultivation of Okra (Abelmoschus* spp.*), a Tropical Fruit-Vegetable with Special Reference to the Ivory Coast*, Wageningen, Netherlands, p. 297.

Simon, S.Y., Musa, I. and Nangere, M.G. (2013) *Int. J. Adv. Res.*, **1**: 45-51.

Singh, A.K., Singh, D.K., Singh, N.K., Kushwaha, M.L. and Maurya, S.K. (2020) *Int. J. Chem. Stud.*, **8**: 2767-2770.

Singh, B., Goswami, A. and Kumar, M. (2013) *Indian J. Hortic.* **70**: 595-598.

Singh, B., Goswami, A. and Sharma, A.K. (2016) *Indian J. Agric. Sci.*, **86**: 1613-1620.

Singh, B., Rai, M., Kalloo, G., Satpathy, S., and Pandey, K. K. (2007b) *Acta Hortic.*, **752**: 323-328.

Singh, B.P. (1987) *HortScience*, **22**: 879-880.

Singh, D.K., Lal, G., Jain, S.K. and Lal, G. (1999) *Scientific Horticulture*, **6**: 121-124.

Singh, H.B. and Bhagchandani, P.M. (1967) *Bhindi Cultivation in India*, ICAR, New Delhi.

Singh, H.B. and Bhatnagar, A. (1975) *Indian J. Genet. Plant Breed.*, **36**: 26-27.

Singh, H.B. and Sikka, S.M. (1955) *Indian Fmg.*, **5**: 27-30.

Singh, H.B., Bhagchandani, P.M. and Thakur, M.R. (1964) *Indian J. Hort.*, **21**: 221-231.

Singh, H.B., Joshi, B.S., Khanna, P.P. and Gupta, P.S. (1962) *Indian J. Genet. Plant Breed.*, **22**: 137-144.

Singh, H.B., Swarup, V. and Singh, B. (1975) *Three Decades of Vegetable Research in India*, ICAR Tech. Bull., New Delhi.

Singh, H.M., Mishra U.S. and Mishra, T.S. (2018) *J. Krishi Vigyan*, **6**: 78-80.

Singh, K. and Singh, R.P. (1965a) *Horticulturist* (Srinagar), **1**: 76-80.

Singh, K. and Singh, R.P. (1965b) *Indian J. Hort.*, **22**: 340-344.

Singh, K.P. and Pandita, M.L. (1981) *Haryana J. Hort. Sci.*, **10**: 136-140.

Singh, Kirti, Nauriyal, J.P. and Singh, A. (1967) *Proc. Int. Symp. Subtrop. and Trop. Hort.*, New Delhi, pp. 501-509.

Singh, N.B., Sharma, H.K. and Srivastava, K.J. (1998) *News Letter, Nat. Hort. Res. and Dev. Foundation*, **18**: 1-6.

Singh, N.P. (1979) *Prog. Hort.*, **10**: 21-30.

Singh, P.V., Kumar, J.R. and Kumar, J. (1998) *Indian J. Agric. Res.*, **32**: 31-36.

Singh, R.N. and Singh, J. (1999) *Shashpa*, **6**: 85-88.

Singh, R.P. and Singh, L.N. (1986) *J. Indian Bot. Sci.*, **65**: 23.

Singh, S., Mandal, A.B. and Ram, T. (1990) *Indian J. Plant Physiol.*, **33**: 266-268.

Singh, S., Singh, B. and Singh, A.P. (2014) *Indian J. Nematol.*, **44**: 172-178.

Singh, S.J. (1996) *Indian J. Virol.*, **12**: 51-52.

Singh, S.J. and Dutta, O.P. (1986) *Indian J. Virol.*, **2**: 114-117.

Singh, S.K., Singh, A.K. and Singh, H.M. (2007a) *J. App. Zool. Res.*, **18**: 121-123.

Singh, V. and Kumar, V. (2012) *Adv. Appl. Sci. Res.*, **3**(3): 1809-1813.

Singh, V. and Singh, V. (1999) *Adv. Plant Sci.*, **12**: 199-202.

Sistrunk, W.A., Jones, L.G. and Miller, J.C. (1960) *Proc. Amer. Soc. Hort. Sci.*, **76**: 486-491.

Sivanappan, R.K., Muthukrishnan, C.R., Natrajan, P. and Siemanthini, Ramdas (1974) *South Indian Hort.*, **22**: 98-100.

Sogalad, A. (2008) *M.Sc. Thesis*, University of Agricultural Sciences, Dharwad, India.

Sohi, H.S. and Sokhi, S.S. (1974) *Indian Phytopath.*, **27**: 91-92.

Solangi, M., Tagar, A.A., Solangi, A.M., Siyal, A.G. Soothar, R.K. and Shah, A.R. (2015) *Sci. Int.* (Lahore), **27**: 4327-4331.

Somkumar, R.G., Mahakal, K.G. and Kale, P.B. (1997) *PKV Res. J.*, **21**: 202-204.

Sood, S. and Kalia, P. (2001) *Haryana J. Hort. Sci.*, **30:** 92-94.

Sridhar, T.S. and Sinha, P. (1989) *Indian J. Agric. Sci.*, **59**: 606-607.

Srivastava, L.S. and Sachan, S.C. (1963) *Allahabad Fmr.*, **47**: 63-65.

Srivastava, V.K. (1964) *Indian J. Hort.*, **21**: 165-169.

Stephan, D., Siddiqua, M., Hoang, A.T., Engelmann, J., Winter, S. and Maiss, E. (2008) *Virus Genes*, **36**: 231-240.

Sulikeri, G.S. and Rao, T.S. (1972) *Prog. Hort.*, **4**: 71-80.

Sundaraj, D.D., Balasubramanian, G. and Soundrapandia, G. (1965) *Madras Agric. J.*, **52**: 10-12.

Surendran, S. and Udayan, P.S. (2020) *Curr. Bot.*, **11**: 38-42.

Swamy Rao, T. and Ramu, P.M. (1975) *Curr. Res.*, **8**: 135-137.

Swanson, M.M. and Harrison, B.D. (1993) *Biochimie*, **75**: 707-711.

Swarup, V. (1977) *Breeding Procedures for Cross-pollinated Vegetable Crops*, ICAR, New Delhi.

Tahir, S.M., Samaila, A.A. and Abdulrahman, M.D. (2019) *Sci. World J.*, **14**: 119-128.

Talukder, M.A.H., Mannaf, M.A., Alam, M.K., Salam, M.A. and Amin, M.M.U. (2003) *Pak. J. Biol. Sci.*, **6:** 1626-1630.

Tanimola, A.A. and Akarekor, C. (2014) *World J. Agric. Sci.*, **10**: 185-193.

Temam, N., Mohamed, W. and Aklilu, S. (2020) *MOJ Eco. Environ. Sci.*, **5**: 80-87.

Ter Avenesian, D.V. (1978) *Theor. Appl. Genet.*, **52**: 77-79.

Terrel, E.E. and Winters, H.F. (1974) *Hortic. Sci.*, **9**: 324-325.

Thakur, M.R. (1976) *SABRAO J.*, **8**: 69-73.

Thombre, M.V. and Deshmukh, S.U. (2006) *Indian J. Genet.*, **66**: 353-354.

Tiwari, K.N., Mal, P.K., Singh, R.M. and Chattopadhyay, A. (1998) *Agric. Water Manag.*, **38**: 91-102.

Tripathi, K., Govila, O., Warrier, R. and Ahuja, V. (2011) *Biology of Abelmoschus esculentus L.* (*Okra*). Govt. of India, Department of Biotechnology, New Delhi.

Trivedi, R.K. and Gunasekaran, M. (2013) *Indian Minimum Seed Certification Standards*, The Central Seed Certification Board, Department of Agriculture and Co-operation, Ministry of Agriculture, Government of India, New Delhi.

Uppal, B.N., Varma, P.M. and Capoor, S.P. (1940) *Curr. Sci.*, **9**: 227-228.

Ushakumari, K., Prabhakumari, P. and Padmaja, P. (1999) *J. Trop. Agric.*, **37**: 87-88.

Van Borssum Waalkes, T. (1966) *Blumea*, **14**: 1-251.

Varma, P.M. (1952) *Indian J. Agric. Sci.*, **22**: 75-91.

Vashisht, V.K., Sharma, B.R. and Dhillon, G.S. (2001) *Crop Improv.*, **28**: 18-25.

Veeraragavathatham, D. and Irulappan, I. (1990) *South Indian Hort.*, **38**: 75-82.

Venkataravanappa, V., Loganathan, M., Saha, S., Rai, A.B. and Singh, B. (2011) *J. Interacademicia*, **15**: 349-356.

Venkataravanappa, V., Reddy, C.N.L. and Reddy, M.K. (2013) *3 Biotech*, **3**: 461-470.

Venkataravanappa, V., Reddy, C.N.L., Jalali, S. and Reddy, M.K. (2012) *Virus Genes*, **44**: 522-535.

Venkataravanappa, V., Reddy, C.N.L., Jalali, S., Briddon, R.W. and Reddy, M.K. (2014) *Eur. J. Plant. Pathol.*, **141**: 217-235.

Verma, J.P., Rathore, S.V.S. and Kushwaha, C.S. (1974) *Prog. Hort.*, **5**: 77-80.

Verma, V.K., Pundrik, K.C. and Chauhan, K.S. (1970) *Punjab Hort. J.*, **10**: 130-136.

Vethamoni, P.I. and Balakrishnan, R. (1990) *Indian J. Hort.*, **47**: 233-238.

Vidhate, A.B., Badgujar, S.L. and Wagh, S.S. (2014) *Trends Biosci.*, **7**: 3010-3012.

Vijay, O.P. and Manohar, M.S. (1986a) *Indian J. Hort.*, **43**: 133-139.

Vijay, O.P. and Manohar, M.S. (1986b) *Indian J. Hort.*, **43**: 252-259.

Walling, N., Kanaujia, S.P., Alila, P., Sharma, M.B. and Ozukum, C. (2020) *Int. J. Recent Sci. Res.*, **11**: 37651-37654.

Wani, A.H., Alam, M. and Alam, M. (1999) *Tests of Agrochemicals and cultivars*, No. 20, pp. 8-9.

Wankhade, R.V., Kale, P.B. and Dod, V.N. (1996) *RKV Res. J.*, **21**: 16-21.

Waychal, G.U., Game, B.C. and Jagtap, S.D. (2018) *Int. J. Chem. Stud.*, **6**: 325-327.

Weerasekara, D., Jagadeesha, R.C., Wali, M.C., Salimath, P.M., Hosamani, R.M. and Kalappanawar, I.K. (2008) *Karnataka J. Agric. Sci.*, **21**: 578-579.

Whitehead, W.F. and Singh, B.P. (2000) *HortScience*, **35**: 849-852.

Woolfe, M.L., Martin, F.C. and Otchere, G. (1977) *J. Sci. Food Agric.*, **28**: 519-529.

Yadav, D.S. (1986) *Indian J. Hort.*, **43**: 274-277.

Yadav, S.R., Sutar, S.P., Bhat, K.V., Joseph, J.K., Latha, M., Scariah, S., Nissar, V.A.M., Krishnan, G., Malik, S.K., Rao, S.R. and Merita, K. (2014) In: *Genus Abelmoschus Medik. in India- an Illustrated Guide for Species Identification*, Indian Council of Agricultural Research, New Delhi, pp. 38-42.

Yadav, Y., Maurya, P.K., Banerjee, S., Bhattacharjee, T., Jamir, I., Chatterjee, S., Mandal, A.K., Dutta, S. and Chattopadhyay, A. (2018a) *Int. J. Veg. Sci.*, DOI: 10.1080/19315260.2018.1540025

Yadav, Y., Maurya, P.K., Bhattacharjee, T., Banerjee, S., Jamir, I., Mandal, A.K., Dutta, S. and Chattopadhyay, A. (2018b) *Agric. Res. Tech.*, **18**: 556071.

Yadav, Y., Maurya, P.K., Devi, P. A., Jamir, I., Bhattacharjee, T., Banerjee, S., Dutta, S., Debnath, D., Mandal, A.K, Dutta, S. and Chattopadhyay, A. (2018) *J. Pharmacog. Phytochem.*, **7**(2): 3795-3802.

Yawalkar, K.S., Agarwal, J.P. and Bokde, S. (1962) *Manures and Fertilizer*, Agri-Horticultural Publishing House, Nagpur, India.

Yildiz, M., Koçak, M. and Baloch, F.S. (2015) *Genet. Mol. Res.*, **14**: 10588-10602.

Younis, R.A.A., Hassan, S.M.K. and Itriby, H.A.E. (2015) *Global J. Biol. Agric. Health Sci.*, **4**: 117-128.

Yuan, C.Y., Zhang, C., Wang, P., Hu, S., Chang, H.P., Xiao, W.J., Lu, X.T., Jiang, S.B., Ye, J.Z. and Guo, X.H. (2014) *Genet. Mol. Res.*, **13**: 165-175.

Zaharuddin, N.D., Noordin, M.I. and Kadivar, A. (2014) *Biomed. Res. Int.*, Article ID 735891.

CASSAVA

T.V.R. Nayar and V. A. Parthasarathy , S. Mitra and J. Tarafdar

1.0 Introduction

Cassava (*Manihot esculenta* Crantz) popularly known as tapioca, also known as yuca, manioc or indioca is a perennial shrub cultivated in the tropics for its starchy tuberous roots and is used as subsidiary food for over 750 million people. The important characteristics that influence the role of cassava within the tropical farming systems are high carbohydrate yields per unit land and labour, adaptation to poor soils and/or stress compatibility with a variety of crops in association, an indeterminate harvestability and tolerance to dominant pests and diseases. In the past, the importance and potential of cassava were not adequately recognized. According to estimates of FAO, on average, farmers produce about 10 tonnes of cassava per hectare, but yields can reach as high as 40 tonnes (www.bphre.com/phindustry/cassava.htm). It is estimated that the introduction of high-yielding varieties, improved pest and disease control and better processing methods could increase cassava production in Africa by 150 per cent. Thailand is engaged in exporting huge cassava pellets, mostly to the European Union. More than 30 per cent of the cassava produced in Latin America is used for domestic animal feed, compared to less than 2 per cent in Africa. Research in Cameroon has shown that poultry breeders could lower their production costs by 40 per cent by incorporating cassava into their chicken feed. Asia leads the way in the production of starches derived from cassava.

2.0 Composition and Uses

2.1.0 Composition

The dry matter content and composition of cassava tubers are influenced by the environmental factors, the variety and the stage of harvest. In general, the tubers contain 30-40 per cent carbohydrate, 1-2 per cent crude protein, 0.2-0.6 per cent other extractives and 55-60 per cent moisture. The carbohydrate fraction is primarily composed of starch; approximately 90 per cent. Cassava starch has 20 per cent amylose and 80 per cent amylopectin. The digestibility of cassava starch is 48 per cent in raw state and 78 per cent when cooked. The proteins present in tubers contain appreciable amounts of all the essential amino acids except methionine and tryptophan. The protein accounts for only 40-50 per cent of the total nitrogen in the tuber and the biological value of the protein is 48 per cent. The tubers also contain significant quantities of vitamin C, thiamine, riboflavin, niacin and smaller amounts of calcium, phosphorus and iron (Table 1). The metabolizable energy of dry cassava is 3500-4000 kcal/kg and so considered as a good source of energy. According to Mitra and Tarafdar (2016), among the biochemical parameters of cassava tubers, dry matter ranged from 34.7-37.6 per cent, total starch ranged between 23.7 and 24.4 per cent in different genotypes, protein ranged from 0.91 to 1.11 per cent, total sugar ranged between 1.11 and 1.31 per cent, beta-carotene content ranged between 1.73 and 2.61 mg/100 g and vit. C was found to vary from 4.36 to 14.27 mg/100 g.

The foliage of cassava is a good source of vegetable protein, since the leaves on dry weight basis contain 26-39 per cent protein. The leaf protein, except for methionine and tryptophan, is rich in all essential amino acids and the digestibility of the same is 80 per cent. Cassava leaves are also enriched with significant amounts of calcium, phosphorus, iron and β-carotene (Hrishi and Nair, 1978).

Table 1: Nutrition Value of Fresh Cassava Tuber (100 g)*

Composition	*Nutrient Value*	*Percentage of RDA*
Energy	160 Kcal	8 per cent
Carbohydrates	38.06 g	29 per cent
Protein	1.36 g	2.5 per cent
Total Fat	0.28 g	1 per cent
Cholesterol	0 mg	0 per cent
Dietary Fibre	1.8 g	4 per cent
Folates	27 µg	7 per cent
Niacin	0.854 mg	5 per cent
Pyridoxine	0.088 mg	7 per cent
Riboflavin	0.048 mg	4 per cent
Thiamin	0.087 mg	7 per cent
Vitamin A	13 IU	<1 per cent
Vitamin C	20.6 mg	34 per cent
Vitamin E	0.19 mg	1 per cent
Vitamin K	1.9 µg	1.5 per cent
Electrolytes		
Sodium	14 mg	1 per cent
Potassium	271 mg	6 per cent
Calcium	16 mg	1.6 per cent
Iron	0.27 mg	3 per cent
Magnesium	21 mg	5 per cent
Manganese	0.383 mg	1.5 per cent
Phosphorus	27 µg	4 per cent
Zinc	0.34 mg	3 per cent

*USDA National Nutrient data base (2019)

Like many other tropical plants cassava also contains cyanide, the toxic principle which limits the food and feed value. In cassava the glycosides linamarin and lataustralin are converted to hydrocyanic acid or prussic acid, a potent toxin, when they come in contact with linamarase, an enzyme that is released on rupture of cassava cells. The distribution and the level of the glycosides differ from part to part in the plant and are influenced by environmental factors, fertilization practices and the varieties. The leaves, especially younger leaves contain more cyanogenic glycosides than the other parts. In the tuber, the peel contains significantly higher amounts of glycosides than the pulp. The activity of the endogenous enzyme linamarase is also higher in the peel. Evaluation of 30 cultivars showed that the mean value of hydrocyanic acid in the peel was 559 mg/kg as against 41 mg/kg in the fresh edible pulp of the tuber (Thampan, 1979). In spite of the toxic principles present in cassava, traditional methods of processing render the same safe for human consumption and livestock feed. Cooking in water for about 5 minutes removes more than 80 per cent of the cyanide. Normal cooking process reduces the cyanide content still further. Sun drying of the tuber slices destroys about 75 per cent of the total cyanide. The cyanide content in leaves also can be reduced to safer limits by cooking, steaming the sun drying.

2.2.0 Uses

Use of cassava varies greatly from region to region. In India, cassava is used as human food, industrial raw material and animal feed. In some states in India, ninety per cent of the harvested cassava tubers are used as food. It is consumed after cooking in different forms of dishes. Almost 70 per cent of the product in Kerala goes for human consumption, while in other states cassava is utilised mainly as industrial raw material. Some value added products prepared from cassava are flour, sago, semolina, macaroni, instant noodles, vermicelli *etc.* (Mitra, 2017). Cassava starch is used as thickener, filler, binder, stabilizer in different preparations. A national sample survey conducted in 1977-78 indicated that average monthly consumption of cassava in Kerala was 5.55 kg in rural and 2.59 in urban areas (Padmaja *et al.*, 1992). Globally 60 per cent of the cassava production is used for human food, 27.5 per cent for animal feed and the rest for starch and starch based industries (Bottema and Henry, 1992). In Africa, nearly all the cassava is used for human consumption. It is estimated that in tropical Africa cassava provides an average of 230 calories per person per day. Forty per cent of the South American cassava production is destined for human food and it provides 150 to 160 calories per person per day. In contrast to Asia, where cassava is dried before feeding to animals, in Latin America, large quantities of fresh tubers are fed to animals. About 50 per cent of the cassava produced in Asia is used for direct human consumption. The remainder is converted to dry chips, pellets or starch, either for export or for use in local industries. Most of the cassava produced in Thailand is pelleted (92 per cent) and converted into chips and starch, and exported to developed countries especially EU countries for use in animal feed (Atthasampunna, 1992, Limsila *et al.*, 1998). In Indonesia about 59 per cent of the cassava is used for human food, 17 per cent as industrial raw material and the rest for other purposes (Damardjati *et al.*, 1992). Bulk of the cassava in China is used for the industrial production of starch and other refined chemicals like sorbitol, mannitol, oxalic acid, glyconic acid and alcohol (Shuren, 1992). The Philippines use 35 per cent of the cassava production for human food, 31 per cent in industry and 33 per cent in animal feed (Van Den *et al.*, 1992). The traditional mode of consumption of cassava in India is usually as boiled or baked tubers. The boiled tubers are eaten with fish curry or taken as side dish with rice. The tubers are also chipped and deep fried in oil for taking as a snack. Cassava tubers are also fed to cattle in southern India either directly or after chipping and drying. The outer peel is used as swine feed in parts of Tamil Nadu.

Cassava leaves are also used as a base in making sauces and soups particularly in the Congo Basin, Tanzania and parts of West Africa. Consumption of leaves in

Uses of Cassava

Different Value Added Products of Cassava

human food has been estimated as high as 500 g/day in Zaire, 40-170 g/day in Congo and 100 g/day in Cameroon (Cock, 1985). In Indonesia also the leaves are used as a vegetable. However, the major use of cassava leaves is as livestock feed in many producing countries. The cassava leaves can also be utilized for indoor rearing of Eri silkworm (*Philosannia ricini*) (Anon., 1983).

3.0 Area and Production

World production of cassava tubers in 2017 was 291.99 million tonnes from an area of 26.34 million hectares (FAOSTAT, 2019). In the year 2017, world leading cassava producing countries like Nigeria, Congo DR and Thailand ranked 1st, 2nd and 3rd respectively in cassava production. Other important cassava producing countries in the world are Indonesia, Brazil, Uganda, Angola, Ghana, Mozambique, Tanzania, Vietnam, Philippines, India, China and Cambodia (Mitra, 2020).

In the world most of the cassava is produced in Africa, South America and Asia. Africa accounts for more than 50 per cent of the production. Nigeria, Congo and Ghana are the major cassava growing countries in Africa. Over the past decade there was substantial increase in cassava production in the African continent, mainly due to the area expansion. In South America, Brazil dominates in area and production of cassava. Cassava is grown throughout in Brazil in the hot humid Amazon basin, in the dry areas of north-east and in the cooler southern areas, where there are frost and even occasional snow (Cock, 1985). Colombia and Paraguay are the other important cassava producing countries in Latin America. Although Brazil, Colombia and Paraguay produce more than 90 per cent of the cassava in South America, the crop is for greater importance in several smaller countries of South and Central America. Asia contributes to 30 per cent of the world's cassava production from 27 per cent of the area under the crop. The following table shows global cassava production.

Table 2: Area, Production and Productivity of Cassava in the World in 2017*

Country	*Area (000' ha)*	*Production (000' tonnes)*	*Productivity (t/ha)*
Nigeria	6,792.35	59,485.95	8.76
Congo, DR	3,877.94	31,596.05	8.15
Thailand	1,342.40	30,973.29	23.07
Brazil	1,314.85	18,876.47	14.36
Angola	1,011.62	11,747.94	11.61
Ghana	965.52	18,470.76	19.13
Rest of the world	11,037.65	120,842.24	10.27
World	26,342.33	291,992.65	11.08

*Mitra (2020)

The Asian continent produces about 78.75 million tonnes of cassava from an area of 18.69 million hectares (FAOSTAT, 2010). It produces 34 per cent of global cassava tubers, and secure first place in productivity (19.85 t/ha). The principal cassava growing countries in Asia are Thailand, Indonesia, Vietnam, India, China,

Cambodia, Philippines, Malaysia, Sri Lanka and Laos (Mitra, 2020). Two third of total production originates from Thailand and Indonesia.

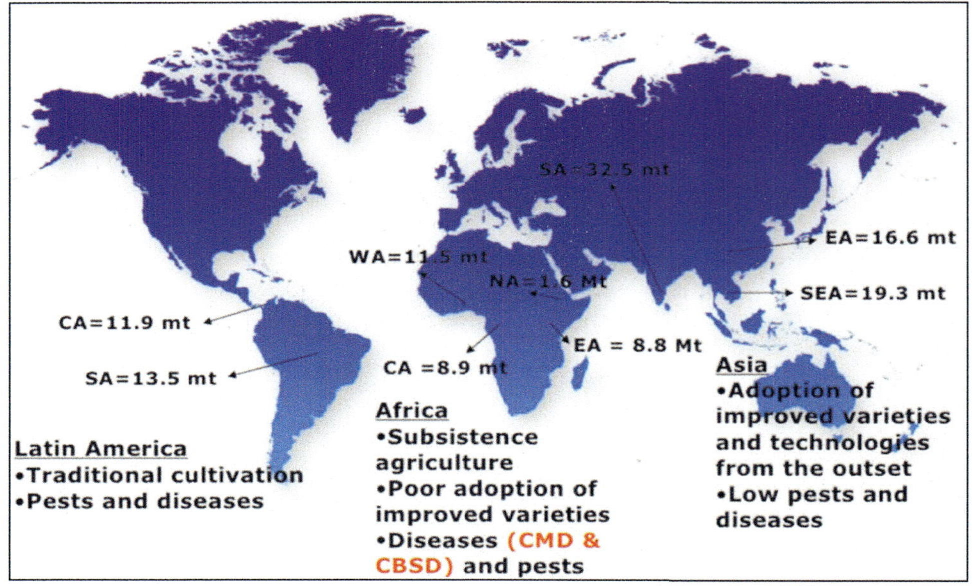

Productivity of Cassava in different Countries

In India, cassava cultivation is concentrated mainly in the state of Tamil Nadu, Kerala, Andhra Pradesh, Meghalaya, Assam, Karnataka and Nagaland. The production of fresh cassava tubers in India in 2012 was 5.484 million tonnes from an area of 1.77 lakh hectares (Anon., 2019). The productivity of the crop in India is about 30.98 t/ha claiming to be the highest in the world (Mitra, 2020).

Table 3: Trend in Area, Production and Productivity of Cassava in India from 2005-2012*

Year	*Area ('000 ha)*	*Production ('000 t)*	*Productivity (t/ha)*
2005	244.60	7854.90	32.10
2006	255.00	8232.00	32.30
2007	270.00	9036.00	33.50
2008	280.00	9623.00	34.40
2009	231.00	8059.90	34.80
2010	221.00	8076.00	36.50
2011	226.70	8746.50	38.60
2012	207.00	7236.60	35.00

*George and Sunitha (2018)

4.0 Origin and Distribution

4.1.0 Origin

Cassava is a native of Brazil in Latin America and was introduced to India by the Portuguese in the 17th century (George and Sunitha, 2018). The evolutionary history of cassava is somewhat speculative due to the complex taxonomy and rare survival of fleshy plant parts, for subsequent archaeological disco-very. Studies of processing artifacts from Colombia and Venezuela provide evidence of cassava cultivation as early as 3000 to 7000 years ago (Rouse and Cruxeet, 1963). According to Harlon (1971) cassava is non-centric, that is with no obvious centre of origin and diversity. Renoize (1973) put forward evidence for multiple origins of cassava. Spath (1973) also suggested four separate areas of origin – Guatemala and Mexico; the north-western coastal savannas of Latin America; eastern Bolivia and north-western Argentina; and eastern Brazil. Box and Doorman (1982) postulated that high and low HCN types might have originated independently in Meso and South America. From the accumulated evidence it may be concluded that cassava is a new world crop with origin in the low land tropics.

4.2.0 Distribution

Cassava was widely distributed in the low land tropics of south and central Americas by the time of Spanish colonization of the continent in the 15th century. The species spread eastward in the post Colombian times (Jennings, 1976). It was taken by the Portuguese to the west coast of Africa in the late 16th century. Evidence for the exact time of introduction of cassava into Asia is not well-documented. It was presumably taken by the Spanish from Mexico to the Philippines in the 17th century with the latter spread to Indonesia and the Asian mainland. To the Indian sub-continent, the crop was introduced by the Portuguese during the early 18th century. Sri Lanka, officially imported cassava cuttings from Mauritius in 1786 and India made importations directly from South America in 1974 and from West Indies in 1840 (Cock, 1985). By the nineteenth century cassava was established as a major crop in the tropics. Tamil Nadu, Kerala, Meghalaya, Assam, Andhra Pradesh, Karnataka and Nagaland are the main cassava growing states in India.

5.0 Taxonomy and Morphology

5.1.0 Taxonomy

Cassava (*Manihot esculenta* Crantz) is a member of the family Euphorbiaceae, tribe Manihotae, sub-family Crotonoideae (Viegs, 1976). In the older literature it is referred as *M. utilissima*. *M. esculenta* never occurs in wild and exists only as a cultivated species, but there are large numbers of wild *Manihot* spp. The only other species of economic importance is *Manihot glaziovii*, cerea rubber. Rogers and Appan (1973) applying taximetric methods defined 98 species within the genus *Manihot*, widely distributed throughout the low and tropics of the Americas. They also suggested that the genus *Manihot* is relatively recent in origin and that the species are still rapidly evolving. There appears to be no sharp genetic delimitation among species. All have 36 chromosomes and show regular trivalent pairing, but there is evidence of polyploidy (Jennings, 1976). Magoon *et al.* (1969) suggested that *Manihot* species are segmental allotetraploids. Nassar (1978) defined four centres of diversity of the wild species; western Mexico with 19 species; central Brazil with 38 species and two lesser centres are in northeast Brazil and in eastern Bolivia. The three Brazilian centres have some species in common, notably from the heterophyla section (Rogers and Appan, 1973). The closest relatives appear to be *M. aesculifolia, M. rubricaulis* and *M. pringlei*. Selection pressure during the domestication of cassava was for large storage roots, erect and less branched stem with ability to sprout and establish from cuttings (Jennings, 1976).

5.2.0 Morphology

Cassava is a perennial woody shrub growing to a height of 3-4 m with erect, branching stems and latex in all parts. The height, branching habit, flowering and other plant characteristics vary greatly among cultivars. In some cultivars branches are from the base of the stem to give an erect nature, while some others which branch profusely produce widely spreading plants. Colour of the stem varies from silvery-grey and different shades of reddish brown to dark brown. The young stem may show 3-4 ribs, but the mature stem is cylindrical and straight. The internal structure conforms to that of dicotyledons. The large deeply lobed leaves are arranged spirally and have long petioles, subtended by small deciduous stipules. Each leaf has five to nine obvate or lanceolate lobes which are broadest about one-third from the tip and taper gradually to their base. The leaves are usually dark green in colour, but red, yellow and various shades of purple pigments occur especially on the young foliage. The leaves tend to be clustered towards the top of the stem, as those below are shed leaving prominent leaf scars.

The stem cuttings used for planting produce extensive fibrous adventitious root system. The root tubers develop as a result of secondary thickening of adventitious roots. In a well-grown plant at the base of the stem five to ten tubers are produced, each having a length of 30-50 cm and thickness of 7-10 cm. The outer periderm colour may be white, pink, red or brown depending on the cultivar. It encloses a thin cortex usually white, sometimes with outer pink or brown streaks. The central portion of the tuber, usually called the pith, constitutes fleshy edible portion made up of secondary xylem and parenchymatous cells filled with starch grains. When cassava is grown from true seed the radical develops into a tap root system typical to other dicots. The tap root and one or two adventitious roots of such plants also become tuberous, but they are rather woody and poor in quality.

Cassava Plant (Left), Flower (Middle) and Tuber (Right)

6.0 Growth and Development

6.1.0 Stem

Cassava is commercially propagated from stem cuttings of 20-30 cm length. When stem cuttings are planted in moist soil, the buds at the distal nodes sprout and develop as shoots. Sprouting is extremely sensitive to temperature and the ideal range is 28-30°C (Keating and Evenson, 1979). Shoot growth is greatest at 30-32°C (Wholey, 1974). The shoots formed show marked apical dominance and new leaves are produced in sequence on the main stem. Once, the apex becomes reproductive, branching takes place depending upon the variety and the environmental factors. Some clones branch early whilst others never branch. Fewer branches are produced at lower fertility, moisture stress and at extremes of temperatures.

6.2.0 Leaf

The first leaves in cassava appear about ten days after planting the stem cutting. Initially the leaves emerge folded along the margin and fully expanded by 10-15 days after appearance. The size of the expanded leaves increase with age up to four months and then declines. The leaf size is greatly reduced by water stress, shading very low temperature and inadequate nutrient supply. The life of individual leaf usually ranges from 60 to 120 days depending upon environmental factors. The leaf area of cassava is influenced by rate of formation of leaves, the size of individual leaf and their longevity. In non-branching types, the leaf area index (LAI) increases in the first 4-6 months and thereafter the LAI declines. The growth and development of cassava plant is dominated by the relation of LAI to total biomass production and distribution, which determines the final yield (Cock and El-Sharkawy, 1988). As LAI increases the crop growth rate tends to increase up to a plateau level at an LAI of 4-5. This LAI corresponds to a light interception of more than 95 per cent and hence it is unlikely that any further increase in LAI will lead to further increase in biomass production without substantial changes in canopy characteristics.

Research on the photosynthesis of cassava indicates that it is intermediate between C_3 and C_4-pathways (El-Sharkawy and Cock, 1987). In general, plants with C_4 pathway produce more biomass per unit area and use water more efficiently. Cassava has the enzyme system of C_4 photosynthesis, though not active as in a typical C_4 plant. Further, cassava does not possess the Krantz anatomy typical of the true C_4 plants. Selection of cassava with higher C_4 enzyme activity and incorporation of *Krantz anatomy* using the modern breeding techniques may pave the way for substantial increase in the photosynthetic capacity of the plant.

6.3.0 Root

Root formation in cassava takes place both from the nodes and the stem cutting used for propagation. Under ideal conditions the first roots emerge within 5 days after planting. After a few weeks growth, some roots exhibit secondary growth in the vascular regions and starch granules can be found in the xylem parenchyma. Indira and Kurian (1977) noted that secondary thickening began 21 days after planting. From 6 weeks after planting root thickening occur rapidly, with formation of large quantities of xylem parenchyma, that are packed with starch granules. According to Tan and Cock (1979) root thickening begins when the supply of carbohydrates exceeds the requirement for growth of the stem and leaves. The number of roots that begin secondary thickening and tuber size may be determined by the genotype, carbohydrate supply and possibly by hormonal factors (Thampan, 1979). Root elongation shows down once the secondary thickening commences. The root bulking or enlargement increase with time and the maximum rate is observed during 5-7 months, in most of the cultivars under rainfed conditions in Kerala. Starch content also increases, with the enlargement of the tubers reaching the peak between 7th and 10th month after planting. Thereafter the starch content decreases and fibre content increases in most of the cultivars grown in India.

6.4.0 Flowering

Cassava is a monocecious species producing a few large basal pistillate and numerous smaller apical staminate flowers on the same racemose inflorescence. Most of the branching cultivars may flower as early as three months after planting, while in non-branching types flowering is very much delayed and sometimes absent. Most genotypes flower preferentially during short days (Harshey and Amaya, 1980). In northern hemisphere flowering occurs from July to January with a peak from October to December. In the southern hemisphere flowering takes place from January to July with a peak in April to June (Hahn *et al.*, 1979). There seems to be a concentration of blooming around mid-way between 11 a.m. and 2 p.m. Pistillate flowers have five petals and locular ovary. The stigma is sticky and receptive at the time of opening to 24 hours. Staminate flowers have ten stamens arranged in two rings of five and do not open until the last female flower of the inflorescence bloomed. Anthers start dehiscence 2-2½ hours before the opening of the male flowers, but complete it only by the time of opening. One male flower produces about 1600 pollen grains of which only 50 per cent are viable. Male sterility has been reported in cassava (Byrne, 1984) due to mechanisms like abscission of staminate buds, no pollen formation, non-dehiscent anthers and chromosomal pairing aberrations. Pollination takes place either through wind or by insects, especially bees, which are attracted by the nectar present in the flowers. The floral biology and pollination habit of cassava make it predominantly allogamous but considerable selfing may occur in profusely flowering genotypes (Kawano *et al.*, 1978).

The interval between pollination and fertilization is about 8 hours, but may extend up to 19 hours. The fertilized ovary grows into a tricolour capsule. Each locule contains a seed. The endosperm contains about 47 per cent lipid and 34 per cent protein and so is a potential source of fat and protein for industrial purposes.

7.0 Cultivars

A large number of traditional and improved cultivars of cassava are grown in different countries. Purseglove (1968) distinguished two major groups of cultivars based on maturity time and the cyanide (HCN) content of the tubers. (i) Short-duration types and matures in 6-11 months and which cannot be kept in the ground longer without serious deterioration in quality. These are often sweet cassavas with low HCN content in the flesh and most of the HCN is confined to the phelloderm.

(ii) Long season types that require at least 12 months to mature and have better ground storage capacity. These tend to be bitter cassavas with high HCN content throughout the tuber. But Cock (1985) stated that there is no sharp distinction between bitter and sweet cassavas, though sweet types usually have lower levels of cyanide than the bitter types. Farmer's choice for the cultivar depends on the end use, *i.e.*, for edible purpose or for industrial use.

The salient features of some of the widely grown cultivars and hybrid varieties released from Central Tuber Crops Research Institute, Trivandrum, Kerala, India (ICAR-CTCRI) for large-scale cultivation in India are provided below:

Table 4: Caltivars of Cassava Released in India*

Sl. No.	*Year of Release*	*Name of Originating Centre*	*Name of Cultivar*	*Area suitable*
1	1976	TNAU	CO 1	Tamil Nadu
2	1977	CTCRI TVM	Sree Sahya	Kerala, Assam, Andhra Pradesh, Karnataka, Tamil Nadu and North Eastern Region
3	1977	CTCRI TVM	Sree Visakham	Kerala, Tamil Nadu, Karnataka, Andhra Pradesh, Maharashtra, Assam and North Eastern Region
4	1985	TNAU	CO 2	Tamil Nadu
5	1987	CTCRI TVM	Sree Prakash	Kerala
6	1992	TNAU	CO 2	Tamil Nadu
7	1996	CTCRI TVM	Sree Harsha	Kerala and Tamil Nadu
8	1998	CTCRI TVM	Sree Jaya	Kerala, Tamil Nadu
9	1998	CTCRI TVM	Sree Vijaya	Kerala, Tamil Nadu
10	2000	CTCRI TVM	Sree Prabha	Kerala, Tamil Nadu, Karnataka, Andhra Pradesh
11	2000	CTCRI TVM	Sree Rekha	Kerala, Tamil Nadu, Karnataka, Andhra Pradesh
12	2002	TNAU	CO (TP) 4	Tamil Nadu
13	2006	CTCRI TVM	Sree Padmanabha	Kerala, Tamil Nadu
14	2012	TNAU	Tapioca YTP – 1	Tamil Nadu
15	2013	CTCRI TVM	Sree Athulya	Plains of Tamil Nadu
16	2013	CTCRI TVM	Sree Apoorva	Plains of Tamil Nadu
17	2014	IGKV	Chhattisgarh Cassava-1	Chhattisgarh (Bastar Region)
18	2015	CTCRI TVM	Sree Swarna	Kerala
19	2016	CTCRI TVM	Sree Reksha	Kerala, Tamil Nadu, Maharashtra and Andhra Pradesh
20	2016	CTCRI TVM	PDP CMR-1	Coastal belts of Andhra Pradesh
21	2016	CTCRI TVM	*TCMS-7	Kerala, Maharashtra, Tamil Nadu and Andhra Pradesh

*George *et al.* (2018)

7.1.0 India

CTCRI has developed many short duration (6-7 months) and long duration (9-10 months) edible varieties of cassava. Some popular short duration varieties are H-165, Sree Jaya, Sree Vijaya, Sree Prakash *etc.*, and some long duration varieties include Sree Visakham, Sree Harsha, Sree Rekha, Sree Prabha *etc.* Along with these varieties, Vellayani Harsha released by Kerala Agricultural University is also a popular variety of cassava.

M-4

A table purpose cultivar introduced from Malaysia and extensively cultivated in the homestead gardens of Kerala. Erect, non-branching plant type producing tubers of excellent culinary quality and very low in HCN. Tuber yield is about 20-25 t/ha in 10 months duration. It has high field tolerance to cassava mosaic disease (CMD), but very susceptible to mites.

Kalikalan

It is an early maturing cultivar with edible tubers of good quality. But it is highly susceptible to CMD. Well adapted to low fertility conditions.

Kuruthakaliyan

It is a spreading type variety with 10 months duration. Produces non-bitter tubers with an average yield of 32 t/ha. But susceptible to CMD, bacterial leaf spot (BLS) and mites.

H-97

It is a hybrid between an indigenous cultivar Manjavella and an exotic selection of Brazilian origin. The variety is characterized by light sepia colour of the emerging leaf and conical medium sized tubers with 27-29 per cent starch on fresh weight basis. Yields about 25-35 t fresh tubers in 10 months duration. This hybrid is tolerant to CMD.

H-165

This is a hybrid between two indigenous cultivars Chadayamangalam Vella and Kalikalan. Generally non-branching with pinkish petiole and light brown emerging leaves. Tuber skin is golden brown and the flesh is creamy with starch content of 23-25 per cent. The hybrid yields about 33-38 t/ha in 8-9 months and tolerant to CMD. This variety is cultivated in large areas of Tamil Nadu and Andhra Pradesh for industrial use.

H-226

It is a cross between M-4 and a local cultivar Ethakkakaruppan. The hybrid is occasionally branching, with characteristic green colour of leaves and tubers with purple rind and white flesh. Due to very high starch content (28-30 per cent), it is very popular for industrial use in Tamil Nadu and Andhra Pradesh.

Sree Visakham (H-1687)

This is a hybrid resulted from crossing an indigenous cultivar and an accession from Madagaskar. The hybrid is non-branching with light brown emerging leaf and tubers having brown rind and cream coloured flesh. The crop period is 8-10 months and the tuber yield is 35-38 t/ha with reasonably good culinary qualities. It is tolerant to CMD.

Sree Sahya (H-2304)

This is a hybrid developed from multiple crossing of five parents of which two are exotic and three indigenous. Plants are all predominantly non-branching, producing long necked tubers. Tuber skin is light brown and the flesh is white with a starch content of 29-31 per cent. Sree Sahya takes 10-11 months to mature and the yield potential is 35-40 t/ha. The tubers are non-bitter and fair in cooking quality.

Sree Prakash (S-856)

It is a short duration cultivar selected from the indigenous collections. The plants are relatively dwarf and non-branching with good leaf retention. The cooking quality of the tuber is fairly good and is non-bitter. Suitable for single cropped rice fields of Kerala with a yield potential of 25-30 t fresh tubers in 6-7 months.

Sree Jaya

It is a short duration, high yielding selection from the indigenous germplasm of cassava (CI-649) from Kottayam, Kerala. It attains maturity in 6 months and is suitable for lowland cultivation as a rotation crop in paddy based cropping system. Average yield is 34 t/ha and potential yield is 58 t/ha.

Sree Vijaya

It is a short duration, high yielding selection from the indigenous germplasm of cassava (CI-731) from Trivandrum, Kerala. It attains maturity in 6 months and is suitable for lowland cultivation as a rotation crop in paddy based cropping system. Average yield is 32 t/ha and potential yield is 51 t/ha. The tubers are more nutritious due to the presence of carotene.

Sree Harsha

A triploid cassava variety obtained by crossing induced tetraploids with some of the cultivated cassava types. They give a higher root yield, dry matter content (above 45 per cent) and recoverable starch content (above 35 per cent).

The cultivars that are identified in some of the important cassava producing countries are:

7.2.0 Indonesia

Adira-1

Adira-1 is a sweet cultivar with only 30 ppm HCN in the tuber and suitable for fresh consumption. It can be harvested in 7-8 months with fresh tuber yield of

25-30 t/ha. Recommended for small scale farms, where cassava is intercropped with other food crops.

Adira-2

It is high yielding cultivar; but the tubers are medium in starch content and high in HCN. Tolerant to cassava bacterial blight (CBB).

Adira-4

A bitter cultivar with approximately 90 ppm HCN but capable of producing high yields and widely adapted. Hence suitable for large plantations and industrial use. The crop can be harvested at 9-10 months with fresh tuber yields of 35-45 t/ha.

Faroka

This cultivar with a yield of 19 t/ha is adapted to drier areas. The tubers are low in HCN with good table qualities.

7.3.0 Thailand

Rayong-1

It is a high yielding (25 t/ha), widely adapted bitter cultivar. Since tubers have high dry matter and starch content it is well accepted as an industrial cultivar.

Rayong-2

It was selected from hybrid seeds of CM 305 introduced from CIAT, Colombia. Its fresh tuber yield, dry matter content and starch content are similar to Rayong-1. This cultivar is table type, good for making fried chips and other convenient foods. Its yellow flesh contains high carotene and very low HCN.

Rayong-3

Rayong-3 is a selection from CM 407 cross which was introduced from CIAT, Colombia. In tuber dry matter and starch content it is superior to Rayong-1. These qualities fetch higher price for Rayong-3 tubers, though the tuber yield of this cultivar is on par with Rayong-1.

Rayong-60

It is a cross between CIAT and Thai clones (M.Co1-1684 x Rayong-1). This cultivar is superior to Rayong-1 in fresh root yield and root dry matter content. It is early maturing and adapted to low fertility conditions with an yield potential of 25-30 t/ha.

Rayong-90

A selection from DOA F_1 hybrid seeds. High dry matter content. Relatively high yielder.

Sriracha 1

A selection from Kasetsart University F_1 hybrid seeds. Have got excellent agronomic traits. Dry matter content is very high.

Kasetsart 50

Selected from KU F_1 hybrid seeds. Dry matter content and yield are high. Well adapted to unfavourable conditions.

Rayong-5

Obtained from a cross between 27-77-10 and Rayong-3 made at Rayong Field Crops Research Centre. Widely adapted by farmers due to its high yield (28 t/ha) and other outstanding features such as ease of harvest due to its good root shape and root formation, good germination and drought tolerance.

7.4.0 The Philippines

Golden Yellow (PR-C 24)

This cultivar gives good yield both under open and partially shaded conditions prevalent in coconut gardens. Tubers are low in HCN. Matures in 10-12 months with an yield of 42 t/ha.

Colombia (PR-C 62)

It is an edible purpose cultivar with high yield potential, 46 t/ha, in 10-12 months duration.

PSB CV-11

Selected by the breeding programme of the Institute of Plant Breeding at the University of the Philippines at Los Banos from the materials introduced from CIAT, Colombia. Used both as a human food and as feed and have a low level of HCN. Gives an average yield of 26 t/ha.

PSB CV-12

This variety is also developed at the University of the Philippines. Used both as a human food and as feed and have a low level of HCN. Average yield is 23 t/ha.

7.5.0 China

SC-201

One of the most popular cultivars in Guanxi region of China. It is tall, broad leaved and the planting material can be stored for long time. The cultivar is adapted to drought and low fertility but susceptible to wind damage. The tubers are bitter and under favourable conditions can yield 30-38 t/ha.

SC-205

Comparatively short narrow leaved cultivar capable of producing 45 t/ha under favourable conditions. It is widely planted in Guanxi and in Hainan Island. It is tolerant to cassava bacterial blight and intermediate in duration.

SC-124

This cultivar is high yielding, cold resistant, drought tolerant and with low HCN content with wide adaptability. The stems can be stored during the winter in the subtropics, ensuring availability of good quality planting material.

SC-8002

This is characterized by a high yield potential and tolerance to cold. It gives an average yield of 34 t/ha.

SC-8013

This is characterized by a high yield and wind resistance. It is suitable for planting in the typhoon affected areas of Hainan and Zhanjiang district of Guangdong. It gives an average yield of 38 t/ha.

7.6.0 Colombia (CIAT)

M. Ven-25

This clone with ideal plant type and high yield (30 t/ha) is suitable for the sandy soils of Media Luna in Colombia. It is an industrial cultivar with high HCN and tolerant to mites.

M. Col-2215

One of the best clones combining high yield (30 t/ha) and good table quality. Adapted to wide soil conditions but susceptible to mite.

M. Bra-12

This clone is resistant to mite and insect pests. It has high yield potential with intermediate dry matter content and adapted to north coast of Colombia.

M. Bra-5

Suitable for the Carimugua region of Colombia. It combines high resistance to CBB and moderate resistance to super-elongation disease. It has got compact plant type and good overall performance.

M. Per-242

This clone also is suitable to the stress conditions prevailing in the Carimagua zone of Colombia. It is capable of producing 30 t/ha with an harvest index of 56. The cultivar is highly resistant to CBB and moderately resistant to super elongation.

M. Col-1684

This clone collected from Lelicia in the Amazon jungle, has extraordinarily high yield potential over a wide range of conditions in Latin America. The yield realised in multi location trials ranged from 35 to 57 t/ha. However, the cultivar performs poorly in cool climates and in areas where thrips are a problem.

7.7.0 Nigeria (IITA, Ibadan)

TMS-30001

Early maturing sweet cassava cultivar, tubers with low HCN and well accepted by consumers. Tuber yield is around 22 t/ha. Resistant to CMD but susceptible to root rot and insect pests.

Cultivars of Cassava

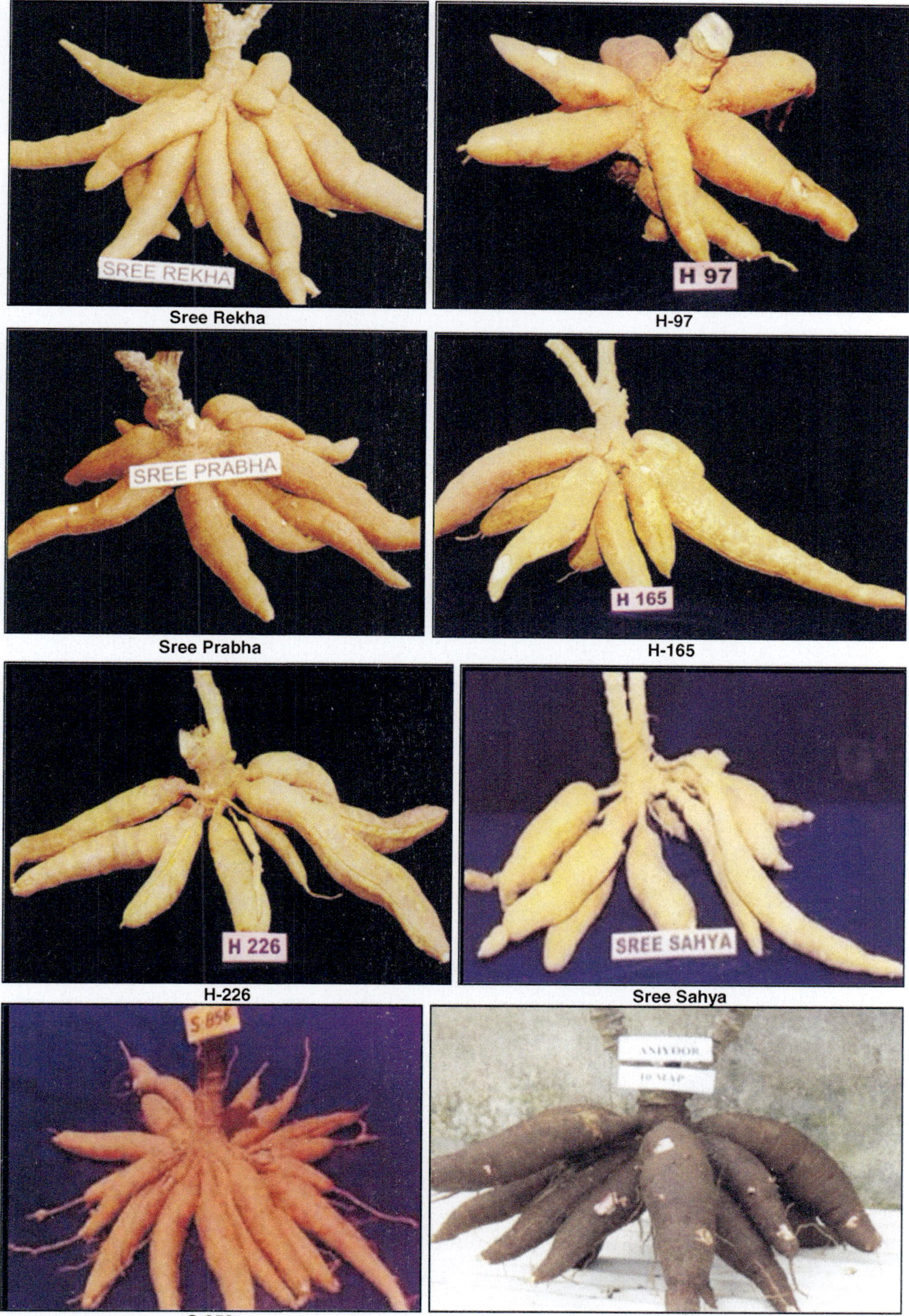

Sree Rekha

H-97

Sree Prabha

H-165

H-226

Sree Sahya

S-856

Aniyoor

TMS-305572

Widely adapted cultivar, due to the high yield (40 t/ha) and quality of the tubers. Resistant to CMD and CBB. Recovers fast, though affected by cassava mealy bug and cassava green mites.

TMS-50395

This cultivar is resistant to CMD, CBB and tolerant to cassava green mite, widely adapted though the tubers are high in HCN. Tuber yield is 33 t/ha.

TMS-30555

High yielding widely adapted cultivar. Resistant to CMD and CBB, with a tuber yield of 40 t/ha.

8.0 Soil and Climate

8.1.0 Soil

Cassava can be grown on a wide range of soils in the tropics and sub-tropics. It is well adapted to low fertility and acidic soils (pH 4.5-7.0), but a well-drained loamy soil of pH 5.5-7.0 is best suited for the crop. It cannot tolerate water logging, high salt concentration and alkalinity, but it can tolerate drought once it is established. Sodium saturation above 2.5 per cent and pH values more than 8 in the soil, drastically reduce the yield of cassava. To obtain maximum productivity, the crop should be raised on well-drained medium to fertile soils with a pH of 5.5 to 7.0. In Africa and South America, cassava is largely cultivated on oxisols and ultisols. It is estimated that 55 per cent of the cassava in Asia is found on ultisols, 18 per cent on inceptisols, 11 per cent on alfisols, 9 per cent on entisols and only 7 per cent on other soil orders combined (Howeler, 1992). In India, most of the cassava is grown on ultisols (central lateritic belt of Kerala) and alfisols (red soils of Thiruvananthapuram, Kanyakumari and South Arcot districts). It is also grown in vertisols (black soils of Salem district) and entisols (sandy soils of coastal belt of Kerala and loamy soils of east Godavari district in Andhra Pradesh).

8.2.0 Climate

Cassava can grow in areas ranging from humid with an annual rainfall more than 2,000 mm to a semi arid condition with an annual rainfall of 700 mm. It is grown up to an altitude of 2000 m near the equator, but performs better at lower altitudes. Ideal temperature for cassava cultivations is between 25°C to 32°C. It is a drought-tolerant crop, but cannot withstand frost and its growth ceases at temperature below 10°C. Cassava is well adapted to annual rainfall ranging between 1000 and 3000 mm, but in areas with an annual rainfall less than 1000 mm, it can be cultivated with supplementary irrigation as in the Salem district of Tamil Nadu.

According to Cock (1985), cassava cultivation is limited to the hotter areas of tropics and sub-tropics, between 30°N and 30°S. It is grown up to an altitude of 2300 m near the equator, but the altitude at which it is found becomes progressively lower away from the equator.

Areas with large seasonal variation in temperature and mean annual temperature lower than 20°C, are not suitable for cassava. However, some special cold-tolerant types are found in the Andean zone where the mean annual temperature is as low as 17°C. In such place frosts may defoliate the plant, but with the rise of temperature plant sprouts from the base and grow normally. Once established, unlike other crops, cassava has no critical period when lack of rainfall may cause crop failure. It is highly tolerant to drought and so even in areas that have dry seasons as long as 6 months, it is a major crop. With the onset of dry weather plant reduces leaf area by producing fewer small sized leaves and shedding the older leaves. At higher temperature the stomata also remain partially closed, thus reducing the transpiration loss of water. If the dry spell continues, the plant remains almost dormant. Once the rains resume or irrigation is given, the plant draws carbohydrate reserves from the stem and roots and produce new leaves. These special attributes of cassava has earned it the reputation as a rustic crop, adapted to a wide range of soil and climatic conditions.

9.0 Propagation and Planting Material

Cassava can be propagated either from the true seed or vegetatively from stem cuttings called stakes or setts. Propagation from true seed is practised only in breeding programmes. Freshly harvested cassava seeds germinate very rapidly and the ideal temperature for germination is between 25°C and 35°C (Ellis and Roberts, 1979). All commercial plantings of cassava are from cuttings of mature stem.

9.1.0 Planting Material

At the time of cassava harvest, mature, healthy and vigorous stems are selected and stored under shade as next season's planting material. Cassava stems of 1 m or more from the basal mature part of the healthy plants treated with a combination of insecticides and fungicides can be stored for a period of about six months under dry, shaded conditions (Ravi and Nair, 1998). Stems kept in vertical position with leaves intact sprout better (Anon., 1983). In areas with prolonged cold weather the mature stems are collected in advance and stored in pits covered with straw or soil to prevent frost damage (Howeler, 1988).

At the time of planting the stems are cut into stakes of 15-20 cm long after discarding the dehydrated base and the sprouted top portion. Such stakes are usually 1.5-2 cm thick and have 5-7 nodes. In some places, as in east Godavari district (Andhra Pradesh), where precipitation is uncertain at the planting season, the stakes are closely planted in a nursery and regularly watered for sprouting. Once the setts strike roots within 15-20 days, they are carefully uprooted and planted in the main field. For the rapid multiplication of cassava planting material, Kamalam *et al.* (1977) developed a technique by inducing sprouting of two nodes, single node and half node cuttings, which could be adopted for the faster multiplication of a new cultivar within a short period.

Osakwe *et al.* (2000) found relationship between length of cuttings and incidence or severity of disease, growth and yield of cassava (cv. TMS 91934) in Nigeria under rainfed conditions. The percentage sprouting was highest (100 per cent) for

Planting Materials, Planting and Crop Establishment of Cassava

15 cm cuttings and lowest (64.43 per cent) for 5 cm cuttings. At one month after planting (MAP), 5 cm cuttings had produced the shortest plants while there was no visible difference in height of plants from the other cuttings. The incidence of cassava mosaic disease at 1 MAP was highest (76.67 per cent) for 25 cm cuttings, and 73.33 per cent for 20 cm cuttings. However, as the plants matured, there was a gradual recovery from cassava mosaic disease; the incidence and severity at 6 MAP were 43.33 per cent and 1.34, respectively, for 25 cm cuttings. With regard to the incidence and severity of cassava bacterial blight (*Xanthomonas axonopodis* pv. *manihotis*), there was a gradual increase in both parameters from 1 MAP until 3 MAP, and then they gradually declined. The highest mean weight of fresh tubers at harvesting was observed for 15 cm cuttings (37.87 t/ha).

10.0 Cultivation

10.1.0 Land Preparation and Planting

The method of land preparation for cassava depends on soil type, topography and farm size. Land preparation, deep ploughing and cultivation before planting increases yield at a lower cost than the use of entomopathogens in controlling the burrowing bug, *Cyrtomenus bergi* (Bellotti *et al.*, 1997).

In Kerala, most of the small scale farmers prepare the land by manual digging, while in Tamil Nadu the usual practice is to provide 3-4 ploughing by animal-drawn implements or with the help of tractor. In small farms of other cassava growing countries the land is prepared with the help of animal-drawn implements, while in commercial plantations tractor is used. Field experiments on tillage requirements of cassava are limited. Loganathan *et al.* (1975) compared four tillage depths *viz.*, 10, 20, 30 and 45 cm and recommended ploughing to a depth of 10-20 cm. In another experiment, 3 systems of planting *viz.*, flat, mound and ridge were evaluated and reported the superiority of ridge system (Shanmugam and Thamburaj, 1975). In a similar study at CTCRI, pit followed by mound system of planting produced higher yield (Anon., 1983). Performance of cassava under different tillage systems such as no tillage, reduced tillage and conventional tillage seems to be site specific (Moreno, 1991). Contour ridges combined with or without live barriers are effective in reducing soil erosion (Anon., 1989; Anon., 1990). Maduakor (1997) conducted a 2-year field trial in an ultisol in the forest zone of SE Nigeria to ascertain if planting of cassava on ridges 35-37 cm high had any significant advantage over flat planting in view of the extra cost of making ridges. The optimum K rates for maximum SRY were 105 kg/ha and 124 kg/ha for flat and ridge planting respectively, the response being linear for plants on ridges. The ridging had no significant advantage over flat planting.

10.1.1 Spacing

The most common plant population for cassava is about 10,000 plants per hectare. Field experiments conducted at CTCRI established the optimum spacing of 90 x 90 cm for branching types and 75 x 75 cm for non-branching types (Anon., 1983). In most of the Asian countries planting density of 10,000-15,000 stakes per hectare is followed for cassava under monocropping (Howeler, 1988). For cassava, intercropped in coconut gardens, the spacing of 90 x 90 cm and a plant population of 8000 was recommended (Nayar and Sadanandan, 1990).

In different cassava growing regions, vertical, horizontal and inclined positions of planting cassava setts are in vogue. Comparison of these positions and different depths of planting showed that, planting 20 cm long stakes to a depth of 5 cm in vertical was better (Anon., 1983). Taro and Atlee (1985) also stated that vertical planting is better when compared to other planting positions. If ridges or mound are prepared the stakes are planted on the upper part, at the recommended spacing. Under normal conditions of planting, most of the stakes sprout within 15-20 days. The unsprouted stakes may be replaced preferably with 40 cm long stem cuttings.

Mechanised planting with tractor-drawn implements, which simultaneously accomplishes furrowing, fertilizer application, horizontal planting, covering the stakes and firming the soil, is practised on large scale in Brazil and to limited extent in Nigeria (Taro and Atlee, 1985).

10.1.2 Planting Season

Cassava can be planted throughout the year under irrigated conditions. April-May with the onset of pre-monsoon showers or August-September, with the onset of north-east monsoon showers is the ideal time of planting as a rainfed crop. Generally cassava planting coincides with the beginning of rainy season. In Tamil Nadu, where the monsoon is from October-November, planting in September gave higher tuber yield (Thampan, 1979). In areas with prolonged winter, cassava planting may be done just after the cold weather. Summer planting is also seen in places where irrigation facilities are available.

10.2.0 Manuring and Fertilization

Cassava, though reputed for its ability to perform well even in poor soils, requires higher levels of fertilizer application for better yields. Moreover, cassava is considered as a soil exhausting crop with its high efficiency in utilizing plant food elements.

Application of FYM or compost @ 10 ha-1 at the time of land preparation and 1 kg manure per pit during filling of the pits is recommended. In addition to FYM, N:P:K @ 100:50:100 kg ha^{-1} for long duration varieties and 75:50:75 kg ha^{-1} for short duration varieties, has been recommended. Green manuring *in situ* with fast growing vegetable cowpea varieties can substitute FYM.

Full dose of P_2O_5 and half dose of Nitrogen and K_2O should be given at the time of sowing, whereas remaining half dose of Nitrogen and K_2O is applied 45-60 days after planting along with first weeding and earthing up. In Gujarat, Maharashtra and Kerala, 100:50:100 kg NPK ha^{-1} fertilizer is recommended.

For acid laterite soils, studies with biofertilizers have been very effective in increasing the yield as well as reducing the use of inorganic fertilizers. In Andhra Pradesh and Tamil Nadu, application of half dose of P_2O_5 (30 kg ha^{-1}) + full N and K_2O (60:60 kg ha^{-1}) + FYM (12.5 t ha^{-1} + phosphorous solubilizing bacterial (PSB) or *Arbuscular mycorrhizal* (AM) fungi (10 kg ha^{-1}) has been recommended for cassava.

In Tamil Nadu, 75:25:75 NPK ha^{-1} is given under rainfed condition and 90:90:240 under irrigated conditions. In Andhra Pradesh 60:60:60 kg NPK ha^{-1} is recommended under rainfed and 100:60:150 NPK ha^{-1} under irrigated condition. For Chhattisgarh 75:50:75 kg NPK ha^{-1} is recommended.

In case of micronutrient deficiency of Mg, Zn and B application of $MgSO_4$ (20 kg ha^{-1}), $ZnSO_4$ (12.5 kg ha^{-1}) and Borax (10 kg ha^{-1}) is recommended. The lime induced chlorsis as seen in Tamil Nadu, can be cured by 1 per cent $ZnSO_4$ and 1 per cent $FeSO_4$ spray.

Experiments on nutrient removal by cassava indicated that, considerable variations existed among cultivars in their nutrient removal pattern. Most of the cultivars under favourable conditions, capable of producing 30 t fresh tubers per hectare, removed 180-200 kg N, 15-22 kg P and 140-160 kg K (Anon., 1983). In addition to the variation due to cultivars, levels of fertilizer application, water management, and cropping system were reported to influence the nutrient removal in cassava. Nayar *et al.* (1986) found that with increase in levels of N and K there was enhancement in the uptake of nutrients both under rainfed and irrigated conditions. However, cassava receiving supplementary irrigation utilized more amounts of N, P and K, respectively of the levels of application.

When cassava is grown in soils with a limited availability of plant nutrients, it grows slowly, remains short and weak, but are quite effective in translocating carbohydrates from the shoots to the roots. This results in a rather low yield in spite of a high harvest index. When grown in a fertile soil or when plant nutrients are added in the form of manures or chemical fertilizers, the greater absorption of essential nutrients results in faster growth, more and bigger leaves, longer life, thicker stems and more and thicker roots. However, if nutrient supply is excessive, especially that of nitrogen, some cassava varieties grown too tall and produce too many leaves resulting in a low harvest index and a reduced root yield. Moreover, excessive supply of phosphorus or potassium can lead to reduced uptake of Fe or Zn and Ca or Mg, respectively. Thus, the maintenance of a balanced nutrient supply is essential for obtaining high yields. General recommended dose under normal package of practice for cassava is NPK @ 100:50:100 kg/ha + FYM @ 10 t/ha. Application of micronutrients along with FYM and NPK in cassava has been found to increase the tuber yield. Among various treatments, application of NPK @ 100:50:100 kg/ha + FYM @ 10 t/ha + $MgSO_4$ @ 20 kg ha^{-1} + $ZnSO_4$ @ 12.5 kg ha^{-1} + Borax @ 10 kg ha^{-1}+ $FeSO_4$ @10 kg ha^{-1} was found to be the most effective in increasing tuber yield of cassava cv. Sree Jaya (28.44 t/ha). Only a slight variation in quality of tubers was observed among different treatments (Anon., 2016).

Maintaining adequate soil fertility through application of manures or judicious use of chemical fertilizers is not only important for the supply of nutrients, but also for maintaining optimum soil physical conditions through the return of organic matter to the soil. This in turn improves the soil's water holding capacity, nutrient retention capacity and stimulates the soil's microflora (Howeler, 1998).

10.2.1 Deficiency Symptoms

Nutritional disorders in cassava especially those of N, P and K are not always very clear under field conditions. However, deficiency symptoms in cassava have been determined in sand and nutrient solution cultures (Asher, 1975) and they can be briefly summarized as follows:

10.2.1.1 Nitrogen

Stunted growth, uniform chlorosis of leaves starting from lower leaves, soon spreading throughout the plant.

10.2.1.2 Phosphorus

Thin stem, short petioles and small pendant leaves; under severe conditions some lower leaves turn yellow, flaccid, necrotic and fall off.

10.2.1.3 Potassium

Small and chlorotic upper leaves, thick stems, under severe cases purple spotting, premature lignifications of upper stem.

10.2.1.4 Calcium

Reduced root growth, border chlorosis of upper leaves, leaf tips and margins bending downwards.

10.2.1.5 Magnesium

Reduction in plant height and interveinal chlorosis of lower leaves.

10.2.1.6 Zinc

Interveinal yellow or white spotting on young leaves, which become narrow and chlorotic at the growing terminal. Necrotic spotting on older leaves and reduced plant growth.

Long-term field experiments at CTCRI showed that withholding application of potassium fertilizers lead to appearance of visual K deficiency symptoms from the 8th year (Kabeerathumma *et al.*, 1988). Typical calcium and magnesium deficiency symptoms also appeared in N and NK treatments. In the absence of clear deficiency symptoms under field conditions, diagnosis of nutritional disorders in cassava may have to depend on soil and/or tissue testing.

10.2.2 Application of Organic Manures

Cassava responds to both bulky and concentrated organic manures. For small-scale cassava production the use of compost, cowdung or chicken manure is recommended. For large scale production these sources are not generally available and not economical. Hence a combination of organic manures and chemical fertilizers are recommended. Studies conducted at CTCRI showed that basal application of farmyard manure (FYM) at 12.5 t/ha was beneficial in enhancing the yield of tubers and maintaining the soil fertility (Mohan Kumar *et al.*, 1976; Pillai *et al.*, 1987). Later experiments showed that FYM application can be substituted by green manuring with cowpea (Nayar and Mohan Kumar, 1989).

Various wastes are generated in enormous amounts in most communities in Nigeria. The management of such wastes pose a serious problem to community hygiene. Poultry manure, cowdung, sawdust, sorted city refuse, shear nut cake and palm kernel cake were collected, composted and enriched with local minerals to produce an organo-mineral fertilizer (OMF), as an alternative nutrient source for Nigerian farmers and an effective waste management strategy. OMF also gave the highest cassava tuber yield of 10.10 t/ha, which was 200 per cent higher than control plots without fertilizer and a 40 per cent increase over mineral fertilizer plots (Ogazi *et al.*, 2000). Gabriel Filho *et al.* (2000b) conducted field experiment in Brazil, with cassava grown with conventional or minimum tillage, and common vetch (*Vicia sativa*) and/or black oat (*Avena nuda*) green manures or no green manure. Soil density and resistance to penetration in the 0-10 cm layer were higher under conventional than minimum tillage. Cassava yield did not differ among tillage or green manure treatments. Florchinger *et al.* (2000) evaluated the impact of 5, 10, 13 and 40 cm topsoil removal on the productivity of cassava in Colombia indicated that organic matter was a key factor for productivity and the yield is dramatically reduced by topsoil loss. Applications of nitrogen, phosphorus, potassium and compound fertilizers with liming could not restore lost productivity. The most important constraint to plant growth was aluminium saturation which increased sharply with topsoil loss.

10.2.3 Application of Chemical Fertilizers

10.2.3.1 Nitrogen

Increase in tuber yield of cassava in response to higher rates of nitrogen application had been reported by many workers. On acid laterite soils of Kerala state, Mandal *et al.* (1971) obtained higher yield with 100 kg N/ha. Saraswat and Chettiar (1976) obtained response up to 150 kg N/ha, but the most economic dose was 100 kg N/ha. Subsequent studies at CTCRI also showed that N application beyond 100 kg/ha were not beneficial. Nair (1982) found that in red loam soils of South Kerala and sandy loam soils of Central Kerala, application of nitrogen at 120 kg/ha produced significantly higher yield in cassava hybrid Sree Sahya. According to Cock (1985) cassava rarely responds to N application above 100 kg/ha. Application of nitrogen at very high levels to cassava promotes vegetative growth at the expense of tuber yield.

10.2.3.2 Phosphorus

The phosphorus requirement of cassava is smaller when compared to that of nitrogen and potassium. Response to applied P by cassava is found to be poor except in the extremely P deficient soils of South America and Africa. Association of mycorrhizal fungi was also found to be beneficial as the fungi promote P uptake in cassava. Long term experiments conducted at CTCRI indicated that exclusion P from the fertilization schedule of cassava grown on acid laterite soils of high P status had no significant adverse effect on the growth and yield (Anon., 1981; Mohan Kumar *et al.*, 1998). Nair *et al.* (1988) investigated the response of cassava to graded levels of phosphorus in acid laterite soil and determined the optimum dose as 50 kg P_2O_5/ha.

10.2.3.3 Potassium

Application of potassium to cassava is essential in sandy and acid low fertility soils, to maintain the yield and quality of tubers. Potassium deficiency causes reduction in dry matter and starch content of the tubers while the HCN content increases. Rajendran *et al.* (1976) found that 100 kg K_2O/ha was optimum for cassava and higher rates resulted in luxury consumption. However, Nair and Sadanandan (1987) recorded yield increase in cassava up to 200 kg K/ha. Howeler (1985) observed that in the initial years, the response of cassava to applied K was poor in Brazil. But in the second and third consecutive cassava crops there were marked response of 100-130 kg K/ha. In general, the response to applied K by cassava depends on the native fertility and mineralogy of the soil. Root HCN contents were significantly reduced by application of K in many cultivars. Application of K fertilizer to low-K soils is warranted to minimize health hazards when fresh cassava is used for human consumption (El Sharkawy and Cadavid, 2000).

10.2.3.4 NPK

For most of local cassava cultivars and M-4, application of FYM at 12.5 t/ha and NPK dosage of 50:50:50 kg/ha were found to be sufficient (Anon., 1986). In the case of high yielding hybrid cassava cultivars *viz.*, H-165, Sree Visakham and Sree Sahya, besides the basal FYM application of 12.5 t/ha, NPK at 100:50:60 kg/ha is recommended (Anon., 1986). However, high yielding cassava hybrids receiving supplementary irrigation responded to an NPK application at 150:100:150 kg/ha (Nayar *et al.*, 1985). Studies on the response of top cross hybrids to different fertilizer levels under upland and lowland conditions indicate that TCH-1 was superior under upland and TCH-3 under lowland conditions at a fertilizer dose of 100:50:100 kg N, P_2O_5 and K_2O per hectare (Anon., 1998).

The relationship between relative yield and the N, P or K concentration in youngest fully expanded leaf (YEEL) blades at 4-5 months after planting (MAP) was discussed by Howeler (1998). There was good relationship between relative yield and the N and K concentration of the leaves, allowing the estimation of a critical concentration (corresponding to 95 per cent of maximum yield) of 5.7 per cent N and 1.9 per cent K. These critical levels are higher than those reported earlier (4.6 per cent N and 1.7 per cent K) (Howeler, 1995), but are within the 'sufficiency' range reported elsewhere for these elements (Howeler, 1996). No clear relationship could be observed between P response and P concentration in YEEL blades, probably due to a lack of serious P deficiency in the four locations for which data are available.

The efficiency and time of application of some of the nitrogen sources like calcium ammonium nitrate (CAN), ammonium sulphate, sodium nitrate and urea were evaluated (Mandal *et al.*, 1971). Application of CAN or urea in two splits, *i.e.*, half as basal and the other half one or two months after planting were found to be superior. With regard to phosphatic fertilizers, basal application of ground rock phosphate was better than single superphosphate (Mohan Kumar, 1978). Potassium also should be applied in two splits along with nitrogen fertilizers.

10.2.3.5 Calcium, Magnesium and Sulphur

Studies on response of cassava on calcium, magnesium and sulphur are limited. Nair and Varghese (1970) found that application of $CaCO_3$ and $MgCO_3$ in equal proportion at 100 kg/ha enhanced cassava yields from 16.7 to 25.5 t/ha. Since cassava is tolerant to soil acidity liming is not recommended except in extreme acidic and peat soils.

In soils low in native sulphur cassava responds to sulphur applied alone or through sulphur containing fertilizers. Mohan Kumar and Nair (1983) reported that application of sulphur at 50 kg/ha resulted in 18 per cent increase in yield over control. Cassava mosaic disease (CMD) severities were unaffected by cropping system or fertilizer (NPK) levels (Ekpe and Chinaka, 2000). Insufficient soil Mg limits crop yield, quality, and use efficiency of N, P and K (Tan *et al.*, 2000)

10.3.0 Irrigation

Cassava is usually raised as a rainfed crop. As a rainfed crop in Kerala, irrigation is normally not advocated, whereas irrigation is common in Tamil Nadu and Andhra Pradesh conditions. Sufficient moisture should be ensured in the field at initial stage *i.e.*, the first 20 days after planting, for proper establishment of cassava plants. Supplementary irrigation during drought period can result in higher yield. Irrigation at approximately 25 per cent available moisture depletion level throughout the growing season could double the tuber yield compared to control. It also improves the starch content and reduces the HCN content.

In parts of Tamil Nadu and in some other non-traditional areas, cassava is cultivated under irrigation. On the day of planting, one irrigation is given followed by two irrigation at an interval of 3-5 days till the plants establish. Further irrigations are scheduled depending upon the distribution of rainfall. In Tamil Nadu irrigation at 8 days interval gave the highest net profit (Shanmugavelu, 1973). At CTCRI the response to supplementary irrigation by high yielding hybrids of cassava was studied (Nayar *et al.*, 1985) and found that cassava responded positively to supplementary irrigation at 20 mm/week, which was approximately equivalent to 70 per cent of the weekly pan evaporation in the locality.

10.4.0 Inter-culture and Weed Control

Earthing up and weeding in the early stage of the crop is essential for proper control of weeds and for improving the soil condition. The first weeding and earthing up should be given 45-60 days after planting and the second a month later.

All the stakes planted may not establish due to poor quality planting material, adverse weather conditions and termite infestation, hence, at the time of planting of stakes in the main field, about 5 per cent of the stakes may be planted separately at a very close spacing of 4 cm x 4 cm in a nursery. At the age of 15-20 days the plants are uprooted from nursery and used for gap filling.

The sprouts emerging from the buds are more vigorous than those emerging from the lower nodes of the stakes. Removal of excess sprouts by nipping at the initial stage of establishment (10-15 days after sprouting) helps to prevent mutual shading and competition between plants. Retention of 2 healthy shoots on opposite sides is better for the production of more number of tubers per plant.

Interculture to suppress weeds in the initial stages of crop growth and to encourage tuber development is an essential requirement in cassava production. Light raking of the soil, followed by earthing up twice at one and two months after planting, had significant beneficial effect on cassava yield (Anon., 1983). Interculture in cassava is usually done with handtools or with the help of animal drawn hoes. In Malaysia, Thailand and in some of the South American countries, on large farms herbicides are used for weed control in cassava. The most widely used herbicide combination for pre-emergent weed control is a mixture of diuron and alachlor in different doses according to the weed floral and soil characteristics. Research on weed control in cassava showed that in addition to one application of selective pre-emergence herbicides, atleast one interculturing was necessary for higher yields (Moreno, 1991). Cassava yield under manual weeding treatment 60 days after planting (20952 kg/ha) was statistically equal to that under the weed-free control treatment (25759 kg/ha) while manual or mechanical weeding 30 days after planting did not result in high crop yields (Gabriel Filho *et al.*, 2000a). Predominant weeds in Venezuela were *Cyperus rotundus, Panicum maximum, Euphorbia heterophyla* and *Convolvulus* sp. Glyphosate and paraquat + oxifluorfen provided the best level of weed control and the lowest weed dry weight. The highest planting distances produced better yields than the lowest (Giron and Alfonzo, 2000).

10.5.0 Mixed Cropping and Rotations

Cassava is suitable as an intercrop under different cropping system. Cassava can be grown well with coconut, *Eucalyptus* and other fruits and forest trees under agri-horti or agri-silviculture system.

In Kerala, cassava is intercropped in 20 per cent of coconut stands and performs well in newly planted coconut gardens up to 5 years. Cassava varieties H-165, H-226, Sree Viskham and Sree Sahya can be grown satisfactorily under different cropping systems. Cassava variety KMC-1 has been reported as a suitable intercrop in coconut gardens of Kuttanad in Kerala.

Cassava varieties of long duration that take 9-10 months for harvest, and take about 3-4 months to develop enough canopy due to slow initial growth, provide a great opportunity to grow short-duration intercrops like ground-nut, vegetable cowpea, French bean, black gram and other legumes in between the plants which are planted at a spacing of 90 x 90 cm. In Tamil Nadu black gram/Bellary onion (*Alliun cepa*), in Andhra Pradesh, green gram/black gram and in Assam soy bean and French bean are recommended as suitable intercrops in cassava.

On most of the small farms cassava is grown as intercrop and in sequential crop patterns. At least 50 per cent the cassava grown in Africa and 40 per cent of the cassava area in tropical America are intercropped with short duration cereals, legumes and vegetable crops (Leihner, 1983). Estimates on extent of intercropping in cassava for Asia are not available. Nevertheless, in small farms cassava is intercropped with legumes like groundnut, cowpea, black gram and onion in India. In Indonesia, dry land cassava follows upland rice and maize as a third crop (Effendi,

1979). In India, Malaysia, the Philippines and Thailand cassava is also intercropped with coconut, oil palm and rubber.

Research on cassava intercropping has shown that land equivalent ratios above one are frequently obtained, when cassava is intercropped with short duration crops such as beans, groundnut, cowpea and some vegetables. This is possible mainly due to the slow initial growth of cassava (Moreno, 1991). When cassava is grown under tree crops in general, the growth and yield of tree crops are not negatively affected, but the productivity of cassava is reduced by shading from the perennials. Ghosh *et al.* (1987) recorded growth, productivity and nutrient uptake by cassava in association with four perennial species *viz.*, coconut, banana, eucalyptus and *Leucaena* and reported that the yield of cassava was significantly reduced when grown in association with eucalyptus. For cassava intercropped in coconut garden a planting density of 8000 cassava and an NPK dosage of 50:50:100 kg/ha were recommended (Nayar, 1986). Investigations at CTCRI identified two sequential cropping systems involving cassava. For the uplands green manure cowpea followed by short duration cassava was found suitable (Nayar and Mohan Kumar, 1989). For the single cropped paddy lands cassava in sequence with rice was suggested (Mohan Kumar *et al.*, 1987).

Cropping system studies conducted at CTCRI have shown that a short duration variety of cassava, Sree Prakash with seven months duration can be grown successfully in a rice based cropping system. A fertilizer dose of 100:25:100 kg/ha N, P_2O_5 and K_2O is recommended for the variety Sree Prakash in a rice based cropping system (Mohan Kumar *et al.*, 1998).

The yield of cassava planted in monoculture decreased yearly without fertilizer application. In contrast, with fertilizer application, the yield was high and constant. Cassava intercropped with upland rice and maize followed by groundnuts or soyabean gave a satisfactory crop yield and maintained soil fertility in upland areas with red yellow podzolic soil in Indonesia (Ardjasa *et al.*, 2001). Experiments conducted in Nigeria indicated that cassava was highly compatible with sunflower although total seed yields of sunflower in sole sunflower and sunflower + cassava were not significantly different (Fagbayide *et al.*, 1997). Kallon *et al.* (1997) studied the effects of different cropping systems and mineral fertilizers on grain yield of rice and dry tuberous root yield of cassava and on nutrient balances and soil chemical properties. Alley cropping without mineral fertilizer application led to a yield decline over time.

Bunch type ground nut varieties (TMV-2 and TMV-7) and French bean (cv. "Contender") are promising and economical intercrops. Sree Prakash and Sree Jaya have been found to be suitable for low land cultivation in a paddy based cropping system *i.e.*, sequential cropping of cassava and paddy. The cultivation details of various intercrops, is given in Table 5.

The gross net income was always higher in the following cassava intercropping systems with legumes/non-legumes than the pure stand of cassava. The gross income increased by 40-82 per cent in a cassava + French bean system, 25 per cent in a cassava + urd bean system, 29 per cent in a cassava + soy bean or mung bean system, 13-20 per cent in a cassava + green gram, or black gram system. Cassava can be successfully relay cropped following maize, peanut, sorghum, selected vegetables and banana) Ravi and Mohankumar, 2004).

Table 5: Cultivation Details of Intercrops in Cassava*

Name of Intercrop	*Cultivar*	*Duration (Days)*	*Spacing (cm)*	*No. of Rows*	*Seed (kg ha^{-1})*	*N:P:K (kg ha^{-1})*	*Yield (kg ha^{-1})*
Groundnut	TMV-2	100	30 x 20	2	40-45 (kernel)	10:20:20	1200 dry pod
Groundnut	TMV-7	100	30 x 20	2	40-45 (kernel)	10:20:20	1200 dry pod
French bean	Contender Top cross-4	70	30 x 20	2	40	20:30:40	2000
Cowpea (grain)	S-488	90	30 x 25	2	20	10:15:10	800
Cowpea (Vegetable)	B-61	65	40 x 20	1	8	10:15:10	3000

*Varma *et al.* (1997)

Sequential cropping of cassava and vegetable cowpea has been recommended for Tamil Nadu with application of P @ 30 kg ha^{-1} along with recommended dose of N and K and for Chhattisgarh, with 5 t ha^{-1} FYM and 50 kg ha^{-1} P_2O_5.

11.0 Harvesting, Yield and Storage of Tubers

11.1.0 Harvesting

Harvesting of cassava depends on the variety, agronomic practices, prevailing weather and the end use. Most of the cultivars are ready for harvest at 8-10 months after planting. Some of the short duration types can be harvested at 6-7th month after planting, while the long-duration types may be retained in the field up to 18 months. For human consumption and starch production the ideal time for harvesting is 8-10 months. When the crop is to be used for direct feeding to animals, harvesting can be done from 7th month onwards depending upon the need. Sometimes, farmers irrigate the crop two to three days before harvesting for avoiding breakage of tubers during harvesting. Cassava is harvested by hand pulling, sometimes using simple tools as lever. Development of harvesting aids and mechanical harvesters are in progress in many places (Taro and Atlee, 1985; Howeler, 1988). Under rainfed conditions, farmers use crowbars or fork for uprooting the tubers.

11.2.0 Yield

The high yielding cultivars are capable of producing 25-30 t fresh tubers per hectare. The average yield of the crop in India is about 19 t/ha. Under good management practices with irrigation facility, tuber yield of 40-50 t/ha, and under rainfed conditions, 20-25 t/ha can be obtained. Some of the high yielding hybrids can produce 55-60 t/ha under good management.

Harvesting of Cassava

11.3.0 Storage

Cassava tubers once harvested are highly perishable due to the physiological and microbiological changes. The harvested tubers should be used within a day or two in the absence of long term storage method. Farmers overcome this problem to some extent either by delaying the harvest or by slicing the tubers and sun drying as dry chips. Keeping the tubers in moist sawdust or moist sand or providing wax coating may prolong the shelf-life by 10-15 days. But these methods are not very effective and costly for the farmer in terms of labour and material inputs. Work at CIAT, Colombia showed that treating the tubers with thiobendazole 0.4 per cent solution and packing them in polythene bags can effectively prevent spoilage up to 3 weeks (Wheatly and Izyuierdo, 1991). The spoilage of fresh cassava can also be prevented by freezing the tubers, which is done on a limited scale in some of the South American countries. However, this method is very expensive.

Cassava tubers harvested 3 weeks after pruning and stored in polyethylene bags or without bags were acceptable in a sensory test after 6 and 12 day in storage (Gonzalez *et al.*, 1991). The major changes observed in cassava storage are high weight loss, vascular streaking (primary deterioration) and starch degradation. High weight loss (7.8 per cent) and presence of vascular streaking (VS) were found only in tubers stored on an earthen floor, the rate of starch degradation was similar in all tubers, irrespective of the storage method (George and Browne, 1994). The occurrence of VS in cassava tubers was primarily related to water stress in the tissue. Oxygen was secondarily involved in the discoloration reaction. Low RH plus an ambient O_2 level enhanced VS development (46 per cent incidence). Reduction of O_2 to 1 per cent at low RH reduced VS development (15 per cent incidence). However, VS was almost completely inhibited (1.4 per cent incidence) at high RH irrespective of the O_2 level (Aracena *et al.*, 1993). High relative humidity (95 per cent) favoured the formation of distinct wound periderm in tubers of cassava cv. HMC 1 and MCol 1684 injured by a 'V' shaped cut or abrasion. In HMC 1 vascular streaking could be prevented by covering the tubers with cling film (Ravi, 1994). When tubers are piled and watered daily to keep them fresh or coated with mud to reduce moisture loss, storage life of tubers can be extended up to 7 days (Mitra and Tarafdar, 2013).

Cassava tubers are stored in Central Africa in the form of 'foufou' (pieces o f tuber which are retted before being sun-dried). 'Foufou' is conducive to the development of numerous stored-products insects, mainly coleoptera. In Congo, 66 species were recorded, 34 of which are believed to feed directly on cassava and which cause severe losses (Delobel, 1992). The main species found in completely dried 'foufou' are *Araecerus fasciculatus, Carpophilus dimidiatus, Sitophilus zeamais, Cryptolestes pusillus, Tribolium castaneum* and *T. confusum*. The stored fermented product, 'dry ball' foofoo, contained *Aspergillus flavus, A. ochraceus* and *P. citrinum*. 'Dry ball' foofoo also contained trace amounts of aflatoxin G_1 (Jonsyn, 1989).

12.0 Diseases and Pests

12.1.0 Pests of Cassava

Among the pests that attack cassava, spidermites, white grubs, termites, thrips and rats, are important and whitefly as a vector of cassava mosaic is significant. The pests of cassava in India can be broadly grouped into (i) pests attacking stems (ii) pests damaging plant setts and young seedlings (iii) pests infesting foliage (iv) pests attacking tubers and (v) storage pests.

12.1.1 Pests Attacking Stems

Scale Insects

Scale insets of different species attack the stem in the field and during postharvest storage of planting material. The most important scale insect is white mussel shaped soft scale (*Aonidomytellus albus* CkII.). The insect sucks and desaps the stem. It multiplies rapidly covering the stem in large numbers and it also attacks the side shoots and at times the petioles. In severe infestation the stems become weak and dry due to the prolonged infestation, resulting in the loss of viability of planting material and poor establishment of the crop. The weakening and breaking of the main stem causes profuse branching which gives bushy appearance. The severity of attack increases during prolonged dry seasons. The problem is perpetuated through planting materials which serves as a source of multiplication and spread. Eggs are laid beneath the scales. The young nymphs which come out in 3-4 days are active and spread over and settle. Once they start feeding they lose their legs and become sedentary. It develops white waxy secretion over which it develops into a scale. In 20-25 days the nymphs become full grown. The male is winged and female is sedentary and wingless.

For the effective control, scale free stems should be collected for storing and planting. Stacking the stems in horizontal position encourage multiplication of scales due to higher temperature and humidity. Therefore healthy stems are to be kept in vertical position under shade to facilitate easy aeration and diffuse day light. As a prophylactic measure the stems may be sprayed with dimethioate 0.05 per cent at the time of storing. If further infestation is observed one more spraying may be done. The infested stems are to be rejected and burned at the time of planting. In case of acute shortage of planting material and the scale attack is mild, them the setts can be given a dip in the above insecticidal solution for 10-15 minutes before planting. Other scale insects observed are *Saissetia nigra* N. and *Lepidosaphes* sp. Larvae and adults of coccenellids – *Chilocorus nigritus* F. *Syemnus quimeti* Muls., *Chilomenus sexmaculatus* F. and *Stethorus pariepunctatus* Kapur prey on the scales (Pillai, 1994, Palaniswami, 1999).

12.1.2 Foliage Feeding Pests

Spider Mites

Among the foliage feeding pests the most serious ones are spider mites. There are two distinct groups of spider mites. One group feeds on the lower surface of the leaves (*Tetranycus cinnabarinus* (Boisd.) and *T. neocaledonicus* (Andre) causing elongated streaks, chlorosis, withering and drying of leaves. The other group which prefer to infest on the upper surface *Eutetranychus orientalis* (Klien) and *Oligonychus biharensis* (Hirst) cause depletion of chlorophyll, resulting in the characteristic bronzing and typical rusted bathery appearance; such leaves begin to curl from margins and later fold into one twisted roll. The terancychid mites usually infest lower surface but during epidemics may cover both sides of the leaf. The predominant species are *T. cinnabarinus* and *E. oreintalis* and are found in cassava growing areas of Kerala and Tamil Nadu. They appear together and multiply rapidly during dry periods. Both groups infest from bottom leaf upwards and in periods of heavy infestation the population density ranged from 200-1000/ leaf in almost all leaves. The lower surface feeding mites causes more severe leaf fall than the other group. The biology and bhinomics of *T. cinnabarinus* and *E. orinetalis* reveal that the total life cycle from egg to adult rangers from 11-20 and 10-14 days respectively (Pillai and Palaniswamai, 1985). The spider mites incidence increases year after year depending upon the climate. The data during 1977 to 1982 revealed that rainfall was the most limiting factor. Heavy rainfall accompanied by wind was deleterious, while a fall in RH and an increase in temperature were highly conductive for rapid multiplication of mites. An outbreak can be expected during rainless summer where temperature is 32°C and above and RH below 75 per cent. The yield loss due to spider mite complex alone ranged from 17.8 to 33.1 per cent in different varieties under natural infestation in Kerala (Pillai and Palaniswami, 1981). Economic threshold levels could be fixed for spider mites; for Sree Vishakham 25 per cent leaf infestation and for Sree Sahya and M4 10 per cent. Spraying of dimethoate or methyldemeton 0.05 per cent during severe infestation (January to April) at monthly interval is highly effective against the pest or spraying water at run off level is also effective in controlling mite damage. Spraying of water can be resorted to where labour is cheap. This has an added advantage of preserving the biotic agents of mites. They include *Coccenellids- Pharoscymnus hornii* Weise, *Stethorus gilvifrons* Muls., thrips – *Scolothrips indicus* P. and phytoseiid mites – *Amblyseius longispinosus* Evans (Pillai and Palaniswami, 1985). Thrips is an effective predator but the thrips population in the field is very low. However even though there are a number of natural enemies their population in general is very low under field conditions. If these predators are multiplied and released they will be effective biological agents for mites. Cassava accessions at CTCRI *viz.*, CE-4, CE-14, CE-38 and CE-139 are found highly resistant to spider mites. The host plant resistance is the most satisfactory and long range solution to control mites on cassava. Foliar application of urea followed by spraying of water at the beginning of mite infestation or spraying of dimethoate @ 0.05 per cent in severe situation suggested to contain the pest outbreak (Pillai and Palaniswami, 1985).

Thrips

Thrips (*Retithrips syriacus* (Mayet) another foliage pest is serious during dry periods when spider mites also occur and accentuate the problem. Infested leaves lose chlorophyll and turn pale with deposits of dark brownish excreta. Infestation starts on the older leaves and spread to the younger ones. The control measures adopted for spider mites will also take care of this pest (Pillai, 1994).

White Fly

Bemisia tabaci (Genn.) is the common white fly on cassava. It is more important as a vector of CMD than as a pest. It is abundant throughout the year Palaniswami *et al.* (2001) Lisha *et al.* (2003) identified two biotypes of *B. tabaci* and reported that only cassava biotype acted as a vector of cassava mosaic virus and not the other biotype *viz.*, sweet potato biotype. There were differences in the gel-banding pattern between the two biotypes. The ability of cassava biotype of transmit ICMV from cassava to cassava was confirmed through nested PCR primers and electron microscopy. *B. tabaci* feeding on cassava leaves induced pathogenesis related proteins (Binu and Palaniswami, 2006a).

Aphelimid parasitoids found often regulating the field population with 6.2 to 12.1 per cent parasitoids (Palaniswami *et al.*, 2001). Parasitoids recorded on *B. tabaci* include *Encarsia transvena* (Timberlake), *E. bimacualata* (Heraty and Polazek), *E. nigriphala* (Dozier), *E. strenua* (Silvestris), *E. haitiensis* (meritoria) (Dozier), *E. Guadeloupe* (Viggiani), *E. azimi* (Hayat), *E. pergandiella* (Howarl), *Eretmocerus mundvls* (Mercet) and *Eretmocerus* sp. (Antony *et al.*, 2006b).

Spiral White Fly

Spiral white fly *Aleurodicus disperses* Russel was found to seriously infest cassava since 1993. It is highly polyphagous in nature. It causes yellowish specks of leaves, resulting in crinkling and curling and sooty mould development. Adults are much larger than *B. tabaci* (2 mm in size). Severity of damage increases during summer. The rainfall had deleterious effect on the pest (Palaniswami, 1999).

12.1.3 Pests Attacking Setts and Seedlings

Termites

Termites (*Odontotermes obesus*) attack the planted cuttings and young seedlings resulting in the poor establishment. The pest problem becomes serious in areas with prolonged dry spell. In certain periods the seedlings mortality ranged from 20 to 40 per cent. It is a sporadic and localized pest. In the case of severe incidence, soil application of insecticides may be resorted to (Palaniswami, 1999).

White Grubs

White grubs (*Leucopholis coneophora* Burm.) are polyphagous and are important in certain localities. The attacked plants become pale, wither and dry up. The grubs feed on the developing roots and on the bank of the planted stems. Eggs are laid in the soil and first instar grubs which emerge out first feed on the organic matter and then they migrate to the cassava roots and cause damage. The pest can be effectively controlled through collection of adult beetles from harbouring plants and trees, deep ploughing to expose the grubs and soil application of insecticides (Palaniswami, 1999).

12.1.4 Pest Attacking Tuber

Rats

Rats are serious and wide spread pest on tubers. The enormity of the damage to tubers is not realized until harvest and attack is often not recognized until lodging occurs because most of the activity is underground. Field rat (*Bandicota bengalensis*), Indian gerbil (*Tatera indica*) and field mouse (*Mus booduga*) are important species noted in the fields, while *Rattus rattus* (house rat) and *Mus muysculus* (house mouse) are problems in the houses and godowns.

Bandicota bengalensis also known as Indian mole rat or lesser bandicoot makes extensive and deep burrows which extent up to 30 m with a depth of 1 m. Each burrow had 1-3 food cum bed chambers with 6-10 outlets. Male and female live separately. Usually only one adult per burrow is observed. They plug the burrow openings with heaps of mud. On an average there are 13 live burrows per ha. The Indian gerbil is nocturnal and live in groups. Burrows are simple but deep (1-2 m) with 3 or more openings and these burrows are usually found in or nearby bunds covered with weeds and bushes. The field mouse lives in shallow burrows with only one or two openings.

The extent and intensity of damage to cassava was assessed and the actual tuber loss was found in the range of 1 to 45 per cent with a mean of 9.4 per cent in upland conditions. The percent of damage done to the plants was found to be in the range of 2 to 90 per cent with a mean of 18 per cent (Rajamma *et al.*, 1989).

12.1.5 Pests of Stored Chips

Among a dozen insets that infest cassava chips in store, the most serious one is the beetle *Araecerus fasciculatus* Da Geer. Others like *Denoderus bifoveolatus, Tribolium castaneum, Lasioderma serricorne* also damage chips. Storing chips in parboiled form is an effective method to protect from these pests (Pillai, 1994).

12.2.0 Diseases of Cassava

White fly transmitted Cassava Mosaic Virus (CMD) is the most common and causes devastation in cassava cultivation worldwide. CMD alone causes heavy loss of the storage roots and Ivory Coast had witnessed annual losses to 500,000t of roots compared to the annual production of 8000, 000t in 1988. The recent pandemic of CMD in 1999 as noticed in Uganda and Kenya and the estimated loss was the worth of US$ 60 and US$ 10 million in Uganda and Kenya respectively. Several strains of CMD from African continent have been notified which manifest variable symptoms. Besides Cassava Mosaic virus, several minor virus diseases also have the evident to attack this crop. The details of the virus diseases are described in the Table 6.

12.2.1 Cassava Mosaic Disease

Cassava mosaic disease is caused by Indian cassava mosaic geminivirus. The symptoms are chlorotic areas intermixing with normal green tissue which gives mosaic pattern. In severe cases leaves are reduced in size, twisted and distorted. It affects the growth of the plant resulting in significant reduction in plant height,

Pests of Cassava

White flies (Left) on Leaves, Pupa and Adults (Right) of *Bemisia tabaci*

Spiraling White fly Infestation on Cassava Leaf

White Scale Infestation on Stem of Cassava

Pests of Cassava

White Peach Scale Infestation on Cassava Stems

Cassava Black Scale Infestation on Leaf and Stem

Mealy Bug Infestation on Cassava Leaf and Stem

Pests of Cassava

Immature Nymph of Thrips and Damage to Leaf

White Grub Larva, Adult Beetle, and Damage to Cassava Tuber

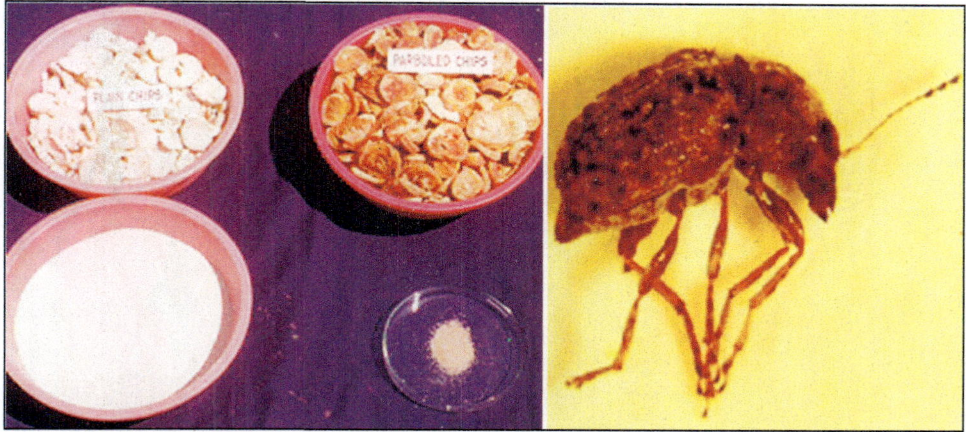

Storage Pests of Cassava

Pests of Cassava

Red Spider Mite Infestation on Leaves of Cassava

Green Mite on Cassava

Termite Nest

stem girth, petiole length and leaf size. It causes 25-80 per cent yield reduction depending up on the varieties and grade of ICMV infection (Thresh *et al.*, 1998).

Table 6: Virus Diseases of Cassava

Africa	
African cassava mosaic virus	Geminiviridae: begomo virus
East African cassava mosaic virus	Geminiviridae: begomo virus
South African cassava mosaic virus	Geminiviridae: begomo virus
Cassava brown streak	Potyviridae: Ipomovirus
Cassava Ivorian bacilliform virus	Unassigned
Cassava kumi virus	Unassigned
Cassava ' Q' virus	Unassigned
Cassava common mosaic virus	Potexvirus
South/Central America	
Cassava common mosaic virus	Potexvirus
Cassava virus X	Potexvirus
Cassava vein mosaic virus	Caulimoviridae
Cassava Colombian symptom less virus	Potex virus
Cassava American latent virus	Comoviridae:Nepovirus
Cassava frog skin virus	
Asia/Pacific	
Cassava common mosaic virus	Potex virus
Indian Cassava mosaic virus	Geminiviridae: begomovirus
Cassava green mottle virus	Comoviridae: Nepovirus

Primary spread of the disease was affected through the indiscriminate use of infected planting material and the secondary spread is through the vector whitefly *Bemisia tabaci* G.-cassava biotype (Lisha *et al.*, 2003). Studies on true seed transmission indicated that the disease was not transmitted through true seeds. Sap transmission of the disease from cassava to cassava could not be achieved. However, the disease could be transmitted through sap from cassava to *Nicotianna benthamiann* and *N. glutinosa*. CMD could be easily transmitted through grafting also.

The disease causing agent (virus) was purified and found to belong to the group Gemini viruses. The particles measured 18-24 nm in diameter. The antisera with adequate titre strength, which can detect successfully even latent infection of the diseases, have also been developed (Thresh *et al.*, 1998; Antony *et al.*, 2006b). Nested primer and few sequencing primers were designed from AUI coat protein (CP) gene of ICMV. This nested primer when used in the diagnosis of ICMV infection in different varieties of cassava was found to be precise and useful (Antony *et al.*, 2006b).

The symptoms of the disease were masked during dry and hot months and make it impossible to identify the diseased plants. Therefore, the disease free planting materials should be selected before the beginning of the dry/hot season.

Natural occurrence of cassava Mosaic Virus in farmers' field of West Bengal was recorded 61-92 per cent incidence and variable symptom with mild chlorotic pattern to severe mosaic and distortion of leaf was observed. The incidence and severity of CMD among eight cassava varieties *viz.* H-5/78, Sree Jaya, Sree vijaya, Sree Prakash, H-118, H-165, CL-590, H-119 was screened and differential response was noticed among the varieties in the experimental farm of ICAR-AICRP on Tuber Crops, BCKV, West Bengal. The nucleotide sequence analysis of PCR product (FN691429) from the isolates of West Bengal showed >90 per cent homology with the Sri Lankan Cassava Mosaic Virus (SLCMV) isolates of India but exhibited somewhat distant relation to SLCMV isolates of Colombo and ICMV (Adhikary *et al.*, 2018).

No cassava variety is immune to CMD. Even the related species *viz.*, *M. glaziovi* is susceptible to artificial inoculation. However, a great variation in the degree of susceptibility was observed. Few field resistant/tolerant lines such as M4, H 97, H 165, H 2304, H 1687, S 1315, S 1310, S 2381, S 2380 MNga^{-1} were identified for cultivation.

The integrated management of CMD including selection of disease free planting material, use of field tolerant varieties like H-97, H-165 and Sree Visakham, selection of disease free meristem derived planting material followed by clonal multiplication with periodical screening and rouging of freshly infected plants, will be useful to raise a good disease free crop. The disease free planting materials can be multiplied in large scale at higher altitude where cassava whitefly population is low. Raising the plants in the nursery at closer spacing prior to transplantation into the main field is an useful step to prevent the primary spread of the disease in the main field. Adherence to strict sanitary practices such as timely harvest and prompt disposal of crop residues, and eradication of self sown plants and weeds which may harbor both the disease and vectors (Thresh *et al.*, 1998).

Development of transgenic plants against CMV through ICMV replicate mediated resistance is in progress. Use of barrier crops around cassava crop and spray of antiviral plant extracts of *Boerhavin diffusa* and *Mirabibis jalapa* were found to be useful in containing the ICMV (Palaniswami *et al.*, 2004).

12.2.2 Cassava Tuber Rot

Cassava tuber rot is caused by *Phytophthora palmivora*. There are no external symptoms seen. Infected tubers show brown discoloration of internal tissues, rotten and exhibit foul smell. Tubers become unfit for consumption or marketing. Causes heavy yield loss when it occurs. Integrated management for tuber rot recommended include the following. Provide proper drainage in the field. Remove infected tubers from field and burn. *Trichoderma viridae* is a good biological control agent and can be incorporated in the problem areas (Thankappan, 1994).

In case of severe incidence, deep ploughing, planting cassava on ridges and incorporation of neem cake (@ 500 kg ha^{-1}) and *Tricoderma viridae* (1 kg culture in 100 kg FYM) with well drained conditions could reduce the tuber rot and increase the yield.

Table 7: Leaf Spot Diseases of Cassava Caused by Several Fungi

Disease	*Causative Organism*
Brown leaf spot	*Cercospora henningsii* Allescher
White leaf spot	*Cercospora caribaea* Chuppa and Ciferri
Snow white leaf spot	*Cercospora manihotae* Viegas
Diffuse leaf spot	*Cercospora vicosae* Muller and Chupp
Phyllosticta leaf spot	*Phyllosticta manihotae* Viegas
Leaf spot	*Periconia shyamala* Roy
Leaf spot	*Epicoccum nigrum* Link

Severe spotting causes premature defoliation resulting in significant loss of leaf area available for photosynthesis. Some of the simple inexpensive cultural practices, if followed at the right time can reduce the severity of the disease and the loss caused by it.

Wider spacing: Cultural practices such as wider spacing would reduce excess humidity in the crop stand, and thereby the multiplication and spread of the disease. Removal of severely infected leaves and burning them with a view to reduce inoculums source has also been recommended. H-97, Sree Visakam and Sree Sahya are moderately resistant, and Sree Prakash is highly resistant to this disease. Spraying of Bordeaux mixture (1 per cent) has been recommended. However, spraying of fungicides should be recommended only when it is absolutely essential and with caution.

Phomopsis leaf spot disease on cassava has been reported from Kanpur (U.P.), Banalore (Karnataka) and Kerala. Two species of *Phomopsis viz., P. manihotis* Swarup, Chauhan and Tripathi, and *P. manihot* (Speg.) Chavaugeon have been recognized as the causal organisms of the disease.

Phomopsis blight is widespread throughout the cooler cassava growing areas at high altitudes or in the lowlands during the rainy season. Large, water soaked spots on young leaves, often at the tips and margins. They are pale green at first, rapidly enlarging and turning brown. Spore-containing pycnidia are formed in concentric rings on the upper leaf surface or on stems. Severe attack leads to defoliation, stem girdling and dieback. It is caused by *Phomopsis manihotis*. Meristem-tip culture eliminates the disease.

12.2.3 Disease Affecting Stem and Tuber

Sett Rot (Diplodia natalensis P.)

Sett rot is an important and wide spread disease of cassava found in the stored stems as well as in the new plantings in drier areas where planting is done after irrigation. Rotting of stems (setts) results in failure of germination and gap formation in the field. Sett rot is caused by the pathogen *Diplodia natalensis* Pole-Evans. (*Botryodiplodia theobromae* Fab.)

Use of healthy planting material is the best means of control. Fungicides, such as benlate (0.1 per cent), bavistin (0.1 per cent), vitavax (0.1 per cent) and thiram (0.2 per cent) have been found to be very effective in checking the disease in field. The cuttings are to be dipped in one of the fungicidal solutions for 15 minutes, dried in shades and then to be planted.

Dry Rot [Fusarium solanii (Mart).*]*

A dry rot of cassava tuber has been found to be prevalent in Tamil Nadu. The infected tubers are usually hard and the colour of the pulp instead of being white varies from brown to reddish brown and dark brown streaks are seen inside the tubers; one or more tubers are affected in each plant and the diseased tubers are unfit for marketing and consumption.

12.3.0 Phytoplasma like Disease

Witches' Broom

Cassava witches' broom is caused by phytoplasma. Disease has been detected in Vietnam, Thailand, Cambodia, Laos, Indonesia, and the Philippines. Three different types of symptoms have been reported: (1) stunting and excessive proliferation of branches; shoots have small leaves and shortened internodes, without distortion or chlorosis; (2) proliferation of shoots from the cutting with generally weak growth; (3) a few weak, stunted shoots germinate from the cutting which never reach normal size (Costa and Kitajima, 1972). Moderate temperatures (between 13 and 20 °C) favour the disease, and at higher temperatures the symptoms disappear.

Frog Skin Disease

Cassava frog skin disease (CFSD) is an important disease affecting cassava roots caused by the phytoplasma. Symptoms disappearance in leaves was obtained by growing cassava-infected plantlets in a chloroxytetracycline solution producing further evidence for the phytoplasma etiology. It is characterized by the thickening of the root epidermis, thin tubers, the irregular swelling and easy breakage of the periderm and thick, cracked, wrinkled skin. The whitefly (*Bemisia tuberculata*) is suspected to be the vector of CFSD. This disease is also transmitted by cuttings, grafting and dodder. Healthy planting material should be used for planting. Thermotherapy followed by meristem tip culture also eliminates the disease.

12.4.0 Root-Knot Nematodes

Meloidogyne incognita, M. javanica, and *M. arenaria* have been reported from cassava from Brazil, Colombia, Dominican Republic, Hawaii, India, Ivory Coast, Nigeria, Rhodesia, Tanzania, Togo, Trinidad, and the United States. Nirula (1963) reported *M. javanica* for the first time from India. *Meloidogyne* spp. feeding produces physically damaging galls with eggs inside them. Galls later merge as the females grow and enlarge, and they interfere with water and nutrient supply (Gapasin, 1980). Cassava roots become tough with age and restrict the movement of the juveniles and the egg release. It is therefore possible that extensive galling can be observed even at low densities following infection (Coyne, 1994).

Diseases of Cassava

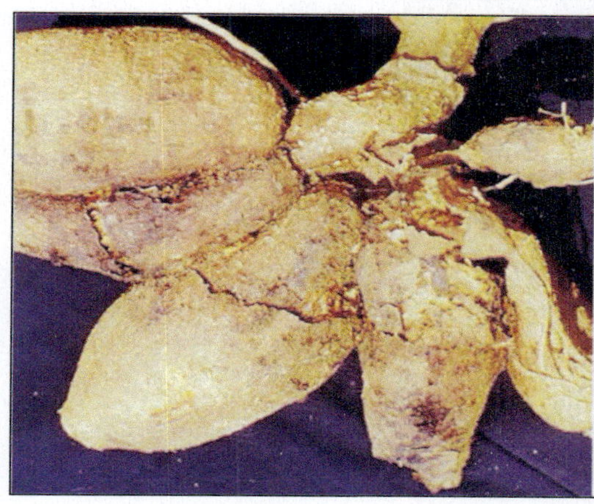

Cassava Tuber Rot

Brown Leaf Spot (Left) and White Leaf Spot (Right) on Leaves

Cassava Plants with Bud Necrosis Lesions on Stem

Cassava Mosaic Disease (CMD)

Symptoms of Cassava Mosaic Disease (CMD) Infection on Leaves (Mosaic and leaf distortion)

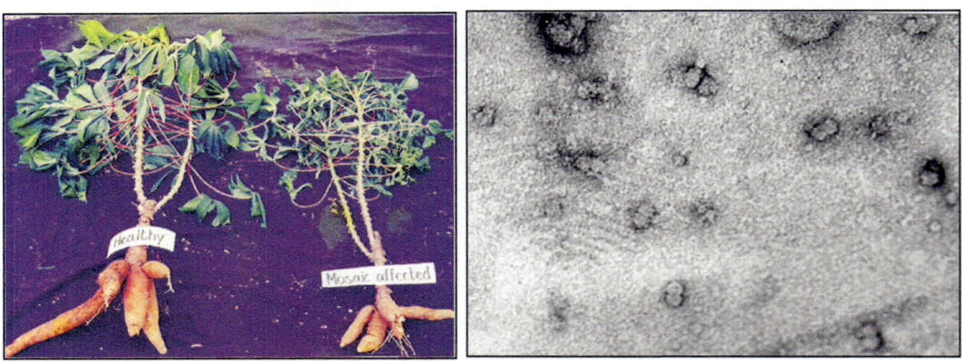

Reduction in Tuber Size Caused by CMD Virus Particles of Indian CMD

Cassava Brown Streak Disease (CBSD)

(a) Dark brown lesions on the green portion of the stem of a severely CBSD-affected cassava plant. (b) Drying and death of axillary buds on woody stem tissue. (c) Dieback of shoot tips. (d) Feathery chlorosis. Irregular chlorosis associated with secondary and tertiary veins. (e) Dry brown necrotic rot in tuberous roots characteristic of CBSD in mature cassava plants. (f) Healthy foliage of a CBSD-resistant cultivar (Legg *et al.,* 2015).

Diseases and Nematodes in Cassava Plants

Cassava Witches' Broom

Frog Skin Disease on Cassava Tubers

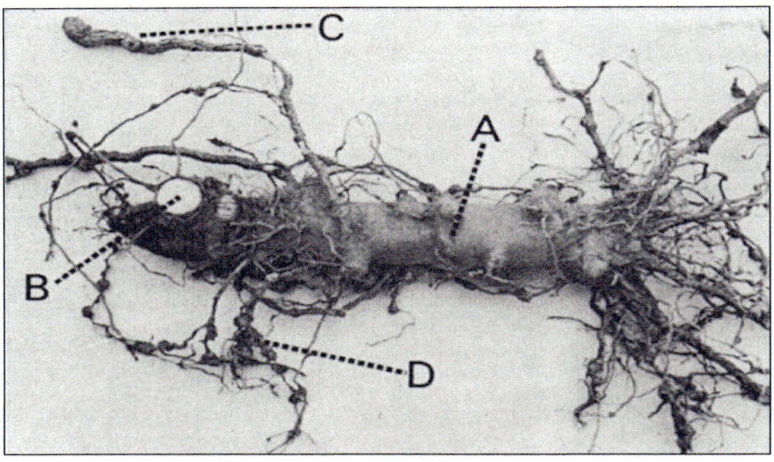

Root-knot Nematode in Cassava

13.0 Crop Improvement

In India, crop improvement work on cassava was taken up at the erstwhile Travan-core University in 1940. Ever since the Central Tuber Crop Research Institute was established in Trivandrum in 1963, significant achievements in varietal improvement of cassava were made. The main cassava breeding methods employed at CTCRI were (1) selection from indigenous and exotic germplasm, and (2) intervarietal hybridization.

13.1.0 Selection from the Germplasm

From a meagre collection of 75 types in 1950, the germplasm has grown to 1655 consisting of 806 indigenous types, 781 exotics, 60 hybrids and 8 wild species. The evaluation of the introduced clones in the earlier years by the Botany Department of the former Travancore University, resulted in the selection of M-4 which is popular even today due to its excellent culinary qualities. Another selection 'Sree Prakash' from the germplasm (accession S-856) was identified and released for its early maturing habit, coupled with good yield and acceptable qualities. Recently two selections, Sree Jaya and Sree Vijaya, were identified and released for their early maturing habit coupled with their suitability for low land cultivation as a rotation in paddy based cropping systems.

13.2.0 Intervarietal Hybridization

As a result of large scale intervarietal hybridization of selected parents from the germplasm, CTCRI produced and released the hybrid clones, H-97, H-165 and H-226 in 1971. Later, two high yielding hybrids Sree Visakham (H-1687) and Sree Sahya (H-2304) were released in 1974 (Jos *et al.*, 1980). The characteristics of these cultivars have been described earlier.

13.3.0 Polyploidy

Production of induced tetraploids and triploids constitute another approach to cassava breeding programme. Though induced tetraploids did not show superior yield performance, the triploids developed by crossing tetraploids with diploids have shown superiority over tetraploids (Josh *et al.*, 1987; Sree Kumari *et al.*, 1996). Work in the triploidy breeding at CTCRI resulted in the recent release of a variety, Sree Harsha, the characteristics of which have been discussed earlier (Sree Kumari *et al.*, 1999).

Many clones of polyploid cassava have been developed at IITA through bilateral and unilateral polyploidization. Some of the polyploids are very vigorous. Several tetraploids have performed well and some triploids out-yielded by over 200 per cent, the best improved varieties, indicating that triploids are more promising than tetraploids (Hahn *et al.*, 1994).

Spontaneous somatic mutants with varying ploidy level occur naturally but they have insufficient vigour for direct use as cultivars, but are of potential use in plant breeding (Hahn *et al.*, 1993). A somatic polyploidization system is proposed for implementation in cassava breeding programmes (Awoleye *et al.*, 1994). Hahn *et al.* (1990) made interspecific crosses of 5 diploid cultivars (2n=36) using 2 related diploid species, *Manihot epruinosa* and *M. glaziovii*. From these interspecific crosses, 4 spontaneous tetraploids (2n=4x=72) and 2 triploids (2n=3x=54) were isolated. These spontaneous polyploids broaden the range of genetic variation and offer opportunities in breeding. With a view to produce higher yielding varieties of cassava, Madeley (1993) crossed the IITA variety TMS30572 with a wild species. Spontaneous polyploids with up to 4 times the normal number of chromosomes were obtained. More than 200 triploids and tetraploids have been generated through open and controlled pollination and 2 spontaneous tetraploids have been adopted. In on-station trials some of the triploids give yields of 50-70 t/ha compared with about 23 t for TMS30572. Nassar (1992) selfed the natural hybrid of *Manihot pseudoglaziovii* and cassava and obtained a triploid from among the progeny and evaluation for agronomic traits showed high productivity of roots (2.13 vs. 1.67 kg plant for cv. Sonoro) and resistance to stem borers under semi-arid conditions.

13.4.0 Mutation

Radiation has been used to increase genetic variability in cassava. As a result of this, short petioled mutant and a chlorophyll-mutant are being evaluated at CTCRI (Nayar and Rajendran, 1985).

Crop Improvement programme on cassava is in progress in some other countries also. At least 28 countries are known to be maintaining cassava germplasm. The largest of these collections are at CIAT, Colombia (5035 accessions), EMBRAPA, Brazil (1960 accessions), IITA, Nigeraia (1286 accessions) and Indonesia (700 accessions). In most of the places, cassava germplasm is maintained in field plots. It has been reported that storing seeds at 5°C and 60 per cent RH can keep the seeds viable up to 7 years (Anon., 1979). Recently, techniques have been developed for the *in vitro* preservation of cassava genetic stocks, mainly by meristem culture. Present day varietal improvement programme in cassava mainly aims at: (1) high yielding ability under low and medium input conditions, (2) higher root dry matter content with low HCN, (3) early harvest ability, (4) tolerance to major pests and diseases, (5) drought resistance, and (6) shade tolerance for intercropping (Nayar *et al.*, 1988; Kawano, 1988). Vasudevan and Jos (1991) used two techniques to propagate 12-13 cm stem cuttings of cv. M-4 that had been exposed to 1-4 kR gamma radiation, the stake (standard) and the new single node cutting (SNC) methods. For the latter, stakes were cut into 7-8 single node cuttings before field planting. After 6 months plants were pruned leaving 4-5 axillary buds which developed secondary shoots (considered to be equivalent to the MIV2). Cut stems were also used to raise an MIV2 generation. Mutations were screened in MIV1 and M1V2 shoots. The SNC method led to the recovery of a total of 54 morphological mutants compared to 8 using the standard method.

14.0 Biotechnology

Unnikrishnan and Sheela (2001) reviewed researches on biotechnology of tropical root and tuber crops which provided comprehensive information in various aspects and achievements reported on the subject.

14.1.0 Tissue Culture

14.1.1 Micropropagation

Viruses and other systemic pathogens restrict the international exchange and conservation of genetic resources. Breeding for resistance to viruses is generally difficult in vegetatively propagated crops and rehabilitation of valuable clones is impossible through conventional methods.

In cassava, Berbee *et al.* (1973) produced *in vitro* cultures and symptom-free plants using explants from shoot-tip. Later, many workers successfully established cassava meristem cultures, multiplied through micropropagation and produced virus-free plants (Kartha and Gamborg, 1975; Smith *et al.*, 1986, Osena *et al.*, 2017). Nodal explants as well as apical buds also could be used for initiation of cultures (Unnikrishnan *et al.*, 1991). Basal medium found effective was that of MS with varying levels of growth regulators, sucrose, thiamine and inositol. Multiple shoot cultures could be produced from meristems by increasing concentration of BA from 2.5 to 5.0 μM (Roca *et al.*, 1984). Multiple shoot cultures could also be produced from nodal explants with NAA (0.25 μM) and BA (1.0 μM). IBA (2.5 μM) was found optimal for inducing roots in shoots grown on multiplication medium (Smith *et al.*, 1986).

14.1.2 Embryo Culture

Fertility and seed viability were found frequently low in cassava breeding programme. Immature zygotic embryos were cultured on half-strength MS medium to recover interspecific hybrid plants of *Manihot esculenta* x *M. aesculifolia*, 35-day-old seeds (after pollination) showed better development and seedling production (Catano *et al.*, 1993).

14.1.3 Anther and Pollen Culture

Callus and root developed on MS medium containing NAA, BA and coconut water when floral buds of 1.5 to 2.5 mm size having microspores of late tetrad to uninucleate stage were inoculated. Chromosome counts of root-tips indicated haploidy (Roca *et al.*, 1984). Callus induction was observed at uninucleate as well as binucleate stage of the pollen and embryogenesis recorded in isolated binucleate pollen culture in 2,4-D incorporated MS medium. Response was found to be influenced by genotype, anther stage and media. Regenerated shoots, however, could not be rooted to realise whole plant (Mukherjee *et al.*, 1989).

14.1.4 Protoplast Culture and Fusion

Techniques for isolation and culture of cassava protoplasts have been described by Szabados *et al.*, in the report of CIAT (Anon., 1987). Protoplasts were isolated from leaf lamina, shoot-tip and somatic embryos. Enzyme mixture of Oriozyka riocellulase, hemicellulase and pectolyase were used. Protoplasts were cultured in higher concentration of sucrose (0.35 M) and CaCl2 (600-900 mg/l). Kao's medium or B5 medium was found suitable. The protoplast-derived FEC was first purified through two rounds of selection of 3 weeks each before being cultured for regeneration of plants. This was done by culturing the protoplast-derived FEC for 11

In vitro Regeneration of Cassava

In vitro **regeneration of cassava using axillary bud explants (A) Induced axillary bud from nodal explant after 10 days of culture on CAM in darkness, (B) Multiple shoots forming from axillary buds after three weeks of culture on CAM, (C) shoots in CBM medium, (D) rooted plantlets in CBM medium after three weeks of culture, (E) acclimatized axillary bud regenerated plant of cultivar TMS60444 in a glasshouse, and (F) acclimatized axillary bud regenerated plant of cultivar Kibandameno in the glasshouse (Osena *et al.,* 2017).**

weeks on maturation medium, yielding a maximum of 184 organized embryos per 10,000 initially cultured protoplasts. Mature secondary somatic embryos developed into shoots in a medium with 1 mg/l BA (Sofiari *et al.*, 1998).

14.1.5 Somatic Embryogenesis

Regeneration through direct somatic embryogenesis has been obtained from cotyledons, embryonic axes of seeds or young leaflets (Li *et al.*, 1996; Szabados *et al.*, 1987; Raemakers *et al.*, 1993). In cassava an improved method for somatic embryogenesis involving optimization of maturation of the embryoids and incorporation of a desiccation stage prior to induction of germination was reported by Mathew *et al.* (1993). Continuous multiple shoot development was obtained from nodal explants grown on MS medium with high levels of 2,4-D (10 mg/l). Somatic embryos developed from explants of immature leaves grown on this medium which later produced plantlets at lower levels of 2,4-D (0.1 mg/l) and GA_3 (1 mg/l) (Cabral *et al.*, 1993). Picloram was more effective than 2,4-D in inducing somatic embryogenesis from leaf-lobes of 15 African cassava genotypes.

14.1.6 Somaclonal Variation

Cassava is tolerant to acid soils. But it is sensitive to high pH and the associated problems of salinity, alkalinity, poor drainage and micronutrient deficiencies (Roca, 1990). It is possible to select stress tolerant lines through cell suspension cultures using selective media. Studies are in progress at the South China Institute of Botany, Gangshu, China, on the utilization of transformed and somaclonal variant cultures for isolating cold tolerant lines in cassava (Li *et al.*, 1996).

14.2.0 Biochemical and Molecular Markers

The varieties could be distinguished by sequential classification on the basis of the isozyme patterns of acid phosphatase, esterase, glutamate oxaloacetate, transaminase and phosphoglucoisomerase. Out of these, esterase was stable isozyme system for cassava (Ramirez *et al.*, 1987).

Amplified fragment length polymorphism (AFLP) was used to study phylogeny of cassava and its taxonomic relationships with the other species of the genus *Manihot*. AFLP evaluation was done with six representative samples of the crops' diversity (core collection) and six wild taxa to estimate genetic relationship within the genus. The studies helped in revealing the unique genetic diversity in *M. esculenta* subsp. *flabellifolia* and *esculenta* subsp. *peruviana* and their genetic similarity to cassava supporting the hypothesis that these two subspecies may be the ancestors of cassava (Roa *et al.*, 1997).

14.3.0 Genetic Transformation

In cassava, transformation of organized type of embryogenic callus cultures with particle bombardment or *Agrobacterium*-mediated method resulted in chimeric transformed embryos. Somatic embryos cocultivated with *Agrobacterium* produced transformed plants through adventitious shoot development (Raemakers *et al.*, 1997). Regeneration of transgenic cassava shoots and plants was achieved via organogenesis of somatic embryos from cotyledon explants after co-cultivation with *A.*

tumefaciens and selection on hygromycin or genticin. The transgenic nature of the shoots were confirmed by GUS activity visually (histochemically) and by Southern blot technique at molecular level (Li *et al.*, 1996).

Recently, Sarria *et al.* (2000) reported successful production of transgenic plants of cassava through *A. tumefaciens* strain CIAT 1182. They used somatic embryo-derived cotyledons as source of explants of the cultivar Mper 183. The transgenics showed resistance to the herbicide, Basta. This is the first report on the successful transfer of *bar* gene in cassava.

14.4.0 *In vitro* Conservation

14.4.1 Slow Growth during Storage

In vitro active gene banks have been set up in cassava using slow growth cultures at various cassava genetic resource centres in Africa, South America and Asia (Unnikrishnan and Sheela, 1998). Studies conducted with different accessions have shown that cultivars responded differently to low temperature. Storage temperature less than 18°C was determined to many cassava varieties. However, culture viability could be maintained at even low temperatures (10-15°C) as long as the illumination was also lowered to less than 500 lux (Unnikrishnan and Sheela, 2001).

Use of osmatic retardants like mannitol has been found effective only if sucrose was also present in the medium (Roca *et al.*, 1984; Unnikrishnan, 1996). Inclusion of activated charcoal reduced shoot growth rate and increased culture viability. Among the growth inhibitors, ABA (2.5 to 10.0 mg/l) was found to slow down growth. Culture growth of cassava was suppressed completely even at the lowest concentration of both 22°C and 30°C.

14.4.2 Cryopreservation

Cassava shoot-tips were frozen in droplets of medium over aluminium foil for the first time and healthy plants recovered after storage (Kartha *et al.*, 1982). Rate of recovery was enhanced by use of low concentration, semi-solid medium and direct immersion in liquid nitrogen (Escobar *et al.*, 1994).

15.0 References

Adhikary, N., Kumar, M. and Tarafdar, J. (2018) *Int. J. Curr. Microbiol. App. Sci.*, **7:** 2887-2893.

Anonymous (1979) *International Institute of Tropical Agriculture, Annual Report*, Ibadan, Nigeria.

Anonymous (1981) *Central Tuber Crops Research Institute, Annual Report*, Trivandrum, India.

Anonymous (1983) *Central Tuber Crops Research Institute*, Two decades of research, Trivandrum, India, p. 81.

Anonymous (1986) *Package of Practices Recommendations*, Kerala Agril. Univ., Trichur, India.

Anonymous (1987) *Techniques of isolation and culture of cassava protoplast*, CIAT, Cali, Colombia.

Anonymous (1989) *Centro International de Agricultura Tropical Cassava Programme, Annual Report*, Cali, Colombia.

Anonymous (1990) *Centro International de Agricultura Tropical Cassava Programme, Annual Report*, Cali, Colombia.

Anonymous (1998) *Annual report 1997-98.* Central Tuber Crops Research Institute, Trivandrum, Kerala, India.

Anonymous (2016) *Annual report.* AICRP on Tuber Crops, Central Tuber Crops Research Institute, Trivandrum, Kerala, India.

Anonymous (2019) http://nhb.gov.in/statistics/State_Level/2018-19 per cent 20(3rd per cent 20Adv.Est_.) per cent 20- per cent 20Website.pdf

Aracena, J.J., Sargent, S.A., Brecht, J.K., Campbell, C.A. and Saltveit, M.E. (1993) *Acta Hortic.*, **343**: 297-299.

Ardjasa, W.S., Ando, H. and Kimura, M. (2001) *Soil Sci. Plant Nut.*, **47**: 101-202.

Asher, C.J. (1975) *Symptoms of Nutritional Disorders in Cassava*, University of Queensland St. Lucia, Australia.

Atthasampunna, P. (1992) In: *Cassava Breeding, Agronomy and Utilization Research in Asia, Proc. III Regional Workshop*, Malang, Indonesia (Ed. R.H. Howeler), Bangkok, Thailand, P. 444.

Awoleye, F., Duren, M. van., Dolezel, J. and Novak, F.J. (1994) *Euphytica*, **76**: 195-202.

Bellotti, A.C., Cardona, C. and Riis, L. (1997) In: *Soil invertebrates in 1997. Proc. 3rd Brisbane Workshop on Soil Invertebrates*, (Eds. P.G. Allsop, D.J. Rogers and L.N. Robertson), pp. 130-133.

Berbee, F.M., Berbee, J.G. and Hilderbrandt, A.C. (1973) *In vitro*, **8**: 421.

Bottema, J.W.T. and Henry, G. (1992) In: *Cassava Breeding, Agronomy and Utilization Research in Asia, Proc. III Regional Workshop*, Malang, Indonesia (Ed. R.H. Howeler), Bangkok, Thailand, P. 444.

Box, L. and Doorman, F. (1982) *Man and Manihot*, Vol.1, Wageningen.

Byrne, D. (1984) In: *Plant Breeding, Review 2* (Ed. J. Janick), Publishing Co. pp. 73-134.

Cabral, C.B., Aragao, F.J.L., Matsumoto, K., Neshich Monte, D.C. and Rech, E.L. (1993) *Proc. First International Scientific Meeting of the CNN*, Catagena, Colombia, August 25-28, 1992, Working document, **123**: 180-184.

Catano, M.L., Mornan, K., Plazas, J. and Roca, W.M. (1993) *Proc. First International Scientific Meeting of CBN*, Cartagena, Colombia, August 25-28, 1992, Working documents, **123**: 149-153.

Cock, J.H. (1985) *Cassava, new potential for a neglected crop*, West view press/Boulder and London, P. 190.

Cock, J.H. and El-Sharkawy, M.H. (1988) *Expt. Agric.*, **24**: 443-448.

Costa, A.S. and Kitajima, E.W. (1972) Studies on virus and mycoplasma diseases of the cassava plant in Brazil. In: Proceedings IDRC/IITA cassava mosaic workshop. International Institute of Tropical Agriculture, Ibadan, Nigeria.

Coyne, D.L. (1994) *Afr. Crop Sci. J.*, **2**: 355–359.

Damardjati, D.S., Widowati, S. and Dimyati, A. (1992) In: *Cassava Breeding, Agronomy and Utilization Research in Asia, Proc. III Regional Workshop*, Malang, Indonesia (Ed. R.H. Howeler), Bangkok, Thailand, P. 444.

Delobel, A. (1992) *J. African Zoo.*, **106**: 17-25.

Effendi, S. (1979) In: *Intercropping with cassava* (Ed. E. Weber) B. Nestel and Champbell, IDRC, Ottawa, pp. 35-36.

Ekpe, E.O. and Chinaka, C.C. (2000) *African Crop Sci. J.*, **8**: 311-316.

El Sharkawy, M.A. and Cadavid, L.F. (2000) *Expt. Agril.*, **36**: 323-334.

Ellis, R.H. and Roberts, E.H. (1979) *Ann. Bot.*, **44**: 677-684.

El-Sharkauy, M.A. and Cock, J.H. (1987) *Photosynthesis Res.*, **12**: 237-241.

El-Sharkauy, M.H., Lopez, Y. and Basidas, G. (1987) *Photosynthesis Research*, **12**: 237-241.

Escobar, R., Mafla, G. and Roca, W.M. (1994) *Proceedings Second International Scientific Meeting CBN II*, August 22-26, 1994, Bogor, Indonesia, Abst. P. 44.

Fagbayide, J.A., Olaiya, A., Gbadamosi, B. and Adetifa, S. (1997) *Rivista di Agricoltura Subtropicalee Tropicale*, **91**: 209-215.

FAO STAT (2019) *Food and Agriculture Organization of the United Nations, Statistical Database*, Statistical Division. Rome.

FAOSTAT (2010) *Food and Agriculture Organization of the United Nations, Statistical Database*, Statistical Division.

FAOSTAT (2013) *Food and Agriculture Organization of the United Nations, Statistical Database*, Statistical Division.

Florchinger, F.A., Leihner, D.E., Stemmuller, N., Muller Samann, K. and El Sharkawy, M.A. (2000) *J. Soil Water Cons., Ankeny*, **55**: 334-339.

Gabriel Filho, A., Pessoa, A.C. dos, S., Staohhaecker, L. and Helmich, J.J. (2000a) *Ciencia Rural.*, **30**: 953-957.

Gabriel Filho, A., Pessoa, A.C. dos. S. and Staohhaecker, L. (2000b) *Acta Scientiarum*, **22**: 1005-1009.

Gapasin, R.M. (1980) *Ann. Trop. Res.*, **2**: 49–53.

George, J.B. and Browne, C.B. (1994) *Tropical Sci.*, **34**: 161-165.

George, J. and Sunitha, S. (2018) *Tropical Tuber Crops*, Westville Publishing House, New Delhi.

George, J. and Sunitha, S. and Suresh Kumar, J. (2018) *Compendium of Tuber crops technologies*, ICAR-AICRP on Tuber Crops, Thiruvanthapuram, Kerala, India, pp. 18-20.

Ghosh, S.P., Nair, G.M., Prabhakar, M., Pillai, N.G., Mohankumar, B., Kabeerathumma, S., Ramanujam, T., Pillai, K.S., Thankappan, M., Lakshmi, K.R. and Pal, T.K. (1987) *Cassava based multiple cropping pattern system. Tech. Bulletin*, 6, CTCRI, Trivandrum, India.

Giron, C. and Alfonzo, E. (2000) *Agronomia Tropical Maracay*, **50**: 31-40.

Gonzalez, A., Ramos, D., Caloni, I.B. de and Figueroa, A. (1991) *J. Agril. Univ. Puerto Rico*, **75**: 269-280.

Hahn, S.K., Bai, K.V. and Asiedu, R. (1990) *Theo. Appl. Genet.*, **79**: 433-439.

Hahn, S.K., Bai, K.V. and Asiedu, R. (1993) *IITA Res.*, **7**: 11-14.

Hahn, S.K., Bai, K.V., Chukwuma, Asiedu, R., Dixion, A. and Ng, S.Y. (1994) *Acta Hortic.*, **380**: 102-109.

Hahn, S.K., Terry, E.R., Leusehner, I.O., Okoli, C. and Lal, R. (1979) *Field Crop Res.*, **2**: 193-226.

Harlon, J.R. (1971) *Science*, **174**: 468-474.

Henry, G. and Hershey, C. (1998) In: *CIAT. Cassava Breeding, Agronomy and Farmer Participatory Research in Asia. Proc. Fifth Regional Workshop*, China, November 3-8, 1996, pp. 3-20.

Howeler, R.H. (1985) In: *Cassava, CIAT*, Cali, Colombia.

Howeler, R.H. (1988) In: *Cassava Breeding, Agronomy and Utilization Research in Asia* (Ed. Howeler R.H. and Kawano, K.), Bangkok, Thailand, p. 350.

Howeler, R.H. (1992) In: *Cassava Breeding, Agronomy and Utilization Research in Asia, Proc. III Regional Workshop*, Malang, Indonesia, Bangkok, Thailand, p. 444.

Howeler, R.H. (1995) In: *Cassava Breeding, Agronomy Research and Technology Transfer in Asia. Proc. 4th Regional Workshop* (Ed. R.H. Howeler), Trivandrum, Kerala, India, November 2-6, 1993, pp. 447-464.

Howeler, R.H. (1996) In: *Tropical Tuber Crops, Problems, Prospects and Future Strategies* (Ed. Kurup, G.T.), Oxford and IBH, New Delhi, India, pp. 181-193.

Howeler, R.H. (1998) In: *Cassava Breeding, Agronomy and Farmer Participatory Research in Asia. Proc. Fifth Regional Workshop*, China, November 3-8, 1996, pp. 3-20.

Hrishi, N. and Nair, R.G. (1978) In: *Cassava Production Technology*, CTCRI, Trivandrum, Kerala, India, P. 54.

Indira, P. and Kurian, T. (1977) *J. Root Crops*, **3**: 29-32.

Jennings, D.L. (1976) In: *Evolution of Crop Plants* (Ed. N.W. Simmonds), Longman, London and New York, p. 814.

Jonsyn, F.E. (1989) *MIRCEN J. Appl. Microbiol. Biotechnol.*, **5**: 457-462.

Jos, J.S., Hrishi, N., Maini, S.B. and Nair, R.G. (1980) *J. Root Crops*, **6**: 1-6.

Josh, J.S., Vijaya, Bai, K. and Sreekumari, M.T. (1987) In: *Proc. Nat. Symp. Production and Utilization of Tropical Tuber Crops*, CTCRI, Trivandrum, India, pp. 7-19.

Kabeerathumma, S., Mohan Kumar, B., Mohan Kumar, C.R., Nair, P.G. and Pillai, N.G. (1988) In: *Proc. VIII Sym. ISTRC*, Bangkok, pp. 259-268.

Kallon, J.M., George, J.B., Amara, D.S., Gichuru, M.P., Nolte, C. and Ekanayake, I.J. (1997) In: *Soil fertility management in West African land use systems. Proc. Regional Workshop, Univ. Hohenheim, ICRISAT Sahelian Centre and INRAN*, Niamey, Niger, 4-8 March, 1997, pp. 259-265.

Kamalam, P., Rajendran, P.G. and Hrishi, N. (1977) *Trop. Agric.*, **54**: 213-217.

Kartha, K.K. and Gamborg, O.L. (1975) *Phytopathol.*, **66**: 826-828.

Kartha, K.K., Leung, N.L. and Mroginski (1982) *Z. Pflanzenphysiol*, **107**: 133-140.

Kawano, K. (1988) In: *Cassava Breeding and Agronomy Research in Asia* (Eds. R.H. Howeler and K. Kawano), Thailand, Bangkok, p. 350.

Kawano, K., Amaya, A., Daza, P. and Rios, M. (1978) *Crop Sci.*, **18**: 373-376.

Keating, B.A. and Evenson, J.P. (1979) *Field Crops Res.*, **2**: 241-251.

Legg, J.P., Kumar, L.P. and Kanju, E.E. (2015) In: *Virus Diseases of Tropical* and Subtropical Crops (Eds. P. Tennant and G. Fermin),CAB International, Wallingford, USA, pp.42-55.

Leihner, D. (1983) *Management and Evaluation of Intercropping Systems with Cassava*, CIAT, Cali, Colombia, p. 70.

Li, H.Q., Ma, G.H., Shu, J. and Gua, J.Y. (1996) *Fifth International Plant Cold Hardiness Seminar*, August 5-8, 1996, Cornvalis, Oregon, USA.

Limsila, J., Limsila, A., Watananonta, W. and Kawano, K. (1998) In: *CIAT. Cassava Breeding, Agronomy and Farmer Participatory Research in Asia. Proc. Fifth Regional Workshop*, China, November 3-8, 1996, pp. 48-54.

Loganathan, S., Ravi Kumar, V. and Krishnamoorthy, K.K. (1975) *Mod. Agri. J.*, **62**: 248-252.

Madeley, J. (1993) *Ceres Rome*, **25**: 5-6.

Maduakor, H.O. (1997) *Soil Tillage Res.*, **41**: 149-156.

Magoon, M.L., Krishnan, R. and Bai, K.V. (1969) *Cytologia*, **34**: 612-624.

Mandal, R.C., Singh, K.D. and Magoon, M.K. (1971) *Indian J. Agron.*, **16**: 449-452.

Mathew, Helena, Schopke, C., Carcamo, R., Chavarriaga, P., Fauquet, C. and Beachy, R.N. (1993) *Plant Cell Rep.*, **12**: 328-333.

Mitra, S. (2017) *Proceeding of the International Seminar on 'Agriculture and Food for Inclusive Growth and Development*, Lucknow, U.P., India.

Mitra, S. (2020) *A Compiled Report on Cassava* using data from FAOSTAT and other sources.

Mitra, S. and Tarafdar, J. (2013) In: *Postharvest Management of Tropical Tubers, BCKV, West Bengal*, India.

Mitra, S. and Tarafdar, J. (2016) *J. Agril. Engineer. Food Tech.*, **3**: 210-214.

Mohan Kumar, B. (1978) In: *Cassava Production Technology*, CTCRI, Trivandrum, pp. 17-20.

Mohan Kumar, B. and Nair, P.G. (1983) *J. Root Crops*, **9**: 16-20.

Mohan Kumar, C.R., Mandal, R.C. and Hrishi, N. (1976) In: *Proc. IV Symp. ISTRC*, Cali, Colombia, IDRC, Ottawa, Canada, pp. 122-124.

Mohan Kumar, C.R., Potty, V.P., Ravindran, C.S., Kabeerathumma, S. and Sudharmai Devi, C.S. (1998) In: *Cassava Breeding, Agronomy and Farmer Participatory Research in Asia. Proc. Fifth Regional Workshop*, China, November 3-8, 1996, pp. 3-20.

Mohan Kumar, C.R., Ravindran, C.S. and Nair, P.G. (1987) In: *Proc. Nat. Symp. Production and Utilization of Tropical Tuber Crops*, CTCRI, Trivandrum, India, pp. 99-102.

Moreno, R.A. (1991) In: *Proc. Expert Consultation on Roots, Tubers, Plantains, Bananas, Animal Feeding*, CIAT, Cali, Colombia.

Mukherjee, A., Unnikrishnan, M., Nair, N.G. and Easwari Amma, C.S. (1989) *J. Root Crops*, **15**: 133-138.

Nair, P.G. and Varghese, T. (1970) *Agric. Res. J. Kerala*, **8**: 14-16.

Nair, P.G., Mohan Kumar, B., Prabhakar, M. and Kabeerathumma, S. (1988) *J. Root Crops*, **14**: 1-9.

Nair, V.M. (1982) *Ph.D. Thesis*, Kerala Agril. Univ., Trichur, Kerala, India.

Nair, V.M. and Sadanandan, N. (1987) In: *Proc. Nat. Symp. on Production and Utilization Tropical Tuber Crops*, CTCRI, Trivandrum, pp. 89-92.

Nassar, N.M.A. (1978) *Econ. Bot.*, **32**: 311-320.

Nassar, N.M.A. (1992) *Field Crops Res.*, **30**: 173-182.

Nayar, G.G. and Rajendran, P.G. (1985) *Mutation Breeding Newsl.*, **30**: 18.

Nayar, G.G., Nair, R.B. and Rajendran, P.G. (1988) In: *Cassava Breeding and Agronomy Research in Asia* (Eds. R.H. Howeler and K. Kawano), Bangkok, Thailand, p. 350.

Nayar, T.V.R. (1986) *Ph. D. Thesis*, Kerala Agril. University, Trichur, Kerala, India.

Nayar, T.V.R. and Mohan Kumar, B. (1989) In: *Silver Jubilee Commemoration Volume 1963-88*, CTCRI, Trivandrum, Kerala, India, pp. 49-51.

Nayar, T.V.R. and Sadanandan, N. (1990) *J. Root Crops*, **16**: 103-109.

Nayar, T.V.R., Mohan Kumar, B. and Pillai, N.G. (1985) *J. Root Crops*, **11**: 37-44.

Nayar, T.V.R., Mohan Kumar, B. and Pillai, N.G. (1986) *J. Root Crops*, **12**: 67-75.

Nirula, K.K. (1963) *Curr. Sci.*, **32**: 221–222.

Ogazi, J.N., Omueti, J.A.I. and Moore, J.A. (2000) *Animal, agricultural and food processing wastes. Proc. Eighth Internatl. Symp.*, Des Moines, Iowa, USA, 9-11 October, 2000, pp. 640-647.

Osakwe, J.A., Ikpe, F.N., Adeniji, M.O. and Folorunso, D.O. (2000) *Indian J. Agril. Sci.*, **70**: 753-756.

Osena, G., Amugune, N. and Nyaboga E. (2017) *J. Plant Sci.*, doi:10.11648/j.jps.20170501.13.

Padmaja, G., Balagopalan, C., Kurup, G.T., Moorthy, S.N. and Nanda, S.K. (1992) In: *Cassava, Breeding, Agronomy and Utilization Research in Asia. Proc. III Regional Workshop* (Ed. R.H. Howeler), Malang, Indonesia, Bangkok, Thailand, p. 44.

Palaniswami,M.S and Tarafdar, J. (2008) In Book: Tuber and Root Crops(Horticultural series-9).Edited by Prof. K.V. Peter. New India Publishing Agency, pp 179-240.

Pillai, N.G., Mohan Kumar, B., Nair, P.G., Kabeerathumma, S. and Mohan Kumar, C.R. (1987) In: *Proc. Nat. Symp. on Production and Utilization of Tropical Tuber Crops*, CTCRI, Trivandrum, pp. 109-111.

Purseglove, J.W. (1968) *Tropical Crops: Dicotyledons*, Longman, London, p. 719.

Raemakers, C.J.J., Sofiari, E., Jacobsen, E. and Visser, R.G.F. (1993) *Euphytica*, **96**: 153-162.

Rajendran, N., Nair, P.G. and Kumar, B.M. (1976) *J. Root Crops*, **2**: 35-37.

Ramirez, H., Hussain, A., Roca, W. and Bushuk, W. (1987) *Euphytica*, **36**: 39-48.

Ravi, V. (1994) *J. Root Crops*, **20**: 85-88.

Ravi, V. and Nair, G.M. (1998) *J. Root Crops*, **24**: 85-90.

Renoize, B.S. (1973) *Econ. Bot.*, **26**: 352-360.

Roa, A.C., Maya, M.M., Duque, M.C., Tohme, J., Allen, A.C. and Bonierbala, M.W. (1997) *Theo. Appl. Genet.*, **95**: 741-750.

Roca, W.M. (1984) In: *Hand Book of Plant Cell Culture*, Vol. 2 (Sharp W.R., Evan, D.A., Ammirato, P.V. and Yamada, Y. eds.), McMillan, New York, pp. 269-301.

Roca, W.M. (1990) *Proc. International Symposium Organized jointly by the Technical Centre for Agricultural and Rural Co-Operation (CTA) and FAO*, held at Luxembourg, June 26-30, 1989, CTA, FAO, pp. 215-219.

Rogers, D.J. and Appan, S.G. (1973) *Monograph 13, Organization for Flora Neotropics*, P. 272.

Rouse, I. and Cruxent, J.M. (1963) *Venezuelan Archeology*, Yale.

Saraswat, V.N. and Chettiar, T.S. (1976) *Indian Agric.*, **20**: 23-26.

Sarria, R., Torres, E., Angel, F. and Chavarriage, P. (2000) *Plant Cell Rep.*, **19**: 339-344.

Shanmugam, A. and Thamburaj, S. (1975) *South Indian Hort.*, **23**: 62-63.

Shanmugavelu, K.G. (1973) *Indian J. Agric. Sci.*, **43**: 789-791.

Shuren, J. (1992) In: *Cassava Breeding Agronomy and Utilization Research Asia. Proc. III Regional Workshop* (Ed. R. H. Howeler), Malang, Indonesia, Bangkok, Thailand, p. 444.

Smith, M.K., Biggs, B.J. and Scott, K.J. (1986) *Plant Cell Tissue Organ Cult.*, **6**: 221-228.

Sofrari, E., Raemakers, C.J.M., Bergervoet, J.E.M., Jacobsen, E. and Visser, R.G.F. (1998) *Plant Cell Rep.*, **18**: 159-165.

Spath, C.D. (1973) *J. Steward Anth. Soc.* 5(1).

Sreekumari, M.T., Jos, J.S. and Kurup, G.T. (1996) In: *Tropical tuber crops: problems, prospects and future strategies*, pp. 95-99.

Sreekumari, M.T., Jos, J.S. and Nair, S.G. (1999) *Euphytica*, **106**: 1-6.

Szabados, L., Hoyos, R. and Roca, W.M. (1987) *Plant Cell Rep.*, **6:** 248-251.

Tan, H.W., Du, C.L. and Zhou, L.Q. (2000) *Better Crops Int.*, **14**: 13-15.

Tan, S.L. and Cock, J.H. (1979) *Field Crops Res.*, **2**: 281-289.

Taro, J.C. and Atlee, C.B. (1985) In: *Cassava*, (Eds. J.H. Cock and J.A. Keyes), CIAT, Cali, Colombia.

Thampan, P.K. (1979) *Cassava*, Kerala Agril. Univ. Press, Trichur, Kerala, India, p. 242.

Unnikrishnan, M. (1996) *International Meet on Tropical Tuber Crops - Tuber Crops in Food Security and Nutrition*, December 9-12, 1996, Trivandrum, Kerala, India, Abst. P. 37.

Unnikrishnan, M. and Sheela, M.N. (1998) *The cassava Biotechnology Network: 4th International Scientific Meeting*, Bahia, Brazil, November 3-7, 1998, Cali, Colombia, CIAT.

Unnikrishnan, M. and Sheela, M.N. (2001) *Biotechnology of Horticultural Crops*, Vol. 2 (Eds. V.A. Parthasarathy, T.K. Bose and P.C. Deka), Naya Prokash, Calcutta, India.

Unnikrishnan, M., Mukherjee, A. and Nair, N.G. (1991) In: *National Symposium Genetics and Biotechnology for Crop Improvement*, Osmania Univeristy Hyderabad, Symp. Abst. P. 61.

USDA (2019) *Cassava, raw. Food Data Central.* U.S. Department of Agriculture, National Nutrient data base.

Van Den, T., Palomar, L.S. and Amestoso, F.J. (1992) In: *Cassava Breeding, Agronomy and Utilization Research in Asia. Proc. III Regional Workshop*, (Ed. R.H. Howeler), Malang, Indonesia, Bangkok, Thailand, P. 444.

Varma, S.P., Suja, G. and Rajendran, P.G. (1997) In: Production Technology of cassava and sweet potato, CTCRI, Trivandrum, P.24.

Vasudevan, K. and Jos, J.S. (1991) *Mutation Breed. Newslett.*, **37**: 9-11.

Viegs, A.P. (1976) *Estudios sobrea Mandioca*, Instituto Agronomico do Estado de Sao Paulo, p. 504.

Wheatley, C. and Izquierdo, D. (1991) In: *Integrated Cassava Products, Working Document*, 78, CIAT, Cali, Colombia, p. 242.

Wholey, D.W. (1974) *Ph.D. Thesis*, University of West Indies, St. Augustine, Trinidad.

SWEET POTATO

S.P. Varma, S.K. Naskar, H. Sen, J. Tarafdar, and S. Mitra

1.0 Introduction

Sweet potato [*Ipomoea batatas* (L.) Lam.] is considered as the seventh most important food crop of the world after wheat, rice, maize, potato, barley and cassava (APEDA, 2007; Jan low *et al.*, 2015; CIP, 2017). Sweet potato is also considered to be the world's most important subsistence crop with high calorie and nutrition yield/unit area and exhibits great diversity in morphological and phenotypic traits, such as habit, leaf shape, and storage root flesh and skin colour. Sweet potato is a dependable and flexible crop grown worldwide under sustainable agriculture and ranks fourth internationally important crop in developing world due to its multipurpose uses ranging from a high energy food source for humans and animals to industrial products especially for starch and alcohol production. Sweet potatoes produce more edible energy (194 MJ ha^{-1} day^{-1}) than any other major food crops (Woolfe, 1992), are more productive within short periods of time on marginal lands, and play an important role in the economy of poor households (Nath *et al.*, 2007). It has proved to yield higher amounts of micronutrients and calories per unit area under low-input regimes than any other crop and designated as 'Sleeping Giant'. It is reported that a hectare of sweet potato can feed more people than a hectare of cereal grains, with less energy input. Sweet potato is a staple food in Africa, while in Asia, in addition to parts of the tuber is consumed; the young sweet potato leaves are also made as a vegetable. For all these reasons sweet potato has great possibilities for enhancing food and nutrition security in developing and underdeveloped countries where most of the farmers' holdings belong to the small and marginal categories (Mukhopadhyay *et al.*, 2011). There is also a sweet potato which is used as an ornamental plant for its beautiful leaves. India is the potential sweet potato growing country and is rich in diversity in its genetic stocks with valued agronomic and other desirable traits. Many heirloom elite sweet potato cultivars are being maintained in the field gene bank of the ICAR- All India Coordinated Research Centres on tuber crops and ICAR-CTCRI, Thiruvananthapuram.

2.0 Composition and Uses

2.1.0 Composition

Sweet potatoes are rich in starch which are small in size, resembles corn starch granules in shape rather than tuber starch granules. The important constituents of fresh tubers are carbohydrate, protein, fat, sugar, appreciable amount of vitamin A, B and C and some minerals (Horton, 1988) (Table 1). Quality of sweet potato is mostly dependent on its sugar content. It varies from 3-6 per cent with cultivars and conditions. Mostly, sucrose and a few reducing sugars are present. Maltose, mannose, galactose and pentose occur only in traces. Sugar percentage in sweet potato tubers increases during storage and cooking.

According to the U.S. Sweet Potato Council Inc., sweet potatoes are cooked with the skin produces more fibre than a serving of oatmeal. It also contains vitamin A, B, C, calcium and potassium which works to relieve inflammation of the stomach.

Based on USDA Nation Nutrient database, it has low calories in every 100 g baked sweet potato (in the skin) containing 90 calories or equal to 376 kJ.

Sweet potatoes are an excellent source of vitamin A (in the form of beta-carotene). They are also a very good source of vitamin C, manganese, copper, pantothenic acid and vitamin B6. Additionally, they are a good source of potassium, dietary fibre, niacin, vitamin B1, vitamin B2 and phosphorus. A 100 g sweet potato storage tuber contains total carbohydrate 20 g, dietary fibre 3 g, sugar 4.2 g and protein 1.6 g. Tuber weight was found to vary significantly from 48.11 g to 240.45 g among the different sweet potato cultivars. TSS and dry matter content in tubers also varied among cultivars ranging from 8° Brix - 10.2° Brix and 23.55 per cent - 30.54 per cent, respectively (Sur *et al.*, 2020).

Table 1: Macronutrients Composition of Sweet Potato (baked 200 g)*

Constituents	*Sweet Potato*	*Constituents*	*Sweet Potato*
Dry matter (per cent FW)	19-35	Amylase (per cent total starch)	8-32
Starch (per cent FW)	18-28	Ca (mg/100 g)	32
Total sugars (per cent FW)	1.5-5.0	P (mg/100 g)	47
Protein (per cent FW)	1.0-2.5	Fe (mg/100 g)	0.7
Fiber (per cent FW)	1.0	Na (mg/100 g)	10
Lipid (g)	0.4	K (kg/100 g)	243
Ash (mg)	1.0	Thiamine (mg/100 g)	0.09
Vitamin A (mg/100 g FW)	900	Riboflavin (mg/100 g)	0.06
Vitamin C (mg/100 g FW)	35	Niacin (mg/100 g)	0.6
Starch grain size (micron)	2-42	Food energy (k cal/100 g)	114

* Horton (1988)

Orange- and yellow-fleshed types have been recognized as good sources of β-carotene (Mitra *et al.*, 2010; Bechoff *et al.*, 2011; Burri, 2011) and both yellow and light fleshed types are good sources of vitamin C (Chattopadhyay *et al.*, 2002a,2002b). The orange-fleshed sweet potatoes (OFSP) are exceedingly rich in β-carotene ranged from 5-18 mg fresh weight basis. Current OFSP varieties contain high concentrations of β-carotene (up to 23,100 μg/100 g FW) and high retinol equivalents (Chattopadhyay *et al.*, 2006a; Bovell-Benjamin, 2007; Kidmose *et al.*, 2007; Ofori *et al.*, 2009), which appear to be more bio-available than those in dark, leafy greens (de Pee *et al.*, 1998). Such varieties could prevent vitamin A deficiency particularly in Sub-Saharan Africa and South Asia, causing over 600,000 deaths per year, mostly of young children and pregnant women (Burri, 2011) and the tribal-dominant areas of India (Burri, 2011; Nath *et al.*, 2005). The purple-fleshed sweet potatoes (PFSP) are outstanding sources of anthocyanins, especially peonidin and cyanidin. The total phenolic content ranged from 313.6 to 1486.7 mg chlorogenic acid equivalent/100 g fresh weight. Both types of sweet potatoes are rich in unique phytonutrients, including polysaccharide-related molecules called batatins and batatosides. Sweet potatoes also include storage proteins called sporamins that have unique antioxidant properties. Leaves of sweet potato also contain lutein, a type of carotenoid, which may help to prevent age-related macular degeneration and cataracts (Isghiguro and Yoshimoto, 2006).

2.2.0 Uses

Sweet potato provides some surprising health benefits and recent advanced research findings on sweet potatoes proved that they have so many unique nutritional benefits to offer. There are a surprising number of nutrient categories responsible for the health benefits of this underappreciated tuber. Among these categories are antioxidants, anti-inflammatory nutrients, and blood sugar-regulating nutrients.

It is eaten usually after boiling, baking and frying and maybe candied with syrup or used as 'puree'. Storage roots are used as the raw material for making liquid glucose, starch, sembei and alcohol in Japan; baby food, weaning food, candies, sauces, snacks, crackers, chips, presto-pi and fruit leather in the USA; transparent noodles in Indonesia; dangmyun, a glass noodle in Korea; Kapek in China, and as an ingredient in pastry filling in Sri Lanka; leather in Malaysia, Japan and China; reconstituted chips in China; and jam, jelly, pickles, sauces, sweets and soft drink in India (Padmaja, 2000; Udensi, 2000; Lee *et al.*, 2006; Reungmaneepaitoon, 2009; Adenuga, 2010). Among different recipes, sweet potato jam prepared with 900gm sugar performed best during storage with high sensory values (Sur *et al.*, 2020). The crop is also used for the production of fuel, textiles, paper, cosmetics, glucose, adhesives and potable alcohol (Wu and Bagby, 1987; Ukom *et al.*, 2009). TSS content in sweet potato sauce was found to remain stable like tomato sauce during storage with a mean value of 30.24° Brix due to standardization of TSS during the processing. Sweet potato sauce can be used as an alternative to conventional sauces available in the market (Devi *et al.*, 2017).

Sweet potato is the cheapest source of calories. In an attempt to increase the economic value of sweet potato the starchy roots have been processed into fruity food products, similar to other fruit and vegetable products (Santhi, 1995). The tender tops and leaves are used as a vegetable in Africa, Indonesia and the Philippines. Apart from a food crop, it is also utilized for feeding livestock. The vines serve as a good source of fodder for livestock.

Commonly available sweet potatoes are white-fleshed with red/pink-red and white coloured skin. Orange-fleshed sweet potatoes and sweet potato-based food products proved not only acceptable to both producers and consumers in terms of appearance, taste, and texture, but contributed to the alleviation of vitamin A deficiency. With high levels of polyphenolic components, purple-fleshed sweet potatoes (PFSP) have been utilized as a healthy food community and source of natural food colourants, where they are used in the production of purple flour and paste for making bread, snacks and noodles (Jusuf *et al.*, 2005; Bovell-Benjamin, 2007). Tai *et al.* (1998) opined that a product from the leaves is effective in raising immune function and controlling hyper cholesterolaemic symptoms. It provides very intriguing nutrient groups especially the PFSP have a unique character of free radical scavenging, antioxidant capacity, age-related degeneration, anti-carcinogenic, anti-ulcerate, and regulate cardiovascular disorder.

Uses of Sweet Potato

Pickles

Jam

French Fries

Squash

Gulab Jamon

Flour

Biscuit

Cookies

3.0 Area and Production

Globally sweet potato is cultivated in 117 countries in an area of 8.62 million ha producing 105.19 million tons with a yield of 12.20 t ha^{-1} (FAO, 2016). Africa is the world largest sweet potato growing region and a majority of its production about 95 per cent comes from developing countries, of which China has the maximum share of 67.09 per cent (FAO, 2016).

In India, it is cultivated in almost all the states but major contribution comes from four states namely Odisha, Kerala, West Bengal and Uttar Pradesh. Odisha is the largest producer of sweet potato in India. The area and production of sweet potato have been increased significantly in most of the sweet potato growing states in India. The other major growing states in India are Chhattisgarh, Karnataka, Maharashtra and Andhra Pradesh. Sweet potato is grown in almost all the States of India except Jammu and Kashmir. The area under sweet potato cultivation in India is 0.13 million ha with a production of 1.47 million tonnes (FAO, 2016).

India exports a significant amount of sweet potato tubers to several countries and the quantity of the export is increased from 2016-2017 (107.58 MT) to 2017-2018 (403.73 MT) and highest import countries are Nepal and Maldives.

4.0 Origin, Taxonomy and Morphology

4.1.0 Origin

Sweet potato is not found in a wild state. South America is considered to be the home of sweet potato. It was used as a food crop in prehistoric times in tropical America and Polynesia (Cooley, 1951; Leon, 1976). Then it spread to Melanesia and New Guinea. It spread into Polynesia by the Spanish which started from Peru in the 16th century (Leon, 1976). Sweet potato was cultivated in the coastal area of Peru where it was introduced possibly from the north, the coastal area of Ecuador and Columbia (Martin *et al.*, 1976). O'Brien (1972) stated its origin in Central America or the North-West of South America about 3000 B.C. The spread of sweet potato into Africa, North America, Europe, India, China, Japan, and the Philippines was the result of Spanish, Portuguese and British trade, exploration and colonization (O'Brien, 1972).

4.2.0 Taxonomy

Sweet potato [*Ipomoea batatas* (L.) Poir] belongs to the family Convolvulaceae. The family Convolvulaceae comprises 45 genera and 1000 species distributed throughout the tropical and warm temperate regions of the world. The genus *Ipomoea* has 650 species which are mostly tropical, few from warm temperate regions. *Ipomoea batatas*, the sweet potato is the only important edible rooted economic plant of the genus. It is a hexaploid with 90 somatic chromosomes. It is probably derived by amphidiploidy from a tetraploid (2n = 60) and a diploid (2n = 30) to produce a triploid (2n = 45), followed by a subsequent doubling of the chromosome to produce hexaploid (2n = 90). The species involved in allopolyploidization are diploid *Ipomoea triloba* (2n=30) and tetraploid *I. tiliacea* and *I. gracilis* (Purseglove, 1977). Sweet potato is thought to have originated from hexaploid *I. trifida* (Nishiyama,

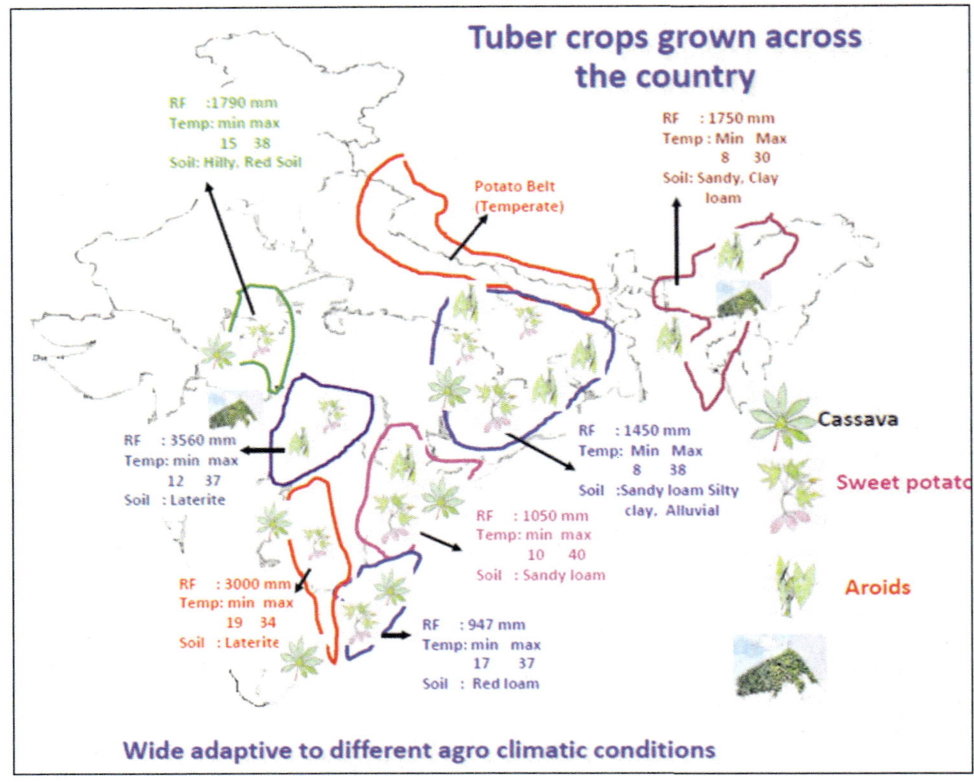

Distribution of Sweet Potato and other Tropical Tuber Crops in India

1961). Purseglove (1965) also considered *I. trifida* as the progenitor of sweet potato. Nishiyama (1971) established *I. trifida* to be the direct ancestor of sweet potato and its diploid predecessors were thought to be *I. leucantha* and *I. litoralis*. Martin (1972) agreed with Nishiyama on *I. trifida* and added two more closely resembled diploid species *I. lacunosa* and *I. trichocarpa*. Purseglove (1977) pointed out that the supposed ancestors do not have edible tubers but some have thickened roots. A hypothesis of the origin of the sweet potato is given with the conclusion that *I. triloba* and *I. tifida* is its closest extant relatives (Daniel *et al.*, 1987). Magoon *et al.* (1970) suggested that the parental genomes are from closely related taxa. According to Nishiyama *et al.* (1973), the evolutionary sequence of sweet potato is *I. leucantha* (BB), *I. litoralis* (BBBB), *I. trifida* (BBBBBB), *I. batatas* (BBBBBB).

4.3.0 Morphology

Sweet potato is a herbaceous perennial but cultivated as annual. The stem is a branching vine with milky juice. The vine is prostrate or ascending and twining. The vines develop roots at the joint during later part of the season. The vine is light green to purple in colour, usually angular in shape and slender to moderate in thickness. The triangular leaf is borne on short to the long petiole. Leaves are

Plant Canopy (Left), Flower (Middle) and Tubers (Right) of Sweet Potato

Types of Sweet Potato

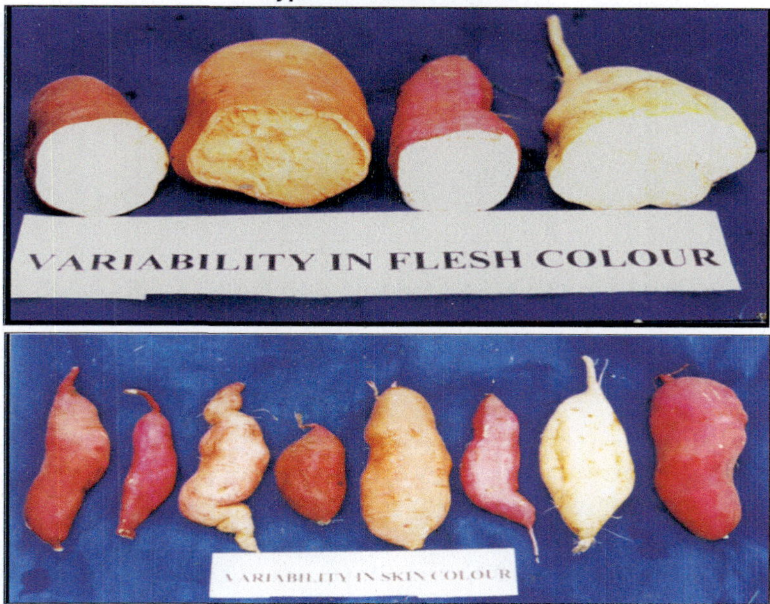

Variability in Tuber Shape, Size, Skin and Flesh Colours

more or less heart-shaped but may be shouldered, toothed, entire, deeply cleft or variously lobed. The surface of the leaf is glabrous or with variable pubescence. The leaves vary in colour from green to purple.

Tubers of different cultivars vary greatly in form, size and skin colour. The size of the tuber is an important attribute from the commercial standpoint. Too large or small a size of the tuber is undesirable to common consumers. The shape of tubers is fusiform and globular with a smooth or ridged surface. The skin colour is commonly white, yellow, orange, yellow-reddish or purple. Root tubers develop on nodes of vine cutting or vine which rests on the soil by the process of secondary thickening of underground adventitious roots. Tubers consist of conducting tissues, parenchymatous storage cells, latex vessels and outer periderm.

The inflorescence of sweet potato is cyme which bears single or few flowers. The corolla is funnel-shaped with 5 lobes and it is purple in colour with deeper throat and pale at the margin. There are five stamens which are attached near the base of the corolla. The stamens are unequal in length. The pistil has a 2 cm long style with 2 lobed stigmas. The seeds of sweet potato are borne on the capsule. A single capsule may contain up to 4 seeds but usually, 1 or 2 develop. The angular seeds are black in colour with hard seed-coat.

5.0 Growth and Development

Growth and development of sweet potato (vine and tuber) are dependent on several factors. Environment plays an important role in growth and development. Mellow (1946) stated that the best growth and yield of tuber were obtained when climate during the growing season was characterized by high day and night temperature. High temperature and plenty of soil moisture during the period of starch accumulation caused excessive vine growth at the cost of root development. In an extensive investigation, Biswas (1981) observed that a range of temperature between 20 and 30°C during the growing season was favourable for tuber formation and high yield. It was also revealed that temperature between 25.6 and 39.8°C during the growing period from May to September caused marked growth of vine but adversely affected tuberization. High rainfall and humidity also stimulated shoot growth.

Sekioka (1971) reported that accumulation of ^{14}C sucrose was highest at the soil and air temperature of 15°C but accumulation in the leaves increased with the rise in temperature. Sink growth of tubers was highest at a soil temperature of 24°C under air temperature of 28°C with a relative humidity of 70 per cent (Eguichi *et al.*, 1994).

Sweet Potato Computer Simulation (SPOTCOMS) has been developed to stimulate growth and yield model of sweet potato (Mithra *et al.*, 2008) and the output of the model is used for optimizing best management practices for its cultivation. SPOTCOMS can stimulate growth and yield of sweet potato irrespective of the agro-climatic conditions (Mithra *et al.*, 2019).

The role of photoperiod on the formation of storage organs has been reported by several workers. In an experiment, long day (16 hour light) was found to stimulate vine length and leaf number while tuberization was promoted under short day (9 hour light) condition. The nitrogen fractions including total, soluble and protein nitrogen and also sugars and reserve polysaccharides in leaves, stem and root were higher under a short day at different stages of growth (Biswas, 1981). Campbell *et al.* (1963) also observed that flowering decreased with increasing day length from 10 to 16 hours and the suppression of flowering was due to continued elongation of shoot under a long day. In another experiment on the effect of light intensity on growth and development of sweet potato, Biswas (1981) concluded that low light intensity (5000-10000 lux) caused greater vine length and leaf number compared to that under full sunlight (18000-40000 lux). Higher light intensity, on the other hand, showed a marked increase in tuber formation and dry weight of both vines and roots.

The cultivars also show a marked variation in growth of tuber in sweet potato. Wide variation in physico-chemical characteristics was observed among the cultivars of sweet potato. The cultivars showed positive correlation between tuber yield and a number of yield attributing and vegetative characters, more significantly to the tuber weight, tuber girth and number of tubers per plant (Reddy *et al.*, 2018).

Enyi (1977) conducted an experiment with 8 sweet potato cultivars and found that tuber yield ranged from 13.2 to 14.8 t/ha in cv. Fadenwaena and from 44.7 to 51.3 t/ha in cv. Loloki No. 1. He also reported that bulking rate and tuber weight was positively correlated with tuber yield. The tuber sink potential had a positive correlation with total dry matter (DM) production, average tuber weight, tuber bulking rate and harvest index but had a negative correlation with tuber number per plant and leaf area index (Goswami *et al.*, 1995). Biswas (1981) carried out an experiment on 6 cultivars with tubers having red and white skin and reported that percentages of nitrogen and carbohydrates in shoot and tubers were higher in cultivars with red skin tubers compared to those with white skin. The highest tuber yield was, however, recorded in a white skin cultivar Pusa Safed followed by Cross-4, another white-tuber cultivar. Huett (1976) reported that among the 16 cultivars of sweet potato tested, cv. L8-92, which contained the highest amount of carotene, produced the highest yield of the marketable tuber. There was a significant negative correlation between tuber number and mean tuber weight in 5 out of 6 cultivars and a positive correlation between the yield components and total yield (Lowe and Wilson, 1975).

The leaf has an important contribution to the growth and yield of tuber in sweet potato. Gonzalez *et al.* (1977) observed that topping of vine reduced tuber production and the highest yield was obtained without topping. Goswami *et al.* (1995) reported that high yielding cultivars (>26 t/ha) recorded higher harvest index values (61.9 to 73.4 per cent) at 120 days of harvest.

5.1.0 Flowering

Many of varieties do not flower. The crop is highly cross-pollinated due to self-incompatibility and male sterility. Cross incompatibility has also been reported. By grafting sweet potato on related species of *Ipomea*, flowering has been induced in the crop. It is reported that related species provide 'flower inducing substance' once they are used as rootstock. Though sweet potato is considered to be shy in flowering, there is a drastic variation in the flowering behaviour of its clones. The habit varies from profuse flowering to almost nil flowering in normal habitat. Sweet potato is a short-day plant. So, flowering is stimulated by short days, long nights and high temperature. Sweet potato flowers readily in the tropical climate. In tropics, flowering and seed set depend on environmental conditions and season. Day length of 11 hours is ideal for flowering (McClelland, 1928; Kasam, 1976). Flowers open early in the morning and wilt on the same day. The stigma becomes receptive in the bud from 6-8 p.m. and anthers start to dehisce from 11-12 p.m. (Purseglove, 1977). In a trial in India, it was observed that stamens shed their pollen around 6-7 a.m. with sunrise (Srinivasan, 1977). The pollen remains viable for 24 hours under normal temperature. The stigma may be receptive for a few hours after the wilting of flowers. Sweet potato is self-incompatible (Martin, 1965; Purseglove, 1977). Natural cross-pollination usually occurs by bees.

5.2.0 Effect of Growth Substances

Treatment with growth substances has proved effective in increasing the tuber yield in sweet potato. Foliar application of CCC at 250, 500 and 1000 ppm increased the yield (El Fouly *et al.*, 1977). Nambiar *et al.* (1976) also reported higher tuber yield by treatment with 50 to 150 ppm CCC. Increase in starch content by treatment with the growth retardant was also recorded by them (1977).

Shanmugam and Srinivasan (1973) obtained a larger number of tubers and higher yield by the application of ethephon at 250 ppm. In China, the application of SADH and CCC each at 300 and 500 ppm after 45 and 75 days from planting resulted in inhibition of vine growth and highest tuber yield (Shee, 1983). Foliar application of Kinetin at 50 and 100 ppm caused a higher tuber yield in terms of number, fresh and dry weight. CCC and GA_3 also showed a significant improvement in yield. McDavid and Alamu (1980a, 1980b) reported inhibition of tuberization by using GA while Kinetin enhanced tuber yield in sweet potato but the effect of ethrel was less pronounced in this respect (Biswas, 1981). Spraying of 1000 ppm CCC at 30th and 45th day after planting increased the tuber yield of sweet potato (Sen *et al.*, 1984; Anon., 1997). Regulating plant growth through the application of CCC at 200 ppm at 30, 50 and 70 days from planting resulted in the highest tuber yield (Misra *et al.*, 1992).

6.0 Cultivars

Most of the existing cultivars are from clonal selection and are self-incompatible. Several cultivars have been obtained from mutation as a chance seedling or deliberate hybridization. There are large variations in the skin and flour colour, shape and size of the tubers, the morphological characters, the depth and period of rooting amongst the cultivars. Sweet potato cultivars grown in India are the most popular local cultivars. They are usually grouped either as white or red type. Few cultivars are currently used for commercial production.

In the last few decades, several traditional and local cultivars of sweet potato were cultivated in India which was white skin with white-fleshed sweet potatoes. There are more than 1000 different sweet potato varieties in the world with different skin and flesh colours. Orange fleshed varieties are most common in the USA

Cultivars of Sweet Potato

Kalinga Sree

Arun

Gouri Sree

Vardhini

BCSP-14

Kamala Sundari

Cultivars of Sweet Potato

Sree Bhadra

Sree Nandini

Sree Kanaka

Sree Varun

ST-10

ST-13

and Europe and white-fleshed sweet potato is more popular in Asian countries (ASPMI). In the USA the most important Centennial varieties are New Jewel, Red Jewel, Oriental, Georgia Jet, Resisto which are deep orange with a marvellous taste. Recently orange-fleshed sweet potato is gaining high importance in Asian countries, the Middle East and India as well. Several orange-fleshed sweet potato varieties have developed in Sub- Saharan Africa (SSA) and catalogued.

Some of the varieties are landraces from African countries while others are introduced germplasms from the USA, South America, and Asia, and be adapted to particular environments in SSA (Kapinga *et al.*, 2010) (Table 2).

In the USA, sweet potatoes are classified as food types and feed types. Usually, the feed type produces higher tuber yield and high starch. Boswell (1950) divided food types into firm-fleshed and soft-fleshed based on the characteristics of cooked sweet potato. According to Thompson and Kelly (1979), soft-fleshed types have greater commercial importance than firm-fleshed types. The most important soft-fleshed cultivars are Puerto Rico, Unit I and Cleitt Bunch Puerto Rico. Other important cultivars in this group are Nancy Hall (deep orange-fleshed), Ranger, Early Port, Goldrush, Heart-O-Gold, Allgold, Australian Canner and Triumph. Cultivar Julian has high carotene, low fibre, firmness, high production of medium-sized roots and resistance to internal cork and stem rot. There are many old and new important firm-fleshed cultivars grown in the USA like Big Stem Jersey and Yellow Jersey, Maryland Golden, Orlisand Jersey Orange. Pelican Processor and Whitestar are two important non-food cultivars.

In Hawaii, there are dozens of good sweet potato cultivars (Tanaka and Sekioka, 1976). The important high yielding cultivars are Cona B, Iliula, Onolena, Waimanalo Red, HSPA-3, Miyashiro and Kaneohe Red. There are many Cuban sweet potatoes grown in Florida.

The cultivar Santa Barbara or Amrillo de Cuba is early maturing, highly productive, with good cooking quality and has well-shaped roots (O'Hair *et al.*, 1983). In Puerto Rico, the highest yielding white-fleshed cultivar is Miguela and Gem is the highest yielding orange-fleshed cultivar. The important new selections are Trompo Negro, Wart, Sunny, Bomara, Papota and Ninety-nine. Cano is considered as the best cultivar to taste in the Dominican Republic (Martin, 1984).

The important cultivars grown in Japan for industrial use are Koganesengan and Minamiyutaka. Both produce high yields with high starch content. Sujuki (1989) recorded that *I. batatas* cv. Fusabeni derived from Kanto 85 x Chikei 7238-19 has a uniform, spindle-shaped tubers with deep red-purple skin, and produces higher yields than Beniko-machi and Beniaka.

There are several improved yellows, orange and purple-fleshed sweet potato recommended for growing in different countries, having better yield, wider adaptability and early maturity and high beta-carotene content (Table 2).

India is the potential sweet potato growing country and is rich in diversity in its genetic stocks with valued agronomic and other desirable traits. Many heirloom elites sweet potato cultivars are being maintained in the field gene bank of the Co-ordinated Research Centres and CTCRI, ICAR and being evaluated. In India from the wide range of germplasms collected from different agro-climatic regions of the country and conserved and characterized at ICAR-AICRP, India centres paved way for the release of 40 improved varieties of sweet potato (Table 3). These varieties have unique and gorgeous flesh colour, rich scarlet red, purple, orange skin and yellow, orange and purple flesh that maintains its colour even when cooked orange, yellow and purple-fleshed colour.

Table 2: The Most Popular Sweetpotato cultivars in USA, Sub-Saharan Africa and other Countries*

Sl.No.	*Name of Cultivar*	*Country of Origin*	*Root Characteristics*
1	CARROTC	Tanzania Pedigree: Landrace	Shape-Long irregular, Skin colour Cream, Dry matter 33.0 per cent, Flesh colour, Deep orange, β-carotene 12390 14370 μg/100g fwb content
2	EJUMULA	Uganda Pedigree: Landrace	Shape-Long irregular, Skin colour Cream, Dry matter - 33.0 per cent, Flesh colour, Deep orange, β-carotene-7760 - 14370 μg/100g fwb
3	JEWEL (CIP440031)	USA Pedigree: Centennial x nugget	Shape Round elliptic, Skin colour Copper brown, Dry matter 28.0 per cent, Flesh colour Orange, (28A: 29A) (CIP colour chart), β-carotene 11030 μg/100g fwb content
4	KAKAMEGA (CIP441768)	Kenya Pedigree: Landrace (SPK004)	Shape Long irregular\, Skin colour Purple red, Dry matter 32.0 per cent, Flesh colour Intermediate orange, (28C: 18B), (CIP colour chart), β-carotene 376.0-3760.0 μg/100g fwb content
5	K566632	Kenya Pedigree: Unknown but SSR analysis suggests closely related to Resisto	Shape Round elliptic, Skin colour Intermediate pink, Dry matter 25.0-26.0 per cent, Flesh colour Deep orange, (29A: 28D) (CIP colour chart), β-carotene 700.0-800.0 μg/100g fwb content
6	MAYAI	Tanzania Pedigree: Landrace	Shape Long elliptic, Skin colour Cream, Dry matter 32.5 per cent, Flesh colour Intermediate orange, (29A: 28C), (CIP colour chart), β-carotene 11030 μg/100g fwb content
7	CN-1424-9 (CIP440245, CN1424-9)	Taiwan Pedigree: Unknown	Shape Long elliptic, Skin colour Brown, Dry matter 27.0 per cent, Flesh colour Orange, (29A: 28C), (CIP colour chart), β-carotene 11030 μg/100g fwb, content
8	RESISTO (CIP440001)	USA Pedigree: W72 x OP	Shape Ovate, Skin colour Pink, Dry matter 24.0 per cent, Flesh colour Deep orange, (30D: 29B), (CIP colour chart), β-carotene 24900 μg/100g fwb, content
9	NASPOT 9 O (VITA)	Uganda Pedigree: SPK004 OP	Shape Obovate with longitudinal grooves, Skin colour Purple red, Dry matter 30.1 per cent, Flesh colour Deep orange, (28A: 29A) (CIP colour chart), β-carotene 11030 μg/100g fwb content
10	NASPOT 10 O (KABODE) OP	Uganda Pedigree: SPK004	Shape Long irregular, Skin colour Purple red, Dry matter 30.5 per cent, Flesh colour Deep orange, (28A: 29A), (CIP colour chart) β-carotene 11030 μg/100g fwb content

Sl.No.	Name of Cultivar	Country of Origin	Root Characteristics
11	TAINUNG 64 (CIP440189)	Taiwan Pedigree: Unknown	Shape Long irregular, Skin colour cream, Dry matter 23.0 per cent, Flesh colour Orange, (29A: 28D), (CIP colour chart), β-carotene 3760-7230 μg/100g fwb content
12	W-151 (CIP440005)	USA Pedigree: Unknown	Shape Round elliptic, Skin colour Purple red, Dry matter 28.0 per cent, Flesh colour Deep orange, (30D: 29B), (CIP colour chart), β-carotene 10500-14370 μg/100g fwb, content
13	ZAMBEZI	Zambia Pedigree: TIS2537 x OP	Shape Round elliptic, Skin colour Pink, Dry matter 28.5 per cent, Flesh colour Deep orange, (29A: 28D), (CIP colour chart), β-carotene 10900 μg/100g fwb content
14	102027.02	Peru Pedigree: CIP breeding population	Shape Round elliptic, Skin colour Cream, Dry matter 25.0 per cent, Flesh colour Dark orange, (28A: 29A), (CIP colour chart), β-carotene 11030 μg/100g fwb content
15	102022.7	Peru Pedigree: CIP breeding population	Shape Round elliptic, Skin colour Orange, Dry matter 25.0 per cent, Flesh colour Orange, (29A: 28D), (CIP colour chart), β-carotene 3760-7230 μg/100g fwb content
16	10300.152	Peru Pedigree: CIP breeding population	Maturity period 4 months, Root yields 25.0 t/ha, Resistance-Moderate to sweet potato to pests weevils, Resistance- Low to sweet potato virus disease
17	CAROMEX (CIP440136)	USA Pedigree: NC 228 x NC 234	Shape Long elliptic, Skin colour Purple red, Dry matter 22.7 per cent, Flesh colour Dark orange, (28A: 29A), (CIP colour chart), β-carotene 11030 μg/100g fwb content
18	CN 1448-49 (CIP440181)	Taiwan Pedigree: Unknown	Shape Ovate, Skin colour Pale purple, Dry matter 22.7 per cent, Flesh colour Intermediate orange (28D; 28C), (CIP colour chart), β-carotene 4470-4920 μg/100g fwb content
19	GABA GABA	Mozambique Pedigree: CIP breeding line	Shape Long elliptic, Skin colour Purple red, Dry matter 23.9 per cent, Flesh colour Deep orange, (28A: 29A), (CIP colour chart), β-carotene 11030 μg/100g fwb content
20	KANDEE (CIP440140)	USA Pedigree: (Yellow Yam x Nancy Hall) x Porta Rico	Shape Round elliptic, Skin colour Light purple, Dry matter 25.3 per cent, Flesh colour Orange, (28A: 29A), (CIP colour chart), β-carotene 11030 μg/100g fwb content
21	JAPONES TRESMESINO SELECT (CIP420009)	Peru Pedigree: Japones Tres Mesino OP	Shape Round elliptic, Skin colour Light purple, Dry matter 21.6 per cent, Flesh colour Light orange, (29A: 28C) (CIP colour chart), β-carotene 3760—7230 μg/100g fwb content
22	LO-323 (CIP440185)	USA Pedigree: Unknown	Shape Ovate, Skin colour Pale purple, Dry matter 21.0 per cent, Flesh colour Intermediate orange, (28C; 18D), (CIP colour chart), β-carotene 5490 μg/100g fwb content

Sl.No.	Name of Cultivar	Country of Origin	Root Characteristics
23	PERSISTENTE (MGCL01)	Mozambique Pedigree: Landrace	Shape Long irregular, Skin colour Cream, Dry matter 37.0 per cent, Flesh colour Dark orange, (28A:29A), (CIP colour chart), β-carotene 11030 μg/100g fwb content
24	CORDNER	USA Pedigree: Unknown	Shape Elliptic, Skin colour Orange brown, Dry matter 25.0 per cent, Flesh colour Deep orange (29A: 28D) and cream, (CIP colour chart) secondary colour, β-carotene 3760-7230 μg/100g fwb content
25	199062.1	Peru Pedigree: SPV 78.001.3 x OP	Shape Ovate/Obovate, Skin colour Pale purple, Dry matter 24.0 per cent, Flesh colour Intermediate orange, (29A; 28D) on (CIP colour chart) new colour chart, β-carotene 3760-7230 μg/100g fwb content
26	CRI-Apomuden (CIP440254)	Bangladesh Pedigree: Unknown	Shape Long irregular, Skin colour Red orange, Dry matter 21.0 per cent, Flesh colour Orange with slight yellow stripes, β-carotene 2000-4000 μg/100g fwb content
27	IMPILO	South Africa Pedigree: Unknown (Bred by ARC)	Shape Round elliptic to elliptic, Skin colour-Pale yellow-orange, Dry matter 21.4 per cent, Flesh colour Pale orange, (29A: 28D), (CIP colour chart), β-carotene 5091 μg/100g fwb content
28	KHANO	South Africa Pedigree: Unknown (Bred by ARC)	Shape Long elliptic to elliptic, Skin colour Pale red Purple, Dry matter 18.2 per cent, Flesh colour Deep Orange, (30D: 29B), (CIP colour chart), β-carotene 14036 μg/100g fwb content
29	W-119 (440004)	USA Pedigree: Unknown	Shape Long elliptic, Skin colour Purple, Dry matter 25.8 per cent, Flesh colour Orange, (30D: 29B), (CIP colour chart), β-carotene 10464 μg/100g fwb content

* Catalogue of orange-fleshed sweet potato varieties for Sub-Saharan Africa. International Potato Center (CIP), Lima, Peru.

Table 3: Sweet potato Cultivars Released in India

Sl.No.	Year of Release	Name of Originating Centre	Name of Cultivar	Area Suitable
1	1976	TNAU	CO 1	Tamil Nadu
2	1980	TNAU	CO 2	Tamil Nadu
3	1980	RAU	Rajendra Sakarkand-5	Plains of Bihar, Jharkhand, Uttar Pradesh
4	1982	TNAU	CO 3	Tamil Nadu
5	1983	CTCRI TVM	Varsha	Konkan region of Maharashtra
6	1987	CTCRI TVM	Sree Nandhini	Kerala
7	1987	CTCRI TVM	Sree Vardhini	Kerala
8	1987	APHU	Samrat	Andhra Pradesh
9	1990	RAU	Rajendra Sakarkand-35	Bihar, Assam, Karnataka, West Bengal and Jharkhand

Sl.No.	*Year of Release*	*Name of Originating Centre*	*Name of Cultivar*	*Area Suitable*
10	1990	RAU	Rajendra Sakarkand 43	Bihar, Assam, Karnataka, Andhra Pradesh, Maharashtra and Jharkhand
11	1993	RAU	Rajendra Sakarkand-47	Bihar, Jharkhand, West Bengal and Uttar Pradesh
12	1994	APHU	Kiran	Andhra Pradesh, Bihar
13	1996	CTCRI TVM	Sree Bhadra	Kerala, Maharashtra, Bihar and Madhya Pradesh
14	1996	CTCRI TVM	Sree Rethna	Kerala
15	1998	CTCRI BBSR	Gouri	Odisha
16	1998	CTCRI BBSR	Sankar	Odisha
17	1999	TNAU	COCIP-1	Tamil Nadu
18	2000	Dr. BSKKV	Konkan Ashwini	Konkan region and adjoining area
19	2001	NDUAT	Narendra Shakarkand-9	Uttar Pradesh
20	2001	RAU	Rajendra Sakarkand-92	Plains of Bihar
21	2004	CTCRI BBSR	Kalinga	Odisha, Bihar, Jharkhand and West Bengal
22	2005	CTCRI BBSR	Goutam	Odisha, suitable in hilly coastal areas of Odisha
23	2005	CTCRI BBSR	Kishan	Odisha
24	2005	CTCRI BBSR	Sourin	Odisha
25	2006	IGKV	Indira Madhur	Chhattisgarh
26	2006	IGKV	Indira Naveen	Chhattisgarh
27	2007	IGKV	Indira Nandini	Chhattisgarh
28	2008	BCKV	Kamala Sundari	West Bengal
29	2011	CTCRI TVM	CO 3-4	Kerala and Odisha
30	2011	IGKV	Indira Priya	Chhattisgarh, Tamil Nadu
31	2011	IGKV	Indira Narangi	Chhattisgarh, Andhra Pradesh
32	2011	TNAU	IGSP-14	Tamil Nadu
33	2011	BCKV	90-101 (Bidhan Jagannath)	West Bengal
34	2013	TNAU	CO-5	Tamil Nadu (Coimbatore, Erode, Thirupur, Namakkal, Salem districts)
35	2015	CAU	NFSP-1	North East Region
36	2017	CTCRI BBSR	Bhu Kanti	Odisha
37	2017	CTCRI BBSR	Bhu Ja	Odisha
38	2017	CTCRI BBSR	Bhu Swami	Odisha
39	2017	CTCRI BBSR	Bhu Sona	Odisha
40	2017	CTCRI BBSR	Bhu Krishna	Odisha

7.0 Soil and Climate

7.1.0 Soil

The best soil for sweet potato is sandy loam and clay subsoil (Key, 1973; Martin *et al.*, 1976). It can be grown in loamy to clay loam soil also. Laterite soils of good depth and better fertility can also be considered suitable for its cultivation. Heavy clay soil which becomes hard after drying checks the development of the tubers and highly sandy soil causes the development of long, cylindrical and thin tubers. In compact soils, tuber yield is low due to poor aeration (Watanabe, 1969). The soil pH between 5.2 and 6.7 is appropriate for sweet potato. Root growth was inhibited at pH 3.5 to 4.5 and salinity reduced the growth of stem and roots (Eguichi *et al.*, 1994). Neutral and high pH may cause pox and scurf diseases (Steinbauer and Kushman, 1971; Key, 1973). Scurf injury was found to be reduced in slightly acid soils (Martin *et al.*, 1976). Highly fertile soil favours luxuriant growth of vines and is unfavourable for tuber development. Heavy clay is undesirable and causes harvest problem. The soil must be well-drained. Sweet potatoes are relatively sensitive to saline and alkaline soil condition.

7.2.0 Climate

The cultivation of sweet potato should be done where moderately warm climate prevails during the growing period of 4 months and the temperature ranges between 21.0-26.7°C (Watts and Watts, 1944; Martin *et al.*, 1976). It grows well at a temperature above 22.5°C (Nair *et al.*, 1984). Temperature less than 20°C were found critical for tuberization in sweet potato. Spence and Humphries (1972) observed early fibrous roots at 15°C and more tubers were formed only at 25 to 30°C. The incidence of even slight frost kills the foliage and inhibits tuber development. Sweet potato requires plenty of sunshine and moderate rainfall. A well-distributed annual rainfall between 75 and 150 cm is considered favourable. If there is drought at the initial stage of the growing period, the tuber yield is reduced considerably (Nair *et al.*, 1984). High rainfall and long photoperiod promote vine growth and reduce the tuber yield. Long day (16-h light) stimulated length and yield of the vine while tuber yield, number of tubers and total biological yield were promoted by 9-h light (Biswas *et al.*, 1989). Low light intensity (5000-10000 lux) resulted in greater vine length and higher light intensity resulted in a higher yield of tuber (Biswas *et al.*, 1990). Sweet potato can be grown from 40°N to 32°S and on equator up to an altitude of 2200 meters (Key, 1973).

Sweet potato performs well in well-drained, fertile, sandy loam soils having high organic matter. Compact and heavy soil reduces the size of tubers due to improper aeration whereas high sandy soil results in long cylindrical pencil-like tubers. The optimum pH range of the soil should be 5.2 to 6.7. Water logging is not good for its cultivation. Sweet potato vines require long warm growing season. A temperature range of 20-27°C is best suited to vine growth, and for tuberization, the ideal temperature is 20-25°C. The areas with an average day temperature of more than 24°C and an average annual rainfall of 750 mm or more are suitable for its cultivation. Not tuber production takes place at very high temperature (35-40°C) and temperature below 15°C, and plants are damaged by temperature below 10°C. A short day with low light intensity promotes root development while an excess of rainfall and long photoperiod encourages vine growth and reduces tuber yield.

Sweet potatoes require at least 500 mm rainfall during the growing season and an annual rainfall of 750-1000 mm is considered the best. They can tolerate considerable periods of drought but yields are very much reduced if water shortage occurs at the time of tuber initiation. Sweet potatoes are grown under high rainfall frequently produce vigorous vine growth but poor tuber yield (Moitra and Tarafdar, 2009). Dry and compact soil favours lignifications leading to the formation of pencil-like roots and more vine growth. For its cultivation pH should be between 5.6 and 6.6. Being an acid-tolerant crop, it cannot withstand salinity and alkalinity. Salinity significantly reduces the growth of stems and roots resulting lateral rolling of leaf lobules, reduction in leaf size and necrosis of older leaves.

8.0 Propagation

Sweet potato is generally propagated vegetatively and rarely through seed. The planting materials are obtained from sprouts or draw produced from tubers and vine cuttings. Vine cuttings, obtained from harvested plants or nursery. The vines obtained from the nursery are found to be healthy and vigorous and give maximum tuber yield while the vines from a freshly harvested mature crop, are commonly used in areas where two crops are raised.

8.1.0 Seed Propagation

Sweet potato is polycrossed in nature and seed propagation of sweet potato is used only for the interest of crop improvement. The seeds are small and covered with a tough coat, which delays germination and therefore, scarification is suggested before sowing. Plants grown from seeds have poorly developed top growth and tuberous root. Potentiality for seed formation in most of the cultivars/varieties of sweet potato is very low, although it is poly cross in nature, compatibility between the selected varieties for breeding strategies is a major constraint. Many of the varieties do not flower.

8.2.0 Vegetative Propagation

To get a good yield, it is advisable to plant selected tubers in the nursery and propagate by sprouts as practised in regions with a short growing season. Medium size healthy tubers free from diseases, pests and bruises are selected for planting in the nursery. Usually, two nurseries, namely primary and secondary are needed for producing planting materials. A well prepared and manured nursery having good drainage helps in producing better planting materials. Planting in the nursery is done 3 months ahead of planting in the main field. The selected tubers are planted 5 to 10 cm deep at a spacing of 25 cm in rows, 60 cm apart. To promote the growth of vines, the nursery is maintained properly and the beds irrigated when necessary. The sprouts are cut after about 40-45 days and planted in the secondary nursery for further growth. The cuttings, 20-30 cm in length are planted in this nursery at a distance of 60 cm between rows and 25 cm in the row. Ultimately when the nursery vines reach a sufficient length, the cuttings are made and planted in the field (Choudhury, 1976).

8.2.1 Nursery Preparation

a) Primary Nursery

The nursery is prepared about 2 months before planting in the main field. About 100 m^2 primary nursery area and 100 kg of medium-sized weevil free seed tubers, weighing 125-150 g each, are enough for planting one hectare of land. The tubers are planted 25 cm apart on ridges formed at a spacing of 60 cm. Urea may be applied 15 days after planting @ 1.5 kg urea/100 m^2 to ensure quick growth of vines and the nursery may be irrigated on every alternate day for the first 10 days and thrice a week thereafter. The vines are clipped off to a length of 20-30 cm after 40-45 days of planting, for replanting in the secondary nursery.

b) Secondary Nursery

The vines obtained from the primary nursery are further multiplied in a secondary nursery of area 500 m^2, which produces enough planting material to plant one hectare of land. About 500 kg FYM or compost may be applied at the time of preparation of the nursery. The vines, obtained from the primary nursery are planted at a spacing of 20 cm on ridges formed 60 cm apart. Urea @ 1 kg/100 m^2 may be applied in two split doses at 15th and 30th day after planting. After 45 days, vines are ready for planting in the main field.

8.2.2 Selection of Planting Material

Vines of 20-30 cm length with at least 3-4 nodes are found to be ideal as planting material. The cuttings obtained from the apical and middle portion of vines are preferable to get higher sprouting percentages and better tuber yields. Mukhopadhyay *et al.* (1990) reported that top and middle vines were equally efficient to produce more tubers than bottom vines. Mohankumar and Mandal (1972) reported the highest yield by using cutting from the apical portion. In North Bihar, 30-40 cm long cuttings having 5-7 nodes are used while in Tamil Nadu, apical cuttings 20-30 cm in length produce a higher yield by 28.6 per cent as compared to that obtained from the middle portion of the vine. In Uttar Pradesh, vines obtained from tubers and grown in the nursery when planted in the field showed a higher yield with large size tubers (Varma, 1984). Esonico *et al.* (1981) also observed higher leaf area index, marketable tuber and total yield by planting apical cuttings than with middle and basal cuttings.

8.2.3 Preparation of Vines

The cuttings with intact leaves are stored under shade for two days before planting in the main field to promote better root initiation, early establishment and higher yield.

9.0 Cultivation

9.1.0 Land Preparation and Planting Methods

The land is ploughed or dug to a depth of 20 cm and harrowed to pulverize the soil. Mounds, ridges and furrows and flatbed methods of planting are being practised in different locations. It is preferable to plant sweet potato on mounds in areas experiencing problems of drainage. Ridges formed across the slope are recommended in sloppy lands to prevent soil erosion. Among the different methods of planting, the highest tuber yield was obtained when planted on mounds. The higher yield obtained in the mound system of planting is attributed to better soil aeration and less tendency for soil compaction.

In Japan, deep cultivation (up to 60 cm) reduced growth and tuber yield and quality because of reduction in soil N and moisture content due to mixing of topsoil and subsoil. Pan-breaking gave higher total and good quality tuber yields than deep cultivation (Ino *et al.*, 1991). Borroto *et al.* (1990) determined the effects of 3 tillage systems on soil salinity and physical properties on yields of sweet potatoes in the saline soil of Guantanamo Valley, Cuba. The 3 systems were: control plot, sub-soiling plus harrowing, multi-plough plus vibrating tiller. The plot which underwent cultivation with the multi-plough and tiller produced the highest yields and was most cost-effective. Salinity was not affected by the systems.

9.2.0 Planting

9.2.1 Planting Time

The time of planting vine cutting affects the yield of sweet potato. In the tropics and subtropics, sweet potato is traditionally planted at the beginning or end of the rainy season. In Cuba, December planted crop yields good (Norman *et al.*, 1989) whereas; in Thailand, March planted crop yields more (Sajjapongse and Wu, 1989). Badillo-Feliciano and Lugo-Lopez (1976) reported that tuber yield was maximum when planting was done in November while the yield was lowest when planted in March or May. Menezes *et al.* (1974) in a similar experiment observed the highest tuber yield by planting in autumn. In Bangladesh, planting in November gave the highest tuber yield (Ahmed and Samad, 1976).In India, planting generally depends on the availability of soil moisture. In North Bihar and parts of Uttar Pradesh, it is planted in two seasons, *i.e.*, spring February-March) and autumn September-October). In Tamil Nadu, planting is usually done in September and May while in Kerala the best time for planting is from September to December (Singh and Mandal, 1970; Mohankumar and Mandal, 1972; Anon., 1983).In the Gangetic alluvium of West Bengal, the highest tuber yield was recorded by planting vine cuttings in September-October (Biswas, 1981), and then declined from the second fortnight of November and in subsequent plantings (Mukhopadhyay *et al.*, 1991a). Sweet potato is mainly grown as a rainfed crop and with irrigation as a summer crop.

9.2.2 Method of Planting and Spacing

The different planting methods, which are practiced in sweet potato, are mound, ridge and furrow, bed and flat method. Among the different methods, the ridge-

Nursery Raising and Planting of Sweet Potato

Nursery Raising of Sweet Potato Vines

Vine Cuttings Ready for Planting

Planting (Left) and Establishment (Right) of Sweet Potato Vines

and-furrow method is mostly practised which gives better yield and is helpful for convenient cultural operations, earthing up and economical water requirements. The land is ploughed to a depth of 15-20 cm to bring a fine tilth. The vines are planted at a spacing of 20 cm on ridges of 20 cm height made 60 cm apart. Only the middle of the vine with nodes is buried at 5-10 cm depth keeping both the ends exposed. Vertical planting on the ridge for sweet potato in sodic soil is recommended for UP. For Bihar, the flat method of planting and for Andhra Pradesh and North-Eastern India, the ridge method of planting is recommended (Palaniswami and Shirly, 2008). Biswas and Singh (1990) planted apical, middle or basal cuttings of sweet potatoes, cv. Pusa Sufaid horizontally or vertically in a field ploughed to 10, 15 or 20 cm depth. Tuber starch and non-reducing, reducing and total sugar contents were highest in plants grown from apical cuttings in soil ploughed to 20 cm depth. Tuber DM percentage was highest from plants grown from apical cutting in soil ploughed to 10 cm depth. Mandal *et al.*, 45 cm was reported to be superior to closer or wider spacing for cv. H-42 (Singh and Mandal, 1970). Nair *et al.* (1971) reported the results of the planting of cv. H-41 at 15, 30 and 45 cm keeping row distance at 60 cm on tuber yield. From another trial, Nair *et al.* (1984) recommended a planting distance of 60 x 20 cm, which will accommodate about 83000 cuttings per hectare, for optimum yield.

9.3.0 Manuring and Fertilization

The nutrient requirements of sweet potato are fairly high and depend mainly on the type of soil and its fertility status. Studies on the uptake of NPK and Ca at different stages of growth revealed that the above nutrients are utilized more by the plants at the early stage, up to about 60 days after planting than at a later stage. Fertilization with N, P and K has been found to increase the yield of sweet potato; N and K are more effective than P. Organic manures in the form of farmyard manure or compost helps to improve the soil condition for proper development of tubers.

FYM or compost @10 ha-1 should be applied at the time of field preparation. In addition to FYM, fertilizers may be applied as recommended doses in major sweet potato growing state of India (Table 4).

Table 4: Recommended Fertilizer Doses of Sweet Potato in different State of India

States	Nitrogen (kg ha^{-1})	Phosphorus (kg ha^{-1})	Potash (kg ha^{-1})
Bihar, West Bengal and Assam	75	60	75
Andhra Pradesh	60	60	60
Karnataka	60	90	60
Kerala	75	50	75

The full dose of P_2O_5 and K_2O and half dose of nitrogen should be given at the time of planting, whereas the remaining half dose of nitrogen should be applied one month after planting along with first weeding and earthing up (Anon., 2017).

Application of recommended level of N along with *Azospirillum* vine dipping @ 2 kg ha^{-1} and soil application of 10 kg ha^{-1} was recommended for Tamil Nadu, Assam, Bihar, West Bengal and Kerala.

In case of micronutrients deficiencies in sweet potato $ZnSO_4$ (1-2 per cent), Ferric sulphate (1-2 per cent) and $MnSO_4$ (0.1 per cent) spray are recommended in several states of India (Anon., 2018).

9.3.1 N, P and K

Sweet potato generally responds to small doses of nitrogen application. Excessive nitrogen application results in profuse leaf production at the expense of root yield. Nitrogen deficiency is usually noticed in sandy soils and soils low in organic matter content. Phosphorus deficiency and response to P application are most common in acid soils, especially in laterite and red soils, which contain high levels of Fe and Al such as Oxisols, Ultisols, Inceptisols, *etc.* Rock phosphate maintains or slightly improves the availability of P in soil. Since sweet potatoes do not require very large quantities of phosphate for root development, a dose of 25-50 kg ha^{-1} P_2O_5 is optimum. Differential response to applied potassium has been reported from various parts of India. Soils high in available potassium status do not respond to added potash in some locations while potassium fertilizer increased the yield of sweet potato by increasing the number of tubers and the ratio of large to small tubers at other locations. A general recommendation of 40-60 kg/ha of N and K_2O and 50 Kg/ha of P_2O_5 is recommended for the lowland conditions. The full dose of P and half of N and K as basal dose at the time of planting and remaining half 30 days after planting along with weeding and earthing up is recommended.

Though the response to phosphorus application is less pronounced in comparison to N and K, about 50 kg P_2O_5/ha or less has been recommended for the cultivation of sweet potato in the several states of India. In China, Chen and Chen (1987) observed 18.2-22.3 per cent higher yield by the application of P fertilizers to newly reclaimed soils at 300 kg/mu (1 mu = 0.067 ha). Pandey and Jauhari (1970) obtained the highest tuber yield and improvement in quality by the application of 90 kg each of N, P and K. From the results of the experiments high yield of sweet potato was recorded by using 50-70 kg N, 40-60 kg P and 60-100 kg K/ha (Varma, 1984). Mohankumar *et al.* (1984) recommended the application of 10 tonnes of FYM and 75 kg each of N and K and 50 kg P/ha. In South Korea, the optimum fertilizer rates for sweet potato were determined to be 27 kg N + 68 kg P_2O_5 + 198 kg K_2O/ha (Nam *et al.*, 1996).

9.3.2 Biofertilizers and other Nutrients

The effects of biofertilizers like *Azotobacter* (AZT) and *Azospirillum* (AZP) on the nitrogen use efficiency as well as growth and yield of sweet potato were found positive. Organic production of sweet potato with soil amendment of Poultry manure @ 5 t ha^{-1} + ash @ 5 t ha^{-1}, Vermicompost @ 5 t ha^{-1} + *Azospirillum* @ 5 Kg ha^{-1} + AM fungi @ 5 Kg ha^{-1} + ash @ 5 t ha^{-1} is highly effective for enhancement of tuber yield both for white and orange-fleshed sweet potato (Anon., 2017). Vine dipping with bio-fertilizer *viz. Azospirillum bracillence* at 2 kg ha^{-1} following application of same @ 10kg/ha and 2/3rd recommended dose of nitrogen (75kg/ha) improved the storage root yield of sweet potato (Chattopadhyay *et al.*, 2005a; Anon., 2018).

Combined inoculations of AZT + AZP gave significantly higher yield than the single application (Jadav *et al.*, 1998). Inoculated crop showed higher N content in leaf and tuber than plants fertilized at optimum rates (Sood *et al.*, 1999). The beneficial response of sweet potato to the inoculation of VAM fungi was also observed by Kandaswamy *et al.* (1988). Root infection by VAM in sweet potato cultivars varied from 13.89 to 46.43 per cent. Xie *et al.* (1982) found that when zinc concentration in soil is 0.24 ppm, application of Zn with P and K proved effective in increasing the yield. Recommended dose of FYM and NPK + $MgSO_4$ @ (20 kg/ha) + Borax @1.5 kg/ha + 2-4 per cent $ZnSO_4$ is best effective treatment for improvement of yield of sweet potato (Anon., 2018).

9.4.0 Irrigation

Light irrigation after planting, to maintain sufficient moisture, gives proper establishment of the crop. Sweet potato is tolerant to drought but a continuous long phase of drought reduces the tuber yield. If there is insufficient moisture, irrigation should be given at 10-15 days interval. The critical stage of moisture supply is 40 days after planting and water stress during this period reduces the marketable tuber yield.

Indira and Kabeerathuma (1990) observed that the tuber yield of sweet potato was significantly reduced by water stress at the initial phase (10 to 30 DAP) but stress during the tuber development phase (30 to 50 DAP) resulted in a slight increase in yield, whereas, stress during the tuber maturity phase (75 to 95 DAP) showed yield reduction in sweet potato. They also observed that the starch and dry matter in tubers were reduced by water stress. Rice straw mulching increased the tuber yield by 23 per cent under a rainfed situation. Application of antitranspirant (kaoline) had a positive effect on yield improvement (Goswami *et al.*, 1995). Black polythene mulch as row cover accelerated the vine growth and increased the tuber yield (Brown *et al.*, 1998).

In the Gangetic alluvial soils, irrigation at monthly interval increased the dry matter accumulation progressively. LAI, CGR, NAR and RGR were also influenced by irrigation (Chakraborty *et al.*, 1993). In lateritic soils, CGR, and yield increased with an increase in irrigation frequency from IW/CPE = 0.2 to 1.0 corresponding to 1-7 irrigations, respectively (Choudhury and Varma, 1997). Jones (1961) obtained the highest tuber yield by irrigating the crop at 45 per cent depletion of available soil moisture. Biswas *et al.* (1980) reported higher tuber yield and high consumptive use efficiency when irrigation was scheduled at 0.70 atm. tension (15-20 days interval) along with 60 kg N/ha and about 60-70 per cent was extracted from 0-30 cm soil layer. Under the irrigated condition, more moisture utilization was from the upper layer whereas, under the rainfed situation, relatively more moisture was utilized from lower layers than irrigated ones.

9.5.0 Interculture, Mulching and Weed Control

The intercultural operation to suppress weeds at an early stage of growth is an

Cultivation Techniques of Sweet Potato

Sweet Potato Cultivation under Protected Cover (Left) and Open Field (Right)

Re-ridging of Vines (left), and Ridge and Furrow Method of Production (Right)

Poly-mulching in Sweet Potato

essential requirement for obtaining high yield. Earthing up of the soil is helpful to suppress weeds besides improving its physical condition. In a trial at the CTCRI, India, Ravindran and Mohankumar (1987) observed that earthing up at 15 or 30 days after planting resulted in tuber yield of 12.68 and 16.55 t/ha, respectively compared to 12.84 t/ha under control (no weeding and earthing up). In China, among the various treatments, the highest yields were obtained with polyethene film mulch (2.6 t/mu). Yields with manual weeding were 1.87t/mu and were not significantly different from those with the application of herbicides but no mulch (1.9 t/mu). The minimum tillage technique with mulching increased soil temperature and water content and improved growth and also decreased the labour requirement (Yang, 1990).

The intercultural operation 30 days after planting along with weeding and earthing-up suppress weeds in the early stages of growth and improves the physical condition of the soil. The second earthing-up should be done 50 days after planting. Being a quick growing crop, it covers soil quickly and suppresses most of the weeds.

9.6.0 Cropping Systems

Sweet potato has been found to grow under different cropping sequences in the eastern and southern region of India. The ideal cropping sequences in eastern India are mentioned below:

Table 5: Sweet Potato Based Cropping System in different States of India

States	*Cropping Sequences*
North Bihar	Maize (Kharif)- Sweet potato (Autumn)- Wheat (Rabi)-Moong (summer)
	Maize (Kharif-Sweet potato- Onion (Rainy)
West Bengal	Moong (summer)-*Colocasia* (Pre-Kharif)- Sweet Potato (Autumn)
	Moong- Elephant foot yam- Sweet potato
Odisha	Mid-season paddy (Rainy)-Sweet potato-Fallow
	Maize-Sweet potato-Fallow
Kerala (low land)	Paddy (1st crop)-Paddy (2nd crop)-Sweet potato
	Paddy-Sweet potato-Fallow
Tamil Nadu	Sweet potato-Cereals
Andhra Pradesh	Maize-Sweet potato-Vegetables
Madhya Pradesh	Vegetable cowpea + Sweet potato
Uttar Pradesh	Vegetable cowpea + Sweet potato
Chhattisgarh	Vegetable cowpea + Sweet potato
Maharashtra	Vegetable cowpea + Sweet potato

Source: Tuber and Root Crops (2009)

10.0 Harvesting and Postharvest Management

10.1.0 Harvesting

The crop matures in 100-120 days after planting depending on variety and environmental conditions. Harvesting should start at 90 days after planting (DAP) for early varieties, 110 DAP for medium duration varieties and 120 DAP onwards for late maturing varieties (Mukhopadhyay *et al.*, 2011). Yield can be increased if the crop remains in the field for longer periods but the tubers become fibrous and less palatable due to damage by weevils (Mohankumar 2000).

Single harvesting and double harvesting (progressive harvesting) are also practised in sweet potato. In Sub- Saharan Africa, farmers harvest piecemeal in order to maintain a supply of fresh roots for as long as possible (Smit, 1997; Hall *et al.*, 1998). In Papua New Guinea, nonmarketable tubers constituted a greater proportion when harvested progressively compared to single harvesting (Mohankumar, 2000). In North Bihar, India a comparison of tuber yields harvested at 150 DAP and at 90 and 150 DAP as well as at intervals of 15 days between 90 and 150 DAP demonstrated that the last resulted in significantly higher tuber yield compared to the single or double harvesting dates (at 150 DAP and 60+150 DAP, respectively; Mishra *et al.*, 1985).

In many areas, small farmers start to harvest early and subsequently at intervals. Before harvest, the vines are cut and tubers are dug up by using spade or fork. Sweet potato is usually harvested when the leaves turn yellow and begin to shed. Maturity can be ascertained by cutting tuber and the latex from mature tuber dries up without turning black (Purseglove, 1977; Key, 1973). Irrigating the field 4-6 days before harvest is helpful when soils become compact. While digging out tubers, care should be taken to avoid mechanical injuries which would affect the storage and quality (Kushman, 1967). In a trial on maturity, Ramanujam and Indira (1979) noticed significant variations among the cultivars for a bulking rate which is considered a reliable parameter for fixing the maturity period. They also observed that H-633 is an early maturing cultivar which showed a higher bulking rate at 75th day. In a trial to determine the optimum stage of harvest, Nawale (1981) reported that tubers of cv. H-268 should be harvested at 120 days. In Bangladesh, Toib and Rashid (1976) suggested that harvesting should be done after 150 days. Garcia-Espinosa (1987) observed that storage root quality, particularly starch, sugar, protein and lipid contents, was unaffected by ploughing depth or planting date when harvested after 5 months. Chattopadhyay *et al.* (2006a) and Mitra *et al.* (2010) observed that β-carotene content of most OFSP cultivars increased up to 105 days and, considering productivity, nutritional quality and consumer acceptability, they suggested that the crop should be harvested at this time.

10.2.0 Yield

The yield of sweet potato tubers varies from 10-40 t/ha depending on varieties, the season of planting, soil types, fertility and cultural practices adopted. Under good management practices, yield can range from 35-40 t/ha in irrigated and 8-10 t/ha under rainfed conditions. Under the Gangetic plains of West Bengal, India the tuber yields varied from 20 to 40 t/ha depending upon the cultivars, and weevil-free good quality tubers can be harvested within 120 days when grown in October to February (Mukhopadhyay *et al.*, 1990, 1991a; Chattopadhyay *et al.*, 2005b; Nath *et al.*, 2005; Anon., 2015).

Harvesting of Sweet Potato Tubers

Sorting, Grading, Packaging and Selling of Sweet Potato

10.3.0 Storage

Sweet potato matures three-and-a-half to four-and-a-half months after planting. Harvesting sweet potato 120 days after planting is normally recommended. Care should be taken to avoid injuries and bruises on tubers. For marketing of fresh tubers, cleaning and grading should be done to get better prices. After harvest, tubers are spread in partial shade for 5-6 days, for healing and curing. In some parts of the country, tubers are stored in a layer of dry sand/soil after curing under ambient conditions. For storing, graded tubers free from sweet potato weevil and bruises should be selected. Farmers store the graded tubers by keeping in a shaded pit and covering it with paddy straw. Finally, the heap is plastered with mud or cow dung slurry.

Roots decline rapidly in quality after harvest because of weight loss, physical damage, sprouting, pathological decay and weevil infestation (Chattopadhyay *et al.*, 2006b; Abano *et al.*, 2011). The weight loss of sweet potato roots is maximum under exposed conditions as compared to other storage methods (Chattopadhyay *et al.*, 2006b; Abano *et al.*, 2011) but least in sand (Chattopadhyay *et al.*, 2006b). Intact roots do not lose weight appreciably under low humidity (60–70 per cent; Shima *et al.*, 1996). Storage temperature has a marked effect upon sprouting (Cooley and Kushman, 1947). Sprouting is a common occurrence in saw dust media (Karuri and Hagenimana, 1995) or in evaporative cooling barns (Abano *et al.*, 2011). However, sprouting may be checked for up to 102 days by immersing tubers in sodium hypochlorite solution at 9.0 per cent by volume for a period of 20 min (Lewthwaite and Triggs, 1995). Soft rot, caused by *Rhizopus stolonifer* (Thankappan, 1989) and weevil infestation (*Cylas formicarius*) are great menaces to effective storage (Abano *et al.*, 2011).

Highly perishable nature of tubers of sweet potato does not permit it to be stored for a long period in the tropics (Key, 1973). In the USA, sweet potato tubers are usually stored after curing. Uncured tubers get spoiled quickly at temperatures of 10°C or below (Lutz and Simons, 1972). A temperature of 29-32°C and relative humidity of 85-90 per cent for 4-7 days are maintained for proper curing (Kushman and Wright, 1969). Sweet potatoes were successfully cured at a relative humidity of 80-90 per cent and 29-40°C for 5-7 days. Curing healed wounds inflicted during harvest (Walter and Schadel, 1982). After curing, sweet potato may be stored at 10-12°C with a relative humidity of 85-90 per cent. At CTCRI, the wounds of harvested tubers were found to heal when stored for 3-4 days at 45-50°C at 85 per cent relative humidity (Anon., 1983). Sweet potato tubers could be stored for 4-12 weeks at a temperature of 15-20°C with 70 per cent relative humidity. The quality of tubers stored in soil or sawdust was as good as those stored at 15°C (Karuri *et al.*, 1994). Satisfactory storage by using copper oxychloride was reported by Gonzalez *et al.* (1983). The percentage of weight loss was 2-3 times higher in open storage than in sand and no rotting was observed up to 30 days (Dayal *et al.*, 1990).

For long-term storage, roots are stored in different structures such as pit storage (Hall and Devereau, 2000); trench storage (Onwueme, 1978); well-ventilated storehouses on raised platforms, in heaps on the floor, in baskets or in roof spaces (Olurunda, 1979); clamps (heaps), bamboo baskets and jute sacks (Gautam, 1991);

Machan (raised platform made of bamboo slices) and Dhol (made of bamboo thatched Chatai) (Ali *et al.*, 1991) and evaporative cooling barns (Abano *et al.*, 2011). Roots also store well in different media like rice or millet (Gautam, 1991); wet sawdust mixed with sand (Ali *et al.*, 1991); dry sand (Samarasinghe, 1991); sawdust (Karuri and Hagenimana, 1995); alternate layers of sand and roots (Chattopadhyay *et al.*, 2006b), earth or sawdust (Mukhopadhyay *et al.*, 1991b), wood ash, earth and straw (Ghosh *et al.*, 1988) or in well-ventilated rooms (Kurup and Balagopalan, 1991). Several other techniques such as cold storage (15°C temperature and 85–95 per cent RH), chemical treatment (postharvest dip in thiabendazole (TBZ) and benomyl for 1 min, especially with hot water at 43 to 53°C for the control of black rot), light exposure (diffused light intensities of 14.30 lux reduced sprout growth) and irradiation (UV radiation [3.6 to 4.8x10^{-4} erg/mm^2]; gamma radiation (0.075 kGy) and X-rays (40–300 krad) significantly lower decay and weight loss) and have been reviewed (Ravi *et al.*, 1996).

11.0 Diseases, Pests and Disorder

11.1.0 Diseases

Several viral, fungal, bacterial pathogens and other microorganisms attack sweet potato and reduce the crop yield and quality. Several pathogenic organisms affect sweet potato including virus, fungal and bacterial diseases most appear to be widespread, but damage levels vary. Although the symptoms of virus diseases appear in the foliage, these have been accorded a separate section because of their importance. Of all the sweet potato pathogens, viruses appear to contribute the most to yield losses (**Ames 1997**). Some diseases occur in the field; others occur during storage, handling, and marketing. Among these viral diseases is the most important and about 20 virus diseases have been reported to infect sweet potato. Among the virus diseases, Sweet potato Feathery Mottle Virus (SPFMV) (Aphid-transmitted potyvirus), whitefly transmitted Sweet potato leaf curl virus (SPCLV), Sweet potato Sunken Vein Virus (SPSVV) (Whitefly-transmitted closterovirus), Sweet potato Virus Disease (SPVD), Sweet potato Mild Mottle Virus (SPMMV) (Whitefly-transmitted potyvirus) are most common and prevalent worldwide. Variable symptoms in sweet potato plants with foliar mosaic, stunting, leaf malformation, chlorosis, and rings spot symptoms were observed in several farms in West Bengal and NCM-ELISA gave many positive results of sweet potato chlorotic stunt virus (SPCSV), C-6 (unknown virus) and seldom occurrence of sweet potato cauliflower-like virus (SPCaLV). Plants obtained from vine cuttings and tuberlets collected from SPFMV, leaf curl, severe mosaic symptomatic plants of different varieties/cultivars in West Bengal gave identical symptoms.

11.1.1 Virus Diseases

Sweet Potato Feathery Mottle Virus (SPFMV) is the most important and most commonly occurring virus. SPFMV alone cannot damage the crop much but mixed infection with other minor viruses inflict sweet potato virus disease (SPVD) syndrome which causes severe damage to the crop. Sweet potato chlorotic stunt virus (SPCSV) (family Closteroviridae) and sweet potato feathery mottle virus (SPFMV) (family Potyviridae) are whitefly-and aphid-transmitted, respectively, which in double infection cause sweet potato virus disease (SPVD) in Africa, a serious sweet potato disease. Sweet potato plants showing symptoms similar to SPVD have been observed in most areas of India. Nevertheless, not much information is available about the identity of the viruses infecting this crop in India. NCM-ELISA, DAS-ELISA and PCR detection of viruses revealed the predominance of SPFMV in the states, Kerala and West Bengal in India.

Symptoms of virus infection vary between varieties and by plant age and growing conditions. Irregular mottled spots or feathery purple discolouration often appear on the leaves, along the veins in some varieties surrounded by a purplish halo. Sometimes leaves develop a feathery discolouration along the veins. Sometimes the SPFMV infected leaves show water-soaked spots with purplish border and coalesce and the leaves become mottling symptoms.

Vigour is reduced and storage roots tend to be thin and elongated. Severe infection can result in large numbers of small cracks over the root surface and pale flesh, particularly in orange varieties. Some infected cultivars express internal (internal cork) or external (russet crack) necrosis on the roots. The disease is non-persistently transmitted by a large number of aphid species and also by vegetative propagation. A common and widespread virus which is primarily spread through infected planting material and aphid (Aphid spp.) also can transmit this virus in a sweet potato field. But infected vine cuttings of sweet potato are the major source of spread of SPFMV in the field. The virus is a flexuous rod with helical morphology, about 13-850 nm (potyvirus group). This disease can be contained using virus-free planting material and roughing out of infected material from the field. Diagnosis can be confirmed by serological tests and graft transmission on *I. setosa*. Meristem tip culture is the method for obtaining disease-free material.

Sweet potato Leaf Curl Virus (SPLCV) is distributed worldwide and causes severe damage to the sweet potato plants to cause yield loss. SPLCV belongs to the family begomovirus of Geminiviridae with monopartite SSDNA genome of Old world group. The name sweepo viruses have been suggested for the sweet potato begomoviruses (Fauquet and Stanley, 2003). Sweet potato leaf curl virus Genus Begomovirus (SPLCV) has been reported from the United States, Taiwan, Korea, Argentina, and Japan. Infected plants show upward leaf curling (Gad Loebenstein, 2015). Three novel species of sweepo viruses have been described in Spain: Sweet potato leaf curl Lanzarote virus (SPLCLaV), Sweet potato leaf curl Spain virus (SPLCSV) and Sweet potato leaf curl Canary virus (SPLCCaV). SPLCLaV was first isolated from a sweet potato plant from Lanzarote (Canary Islands, Spain) and later found in the Malaga province of continental Spain (Lozano *et al.*, 2009). The occurrence of sweet potato viruses in West Bengal, India was studied (Sinha and Tarafdar, 2007) and whole-genome characterization of SPLCV infecting sweet potato in West Bengal, India has first time been done (Kumar *et al.*, 2020).

Sweet potato chlorotic leaf spot virus, sweet potato chlorotic stunt virus, sweet potato leaf curl virus, C-6 virus, sweet potato caulimo virus, sweet potato mild mottle virus, sweet potato sunken vein virus, sweet potato vein cleaning virus, sweet potato vein mosaic virus, sweet potato yellow dwarf virus, sweet potato virus A are other viruses reported on sweet potato. Sweet potato mild mottle virus and sweet potato leaf curl virus are found in Kerala, Bihar, Orissa and West Bengal (Tarafdar and Sarkar, 2006; Palaniswami and Shirly, 2006).

Sweet potato latent virus (SPLV) belongs to potyvirus group. It causes latent or mild chlorosis in sweet potato. There is no known vector. The virus is a flexuous rod with helical morphology, about 13-750 nm. Graft transmission to *I. setosa*. Diagnosis can be confirmed by serological tests of *I. setosa*.

Sweet potato mild mottle virus (SPMMV) infected sweet potatoes exhibiting stunting and leaf mottling. The disease is efficiently transmitted by the whitefly *Bemisia tabaci* and by vegetative propagation material. The virus is flexuous rod-shaped particles, 800-900 nm in length (potyvirus group). Graft transmission to *I. setosa*. Diagnosis can be confirmed by serological tests of *I. setosa*.

11.1.2 Russet Crack Syndrome

The sweet potato plants infected with virus diseases (SPVD) showed mixed symptoms on the foliage. The severely infected plants produced deformed storage root, splitting and internal cork tissues. The flesh of the roots produces dark, hard, corky, a patch with dark brown to black. The central portion of corky tissues produces hollow zone with shrinkage of tissues. Sometimes the roots do not show external symptoms and occasionally produce tiny corky growth on the skin. Production and planting of healthy virus-free vine cuttings are the best options for controlling the disease. The leaves on the vine cuttings showing any symptoms like SPFMV the vine may be destroyed. Supervision of the sweet potato field may be conducted in regular basis and the SPFMV infected plants may be uprooted and destroyed to avoid the secondary spread of the virus through insect vector as well as to minimize the source of infection in the vine cuttings. During the growing period, the suitable insecticide or organics may be sprayed to control the aphid population in the sweet potato field. Intercropping of sweet potato with coriander or garlic is effective for the management of SPFMV. In some cases severe symptoms of SPMFV may be confused with phosphorous deficiency, however, critical observation is needed. Aphid control is not economically feasible. The main controls are avoidance of the use of diseased plants for cutting material, sanitation, and use of resistant varieties. The main controls are avoidance of diseased plants as sources of planting material and use of resistant varieties. Farmers usually avoid diseased planting material because symptoms are so severe.

11.1.3 Stem Rot or Fusarium Wilt (*Fusarium oxysporum f. batatas*)

This disease is more prevalent in the nursery beds during vine multiplication. The infected plants become very weak, thin and leave become yellow. Initially wilting appears in a patch and infected plants gradually wilt and dry. Stems are blackened and the roots are dried.

This disease is caused by *Fusarium lateritium* Nees Fr. The prominent symptoms are bright chlorosis on the young leaves, twisting or deformation of leaves in an advanced stage, white powdery growth of fungal structures on the leaf surface and margin or leaf blade. It affects the leaf opening and reduces the chlorophyll content. Spraying of dipotassium hydrogen phosphate or disodium hydrogen phosphate @ 0.001 per cent for 3-4 times control the disease (Thankappan, 1994).

Disease of Sweet Potato

Fasciated stems

Russet Crack of Sweetpotato
caused by SPVD

Disease of Sweet Potato

Leaf Symptoms of Sweet Potato Feathery Mottle Virus Disease

11.1.4 Bacterial Wilt and Root Rot (*Erwinia chrysanthemi*)

This disease is also referred to as bacterial soft rot. The disease manifests with yellow leaves and water-soaked lesions on stems and petioles. The infections occur in a more localized area of the plant and most cases only one or two branches collapse and die. In advanced condition, the leaves and petiole turn black and the whole plant dies. Roots decompose completely. The disease is more prevalent in rainy season mainly during vine multiplication. The heavy rainfall with poor drainage and waterlogged condition predisposes this disease.

11.1.5 Leaf Spots

Two species of *Cercospora* have been reported to cause leaf spot disease on sweet potato. *Cercospora batatae* Zimm., a synonym of *C. timorensis* Cooke, reported by Sydow and Mc. Rae in 1929 (from Tamil Nadu and Bihar), Salam and Rao in 1957 (from Hyderabad) and Chiddarwar in 1959 (from Maharashtra). Later in 1982, *C. ipomoeae* Wint. was also reported to cause leaf spot disease on sweet potato in Bihar (Sinha and Mehta, 1982).

Other leaf spots are *Cercospora batateae* Zimm., *Cercospora timorensis* Cooke, *Cercospora ipomoee* Wint. *Alternaria* sp., *Alternaria capsici-* annui Saval and Sandhu, *Colletotrichum capsici* (Syd.) Butler and Bisby, *Helminthosporium euphorbiae* Hans.

Use of field resistant varieties alone can give satisfactory control for this disease. Fifteen sweet potato lines out of ninety-five tested have been identified as resistant at Dr. Rajendra Prasad Central Agricultural University, Bihar.

11.1.6 Scurf

Scurf is caused by the fungus *Monilochaetes infuscans* and occurs in the USA, Hawaii, Japan and Australia. The disease produces black spots of varying sizes and shapes on the tubers. Practising long rotation or applying organic matter at least 1 year before planting is recommended. Good drainage and application of sulphur resulted in the reduction of disease (Chupp and Sherf, 1960). Disease-free root selection and routine chemical treatment of roots are useful.

11.1.7 Vine Fasciation and other Diseases

Striking diagnostic symptom of unknown etiology in the appearance of the thick flattened transformation of vines which remain green and grow till harvest is occurring sporadically (Palaniswami and Tarafdar, 2008).

Other less important field diseases of sweet potatoes are Texas root rot (*Phymato-trichum omnivorum*), mottle necrosis (*Phythium* sp.), *Phyllosticta* leaf blight (*Phyllosticta batatas*), and septoria leaf spot (*Septoria bataticola*) and white rust (*Albugo ipomoeae-pandurata*).

11.1.8 Other Storage Rots

Other storage rots affecting sweet potatoes are black rot (*Ceratocystis fimbriata*), surface rot (*Fusarium oxysporum*), dry rot (*Diaporthe phaseolorum* var. *batatatis*), Java black rot (*Diplodia theobromae*), and charcoal rot (*Macrophomina phaseolina, Sclerotium bataticola*).

Black Rot (caused by fungus: *Ceratocystis fimbriata*) can cause severe loss in storage, in the plant bed, and the field. The infected storage root of sweet potato shows black, sunken circular spots that often cover the surface. The infected tissues may spoil and make it non-marketable.

Rhizopus Soft Rot (caused by fungus: *Rhizopus stolonifer*) is found in the storage and is the most destructive disease. Once the disease appears it spread very quickly in the storage and may cause a heavy toll to the harvested tubers. The infected sweet potato becomes soft, watery with offensive odour and producing fluffy white fungal growth. The disease is more prevalent in high humidity with poor ventilation in the storage.

Scurf (caused by fungus: *Monilochaetes infuscans*) turns the skin of the sweet potato blackish patch, sunken and shrivelled later. The disease is usually worse during wet weather and late harvesting during summer scurf occurs in the storage and whole marketable roots become hard and shrivelled. The disease spreads when healthy sweet potatoes come in contact with diseased roots (Tarafdar, 2006).

11.2.0 Pests

More than 80 species of insect pests and few species of mites and nematodes cause damage by feeding on roots, tubers and foliage (Pillai and Lal, 1978).

11.2.1 Sweet Potato Weevil

Among the pests that attack the crop, the most serious and everywhere one is the weevil (*Cylas formicarius* F.) which originated in India and was first recorded in 1792 near Pondicherry. Even though many pests are recorded on this crop, none of them will come near to the weevil in inflicting direct damage, yield loss and at times destruction of the crop. Both adults and grubs cause damage to the crop both in the field and in-store. Adult weevil feeds on all parts of the plant while the attack of the grubs is restricted to vines and tubers. Economic damage is caused by grubs. The feeding causes a characteristic terpenoid odour and a bitter taste on cooking, thus rendering the tubers unsuitable for human consumption. Other related species like *Cylas brunneus* F., *C. puncticollis* Boh. and the West Indian weevil *Eusceps postfasiatus* De Geer, which are considered to be important elsewhere, have so far not been recorded in India (Jannson and Raman, 1991). The weevils lay eggs on vines and maintain its population in the vines before tuber formation. The grubs bore and feed by making irregular tunnels. As a result of feeding by several grubs, the collar region usually bulges out due to hypertrophy and proliferation of tissues. At times of severe damage, the region breaks off from the root zone. The emerging adult weevils get access into the tubers through the cracks and crevices formed in the soil during bulking. The pre oviposition, oviposition and post oviposition periods were 7.8, 8.1 and 4.0 days, respectively. A single female on an average laid 180 eggs in its life span. The adult longevity is 95.8 days for female and 112.6 days for male. According to Jayaramaiah (1975), it was 90.5 and 91.1 days, respectively. Subramaniam *et al.* (1977) observed 61 to 120 days of adult longevity for males and 81 to 107 days for females. However, longevity was found higher in males than in females. Observed sex ratio was 1.5: 1 for a male to female (Jayaramaiah, 1975). Seven to nine overlapping generations are completed in a year.

Disease of Sweet Potato

Symptoms of Sweet Potato Viral Diseases

Disease of Sweet Potato

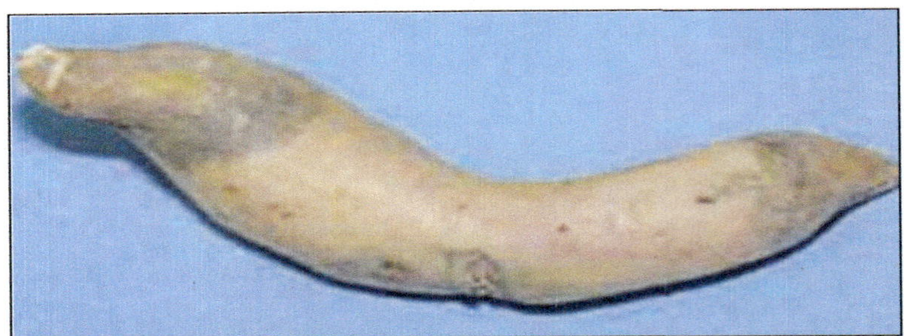

Cercospora Leaf Spot

Sweet Potato Scab

Browning of Vascular Tissue Due to *Fusarium* Wilt

Symptoms of Chlorotic Leaf Distortion Disease on Sweet Potato

In India, the yield loss assessment due to weevil damage ranged from 17 to 70 per cent at Coimbatore (Subramaniam *et al.*, 1977), 16 to 40 per cent in Bangalore (Jayaramaiah, 1975) and 27 to 82 per cent in Trivandrum (Rajamma *et al.*, 1993). In the farmer's field in Kerala, the damage ranged from 0 to 22 in lowland and 4 to 50 per cent in upland. The annual yield loss due to weevil damage in Kerala is estimated at Rs. 96 lakhs (Palaniswami *et al.*, 1991). The damage in Orissa, Bihar, U.P. and West Bengal varied from 25 to 45, 5 to 50, 7 and 5 to 11 per cent, respectively. In Odisha, the tribal people harvested tubers after five months and in such condition, about 50 per cent tuber damage was observed. In some villages of Bihar, vine damage was more than 50 per cent while the tuber damage was less than 5 per cent indicating a lack of relationship between tuber damage and vine damage. Yield loss due to weevil damage also varied from season to season, soil to soil, age of the crop, moisture and temperature of the soil. The population build-up of weevil and tuber damage at different ages of the crop during the three seasons revealed that mean population increased from 2.5 at 30 days to 22.4 per plant at 120 days maturity in June to September season. The tuber damage at 105 days was 50 per cent which increased to 72 per cent at 120 days. In October-January season the weevil population increased from 2 per plant at 30 days to 27.3 at 120 days and the tuber damage 61 per cent at 105 days and 82 per cent at 120 days maturity (Rajamma *et al.*, 1993). Generally, the weevil damage was higher in summer crop. There was a progressive increase in the weevil incidence and damage as the age of the crop increased. The crop is grown in red loam and clay loam had a higher incidence than that of sandy loam probably due to cracking nature of the soil (Palaniswami *et al.*, 1991). In North India, the crop raised in Kharif and Rabi seasons showed varying weevil damage. In Rabi, the weevil damage was lower while it was higher in the Kharif; this phenomenon can be mainly attributed to low soil temperature.

Chemical Control

The insecticidal application was the most widely recommended control measure against sweet potato weevil. Spraying the crop at monthly intervals with fenitrothion 0.05 per cent or methyl demeton 0.05 per cent was highly effective to reduce weevil on sweet potato. There was 80-86 per cent increased production of marketable tubers in fenitrothion treatment (Palaniswami, 1999).

Elaborate field experiments were done at ICAR-CTCRI using eleven insecticides (E.C and Granules) and four soil amendments. Spraying the crop first after a month of planting and subsequently at an interval of three weeks with 0.05 per cent fenthion or fenitrothion (500 g a.i./ha) was found highly effective in reducing weevil infestation and to increase the marketable tuber yield. The residue analysis of the tubers after harvest showed that the residue was 0.09 ppm in fenitrothion and 0.07 ppm in fenthion which were far below the tolerance limit. Soil application of phorate or carbofuran (1.5 kg a.i./ha) at 45th day of planting also reduced weevil infestations. Soil drenching with any one of the chemicals *viz.*, fenthion, endosulfan, quinalphos, monocrotophos and fenitrothion 0.05 per cent at 50 and 70 days after planting was found effective to reduce weevil infestation. The tuber damage was brought down from 44.4 per cent in the control to 14.4-19.5 per cent in the above treatments.

The cost-benefit ratio was highest in fenthion (1:5.7) followed by endosulfan and fenitrothion. Residues of the above chemicals were below the tolerance limit.

Different insecticides reported as effective for the control of weevil in various coordinated centres were selected for final evaluation and also to find out the cost-benefit ratio. Accordingly fenthion, endosulfan, fenitrothion and quinalphos each at 500 g a.i./ha (per spray) were tried against weevil for three seasons in 1983-84 adopting four sprayings. The mean tuber yield in different chemical treatments ranged from 12 to 17.4 t/ha against 7.4 t/ha in the control.

Sweet potato intercropped with garlic showed promising for reduction of weevil infestation. Intercropping confirmed that planting of sweet potato with garlic 1:1 ratio could reduce the weevil infestation and the result was very close to the same design 2:1 ratio (Anon., 2014). As sweet potato was intercropped, the total plant population was reduced than sole crop, hence average marketable tuber yield was little lower but non-marketable tuber yield was very high in the sole crop. Moreover, in intercropped plots with garlic, coriander and chilli, additional return increased the net income than a sole crop.

Resistant Varieties

Isolation of clones that are resistant or less susceptible to weevil has been one of the important lines of work. Clones with a long neck and deep rooting nature escape severe damage. Some of the clones with a moderately long stalk and deep rooting nature *viz.* S 3, S 13, S 234, S 238 and S 248 showed lower susceptibility to weevil. The resistance can be considered only in relative terms in sweet potato as purely resistant genotypes are not available in India or elsewhere. The cultivation of less susceptible or relatively resistant clones under good agronomic practices will help to reduce weevil damage leading to increased production (Palaniswami and Mohandas, 1992; Rajamma, *et al.*, 1993).

Biological Control

Different parasites, predators and pathogens have been recorded on different stages of weevil in India. Two hymenopteran ecoparasitoids (*Rhaconotus* sp and *Bracon* sp) and dipteran predator (*Drapetis* sp) were identified on grubs. Forty per cent of the grubs collected from aerial vines were found parasitized by *Rhaconotus* sp and *Bracon* sp. *Metarhizium anisopliae*, a commonly prevalent fungus in the field was found to infect the adults. Weevil could be infected by *Bacillus thuringiensis* (60-64 per cent). Among the biotic agents attacking the weevil, larval mortality factors were maximum and the parasitoids itself contributed up to 45 per cent, while others contributed only 5 per cent. *Rhaconothus menippus* and *Metarhizium anisopliae* were identified as potential ones in the biological control of *Cylas formicarius*. An entomopathogenic nematode was found frequently occurring in the soil which infects the weevil in the field. The nematode was identified as *Heterorhabditis indicus* (Palaniswami, 1994, 1999, 2000).

Sex Pheromone

The synthetic sex pheromone for the control of SPW was tried and the semiochemical, Z(3)-dodecen 1-ol-E(2) butenoate pheromone attracting males was

Disease of Sweet Potato

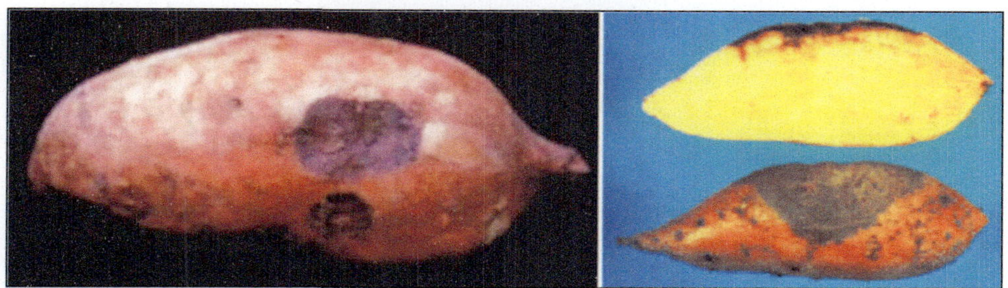

Black Rot of Sweet Potato Tubers

Root Rot at the Tips   Anthracnose on Stem

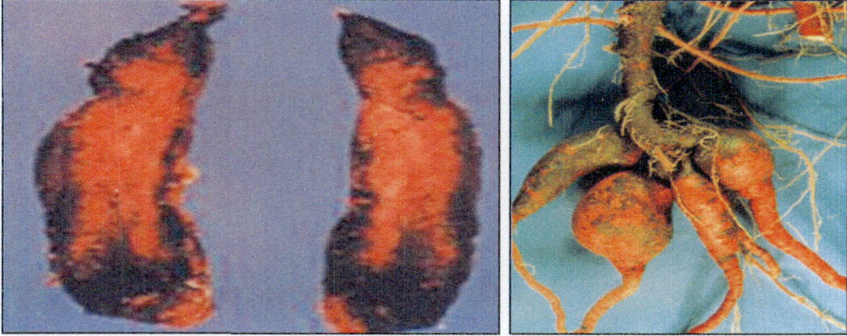

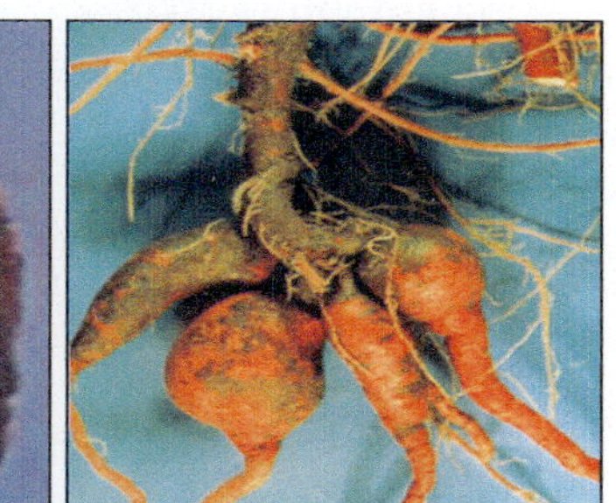

Charcoal Rot-Infected Sweet Potato Tubers (Left) and a Hill of Sweet Potatoes Showing Scurf Lesions (Right) that had Spread from the Stem of the Mother Plant onto the Daughter Roots

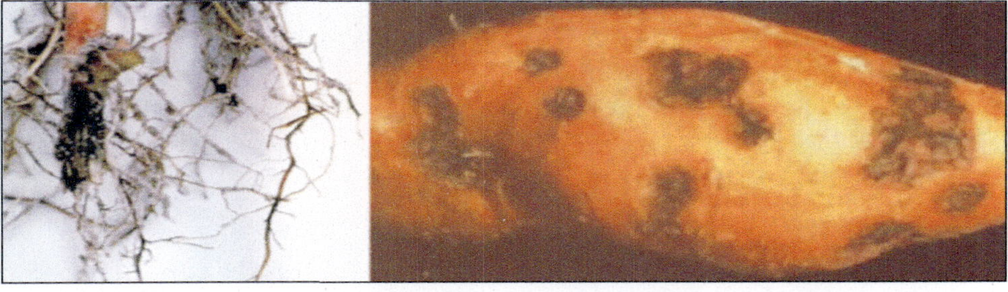

Rootlet Rot (Left) and Scab Lesions with Irregular Margins (Right) on Tuber

standardized. One mg was useful for mass trapping of male adult weevils. 10 mg was enough for monitoring and the results were spectacular. Field experiments were conducted for two crop seasons (June-September and November-February) and the pheromone traps were installed @ 1 trap/100 m^2 area from the first day of planting till harvest and thereafter for 1 month to collect the residual population (Anon., 2012). The male weevils were collected and killed in alternate days from the traps. During June-September season nearly 12,000 male weevils were collected from the experimental field of 400 m^2 (within 15 weeks) and 8300 weevils in November-February season. The residual populations in June-September and November-February were 6300 and 1500 respectively. In the experimental fields, the tuber damage was less than 10 per cent in both the seasons while in the control the damage ranged from 33 and 39 per cent respectively. The population build-up of the weevil was drastically reduced in the experimental field by continuous elimination of male weevil. The mean weevil population/stem and tuber were 1.3 in the treatment and 9.5 in the control. Thus synthetic sex pheromone has been found as an efficient tool to contain the population build-up of weevil and consequent damage of the crop (Rajamma *et al.*, 1993; Palaniswami and Chattopadhyay, 2006, Tarafdar and Sarkar, 2006).

Isolation of clones that are resistant or less susceptible to weevil has been one of the important lines of work. Clones with a long neck and deep rooting nature escape severe damage. Some of the clones with a moderately long stalk and deep rooting nature *viz.* S 3, S 13, S 234, S 238 and S 248 showed lower susceptibility to weevil. The resistance can be considered only in relative terms in sweet potato as purely resistant genotypes are not available in India or elsewhere. The cultivation of less susceptible or relatively resistant clones under good agronomic practices will help to reduce weevil damage leading to increased production (Palaniswami and Mohandas, 1992; Rajamma, *et al.*, 1993).

Integrated Pest Management (IPM)

Sweet potato weevil incidence is a complex phenomenon influenced by several factors such as seasons, type of soil, soil moisture, age of the crop, varietal susceptibility and cultural practices including crop sanitation, crop rotation *etc.* Therefore, a proper technology of integrated pest management utilizing mechanical, cultural, chemical and biological methods systematically would be an ideal one for the pest management. Thus we can integrate the effective mechanical and cultural practices, use of resistant/tolerant variety and or chemical control with due consideration to the parasitoids, predators and pathogens of the weevil. An integrated approach for the control of weevil using a new sex phenomenon developed at BARC was found to be cost-effective and sustainable.

1. Collection and Removal of *Ipomoea* species weeds which will otherwise serve as alternate hosts.
2. Selection of pest-free planting materials from the nursery.
3. Proper earthing up whenever there is soil crack, especially after tuber bulking.

Integrated Pest Management against Sweet Potato Weevil

Installation of Pheromone Trap in Sweet Potato Field

Pests of Sweet Potato

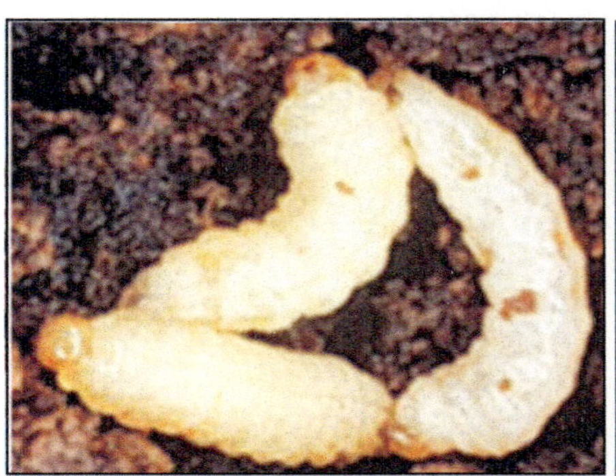

Adult of Sweet Potato Weevil

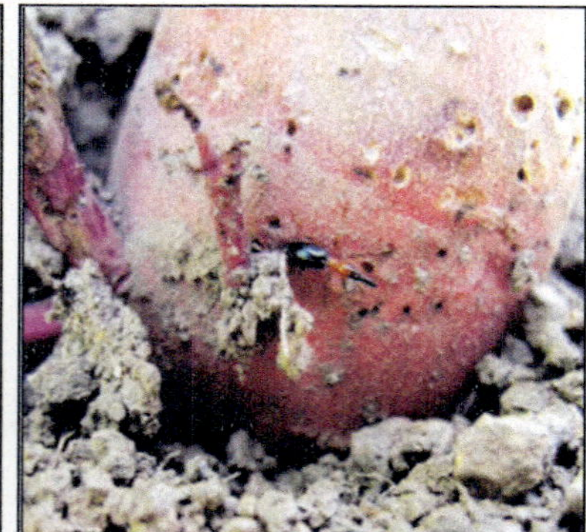

Sweet Potato Weevil Infested Tubers

Pest of Sweet Potato

Fold Produced by Green Leaf Folder with Web Visible through Feeding Hole (Left)
Browning and Holes Inside the Folds Produced by Brown Leaf Folder (Right)

Hornworm Feeding on Leaf (Left) and Fields Severely Damaged by Leaf Miner (Right)

Stem Borer Damage on Sweet Potato

Pests and Rat Damage on Sweet Potato Tuber

White Grub Damage on Sweet Potato Tubers (Left); Grubs in Soil (Upper Right) and Sweet Potato Roots Injured by Cucumber Beetles (Bottom Right)

White flies on Sweet Potato Leaf

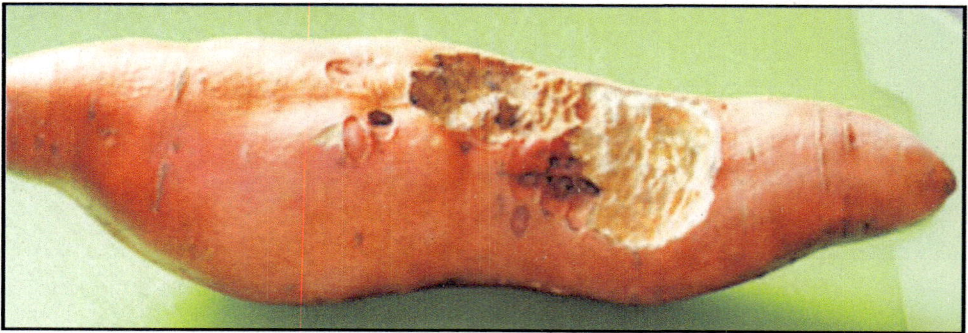

Rat Damage on Sweet Potato Tuber

Nematode Infestation in Sweet Potato Tubers

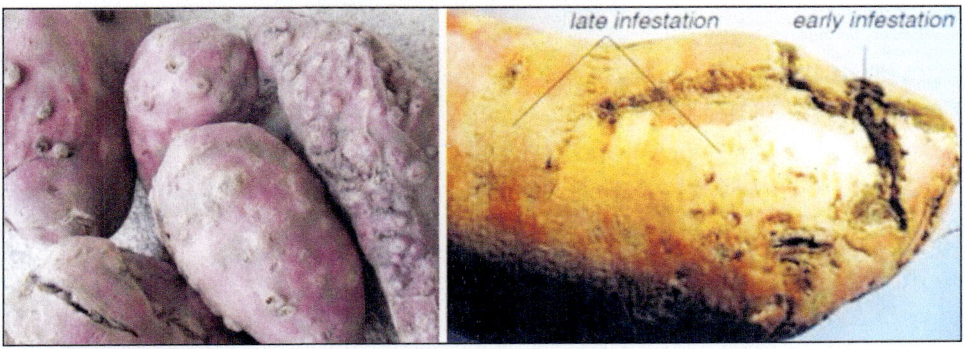

Root-knot Galls on Sweet Potato Tubers (Left), Cracks in Sweet Potato Tuber Due to root-knot Nematode Infection (Right)

Symptoms of Reniform Nematode Infection

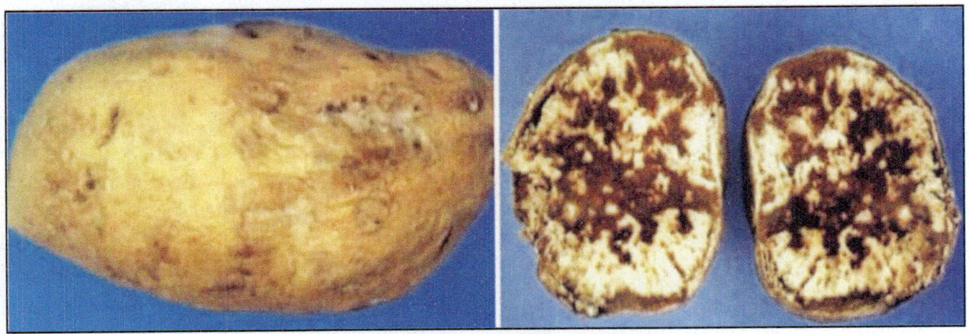

Symptoms of Brown Ring Nematode

4. Basal application of mahua cake or *Calophyllum* cake @ 2500 kg/ha or leaves of *Clerodendron* or *Chromolaena* @ 5000 kg/ha.
5. Adoption of crop rotation utilizing only one sweet potato crop in the sequence.
6. Retention of soil moisture in the field.
7. Use of resistant/tolerant varieties.
8. Encouraging the naturally occurring biotic agents.
9. Installation of pheromone traps in the field from first DAP to trap and kill the male weevil at a distance of 10 x 10 m using BARC sex pheromone. Continue trapping till a fortnight after harvest.
10. Harvesting at 110 days.
11. Destruction of weevil infested crop residues.

This IPM programme resulted in the increased production of marketable tubers (over 53 per cent) (Palaniswami, 2000; Palaniswami and Chattopadhyay, 2006). On-farm trials of IPM using the suggested components indicated a drastic reduction of sweet potato weevil damage in tubers and enhanced marketable tuber yield. Yam bean has been identified as a barrier crop to the weevil. Yam bean as border crop in sweet potato or as an intercrop in between two rows of sweet potato has helped in the reduction of weevil damage in the crop (Anon., 2012).

11.2.2 Vine Borer (*Omphisa anastomosalis*)

Generally, the pest is widespread and sporadic. White stout caterpillars bore into the vines causing terminal drying and wilting. Larvae occasionally bore into tubers. The larval period is about 28-50 days. Egg and pupal periods are about one and two weeks, respectively. The adult moth has straw-coloured wings with wavy markings. Crop hygiene and crop rotation reduce incidence. Spraying 0.05 per cent fenthion, fenitrothion or endosulfan in the case of a severe infestation is advocated. Another borer (*Metasis coniotalis*) is also found in India (Pillai and Palaniswami, 1984).

11.2.3 Other Pests

The other pests of minor importance are stem borer (*Omphisa anastamosalis*), leaf feeding beetles (*Aspidomorpha miliaris, Chirida sexnotata* and *Metriona circumdata*), leaf-feeding caterpillars (*Diacrisia obliqua*), brown hairy caterpillar (*Euchromia polymena*), sweet potato sphingid (*Herse convolvuli*), sweet potato horn worm (*Agrius cingulatus*), southern armyworm (*Prodenia eridania*) and the spiny butterfly caterpillar (*Grecis orithiya*). These leaf feeders can be controlled by spraying monocrotophos or endosulfan at 0.05 per cent. Besides these insect pests, red mite (*Tetranychus telarius*) and rust mite (*Epitremerus* sp.) also cause damage.

Plant-parasitic nematodes like *Meloidogyne incognita, Rotylenchulus reniformis, Ditylenchus destructor, Helicotylenchus* sp., *Tylenchorhynchus* sp. and *Dorylaimus* are also observed. Rats also damage sweet potato tuber in the field. For controlling rats, live burrows in the field or immediate vicinity should be fumigated with aluminium phosphide or with cynogas (Anon., 2001).

11.3.0 Disorders

Tuber cracking at maturity in some cultivars is often a problem in sweet potato cultivation. This disorder generally occurs in excess or shortage of soil moisture or delay in harvest. Application of potassium reduces the tendency of tuber cracking.

The tuber skin disorder of sweet potato was reported by Kawashiro and Takeda (1991). In pot trials, the incidence of the disorder increased both in extent and severity as the water supply increased. In field trials, the severity of the disorder was decreased by increasing the depth of rotary cultivation and the incidence varied with cultivar.

12.0 Crop Improvement

Sweet potato is usually propagated vegetatively. Only for breeding works, it is propagated by seeds. Sexual reproduction in this crop is limited for two reasons – (i) the non-blooming habit, and (ii) low fertility of blossoms. There is a wide range of variability found amongst cultivars due to its highly heterozygous nature. Presence of self- and cross-incompatibility between some cultivars hinders breeding work (Williams, 1964; Martin, 1965). Incompatibility in sweet potato is sporophytic (Martin, 1965). Seed setting is very poor even in the most compatible combination in sweet potato. Even controlled pollination produces less number of seeds. High seed set may indicate good physiological balance and lead to long term yield improvement (Jones and Dukes, 1976). In light of the above factors, the improvement of sweet potato is done principally through introduction, selection and hybridization. X-ray irradiation has been used for creating new variability. Use of polyploidy may be beneficial as an aid in inter-specific hybridization. Anther culture also has been used recently in the sweet potato improvement programme. Introduction and collection of both indigenous and exotic genetic materials have been made.

Fuglie (2007) reported that more than 7000 indigenous accessions are being maintained in the gene banks of sweet potato research programmes in Asian countries. In recent years, most countries in Asia, in collaboration with International Potato Centre (CIP), have introduced non-indigenous material into their breeding efforts. Indigenous genetic resources together with introduced clones from Japan and USA have been the source of parental material for breeding and selecting new varieties. By the mid-1990s, global ex situ gene bank collections contained more than 32,000 accessions, with ~21 per cent held by CIP (FAO, 1997). In India, CTCRI and All India Coordinated Research Improvement Project (Tuber Crops) are mainly responsible for a large collection of germplasms of sweet potato. CTCRI is maintaining over 1796 genetic stocks (George *et al.*, 2018) of sweet potato obtained from different parts of India.

In India, Rajendra Prasad Agricultural University, Bidhan Chandra Krishi Viswavidyalaya and Sri Konda Laxman Telangana State Agricultural University are maintaining a large number of indigenous sweet potato germplasms in the field gene bank (George *et al.*, 2018). Field gene banks are normally maintained by replanting every quarter (Naskar *et al.*, 1996) but this is difficult owing to the high genetic variability of the plant, possible loss of accessions, chances of misidentification and requirement for heavy inputs (Edison *et al.*, 2006). In India guidelines for Distinctiveness, Uniformity and Stability (DUS) of sweet potato has been developed and published by Protection of Plant Varieties and Farmers' Rights Authority (PPV and FRA), Government of India (2018).

At the beginning of the improvement programme, Pusa Safed and Pusa Red cultivars were developed from selections. As the crop is highly heterozygous, screening from open-pollinated seeds was beneficial to find out desirable phenotypes. A large number of promising lines: 75-OP-1, 75-OP-2, 75-OP-5, 75-OP-6, 75-OP-9, 75-OP-59, 76-OP-217 and 77-OP-34 were developed through screening at CTCRI (Anon., 1983). In the USA, most of the older cultivars were evolved from root selection and later from controlled crosses and bud mutations (Steinbauer, 1962). Hybridization has been used to combine desirable plant characters into the hybrid. Most of the cultivars are self-sterile and self-incompatible (Purseglove, 1977; Martin, 1984). It is, therefore, not necessary to emasculate the flowers, but in case of self-fertile cultivars, emasculation is done on the day before anthesis by cutting the corolla tubes at the mid-point and removing the adhering stamens. The emasculated flower buds and flowers to be used as a source of pollens are bagged. Pollination is done between 6 a.m. on the day following emasculation by placing the pollen of the desired male to the stigma of the female parent. Pollen grains are generally released over the stigma by splitting open the anther with small fine-pointed forceps. After pollination, the flowers are again bagged and kept until berries are formed (Naskar, 1984). As sweet potato is a vegetatively propagated crop, segregation of character is found in F_1 generation following hybridization. Therefore, selection work starts in the F1 generation itself and cultivars are evolved through continuous selection.

In Japan, Minamiyutaka cultivar which has high yield and resistance to some diseases and insects was developed through interspecific hybridization (Kobayashi and Miyazaki, 1976). Interspecific transfer of resistance genes of *Cylas formicarius* into *I. batatas* was a major objective and mass selection was practised for resistance to insects and diseases and horticultural characteristics (Jones *et al.*, 1976). The relationship between root tuber characteristics of immature sweet potato and earliness was studied by Fang *et al.* (1992) in 1680 plants derived from 25 crosses. Tuberosity, bulking and mini-tuber production was closely related to earliness. Tuberosity evaluation at 25-30 days allowed the selection of 76.4 per cent of early-maturing clones. It was suggested that when using mini-tuber evaluation to advance generations, the first clonal generation plants can be obtained in the current year. Mutation breeding using X-rays and gamma-rays has been used for evolving cold-tolerant type (Mie, 1973; Mie *et al.*, 1973) and other characters (Kukimara, 1971). Kamalam *et al.* (1977) observed that tuber yield was positively correlated with the number of tubers but negatively correlated with vine length and vine weight. In inheritance studies, Constantine (1965) reported that skin colour is controlled by complementary genes and colour, sweetness and flavour are positively correlated with baking qualities.

Disease and pest resistance and nutritive value are two areas in which more attention should be paid for improving this crop. In areas where sweet potato forms a dominant part of the diet and protein intake is low, Oomen *et al.* (1961) suggested the upgrading of amino-acid content in sweet potato by breeding, since cultivars exhibit considerable variability in this property. Besides these, suitable type for mechanical harvesting and the use of tubers in processed foods are other objectives in improving this crop which is likely to have a considerable impact (Yen, 1976).

13.0 Biotechnology

Application of biotechnology has proved very useful in crop improvement and utilization programmes of tropical tuber crops.

13.1.0 Micropropagation

A comprehensive review by Unnikrishnan and Sheela (2001) on biotechnology of sweet potato provides scientific and applied aspects including advances in micro-propagation. Shoot-tips along with a leaf primordium as well as meristem, cultured on MS medium with GA_3 and kinetin produced plantlets which were tested as virus-free. GA_3 and kinetin or GA_3 alone was effective for faster development of meristem culture (Unnikrishnan *et al.*, 1991). *In vitro* cultures could be established in sweet potato using meristem-tip (0.1 mm) excised from shoots derived from tuber sprouts (Zamora and Gruezo, 1993). Sweet potato cultures could be initiated from nodal cultures as well as internodal callus cultures. Growth of *in vitro* cultures of sweet potato was enhanced under optimal photoautotrophic micropropagation conditions (without sugar in the medium and enriched CO_2 concentration) (Kozai *et al.*, 1996). A two-stage protocol was found better for micro-propagation of 27 sweet potato genotypes. First of all, leaf explants were grown on MS medium with 2,4-D (0.1 mg/l) and zeatin (0.2 mg/l) until the base of the petiole began to swell (2-4 days). Then they were transferred to medium with zeatin (0.8 mg/l) where high frequency of shoot regeneration was found to occur (Porobo Desai *et al.*, 1995).

13.1.1 Embryo Culture

A good number of reports on tissue culture of sweet potato (*Ipomoea batatas* L.) using different tissues of various cultivars with varying level of efficiency and reproducibility are available but callus induction and plantlet regeneration are recalcitrant and limited to a few genotypes in response to different treatments in vitro. Xue (1988) also reported plantlet regeneration from unfertilized ovules of *Ipomoea* species. Embryos excised from fertilized ovules of 6- to 12-day-old capsules were cultured on MS medium supplemented with NAA, BA, GA3 separately and in combinations. Seedlings were recovered from late globular stage onwards but best recovery was from advanced embryo stages. GA3 was found essential for initial morphogenesis of globular and heart stages. Differentiated embryos produced multiple shoots on MS medium with NAA (1.0 μM), BA (2.0 μM) and GA3 (0.5 μM) (Mukherjee *et al.*, 1991a). A large number of embryogenic calli that are capable of regenerating whole plantlets on all the different types of explants tested, however, attempts made in plantlet regeneration have failed and discusses ways in which improvement of the regeneration media tested can be made for plant regeneration (Michael 2009). An efficient method of embryogenic calli formation from leaf explants and regeneration of plant and shoot tip culture have been developed using the level of BAP and Thidiazuron (TDZ) exhibits the unique property for shoot regeneration in sweet potato (Murthy *et al.*, 1998, Moitra and Tarafdar, 2009). Three sweet potato cultivars that are widely grown in Kenya; KSP36, Kemb36 and Mweu mutheke along with an exotic model cultivar Jewel were exhibited maximum somatic embryogenic induction, at 96.72 per cent, which was obtained from explants cultured on Linsmaier and Skoog salts and vitamins medium supplemented with 0.5 mg/l dichlorophenoxyacetic acid and 0.2 mg/l zeatin riboside (Mbinda *et al.*, 2016).

Compact and organized callus, as well as friable callus, was obtained. The compact callus gave rise to embryoids on transfer to growth regulator-free medium which later developed into plantlets. Direct shoot regeneration was obtained from roots of sweet potato leaves grown *in vitro* and incubated at 12-h photoperiod (Mukherjee *et al.*, 1993). Increased levels of picolinic acid (2 to 3 mg/l) in combination with kinetin (0.25 and 1.0 mg/l) induced direct somatic embryogenesis from apical meristem-tips which later showed secondary embryogenesis as well as development into whole plants (Desamero *et al.*, 1994).

13.1.2 Anther and Pollen Culture

Response to anther culture in sweet potato also depended on genotypes as well as on media (Mukherjee *et al.*, 1991b). Callus induction was obtained on MS medium with NAA and cytokinins (BA or kinetin or 2iP). Regeneration was achieved on a subsequent culture of anthers on MS medium with 2,4-D and MS basal medium without growth regulators. The plants could be transferred and established. They showed heteroploidy (chromosome number (70 to 80) with anomalies like lagging chromosome and disturbed polarity of metaphase plates. Lower ploidy (2n=51) was also noticed.

13.1.3 Protoplast Culture and Fusion

Regeneration of plantlet from protoplast of petiole was reported by Murata *et al.* (1987) and Dhir *et al.* (1998). Murata *et al.* (1989) also isolated of protoplasts from young petioles of Chugoku 25 plated onto MS medium containing various combinations of growth regulators. Shoot formation occurred on media supplemented with glutamic acid, asparagine and proline. Distant hybridization of sweet potato (*I. batatas*) and two wild species (*I. trifida* and *I. lacunosa*) were obtained through protoplast fusion and regeneration. Electrofusion was more effective than polyethylene glycol (PEG) treatment as this alone induced regeneration from fused calli (Belarmino *et al.*, 1996). Protoplasts of sweet potato cv. Kokei No. 14 and *I. lacunosa* were fused by the PEG method. Fused protoplasts were cultured in a modified MS medium containing 0.05 mg/l 2,4-D and 0.5 mg/l kinetin. First cell division occurred within 4-5 days. At 12 weeks after plating small calli up to 1-2 mm in diameter were transferred to MS medium supplemented with 0.05 mg/l, 2,4-D and 0.5 mg/l kinetin resulting in rapid callus proliferation. Calli were cultured on MS medium supplemented with 3.0 mg/l BAP and 9 regenerated plants were obtained. Analysis of peroxidase and esterase isozymes and RAPDs indicated that 2 or the 9 regenerated plants (KL1 and KL3) were interspecific somatic hybrids of Kokei No. 14 and *I. lacunosa* (Liu *et al.*, 1998).

13.2.0 Molecular Markers

Starch gel electrophoregrams of nine enzymes that expressed polymorphism was used to identify sweet potato cultivars (Kennedy and Thompson, 1991).

Morphological descriptors are potentially useful for clonal identification and genetic distance estimation. Genotypic and environmental variance components of each character should also be estimated to design breeding strategies for character improvement. On the other hand, environmental influences on phenotypically plastic genotypes make the process of phenotypic evaluation complex (Lin *et al.*, 2007).

Comparative studies in plants have shown that SSR markers, which are single-locus markers with multiple alleles, are more variable than other markers and provide an effective means for discriminating between genotypes (Powell *et al.*, 1996). It is possible as the microsatellites show a high level of polymorphism (Akagi *et al.*, 1997) and co-dominant inheritance. Molecular markers have also shown important and critical applications in the assessment and conservation of genetic variation.

In recent years it has become well recognized that PCR-based DNA markers make powerful tools for genetic analysis and for breeding programs because of their simplicity and ease of handling. Primers designed to flank simple sequence repeats (SSRs) loci were developed (Jarret and Bowen, 1994) and used to genotype sweet potato cultivars (Hu *et al.*, 2004), determine genetic relationships between elite sweet potato cultivars and conclude that poly cross-derived cultivars originated from various sources (Hwang *et al.*, 2002), for analyzing the genetic diversity of Latin American and East African sweet potato landraces (Zhang *et al.*, 2001; Gichuru *et al.*, 2006) as well as for analyzing the genetic diversity of sweet potato landraces (Veasey *et al.*, 2008). Differences in sequences of SSR products were also studied to determine levels of polymorphism among Sweet potato cultivars (Butler *et al.*, 1999).

Genetic diversity and relationships of Indian 82 cultivars of sweet potato representing white, pink, red, orange and purple coloured root skin and white, yellow, orange flesh of the sweet potato, were examined using Simple Sequence Repeats (SSR) at ICAR-AICRP on Tuber Crops. A total of 40 SSR primers were tested among the 82 cultivars. Most of the primers contained di/tri/tetra nucleotide repeat motifs. 3552 fragments were generated with the selected primers in these cultivars and an average of 43.32 bands per cultivar was amplified. The SSR fragments were found to be highly polymorphic (90.66 per cent) among the 82 sweet potato lines. Levels of polymorphism with the SSR markers had a total of 75 loci, all were polymorphic between 1 and 4 loci per primer. The polymorphic SSR loci revealed a diverse relationship among the sweet potato cultivars of which reflected in eight major clusters (Anonymous, 2015).

To assess the genetic diversity and interrelationship of sweet potato cultivars from Oceania and Latin America and to test the hypothesis of human transfer of this crop to the Pacific Islands in prehistoric time, seventy-six sweet potato cultivars from Peru-Ecuador, the Philippines and eight Oceania countries were analyzed using amplified fragment length polymorphism (AFLP) (Rossel *et al.*, 2001).

Multidimensional scaling (MDS) and analysis of molecular variance (ANOVA) revealed wide genetic variation in the Oceania gene pool, greater than that of Peru-Ecuador. There was a significant sweet potato 'gene flow' from Mexico to Oceania. In contrast, there is little association between the Peru-Ecuador germplasms and that of Oceania.

13.3.0 Genetic Transformation

Although several high throughputs advanced technologies developed for plant genetic transformation, little success has been achieved in genetic engineering of sweet potato due to fewer difficulties in an efficient and reproducible system in plant regeneration from transformed explants of sweet potato. Okado *et al.* (2001) reported development of transgenic plant having resistant to Russet Crack disease caused by Sweet potato Feathery Mottle Virus (SPFMV). They developed an expression vector carrying coat protein and selectable marker genes driven by CuMV 35S promoter.

Agrobacterium vectors and particle bombardment approaches were being employed to introduce marker gene (gusA, nptII and hpt) into sweet potato (Otani *et al.*, 1998). The disease resistance geno (Shiva) is encoded for a lytic peptide which is highly antimicrobial and the artificial storage protein (*aspI*) gene is encoded for essential amino acids. The expression of these genes were being selectively targeted to the storage food by fusing the sporamin promoter to the aspI coding sequence (Prakash *et al.*, 1996).

Clones of sweet potato transgenic plants (cv. Jewel), carrying the *cryIIA* gene from *Bacillus thuringiensis* subsp. *tenebrionis* (*Bt.t*) were obtained. Molecular tests and biological activity experiments against sweet potato weevil (*Cylas formicarius*) were performed. The *cryIIIA* delta-endotoxin gene was cloned under the regulation of the 35S CaMV promoter and the TMV Omega fragment (Moran *et al.*, 1999).

Mary *et al.* (2016) reviewed the genetic transformation of sweet potato and its success story against several biotic and abiotic stresses. Sweet potato expressing the endotoxin cry8Db, cry7A1 and cry3Ca genes showed lower sweet potato weevil infestation than non-transformed lines. Transgenic cultivar 'Xushu18' expressing the oryzacystatin-1 (OC1) gene showed enhanced resistance to sweet potato stem nematodes. Sweet potato line 'Chikei 682-11' expressing the coat protein (CP) exhibited resistance to the sweet potato feathery mottle virus (SPFMV). Transgenic expressing the rice cysteine inhibitor gene oryzacystatin-1 (OC1) also exhibited resistance to the SPFMV. Besides, effective and efficient transmission of genes of interest from selected parents to their progeny needs comprehensive knowledge on gene action (Mwije *et al.*, 2014).

Transformation for weevil resistance is among the first genetic transformations that were attempted in sweet potato (Mwanga *et al.*, 2011). This involved the use of proteins that reduced the digestibility of sweet potato for the insects. Sweet potato cultivar KB1 was transformed with a delta endotoxin gene cry8Db from *Bacillus thuringiensis* via *Agrobacterium* mediated transformation. The cry8Db gene codes for an insecticidal protein that has been shown to kill the larvae of many insects. Twenty-one transgenic lines were regenerated from transformed calli. The transgenic plants were shown to have a lower sweet potato infestation level than the untransformed lines (Ngoc *et al.*, 2015). Most of the existing transformation and regeneration protocols for sweet potato are confined to particular cultivars which further limits the development of new varieties (Ngoc *et al.*, 2015; Ogero *et al.*, 2012). Recently the structural and functional genes for anthocyanin biotransformation in purple-fleshed sweet potato were characterized in ICAR-AICRP on Tuber Crops, BCKV. The major up- and down-regulated genes for an accumulation of anthocyanin in sweet potato *viz.* IT4 partial gene (Genbank accession No. HE980452) responsible for the regulatory gene for Sporamin A precursor molecule expresses itself predominantly in the flesh of tuberous root), 3GT partial gene (Genebank accession No.HE978836) encodes UDPG-flavonoid-3-0-glucosyltransferase (UF3GT) enzyme coding gene 3GT and IT666 (Genbank accession No. HE980451) and IT666 gene (Genbank accession No. HF937132) for R2R3-type MYB genes in tissue and cultivar-specific expression of anthocyanin were characterized (Chowdhury *et al.*, 2019) which would be useful for future molecular breeding of sweet potato.

14.0 References

Abano, E.E., Teye, E., Amoah, R.S. and Tetteh, J.P. (2011) *Asian J. Agric. Res.*, **5**: 1-12.

Adenuga, W. (2010) *J. Food Tech.*, **8**: 223-228.

Ahmed, K. and Samad, M.A. (1976) *Bangladesh Hort.*, **4:** 32-33.

Ahn, Y.S., Jeong, B.C., Oh, Y.B., Cho, S.Y., Min, K.S., Lee, H.B. and Kim, T.G. (1998) *J. Crop Science-II*, **40:** 134-141.

Ali, M.S., Bhuiyan, M.K.R., Mannan, M.A. and Rashid, M.M. (1991) In: Sweet potato in South Asia: Postharvest, Handling, Storage, Processing and Use. (Eds. T. R. Dayal, G. J. Scott, G. T. Kurup, C. Balagopalan), Proceedings of a workshop held at CICRI, Trivandrum, 9-13 September, pp 13-22.

Anonymous (2013) *Annual Report*, AICRP on Tuber Crops, BCKV, Kalyani, West Bengal, India.

Anonymous (2014) *Annual Report*, AICRP on Tuber Crops, BCKV, Kalyani, West Bengal, India.

Anonymous (2015) *Annual Report*, AICRP on Tuber Crops, BCKV, Kalyani, West Bengal, India.

Anonymous (2016) *Annual Report*, AICRP on Tuber Crops, BCKV, Kalyani, West Bengal, India.

Anonymous (2017) *Annual Report*, AICRP on Tuber Crops, BCKV, Kalyani, West Bengal, India.

Anonymous (2018) *Annual Report*, AICRP on Tuber Crops, BCKV, Kalyani, West Bengal, India.

Anonymous (2019) *Annual Report*, AICRP on Tuber Crops, BCKV, Kalyani, West Bengal, India.

APEDA -Agricultural and Processed Food Products Export Development Authority (2017) Ministry of Commerce and Industry, Government of India.

Badillo-Feliciano, J. and Lugo-Lopez, M.A. (1976) *J. Agric. Univ., Puerto Rico*, **60**: 597-605.

Bechoff, A., Tomlins, K., Dhuique-Mayer, C., Dove, R. and Westby, A. (2011) *Int. J. Food Sci. Tech.*, **46**: 52-60.

Belarmino, M.M., Abe, T. and Sasahara, T. (1996) *Plant Cell Tissue Organ Cult.*, **46**: 195-202.

Biswas, D., Sen, H., Som, M.G. and Jana, P.K. (1980) *Veg. Sci.*, **7**: 1-5.

Biswas, J. (1981) *Ph.D. Thesis*, Bidhan Chandra Krishi Viswavidyalaya, West Bengal.

Biswas, J., Sen, H. and Mukhopadhyay, S.K. (1989) *J. Root Crops*, **15**: 123-125.

Biswas, J., Sen, H. and Mukhopadhyay, S.K. (1990) *J. Root Crops*, **16**: 1-3.

Biswas, V.R. and Singh, I.J. (1990) *Veg. Sci.*, **17**: 82-86.

Borroto, M., Borges, O., Toledano, M., Gell, P., Acosta, R., Saiz, J., Lobaina, P. and Mendez, A. (1990) *Ciencias del Suelo, Riego y Mecanizacion*, **2**: 37-46.

Bosewell, V.R. (1950) *USDA Farmers Bull.*, 2020.

Bourke, R.M. (1985) *Field Crops Research*, **12**: 363-375.

Bovell-Benjamin, A.C. (2007) *Adv Food Nutr. Res.*, **52**: 1-59.

Brown, J.E., Woods, E.M. and Channell, B.C. (1998) *J. Veg. Crop Production*, **4**: 49-54.

Burri, J.B. (2011) *Comp. Rev. Food Sci. Food Safety*, **10**: 118-130.

Campbell, G.M., Hernandez, T.P. and Miller, J.C. (1963) *Proc. Amer. Soc. Hort. Sci.*, **83**: 618-622.

Chakraborty, A., Sen, H. and Goswami, S.B. (1993) *J. Potassium Res.*, **9**: 55-61.

Chatterjee, B.N. and Mondal, N.C. (1976) *J. Root Crops*, **2:** 25-28.

Chattopadhyay, A., Chakraborty, I. and Sen, H. (2002a) *The Hort J.*, **15**: 45–50.

Chattopadhyay, A., Sen, H., Chakraborty, I., Satapathy, M., Prakasa Rao, L.L.T. and Dutta, A. (2002b) *Indian J. Plant Genet. Resour.*, **15**: 223-226.

Chattopadhyay, A., Sen, H., Dutta, A., Soumik, M. and Satapathy, M.R. (2005) *The Hort. J.*, **18**: 59-63.

Chattopadhyay, A., Chakraborty, I., Mukhopadhyay, S.K., Kumar, P. R. and Sen, H. (2006a) *Acta Hortic.*, **703:** 211-218.

Chattopadhyay, A., Chakraborty, I., Kumar, P.R., Nanda, M.K. and Sen, H. (2006b) *J. Food Sci. Technol.*, **43**: 41-45.

Choudhury, B. (1976) In: *Vegetables*, National Book Trust, India, New Delhi.

Choudhury, N., Tarafdar, J. and Sur, P.P. (2019) *J. Crops Weeds*, **15**(1):1-16

Chowdhury, S.R. and Varma, S.P. (1997) *Indian J. Plant Physiol.*, **2**: 29-31.

Chupp, C. and Sherf, A.F. (1960) In: *Vegetable Diseases and their control*, The Ronald Press Company, New York.

Constantine, R.J. (1965) *Diss. Abstr.*, **25**: Order No. 65-3372, 8477.

Cooley, J.S. and Kushman, L.J. (1947) *Proc. Am. Soc. Hortic. Sci.*, **50**: 311-316.

Cooley, J.S. (1951) *Econ. Bot.*, **5**: 378-386.

Dayal, T.R., Upadhya, M.D. and Mehra, S.K. (1990) *Abst. Natioal Symp. on Tropical Tuber Crops*, p. 11.

Desamero, N.V., Rhodes, B.B., Decoteaw, D.R. and Bridges, W.C. (1994) *Plant Cell Tissue Organ Cult.*, **37**: 102-111.

de Pee, S., Bloem, M.W., Gorstein, J., Sari, M., Satoto, Y.R. and Shrimpton, R. (1998) *Amer. J. Clin. Nutr.*, **68**: 1068-1074.

Devi, C., Hynniewta, L. and Mitra, S. (2017) *Int.J.Curr.Microbiol.Appl.Sci*, **6**(10): 3257-3265.

Dhir, S.K., Oglesby, J. and Bhagsari, A.S. (1998) *Plant Cell Rep.*, **17**: 665-669.

DUS Guide lines of Sweet potato (2018) Protection of Plant Varieties and Farmers' Rights Authority (PPV and FRA) Government of India.

Edison, S. (2000) *Souvenir International Symposium on Tropical Root and Tuber Crops*, pp. 28-33.

Edison,S., Unnikrishnan, M., Vimala, B., Pillai Santha, V., Sheela, M.N. and Sreekumari, M.T. (2006) Biodiversity of tropical tuber crops in India. NBA scientific bulletin number - 7, National Biodiversity Authority, Chennai, Tamil Nadu, p. 60.

Eguichi, T., Kitano, M. and Eguichi, H. (1994) *Biotronics*, 23: 75-80.

El-Fouly, M.M., Masoun, Y.A. and El-Hindi, M.H. (1977) *World Crops*, **23**: 133.

Enyi, B.A.C. (1977) *J. Agric. Sci.*, U.K., **88**: 421-430.

Esonico, C.A., Escalada, R.G. and Trenucla, R.M. (1981) *Annals Tropical Res.*, **3**: 144-149.

Fang, Z.Y., Ye, Y.F., Li, B.Q. and Qian, L.S. (1992) *Acta Hortic.*, **18**: 96-102.

FAO (2018) *Production Year Book.*

Food and Agriculture Organization (FAO) (1997) FAO Statistics. In http://apps.fao.org.

Fauquet, C.M. and Stanley, J. (2003) *Ann. App. Biol.*, **142**: 165-189.

Fuglie, K.O. (2007) *HortScience*, **42**: 1200-1206.

Fujise, K. and Tsuno, Y. (1962) *Proc. Crop. Sci. Soc.*, Japan, **13**: 145-149.

Galze, N.C. and Hall, M.R. (1990) *Weed Technol.*, **4**: 518-523.

Gao, F., Gong, Y.F. and Zhang, P.B. (2000) *Crop Prot.*, **19**: 105-111.

Garcia-Espinosa, G. (1987) Documentos-de-Ciencia-y-Tecnica-Cien-cias-Agropecuarias-ISCAH, **3**: 72-84.

Garner, L.C., Mortley, D.G., Trotman, A.A. and David, P.P. (1995) *J. Root Crops*, **21**: 36-38.

Gautam, D. M. (1991) In: *Sweet potato in South Asia: Postharvest Handling, Storage, Processing and Use*. (Eds. T. R. Dayal, G. J. Scott, G. T. Kurup, C. Balagopalan), Proceedings of a workshop held at CICRI, Trivandrum, 9-13 September, pp. 23-28.

George, J., Sunitha, S. and Suresh Kumar, J. (2018) Compendium of Tuber Crops Technologies, ICAR-AICRP on Tuber Crops, Thiruvanthapuram, Kerala India, pp 18-20.

Ghosh, S.P. and Nair, R.G. (1984) *Indian Fmg.*, **53**: 5-8.

Gonzalez, F.R., Cadiz, T.G. and Bugawan, M.S. (1977) *Philippine J. Crop. Sci.*, **2**: 97-102.

Gonzalez, T.M., Aiazii, M.T., Perser, R.A. and Racca, R.W. (1983) *Revista de ciencias Agropecuarias*.

Goswami, S.B., Sen, H. and Jana, P.K. (1995) *J. Root Crops*, **21**: 77-81.

Gowda, P.H.R., Lingaiah, H.B., Seenappa, K. and Shivashankara, K.T. (1990) *J. Potassium Res.*, **6**: 79-81.

Hahn, S.K. (1977) *Acad. Press*, New York, pp. 237-248.

Hall, A.J. and Devereau, A.D. (2000) *Outlook on Agric.*, **29**: 275-282.

Hernandez, Teme, P. (1970) *Southern Co-op. Ser. Bull.*, 159, pp. 25-32.

Hill, W.A., Dodo, H., Hahn, S.K., Mulongoy, K. and Adeyeye, S.O. (1990) *Agron. J.*, **82**: 1120-1122.

Horticultural Statistics at a Glance (2018) Horticulture Statistics Division, Department of Agriculture, Cooperation and Farmers' Welfare Ministry of Agriculture and Farmers' Welfare Government of India.

Horton, D. (1988) Underground Crops. Long-term trends in production of roots and tubers. Morriton, A.R.USA: Winrock International Institute of Agricultural development.

Hossain, M.M., Siddique, M.A. and Choudhury, B. (1987) *Bangladesh J. Agric.*, **12**: 181-188.

Huang, J.C. and Sun, M. (1999) *Biotech. Techniques*, **13**: 277-278.

Huett, D.O. (1976) *Exp. Agric.*, **12**: 9-16.

Isghiguro, K., and Yoshimoto, M. (2006) In: *2nd International Symposium on sweetpotato and cassava*, held on 14-17 June, 2005, Kuala Lumpur, Malaysia, pp. 213-214.

Indira, P. and Kabeerathuma, S. (1990) *J. Root Crops*, **16**: 28-32.

Ino, M., Shinohara, S. and Yashiki, T. (1991) *Bull. Chiba Prefectural Agricultural Experiment Station*, **32**: 81-89.

International Potato Centre (2017) Accessed on 18.04.2018 from https://cipotato.org/research/sweetpotato-in-africa/.

Jadhav, A.C., Memane, S.A. and Konde, B.K. (1998) *J. Maharashtra Agric. Univ.*, **23**: 80-82.

Jan Low., Moses Nyongesa., Sara Quinn and Monica Parker (2015) http://www.rtb.cgiar.org/blog/2015/12/12/potato-and-sweetpotato-in-africa-transforming-the-value-chains-for-food-and-nutrition-security/.

Jones, A. and Dukes, D.D. (1976) *J. Amer. Soc. Hort. Sci.*, **101**: 3585-3588.

Jones, A., Dukes, P.D. and Cuthbert, F.P.Jr. (1976) *J. Amer. Soc. Hort. Sci.*, **101**: 707-714.

Jones, S.T. (1961) *Proc. Amer. Soc. Hort. Sci.*, **77**: 458-462.

Jusuf, M., Hilman, Y. and Setiawan, A. (2005) In: *2nd International Symposium on sweet potato and cassava*, held on 14-17 June, 2005, Kuala Lumpur, Malaysia, pp. 40-41.

Kamalam, P., Biradar, R.S., Hrishi, N. and Rajendran, P.G. (1977) *J. Root Crops*, **3**: 5-11.

Kandaswamy, D., Palaniswami, D. and Oblisami, G. (1988) *J. Root Crops*, **14**: 37-42.

Kapinga, R., Tumwegamire, S., Ndunguru, J., Andrade, M.I., Agili, S., Mwanga, R.O., Laurie, S. and Dapaah, H. (2010) Catalogue of orange-fleshed sweet potato varieties for Sub-Saharan Africa. International Potato Center (CIP), Lima, Peru, p. 40.

Karuri, E.G. and Hagenimana, V. (1995) *Zimbabwe J. Agric. Res.*, **33**: 83-91.

Karuri, E.G., Ojijo, N.K.O. and Sass, P. (1994) *Acta Hortic.*, **368**: 441-452.

Kasam, A.H. (1976) *Crops of the West African Semi-Arid Tropics*, ICRISAT, Hyderabad, p. 65.

Kennedy, L.S. and Thompson, P.G. (1991) *HortScience*, **26**: 300-302.

Key, D.E. (1973) *Root Crops, Tropical Products Institute*, London, pp. 144-159.

Kidmose, U., Christensen, L.P., Agili, S.M., and Thilsted, S.H. (2007) *Inn. Food Sci. Emer. Technol.*, **8**: 399-406

Kobayashi, M. and Miyazki, T. (1976) *Proc. 4th Int. Symp. Soc. Trop. Root Crops*, Columbia, 5557.

Komada, H. (1990) *Biological control of soil borne plant pathogens*, pp. 65-75.

Komaki, K., Regmi, H.N., Katayama, K. and Ramiya, S. (1998) *Breed. Sci.*, **48**: 281-286.

Kozai, T., Kitaya, Y., Kubota, C., Kobayashi, N. and Watanabe, S. (1996) *Acta Hortic.*, **440**: 566-569.

Kukimara, H. (1971) *Technical News, Institute of Radiation Breeding*, **8**: 2.

Kuo, K.E. (1972) *Taiwan Agric. Quart.*, **8**: 183-188.

Kurup, G. T., and Balagopalan, C. (1991) In: *Sweet potato in South Asia: Postharvest Handling, storage, Processing and Use* (Eds. T. R. Dayal, G. J. Scott, G. T. Kurup, C. Balagopalan), Proceedings of a workshop held at CTCRI, Trivandrum, 9-13 September, pp. 33–39.

Kushman, L.J. (1967) *USDA Marketing Bull.*, 38.

Kushman, L.J. and Wright, F.S. (1969) *USDA Farmers Bull.*, 1442 (rev.).

Lee, J.S., Kim, H.S., Chung, M.N., Ahn, Y.S., Jeong, B.C. and Bang, J.K. (2006) *Acta Hort.* (ISHS), **703**: 125-132.

Leon, J. (1976) *Proc. 4th Int. Symp. Soc. Trop. Root Crops, Columbia*, pp. 20-36.

Lewthwaite, S.L. and Triggs, C.M. (1995) *New Zealand J. Crop Hortic. Sci.*, **23**: 283-287.

Liu, Q.C., Mi, K.X., Zhou, H.Y., Ma, B. and Zhai, H. (1998) *Acta Agronomica Sinica*, **24**: 529-535.

Loebenstein G (2015) *Dis. Adv. Virus Res.*, **91**: 33-45.

Lowe, S.B. and Wilson, L.A. (1975) *Exp. Agric.*, **11**: 39-48.

Lutz, J.M. and Simons, J.W. (1972) *USDA Prod. Rsh. Rept.*, 147-1-17.

Maddipatla, R. and Mitra, S. (2017) *J. Pharmacog. Phytochem.*, **6**(6): 2313-2317.

Magoon, M.L., Krishnan, R. and Vijaya Bai, K. (1970) *Trop. Root Tuber Crops Newsl.*, **3**: 20-21.

Mandal, R.C. (1976) *Short Specialized Training Course on Sweet Potato Production Tech.*, CTCRI, Trivandrum.

Mandal, R.C., Singh, K.D., Maini, S.B. and Magoon, M.L. (1971) In: *Two Decades of Research*, CTCRI, Trivandrum, 1983, p. 69.

Martin, F.W. (1965) *Econ. Bot.*, **19**: 406-415.

Martin, F.W. (1972) *USDA Prod. Rsh. Rept.*, **147**: 1-17.

Martin, F.W. (1984) *Ann. Rept. Trop. Agric. Res. Sta. Mayaguez*, Puerto Rico, USA.

Martin, J.H., Leonard, W.H. and Stamp, D.L. (1976) *Principles of Field Crop Production* (3rd edn.), Macmillan Publishing Co., Inc., New York.

Mary, C. Imbo, Nancy, Budambula, L. M., Cecilia, M. Mweu, Joshua, Kiilu Muli and Sylvester, E.A. (2016) *Adv. Life Sci. Technol.*, **49**: 66-76.

McClelland, T.B. (1928) *J. Agric. Res.*, **37**: 603-628.

McDavid, C.R. and Alamu, S. (1980a) *Ann. Bot.*, **45**: 363-364.

McDavid, C.R. and Alamu, S. (1980b) *Trop. Agric.*, **57**: 113-119.

Mellow, D.A. (1946) *Ceres.*, **7**: 54-60.

Menezes, D.M.De., Cezar, T.I. and Carvalho, J.F.De. (1974) *Pesquisa Agropecuaria Brasileira*, Agronomia, **9**: 145-149.

Mie, T.S. (1973) *Genetics*, **74**: 183.

Mie, T.S., Yu, C.Y. and Chang, C.S. (1973) *Taiwan Agric. Quart.*, **9**: 28-40.

Mishra, S., Misra, S.S. and Sinha, U.P. (1992) *J. Root Crops*, **18**: 6-9.

Mishra, S., Mishra, S.S., Ray, K. and Singh, S.J. (1985) Proc. Nat. Symp. On Production and Utilization of Tropical Tuber Crops, 27-29 November, Trivandrum, India.

Mithra, S., V. S. and Somasundharam, (2008) *World Appl. Sci. J.*, **4**(4): 568-577.

Mitra, S., Tarafdar, J. and Palaniswami, M.S. (2010) *Acta Hortic.*, **858:** 205-212.

Mohankumar, B., Kabeerathuma, S. and Nair, P.G. (1984) *Indian Fmg.*, **33**: 35-37.

Mohankumar, C.R. and Mandal, R.C. (1972) In: *Two Decades of Research*, CTCRI, Trivandrum.

Mohankumar, C.R. (2000) Fourth International Training Course on Integrated production and Processing Technologies for sweetpotato, 21-28 November, Trivandrum, Kerala, India, pp. 59-69.

Moitra, D. and J. Tarafdar. (2009) *Adv.Plant Sci.*, **22**(II):377-380

Morales, A. (1981) *Cienca y Tecnica en la Agricultura, Vindas Hortalizas*, Y. Granos, **4**: 47-53.

Moran, R., Garcia, R., Lopez, A., Zaldua, Z., Mena, J., Garcia, M., Armas, R., Somonte, D., Rodriguez, J., Gomez, M. and Pimentel, E. (1999) *Plant Sci. Limer.*, **139**: 175-184.

Morita, T. (1967) *J. Japanese Soc. Hort. Sci.*, 36: 421-426.

Motoyasu Otani, Takiko Shimada, Takashi Kimura and Akira Saito (1998) *Plant Biotechnology*, **15** (1):11-16

Mukherjee, A., Nair, N.G. and Jos, J.S. (1993) *J. Root Crops*, **19**: 127-130.

Mukherjee, A., Unnikrishnan, M. and Nair, N.G. (1991a) *Plant Cell Tissue Organ Cult.*, **26**: 97-99.

Mukherjee, A., Unnikrishnan, M. and Nair, N.G. (1991b) *J. Root Crops: ISRC Nat. Sym.*, 1990, Special: 302-304.

Mukhopadhyay, S.K., Chattopadhyay, A., Chakraborty, I. and Bhattacharya, I. (2011) *Food Security*, **3** (3): 283-305.

Mukhopadhyay, S.K., Sen, H. and Jana, P.K. (1990) *J. Potash Res.*, **6**: 114-118.

Mukhopadhyay, S.K., Sen, H. and Jana, P.K. (1991a) *J. Root Crops*, **17**: 19-25.

Mukhopadhyay, S.K., Sen, H. and Jana, P.K. (1991b) *J. Root Crops*, **17**: 71-72.

Mukhopadhyay, S.K., Sen, H. and Jana, P.K. (1992) *J. Root Crops*, **18**: 10-14.

Murata, T., Hoshino, K. and Miyaji, Y. (1987) *Japanese J. Breed.*, **37**: 291-298.

Murata, T., Hoshino, K., Fukuoka, H. and Miyaji, Y. (1989) *Proceedings of Faculty of Agriculture, Kyushu Tokai Univ.*, **8**: 23-27.

Murthy, B. N. S., Murch, S. J. and Saxena, Praveen K. (1998) *Planta*, **34**: 267

Mwanga, R. O. M., Ghislain, M., Kreuze, J., Ssemakula, G. N., and Yencho, C. (2011) In: Agro-*Biotechnology, Biosafety and Seed Systems in Developing Countries*, pp. 25-31.

Mwije, A., Mukasa, S. B., Gibson, P. and Kyamanywa, S. (2014) *African Crop Sci. J.*, **22**: 79-87.

Naskar, S.K., Eswari Amma, C.S. and Nair, S.G. (1996) Proceedings of the workshop on the formation of a Network for the Conservation of Sweetpotato Biodiversity (Ed. V. Anatha Rao), 1–5 May, 1996, Bogor, Indonesia, pp. 57-80.

Nair, G.M. (1972) *M.Sc. (Ag.) thesis*, Kerala University.

Nair, G.M. and Nair, V.M. (1995) *J. Root Crops*, **21**: 17-23.

Nair, G.M., Ravindran, C.S. and Prasad, M. (1984) *Indian Fmg.*, **33**: 29-33.

Nam, S.Y., Jong, S.K., Rho, C.W. and Kim, I.J. (1996) *RDA J. Agri. Sci., Upland and Industrial Crops*, **38**: 135-142.

Nam, S.Y., Jong, S.K., Rho, C.W., Kim, I.J. and Park, S.K. (1996) *J. Agric. Sci. Soil Fert.*, **38**: 396-401.

Nambiar, I.P.S., Sadanandan, N. and Kunju, U.M. (1976) *Agric. Res. J., Kerala*, **14**: 189-190.

Nambiar, I.P.S., Sadanandan, N. and Kunju, U.M. (1977) *Agric. Res. J., Kerala*, **15**: 93-94.

Naskar, S.K. (1984) *Training-cum-Discussion Seminar on Sweet Potato and Tapioca held at RCCTCRI*, Bhubaneswar.

Nath, P.S. Ramdas Mandi and Tarafdar, J (2003) *J. Mycopathol. Res.*, **41**(2): 245-246.

Nath, R., Kundu, C.K., Majumder, A., Islam, Sk. J., Gunri, S. and Chattopadhyay, A. (2005) In: *2nd International Symposium on Sweet potato and Cassava*, held on 14–17 June, 2005, Kuala Lumpur, Malaysia, pp. 197-198.

Nath, R., Chattopadhyay, A., Kundu, C.K., Majumder, A., Islam Sk, J. and Gunri, S. (2007) *J. Root Crops*, **33:** 97-103.

Nawale, R.N. (1981) *J. Root Crops*, **7**: 29-31.

Nicholaides, J.J., Chancy, H.F., Mascagni, H.J., Wilson, L.G. and Caddy, D.W. (1985) *Agron. J.*, **77**: 466-470.

Nishiyama, I. (1961) *Proc. 10th Pacific Science Congress*, Hawaii, pp. 119-228.

Nishiyama, I. (1971) *Bot. Mag.*, **84**: 377-387.

Nishiyama, U., Miyazuki, T. and Sakamito, S. (1973) *Genetics*, **74:** 197.

Norman, M.C., Rodrjgues, N. and Cuadra, A. (1989) *Ciencia y Tecnica en la Agricultura*, **12**: 49-57.

O'Brien, P.J. (1972) *American Anthropologist*, **74:** 342-365.

O'Hair, S.K., McSorley, R., Parrqado, J.L. and Mathews, R.F. (1983) *Proc. Amer. Soc. Hort. Sci. Trop. Region*, **27**: 72-75.

Ochoa-Alejo, N. and Lopez-Gutierrez, F. (1987) *Ann. Bot.*, **59**: 595-597.

Ogero, K.O., Mburugu, G.N., Mwangi, M., Ngugi, M. M. and Ombori, O. (2012) *Res. J. Bio.*, **2**: 51-58.

Ofori, G., Oduro, I., Ellis, W.O. and Dapaah, H.K. (2009) *African J. Food Sci.*, **3**: 184-192.

Olurunda, M. (1979) In: *Small scale processing and storage of tropical root crops*, (Ed. D. L. Plucknett), Boulder: Westview.

Onwueme, I.C. (1978) In: *The tropical Tuber Crops*, Willey, New York, pp. 168–171.

Oomen, H.A.P.C., Spoon, W., Heesterman, J.E., Ruinard, J., Luyken, R. and Slump, P. (1961) *Trop. Georg. Medicine*, **13**: 55-56.

Padmaja, G. and Rajamma, P. (1982) *J. Food Sci. Tech.*, **19**: 162-163.

Palaniswami, H.S., Mohandas, N. and Visalakshi, K. (1992) *J. Res. Crops*, **18**: 113-119.

Palaniswami, M.S. and Mohandas, N. (1994) *J. Root Crops*, **20**: 101-105.

Palaniswami, M.S. and Shirly Raichal, A. (2008) In: *Tuber and Root Crops* (Horticultural series-9) (Ed. K.V. Peter), New India Publishing Agency, pp. 85-118.

Palaniswami, M.S. and Tarafdar, J. (2008) In: *Tuber and Root Crops* (Horticultural series-9) (Ed. K.V. Peter), New India Publishing Agency, pp. 179-240.

Park, R.K., Jung, J.W., Cha, Y.H. and Lee, H.B. (1991) *Research Reports of the Rural Development Administration, Upland and Industrial Crops, 1991*, **33**: 22-28.

Patrick, S.M. (2009) In 'Food, Health and Fuel: Plants for the Future At: Selangor, Malaysia.

Pham, Bich Ngoc, Vu, Thi Lan, Tran, Thu Trang, Nguyen, Hoai Thuong, Le, Thu Ngoc, Chu, Hoang Ha, and Le, Tran Binh. (2015) *J. Life Sci.*, **10**: 262-271.

Pillai, K.S. and Lal, S.S. (1978) *J. Root Crops*, **4:** 45-47.

Pillai, K.S., Lal, S.S. and Palaniswami, M.S. (1981) *Entomon*, **6:** 69-72.

Porobo Desai, A., Gosukonda, R.M., Blay, E., Korsidumeyo, C., Medina Bolivar, F. and Prakash, C.S. (1995) *Sci. Hortic.*, **62**: 217-224.

Prakash, C.S., Porobo Desai, A., Ramana Murthy, G., Korsi Dumenyo, C., He, G., Zheng, Q., Egnin, M. and Kanyand, M. (1996) In: *Tropical Tuber Crops: Problems, Prospects and Future Strategies* (Eds. G.T. Kurup, M.S. Palaniswami, V.P. Potty, G. Padmaja, S. Kabeera-thumma and V.xPillai, Santha), Oxford and IBH, New Delhi, pp. 27-36.

Prasad, U.K., Prasad, T.N., Narayana, S. and Kumar, A. (1997) *J. Potassium Res.*, **13**: 283-289.

Purseglove, J.W. (1977) *Tropical Crops-Dicotyledons*, Longaman Group Ltd., London, pp. 78-88.

Rajamma, P. and Premkumar (1984) *Indian Fmg.*, **33**: 62-63.

Rajendran, P.G. and Sreekumar, J. (2001) In: *Textbook of Vegetables, Tuber Crops and Spices* (Eds. T. Thamburaj, and N. Singh), ICAR, New Delhi, India, pp. 348.

Ramanujam, T. and Indira, P. (1979) *J. Root Crops*, **5**: 43-45.

Ravi,V., Aked, J. and Balagopalan, C. (1996) *Cri. Rev. Food Sci. Nut.*, **36**: 661-709.

Ravindran, C.S. and Mohankumar, C.R. (1987) In: *Cultural Practices of Sweet Potato*, CTCRI, Technical Bulletin, Trivandrum.

Reddy, R., Soibam, H., Ayam, V.S., Panja, P. and Mitra, S. (2018) *Indian J. Agric. Res.*, **52** (1): 46-50.

Rossel, G., Kriegner, A. and Zhang, D.P. (2001) *Scientist and farmers: partners in research for the 21st Century*, Program Report, 1999-2000, pp. 315-321.

Reungmaneepaitoon, S. (2009) *Kasetsart J. Nat. Sci.*, **43**: 768-779.

Sajjapongse, A. and Roan, Y.C. (1982) *Sweet Potato*, pp. 203-208.

Sajjapongse, A. and Wu, M.H. (1989) *J. Natural Sci.*, **23:** 86-92.

Salgado-Garciglia, R., Lopez-Gutierrez, F. and Ochoa-Alejo, N. (1985) *Plant Cell Organ. Cult.*, **5**: 9-12.

Samarasinghe, S.H.M.C. (1991). In: *Sweet potato in South Asia: Postharvest Handling, Storage, Processing and Use* (Eds. T. R. Dayal, G. J. Scott, G.T. Kurup, and C. Balagopalan), Proceedings of a workshop held at CICRI, Trivandrum, 9-13 September, pp. 29–32.

Santhi, B. (1995) *J. Root Crops*, **2:** 107-110.

Santhosh Mithra V. S., Raji Pushpalatha,S.Sunitha, James G oerge, P. P. S ing, R. S. Singh, J. Tarafdar, Mitra Surajit, Chandra Deo, Sunil Pareek, K. M. Lakshmi,R. Shiny and Byju, G. (2019) *Curr. Sci.*, **116**: 1-5.

Satistica (2018) https://www.statista.com/.

Sekioka, H. (1971) *Trop. Root and Tuber Crops Tomorrow*, **1**: 37-40.

Sen, H., Bose, T.K. and Jana, P.K. (1981) *Indian Pot. J.*, **6**: 2-5.

Sen, H., Bose, T.K. and Jana,P.K. (1981) *Indian Potash J.*, **6**: 1-4.

Sen, H., Mukhopadhyay, S.K. and Goswami, S.B. (1990) *J. Root Crops*, **16**: 18-21.

Sen, H., Roychoudhury, N. and Bose, T.K. (1984) *The Lalbaugh J.*, **29**: 1-3.

Shanmugam, A. and Srinivasan, C. (1973) *Hort. Res.*, **13:** 143-145.

Shee, B.W. (1983) *J. Agric. Assoc.*, China, **122:** 24-32.

Shima, Y., Nagahama, T., Suganuma, T. and Kitahara, K. (1996) *Jap. J. Trop. Agr.*, **40**: 204-212.

Shinohara, S., Ino, M., Kimura, N. and Katsukita, H. (1989) *Bull. Chiba Prefectural Agric. Exp. Stat.*, **30**: 51-60.

Singh, K.D. and Mandal, R.C. (1970) In: *Two Decades of Research*, CTCRI, Trivandrum, 1983, India.

Smit, N.E.J.M. (1997) *Agric. Ecosyst. Env.*, **64**: 191-200.

Sood, M.S., Sbuddin, A.S.A., Anus, A.G. and Shamsuddin, Z.H. (1999) *Soil Sci. Plant Analysis*, **30**: 11-12.

Speights, D.E., Burns, E.E., Patterson, D.R. and Thomas, W.H. (1967) *Proc. Amer. Soc. Hort. Sci.*, **91**: 478-485.

Spence, J.A. and Humphries, E.C. (1972) *Ann. Bot.*, **36**: 115-121.

Srinivasan, G. (1977) *J. Root Crops*, **3**: 55-57.

Steinbauer, C.E. (1962) *Horticultural News*, 43.

Steinbauer, C.E. and Kushman, L.J. (1971) *Sweet Potato Culture and Diseases*, USDA Hand Book, 388, pp. 1-74.

Sur, R., Mahato, A., Mani, A. and Mitra, S. (2020) *Intl. J. Chem. Stud.*, **8**(1): 2830-2834.

Sur, R., Paik, C., Mitra, S. and Chakraborty, I. (2020) *J. Pharmacog.Phytochem.*, **9**(2): 580-584.

Suzuki, M. (1989) *Japanese J. Breed.*, **39**: 499-506.

Tai, J.X., Hua, X.X., Wang, J.W., Niu, F.X., Huang, G.R. and Fu, Q. (1998) *Acta Agronomica Sinica*, **24**: 161-167.

Talbert, R.E. (1967) *Arkansas Farm Res.*, **16**: 13.

Tanaka, J.S. and Sekioka, T.S. (1976) *Proc. 4th Symp. Int. Soc. Trop. Root Crops*, Columbia, pp. 150-151.

Tarafdar, J. and Sarkar, M.A. (2006) *Acta Hort.*, **703:** 189-196.

Thankappan, M. (1989) International Seminar on Pest and Diseases of Sweet potato in Asia, held on 7-8 August at CTCRI, Trivandrum, India.

Thakur, C. (1984) *Weed Science*, Metropolitan Book Co., Pvt. Ltd., New Delhi, p. 122.

Thompson, C. and Kelly, W.C. (1979) In: *Vegetable Crops*, Tata McGraw-Hill Publishing Co. Ltd., New Delhi, pp. 405-430.

Toib, M.A. and Rashid, M.M. (1976) *Bangladesh Hort.*, **4:** 1-6.

Trust, I. (1986) CIP Annual Report, CIP, Lima, Peru, p. 51.

Tsuno, Y. and Fujise, K. (1968) *Proc. Crop. Sci.*, Japan, **37**: 3-21.

Udensi, E.A. (2000) In *Sustainable agro-allied projects with great economic potentials for Nigeria* F.I. Onyenobi and W. Rose (Eds.), Abakaliki: Appleseed Publishing Coy, pp. 142-149.

Ukom, A.N., Ojimelukwe, P.C. and Okpara, D.A. (2009) *Pakistan J. Nutr.*, **8**: 1791-1795.

Unnikrishnan, M. and Sheela, M.N. (2001) In: *Biotechnology of Horticultural Crops* Vol. 2 (Eds. V.A. Parthasarathy, T.K. Bose and P. C. Deka), Naya Prokash, Calcutta, India.

Unnikrishnan, M., Mukherjee, A. and Nair, N.G. (1991) In: *National Symposium on Genetics and Biotechnology for Crop Improvement*, Osmania University Hyderabad, Symp. Abst. p. 61.

Walter, W.M.Jr. and Schadel, W.E. (1982) *J. Amer. Soc. Hor. Sci.*, **107**: 1129-1138.

Watanabe, K. (1969) *Proc. Crop Sci. Soc., Japan*, **38**: 652-56.

Williams, O.B. (1964) *J. Agric. Soc.*, Trin., Tobago, 89, pp. 419-428.

Wilton, M.,Sylvester, A., Omwoyo O., Christina, D. and Richard, O. (2016) *J. Tiss. Sci. Eng.*, **7**: 2-6.

Woolfe, J.A. (1992) Cambridge University Press, UK.

Wu, Y.V. and Bagby, M.O. (1987) *J. Agric. Food Chem.*, **35**: 321-325.

Xie, R.Y., Peng, F. and Cao, L. (1982) *Soils* (Turang), **14:** 133-136.

Xue, Q.H. (1988) *J. Agric. Sci.*, **4**: 56-62.

Yang, Z.G. (1990) *J. South China Agri. Univ.*, **11**: 64-71.

Yen, D.E. (1976) *Evolution of Crop Plants*, (Ed. N.W. Simmonds), Longman, London and New York.

Zamora, A.B. and Gruezo, S.S. (1993) *Plant Genet. Resour. Newsl.*, **91-92**: 25-28.

YAM

T.K. Bose, T. Maharana , H. Sen, S. Mitra and J. Tarafdar

1.0 Introduction

Yams (*Dioscorea* spp.), though considered as under-utilized tuber crops, are cultivated widely in the tropics and are important food crops in West Africa, West Indies, Tropical America and South-East Asia. Though, these herbaceous vines are perennial in nature, cultivated mainly as a seasonal crop restricting its availability in a specific season of the year (Mitra and Tarafdar, 2013). Yam is an economically important food in many tropical countries particularly in West Africa and the Caribbean having a social and cultural importance. It is essentially a crop of subsistence agriculture.

Worldwide production of yams is about 73.0 million tonnes, led by Nigeria with 66 per cent of the global total production (FAOSTAT, 2017). Nigeria farmed yams on 5.9 million hectares, 70 per cent of the world land area of 8.6 million hectares devoted to yam farming (Obi-Young, 2020). The world average annual yield of yams was 8.8 tonnes per hectare in 2017, with Ethiopia having the most productive farms of 29.2 tonnes per hectare (FAOSTAT, 2017).

In Asia, along with the use as vegetable, the yam is also used for medicinal purposes. In India, yams are grown in many States but the major yam producing States are Kerala, Orissa, West Bengal, Bihar, Tamil Nadu, Assam, Rajasthan, Gujarat and Maharashtra.

2.0 Composition and Uses

2.1.0 Composition

The tubers contain nearly all the essential nutrients including minerals and vitamins and have several medicinal properties. Yam is considered as one of the important staples in the diets of many tropical countries because of its high carbohydrate content. The tubers are found to have high amounts of protein as compared to the protein content in cassava and sweet potato (Soibam *et al.*, 2017). Apart from food, yams are also a source of pharmaceutical compounds like saponins and sapogenins (Higdon *et al.*, 2001), which are precursors of cortisone, a hormone from the adrenal gland used in the treatment of arthritis and some allergies. Some yams are also used as medicines in oriental countries to prevent diarrhoea and diabetes. Modern researches have shown that yam extracts can reduce blood sugar (Undie and Akubue, 1986), blood lipid (Araghiniknam *et al.*, 1996), inhibit microbial activity (Kelmanson *et al.*, 2000) and shows antioxidative activity (Chan *et al.*, 2004). Thus, yams are considered to be helpful to human health. The species *Dioscorea alata* shows higher antioxidant activity than *D. rotundata.*

The higher antioxidant percentage of *D. alata* correlates with the presence of their higher phenol, vitamin C and β-carotene content (Soibam *et al.*, 2017).

However, historical reports of these yam tubers also suggest that tubers are acrid, causing irritation and inflammation and resulting in gastrointestinal disturbances, vomiting and diarrhoea when consumed in large amounts. Yams are also known to contain some antinutritional components that may have adverse effects on human nutrition. These are mainly tannins, phenols and phytic acid. However, bitter principles may be polyphenols or tannin like compounds (Coursey, 1983) while phytic acid (inositol hexaphosphate) is an organic acid found in plant materials (Heldt, 1997) which combines with some essential elements such as iron, calcium, zinc and phosphorus to form insoluble salts called phytate which are not absorbed by the body thereby reducing the bioavailability of these elements. It is an important dietary staple for low income groups and consumed mainly in boiled and roasted forms. Processing and value addition in yams could be helpful in reducing anti-nutritional components and promoting consumer preferences enabling the producers to increase their income (Mitra, 2011).

According to Patel *et al.* (2019), among the biochemical parameters of yam, total carbohydrate ranged between 51.87 and 87.85 per cent in different genotypes, starch ranged from 47 to 80.67 per cent, fat ranged from 0.6 to 2.32 per cent, crude fibre ranged between 1.10 and 4.09 per cent, anthocyanin content ranged from 1.01 to 3.25 mg/g, beta-carotene content ranged between 0.97 and 1.88 µg/g.

The chemical composition of fresh yam tuber is given in Table 1.

Table 1: Chemical Composition of Three Edible Species of Yam*

Species	*Moisture (Per cent)*	*Carbohydrate (Per cent)*	*Fat (Per cent)*	*Crude Protein (Per cent)*	*Vit.C (mg/100g)*	*Antioxidant (Per cent)*
Dioscorea alata	68.3-83.5	23-27	0.33-0.46	1.50-2.70	7.6-9.4	78.6-85.7
D. rotundata	63.9-67.5	25-28	0.22-0.28	1.20-2.40	6.5-7.4	78.9-80.4
D. esculenta	67.4-78.3	21-26	0.19-0.31	1.31-2.26	6.2-81	76.2-81.3

**Source:* Soibam *et al.* (2017), Mitra (2011)

2.2.0 Uses

Yam tuber is consumed after roasting, boiling or with other vegetables. It is also used as chips, flakes, and flour. Yam chips are a stabilized product, made from small peeled, pre-cooked and sun-dried tubers. Yam chips dehydrated at 50 °C, (12±2) hrs. were found to retain high nutrient and low antinutrient values with desirable sensory attributes and minimal microbial counts for 3 months. Therefore, yam chips can be safely stored and used as reliable and energy security food if dehydrated at the specified temperature and duration (Soibam *et al.*, 2017). They are mainly consumed in the form of a paste (amala or telubo) prepared from the flour obtained by grinding the chips (Bricas *et al.*, 1997). Some poisonous types like *Dioscorea dumetorum* and *D. hispida* are eaten during scarcity after clarification or as such used as an aid in hunting, fishing, arrow poisoning and for insecticidal purpose. Many species of yam contain small amounts of sapogenins and alkaloids for various uses. The main sapogenins present is diosgenin which is a starting point of several corticosteroid drugs. Of the 80 Chinese species of *Dioscorea*, tested by Zhao and Xiao (1989), 17 contained diosgenin. Those with high contents and a prospect of medicinal utilization are: *D. althaeoides, D. collettii, D. colletti* var. *hypoglauca, D.*

Uses of *Dioscorea*

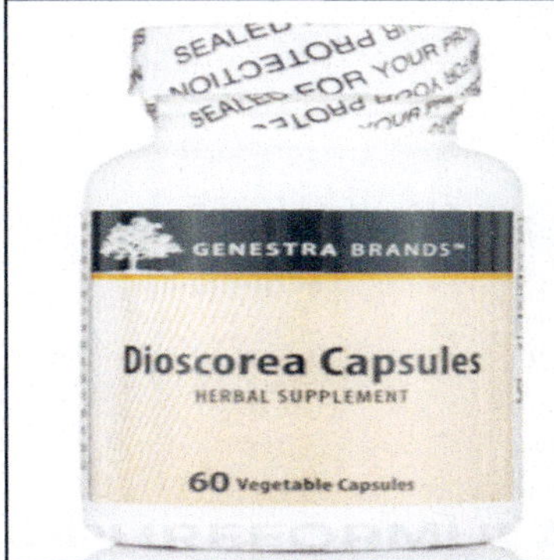

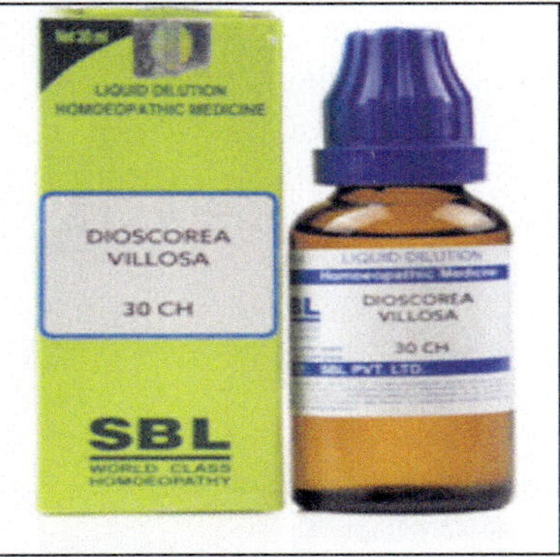

Different Value added Products of Yam

deltoides, D. nipponica, D. panthaica, D. zingiberensis, D. futschauensis, D. gracillima and *D. tencipes.*

3.0 Origin, Distribution and Taxonomy

3.1.0 Origin and Distribution

The family Dioscoreaceae shows some primitive characters of both monocot and dicot and they were considered as earlier Angiosperms. It was suggested by Axelrod (1970), cited by Coursey (1975) that the ancestral Dioscoreaceae might have originated in Southeast Asia. Burkill (1960) also indicated Southeast Asia as the origin of early Dioscoreaceae.

In Asia, *D. alata* seems to have developed as a result of human selection from wild form related to *D. hamiltonii* and *D. persimilis* in the Southeast Asian peninsula. *D. esculenta* indigenous to the same area has also been developed by selection (Alexander and Coursey, 1969; Coursey, 1975). The major centre of diversity of cultivars of both species is in Papua New Guinea (Coursey, 1976). The evolution of African yams as crop plant is mainly due to the identification of *D. cayenensis* (Ayensu and Coursey, 1972). Of the two major species, *D. cayenensis* is indigenous to forest zone of Africa, while *D. rotundata*, a cultigen close to *D. cayenensis* is considered as a subspecies (Miege, 1968, cited by Ayensu and Coursey, 1972). *D. trifida* is an Amerindian domesticate and the presence of cultivars suggests domestication on the border of Brazil and Guyana, followed by spread through the Caribbean (Ayensu and Coursey, 1972). *D. bulbifera* is the only edible yam species that is native to both Asia and Africa.

Using IPGRI's descriptors, 560 accessions of cultivated yams (*D. cayenensis/D. rotundata* complex) collected throughout Benin, were characterized by Dansi *et al.* (1999). Ninety morphotypes were identified and further classified into 26 cultivar groups based on their morphological similarities. An identification key is proposed for the cultivar groups. The geographical distribution of cultivar groups is also presented and two centres of diversity are identified. A comparative study between the cultivated forms and their wild relatives (*D. abyssinica, D. praehensilis* and *D. burkilliana*) has allowed the establishment of linkage between the identified morphotypes and wild species.

Collecting missions of yams were made in Ethiopia, and 20 accessions were collected and planted at Jima for evaluation. The species collected included *D. aleta* (*D. alata*), *D. esculenta, D. bulbifera, D. abyssinica* and other unidentified types. Evaluation over 5 years showed that accession Yam 2/81 gave a tuber yield of 60.70 t/ha, followed by Yam 56/76 (44.82 t/ha). A total of 81 cultivated yam accessions were collected in Guinea, most of which were *D. rotundata* and *D. cayenensis* (Mignouna *et al.*, 1997).

Pedralli (1994) collected 81 accessions of endangered wild relatives and cultivated species of yam from the Espinhaco range, Brazil. Morphological analysis of tubers, rhizomes, seedlings, flowers, fruits and seeds showed that *D. multiflora, D. regnellii, D. sincorensis* and *D. sinuata* var. *sinuata* were the wild species with the most variations. Of the cultivated species *D. dodecaneura* and *D. alata* showed the widest variation for tuber morphology. *D. kunthiana* and *D. microbotrya* were discovered for the first time in the national reserve of Serra do Cipo.

N'-kounkou *et al.* (1993) described the origin, distribution and uses of *Dioscorea* in Congo. They reported that the family *Dioscoreaceae* was represented by the single genus *Dioscorea*, with 18 species, 1 subspecies, 3 varieties and 4 cultivars. Information on the medicinal value of *D. bulbifera, D. dumetorum, D. praehenzilis* and *D. smilacifolia* were also included. *D. schimperiana, D. hirtiflora, D. pynaertii, D. liebrechtsiana* and *D. cayenensis* were recorded for the first time in Congo. *D. lecardii* was reduced to a variety as *D. sagittifolia* var. *lecardii*, and *D. lecardii* var. *chevalieri* and *D. lecardii* var. *vera* were considered synonyms with this variety. *D. odoratissima* was reduced to synonymy with *D. praehensilis, D. cameruensis* was identical to *D. cayenensis* and *D. liebrechtsiana* was reinstated as a species. N'-kounkou in 1994 observed that the genus Dioscorea consists of 27 species, in eight sections, in central Africa (Congo, Zaire, Rwanda, and Burundi). Data on 28 morphological characters were subjected to phylogenetic analyses and these results suggested that the closely related family Menispermaceae may be a hypothetical ancestor of the *Dioscoreaceae*.

Missions for collecting germplasm were undertaken in West Africa. A total of 177 samples of tubers, bulbs or fruit representing 11 *Dioscorea* spp. were collected from forest, savannah and forest-savannah transition zones in Benin, Cameroon, and Cote d'Ivorie (Zoundjihekpon and Tio-Toure, 1992). Hamon and Ahoussou (1987) reported that in Ivory Coast three missions, each to a different region of the country, collected a total of 1359 accessions from 13 species. The indigenous species *D. dumetorum, D. bulbifera* and *D. cayenensis-rotundata* [*D. cayenensis, D. rotundata*], and also *D. alata* (introduced), are widely grown while several others are grown on a small scale or may be useful as genetic resources even if not edible.

Tellez-Valdes and Schubert (1990) described new species *D. carpomaculata* and *D. sumiderensis* in Central America.

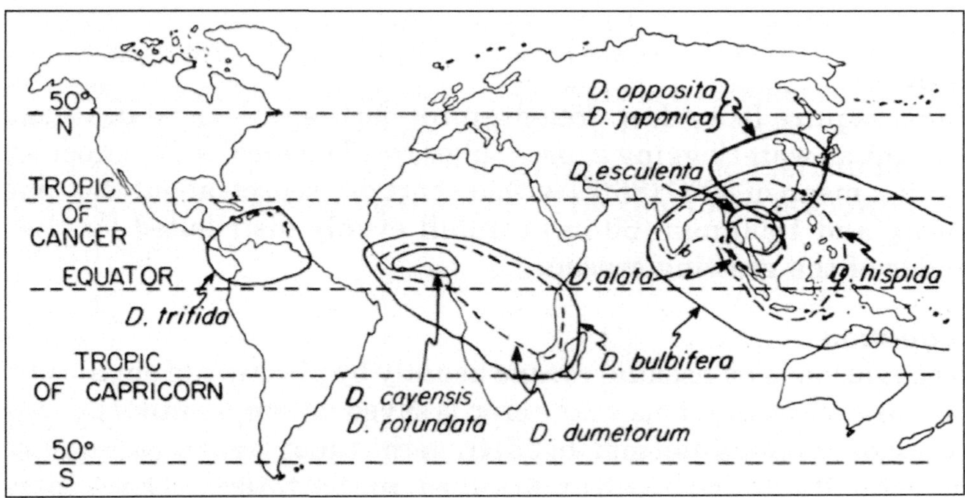

Yam Producing Regions of the World (Coursey, 1967)

3.2.0 Taxonomy

The yam belongs to the family Dioscoreaceae and genus *Dioscorea* which contains about 850 species. Of the total species, about 12 are known for edible purposes. The species *D. alata* L., *D. rotundata* (L) Poir., *D. esculenta* (Lour.) Burk., *D. bulbifera* L. and *D. trifida* L. are the principal yams of the tropics. However, the former 3 are cultivated on large scale for edible purposes. In India, *D. alata, D. esculenta* are extensively cultivated in the different regions as a subsidiary food crop. The evolu-tionary difference of old and new world groups has been observed in their cytology. The old world species in most cases have chromosome number based on x =10. The primitive section Stenophora has 2n = 2x = 20, but many tropical species used as food have 2n = 4x = 40 (Burkill, 1960; Coursey, 1967).

The new world *Dioscorea* have x = 9, most being tetraploids or hexaploids (2n = 36, 54) (Martin and Ortiz, 1963, 1966, Coursey, 1967; Ayensu and Coursey, 1972). Among the old world yams, high degrees of polyploidy were found to occur and chromosome counts of 50, 60, 70, 80 and up to 140 have been reported by Coursey (1967). Using morphological descriptors and starch gel electrophoresis, 393 West African cultivated yam accessions representing 20 previously described varietal groups, Hamon and Toure, (1990a) classified them into 2 distinct clusters which differed in tuber flesh pigmentation, duration of the aerial vegetative cycle and electromorphs of 3 enzyme systems. Fifteen varietal groups were included in one cluster and 3 in the other; the remainder were thought to represent inter-cluster hybrids.

Gamiette *et al.* (1999) used chromosomes counting and flow cytometry for assessing the ploidy level of various yam accessions of the INRA collection located in Guade-loupe. About 85 different clones were evaluated in *D. alata* and *D. cayenensis-rotundata* and four other wild species related to *D. cayenensis-rotundata* (*D. mangenotiana, D. abyssinica, D. praehensilis* and *D. burkilliana*). All the studied clones fitted ploidy levels of 4x, 6x or 8x; no diploid clones were found. Chromosome counts and flow cytometry data led to the same results.

Dansi *et al.* (2001) determined the ploidy levels of twenty-two yam (*D. cayenensis/D. rotundata* complex) cultivars within germplasm Cameroon Guinea yam by flow cytometry. Three different ploidy levels (4x, 6x, 8x) were detected within the samples analysed. Fifteen cultivars were tetraploids, five were hexaploids and two were octoploids. The cultivar group ECHOED showed a high level of ploidy variation with tetraploid, hexaploid and octoploid cultivars. The hexaploid nature of cultivars Dobnawo and Bilougnou supported the hypothesis that they are hybrids between cultivars of the ECHOED group and either the KPE or BAKOKAE groups. The ploidy levels of ninety Guinea yam (*D. cayenensis/D. rotundata* complex) cultivars determined cytometry. Three different ploidy levels (4x, 6x, 8x) were detected among the samples. Eighty cultivars were tetraploid, five were hexaploid, three were octoploid and two cultivars, Tam-Sam and Yoube were mixoploid with both 4x and 8x ploidy levels. Chromosome counts were in agreement with data from flow cytometry that provided an easier assay for ploidy analysis. Flow cytometry was found to be a reliable tool for rapid determination of ploidy level in yam.

Different Types of Yam

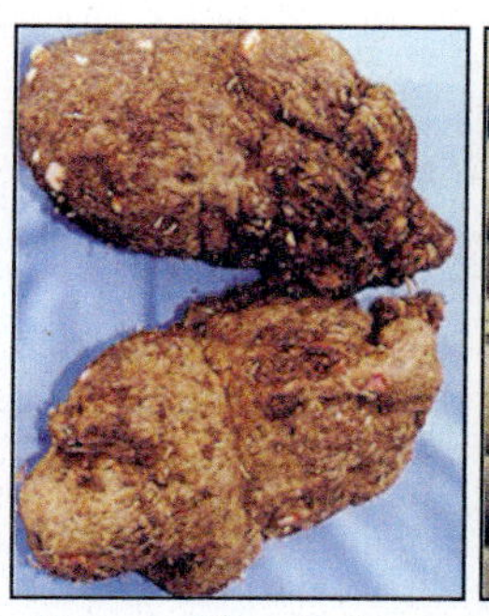
Greater Yam Tuber

White Yam Tuber

Lesser Yam Tuber

Purple Yam

Yellow Yam

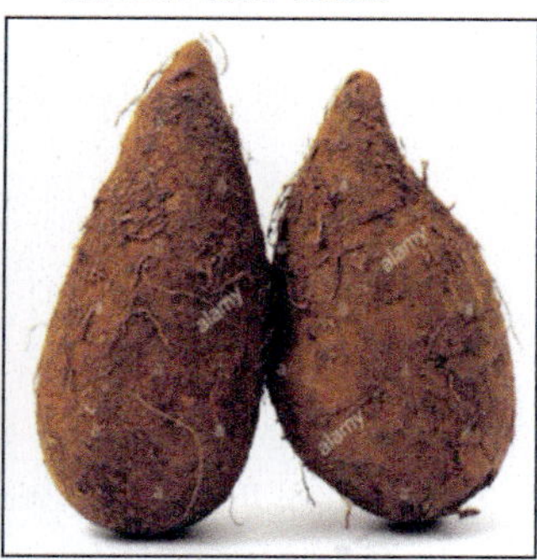
Cush-cush Yam

Chinese Yam

Japanese Mountain Yam

Bulbils of Aerial Yam

Variations among 326 accessions of 3 species of Dioscorea in the NPGRL collection were analyzed using 113 agro-morphological characters (Gruz *et al.*, 1999). Cluster analysis and ordination methods of 202 *D. alata*, 91 *D. esculenta* and 33 *D. hispida* accessions revealed that each species formed distinct and tight clusters. Of the accessions, 8 were identified as not belonging to *D. alata* and 6 were identified as *D. bulbifera*. Geographical distribution of the 3 species encompassed 40 provinces of the Philippines in 12 political regions.

The morphological characters of three important species of edible yam are given below:

***D. alata*:** It is also called the greater yam which has winged spineless stems twinning to the right, leaves are large and arranged in opposite manner. The tubers of cultivated type are normally round, oval or irregular in shape. The skin of the tuber is black or brown white and flesh is white, yellowish or purplish. Plants produce usually 1-3 tubers at the end of a stolon.

***D. esculenta*:** This is of short duration (7-8 months) type and is called lesser yam. It is characterized by thin, pubescent and spiny stems which twine to the left. Leaves are small, cordate and alternate in arrangements. Tubers are smaller as compared to other yams and they are produced in clusters of 10-15 tubers. They have thin yellowish brown skin and white flesh.

***D. rotundata*:** This species is mainly cultivated in West Africa and is known as white yam. It is characterized by sturdy, spiny and roughly circular stem which twines to the right. Leaves are small, simple, cordate and opposite or alternate in arrangement. Tubers are normally cylindrical having brown skin and white flesh.

Commercial exploitation of yams for diosgenin occurs most prevalently in Mexico where *D. mexicana, D. floribunda* and *D. composita* are grown for the purpose (Onwueme, 1978a).

4.0 Botany, Flowering and Floral Biology

4.1.0 Botany

Dioscorea are deciduous perennial plants. The root system develops from the end of tuber, thick, unbranched at the beginning, become fibrous and branched later. Most species produce underground stem tubers, few wild species are rhizomatous. The tubers vary in number, size and forms, may be globular, elongated or flattened. Some species produce bulbils in leaf axils.

An annual stem arises from the top of the tuber. Stem climbs by twining, in the section Enantiophyllum to which *D. alata, D. rotundata* and *D. cayenensis* belong, stems twine to right while in other edible yams stems twine to left (Purseglove, 1968). Stem may be spiny or winged or with longitudinal ridges. Leaves are usually simple, cordate, acuminate having main veins from the base of the lamina, subsequently reticulate. Small glands are present in several species and the secretion from these glands is known to attract insects.

4.2.0 Flowering and Floral Biology

In *Dioscorea*, male and female flowers are borne on separate plants; male flowers arise on long axillary panicle, while female flowers are borne on axillary spikes. Flowers small, greenish or cream in colour, perianth in 2 whorls of 3 segments each of similar size, stamens 6 in two whorls, female flowers have inferior ovary, 3 locular and each loculus has 2 ovules, fruit a dehiscent trilocular capsule, winged or angled, seeds winged.

Segnou *et al.* (1992) investigated natural pollinators, flowering pattern and hybridization of *D. rotundata* at the International Institute of Tropical Agriculture, Ibadan, Nigeria. Insects belonging to the Coleoptera (48.0 per cent), Diptera (20.4 per cent), Hymenoptera (20.0 per cent), Hemiptera (5.8 per cent) and Thysanoptera (5.8 per cent) were caught around yam plants and presumed to be pollinators: they entered open-receptive-flowers and their presence on yam plants coincided with the duration of the flowering period. Studies on the flowering pattern revealed that (1) male and female clones differed in their periods of flowers initiation as well as the flowering duration; and (2) seed production and conventional breeding methods are feasible if staggered planting, water and soil fertility could be managed to ensure that flowering periods of males and females nick and also extend flowering. Artificial pollination with white yam and between white yam and yellow yam (*D. cayenensis*) using a camel hair brush method were successful, but resulted in a relatively low fruit set (31.8 per cent maximum). This method was more effective than natural pollination (19.2 per cent) and can thus supplement it for the production of hybrid seeds for introducing genetic variation in yam breeding populations.

Observations on the floral morphology of *D. bulbifera, D. dumetorum, D. hirtiflora, D. minutiflora, D. preussii, D. odoratissima, D. rotundata* and *D. smilacifolia* were reported by Edeoga *et al.* (1998). The most distinctive features of the taxa studied were the presence of large fruits and raceme inflorescences in *D. preussii*, the absence of pendulous inflorescences in *D. dumetorum*, and the absence of stalks in female flowers of *D. bulbifera*.

5.0 Growth and Development

A study was carried out by Siwachi *et al.* (1997) to analyze the ecological characteristics of water yam (*D. alata*), Chinese yam (*D. opposita*), jinen-jo (*D. japonica*) and their strains by comparing their growth characteristics. The growth periods of water yam and jinen-jo were longer than that of Chinese yam in Kagoshima, Japan. In the early-maturing type of water yam tubers started to enlarge before August and became mature at the beginning of December. However, in the middle and/or late-maturing types, the tubers began to enlarge later than in the early-maturing type and the tubers continued to enlarge until harvest. In the extremely early-maturing type originated from southern Kyushu, the tubers began to enlarge and they became mature earlier than those of the early-maturing type. Thus, by introducing and selecting the extremely early-maturing and early-maturing types, it may be possible to extend water yam cultivation to warmer areas in temperate regions. In Chinese yam, the growth pattern of tubers, stems and leaves differed depending on the strains or groups. In the Naga-imo and Icho-imo groups, new tubers were formed at the budding stage of young plants and began to grow immediately. However, in the Tsukune-imo group, after the new tubers were formed they did not enlarge for a month and then began to enlarge rapidly. In this group tubers reached their maximum weight one month later than those in the Naga-imo and Icho-imo groups. Thus the differences in the period during which tubers did not develop affected the cultivation period of the three groups of Chinese yam. Growth pattern of tubers in the Tsukune-imo group was similar to that in the extremely early-maturing type of water yam.

6.0 Cultivars

Some high yielding varieties of *Dioscorea alata, D. rotundata* and *D. esculenta* developed at Central Tuber Crops Research Institute, Thiruvananthapuram, India are described below.

Dioscorea alata

Sree Keerthi

It is a high yielding variety with large sized tubers, good cooking quality and excellent taste. It is a clonal selection from indigenous genotype. It climbs to a height of 4-5 meter and matures at 9-10 months after planting. Yield ranges from 25 to 30 t/ha and starch content is about 22 per cent.

Sree Swathy

It is a variety of greater yam having high yield of about 30 t/ha, good culinary and nutritive quality and moderate tolerance to anthracnose disease.

Sree Neelima

It is a high yielding yam variety with an average yield of 35 t/ha. Tubers are light purple in colour having good culinary and nutritive quality.

Sree Roopa

It is a high yielding variety of greater yam with medium sized tubers and excellent cooking quality. It is a clonal selection from indigenous genotype. Vines climb to a height of 4-5 meter and tubers mature after 9-10 months of planting. Average yield is 25-30 t/ha. Starch and protein content varies from 16-18 per cent and 1-2 per cent, respectively.

Sree Karthika

It is a high yielding greater yam variety with long, oval tubers having excellent cooking quality. Maturity period is about 9 months and average yield is 30 t/ha.

Sree Shilpa

It is the first hybrid released in greater yam with medium sized, oval tubers and good cooking quality. It is an inter-varietal hybrid between Da-140 and Sree Keerthi. It has cordate leaves with overlapping lobes, lamina slightly cupping and petiole purple at both ends. Swollen oval tubers with white flesh are easily harvestable. Average yield is 28 t/ha with a yield potential of 40 t/ha.

Orissa Elite

It is a clonal selection of greater yam with long cylindrical tubers having excellent cooking quality. Average yield is 25 t/ha. Plants are field tolerant to leaf spot, scales and mealy bugs.

Konkan Ghorkhand

It is high yielding and erect growing vine produces aerial tubers having purplish green colour stem and simple leaves. Crop duration is 6 months. Average yield is 16 t/ha.

Dioscorea rotundata

Sree Priya

It is a seedling selection from the African variety "Umidika" with cylindrical tubers having excellent cooking quality and pleasant flavour. Plants are drought tolerant, and are suitable for intercropping in mature coconut gardens and banana orchards. Maturity duration is 9–10 months and average yield is 35–40 t/ha. Starch and protein content varies from 20–21 per cent and 2–3 per cent, respectively.

Sree Subhra

It is a high yielding variety of white yam having tubers of brown skin with excellent cooking quality. It is a selection from the African variety "lwo". Tubers are partially hairy with cylindrical shape. Plants are drought tolerant. Duration for maturity is 9–10 months and average yield is 35–40 t/ha. Starch content is 21–23 per cent.

Sree Dhanya

It is a high yielding variety of white yam with cylindrical tubers. Duration for maturity is 9 months. It is the first dwarf bushy variety and it does not require staking and can be accommodated in closer spacing. Average yield is 20 t/ha.

Dioscorea esculenta

Sree Latha

Sree Latha is a high yielding and widely adaptable variety of lesser yam. It is a selection from indigenous germplasm. Plants climb to a height of 2.5–3 m. Tubers mature after 7.5–8 months of planting and average yield is 25 t/ha. Cooking quality is good having pleasant flavour and acceptable taste. Average starch content is 24-25 per cent and shelf life of tuber is 2.5–3 months.

Sree Kala

It is an early maturing variety of lesser yam with oval tuber shape and excellent culinary quality. It is a selection from exotic germplasm named Kombi. Tubers mature at 7.5 months after planting. Average yield is 20 t/ha.

Konkan Kanchan

Tuber shape is Uniform, non-branching, long cylindrical with blunt ends. Tuber skin colour is yellowish light brown skin. Tuber flesh colour is white having non bitter and acceptable taste. Average yield is 20-22 t/ha.

Childer *et al.* (1950) mentioned that Guniea, a white flesh cultivar of *D. cayenensis* was very popular in Puerto Rico. The cultivars Congo Yellow or Guniea Yellow are also grown. Patato and Tango (*D. aculeata*) and Agua (*D. alata*) are white fleshed cultivars. Purple Ceylon is a good purple cultivar (Thompson and Kelly, 1957). Martin (1974) has described several promising cultivars of *D. alata* including Florida, Forastero, Gamelos, Leone Globe, Veveen, *etc.* There is a large number of cultivars of *D. rotundata*, especially in West Africa where it originated (Waitt, 1961).

The Central Tuber Crops Research Institute, Trivandrum (CTCRI) has a large collection of germplasm of *D. alata* and *D. esculenta* mostly from indigenous sources. As a result of the systematic screening and testing 3 superior types of *D. alata* have been identified. They are Da 60, Da 80 and Da 122. Da 60 has large conical shaped tubers, Da 80 has medium sized irregular tubers and Da 122 has medium sized, round to oval tubers. These selections give high yield of about 30-35 tonnes of quality tubers per hectare.

As a result of selections in *D. esculenta*, a high yielding superior cultivar (De 11) namely, 'Sree Latha' has been released for general cultivation in Kerala. This is the first cultivar of an edible yam released in India. The cultivar produces about 20-25 medium to large tubers per plant. The average tuber length is about 13.5 cm and girth 12.2 cm. Tubers are oblong to fusiform with greyish brown skin which is partly hairy. They have good cooking quality and taste. The yield of the cultivar is about 25 tonnes of tubers per hectare (Anon., 1985).

Amma *et al.* (1989) evaluated 192 indigenous and 18 exotic genotypes of *D. alata*. The high yielding selection, designated Sree Keerthi (SK) and Sree Roopa (SR), were isolated and released in 1986. SK is a male clone, plants producing 1-2 tubers with a mean size of 32 x 44 cm, brown skin and a conical shape. SR is a female clone producing 2-3 tubers/plant, tubers having a mean size of 24 x 38 cm, black skin and a digitate shape. Dry matter, starch and protein content of SR and SR tubers is 30.4 and 23.2 per cent, 21.0 and 17.8 per cent, and 1.8 and 1.6 per cent, respectively.

A total of 13 *Dioscorea* genotypes were cultivated, analyzed and evaluated at Taiwan Agricultural Research Institute (TARI). Comparison of the varietal characters, yield potential and phytochemical properties of these genotypes indicated that TNG2, 70W42, 70W47, 70W35 and 80W02 might be the most promising lines due to their high yield potential (over 34 t/ha) (Liu *et al.*, 1999). The new yam (*D. alata*) cultivar, Tainung 2, from Nantou, Taiwan, was reported by Lin *et al.* (1996). After a series of trials, Tainung 2 was considered as superior to Tainung 1 and other native and introduced yam lines.

Several water yam (*D. alata*) strains were introduced from high altitude areas (600 to 1800 m) in Nepal. Some high altitude strains (Arata, Osato No. 1, Rato, Seto, Zate and 5 unknown strains) were grown in Kagoshima, Japan, and their growth and tuber enlargement were investigated by Shiwachi *et al.* (2001). All the tested strains were early-maturing types and some extremely early-maturing strains were

Cultivars of Greater Yam

Konkan Ghorkand

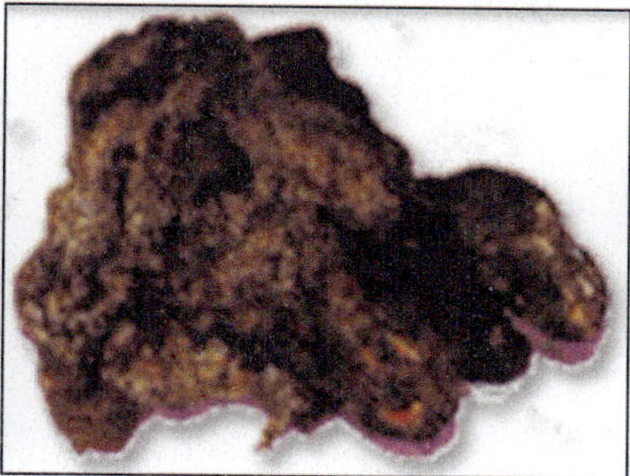
Sree Keerthi

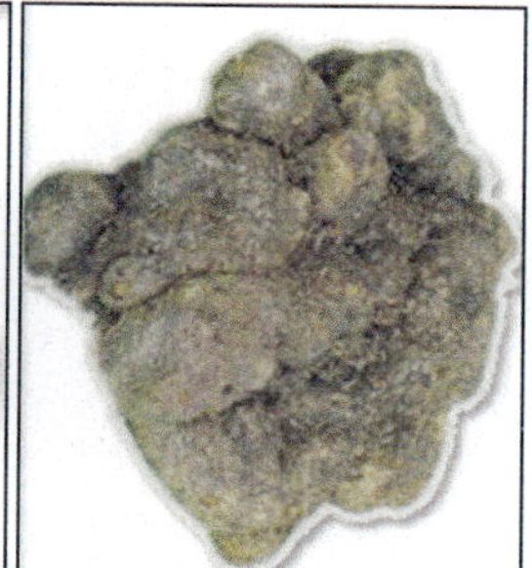
Sree Silpa

Cultivars of White Yam

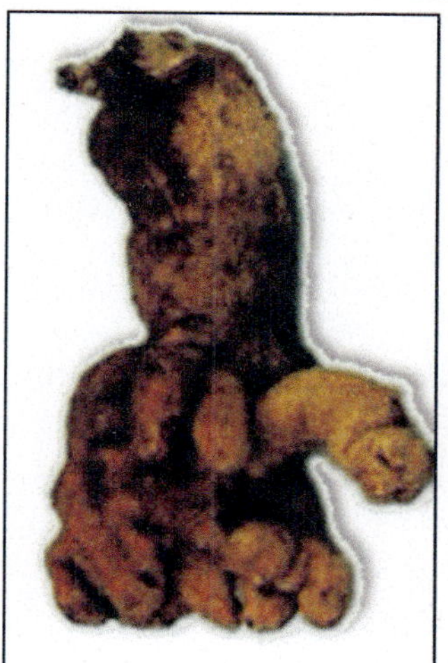
Sree Dhanya

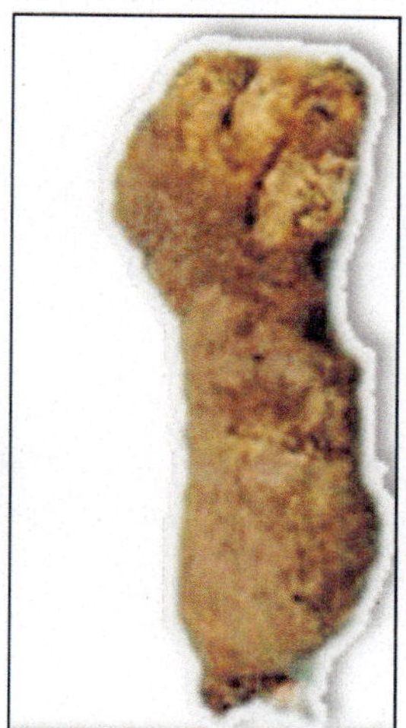
Sree Priya

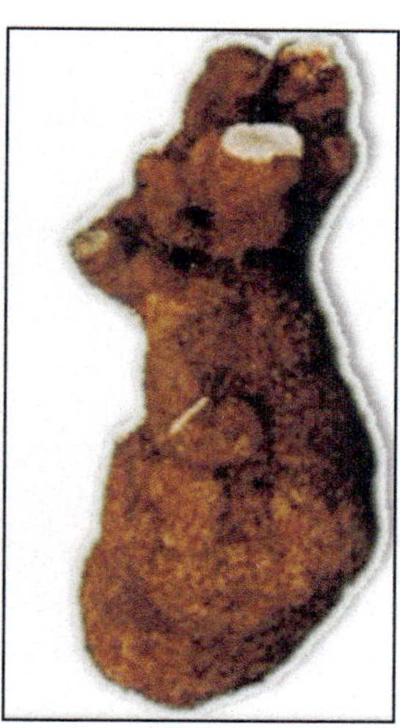
Sree Subhra

Cultivars of Lesser Yam

Sree Latha

Konkan Kanchan

detected in this group. Of this type, the tubers began to enlarge and mature earlier than extremely early-maturing strains originating from Kagoshima.

7.0 Soil and Climate

7.1.0 Soil

Yam can be grown in different types of soil but it grows well in sandy loam soil with a pH of 5.5-7.5 (Mitra and Tarafdar, 2013).

Tuber growth and development largely depend on the texture and preparation of the soil. Some yams go deep into the soil and some spread on the upper strata. Well drained, loose friable soil containing good amount of organic matter is preferred for yam cultivation. In sandy soil, plant declines and in stiff clay soil, pressed and deformed tubers are formed with poor storage quality.

7.2.0 Climate

Plants are mainly grown in the tropic and sub-tropics of the world. However, major yam growing areas are in the tropical region. The optimum range of temperature is between 25 and 30°C. Yams cannot tolerate frost and growth is affected at a temperature below 20°C. High rainfall and warm climate during the growing season promote in vegetative growth and tuber bulking. The white yam (*D. rotundata*) can tolerate a long dry season and optimum yield is obtained under 100 to 150 cm rainfall evenly distributed over 6 to 7 months. Low moisture supply at the early stage may kill the young shoots which have exhausted reserve food of the old sett. Onwueme (1975) found that moisture stress in the early stage of growth resulted in delay in the onset of tubers. Day length more than 12 hours at the early stage promotes vine growth, while short photoperiod favours tuber development.

8.0 Propagation

8.1.0 Vegetative Propagation

8.1.1 Tubers

Yams are mainly propagated vegetatively and the material commonly used for planting is tuber piece or small whole tubers. The whole tuber is cut into seed pieces consisting of head, middle and tail. The size and weight of tuber pieces at planting influence the yield. The relationship between sett weight and yield was confirmed by Onwueme (1972) and Nwoke *et al.* (1973). Among the portions of tuber, heads are the best, which sprout readily followed by tails and middles. Although large sized seed tubers give higher yield, a seed tuber weighing 200-250 g is ideal for optimum production in *D. alata*. Nwoke *et al.* (1984) found that larger sett gave higher tuber yields than smaller ones of *D. rotundata*, the former planting materials showed vigorous initial growth, an advantage that was maintained throughout the growing season leading to higher final yield. Tuber yield per plant decreased but the tuber multiplication increased with decreasing sett weight. High yield from small head setts was almost as good as from large setts due to longer duration of early and rapid tuber bulking (Ferguson *et al.*, 1983). Gyansa-Easmon (2000) in an experiment on planting material reported significantly high fresh tuber yield of 19.3 t/ha from 'pre sprouted' slip (leafless sprouts from stored tuber after the dormancy) compared with 8.8 t from the unsprouted slips. Direct planting produced significantly greater tuber size of 362.6 g relative to the pre-sprouted or pre-rooted slips, although tuber numbers/ha at 24785 were significantly lower than the 103910 with pre-sprouting. The directly-planted and pre-sprouted slips produced tubers largely in the conventional seed yam range (200-1000 g). This was an indication that the etiolated slips (the pre-sprouted form), could be a new source of planting material for commercial seed yam production. The plants exhibited'bushy' canopy architecture rather than viny and thus staking may not be necessary.

Multiplication of yams by vine cuttings is also possible but tuber production by this method is slow. Production of successful cutting has been established in *D. floribunda, D. dumetorum* and *D. rotundata*. Propagation is also possible through seed and bulbils. In *D. pentaphylla* and *D. alata*, small bulbils are produced in abundance, whereas in *D. hispida* bulbils are not produced but fertile seeds are formed. In *D. esculenta* and *D. trifida*, small tubers are produced in clusters which are utilized for the purpose of propagation.

Yam tubers undergo a period of dormancy at the end of the season, usually for 2-3 months. Early sprouting causes considerable shrinkage and damage to seed tubers, hence sprouts are frequently removed. Treatment with 4-8 per cent solution of ethylene chlorhydrin by quick-dip breaks the dormancy of yam. Soaking of tubers in 1 per cent solution of maleic hydrozide (MH), on the other hand, delayed the sprouting in storage, pre-harvest application of MH also proved effective in this respect (Hayward and Walker, 1961).

8.1.2 Tubers from Minisett and Slip

George and Suja (1995) reviewed the research on yam production from minisett and concluded that it is a cheap and rapid means of multiplication of yams, as only about 4 per cent of harvested yams need to be stored as planting material. Kalu (1989) carried out field trials on *D. rotundata* cv. Dan Onitcha, *D. alata* cv. Obuna and *D. cayenensis* cv. Oku to evaluate their suitability for use as minisetts for rapid seed yam production. For each species, minisetts weighing 20, 25, 30, 35 or 40 g were each cut from the head, middle and tail regions of mother tubers. Sets were treated with Aldrex (thiram + aldrin), cured overnight and planted directly at 100 cm x 25 cm spacing on ridges. Heavier tubers at harvest were obtained from *D. alata* (285-630 g) than from *D. rotundata* (105-450 g) and *D. cayenensis* (40-230 g). The potential value of the use of the minisett technique appeared more promising for *D. alata* and *D. rotundata* than for *D. cayenensis*.

Gyansa-Ameyaw *et al.* (1994) reported that yields of 5.01-10 g sets of *D. rotundata* were not significantly different from those of the 10.01-20 or 2-5 g classes. Use of the 5.01-10 g class for seed yam production was recommended. In another investigation the traditional practice of using 100-200 g sets of *D. rotundata* cv. Dan Onitcha, *D. alata* cv. Obuna and *D. cayenensis* cv. Oku was compared with the use of minisetts weighing 25, 20, 15 or 10 g, in field experiments (Kalu *et al.*, 1989). The minisett technique required only 6-33 per cent of the number of mother tubers needed for the traditional planting system. Number and total weight of harvested tubers were similar for traditional sets and 25 g minisetts in all species in all years. The number of tubers produced decreased with decreasing size of minisett and overall tuber weight was lower for *D. cayenensis* than the other 2 species. The proportion of seed tubers <100 g was greater from minisetts than from traditional sets but total seed yam (<400 g) yield was similar for traditional sets and minisetts >20 g in *D. cayenensis* and >15 g in *D. rotundata* and *D. alata.*

In an experiment to determine the optimum minisett size for rapid multiplication of seed yams (*D. rotundata* cv. Sree Priya) four sizes on minisetts (15, 30, 45 and 60 g) were used (George *et al.*, 1996). A 30 g minisett was best for rapid seed yam production. Although 60 g minisetts gave higher yields (21.7 t/ha), multiplication ratio was only 18:1. Minisetts of 30 g size were on par with 45 g minisetts for percentage sprouting. Establishment percentage, however, was higher in the former (86.8 per cent) than in the latter (82.6 per cent). Yield obtained from 30 g minisetts (15.5 t/ha) was also on par with that of 45 g minisetts (17.3 t/ha).

Sprouting of 15, 30, 45 and 60 g minisetts (small yam pieces) of *D. alata* cv. Sree Roopa and *D. rotundata* cv. Sree Priya in nursery beds of sawdust, sea sand, leaf litter or compost was assessed by George (1990). Sprouting occurred after 7-14 days in *D. alata* and 20-35 days in *D. rotundata*. Sprouting percentage in both species increased with increasing minisett size. Sprouting percentage was highest (68.1 and 68.6 per cent) in sawdust and sand in *D. alata,* and in compost (29.2 per cent) and sawdust (28.4 per cent) in *D. rotundata.* In on-farm field experiments in Kerala, India minisetts of *D. alata* of 30 g size were planted in nursery beds made of sawdust, leaf litter or river sand. In the saw dust and river sand media, around 80 per cent of the minisetts sprouted compared with 51-53 per cent in leaf litter. In another experiment, sprouting percentage and tuber yields were significantly greater from *D. alata* raised from minisetts than in *D. rotundata* (Sreekanta *et al.*, 1995). Osai and Ikotun (1994) recorded that sawdust and garden soil employed in minisett propagation were the main sources of microbial rot. Other inoculum sources also infected yam tubers.

In Njombe, Cameroon, studies were undertaken by Segnou (1999) to investigate different media which could replace sawdust for germination of yam minisetts and the effect of different types of mulch, plant staking and different plant densities on seed yam yields, using 20-30 g minisetts as planting material. Topsoil, sand, coffee husks and volcanic gravel could replace sawdust for germination of yam minisetts. Mulching with black and white polyethylene, with the white side up, and plant staking increased seed yam yields, with a combination of the 2 practices gave the greatest yields. These treatments also provided the best conditions for the production of abundant high quality planting material. Yields increased with increasing plant density, while tuber size decreased. The proportion of marketable tubers (>200 g) was greater with mulch.

Wilson *et al.* (1998) suggested that development of improved techniques for combination of minisett production and GA_3-induced enhancement of tuber storage life could lead to recognition of minitubers as a new staple tropical product.

8.1.3 Tissue Culture

Unnikrishnan and Sheela (2001) have reviewed the results of investigations on biotechnology of root and tuber crops.

8.1.3.1 Micropropagation

Tissue culture has been used for clonal propagation of yams. A technique of micropropagation has been developed for *D. esculenta* at the Central Tuber Crops Research Institute (CTCRI), Trivandrum, India (Nair, 1984). The three edible species of *Dioscorea viz., D. alata, D. rotundata, D. esculenta* and *D. bulbifera* were brought into *in vitro* cultures using meristem tip or nodal explants of vines or tuber sprouts. *Dioscorea esculenta* cultures produced shoots in presence of NAA (1 µM) and BA (10 µM), *D. alata, D. rotundata* nodal segments developed shoots in presence of NAA (1 µM) and BA (2 µM). These shoots on transfer to medium containing 1 µM each of NAA and BA produced whole plants (Nair and Chandrababu, 1996). Nodal segments of *D. alata* and *D. rotundata* grown under long days (16-h light) developed into plantlets on MS basal medium without growth regulators (Mantell *et al.*, 1996). A micropropagation system using mericlones obtained from immature leaves cultured on MS medium with (8.9 M) was developed for *D. opposita* (Kohmura *et al.*, 1995). *In vitro* cultures of Jamaican yams (*D. cayenensis* and *D. trifida*) were established by nodal cultures on MS medium with NH_4NO_3 (800 mg/l) and BA (0.5 mg/l) (Mitchell *et al.*, 1978).

Mix-Wagner (1993) reported regeneration of plantlets with well-developed root systems from axillary buds and stem apices of *D. rotundata*. Shoot regeneration was only about 9 per cent in the absence of growth regulators, but increased to 71-80 per cent when GA and NAA were added. On all media the use of rockwool instead of agar increased shoot regeneration. Saborio *et al.* (1998) observed that production of clean asexual planting stock of *D. trifida, D. alata* in Costa Rica. The strategy used was to produce 205 000 *in vitro* plantlets of these crops, and then further propagate them by establishing mother blocks, under isolated conditions, in the field.

Plantlets with well-developed root systems were regenerated from axillary buds and stem spices of *D. rotundata* (Mix-Wagner, 1993). Shoot regeneration was only about 9 per cent in the absence of growth regulators, but increased to 71-80 per cent when BA and NAA were added. On all media the use of rockwool instead of agar increased shoot regeneration. Addition of CCC to a medium with 2 mg BA + 2 mg NAA + 80 g sucrose/l led to 85 per cent regeneration with an average of 15 shoots/plant.

Multiple buds were formed from node of shoots from bulbils *in vitro* cultured in the MS medium with or without 10 mg ancymidol/l in a rotary or shaking culture for 40 days. Multiple bud multiplication was particularly effective in shaking culture with ancymidol. Seven cycles of subculture in aqueous basal medium with 10 mg ancy-midol/l and 2 per cent sucrose furnished numerous buds (multiplication rate 30 per subculture) of uniform size (Matsubara *et al.*, 1992). *In vitro* morphological development of meristem tips, excised from microshoots was reported by Malaurie *et al.* (1995) for one clone (Grosse Caille) of *D. cayenensis/D. rotundata* complex and for one clone of *D. praehensilis*. Larger meristem tips had the best rooting and plantlet formation, while the smaller size gave the best bud production. Use of antiviral chemicals like DHT (2,4-dioxohexahydro 1,3,5 triazin) and diacetyl DHT were studied in eliminating mosaic disease in *D. esculenta*. But they were found to affect growth of *in vitro* plants (Nair *et al.*, 1994). Shoot apex cultures were used to clean yam mosaic virus (YMV) infection in *D. trifida* (Saleil, 1990).

8.1.3.2 Somatic Embryogenesis

Viana and Mantell (1989) obtained callus and regeneration from embryo explants of *D. cayenensis*, on MS medium supplemented with 2, 4-D (18 µM), zeatin (0.1 µM) and glutamine (3.3 mM). In *D. alata*, somatic embryogenesis was obtained from calli derived from leaf and nodal explants of *in vitro* plants. Microcalli was obtained from protoplasts isolated from young lamina cells (Cedric *et al.*, 1988). Regeneration from callus cultures of *D. alata* and *D. esculenta* was also obtained on MS medium (Nair and Chandrababu, 1996). *D. rotundata* embryos showed faster germination on modified MS medium and LS medium (Okezie *et al.*, 1984).

Somatic embryos were produced by the callus cultures of 2 rare endemic *Dioscorea* species, *D. caucasica* and *D. balcanica* (Chulafich *et al.*, 1994). Seedlings developed (initially roots, then shoots) following transfer of embryos to agar-solidified medium, supplemented with IAA at 0.18 mg/l and BA at 2.25 mg/l. After 8 weeks, the plantlets were established in pots of soil for acclimation and transferred to soil outdoors. After one year, the regenerated plants flowered.

In vitro-grown yam plantlets belonging to four species, *Dioscorea rotundata, D. cayenensis, D. alata* and *D. trifida* were acclimatized in a shaded house (75 per cent shade cloth) over a 4-week period using a simple, cost-effective and efficient procedure which gave 99 per cent survival (Asemota *et al.*, 1997).

8.1.3.3 Tuberization in vitro

Morphogenetic and physiological responses in vitro

In vitro tuberization was reported in *D. alata* by Ammirato (1984) and in *D. bulbifera* by Forsyth and Van Staden (1984). Tuberization occurred in nodal cultures grown on MS medium and was found to be influenced by the sucrose and kinetin content of the medium (Forsyth and Van Staden, 1984). In *D. rotundata* microtuber production was influenced by sucrose and day length. The frequency of tuber isolation was highest at 5 per cent sucrose under 12 to 16 hour day length (Ng, 1990).

The effects of photoperiod (8, 12 or 16 h), mineral medium strength (dilutions of a tuberization medium, T medium), sucrose (0, 2, 4, 8 per cent w/v) and kinetin (0, 2.5 µM) on organ development in shoot cultures were evaluated by Mantell and Hugo (1989). The effect of high and low inorganic ammonium containing media on microtuberization indicated that ammonium ions inhibited microtuber induction in *D. alata* but not *D. bulbifera*. Microtubers of *D. alata* were only formed under 8-h days; *D. bulbifera* cultures produced microtubers under this photoperiod as well as under 16-h photo-periods, provided that kinetin was present in the media. Nodal cuttings were cultured *in vitro* to assess the influence of NAA on the production of microtubers (Jean and Cappadocia, 1992). In *D. alata*, high concentrations of NAA

(27 and 54 µM) favoured the production of large microtubers, whereas the highest number of microtubers was obtained with 2.7 µM. *D. abyssinica* was more sensitive to NAA since concentrations higher than 0.27 µM promoted the growth of callus on the root system. In this species, the production of the largest microtubers was obtained at 2.7 µM whereas the number of microtubers was not affected by any concentration tested.

The effects of 0, 1.0, 2.5 or 5.0 µM kinetin on *D. alata* cv. Oriental Lisbon shoot cultures grown in T medium (Heller's medium microsalts) with or without ABA or in MS medium were studied by Alhassan and Mantell (1994). Supplementing T medium with 1.0 µM kinetin improved microtuber frequencies. Interaction of kinetin at all levels with a set level of 1 µM ABA reduced microtuber frequency by 14 per cent and increased individual microtuber weights by 6 per cent. Passam (1995) reported that single-node vine cuttings of *Dioscorea rotunda* and *D. bulbifera* rooted in moist peat or perlite yielded microtubers which could be stored satisfactorily for 50 and 100 days, respectively and on subsequent planting produced tubers of a satisfactory seed size.

The effects of plant growth regulators, light intensity and end-of-day (EOD) light quality treatments on node and microtuber induction (per cent of cultures with microtubers) and development (fresh weight of microtubers) in yam cultures were studied by John *et al.* (1993). Nodal segments were excised from plantlets cultured on tuberization medium containing growth regulators and exposed to various light treatments. Abscisic acid (1 µM) stimulated and cytokinins (2.5 µM) inhibited microtuber development from nodal segments cultured on Mantell's and Hugo's full-strength tuberization medium less than 8 h photoperiods. EOD far-red (FR) light inhibited microtuber induction and development and enhanced node formation. EOD FR light effects were nullified by immediately following the FR treatment with red light. This suggested the involvement of phytochrome in these processes.

8.2.0 Seed Propagation

The seeds of *D. abyssinica, D. bulbifera, D. dumetorum, D. preusii* (*D. preussii*), *D. schimperiana* and *D. togoensis* displayed rapid germination at temperatures around 26°C and high-temperature inhibited germination. The lowest temperature for germination was 20°C in all species (Okagami *et al.*, 1999).

Abraham (1992) studied the dormancy and viability of seed of *D. alata* and sun dried seeds were germinated in a sand-soil mixture immediately after harvest or stored for 1 to 10 months and tested for germination at monthly intervals. The highest germination percentage was observed in seeds sown immediately after harvest or stored for 1 month. Germination percentage fell to between 51 and 54 per cent after 2-5 months storage and to 49 per cent in months 6 and 7. Seeds stored for 10 months did not germinate. Viable seeds of all ages started to germinate after 12 to 17 days and 50 per cent germination of viable seeds was observed within 14 to 24 days after sowing.

Danial *et al.* (1999) reported the highest seed storage life of 21 years was derived for seed dried to 5 per cent moisture content and stored at –20°C. This was followed by 19 years with the seeds dried to 5 per cent moisture content and stored at 5°C.

The shortest seed storage life of about seven years was estimated for seeds of 14.5 per cent moisture content at 25°C.

9.0 Cultivation

9.1.0 Land Preparation and Planting

The soil should be deeply ploughed, pulverized and levelled. Yams are planted in flat or raised beds or on mounds formed over pits. It is vegetative propagated mainly by small whole seed tubers or cut tuber pieces of 200-500 g. Pits of the size 45 x 45 x 45 cm are dug, about 3/4th size of these pits are filled with dry farm yard manure and top soil, then whole seed corms or yam setts are planted in pits and covered with soil. Staking in yam is essential for increased tuber production (Mitra and Tarafdar, 2013).

The seed tubers are planted at a depth of 10-15 cm in the soil from March-May before the onset of monsoon. In Puerto Rico, Irizarry and Rivera (1994) recorded that the best planting date of *D. alata* was March, with a marketable yield of 70.59 t/ha. About 88.2 per cent (62.25 t/ha) was highly graded tubers. The delay in planting from March to July significantly reduced yields by 44.8 per cent.

The weight of seed tuber for optimum yield is 200-250 g in *D. alata* and 100-125 g in *D. esculenta*. Kayode (1984) observed that in *D. rotundata*, sets weighing about 400 g gave the best tuber yield and larger planting material was not economical. Cut tubers are suberized or smeared with wood ash before planting. Treatment of benlate powder on the cut surface prevents soft rot.

The effect of the seed tuber size on the growth and development of yam was investigated by Toyohara *et al.* (1998) using two cultivars: Lisiteye introduced from Papua New Guinea and a local cultivar from Africa. Tubers of these cultivars were cut into fragments with five sizes (20, 50, 80, 110 and 140 g). These planting materials were grown in the experimental field of Tokyo University of Agriculture. The effects of seed tuber size of both cultivars on growth were evident from the early growth stages, with increases in stem length, in number of branches and in total leaf area in proportion to the seed tuber weight. The length and weight of the new tubers were also positively correlated with the seed tuber size. There were differences between cultivars in the effect of seed tuber size. Seed tubers weighing 20 g were considered to be too small for practical cultivation. In another experiment they also reported that among the three types of tuber fragments weighing 20, 50 and 80 g planted in pots filled with soil given no fertilizers, larger seed tuber fragments showed conspicuous reduction in tuber weight, resulting in larger initial growth, compared with the smaller ones.

In field trials in Brazil, tips or portions of large or small rhizomes of 5 cultivars of *Dioscorea* sp. were used as planting stock (Zarate *et al.*, 1998). A significant interaction was observed between cultivars and type of planting stock, the greatest difference being observed in cv. Roxo where portions of large rhizomes resulted in fresh yields 51.4 and 69.2 per cent higher than those given by tips and small rhizome sections. Use of large rhizomes gave highest rhizome yields for all except cv. Minoso.

In a field experiment in India, seed tuber pieces of *D. alata* weighing 50, 100 or 200 g were planted at intra-row spacings of 0.5, 1 or 1.5 m and inter-row spacing of 1 m. The intra-row spacing of 0.5 m and the use of 200 g seed tuber pieces produced the highest mean tuber yield 239.8 t/ha (Singh *et al.*, 1994).

In yam species, namely *D. bulbifera* and some types of *D. alata*, bulbils or aerial tubers are planted to obtain small-sized tubers in the first season, which in the subsequent season are used as setts for planting.

9.2.0 Spacing

Plant population per unit area depends upon the growth habit of the species and in general, cultivars having broader leaves and vigorous growth require wider spacing. A spacing of 75 x 75 cm for *D. esculenta* and 90 x 90 cm for *D. alata* may be followed for the optimum yield of tubers (Anon., 1985). Onwueme (1978a) reported that yams are planted at a spacing of 1 m between rows and tubers.

An experiment was conducted in Brazil by Heredia *et al.* (2000) to study the yield of rhizomes of yam clones (Caramujo, Pezao, Roxo, Florida and Mimoso) under plant densities of 4000, 8000, 12000 and 16000 per hectare. Harvesting was carried out when aerial parts from 3 of the 5 clones were in advanced stages of senescence. The evaluated parameters were tuber and rhizome total yield. The total average yield of Caramujo (65.79 t/ha) was greater by 12.06, 57.96, 83.98 and 85.79 per cent than that of Pezao, Florida, Roxo and Minoso, respectively. Highest and lowest rhizome yields were recorded from Florida (41.65 t/ha in average) and Pezao (28.65 t/ha), respectively. The optimum plant density for rhizome and tuber yield was 16000 plants/ha, except for Mimmoso and Roxo (12000 plants/ha).

Four planting rates (1, 3, 5 and 7 t/ha) and four plant populations (20, 45, 70 and 95 x 103 plants/ha) of the 6322 accessions of greater yam (*D. alata*) were combined in a two-factorial experiment conducted in Costa Rica (Rodriguez-Montero *et al.*, 2001). The planting rate strongly affected the yield components, whereas the effect of plant population was only significant for the commercial tuber yield and the number of non-commercial tubers. Commercial tuber yield tended to increase asymptotically with increasing sett size. For total yield, however, tuber production was a function of planting rate and did not respond to plant population, except at the lowest planting rate.

Enyinnay *et al.* (1987) concluded that planting depth of 9-12 cm was best for minisett planting of *D. rotundata* and *D. alata*. This depth minimized exposure by rainfall, pest damage and sun-induced rot.

9.3.0 Trailing of Vines

Trailing of vines is an important operation to expose the leaves to sunlight. Staking to provide support to the vine has been found essential for better growth, development and yield of yams (Lyonga *et al.*, 1973). The emerging shoots should be immediately provided with supports to avoid any injury to the tender shoots. The species *D. alata* requires trailing to a greater height as compared to *D. esculenta* due to its bushy growth. Trailing is done within 15 days of sprouting by coir rope attached to artificial supports or trees in the growing areas. The most effective method is pandal system which produced a yield of 22.0 tonnes, while the control plants (without trailing) showed a yield of 14.7 t/ha (Abrahman and Easwariamma, 1984).

Yams are also staked on to living or dead trees or other quick growing plants. Branched stakes are found to be superior over unbranched stakes (Haynes and Coursey, 1969). In Trinidad, Chapman (1965) reported a reduction in yield by more than 50 per cent in the unstaked crop.

Igwilo (1989) reported that staking increased fresh tuber yield of *D. rotundata* cv. Nwopoko from 5.5 to 8.6 t/ha and from 14.4 and 22.8 t to 18.6 and 29.7 t in cv. Obioaturugo and Um 580, respectively. NPK application increased the tuber yield of Um 680 in both staked and unstaked crops by increasing the LAI. NPK application had no significant effect on the staked plants of cvs. Nwopoko and Obioaturugo, and reduced the yield of unstaked plants of these cultivars. Cultivar differences in response to staking could arise from a difference in canopy structure. Heredia *et al.* (1999) reported that when *D. cayenensis* cv. Caramujo was staked, it produced 92.06 per cent rhizomes and 7.94 per cent tubers, while on the ground it produced 46.12 per cent rhizomes and 53.88 per cent tubers. The Liso (*D. alata*) rhizomes had the best nutritive value, while the rhizomes of Caramujo had a better nutritive value than the tubers. Yield of Caramujo (50.3 t/ha) was significantly higher than that of Liso (22.7 t/ha). Yields of both species were higher without staking.

Budelman (1990a) concluded that *Gliricidia sepium* as support and the mulch derived from its leaves created more favourable conditions for development of the yam plant (*D. alata*). Budelman (1990b) reported of 3 leguminous perennials (*Leucaena leucocephala, Flemingia macrophylla* and *Gliricidia sepium*) as live support systems in yam (*D. alata* cv. Brazo Fuerte) cultivation. Tuber yields per hectare were 3.4 (leucaena), 5.3 (flemingia) and 10.1 (gliricidia) tonnes fresh weight. The significantly higher tuber yield of the yam crop growth with *G. sepium* is concluded to be a function of specific properties of the tree species: low leaf productivity, a relatively weakly developed root system, and an open architecture, allowing sufficient space for a yam crop grown in association. The effects of staking with live *G. sepium* planted at 2500 plants/ha or bamboo canes on yield of *D. rotundata* cv. TDr 131 (improved) or Ikokoro (local) were also studied (Otu and Agboola, 1994). The live stakes were pruned 3 times giving DM yields of 3.7, 2.7 and 2.1 t/ha at the 1st, 2nd and 3rd pruning, respectively. After decomposition the nutrient contribution from the pruning was 354 kg N, 22 kg P, 134 kg K, 121 kg Ca and 38 kg Mg/ha per year. Usage of *G. sepium* as mulch decreased soil bulk density by 14 per cent, increased soil moisture content by 16 per cent, decreased weed population by 14 per cent and decreased weed DM by 21 per cent. In *D. rotundata* tuber yield was not significantly affected by stake. After 3 pruning, tuber yield under the live stake was 16 per cent higher than that with bamboo stakes.

Supporting *D. opposita* with nets improved stem and leaf growth, light interception, leaf area index and tuber weight/plant compared with unsupported plants or conventional support. Yield was more than double from tuber planting compared with bulbil planting (Kwon *et al.*, 1996).

Cultivation Practices of Yams

9.4.0 Manuring and Fertilization

Yams require plenty of organic manure. Well rotten FYM at the rate of about 10 t/ha is recommended. Among different treatments under validation programme of organic farming system in greater yam, the highest yield of 25.13 tha^{-1} was recorded with the organic and biofertilizer treatment combinations (FYM @ 15 t ha^{-1} + green manure @ 15-20 t ha^{-1} + neem cake @ 1 t ha^{-1}+ ash @ 1.5 t $ha^{-1)}$ + biofertilizers (*Azospirillum* @ 3 kg ha^{-1}, mycorrhiza @ 5 kg $ha^{-1}$and phosphobacteria @ 3 kg ha^{-1}) followed by 21.96 t ha^{-1} with the treatment- conventional (FYM @ 10 t ha^{-1} + NPK @ 80:60:80 kg ha^{-1}). There was slight difference in nutritional parameters in tubers under different treatments (Anon., 2020).

The results of the experiments on the effect of fertilizers on the yield of tubers clearly indicated that yams are very responsive to the application of inorganic fertilizers along with organic manures. Application of NPK at the rate of 80:60:80 kg/ha was recommended for *D. esculenta* and *D. alata* (Mohan Kumar *et al.*, 1984). In an experiment on *D. esculenta,* the tuber yield increased up to 20.6 t/ha at 80 kg N and 21.2 t/ha at 120 kg/ha. Further increase caused a decline in yield (Singh *et al.*, 1973). In *D. alata,* they concluded that NPK @ 120:80:80 kg/ha was optimum for obtaining high yield of good quality tuber. Results of the experiments also indicated that the response to P was low (Lyonga *et al.*, 1973). Onwueme (1978b) recommended fertilization one month after shoot emergence, whereas Obigbesan and Agboola (1978) suggested waiting till yam vines are established. Fertilization in white yam is effective any time from 8 to 12 weeks after planting (Nwinyi, 1984). Application of nitrogen as ammonium sulphate at planting depressed the tuber yield while it was beneficial when applied after 3 months (Chapman, 1965).

Trials conducted at IITA, Nigeria, indicated that *D. rotundata* responded significantly to high rates of N (90 kg), low rates of P (30 kg/ha) in soil low in these elements. However, a fertilizer dose of N 90, P 25, K 30 was considered optimum (Kpeglo *et al.*, 1980). If the soil moisture is not adequate the plants should be irrigated after fertilization. Half dose of nitrogen and potassium along with full dose of phosphorus should be applied at the time of first inter-culturing. The remaining half dose of nitrogen and potassium should be applied a month later along with second inter-culturing.

Fertilizer application (0 or 100 kg N + 0 or 60 kg P_2O_5 + 0 or 200 kg K_2O/ha) significantly increased leaf and tuber N and crude protein (CP) contents 5 months after planting. Utilization of N, P and K was 40.1, 17.7 and 24.9 per cent, respectively. For each tonne of tubers harvested after 11 months, uptake of N, P and K was greatest (4.0, 1.06 and 5.94 kg, respectively) with 100: 60: 200 kg N: P: K. Tuber yield ranged from 14.97 t/ha without fertilizer to 28.64 at highest NPK rates (Castillo *et al.*, 1989). Liu (1989) standardized cultivation technique for productivity of *D. alata* cv. Coconut-Lesbon (a) and a Tainung cultivar (b). Tuber yields were highest with 90 kg N + 60 kg each of P_2O_5 and K_2O in (a) and 90 kg N + 60 kg P_2O_5 + 120 kg K_2O in (b).

The effect of NPK on the tuber yield *D. deltoidea* (a medicinal plant yielding diosgenin, used for corticisteroid production) on a loamy soil showed that the interactions K x N, N x P and K x N x P were significant, and that the maximum tuber yield/ha was obtained with the low level of nitrogen and high levels of potassium and phosphorus. There was not much significant difference in the percentage of diosgenin yield with treatment (Rishi *et al.*, 1994). Application of N rates of 0, 50, 150, 250 or 350 kg/ha, with fixed levels of P_2O_5 (50 kg/ha) and K_2O (150 kg/h a), to soils with (12 q/ha) or without lime. Leaf area index, dry matter, net assimilation rate, crop weight and diosgenin yield of *D. floribunda* increased with increasing N, particularly on limed soil. The best economic return was with 350 kg/ha N in split doses (Saha *et al.*, 1991).

N, P and K rates of 150 kg/ha; foliar applications of Mg + Mn + B; GA or IAA at 20 mg/l; and removal of flowers or fruits enhanced tuberization of *D. composita* and *D. floribunda* and diosgenic content was increased by P alone at 150 kg/ha, foliar application of Mg + Mn; GA or IAA at 200 mg/l; and removal of flowers of fruits (Nandi and Chatterjee, 1992).

The effect of organic and inorganic fertilizers on growth and yield of planting stock of *D. doryophora*, a good source of steroids, was investigated in China (Sheu *et al.*, 2000). Organic fertilizer in the form of goat manure was applied at a rate of 20,000 kg/ha. Inorganic N, P, K fertilizers acting as the control, were applied separately at rates of 147, 144 and 120 kg/ha. The plants were harvested 8.5 months after planting. The results showed that growth and yield (23 t/ha) were similar regardless of fertilizer source.

Plantain + yam or cocoyam intercropping systems fertilized with 320 kg N/ha were recommended by Obiefuna (1989) because of improved plantain establishment and increased combined crop yield. Nnoke *et al.* (1987) determined the optimum rate and date of fertilizer application in *Dioscorea* sp./maize/cassava intercrops. NPK fertilizer (15:15:15) at 0, 200, 400, 600 or 800 kg/ha was applied to the crop mixture 3, 4 or 8 weeks after planting (WAP). Results showed that yields of maize grain, *Dioscorea* tubers, and cassava roots increased significantly with increasing fertilizer rates, reaching a maximum at 800 kg NPK/ha. Although fertilizer application at 3 WAP gave the highest yields in all cases, the differences were not significant.

The nutrient uptake in several species of *Dioscorea* varied between 116.1 and 168.0 kg N, 25.5 and 30.0 kg P and 108 and 146 kg K per hectare (Mohan Kumar *et al.*, 1984). The nutrient removal by yam is considered high. The uptake of nutrients by different species of *Dioscorea* studied at CTCRI, Trivandrum clearly showed the preference for both N and K (Table 2).

Table 2: Yield and Nutrient Removal of Three Edible Yam Species*

Species	*Yield (t/ha)*	*Nutrient Removal (kg/ha)*		
		N	*P*	*K*
D. esculenta	19	116.1	26.2	108.3
D. alata	25	163.0	23.5	126.0
D. rotunda	27	168.0	30.0	146.0

*Anon. (1983)

In field trials on *D. alata* cv. Gunung, Irizarry *et al.* (1995) reported that tuber dry matter yield was not significantly affected by the fertilizer treatment. N, K and Ca uptake peaks occurred about 5 months after planting, while P and Mg uptake increased up to the end of the 7-month cropping cycle. The dry matter production peak occurred at the completion of the 7-month cropping cycle. The plants used 24.7 kg N, 2.2 kg P, 25.7 kg K, 11.0 kg Ca and 1.0 kg Mg/ha for every tonne of edible dry matter production.

9.5.0 Intercropping and Weed Control

Adequate fertilizer and moist soil are conducive conditions for the rapid growth of numerous monocot and dicot weeds. Control of weeds is very important for the proper growth of the plants and normally 2-3 intercultural operations would be sufficient to check the growth of weeds. The first interculturing which must be sufficiently deep is done one week after sprouting of 50 per cent tubers and the second and third ones at an interval of 15 days to 1 month depending on the weed growth.

Orkwor *et al.* (1995) conducted experiments to determine: the critical period of weed interference in yam/maize/okra/sweet potato intercrops; assess crop yield losses due to uncontrolled associated weeds; and assess the timing of weed removal. The critical period of weed interference in the intercrop was between 3 and 16 weeks after planting (WAP). Weed interference in the mixture reduced yam tuber yield by 35 per cent, maize by 60 per cent, okra by 79 per cent and sweet potato roots by 80 per cent. Yam was significantly affected by weed competition in the intercrop 3 WAP and weeds continued to depress yam yields up to 16 WAP as in sole cropping.

Kolo (1995) observed that yam (*D. rotundata*) intercropped with melon (*Citrullus lanatus*) significantly ($P<0.05$) reduced weed growth by about 50 per cent compared with monocropped yam in both years of study, irrespective of the staking method. The individual staking method resulted in a higher weed infestation than other treatments with or without intercropping.

Anuebunwa (2000) reported that the prevalent grass weeds were suppressed by egusi (watermelon) and sweet potato although they later reappeared sparsely. The egusi acted as the early weed suppressant during the initial growth of the component crops. Sweet potato continued the ground cover provided by egusi, but to a different extent. The yield of yam and cassava was higher where sweet potato was introduced at 10 weeks after planting (9.4 per cent and 7.6 per cent, respectively) than with two hand weeding.

In Nigeria, across agro-ecological zones, speargrass (*Imperata cylindrica*) was ranked by 55 per cent of respondents as the most serious weed, infesting the most important crops of the region (maize, cassava, yam, cowpea, groundnut, cotton and sorghum). Besides speargrass, *Commelina benghalensis, Digitaria nuda, Cyperus rotundus, Tridax procumbens, Rottboellia cochinchinensis, Digitaria horizontalis, Striga* spp., and *Euphorbia heterophylla* were also considered as important weeds in major cropping systems (Chikoye *et al.*, 2000).

Renaut and Merlier (1973) found that Diuron at the rate of 3-3.5 kg/ha was effective in pre-emergence control of weeds in yam plots in Ivory Coast. Lagoke *et al.* (1987) found that ametryn at 2 kg/ha + alachlor at 2 kg provided effective early weed control and increased yam tuber yield compared with manual weeding, 16 weeks after planting. Fluometuron at 2 kg consistently gave higher tuber yields but did not reduce weed infestation significantly. Prometryn at 1 kg + metolachlor at 3 kg resulted in high tuber yields.

The results of field trials conducted in Central Cote d'Ivoire to assess the efficacy of some 20 herbicides used alone or in combination for weed control in yam (*D. alata* and *D. cayenensis*) grown in monoculture or intercropped were summarized by Marnotte and Tehia-Kouakou (1995). Best weed control and economic benefit in the yam monocrops was obtained with 700 g/ha metribuzin + 800 g diuron or 240 g clomazone. When yam was intercropped with maize, best control and crop selectivity were recorded with atrazine + metolachlor, atrazine + pendimethalin or terbutryn + metolachlor mixtures. Liu *et al.* (1990) reported that ametryn at 4.48 kg/ha pre-emergence followed by fluazifop-P at 0.56 kg post-emergence was highly effective against broad leaved weeds in yams cv. Guinea Negro. Oxyfluorfen at 2.4 kg pre-emergence followed by paraquat at 2.34 litres post-emergence was also highly effective, but both treatments caused slight to moderate crop injury. Cinmethylin alone at 2.05 and 4.10 l/ha was poor against broadleaved weeds, but showed improved results when followed by paraquat post-emergence. Metolachlor at 3.54 l/ha followed by paraquat also provided good weed control.

Monoculture and intercropping systems for yam and sweet potato production were evaluated in field trials over 2 planting seasons (Roman *et al.*, 1991). In the first experiment, yam cv. Guinea Blanco and sweet potato cv. Dominicana were planted on the same date, or else sweet potatoes were planted 6 weeks before or after yams in February, April and May. In the second experiment, Guinea Negro was used and the same cropping treatments were evaluated, but planting took place in October and December. The results revealed that in the first experiment, sweet potatoes planted in monoculture or intercropped suppressed weed growth. However, yam competition significantly reduced sweet potato yields. Although weeds grew freely in crop mono-cultures, yields were not significantly reduced. In the second experiment no significant differences were recorded among cropping systems with regard to weed control. However, planting in December and January reduced weed populations in both systems. When yams and sweet potatoes were intercropped during these months, yam production was not affected. The highest yield of combined crops (59,014 kg/ha) and yams (36,738) occurred on intercropped plots planted in December. This planting date also resulted in the highest sweet potato yield (36,559 kg/ha), providing that the crop was grown as monoculture.

9.6.0 Mulching

Mulching, after the planting of set or tuber is very useful for increasing the yield. It provides protection from excessive temperature, conserves soil moisture, ensures quick and uniform sprouting of the tubers and suppression of weeds. Manu (1997) also reported that mulching with guinea grass, coconut coir dust and mature coconut fronds increased yam fresh tuber yields to 43.5-45.0 t/ha compared with 38.0 t/ha in unmulched controls. There were no significant differences in yield between the mulching treatments. Olasantan (1999) in an investigation on mulching concluded that yam planted in October, just before the rain stops, can be mulched with dry *Enpatorium odoratum* in January or February without detrimental effect on emergence, growth or tuber yield. Mulching with leaves of *Gliricidia sepium* shortened the time taken for the yam sets to sprout, improved the leaf nutrient contents of the yam crop and resulted in significantly higher tuber yields. Over a tuber yield range up to approximately 15 t/ha each additional tonne dry matter of mulch applied resulted in a yield increment of about 2 t of yam tuber (Budelman, 1989).

A total of 25 yam varieties from Africa, Japan and Papua New Guinea were compared with agronomic characters and tuber characteristics when grown with or without a mulch treatment (vinyl sheet 0.2 mm thick) (Toyohara *et al.*, 1997). Mulch treatment resulted in increased soil moisture and soil temperature during the early growing season. There were varietal differences for most of the characters measured. Effects of mulching became evident at around 2 months after planting, resulting in an increased number of branches and leaves, and increased leaf length and width, compared to plants without mulch treatment. Total top dry weight, leaf area, tuber weight and number of tubers at harvest were generally larger under mulched condition. A large positive correlation was found between the magnitude of increase induced by mulching for total top dry weight and tuber weight, indicating that varieties with higher responses of top weight to mulching tended to show higher responses of tuber weight.

10.0 Harvesting, Yield and Storage

10.1.0 Harvesting

The period from planting to maturity varies depending on the species. Cultivated species are generally harvested at 7 to 11 months after planting. It takes 7-8 months after planting to produce mature tubers of desirable size. When the leaves turn yellow and the vines completely dry up, the crop is ready for lifting, which will be in about 8-9 months after planting. *D. esculenta* matures early as compared to other species. There is a practice of double harvesting, *i.e.*, removing mother tubers after two months of growth and allowing subsequent production of side tubers. However, double harvesting is not economical as compared to single harvesting. While harvesting care should be taken so that tubers are not cut or damaged as this causes rotting of tubers.

10.2.0 Yield

The tuber yield depends on species and cultural practices adopted. *D. alata* and *D. rotundata* have been found to give a yield of 20-40 tonnes of tubers per hectare depending on sett-sizes used, whereas *D. esculenta* yields 10-30 tonnes per hectare under field conditions in one season. Sanchez *et al.* (1987) reported that the mean commercial root yields from brown, alluvial and ferralitic soils in Cuba were 25.9, 18.1 and 8.4 t/ha, respectively, and from 90 cm ridges, furrows and 120 cm ridges were 18.7 and 20.5, 13.8 and 17.3, and 16.6 and 18.0 t/ha when planted with cuts facing upwards or downwards, respectively.

In seven yam (*Dioscorea* spp.) cultivars belonging to three different species, *D. rotundata, D. alata* and *D. esculenta*, the yield was highest in one of the 2 *D. rotundata* cultivars (Sree Priya), while both *D. esculenta* cultivars gave lower yields than any cultivar of the other 2 species. The rate of dry matter (DM) production (mean crop growth rate) was not significantly associated with tuber yield, whereas yield was closely positively correlated with harvest index in each species. Mean harvest index was higher in *D. rotundata* (Chowdhury, 1998). Akoroda (1987) concluded that higher tuber yields were be associated with systems that maximize low inputs of sett, stake, soil nutrients and water within a specified site and crop season.

10.3.0 Storage

Like other tuber crops, curing before storage allows suberization of surface injuries occurred during harvesting and handling of yam tubers, and helps in reducing weight loss and rotting. Curing of yam tubers should be done at 30-32°C at 90 per cent to 95 per cent relative humidity for 5 to 7 days (Mitra and Tarafdar, 2018). Yams are stored in the tropical countries at a temperature of 30°C. Tubers may suffer from chilling injury when stored below 12°C. Several storage methods are followed in different yam growing countries. The most common practice is to leave yam as such in the field and harvest whenever required but in this method the quality of tubers deteriorates after the onset of rains. In some regions the tuber are kept in shallow pit and covered with soil. In some cases they are heaped in a pyramid shape and covered with moist soil. Yam barns are constructed consisting of vertical wooden frame on which yams are tied. In Oceania, specially constructed huts are used for stacking the yams. In some parts of India, yams are stored in the layers of sand. Dipping of tubers in fungicides like benlate and captan (Adesuyi, 1973) and in 250 ppm thiabendazole (Ricci and Arnolin, 1973) are effective in controlling storage rots.

Cush-cush yam (*D. trifida*) seed tubers were stored at 8, 15 or 30°C and relative humidities of <60, 76 or 98 per cent. Tuber storage as assessed by percentage of non-sprouted and non-rotted tubers reached an optimum of about 90 per cent healthy tubers at 15°C and <60 per cent RH (Solano *et al.*, 1996).

Girardin *et al.* (1998) studied the storage system and postharvest removal of sprouts on fresh weight losses of tubers. The pits allowed the daily range of variation in temperature to be reduced while maintaining a high relative humidity (RH), whereas the temperature and RH between the storage systems had little effect on the fresh weight losses of the tubers. Sprout removal reduced fresh weight losses equally over 5 and 6 months' storage. The reduction in fresh weight loss was 12 per cent for cv. Gnan (*D. cayenensis rotundata*) during 6 months' storage. Monthly sprout removal gave an equal benefit in pits, in sheds and on yam barns.

The problems associated with postharvest handling and marketing of yam in Ghana were described by Bancroft *et al.* (1998). Observations suggest that, during the early season, the loss of yam quality is associated with certain pre-harvest infestations of nematodes and termites as well as a systemic internal browning;

postharvest tissue damage associated with the stacking of tubers, various rots and the prolonged exposure of tubers to intense sunlight in the market place are subsequent factors. Such deterioration may lead to price discounting of 25-63 per cent and absolute biological losses of 10 per cent.

It was found that treatment of tubers with insecticide dust (actellic 2 per cent dust; pirimiphos-methyl) significantly reduced fungal infections (caused by *Fusarium* spp., *Penicillium* spp., *Aspergillus* spp., *Curvularia* spp., *Epicoccum* spp. and *Helminthosporium* spp.) resulting from insect attack (mainly by storage beetles Coleoptera), mealy bug (*Planococcus citri*) and scale insect (*Aspidiella hartii*) during storage (Morse *et al.*, 2000).

One of the limiting factors for the production of cush-cush yam (*D. trifida*) in the Atlantic region of Costa Rica is the non-availability of high quality planting material. Severe losses may occur during storage by sprouting and rotting and after planting if large tubers have to be cut into pieces facilitating pathogenic infection. Solano *et al.* (1996) reported that storability was best after pre-harvest spraying with a 0.1 per cent solution of 2,4-D, maintaining 75 per cent of tubers suitable for planting after 3 months as compared with 48 per cent in the unsprayed control. In the field, the largest seed tuber number of 463500/ha was harvested in mulched plots of non-disinfected soil.

The investigation on fertilizer application on yam (*D. rotundata*) by Okwuowulu *et al.* (1995) revealed that neither fertilizer application (recommended levels of NPK) nor its timing (6, 8 or 10 weeks after planting) significantly affected loss in weight or rotting of tubers during storage but sprouting was affected by the timing of application. Age at harvest significantly affected storability. Tubers harvested 6 or 7 months after planting stored best. Weight losses were lower in the traditional barn than in the modified store (Okwuowulu *et al.*, 1997).

Studies were conducted on the chemical composition, as well as the biochemical and textural changes occurring during storage of 2 cultivars of *D. dumetorum* yam tubers (Afoaka and Sefa-Dedeh, 2001). The tubers were harvested and stored under tropical ambient (28°C) and cold room conditions (4°C) for 24, 48 and 72 h. The samples were evaluated for moisture, starch, sugars, fibres and textural properties. Chemical analysis showed no significant differences between the means obtained for moisture, ash, protein, fibre, fat, carbohydrate, calcium, phosphorus, iron, vitamin A and vitamin C contents of the cultivars (white and yellow) studied. During storage, moisture content decreased by approximately 6-10 per cent after 72 h of harvest due to dehydration. Starch levels declined from 70.5-66.5 g/100 g, while sugar and fibre contents increased slightly in the samples. Textural properties (peak force and curve areas) also increased considerably during storage. Peak force (hardness) increased from 1143.0 to 5711.5 g in the white cultivars, whilst the yellow cultivars increased from 2130.5 to 4194.1 g.

Chang (1998) studied the biochemical changes in yam tubers after harvest and reported decreases in fresh weight, water content and total solids in Keelong yam (*D. japonica* var. *pseudojaponica*) tubers stored at room temperature. Starch content decreased gradually while sucrose and reducing sugar contents increased. Activities

Transportation and Selling of Yams in the Market

of starch degradation enzymes, amylases and starch phosphorylase fluctuated during storage while a key enzyme of glycolysis, glyceraldehyde-3-phosphate dehydrogenase, increased gradually. Browning and defense-related enzymes, polyphenol oxidase (catechol oxidase), peroxidase and chitinase increased during early storage and then decreased or fluctuated during prolonged storage. It is suggested that loss of fresh weight of yam tubers during storage might be due to transpiration and respiration. The decrease in the starch content might be due to respiration and conversion to sucrose and reducing sugars. Tubers of *D. esculenta* cv. Sree Latha stored at room temperature (30-32°C) and 80-85 per cent RH in the dark sprouted 70 to 80 days after harvest. In the apical and basal regions, the starch content decreased while the sugars and α-amylase activity increased (Muthukumarasamy and Panneerselvam, 2000).

Mozie (1986-87) reported that the breakage of dormancy of yam tubers (*D. rotundata*) was characterized by the appearance of fructose spots on chromatograms. Storage at 15°C suppressed the appearance of fructose in yam tissues and also suppressed sprouting for over 9 months after harvest. Storage in the conventional barn and under controlled temperature of 20° and 25°C showed an unsynchronized pattern of dormancy breakage. All tubers stored at 5° exhibited soft rotting and putrefaction after 3 months of storage and neither fructose spots nor sprouts were observed in these tubers throughout the storage period.

Treatment of tubers of 2 local *D. trifida* cultivars immediately after harvest by submergence for 22 h in water containing 150 ppm GA_3 extended the dormancy period (100 per cent sprouting) to 20 weeks and decreased tuber weight loss to 17 per cent compared with 13 weeks and 26 per cent in untreated controls (Wickham, 1988).

11.0 Diseases and Pests

11.1.0 Diseases

Fungal Diseases

Two major fungal diseases, the blight and dieback caused by *Cercospora* sp. and *Colletotrichum* sp. are found in *D. esculenta* and *D. alata*. The symptoms start as brown to black leaf spots at the beginning of the rainy season. The spots gradually coalesce, blight the leaves. The diseases cause defoliation and plants die in severe infection. Spraying of cercobin or MBC (0.2 per cent) at monthly interval controlled the disease and increased the tuber yield by 29 to 34 per cent (Anon., 1985).

Yam tubers are also infected with wet rot, soft rot or brown rot during storage (Adeneji, 1970). These diseases are caused by *Botrydiplodia* sp., *Penicillium* sp., *Aspergillus* spp. *Fusarium* spp., *etc.* Storage rot diseases are controlled to some extent by treatment with copper fungicides and lime washing.

The response of yam (*Dioscorea*) cultivars to artificial and natural infections with minisett rot pathogens was investigated by Osai *et al.* (1996). None of the 20 yam cultivars were resistant to *Sclerotium rolfsii* [*Corticium rolfsii*] or *Trichoderma longibrachyatum*. Cultivars TDa 84/093, 85/236, 85/257, TDr 608, 745, 81/453 and Efuru were not susceptible to *Aspergillus niger*. Cultivar TDa 382 was not susceptible

Diseases of Yams

Symptoms of Anthracnose on Yam Leaves and Shoots

Cercospora Leaf Spot *Curvularia* Leaf Spot

Tuber Rot Infection

Large, Brown to Black Leaf Spots (Left) with well-defined Diffuse Margins; Veinal Necrosis of Yam Leaves (Right) Associated with Rust Disease

Diseases of Yams

"Bumpy" Leaf Spot Symptom (Left) Associated with Yam Rust Disease; Defoliation (Right) Due to Yam Rust Disease

Yam Plants Infected with Yam Mosaic Virus Showing Mottling Symptom in *Dioscorea esculenta* (Left) and Mosaic Symptom in *D. alata* (Right)

Leaves of Yam: Chlorotic Mosaic and Leaf Curling (YMV) (Left); Vein Clearing and Necrotic Mosaic (YMV) (Middle); Shoe Stringing and Stunting (YMV, badnavirus) (Right)

to *Penicillium oxalicum* and TDa 84/093, 176, 382 and TDr 608 and 81/453 were mildly susceptible to *Botryodiplodia theobromae*.

Viral Diseases

Mosaic-like symptoms are observed in *D. alata*, *D. esculenta* and *D. rotundata*. The symptoms on *D. alata* include leaf puckering, chlorosis and necrosis of veins on the surface. Leaf mottling is also noticed. In *D. esculenta* the symptoms are dark and light green areas on leaves, green vein-banding, crinkling of leaves and stunting of plants.

A disease was also noticed on some cultivars of *D. rotundata* introduced from Nigeria. The symptoms include green vein-banding, curling and puckering of leaves and stunting of severely affected plants. The disease was mechanically transmitted to *D. rotundata* but the percentage was very low. By tuber grafting about 80 per cent of the inoculated plants were diseased. The affected plants showed reduction in growth rate, root formation and yield (Anon., 1983).

11.2.0 Pests

Yams are attacked by more than a dozen pests in the field and in storage. The important pests include the scale insects, white grubs, termites, chrysomelids and hairy caterpillars. Among these, scale insects are the most serious (Pillai and Palaniswami, 1984). White scale insect (*Aspidiella hartii*) is a serious pest on *D. esculenta* and *D. alata*. The insects attack the base of vines and tubers and plants wither when infestation is severe. The attacked tubers shrivel and lose the quality and marketability. The pest spreads through infested seed tuber and they migrate in the tuber during storage (Palaniswami *et al.*, 1979). The pest has two aphelinid parasites. Soon after the harvest infested tubers are removed and destroyed. The seed tubers should be dipped in 0.1 per cent monocrotophos before planting.

White grub (*Leucopholis coneophora*) is a polyphagus pest that attacks a number of plants including yam. The grubs attack the tubers of *D. esculenta* and *D. alata* and shallow round pits are formed. Application of 20-25 kg/ha sevidol has proved effective. Other pests on yams are chrysomelid (*Lema laecordaeiri*) and hairy caterpillar, termites and spider mites.

In Nigeria, field experiments were conducted to evaluate the effectiveness of aldrin 2.5 WP (at 34 kg/ha) and furadon 3 G (carbofuran) at 47 kg/ha in controlling yam tuber beetles (*Heteroligus meles* and *H. appius*). Umeozor (1997-1998) reported that furadon significantly reduced the amount of damage. In 1993 and 1994 the mean numbers of beetle feeding holes were 9.8, 4.2 and 9.1, and 25.0, 9.2 and 29.9 in the aldrin, furadon and control treatments, respectively.

Several nematode species attack yam tubers. These are *Scutellonema bradys*, *Meloidogyne* spp. and *Pratylenchus* spp. Setts with nematode infestation should be avoided and crop rotation is useful, fumigation with nemagon has been partially successful and dipping of infested setts in hot water (50°C) reduced nematode infestation. To control *Pratylenchus coffeae* and *Scutellonema bradys* on yams, Castagnone-Sereno (1988) recommended the use of hot water treatment (immersion of tubers for 40 min in water at 50°C) and nematicides (oxamyl) was recommended during planting of the crop. Different methods of oxamyl application evaluated by Ijenoma (1999) to protect microseed yams (45-50 g) against *M. incognita* during cultivation did not inhibit the sprouting potentials of the seed yams. At harvest of the yam tubers 28 weeks after planting, oxamyl application by drench at 1000 mg a.i./ml/plant + side dressing after 6 weeks at 2 kg a.i./ha; and side dressing (2 kg a.i./ha) at planting gave the highest total weight of yam tuber yield and weight of gall-free tubers compared with the untreated control.

Rajendran and Vadivelu (1991) screened seven cultivars of *D. alata*: DA6, DA7, DA8, DA42, DA60, DA80 and DA122 against *M. incognita* and all of them were highly susceptible with root knot index of 5. *D. esculenta* cultivars DE1, DE3, DE4, DE11 and DE17 also showed severe susceptibility expressing a similar root knot index. Out of a total of 33 yam accessions tested against *M. incognita* in pot experiments, 5 were immune, 9 resistant, 5 moderately susceptible and 14 were susceptible (Mohandas *et al.*, 1998).

12.0 Crop Improvement

In spite of great potentiality of yam crop as staple food in many tropical countries, very little attention has been paid for its improvement. Objectives for yam improvement are many and some of them are – i) greater efficiency in translocation of assimilates, ii) shortening of pre-bulking phase, iii) faster tuber maturation, iv) rapid drying of top and v) higher nutritive value. Besides these, plan for the improvement of food yams should be aimed to develop improved clones for growing under virtually horticultural conditions and for mechanized cultivation. Ayensu and Coursey (1972) emphasized to develop cultivars which can be grown without staking in order to reduce the cost of cultivation.

The variation already known among yam cultivars is so vast that much can be done simply by selection and dissemination of the best existing clones. Even apparently inferior clones may have value for use in different ecosystems or for particular breeding characteristics. Classification of *Dioscorea* germplasm, comprising several species, was studied by undertaking characterization using morphological descriptors and by subjecting the data to classification using principal component analysis (Albuquerque-Melo *et al.*, 2000). It was observed that leaf length and width provided high contribution to discrimination of the accessions studied. The accumulated variance on the first and second principal components accounted for 89 per cent of all the variation noted. Six groups could be distinguished representing the *Dioscorea* accessions studied.

Clonal selection in the Chinese yam (*D. opposita*) cv. Kisa 1 resulted in the new cv. Hirokei 1 (Kohmura *et al.*, 1998). Tubers of this new cultivar sprout earlier than those of Kisa 1, Takashiro or Aoyama and most are spherical, but in some years tubers with a neck are produced. Average tuber yields are greater in Hirokei 1 than in Kisa 1, Takashiro or Aoyama, and tuber weight is averaged 459 g in Hirokei 1 and 304-423 g in the other cultivars.

In most of the species, seed-set is poor due to dioecism, small flowers, sticky nature of pollen, inter- and intra-species barrier in fertilization, and moreover there is pre- and post-ovular breakdown and embryo abortion (Rao *et al.*, 1973).

Pests of Yam

Yam Tubers Infected with Scale Insects

Adult Yam Moth

Red Spider Mite Adult

Sawfly on Yam Leaf and its Damage on Leaves

Pests of Yam

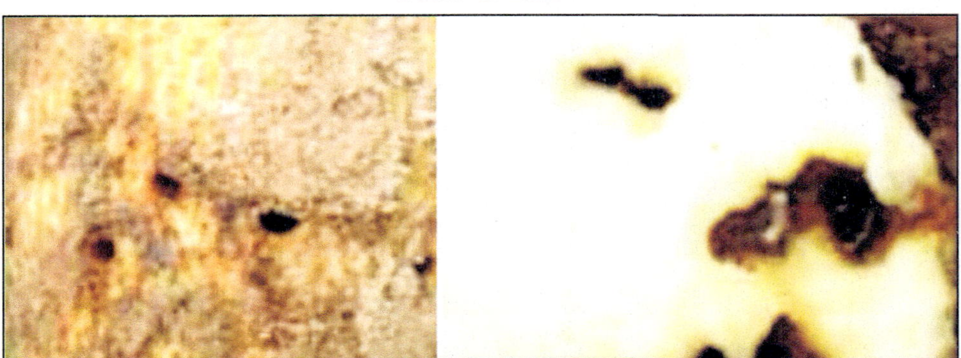

External and Internal Termite Damage of Yam Tubers

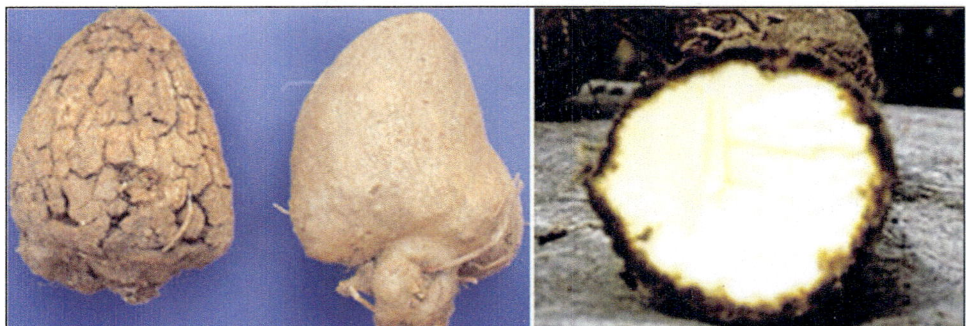

Nematode-Infested and Healthy Tubers (Left);
Dry Rot Nematode on Tubers (Right)

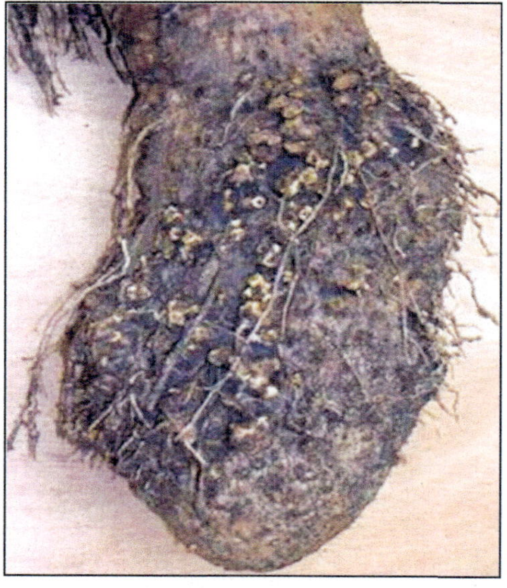

Root-knot Nematode-Infested Yam Tuber

However, success has been obtained in *D. rotundata* (Sadik, 1976). Selection of seedling segregates is another method for improvement of this crop. Many of the important yam species produce one or few large tubers, suitable to the manual farmer. Others, such as *D. esculenta* and *D. trifida* which form many smaller tubers, are more adaptable to industrial agriculture; the latter species is sexually highly fertile. A much larger proportion of the variability within the genus should be exploited either by direct introduction or by hybridization with cultivated forms (Coursey, 1976). Thus for radical improvement, hybridization programmes will be required to overcome low fertility of many species. Occasional attempts to breed food yam have been made in Nigeria (Waitt, 1960), Puerto Rico (Martin, 1974) and Petif Bourg (Degras, 1976).

Three cultivars, Nagaimo, Ichoimo and Tsukuneimo, of *D. opposita* and 9 lines of *D. japonica* were artificially pollinated to observe capsule development and hybrid seed formation. A large number of normal seeds were produced in 7 combinations in *D. japonica*. Almost all of them contained an embryo (larger than 0.8 mm) which developed into a seedling when sown in soil or cultured *in vitro*. However, cross incompatibility existed among 4 combinations in *D. japonica*, and no normal seed was produced. In the crosses Ichoimo x *D. japonica* and Tsukuneimo x *D. japonica* there was a difference in the seed-forming capacity depending on the *D. japonica* line used as the pollen parent. Only a few seeds containing an immature embryo were produced in Ichoimo x Nagaimo, Ichoimo x *D. japonica*, Tsukuneimo x Nagaimo, Tsukuneimo x *D. japonica* and *D. japonica* x Nagaimo. However, in these 5 combinations, some seedlings were obtained by embryo culture.

Colletotrichum gloeosporioides (*Glomerella cingulata*) causes anthracnose, the most severe foliar disease of field-grown water yam (*D. alata*). Mignouna *et al.* (2001) studied the inheritance of resistance to a moderately virulent (FGS) strain of the pathogen was investigated in crosses between tetraploid *D. alata* genotypes: cross A: TDa 95/00328 (resistant) x TDa 95-310 (susceptible), and cross B: TDa 85/00257 (resistant) x TDa 92-2 (susceptible). Segregation of F_1 progeny fitted genetic ratios of 3:1, 5:1 (crosses A and B) and 7:1 (cross A) resistant: susceptible when inoculated with the FGS strain (Cg33), indicating that resistance in dominantly inherited and suggesting that more than one gene controls the inheritance of resistance to this strain in the accessions studied.

The polyploidy of many preferred cultivars may also contribute to infertility which appears most extreme in the Asiatic *D. alata* and *D. esculenta* where seed set is almost unknown.

Selection of somatic mutants is another effective method for its improvement and it has contributed to the ancestry of some yams. Yam improvement through mutation breeding requires special attention.

13.0 Biotechnology

Tissue culture in yam has been summarized under the sub-heading propagation. Results of investigations on protoplast culture, biochemical and molecular markers, genetic transformation and *in vitro* conservation are given below:

13.1.0 Protoplast Culture and Fusion

In *D. alata* and other *Dioscorea* species, protoplasts have been successfully isolated from unexpanded leaves and cultured on agarose droplet culture to develop microcalli consisting of upt o 100 cells (Twyford and Mantell, 1988; Tardieu and Malaurie, 1988). Leaf mesophyll protoplasts from immature leaves of *in vitro* shoot cultures of a range of cultivars of three species of food yam (*D. alata, D. bulbifera* and *D. cayenensis-rotundata*) were isolated and their responses to culture in agarose-solidified media compared by Tor *et al.* (1998). Formation of cell colonies to the 50-cell stage was observed in protoplast cultures in five of ten cultivars of *D. alata* and to the 30-cell stage in two cultivars of *D. cayenensis-rotundata* but not in cultures of *D. bulbifera.*

13.2.0 Biochemical and Molecular Markers

13.2.1 Phylogenetic Relationship

Investigations have been carried out to study phylogenetic relation in different species and varieties of *Dioscorea* using biochemical and molecular markers.

A method was developed for the isoelectric focusing of soluble peroxidase and acid phosphatase isoenzymes of yams. It was found suitable for the positive identification of shoot culture materials of the four yam species *D. alata, D. cayenensis/D. rotundata, D. esculenta* and *D. bulbifera*. It was sufficiently sensitive to detect differences between four clones of *D. alata* which were otherwise difficult to distinguish with certainty on the basis of morphological characters alone. The method was suitable for identification of clones of all the four yam species when used in protoplast fusion and regeneration of cybrid plants from microcalli (Twyford *et al.*, 1990).

Isoenzyme pattern studies were found useful in the characterization of varieties of *D. cayenensis-rotundata* complex.

Hamon and Toure (1990a) described cultivated and wild West African yams and the relationship between the 2 groups was examined. Morphological and isoenzymatic data indicated that of the 20 varietal groups of cultivated yam, 15 and related to the annual wild species *D. abyssinica* and *D. praehensilis*, 1-2 are closely related to the semi-perennial *D. mangenotiana*, while the other 3 appear to be hybrids between perennial and annual wild species. In another investigation, from a total of 453 accessions of cultivated yams collected in West Africa, sixty-two of the accessions were identifiable by their isoenzyme patterns for malate dehydrogenase, isocitrate dehydrogenase, 6-phosphogluconate dehydrogenase, shikimate dehydrogenase and phosphoglucoseisomerase. Twelve cultivar groups could easily be distinguished. The results showed a good agreement between morphological polymorphism and enzymatic variation (Hamon and Toure, 1990b). Lebot *et al.* (1998) studied isoenzyme variation to determine relationship among 269 cultivars of *D. alata* originating from the South Pacific, Asia, Africa and the Caribbean.

Some 467 accessions of cultivated Guinea yam (*Dioscorea cayenensis/Dioscorea rotundata* complex) collected from different localities of Benin Republic were analysed to study isoenzymatic variability in seven enzyme system:aspartate aminotransferase, esterase, glucose-6-phosphate dehydrogenase, isocitrate dehydrogenase, phosphoglucomutase, phosphoglucoisomerase [glucose-6-phosphate isomerase] and shikimate dehydrogenase using starch gel electrophoresis (Dansi *et al.*, 2000). Different combinations of banding patterns of these systems led to identification of 227 different cultivars within the 467 accessions analysed. Cluster analysis (UPGMA) produced a most likely division of the accessions into two groups corresponding to *D. rotundata* and *D. cayenensis*, supporting the concept that the two forms of Guinea yam represent different genetic entities.

RAPD analysis was used to characterize eleven cultivars of the five economically most important yam species grown in Jamaica (*Dioscorea alata, D. cayenensis, D. rotundata, D. trifida* and *D. esculenta*) (Asemota *et al.*, 1996). While no variation was observed among plants belonging to the same cultivar, a large number of inter-varietal and interspecific polymorphisms enabled the reliable discrimination of all Jamaican cultivars investigated. Ramser *et al.* (1996) also used RAPD markers to assess intraspecific variability and relationships in aerial yam (*Dioscorea bulbifera*). A total of 23 accessions from different geographical locations in Africa, Asia and Polynesia were analysed by 10 arbitrarily chosen GC-rich decamer primers. All methods of data evaluation resulted in similar groupings that reflected the geographical origin of the samples. The African accessions formed a distinct isolated group, whereas Asian and Polynesian accessions proved to be more heterogeneous.

13.2.2 Screening for Disease Resistance

Aiming at developing a suitable *in vitro* method for screening anthracnose-resistant yam (*Dioscorea*) germplasm, a study was initially conducted by Moura-Costa and Mantell (1993) on the effects of temperature (15, 19, 25 and 29°C), culture media (PDA and Czapek Dox agar) and light on the sporulation of a Caribbean pathovar of *Colletotrichum gloeosporioides* f. sp. *dioscoreae*. Methods for inoculation of yams with conidiospores of *C. gloeosporioides* were then developed at the whole plant, plantlet (shoot) and cell (protoplast) levels. *In vitro* shoot culture clones showed a wide range of reactions to the pathovar, similar to those of greenhouse-grown plants. *D. alata* showed marked sensitivity to the disease while *D. esculenta* was the most tolerant species, followed by *D. composita* and *D. cayenensis*. Protoplasts of *Dioscorea* responded differently when cultured in the presence of mycelial culture filtrates and Czepek Dox liquid medium. In contrast to *in vitro*-plantlets, greenhouse-raised plants developed anthracnose symptoms only after wounding.

13.3.0 Genetic Transformation

Successful transfer and expression of specific tumour inducing t-DNA region of *A. tumefaciens* into nuclear genome of *Dioscorea bulbifera* was affected using *Agrobacterium* treated with wound substances from a dicot plant (Shafer *et al.*, 1987).

13.4.0 *In vitro* Conservation

Different species of *Dioscorea* (10 wild and 5 edible) were stored under slow growth. Node culture of *D. alata, D. rotundata* and *D. esculenta* could be conserved up to 12 months (without subculturing) on MS basal medium containing 3 per cent mannitol and no growth regulators (Nair and Chandrababu, 1994). Use of mannitol, low light intensity and reduced temperature to prolong the storage period beyond 6 months (28°C incubation) were studied at CATIE, Costa Rica (Mandal and Chandel, 1996).

14.0 References

Abraham, K. (1992) *J. Root Crops*, **18**: 142-143.

Abraham, K. and Easwariamma, C.S. (1984) *Indian Fmg.*, **33**: 17-19.

Adeneji, M.O. (1970) *Phytopathology*, **60**: 590-592.

Adesuyi, S.A. (1973) *Proc. 3rd Symp. Trop. Root Crops*, Ibdan, Nigeria.

Afoakwa, E.O. and Sefa-Dedeh, S. (2001) *Food Chem.*, **75**: 85-91.

Akoroda, M.O. (1987) *Proceedings of the third triennial symposium of the International Society for Tropical Root Crops - Africa branch.* p. 61.

Albuquerque-Melo-Filho, P. de., Santos, R.C. dos, Mafra, R.C., Santos, J.W. dos and Anunciacao-Filho, C.J. da (2000) *Ciencia Rural*, **30**: 619-623.

Alexander, J. and Coursey, D.G. (1969) In: *The Decades of Research*, CTCRI, Trivandrum, pp. 117-118.

Alhassan, A.Y. and Mantell, S.H. (1994) *Acta Hortic.*, **380**: 342-348.

Amma, C.S.E., Abraham, K. and Nair, S.G. (1989) *J. Root Crops*, **15**: 19-23.

Ammirato, P.V. (1984) In: *Handbook of Plant Cell Culture, Crop species,* Vol. 3 (Ammirato, P. V., Evans, D. A., Sharp, W. R. and Yamada, Y., Eds.), Macmillan, New York, pp. 113-122.

Anonymous (1983) *The Decades of Research*, CTCRI, Trivandrum, pp. 117-118.

Anonymous (1985) *Edible Yams,* Bull. No. 7, CTCRI, Trivandrum.

Anonymous (2020) *Annual Report,* AICRP on Tuber Crops, CTCRI, Trivandrum, Kerala.

Anuebunwa, F.O. (2000) *Biological Agriculture and Horticulture*, **18**: 95-102.

Araghinkinam, M., Chung, S., White, T.N., Eskelson, C. and Watson, R.R. (1996) *Life Sciences*, **59** (11): 147–157.

Asemota, H.N., Iyare, O.A., Wheatley, A.O. and Ahmad, M.H. (1997) *Trop. Agric.*, **74**: 243-247.

Asemota, H.N., Ramser, J., Lopez-Peralta, C., Weising, K. and Kahl, G. (1996) *Euphytica*, **92**: 341-351.

Ayensu, E.S. and Coursey, D.G. (1972) *Econ. Bot.*, **26**: 301-318.

Bancroft, R., Gray, A., Gallat, S., Crentsil, D. and Gogoe, S. (1998) *Tropical Agriculture*, **75**: 115-119.

Bricas, N., Vernier, P., Ategbo, E., Hounouigan, J., Mitchikpe, E., N'-Kpenu, K.E. and Orkwor, G. (1997) *Cahiers de la Recherche Developpement*, **44**: 100-114.

Budelman, A. (1989) *Agroforestry Systems*, **8**: 243-256.

Budelman, A. (1990a) *Agroforestry Systems*, **10**: 61-69.

Budelman, A. (1990b) *Agroforestry Systems*, **10**: 47-59.

Burkill, I.H. (1960) *J. Linn. Soc. (Bot.)*, **56**: 319-412.

Castagnone-Sereno, P. (1988) *Turrialba*, **38**: 337-340.

Castillo, R.J., Ruiz-Martinez, L.A., Portieles-Rodriguez, J.M. and Milian-Morales, J.O. (1989) *Ciencia y Tecnica en la Agricultura*, Viandas Tropicales, **12**: 59-73.

Cedric, T., Twyford and Mantell, S. H. (1988) *J. Agric. Univ. Puerto Rico*, **72**: 515-520.

Chan, Y.C.; Hsu, C.K.; Wang, M.F. and Su, T.Y. (2004) *Int. J. Food Sci. Tech.*, **39**: 99-107.

Chang, C.T. (1998) *J. Chinese Agri. Chem. Soc.*, **36**: 547-554.

Chapman, T. (1965) *Trop. Agric.*, Trinidad, **42**: 145-151.

Chikoye, D., Manyong, V.M. and Ekeleme, F. (2000) *Crop Prot.*, **19**: 481-497.

Childers, N.F. (1950) *Feb. Exp. Sta., Puero Rico, Circ.*, 32.

Chowdhury, S.R. (1998) *Tro. Agric.*, **75**: 435-439.

Chulafich, L., Grubishich, D., Vuiichich, R., Volkova, L.A. and Popov, A.S. (1994) *Russain J. Plant Physio.*, **41**: 821-826.

Coursey, D.G. (1983) In: *Handbook of Tropical Foods* (Ed. H.T. Chan), Marcal Dicker, New York, 555-602.

Coursey, D.G. and Walker, H.M. (1960) *Ann. Rept. West Africa Stored Prod. Res. Unit*, pp. 61-64.

Coursey, D.G. (1975) *Origin of African Plant Domestication* (Ed. J.R. Harlan), The Hague.

Coursey, D.G. (1967) *J. Stored Prod. Res.*, **2**: 229-244.

Coursey, D.G. (1976) *Evolution of Crop Plant* (Ed. N.W. Simmonds), Longman, London, pp. 70-74.

Cruz, V.M.V., Altoveros, N.C., Mendioro, M.S. and Ramirez, D.A. (1999) *Plant Gent. Res. Newslet.*, **119**: 7-11.

Daniel, I.O., Ng, N.Q., Tayo, T.O. and Togun, A.O. (1999) *Seed Sci. Tech.*, **27**: 969-975.

Dansi, A., Mignouna, H.D., Pillay, M and Zok, S. (2001) *Euphytica*, **119**: 301-307.

Dansi, A., Mignouna, H.D., Zoundjihekpon, J., Sangare, A., Ahoussou, N. and Asiedu, R. (2000) *Genet Resour Crop Ev.*, **47**: 619-625.

Dansi, A., Mignouna, H.D., Zoundjihekpon, J., Sangare, A., Asiedu, R. and Quin, F.M. (1999) *Genet. Resour. Crop Ev.*, **46**: 371-388.

Degras, L. (1976) *Proc. 4th Int. Symp. Trop. Root Crops*, pp. 58-62.

Edeoga, H.O., Okoli, B.E. and Agba, I. (1998) *Folia Geobotanica*, **33**: 201-205.

Enyinnaya, A.M., Igbokwe, M.C. and Nwankiti, A.O. (1987) *Proceedings of the third triennial symposium of the International Society for Tropical Root Crops - African Branch.*

FAOSTAT (2017) Statistics Division of the UN Food and Agriculture Organization.

Ferguson, T.U., Haynes, P.H. and Spence, J.A. (1983) *Proc. 6th Sym. Int. Soc. Trop. Root Crops*, Lima, Peru, pp. 649-655.

Forsyth, C. and Van Staden, J. (1984) *J. Plant Physiol.*, **115**: 79-83.

Gamiette, F., Bakry, F. and Ano, G. (1999) *Genet. Resour. Crop Evol.*, **46**: 19-21.

George, J. (1990) *J. Root Crops*, **16**: 71-75.

George, J. and Suja, G. (1995) *J. Root Crops*, **21**: 1-6.

George, J., Palaniswami, M.S., Potty, V.P., Padmaja, G., Kabeerathumma, S. and Pillai, S.V. (1996) In: *Tropical tuber crops: problems, prospects and future strategies*, pp. 306-309.

Girardin, O., Nindjin, C., Farah, Z., Escher, F., Stamp, P. and Otokore, D. (1998) *J. Agric. Sci.*, **130**: 329-336.

Gonzalez, M.A. and Reviera, A.C. de (1972) *J. Agric. Univ.*, Peurto Rico, **56**: 46-56.

Gyansa-Ameyaw, C.E., Hahn, S.K., Alvarez, N.M. and Doku, E.V. (1994) *Acta Hortic.*, **380**: 335-341.

Gyansa-Easmon, C.E. (2000) *Dis. Innov.*, **12**: 128-131.

Hamon, P. and Ahoussou, N. (1987) *Plant Gene. Res. Newslet.*, **73**: 20-23.

Hamon, P. and Toure, B. (1990a) *Euphytica*, **46**: 101-107.

Hamon, P. and Toure, B. (1990b) *Euphytica*, **47**: 179-187.

Haynes, P.H. and Coursey, D.G. (1969) *Trop. Sci.*, **11**: 93-96.

Hayward, L.A.W. and Walker, H.M. (1961) *Ann. Rept. West Africa Stored Prod. Res. Unit*, pp. 107-115.

Heldt, H.W. (1997). Plant biochemistry and molecular biology. Oxford University press New York, 153.

Heredia Z-N.A., Vieira, M-do C. and Criep, R. (1999) *Horticultura Brasileira*, **17**: 45-48.

Heredia, Z-N.A., Vieira, M-do-C., Minuzzi, A. and do-C-Vieira, M. (2000) *Ciencia e Agrotecnologia*, **24**: 287-391.

Higdon, K.,Scott, A., Tucci, M., Benghuzzi, H., Tsao, A., Puckett, A., Cason, Z. and Hughes, J. (2001) *Biomedical Scientific Instrument*, **37**: 281-286.

Igwilo, N. (1989) *Trop. Agric.*, **66**: 38-42.

Ijenoma, O.F. (1999) *J. Root Crops*, **25**: 21-27.

Irizarry, H. and Rivera, E. (1994) *J. Agriculture University of Puerto Rico*, **77**: 161-166.

Irizarry, H., Goenaga, R. and Chardon, U. (1995) *J. Agriculture University of Puerto Rico*, **79**: 121-130.

Jean, M. and Cappadocia, M. (1992) *Plant Cell Rep.*, **11**: 34-38.

John, J.L., Courtney, W.H. and Decoteau, D.R. (1993) *Plant Cell Tissue Organ Cult.*, **34**: 245-252.

Kalu, B.A. (1989) *Trop. Agric.*, **66**: 83-86.

Kalu, B.A., Norman, J.C., Pal, U.R. and Adedzwa, D.K. (1989) *Exp. Agric.*, **25:** 181-188.

Kayode, G.O. (1984) *Expl. Agric.*, **20**: 53-57.

Kelmanson, J.E.; Jager, A.K. and Van, S. J. (2000) *J. Ethnopharmacol.*, **69**: 241–246.

Kohmura, H., Araki, H. and Imoto, M. (1995) *Plant Cell Tissue Organ Cult.*, **40**: 271-276.

Kohmura, H., Imoto, M. and Hirao, A. (1998) *Bulletin Hiroshima Prefectural Agriculture Research Centre*, **66**: 25-31.

Kolo, M.G.M. (1995) *Trop. Agric.*, **72**: 102-104.

Kpeglo, K.D., Obligbesan, G.O. and Wilson, J.E. (1980) In: *Trop. Root Crops* (Eds. E.R. Terry, K.A. Oduro and F. Caveness), IDRC, Canada, pp. 198-203.

Kwon, J.K., Kim, J.C., Park, D.K., Lee, J.H. and Son, S.Y. (1996) *RDA J. Agricultural Science*, Upland and Industrial Crops, **38**: 149-157.

Lagoke, S.T.O., Mansfield, R.A. and Chandra-Singh, D.J. (1987) *Samaru J. Agric. Res.*, **5**: 3-12.

Lebot, V., Trilles, B., Noyer, J.L. and Modesto, J. (1998) *Genet. Res. Crop Evol.*, **45**: 499-509.

Liu, L.C., Santiago-Cordova, M. and Gonzalez, A. (1990) *J. Agriculture of the University of Puerto Rico*, **74**: 221-228.

Liu, S.Y. (1989) *J. Agric. Res. China*, **38**: 312-325.

Liu, S.Y., Wang, J.Y., Shyu, T.Y., Hu, M.F., Yang, H.R. and Ho, C.C. (1996) *J. Agric. Res. China*, **45**: 260-284.

Liu, S.Y., Chang, T.W., Lin, Y.K., Chen, S.F., Wang, J.Y., Zu, G.L. and Wang, S.C. (1999) *J. Agric. Res. China*, **48**: 1-22.

Lyonga, S.N., Faymi, A.A. and Agboola, A.A. (1973) *Proc. 3rd Int. Symp. Trop. Root Crops*, Ibadan, Nigeria.

Malaurie, B., Thouvenel, J.C. and Pungu, O. (1995) *Plant Cell Tissue Organ Cult.*, **42**: 215-218.

Mantell, S. H., Haque, S.Q. and Whitehall, A. P. (1978) *J. Hort. Science*, **53**: 95-98.

Mantell, S.H. and Hugo, S.A. (1989) *Plant Cell Tissue Organ Cult.*, **16**: 23-37.

Manu, V.T. (1997) *J. South Pac. Agric.*, **4**: 135-138.

Marnotte, P. and Tehia-Kouakou, E. (1995) *Seizieme conference du COLUMA. Journess internationales sur la lutte contre les mauvaises herbes, Reims, France, 6-8 december, 1995*, Tome 3, 1996, pp. 1355-1363.

Martin, F.W. (1974) *Memo. Red. Expt. Sta. Mayagunez*, Peuro Rico.

Matsubara, S., Ohmori, Y., Komasadomi, T. and Takada, H. (1992) *Scientific Reports of the Faculty of Agriculture, Okayama University*, **79**: 37-44.

Miege, J. (1968) In: *Flora of West Tropical Africa* (Ed. J. Hutchinson), J.M. Dalziel and F.N. Hepper, London.

Mignouna, H.D., Abang, M.M., Green, K.R. and Asiedu, R. (2001) *Theor Appl. Genet.*, **103**: 52-55.

Mignouna, H.D., Ellis, N.T.H., Knox, M.R., Asiedu, R. and Ng, Q.N. (1997) *Trop. Agric.*, **75**: 224-229.

Mitra, S. (2011) *Proceedings of the International Symposium on Minor Fruits and Medicinal Plants for Health and Ecological Security.* Faculty of Horticulture, BCKV, Kalyani, West Bengal, pp. 340-343.

Mitra, S. and Tarafdar, J. (2013) In: *Postharvest Management of Tropical Tuber Crops,* AICRP on Tuber Crops, Directorate of Research, BCKV, Kalyani, West Bengal.

Mitra, S. and Tarafdar, J. (2018) In: *Tropical Tuber Crops,* (Eds. James George and S. Sunitha), Westville Publishing House, Paschim Vihar, New Delhi.

Mitchell, S.A., Asemoto, H.N. and Ahmad, M.H. (1995) *J. Sci. Food Agri.*, **67**: 541-550.

Mix-Wagner, G. (1993) *Landbauforschung Volkenrode*, **43**: 93-100.

Mohan Kumar, B., Kabeerathumma, S. and Nair, P.G. (1984) *Indian Fmg.*, **33: 35-37**

Mohandas, C., Sreeja, P., Wari, S.N., Sheela, M.N. and Kaushal, K.K. (1998) *Proceedings of national symposium on rational approaches in nematode management for sustainable agriculture* (Ed. S.C. Dhawan), Anand, India, 23-25 November, pp. 63-64.

Mondal, B.B. and Chandel, K.P.S. (1996) In: *Tropical Tuber Crops: Problems, Prospects and Future Strategies* (Kurup, G.T., Palaniswami, M.S., Potty, V.P., Padmaja, G., Kabeera-thumma, S. and Pillai, Sandha, V. eds.), Oxford and IBH, New Delhi, pp. 49-54.

Morse, S., Acholo, M., McNamara, N. and Oliver, R. (2000) *J. Stored Products Res.*, **36**: 37-45.

Moura-Costa, P.H. and Mantell, S.H. (1993) *Pesquisa Agropecuaria Brasileira*, **28**: 1297-1306.

Mozie, O. (1986-87) *Trop. Root Tuber Crops Newsl.*, No. 16-17, pp. 39-45.

Muthukumarasamy, M. and Panneerselvam, R. (2000) *Trop. Sci.*, **40**: 63-66.

Ng, S.Y.C. (1990) *IITA Research*, **1**: 11-13.

N'-kounkou, J.S. (1994) *Fragmenta Floristica et Geobotanica*, **39**: 401-416.

N'-kounkou, J.S., Lejoly, J. and Geerinck, D. (1993) *Fragmenta Floristica et Geobotanica*, **1**: 139-182.

Nair, P.G. (1984) *Indian Fmg.*, **33**: 45.

Nair, N.G. and Chandrababu, S.C. (1994) *J. Root Crops*, **20**: 68-69.

Nair, N. G., Unnikrishnan, M. and Mukherjee, A. (1994) In: *Advances in Horticulture, Tuber Crops.* Vol. 8 (Eds. K. L. Chadha, and G. G. Nayar,), Malhotra Publishing House, New Delhi, India, pp. 211-232.

Nair, N.G. and Chandrababu, S.C. (1996) *Tropical tuber crops: problems, prospects and future strategies*, pp. 55-60.

Nandi, R.P. and Chatterjee, S.K. (1992) *Acta Hortic.*, **306**: 346-352.

Nnoke, F.N., Unamma, R.P.a., Ene, L.S.O., Odurukwe, S.O., Terry, E.R. Akoroda, M.O. and Arene, O.B. (1987) *Proc. of the third triennial symposium of the International Society for Tropical Root Crops* -African branch, 1987, pp. 63-69.

Nwinyi, S.C.O. (1984) In: *Trop. Root Crops-Production and uses in Africa*, IDRC, Canada, pp. 136-137.

Nwoke, F.I.O., Njoku, E. and Okonkko (1973) *Proc. 3rd Int. Symp. Trop. Root Crops*, Ibadan, Nigeria.

Nwoke, F.I.O., Njoku, E. and Okonkwo, S.N.C. (1984) *Trop. Agric.*, **61**: 99-101.

Obiefuna, J.C. (1989) *Fertil. Res.*, **18**: 245-250.

Obigbesan, G.O. and Agboola, A.A. (1978) *Expl. Agric.*, **13**: 349-355.

Okagami, N., Terui, K., Araki, H., Kanazawa, T. and Kawai, M. (1999) *Japanese J. Trop. Agric.*, **43**: 265-270.

Okezie, C.E.A., Nwoke, F.I.O. and Oknkwo, S.N.C. (1984) In: *Tropical Root Crops: Problems, Prospects and Future Strategies* (Eds. G.T. Kurup, M.S. Palaniswami, V.P. Potty, G. Padmaja, S. Kabeera-thumma and V. Pillai Sandha), Oxford and IBH, New Delhi, pp. 121-124.

Okwuowulu, P.A., Nnodu, E.C., Udensi, U.A. and Opara, M.U.A. (1995) *Trop. Sci.*, **35**: 58-63.

Olasantan, F.O. (1999) *Soil Tillage Res.*, **50**: 215-221.

Onwueme, I.C. (1972) *Niger. Agric. J.*, **11**: 17-22.

Onwueme, I.C. (1975) *J. Agric. Sci. Camb.*, **84**: 503-505.

Onwueme, I.C. (1978a) *The Tropical Tuber Crops*, John Wiley and Sons, New York.

Onwueme, I.C. (1978b) *J. Agric. Sci., Camb.*, **19**: 317-319.

Orkwor, G.C., Okereke, O.U., Ezedinma, F.O.C., Hahn, S.K., Ezumah, H.C., Akobundu, I.O. and Ofori, F. (1995) *Acta Hortic.*, **380**: 349-354.

Osai, E.O. and Ikotun, T. (1994) *Fitopatologia Brasileira*, **19**: 408-412.

Osai, E.O., Ikotun, T., Asiedu, R. and Hahn, S.K. (1996) *Trop. Sci.*, **36**: 201-205.

Obi-Young, O. (2020) In: *Zaki Biam,Nigeria's Largest Mono-Product Market, Folio Nigeria.*

Otu, O.I. and Agboola, A.A. (1994) *Acta Hortic.*, **380**: 360-366.

Palaniswasmi, M.S., Pillai, K.S. and Abraham, K. (1979) *J. Root Crops*, **5**: 65-66.

Passam, H.C. (1995) *Trop. Sci.*, **35**: 217-219.

Patel, K.S., Karmakar, N., Desai, K.D., Narwade, A.V., Chakravarty, G. and Debnath, M.K. (2019) *Physiol. Mol. Biol. Plants*, **25:** 1235–1249.

Pedralli, G. (1994) *Plant Genet. Resour. Newsl.*, **97**: 51.

Pillai, K.S. and Palaniswami, M.S. (1984) *Indian Fmg.*, **33**: 58-61.

Purseglove, J.W. (1968) *Tropical Crops: Monocotyledons*, Longman, Londan.

Rajendran, G. and Vadivelu, S. (1991) *J. Root Crops*, **17**: 223.

Ramser, J., Lopez-Peralta, C., Wetzel, R., Weising, K. and Kahl, G. (1996) *Genome*, **39**: 17-25.

Rao, U.P., Bammi, R.K. and Randhawa, G.S. (1973) *Ann. Bot.*, **37**: 395-401.

Renaut, G. and Merlier, H. (1973) *Proc. 3rd Int. Symp. Trop. Root. Crops*, Ibadan, Negeria.

Ricci, P. and Arnolin, R. (1973) *Proc. 3rd Int. Symp. Trop. Root. Crops*, Ibadan, Nigeria.

Rishi, A.K., Bhan, M.K., Dhar, P.L., Nasir, Ali and Ali, N. (1994) *Indian J. Forestry*, **17**: 53-55.

Roman, F.M., Beale, A. and Irizarry, H. (1991) *J. Agriculture of the University of Puerto Rico*, **75**: 11-19.

Saborio, F., Torres, S., Gomez, L. and Drew, R.A. (1998) *Acta Hortic.*, **461**: 495-501.

Sadik, S. (1976) *Proc. 4th Int. Symp. Trop. Root Crops*, Ibadan, Nigeria.

Saha, P.K., Chatterjee, S.K. and Raychaudhuri, S.P. (1991) In: *Recent advances in medicinal, aromatic and spice crops, Volume 1. International conference held on 28-31 January, 1989*, New Delhi, India, pp. 119-122.

Saleil, V. (1990) *Agronomie*, **10**: 605-616.

Sanchez, E.E., Ruiz, L., Betancourt, F. and Cuadra, A. (1987) *Ciencia y Tecnica en la Agricultura, Viandas Tropicales*, **10**: 19-28.

Segnou,M., Fatokun, C.A., Akoroda, M.O. and Hahn, S.K. (1992) *Euphytica*, **64**: 197-203.

Segnou, M. (1999) *Cahiers Agricultures*, **8**: 405-408.

Shafer, W., Gorz, A. and Kahil, G. (1987) *Nature*, **237**: 529-532.

Sheu, B.W., Juan, Y.H. and Yan, H.F. (2000) *Bulletin National Pingtung University of Science and Technology*, **9**: 103-107.

Shiwachi, H., Onjo, M. and Hayashi, M. (1999) *Japanese J. Trop. Agric.*, **43**: 149-156.

Shiwachi, H., Onjo, M. and Hayashi, M. (2001) *Japanese J. Trop. Agric.*, **45**: 15-21.

Singh, K.D., Mandal, R.C., Maini, S.B. and Magoon, M.L. (1973) *Indian J. agron.*, **18**: 17-21.

Singh, D.N., Tripathy, P., Sahu, A. and Nandi, A. (1994) *J. Root Crops*, **20**: 81-84.

Soibam, Helen, Singh Ayam, V. and Mitra, Surajit (2017) *Pharm. Innov. J.*, **6**: 465-469.

Solano, X., Navarro, J.R., Leihner, D.E. and Hilger, T.H. (1996) *Angewandte Botanik*, **70**: 215-217.

Sreekantan, L., George, S. and Nair, K.H. (1995) *Indian J. Agron.*, **40**: 149-150.

Tardieu, F. and Malaurie, B. (1988) *ISRTC. Symp.*, 1988, pp. 70-71.

Tellez-Valdes, O. and Schubert, B.G. (1990) *Annals Missouri Botanical Garden*, **77**: 245-253.

Thompson, H.C. and Kelly, W.C. (1957) *Vegetable Crops*, McGraw-Hill Book Company, Inc., New York.

Tor, M., Twyford, C.T., Funes, I., Boccon-Gibod, J., Ainsworth, C.C. and Mantell, S.H. (1998) *Plant Cell Tissue Organ Cult.*, **53**: 113-125.

Toyohara, H., Kikuno, H., Irie, K. and Kikuchi, F. (1997) *Japanese J. Trop. Agri.*, **41**: 74-80.

Toyohara, H., Tanifuji, H., Irie, J., Takahashi, M., Myoda, T. and Kikuchi, F. (1998) *Japanese J. Trop. Agric.*, **42**: 282-287.

Twyford, C.T. and Mantell, S.H. (1988) *ISTRC 8th Symposium*, pp. 74-75.

Twyford, C.T., Viana, A.M., James, A.C. and Mantell, S.H. (1990) *Trop. Agric.*, **67**: 337-341.

Umeozor, O.C. (1997-1998) *Nigerian J. Entomol.*, **14-15**: 100-106.

Undie, A.S. and Akubue, P.I. (1986) *J. Ethnopharmacol.*, **15**: 133-144.

Unnikrishnan, M. and Sheela, M.N. (2001) In: *Biotechnology of Horticultural Crops*, Vol. 2 (Eds. V.A. Parthasarathy, T.K. Bose, P.C. Deka), pp. 455-494.

Viana, A.M. and Mantell, S.H. (1989) *Plant Cell Tissue Organ Cult.*, **16**: 113-122.

Waitt. A.W. (1960) *Ann. Rept. Red. Dept. Agric. Res.*, Negeria.

Waitt, A.W. (1961) *Memo. Fed. Dept. Agric. Res.*, Negeria.

Wickham, L.D. (1998) *Trop. Sci.*, **28**: 75-77.

Wilson, L.A., Wickham, L.D. and Ferguson, T. (1998) *Trop. Agric.*, **75**: 77-83.

Zarate, N.A.H., Vieira, M. do C. and Ortiz, A.C.S. (1998) *SOB Informa*, **17**: 16-17.

Zhao, Y. and Xiao, P. (1989) *Zuowu Pinzhong Ziyuan*, **1**: 23-24.

Zoundjihekpon, J. and Tio-Toure, B. (1992) *Plant Genet. Resour. Newsl.*, **90**: 39-41.

TARO

S. Mitra, J. Tarafdar and V. A. Parthasarathy

1.0 Introduction

Edible aroids belonging to the family Araceae are comprised of many underground root crops grown in several tropical and sub-tropical countries. Taro (*Colocasia esculenta* (L.) Schott), an important member of the family Araceae mainly serves as a vegetable in the Tropical countries but in the Pacific Islands, it is cultivated as a staple food (Mitra and Tarafdar, 2018). It is the most widely cultivated species in the family Araceae. The subfamily Colocasiodeae of family Araceae consists of three edible tuber crops, namely 'taro' (*Colocasia esculenta* Schott), 'tannia' (*Xanthosoma sagittifolium*) and 'Giant taro' (*Alocasia macrorhiza*). Among these crops, taro and tannia are cultivated to a larger extent, while giant taro is not as common as commercial crop like the other two. In general, these are crops of third world countries, particularly grown in Africa and Asia. The world annual production of taro is about 10.64 million tonnes (Anon., 2020).

About 88 per cent of the total world acreage is in Africa which produces about 80 per cent of total production (Onwueme, 1978). Taro is an important crop in Hawaii. Among the three crops, taro is more common in India. It is one of the ancient crops with an interesting history blending with the evolution of agricultural systems.

2.0 Composition and Uses

2.1.0 Composition

The leaves, petioles, corms and cormels are edible. The corm is rich in starch which contains 17-25 per cent amylose. Mitra *et al.* (2007) reported that dry matter, starch and protein content in taro tubers varied significantly among genotypes and dry matter content was found to vary from 23.23 to 29.18 per cent.

The composition of leaves, stem and root are given below:

Table 1: Composition of Colocasia Leaves, Stem and Root (per 100 g of edible portion)*

Composition	Leaves (green)	Stem	Root
Moisture	82.7 g	940 g	73.1 g
Carbohydrates	6.8 g	3.6 g	3.1 g
Protein	3.9 g	0.3 g	21.1 g
Fat	1.5 g	0.3 g	0.1 g
β-carotene	10278 μg	104 μg	24 μg
Thiamine	0.22 mg	0.07 mg	0.09 mg
Riboflavin	0.26 mg	0.07 mg	0.03 mg
Vitamin C	12 mg	3 mg	–
Calcium	227 mg	60 mg	40 mg
Iron	10 mg	0.5 mg	1.7 mg

*Anon. (1980)

2.2.0 Uses

Taro being a rich source of good quality starch finds its use in baby foods and diet of the people allergic to cereals (Mitra *et al.*, 2007). Taro is also recommended for gastric patients. In Hawaii and Polynesia, a fermented product prepared from taro is very popular. The pressure cooked taro corms after being passed through strainer are allowed to ferment giving an acidic product called 'poi'. In Africa, 'fufu', the corm paste prepared from the cooked taro is widely consumed. In India, both the tubers and the leaves are used as vegetable, and the tubers are also used for the preparation of chips.

3.0 Origin and Taxonomy

3.1.0 Origin

It is believed to have originated in South-Central Asia, perhaps India or Malaysia. The origin and dispersal of *Colocasia* make an interesting reading by virtue of its antiquity and wide distribution. Plucknett *et al.* (1970), Plucknett (1976) and Leon (1976) presented comprehensive accounts of the origin and dispersal of aroids. India to Southern Asia is the centre of origin of *Colocasia*. From its home it had travelled east to all the islands of Oceania, Philippines, China and Japan. Based on ethno-botany, Spier (1951) placed India as the home of *Colocasia*. The spread to the west is ill documented (Leon, 1976). Mauny (1953), as quoted by Leon, reported that taro was in cultivation in Gambia and San Thome around A.D. 1500. The confusion between *Xanthosoma* and *Colocasia* made it difficult to establish the arrival of taro in America even though in the 18th century its spread has extended to Brazil through the Caribbean, the check of spread of *Colocasia* being *Xanthosoma* (Leon, 1976). But De Candolle (1885) opined that taro cultivation might have begun independently in Malay Peninsula, Fiji and India. Plucknett *et al.* (1970), citing earlier works, indicated the origin of generic name *Colocasia* from the Egyptian word Colcas or Qolquas. Coates *et al.* (1988) analyzed the karyological data from 107 clonal varieties from the Pacific region as well as from India, Sri Lanka, Thailand and Nepal and found that there were 2 separate lineages within contemporary *C. esculenta* populations. The findings upheld the putative Indian-South-East Asian mainland origin of one form of the species, but also indicated a separate but parallel domestication process of another form in the western Pacific region. Isshiki *et al.* (1999) analysed ploidy levels of the seedlings derived by crossing diploid accessions (2n = 28) of wild taro (*C. esculenta*) from Bangladesh to determine the origin of triploids (2n = 42). Six and two triploids were detected among 1041 hybrids and 354 selfs, respectively, although most hybrid and selfed seedlings were diploid. These results indicated that triploids in taro originated both in hybrid and selfed seedlings of diploids. Isozyme analysis at the *Aat-1* locus of aspartate aminotransferase in the six hybrid triploids revealed that the seed parent was the double genome donor in five triploids, whereas the pollen parent served as the donor in the sixth. Therefore, both seed and pollen parents are able to serve as a double genome donor of the triploids. The Japanese diploid groups seem to have been produced by out crossing, and

Uses of Taro

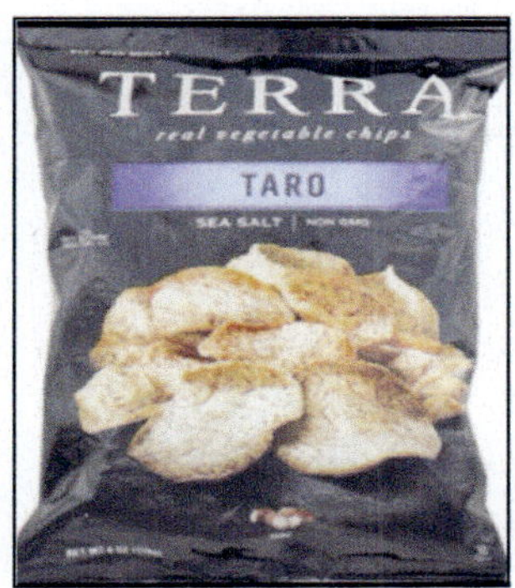

Taro Chips

Different Taro Recipes in Indian Cuisine

accessions within the groups seem to be progenies derived from clonal line, because of their morphological similarity. The Taiwanese groups might have been produced by out crossing with the other groups in Taiwan because of their diploidy and incomplete sexual isolation (Tanimoto, 1990). Zhang and Zhang (1990) investigated the distribution of diploid (2n = 2x = 28) and triploid (2n = 3x = 42) plants using 105 cultivars and 20 wild accessions collected from various parts of China. The results showed that the 2 ploidy levels have different geographical distributions with both diploids and triploids occurring in southern and eastern China, whereas triploid cultivars only were found in central and northern regions. All 14 cultivars collected from the southernmost tropical Hainan province were diploid. A greater percentage of triploids occurred at higher altitudes than at lower altitudes where diploids were more frequent.

Yang and Kong (1994) considered China as a centre of origin for taro (*C. esculenta*) as it has abundant genetic resources. According to Ivancic and Lebot (1999) taro is considered to be an introduced crop in New Caledonia and has been cultivated since its introduction by Melanesian farmers. Wild germplasm exists on the main (continental) island and is represented by 3 easily distinguished morphotypes: one with purple leaves, another with green leaves, and a third with green leaves and a purple vein junction on the lamina. All 3 morphotypes are diploids (2n = 2x = 28), and have well-established wild populations in many valleys and gulches of the main island. The morphotype with purple leaves has all typical traits of a wild genotype (inedible corms; long, thin stolons), while the other two produce edible corms. The purple and the green morphotypes flower produce fertile pollen. The spathes of the green morphotype can be greater than an extremely long appendix a typical for pacific taros characterizes 40 cm long and the spadix. Isoenzyme analysis conducted using 4 enzyme systems (esterase, phosphoglucomutase, shikimic dehydrogenase and peroxidase) indicated that New Caledonian wild taros differ from most widely grown local cultivars, and Pacific cultivated and wild genotypes. Evidence presented here suggests that *C. esculenta* is endemic to New Caledonia. Cultivars were probably introduced as clones from what is now Vanuatu by early Melanesian migrants and were not domesticated locally from existing wild forms, which appear to be genetically distant from other Melanesian wild taros.

The Assam-Bengal region of India or Indo-China is the possible centres of domestication of the primitive *Alocasia mucrorrhiza* (*A. indica*) where another relative *A. cucullata* is also under cultivation (Leon, 1976). According to him, *Alocasia* has spread eastwards to Melanesia and Polynesia and introduced into Brazil as cattle feed. From America, *Xanthosoma* was taken during slave trading era, to Africa where it assumed the name of new cocoyams and now ranks next only to cassava and yams, particularly in West Africa (Plucknett, 1976). Massal and Barrau (1955) attributed the spread and acceptance of *Xanthosoma* in the Pacific and Asia to its resistance to diseases and pests.

3.2.0 Taxonomy

The edible aroids can be classified as follows:

Order: Lasiodeae

Family: *Araceae*

Sub-Family: Colocasioideae

Colocasia, Alocasia, Xanthosoma, Cyrtosperma, Amorphophallus

The species status and taxonomy of the genera in subfamily Colocasioideae is more confusing. A persual of literature would amply support the statement. Bentham and Hooker (1883) described 5 Asian species of *Colocasia,* while the report of Conner (1908) revealed 7 species of *Colocasia* to be native of Indo-China. But subsequent compilation of Fedrov (1974) indicated only 3 species of *Colocasia, viz., antiquorum* Schott, *esculenta* Schott and *gigantea.* While many authorities felt that *C. antiquo-rum* and *C. esculenta* are synonyms. Most of the local varieties found in India belongs to mixed type. Moreover, dasheen of South USA is eddoe of Trinidad. Bailey (1925) opined that Egyptian taros belong to the species *antiquorum,* while *esculenta* encompasses Indian taros. The confusion could not be relegated easily because of the variation in somatic chro-mosome numbers in the same species. This can be substantiated by the fact that 5 somatic chromosome numbers have been reported for *C. antiquorum* alone. Sharma (1956) reported 14 and 42 in the somatic cells, while Kurakubo (1940) and Sharma and Das (1954) reported 28. Other 2n numbers reported are 36 (Rao, 1947) and 48 (Malvesin-Fabre, 1945). Leon (1976) suggested that a general survey of variability in taro is badly needed. He stated that two basic chromosome numbers have been recorded in taro, *i.e.,* x = 12 and x =14. Clones with 2n = 24 and 2n = 28 are reported from India to Japan in one direction and to Timor, New Caledonia and New Zealand in the other.

Thirteen *Alocasia* species have been reported by Fedrov (1974) out of which 3 species, namely, *A. macrorrhiza, A. indica* and *A. cucullata* are edible. Plucknett (1976) stated that confusion exists concerning the taxonomy of edible *Xanthosoma.* Coursey (1968) listed 5 edible species, *viz., X. brasiliense, X. belophyllum, X. caracu, X. jacquini* and *X. mafaffa.* But Plucknett (1976) was of the opinion that *Xanthosoma* should be considered as one polymorphic species (*X. sagittifolium*) since most of them were given species status based on morphology.

Colocasia are herbaceous plants up to 2 m tall and have underground corms. Leaves are large peltate with long petiole clasping at the base. Corms are cylindrical with short internodes and few side tubers. Root system is shallow and fibrous. Inflorescence on stout peduncle; spathe about 20 cm long, pale yellow; spadix cylindrical with basal female flowers, above which is a section of aborted flowers compressed by neck of spathe, a longer section of male flower and a short sterile appendage at the apex. Many cultivars do not flower (Purseglove, 1968).

4.0 Cultivars

Depending on the manner of corm development, *Colocasia* is grouped into two varietal types. These are *Colocasia esculenta* var. *esculenta* (Dasheen or Banda type) which produce a large edible mother corm with few cormels and *C. esculenta* var. *antiquorum* (Eddoe type) which has a small or medium sized mother corm and large number of edible cormels (Table 2). In West Bengal as well as in most of the taro growing areas in India, Eddoe type is of major importance and makes a significant contribution as a vegetable in the diet of people. Several cultivars of taro are grown in different parts of India. However, the performance of a particular cultivar varies in different taro growing areas (Mitra and Tarafdar, 2018).

Table 2: Differences between Eddoe and Dasheen Taro

Eddoe Taro	*Dasheen Taro*
Cormels are edible	Mother corm is edible
It produces 15-20 cormels/plant	It produces 4-5 cormels/plant
It matures within 5-6 months after planting	It matures within 7-9 months after planting
It cannot tolerate flood	It can tolerate flood

From the wide range of germplasm collected from different agro climatic regions of the country and conserved and characterized at ICAR-AICRP centres paved way for the release of many improved varieties in tuber crops. Based on the performance at various stages of the varietal evaluation trials *viz.*, initial evaluation trial, uniform regional trial and finally multi location trials conducted across the centres, AICRP on tuber crops so far could release or recommend a total of 21 in taro, 3 in bunda varieties (Table 3).

Table 3: The Released Cultivars of Eddoe, Dasheen and Swamp Taro in Indiacultivars in India

Year of Release	*Originating Centre*	*Name of Cultivar*	*Area Suitable for Cultivation*
A. Eddoe taro			
1983	APHU, Andhra Pradesh	Satamukhi	Andhra Pradesh, Bihar, West Bengal, Tamil Nadu, Kerala, Bihar
1987	CTCRI, TVM	Sree Pallavi	Kerala
1987	CTCRI, TVM	Sree Rashmi	Kerala
1991	TNAU, Coimbatore	CO 1	Tamil Nadu
1999	NDUAT, Faizabad	Narendra Arvi-1	Uttar Pradesh
1999	NDUAT, Faizabad	Narendra Arvi-2	Uttar Pradesh and Bihar
2000	BCKV, Mohanpur, West Bengal	Bidhan Chitanya	West Bengal
2002	CTCRI, BBSR	Muktakeshi	Odisha, Chhattisgarh and Jharkhand
2003	IGKV, Raipur	Indira Arvi-1	Chhattisgarh and Madhya Pradesh
2005	CTCRI, BBSR	PaniSaru-I	Coastal areas of Odisha
2005	CTCRI BBSR	PaniSaru-II	Odisha
2006	APHU, Andhra Pradesh	Bhavapuri	Coastal region of Andhra Pradesh, Telangana
2008	RAU, Samastipur	Rajendra Arvi-I	Bihar, Jharkhand and Chhattisgarh
2011	CTCRI, BBSR	Jhankri (Bhu Kripa)	All over Odisha
2011	CTCRI, BBSR	Sonajuli (Bhu Sree)	Odisha

Year of Release	*Originating Centre*	*Name of Cultivar*	*Area Suitable for Cultivation*
2011	BCKV, Mohanpur, West Bengal	BCC-17 (Upland Taro)	West Bengal
2013	ICAR RC NEH, Meghalaya	ML-1(Megha Taro-1)	North East Region
2014	Dr. YSRHU, Andhra Pradesh	RNCA-1	Andhra Pradesh
2015	IGAU, Raipur	IGCOL-E-9	Chhattisgarh
2015	RAU, Samastipur	AAU COL -46	Bihar
2017	Dr. YSRHU, Andhra Pradesh	Goodavari Chema	Andhra Pradesh
B. Dasheen taro			
1999	NDUAT, Faizabad	Narendra Bunda- 1	Uttar Pradesh and Arunachal Pradesh
2016	BCKV, Mohanpur, West Bengal	BCB-2	West Bengal
2016	IGKV, Raipur	IGB-5	Chhattisgarh
C. Swamp taro			
2011	BCKV, Mohanpur, West Bengal	Bidhan Latika	West Bengal, Assam

Important characteristics of some eddoe taro varieties grown in different parts of India are given below:

Sree Rashmi

It is a high yielding variety of taro with good cooking quality. It matures after 7-8 months of planting with an average yield of 18 t/ha.

Sree Pallavi

It is a tall variety, and attains a plant height up to 1.5 meter with large number of small sized tubers (20-25 per plant). It matures after 6-7 months of planting with an average yield of 16 t/ha. It has field tolerance to taro leaf blight.

Muktakeshi

The cormels mature after 5-6 months of planting with an average yield of 20 t/ha. The plants are resistant to taro leaf blight under field conditions, and field tolerant to dasheen mosaic, aphids, cut worm and scale insects.

Sree Kiran

It is the first hybrid taro in India. Cormels are round to oval in shape with good cooking quality. Keeping quality of the cormels are also good.

Telia

Though the plants are susceptible to taro leaf blight, cormels are of good shape and size. It matures at 6-7 months after planting with an average yield of 15 t/ha.

Bidhan Chaitanya

This is a high yielding variety, mainly grown in West Bengal and Eastern parts of India. Plants are tall and erect (100-140 cm). It matures after 6-7 months of planting with an average yield of 17 t/ha.

Onwueme (1978) described two botanical varieties, *viz. Colocasia esculenta* var. *antiquorum* (Eddoe types) and *C. esculenta* var. *esculenta* (Dasheen types). But Whitney *et al.* (1939) reported that Dasheens belong to the botanical variety *globulifera*. Eddoes have bigger cormels than the mother corms, while it is *vice-versa* in case of Dasheen types. Hill, cited by Plucknett *et al.* (1970), described 8 botanical varieties namely, var. *nymphaefolia*, var. *globulifera*, var. *aquatilis*, var. *acris*, var. *antiquorum*, var. *euchlora*, var. *fontanesii*, and var. *illustris*. This taro cultivars vary in colour of leaves and petioles and shape of corms and cormels. Parthasarathy *et al.* (1983) stated that the north-eastern region of India where taro has originated, has many cultivated and feral forms of taro. In this region the cultivars are available with the tribals who grow in their *jhum* (shifting cultivation) plots. The taro cultivars are named after the shape of corms such as 'Panch mukhi' (five faced) in Assam; 'Sahasra mukhi' in Maharashtra and Tamarakannan in Kerala and after names of places like Kovvur in Andhra Pradesh. Parthasarathy *et al.* (1983) further reported the cultivar Khasi Bhunga is suitable for Meghalaya. Collections like C-9, C-135, C-266, C-149 at CTCRI, Trivandrum are quite promising. C-9 is a popular cultivar among farmers in Kerala and Sree Rashmi is a high yielding cultivar released by the Institute. C-9 a selection from Tamarakannan possesses good culinary qualities and starch content (16.5 per cent) and has given the yield up to 20 tonnes per hectare. C-266 is another promising selection giving a yield of 10-12 tonnes/ha (Unikrishnan and Sreekumari, 1984). Few collections of *Xanthosoma* spp. namely Xa-31, Xa-33, Xa-40 and Xa-41 were found promising from the yield trials conducted at CTCRI (Unikrishnan and Sreekumari, 1984). Kakakachu, a Jorhat collection showed better yielding ability with high degree of susceptibility to phytophthora blight whereas Panch mukhi and Nadia Local gave satisfactory cormel yields in West Bengal with high degree of disease resistance (Sen and Das, 1991). Rodriguez *et al.* (1994) proposed a new classification based on a study of 42 *C. esculenta* varieties (27 landraces and the rest introduced varieties) in Santo Domingo, Cuba, with 3 main groups (diploid, triploid and aneuploid), each with 4 subgroups according to corm flesh colour (white, pink, yellow and purple). Groupings based on peroxidase and esterase isoenzyme patterns differed from each other, and the groupings based on morphological and karyotypic characters did not agree with the ones based on isoenzyme patterns. These patterns were useful for identifying duplicate varieties (of which there were none in the original collection) and for discriminating among varieties. Sreekumari and Mathew (1991b) observed significant karyomorphological differences between 2 distinct morphotypes of diploid taro suggesting that a taxonomic separation of the morphotypes should be considered. They (Sreekumai and Mathew, 1991a) also reported karyomorphology for 5 taro diploid morphotypes (2n = 28). Morphotypes I to IV were similar with respect to gross karyomorphology, but differed in the finer details of individual chromosomes. The karyotype of morphotype V was markedly different from the others, showing a reduced frequency of m-type chromosomes (3 pairs vs. 8 or more for morphotypes I-IV) and more st-type chromosomes (5 vs. 1-2 pairs).

Rodriguez Manzano *et al.* (1998) proposed another new classification system where the degree of ploidy defines the group and pulp colouration defines the subgroup. Other morphological characters are used to distinguish between

Cultivars of Eddoe Taro

Different Eddoe Taro Cultivars

Bidhan Chaitanya

Bidhan Joydeb

BCC-17 (Eddoe Taro)

Cultivars Resistant to *Phytophthora* Blight

Cultivars of Eddoe Taro

Muktakeshi

Jhankri

Sonajuli

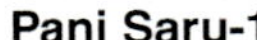

Pani Saru-1

Pani Saru-2

Cultivars of Dasheen (Banda) Taro

Bidhan Banda

BCB-2

EAC-2

clones in subgroups, and peroxidase and esterase isoenzyme patterns are used for distinguishing between morphologically similar clones within one subgroup.

5.0 Growth and Development

Three phases of growth have been reported by Sivan (1979) namely, phase 1– regeneration of mother corm mainly characterized by the emergence of roots and leaves extending up to eight weeks; phase 2 – grand growth period characterized by rapid vegetative growth commencing from the second to sixth month and phase 3 crop maturity and duration which varied with cultivars and growing conditions. During phase 1 and 2 root dry weights increased rapidly with a peak at about 24 weeks after planting and then declined due to dry conditions in upland while in low land it increased continuously according to the availability of moisture and age. The corm development could be noticed from eighth week after planting in upland taro and the corm weight increased continuously reaching a peak value at 40-46 weeks after planting under Fiji conditions. Varieties differ significantly in leaf production. Usually, a plant produces 20-35 leaves retaining 4-7 leaves at a time with longevity of 30-50 days depending upon the cultivar. Leaf number and leaf area index (LAI) in low land taro are always more than those in upland taro due to higher rate of leaf production, longer leaf life and more number of suckers. The optimum LAI for taro in both upland as well as low land conditions was found to be 3.0 (Sivan, 1983).

6.0 Soil and Climate

6.1.0 Soil

Taro can grow on all types of soils, but the deep well drained, sandy loam soil is ideal for its successful cultivation (Mitra and Tarafdar, 2018). Taro is a tropical crop preferring abundant supply of water. In India, taro is cultivated both under low land as well as upland conditions while in Hawaii, it is grown in flooded puddled soils. The soils of low land taro are usually alluvial and rich in fertility while upland taro is normally grown in soils of marginal fertility. All the three species of edible aroids tolerate shade to some extent and are often grown as inter-crop. The optimum soil pH range is between 5.5 and7.0. Some cultivars of taro are also tolerant to salinity.

6.2.0 Climate

Taro requires warm and humid climate of tropical and sub-tropical regions with a mean temperature of 30-35°C. Though taro generally requires warm climate with plenty of moisture, it can also be grown in the warmer temperate regions (Mitra and Tarafdar, 2018).

As observation of the natural habitat of taro along water sources, *viz.*, banks of rivers and channels would reveal that the crop prefers flooded conditions. It is grown on hills and plains of North-Eastern India as rainfed crop where the rainfall is copious. Rainfall in taro growing areas is highly variable. It ranges from 1000 mm in parts of Myanmar to over 5000 mm annually in some regions of India (Purewall and Dargan, 1957). Taro is primarily adapted to moist environment but can be grown under a wide range of moisture regimes ranging from paddy culture of swampy lands and also in upland conditions under irrigation or as rainfed crop (Plucknett *et al.*, 1967). It cannot stand cool climate where frost is a problem but grows even under subtropical climate where temperature averages around 21°C.

7.0 Cultivation

7.1.0 Land Preparation

The land preparation differs from place to place. Plucknett and de la Pena (1971) reported Hawaii to be the only country where taro is cultivated on a scientific way. The land is puddled before planting in Hawaii, while it is planted on ridges and furrows in parts of South India and in beds in North-Eastern India. Ezumah (1973) found that puddling did not increase the yield. Whatever be the preparatory cultivation method, the main objective is to give proper soil condition for tuber development and prevent percolation of impounded water (Onwueme, 1978). For upland taro, farmyard manure is incorporated at the time of land preparation and flat or mound planting is normally practised in India. Flat planting followed by earthing up at 60 to 70 days after planting give better soil coverage to the developing cormels and better support to the crop.

7.2.0 Planting Material and Planting

Taro is vegetatively propagated by mother corms or cormels. Small cormels of 20-25 g are generally planted following ridge and furrow method at a row to row and plant to plant spacing of 60 x 45 cm during March-April in West Bengal, India (Mitra, 2013).

Plucknett *et al.* (1970) reported that 1 cm corm tips with 15-20 cm petiole (called 'huli') are used in Hawaii which yields higher than corms as planting material. Cormels are also commonly used as planting material in many taro growing areas under upland condition. The cormels are preferred in places where winter is severe for the convenience of storing till the next planting season. The cormels are usually planted in shallow pits during the pre-monsoon season and leaf mulching is commonly practised.

Reports on the optimum size of planting materials are contradictory. Hodnett (1958) found that mother corms are better than cormels. In Kerala, Joseph *et al.* (1981) compared 3 cormel sizes, *viz.*, 15-25 g, 34-55 g, and 55-65 g and corms weighing 60-70 g. They found that large sized cormels are better than corms. The results at Central Tuber Crops Research Institute, Trivandrum revealed the absence of any effect of seed size. However, the crop establishment, growth and productivity of edible aroids are largely governed by the size and quality of planting materials.

The depth of planting varies between 5 and 10 cm. In case of 'huli', 15-23 cm of the leaf petioles along with corm tips are inserted into puddled soil (Plucknett *et al.*, 1970). The spacing also varies in different places. Mohan Kumar *et al.* (1976) recommended a spacing of 60 x 45 cm in Kerala (India) while in Cuba, Torres and Vazquez (1981) suggested a spacing of 90 x 30 cm. In Japan, Igbokwe and Ogbonnaya (1981) reported planting at 60 x 100 cm for high yield. Silva *et al.* (1971) in Brazil and Ezumah (1973) in Hawaii found that close spacing increased corm and shoot yield

Planting Materials and Field Growing of Eddoe Taro

Planting Materials of Taro

Planting of Taro

Evaluation of different Eddoe Taro Varieties

per hectare but decreased yield per plant. The spacing can be closer in lowlands than in uplands since water may not be a limiting factor in lowlands (Onwueme, 1978). Ezumah and Plucknett (1981) found that the most suitable plant population depends upon the age at harvest and land preparation. Their experiments showed that in flat culture at 60 cm flat row spacing, population was increased from 18000 to 54000 plants/ha and the yield increased at all ages depending upon harvest stage. The corm yield increased with the increase in population from 12000 to 36000 plants when planted at 60 x 90 cm on ridges. However, on per hill basis, the yield components namely corm weight, cormel weight, the number of cormels increased significantly as the plant density decreased and showed an asymptotic relationship with plant density both in upland and low land taro. In India high yields of taro were reported at plants pacing of 30 x 45 cm (Purewal and Dargan, 1957) and 45 x 60 cm (Mohan Kumar *et al.*, 1976). The planting is normally done during the rainy season and it can be done anytime if irrigation facilities are available. The results obtained at CTCRI indicated that April to June was the ideal time of planting taro under rainfed condition. Drastic reduction in yield was noticed when planting was delayed beyond June (Prabhakar and Nair, 1984). In temperate regions, planting is avoided during winter. The planting is done in beds, pits or ridges and furrows. The method of planting has least influence on yield (Joseph *et al.*, 1981).

7.3.0 Manuring and Fertilization

Like other tuber crops, taro is responsive to fertilizer applications. The requirement varies with soil and cultural conditions. The plots amended with FYM (10 t ha^{-1}) + Mustard cake (10 t ha^{-1}) recorded the highest cormel yield of 15.99 t/ha followed by a dose of NPK @ 80:50:100 Kg ha^{-1} (15.75 t/ha) and vermi compost (2 t ha^{-1}) + 75 per cent of recommended dose of NPK (15.12 t/ha) as against 7.10 t/ha under control (Anon., 2016). de la Pena and Plucknett (1969) found that N and P content of petiole and leaf blade decreased from younger to oldest leaves. Widdowson (1966) reported zinc deficiency in calcareous soils. When taro is grown under calcium deficient condition a disorder known as 'metsubure' occurs (Miyaji *et al.*, 1976). The defective corms have smooth or concave top, slightly brownish in colour and of varying size occurring mostly on secondary corms (Tanabe *et al.*, 1980). Use of lime in such soils can solve the problem. Magnesium is an important element in the structural unit of chlorophyll but there appears to be a lack of awareness of the importance of mangesium in plant production and it has not always being given due recognition in fertilizer application considerations (Kubuabola *et al.*, 2000). In hydroponic culture of cv. Bun Long, Austin *et al.* (1994) observed that dry matter (DM) of leaves, roots, corms, petioles, and total biomass were significantly higher in all treatments with Mg. Kubuabola *et al.* (2000) determined that balance between inputs and outputs of magnesium in a cropping system of 12 months and studied its accumulation and distribution relative to calcium in taro. Magnesium and calcium inputs were measured by crop uptake/removal. A partitioning of the biomass was made to trace magnesium and calcium distribution in the various parts of the plants (leaf, petiole, root and corm) during growth to maturity. During early growth, magnesium and calcium accumulation occurred mostly in the leaves and petioles. Towards maturity there appeared to be a decline in the leaf and petiole content with an increase in the corm suggesting a translocation of magnesium and calcium to the corm. At final harvest, magnesium and calcium removed from the system by taro harvesting was calculated at 6 and 15 kg/ha/year, respectively.

Portieles *et al.* (1986) applied N fertilizer in equal parts at planting and at 70, 100 and 130 days to give a total of 0-450 kg/ha. Commercial yield in the unfertilized control was 17 t/ha which increased to a maximum of 27.1 and 28.9 t/ha with 300 and 200 kg N, respectively. Yields declined with higher rates in a range of 16.6 to 23.2 t/ha and 200 kg N/ha gave the best financial return. Ruiz *et al.* (1989) reported that the average total corm + cormel yield ranged from 24.41 t/ha with no fertilizer to 43.73 t with 350 kg N. N application increased crude protein, K and Ca percentages in corms and cormels but tended to decrease DM and P percentages. Applying 350 kg N gave the greatest economic returns.

Jacobs and Clarke (1993) suggested that N availability in the soil is one of the determinants of biomass production in taro and that management of N nutrition for corm production may need to take account of changes in partitioning of DM and N that occur under varying supplies of N. Ramnanan *et al.* (1995) found that N applied as CO $(NH_2)_2$ at 280 kg N/ha at 14 DAP is optimum for upland taro production under these conditions. Under normal field conditions surface washing may be the most important N loss pathway but when excessive fertilizers are used then gaseous pathways may also contribute to N losses. Purea and Mataora (1996) studied the effects of N fertilizer and Azolla green manure on yields of taro, intercropping taro with grain legumes. Yields of taro cv. Niue Po and Niue Ma, were increased while cv. Veo decreased by application of N.

The highest yield of best grade corms (32,622 lb/acre) and marketable corms (32,912 lb/acre) was produced with the application of 5051 lb triple superphosphate/acre (Sato *et al.*, 1990b). Using prediction equations based on the observed data, 3571 lb superphosphate gave maximum predicted marketable yield (35,546 lb/acre) and 3061 lb superphosphate gave maximum predicted net revenue. A preliminary recommendation of 3061 lb superphosphate/acre as a preplant application with 14520 plants/acre is made for Hilo soils with approximate 46 ppm available P. Halavatau *et al.* (1998) found the placement of P fertilizers and number and timing of N applications increased fertilizer efficiencies.

Taro corm can sustain itself up to 6 leaf stage from the nutrient stored in corm (Hayashi, 1973). It requires more potassium than nitrogen and phosphorus. An optimum level of 103.9, 74.1 and 135.7 kg of N, P and K, respectively was recommended by Mohandas and Sethumadhavan (1980) under Kerala conditions. In Hawaii, satisfactory yield was obtained by applying 250 kg each of N, P and K per hectare for upland and 250 kg each of N and K for lowland taro (de la Pena and Plucknett, 1969). The effect of N, P and K on quality was studied by Pillai (1975). He found that application of phosphorus enhanced dry matter, while potassium had negative influence on dry matter but it proved effective in increasing protein content of corms. On the other hand, tuber and total dry matter production were found to increase up to 80 kg N and 100 kg K_2O/ha but levels of P_2O_5 had no significant effect on these characters (Mohan Kumar and Sadanandan, 1989). Potassium application beyond 40 kg/ha, under rainfed condition in acid laterite soils did not show any increase in yield; however, the economic optimum doses of N and K_2O worked out from quadratic response surface model were 89 and 79 kg/ha, respectively (Ashokan and Nair, 1984). Starch content increased by the application of nitrogen and potassium and recommended application of N and K at the rate of 120 and 60 kg/ha, respectively. From the results of the experiments at CTCRI, Trivandrum, Mohan Kumar *et al.* (1984) recommended 12.5 tonnes of FYM and 80, 50 and 100 kg/ha of N, P and K, respectively. In West Bengal, a marked improvement on cormel yield was recorded with 120 kg N in combination with 80 kg K_2O/ha (15.0 t/ha) which was significantly superior to control (10.3 t/ha) (Sen and Roychoudhury, 1988). Time of application of N and K plays an important role in the production of cormels. Three split applications of nitrogen and potassium gave significantly higher cormel yield over two splits by Mohan Kumar *et al.* (1990), Mohan Kumar and Sadanandan (1990). Kumar *et al.* (1998) found a fertilizer application containing 120 kg N + 80 kg K_2O/ha is optimum for taro cultivation in the agro-climatic conditions in the plateau region of Bihar.

Silva *et al.* (1990) studied the response of taro cv. Bun Long to 12 combinations of N and K on a crop grown for 9 months. N had the greatest effect on weight of No. 1 corms while K had relatively little effect. The highest yields were obtained with the higher rates of N (> 320 lb N/acre) at all rates of K. Predicted net revenue of $16200-17000 may be achieved with applications of 460 and 600 lb N/acre at all rates of K and also with 320 lb N and 600 lb K/acre. A fertilizer recommendation for highest net revenue of 460 lb N and 600 lb K/acre is made. The proposed adequate range for leaf N content at 3 months is 4.3 to 4.5 per cent and for leaf K content is 4.1 to 4.3 per cent. Sato and Silva (1990a) carried out experiment with five plant spacing treatments and 4 fertilizer treatments in taro grown on land previously cropped to *Zingiber officinale*. Total fertilizer was 695 lb urea and 630 lb muriate of potash/acre applied at planting or as a split application. A spacing of 0.5 x 4 ft gave the highest corm yield of 36,145 lb/acre however only 67 per cent were No.1 corms. Spacing of 1 x 4 and 1 x 3 ft gave yields of 28,159 and 28,488 lb/acre, respectively with 87 per cent and 76 per cent No.1 corms, respectively. Splitting fertilizer application had no effect on total corm yield or percentage No.1 corms. Velez and Lopez (1999) evaluated taro cv. Blanca under two plant spacing (0.91 x 0.30 m and 0.91 x 0.60 m) and three fertilizer levels (0, 56, 112 g per plant of a 14-3-13 NPK commercial mixture) in a Corozal clay soil in Puerto Rico. An increase in the planting distance caused a significant reduction in the corm fresh yield but a significant increase in the average corm weight in all of the planting methods. The fertilizer levels had no effect on the corm fresh yield nor on the average corm weight. Fertilizer levels had no significant effect on the foliar concentration of N, P, K, Ca and Mg measured six months after planting. Average foliar concentrations of these elements across treatments were 4.6, 0.36, 4.1, 0.23 and 1.5 per cent, respectively.

Goenaga and Chardon (1995) found that there were no significant differences in total and edible DM content between cultivars. However, cv. Lila absorbed significantly smaller amounts of N, P, K, Ca and Zn than cv. Blanca, suggesting that it had a higher nutrient use efficiency. Fresh corm yields were not significantly different and averaged 20.2 t/ha for both cultivars. Hartemink and Johnston (1998) presented data on nutrient uptake of taro roots in relation to corm yield and shoot biomass on a Typic Tropo-fluvents in the humid lowlands of Papua New Guinea. Fertilized (100: 50: 100 kg NPK/ha) and unfertilized plants were harvested 126 days after planting (DAP) (mid season) and 231 DAP (harvest). Rooting depth at both sampling times was < 0.2 m and a unit soil area equivalent to the planting distance (0.5 m x 0.8 m) was removed from the field where after roots were washed. Root biomass at 126 DAP was 0.26 t/ha (15 per cent of total biomass) in the unfertilized plots and 0.52 t/ha (13 per cent of total) in the fertilized plots, but at 231 DAP root biomass was similar (0.50 t/ha). Root nutrient concentration at 126 DAP was similar in both plot but N, Ca, and S significantly declined in the unfertilized plots at 231 DAP whereas B increased with 18 mg/kg (P<0.01). In the fertilized plot, P, K, Mg, Mn and Cu had decreased at 231 DAP whereas Zn had significantly (P<0.01) increased. Nutrients in the root biomass as fractions of the total nutrient uptake were similar at 126 DAP for both treatments. At 231 DAP; however, the fraction of nutrients in the root biomass was considerably lower in the fertilized plots. There was a high uptake of Mg by taro roots in the unfertilized plots at 126 DAP (38 per cent of total) and at 231 DAP (36 per cent of total). It was concluded that the amount of nutrients taken up by roots of fertilized and unfertilized taro was similar at harvest, but that a much larger proportion of plant nutrients is allocated to the roots under unfertilized conditions. Ni *et al.* (1997) reported that marketable yield of taro increased significantly with K fertilization. K application also lowers NO_{3-} content. Sulphate of potash was more effective with regard to yield and quality, compared with muriate of potash. The slow release K content of the soil in the ploughing horizon (0-20 cm) dropped. This is thought to be due to the decrease in soil pH caused by the different fertilizer treatments. Hartemink *et al.* (1997) studied the nutrient uptake of taro roots in relation to corm yield and aboveground biomass in the humid lowlands of Papua New Guinea. Fertilized (100-50-100 kg NPK/ha) and unfertilized plants were harvested at 126 days after planting (126 DAP = mid-season) and 231 DAP (harvest). Root biomass at 126 DAP was 0.26 t/ha (15 per cent of total biomass) in the unfertilized plots and 0.52 t/ha (13 per cent of total) in the fertilized plots but at 231 DAP root biomass was similar (0.50 t/ha). Root nutrient concentrations at 126 DAP was similar in both plots but N and Ca significantly declined in the unfertilized plots at 231 DAP whereas B increased from 12 to 30 mg/kg (P<0.01). In the fertilized plots P, K, Mg, Mn and Cu had decreased at 231 DAP but Zn had significantly (P<0.01) increased. Nutrients in the root biomass as fractions of the total nutrient uptake were similar at 126 DAP for both treatments. At 231 DAP; however, the fraction of nutrients in the root biomass was considerably lower in the fertilized plots. Nutrient up-take by roots at harvest was 5 kg N, 1 kg P, 25 kg K, 5 kg Ca, and 3 kg Mg/ha for both fertilized and unfertilized conditions. It was concluded that taro roots contain a considerable amount of nutrients but that a much larger proportion of plant nutrients is allocated to the roots under unfertilized conditions.

The requirement of fertilizer also varies with the cultivation methods including spacing, cultivar and type of soil. Igbokwe and Ogbonnaya (1981) found that nitrogen uptake was highest when grown at a distance of 60 x 100 cm and treated with 40kg N/ha. Mohan Kumar and Sadanandan (1990) reported that the total uptake of nitrogen increased significantly only up to 80 kg/ha, but increasing levels of P and K application had significant effect on the uptake of these nutrients at all levels tried.

7.3.1 Organic Manuring

The use of organic manure is becoming more important. Escalada and Ratilla (1998) reported the use of 'ipil-ipil' (*Leucaena leucocephala*) green manure supplemented with inorganic P and K (and compared with inorganic NPK) for growing taro; ipil-ipil green manure alone was tested in later croppings. Application of ipil-ipil green manure (7.23 or 10.84 t/ha) also promoted vigorous stand growth and a higher crop yield. Significantly larger corms were formed, resulting in higher total corm yields and marketability than the control (untreated) and those treated only with 60-39.6-74.7 kg/ha N, P and K from inorganic fertilizer. Increasing the rate of inorganic N applied to plants (90-36.9-74.7 kg/ha N, P, K) during a third cropping, gave comparable yield to treatment involving 7.23 t/ha ipil-ipil supplemented with inorganic P and K. Despite the relatively higher gross income obtained from organic and inorganic treatment combinations, slightly lower net income was obtained compared with inorganic fertilizer-applied plants due to higher production expenses incurred with the former treatments.

Tekle Haimanot and Doku (1995) obtained significantly higher yields from incorpo-ration of azolla and intercropping with azolla under flowing water than from other treatments; the control treatment gave higher yields than the treatments with N and P fertilizers, due to greater amounts of fresh water received by the control.

7.4.0 Irrigation

Irrigation had least influence on sucker initiation (Ezumah and Plucknett, 1981) but corm yields increased with increasing water levels because of very extensive rooting beyond 9 months under flooded conditions (Ezumah and Plucknett, 1982). In most oceanic countries, taro is grown in paddy lands. Experiment in Cuba (Vazquez and Torres, 1981) showed that irrigation throughout the lifecycle of taro resulted in higher total yield of corms and cormels than unirrigated plots which was also confirmed by Ezumah and Plucknett (1982) in Hawaii. Yamasaki (1952) gave an explanation for the tolerance to waterlogged conditions by taro. Moisture stress adversely affects canopy development in taro which in turn restricts the dry matter production and finally the tuber yield (Ravi and Choudhury, 1990). Taro is a traditional and popular staple food in Japan. Japanese taro is more cool tolerant than Pacific Island taro, which prefers a long growing season and high temperatures. It also has a sweeter taste and stickier texture. Scheffer *et al.* (1999) conducted trials to determine the effect of irrigation, size of planting propagule, harvest time, plant density, and nitrogen nutrition on cormel yield and quality of Japanese taro in New Zealand. In cold frames the average total cormel yield from planting propagules irrigated daily for 194 days was 1.9 kg/plant, compared to 1.0 kg/plant produced by planting material irrigated at the same rate for the first 82 days. Large treatment differences were also caused by planting propagule size with small propagules producing 0.6 kg/plant, medium propagules 1.5 kg/plant and large propagules 2.2 kg/plant. In field trials, cormel yields increased each month when harvest was delayed from March until June. A high density planting of 6.7 plants/m_2 produced a total cormel yield of 42 t/ha compared to 28, 30 and 38 t/ha at 2.7, 3.3 and 4.4 plants/m_2 respectively. Plant density did not affect average cormel weight. Nitrogen application up to 150 kg N/ha increased yields with a slight reduction at a higher rate. Small cormels (< 30 g), and cormel cracking and sprouting were identified as major factors adversely affecting product quality. This research has shown that Japanese taro can be successfully grown in New Zealand. Komamura *et al.* (1994) determined the irrigation requirements in greenhouse at Atsugi Central Experimental Farm and Gunma prefecture. Evapotranspiration was determined using pan evaporation and crop coefficients, which averaged 1.2-1.4. The soil water response coefficients of taro were close to 1.0. The relationship between soil water and response coefficient was affected by increasing soil water. The upward moving water supply in the greenhouse was 0.5 mm/day, which was less than in open cultivation. This upward movement was affected by crop water requirement and soil water management. Karimata *et al.* (1994) reported that the effects of drip and perforated pipe irrigation systems combined with soil mulching on yield and water requirements of taro were better.

Ravi and Chowdhury (1993) grew thirty-day-old taro plants in pots at soil moisture contents of 25, 50 or 100 per cent of field capacity (FC). Percentages of stomata open, both on the adaxial and abaxial leaf surfaces, were highest in the unstressed plants and lowest in highly stressed plants (25 per cent FC) during 06.00 to 16.00 h. In all the treatments, the percentage of stomata open was highest at around 09.00 h. Stomatal closure began from 09.00 h and continued until 13.00 h followed by reopening during 14.00 to 15.00 h.

Klee (1999) described an integrated biosystem installed in Fiji which was designed by the United Nations University in Tokyo in collaboration with the Zero Emissions Research Initiative (ZERI) Foundation. It operates on a basic concept where the waste from one process in upgraded and completely utilized in another, resulting in zero emissions from the system. Mushroom growing breaks down bagasse/brewery waste substrate, generating a high-value feed for pigs. Pig manure effluent flows into an anaerobic digester producing biogas used as household fuel. After the effluent is anaerobically treated, it flows into a series of three algae ponds and subsequently flows into two 3000 m^3 fish ponds. The algae is harvested daily and used as feed for livestock. The nutrient rich waters permeate the soils in the fish pond embankments, creating fertile soil for vegetables and crops (including eggplants, cabbage, pumpkins, maize, peppers, sweet potatoes, beans, taro, chilli, bananas, pineapples, and cassava, among others) without irrigation or chemical fertilizers. The integrated biosystem has been designed to grow and adapt to changing environmental, economic or technical needs.

7.5.0 Interculture

Weeding becomes natural under flooded conditions. Keeping the plots free from weeds during initial periods of crop growth helps in better growth. Plucknett *et al.* (1967) reported that herbicides like propanil, prameton and nitrofen are

Intercultural Operations and Irrigation in Taro

Taro Intercropped with Coconut

Interculture (Above) and Irrigation (Below) in Taro Field

effective in wet lands, while for uplands they recommended ametryne. Manual weeding is still the most common practice. Maximum yield reductions due to weed competition were < 80 and > 90 per cent under favourable and unfavourable cropping conditions, respectively (Kurschner and Koch, 1986). Corm shape, rot and DM were also affected by the level of weed infestation. Three applications of paraquat provided slightly higher yields than 2 hand weedings. The herbicide caused a change in species composition from creeping broadleaved to perennial grass weeds (Kurschner and Koch, 1986).

Mulching with dried leaves and other plant materials and organic matter provides nutrients and also prevents rapid depletion of soil moisture and reduces weed growth.

Use of polythene sheet has been recommended by some workers (Hayashi, 1973), while Plucknett *et al.* (1970) felt that growth under these conditions might be satisfactory but shoots of cormel which are important for total yield get entrapped and the yield is affected. Pardales (1985) reported that mulching with straw had the highest yields of corm, cormel and herbage (10.98, 3.32 and 17.63 t/ha, respectively) when compared to unmulched crop (8.83, 1.5 and 8.96 t/ha). Early sprouting and favourable soil temperature were considered to be the major factors for higher yield due to mulching. The plastic mulch provided weed control and the yield in August was 15-30 per cent higher than the yield in July. The yield with mulching was 30 per cent higher than that without mulching (Chen, 1996). Mulching with dry leaves or paddy husk after planting has been found to increase yield by conserving soil moisture and checking weed growth (Mitra and Tarafdar, 2018).

8.0 Harvesting and Yield

8.1.0 Harvesting

The duration to reach maturity varies depending upon the type, cultivar and climate. Taro is harvested at 5-6 months after planting for vegetable purpose, but for seed purpose it is harvested at 8-9 months after planting (Mitra, *et al.*, 2007). Barrett (1928) reported that taro matures in 3 months in Ceylon, while it takes 12-15 months in Hawaii (Plucknett *et al.*, 1970). In India, the crop becomes ready for harvest in 5-7 months. Onwueme (1978) reported that Dasheen type matures in 8-10 months, while Eddoes take only 5-6 months in Trinidad. The crop is harvested by pulling the plants out of soil with the help of hoes or spades. The crop maturity is ascertained by yellowing of leaves.

8.2.0 Yield

Yield of taro largely depends on the cultivars grown and also on the growing region. Averages yield ranges from 14-20 t/ha (Mitra *et al.*, 2007). Mohan Kumar *et al.* (1976) reported a yield of 16-26 t/ha in Kerala, while in the neighbouring State of Tamil Nadu, Shanmugavelu and Vijayakumar (1980) reported a yield of 13 t/ha. The average yields in Malaysia, Philippines, West Africa and in low and upland in Hawaii, compiled by Plucknett *et al.* (1970), are 3.5-40, 10.0, 1.5-3.5, 15-30 and 6-10 t/acre, respectively. Dipping of the corms in 500 ppm captan or benlate before storing in soil pits was effective in reducing storage rots (Nwufo, 1988). The yield also varies with maturity.

8.3.0 Storage

Taro cormels should be stored after proper curing in single layer instead of heaping in ambient condition. If storage environment is maintained at 11-13°C and 85-90 per cent relative humidity, taro can be stored up to 5 months (Mitra and Tarafdar, 2013).

9.0 Diseases and Pests

9.1.0 Diseases

9.1.1 Fungal Diseases

Leaf blight and *Pythium* rot or soft rot are known to be the most destructive. Taro is infected by a number of diseases on leaves and corms.

9.1.1.1 Leaf Blight

The taro leaf blight disease appeared throughout the world. The disease was first reported from Java (Raciborski, 1990). In India disease was first reported by Butler and Kulkarni (1993). Several reports on the worldwide occurrence of the disease have been made by many authors time to time. Leaf blight of taro caused by *Phytophthora colocasiae* is the most destructive amongst the many diseases which attack this crop. The leaf blight disease of taro (*C. esculenta*) was first studied by Raciborski (1900) in Java. He briefly chosen the causal organism which he named *Phytophthora colocasiae* and in chronological sequence it became the eight Species of the genus to be named.

The disease appears on the foliage, first as purple to brownish circular water-soaked lesions 1-2 cm in diameter, usually at the tip, base and margins of the leaves. These lesions gradually enlarge, coalesce and the entire leaf dies. In favourable conditions the infected leaves get killed and the entire crop is blighted in 7 to 10 days time. Losses ranging from 25-50 per cent have been recorded due to this disease. The *Phytophthora* blight is predisposed with continuous high humidity (90-95 per cent) and temperature 30-35°C during July to September. Heavy rainfall helps to spread the sporangia and short phase development of diseases occurs in the field (Palaniswami and Tarafdar, 2009).

The blight pathogen, *Phytophthora colocasiae*, has a complex life cycle and can attack taro and other aroids in several ways. The assault usually begins when airborne spores land on the foliage and either germinate directly or release swimming zoospores. In either case, the pathogen penetrates the leaf or stem and grows into a dark lesion that releases thousands of new spores. Carried by wind or rain-splash, the spores can infect the same or other plants, sometimes miles away. Zoospores can be washed from the foliage to the taro corms in the soil, causing them to rot immediately or in storage. Infected seed tubers can start a new cycle of infection in a following season.

The pattern of disease development in resistant cultivars highly depends on the leaf characters, angle of divergent with petiole. The leaves with narrow angle of divergent cannot hold free water on the leaf surface which is needed for germination of sporangium. The spread of leaf blight of taro is more in the broad ovate leaves with wide angle of divergent. The resistant varieties of taro show Hypersensitive Reaction (HR) due to pathogen invasion and induce extremely fast and plant cells rapidly sacrifice to check fungal growth inside tissue, especially in highly resistant taro against *Phytophthora*, which causes small necrotic spots on the leaves. The cytological evidence in infected resistant cultivars of taro shows containment of hyphae in HR lesion causing trailing necrosis, hyphal coagulation, cytoplasmic granulation with high intensity of stain of Trypan blue due to irreversible membrane damage. Several mesophyll tissue of partial resistant plant is involved in HR. Sometimes, the *Phytophthora* invaded tissues of resistant cultivars induce series of cell death, condensed protoplast which give barrier to the pathogen spread. Penetrated cell of susceptible taro cultivars do not show visible plant response and pathogen hyphae expand through the plant tissue leads to the external symptoms manifestation with massive gum deposition.

Several biochemical changes occur due to pathogen invasion and plant peroxidises have many putative functions including lignifications, and likely to play indispensable roles in plant defense system against leaf blight pathogen attack in taro. Moreover, Super Oxide dismutase, Esterase, Glucose-6 phosphate dehydrogenase exudates may work together as a multi-component defense system to protect plant tissue during pathogenic assault in taro. Isozyme steadily increased in activity during pathogenesis. The diversity of peroxidise, EST in individual cultivar expressed in leaf tissue of taro which is quick migrating ixozyme, was expressed at a slightly higher level in infected plant tissue than in healthy tissues. Nitrate reductase is also an important signalling and defense molecule, plays a key role in activating disease resistance in plants. Hypersensitive is always associated with the resistance response of taro to *Phytophthora colocasiae*; the timing, severity and extent of the HR varied considerably, depending upon the examined cultivars. As observed in taro leaves *P. colocasiae* interactions, various level of pathogen ingress correlated with the different necrotic responses (Mandi *et al.*, 2009).

Several studies have been done on the cytological and biochemical studies on the infection process and defence mechanisms in taro plant. Lipid peroxidation and lipid damage may be partially responsible for some of these cell changes and probably affect membrane function. Around the lesion the phenolic and phytoalexins such as glyceollin (in soybean), and other compounds are synthesized. Callose, lignin are deposited and pathogen related (PR) protein like β-1, 3-glucanase and chitinase are produced. The HR leads the secondary resistance response, the systemic acquired response (SAR) and is characterized by necrotic lesion the infection site. Biochemical changes comprise of generation of active oxygen species (oxidative burst), cell death, over production of lignin related materials and the Induction of certain pathogen related protein. During hypersensitive response a conserved programmed cell death (PCD) mechanism is activated (1998). The programmed cell death (PCD) is a genetically coordinated process of cellular suicide by both endogenous stimuli and environmental stresses (Mandi *et al.*, 2009).

Diseases of Taro

Symptoms of Leaf Blight on Taro Foliage

Symptoms of Brown Leaf Spot on Taro Leaves

Symptoms of Corm Soft Rot on Taro

Leaves Infected by White Spot

Diseases of Taro

Wilting and Leaf Death Associated with Infection by Corm Rot

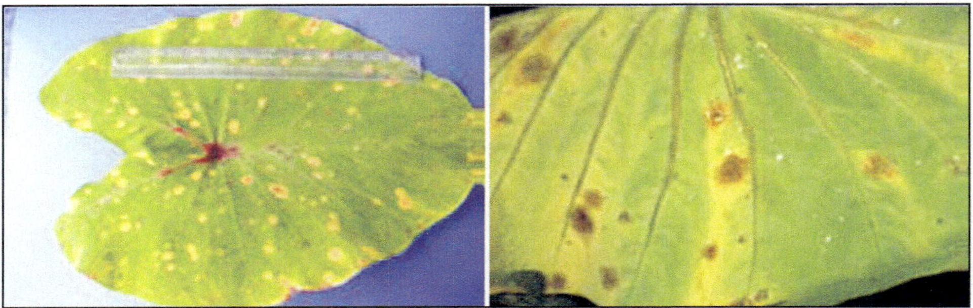

Symptoms of Orange Leaf Spot Disease

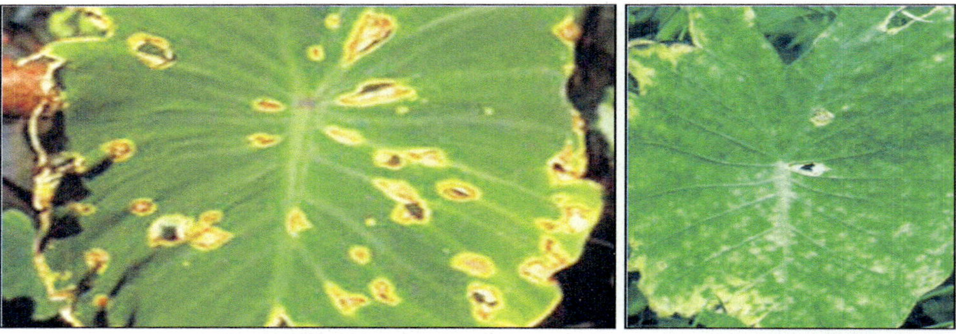

Shot Hole Disease **Leaf Blotch Disease**

Taro Plants Infected with Bacterial Soft Rot; Collapse of Plant on the Right

Viral Diseases of Taro

Symptoms of Dasheen Mosaic Virus

Colocasia bobone (Left) Disease (Distorted, thickened, and brittle leaves) and Taro Badnavirus (Right) (Distinct vein chlorosis)

Symptoms of Taro Vein Chlorosis Virus (Distinct vein chlorosis on the middle leaf and feathering symptoms on right side leaf)

Development of resistant varieties and application of proper chemicals are the only option for management of *Phytophthora* blight in taro. Several varieties have been developed in India *viz.*, Muktakeshi, Bhu Sree, Bhu Kripa, Sree Rashmi, Shree Pallavi and Bidhan Chaitanya are moderate to high resistant to *Phytophthora* blight. Among Taro (Banda) type, two popular varieties BCB-2 and IGB-5 are found resistant to this blight disease. There are several germplasm of taro in India show differential symptoms in field and very low to moderate blight infection of blight (Mandi *et al.*, 2006; Sarkar *et al.*, 2017). Early planting during February-March, especially in Eastern part of India may escape the disease. Application of suitable fungicides like Metalaxyl-M + Mancozeb 72 per cent WP, Cymoxanil + Mancozeb 72 per cent WP @ 2.0 g/lt of water are effective. Seed corm treatment with Metalaxyl-M + Chlorothalonil 440 SC @1gm/lit. of water gives good germination and less infection especially in late sown crop. But in severe case two to three spraying of Metalaxyl-M 4 per cent + Mancozeb 64 per cent WP @ 1000 g/ha or Cymoxanil 22.1 per cent + Famoxadone 16.6 per cent SC @ 500g/ha is highly effective against this disease.

9.1.1.2 Pythium Rot

Several species of *Pythium* which are soil-borne infect the roots and corms. The initial symptoms appear as chlorosis of the leaves and the affected plants gradually wilt. The corms become soft and rot with a foul odour. Field sanitation, crop rotation, use of disease free planting materials and treatment of soil and planting material with fungicides offer control.

9.1.2 Virus Disease

Mosaic (*Dasheen mosaic virus*) causes interveinal yellowing and mosaic resulting in distortion of leaves and acute reduction of chlorophyll content which resulted poor growth of plant and yield. Use of healthy planting material and rouging infected plants is advocated. Foliar mosaic, mottle and 'feathering' symptoms are apparent. Foliar symptoms are intermittently expressed. Transmission by aphids *Pentsalonia negronervosa*, non-persistently, under natural conditions is the vector. Flexuous rod shaped about 11 x 750 nm induces cylindrical inclusions, as do other potyviruses (potyvirus group). Plants free of this virus can easily be obtained through meristem-tip culture when small (0.5 mm or less) meristem-tips are used (Anon., 2018).

Among the species of the genus Potyvirus, Dasheen mosaic virus (DsMV) infects a wide variety of aroid plants worldwide, including at least 16 genera of the family Araceae (Zettler and Hartman, 1986, 1987; Zettler *et al.*, 1978). It causes serious damage to ornamentals like Caladium, Dieffenbachia, and Zantedeschia, and is ubiquitous in commercial plantings of tropical root crops of the genera Colocasia, *Xanthosoma* and *Amorphophallus*. The virus is experimentally transmitted by infected sap and, in nature, by several widely distributed aphid species, including *Myzus persicae* and *Aphis gossypii*.

9.1.3 Other Diseases

Diseases of minor importance reported on taro are *Sclerotium* rot, leaf spots caused by *Cladosporium colocasiae* and *Phyllosticta colocasiophila* in upland taro.

Spraying of a common protective fungicide like dithane M-45 (0.2 per cent) or blitox (03 per cent) mixed with a common sticker may check the build-up of the diseases. Recently, storage rot caused by a number of fungi like *Aspergillus niger*, *Botryodiplodia theobromae*, *Fusarium solani*, *Rhizopus stolonifer* and *Sclerotium rolfsii* has been reported (Nwufo and Fajola, 1981). Mechanical injury during harvesting is known to predispose the corm to the pathogens in storage. As determined in other occasions, dipping of root of taro in solution of difolatan (0.1 per cent) or miltox (02 per cent) may check the rotting in storage.

9.1.4 Storage Diseases

The other fungal pathogens like *Phythium* spp., *Fusarium* spp., *Sclerotium* spp., *Botryodiplodia* spp., *Rhizopus* spp. *etc.* are the important pathogens which attack the plant in the field but mainly in the storage.

Pythium sp.: Infected tissues of corm become pale and soft. The lesion is bounded by irregular but sharp margin. In advance stage decay is accompanied by unpleasant odour. The pathogen, *P. helicoides* was 1st reported from the AICRP Tuber Crops, Kalyani centre in 2008 which causes serious corm rot in the field as well as in the storage.

Fusarium sp.: Decay usually begins at the base or side of the Spongy whitish grey with distinct brown margin. In case of infection with *F. Solani*, the dry corky or powdery rot with white mould develops on the surface.

Sclerotium rolfsii: Caused tuber rotting and the tissue becomes pale brown to pinkish, soft but not watery. In humid condition there is development of white mould which given rise to small spherical sclerotia.

Botryodiplodia theobromae: Showed no external symptoms even when decay is advance. Internal tissue is initially greyish or pinkish and subsequently turn black distinct margin. Flesh growth may turn spongy.

Rhizopus stolonifer: The damage starts at one end of the corm. In dry condition firm flesh growth appear on sunken lesion. In humid condition affected tissue is soft and watery with a pronounce odour. The periderm develops white mycelia (Palaniswami and Tarafdar, 2009).

Aspergillus niger: The tuber tissue is affected with typically fawn or brown colouring. Besides several fungal species *e.g. Pythium aphanidermatum, P. splendens, P. myriotylum, Fusarium solani, F. moniliforme* cause rotting of EFY. Black rot (*Ceratocystis fimbriata*), Bacterial soft rot (*Erwinia carotovora*), Blue mold rot (*Penicillium crustosum*), Cylindrocarpon rot (*Cylindrocarpon lichenicola*), Pink mould rot (*Trichothecium roseum*) and Rhizoctonia rot (*Rhizoctonia solani*) are other storage diseases on corms of aroids (Palaniswami and Tarafdar, 2009).

9.2.0 Pests

Taro leaf hopper (*Tarophagus proserpina*) causes severe losses in most taro growing areas. The Philippine egg sucking bug (*Cyrtorhinus fulvus*) is an effective biological agent to control the hoppers. Spraying any systemic insecticides controls the pest. The other important insect pests are sweet potato hawk moth (*Hippotion celerio*) and aphid (*Aphis gossypii*). The non-insect pests include red spider mites (*Tetranychus cinnabarinus* and *T. neocaledonicus*). The mites can be controlled by spray of acaricides.

Aphids (*Aphis gossypii*) causes yellowish speckling and curling of leaves. In severe cases sooty mould develops, reducing vitality of leaves. Major pests of taro and elephant -foot yam can be controlled by spraying insecticides such as Dimethoate, Quinalphos or Fenthion (0.05 per cent) at 30 and 60 days after planting.

Thrips (*Caliothrips indivus*) causes characteristic silverish white specks with black dots on the leaves and drying foliage.

Leaf-eating caterpillars (*Spodoptera litura*) cause economic defoliation ranging from16-80 per cent. Young caterpillars feed on leaves by scraping gregariously. Older caterpillars feed voraciously on the entire leaf.

Scale insect (*Aspidiella hartii*) causes shrivelling due to desapping and feeding. In case of severe infestation, the plant shows withering and drying, reducing vigour and vitality of plants. The infested tubers lose quality and viability. Use of pest-free seed tubers and dipping seed tubers before planting in 0.1 per cent Dimethoate is advocated.

Arecanut beetle (*Araecerus fasciculatus*) bores into tubers and converts it to powder causing loss in weight and quality. Storing uninfested tubers in white sand or saw dust prevents infestation. Seed tubers can be stored in a mixture of sand and Carbaryl dust (10 per cent) in a 100:1 ratio.

Mealy bugs (*Pseudococcus citriculus; Rhizoecus* sp.) Nymphs and adults and their powdery meal cover the tuber surface. It leads to poor germination and also affects cooking quality. If pest infestation is severe, spray of Dimethoate (0.05 per cent) is effective in checking its further spread.

9.2.1 Nematode

The use of nematode free planting materials and treating the planting materials in hot water (40°C) for an hour have been suggested to control root-knot nematode in taro.

10.0 Crop Improvement

The complex cytology, sterility due to degeneration of the megaspores followed by the abortion of ovules (Banerji, 1937) and rare occurrence of natural fertile seed (Porterfield, 1944) are responsible in making the conventional breeding programmes more difficult in taro. In spite of the difficulties, breeders in India (Venkateswarlu, 1980), Hawaii (Plucknett and de la Pena, 1971) and Japan (Odawara, 1959) have attempted to breed better taro cultivars. Some cultivars which flower naturally can be exploited for hybridization programme. Intervarietal hybridization for induction of resistance to leaf blight is reported in India. First taro hybrid 'Somoa Hybrid' was released from the Research division of the Ministry of Primary Industries, Fiji in 1984. Later the University of South Pacific, Western Samoa released 'Alafua Sunrise' in 1991.

Pests of Taro

Aphid Infestation in Taro

Mealy Bugs on Taro Roots

Adults of Plant Hopper on the Leaf

Spiralling White Fly infestation in Taro

Corm Borer infestation in Taro

Pests and Nematode Infestation in Taro

Armyworm Infestation in Taro

Spider Mite infestation (Left) in Taro and Adult Mite (Right)

Giant African Snail attack in Taro

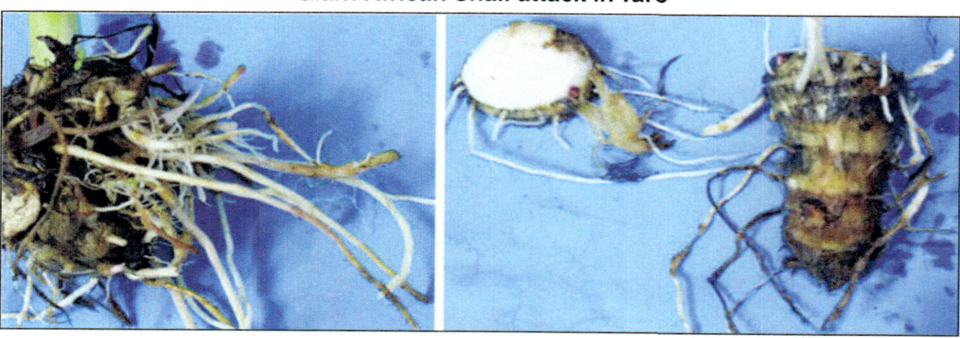

Root-knot Nematode Infestation in Corms and Roots of Taro

The naturally occurring variability is very large, and useful mutants and ecotypes can be isolated from nature. The vegetative propagation offers potentiality for selection of ideal mutants through mutation breeding. Besides, the seedling progeny from the maturally flowering types would present an array of variability for selection and improvement. Johnston and Gendua (1998) grew taro seedlings derived from two naturally pollinated inflorescences (*Ph15* and *Ph 17*) to maturity. The growth performance of the seedlings was equal to, or better than, the vegetatively-propagated parent plants. The seedlings had greater leaf lamina length and width, leaf area, petiole length, and had more leaves and suckers than the parent plants. Guidelines for the Conduct of Test for Distinctiveness, Uniformity and Stability on Taro (*Colocasia* spp.) has been developed and notified in India (DUS, PPV and FRA 2018) and digital morphometrics have been developed (Mandal *et al.*, 2015); these test guidelines shall apply to all varieties, hybrids and parental lines of elephant foot yam for future crop improvement program.

Ivancic and Lebot (2000) brought out a publication on genetics and breeding of taro detailing systematics, origin and centres of diversity, genetic resources, reproductive biology, and breeding. The objective in breeding of taro is mainly to breed tolerant/resistant lines to *Phytophthora* blight. Lebot (1997) emphasised the urgent need to characterize and evaluate existing genetic resources. Taro breeding success depends strongly on hybridization. The main problem in artificial hybridization is the irregularity of flowering and the occurrence of abnormal floral structures. Abnormal inflorescences appear continuously in cultivated and wild taro populations in Papua New Guinea and the Solomon Islands (Ivancic, 1995). The frequency of abnormal inflorescences increases with the application of flower-inducing hormones which are frequently used in breeding programmes. Abnormal and unusual inflorescences are classified into 16 groups. Several types can be efficiently used in hybridization. Some of the genotypes characterized by unusual or abnormal floral traits represent a potential source of future improvements in taro. They may be used as genetic markers, the sources of the genetic variation for ornamental characteristics or for the improvement of flowering ability (Ivancic, 1995).

Gamma rays could be used to induce mutants. Vasudevan *et al.* (1989) noticed variability in tuber storability among induced mutants of taro. Tubers of CM17 in particular retained their quality even after 180 days storage. CM17 tubers content low phenol and sugar content but high DM and starch content, making them suitable for the production of taro chips.

Rodriguez Manzano *et al.* (1999) identified twelve qualitative and quantitative important morphological characters, for the characterization and identification of clones. The relationship between palatability and the pigmentation of buds, roots and pulp of the cormels was noteworthy.

Pandey and Dobhal (1997) studied the genetic divergence of 31 genotypes of *C. escu-lenta* collected from Assam, Meghalaya and Nagaland and grouped them into 11 clusters irrespectively of geographical origin. Genotypes IC-87168 (M), IC-87042 (M) and IC-89548 (A) were distinct and classed as separate clusters. IC-89548 (A) exhibited the highest mean values for yield/plant (933.8 g) and weight of mother cormels (640.5 g). The cluster containing IC-87132 (A) and IC-87163 (M) also exhibited high yields (875.2 g) and cormel weight (449.8 g) combined with greater plant height, leaf length and leaf width than IC-89548(A). Medhi and Parthasarathy (1998) evaluated corm quality and size, and yield of twenty-nine genotypes of taro collected from Meghalaya (13), Nagaland (11) and Assam (5). The genotypes from Meghalaya were distributed evenly among 4 of the 6 clusters obtained suggesting wide divergence in the native germplasms. A genotype from Assam (AS6) with large corms and high yields formed a lone cluster while cluster 6 is predominated by genotypes from Nagaland.

Rekha *et al.* (1997) grouped 180 taro cultivars into apparent amylase inhibitory activity and found that 9 belonged to type I, having < 2000 amylase inhibitor units (AIU)/g tuber dry matter (DM), 35 cultivars belonged to type II, having 2000-4000 AIU/g DM, 87 belonged to type III, with 4001-6000 AIU/g DM, and 49 belonged to type IV, with > 6000 AIU/g DM. Variation in sugar content of different taro cultivars was reported by Ghosh and Hasan (1992). Significant differences for sugar content were observed among the 11 cultivars grown in West Bengal. Kovvur Local gave the highest percentage (9.1 per cent) of total sugar per fresh corm weight. The highest levels of non-reducing and reducing sugars were produced by Nadia Local (7.4 per cent) and C135 (5.7 per cent), respectively.

Path analysis revealed that weight of mother cormels and weight of cormels had the highest direct and indirect effects, respectively, and could therefore be used as selection criteria for higher yielding genotypes. Of the traits studied by Pillai *et al.* (1996), number of cormels/plant had the maximum direct effect on yield, followed by mean cormel weight. Medhi *et al.* (1994) observed that all the yield components had heritability exceeding 60 per cent. The best lines for cultivation in Meghalaya were ML8 and ML1. Mohankumar *et al.* (1990) found yield was most closely positively correlated with LAI 120 days after planting and with mean cormel weight.

In a study of 239 Indian taro accessions for 7 flowering related characteristics, flowering was observed to be rare, erratic and seasonal. Flowering habit was irregular and non-synchronous. The very low frequency and sporadic nature of flowering and the very high frequency of sterile triploids among vegetatively propagated cultivated taro have severely restricted the possibilities of sexual reproduction (Sreekumari and Pillai, 1994).

10.1.0 Resistance Breeding

The taro industry in Samoa was decimated in 1993 with the incidence of taro leaf blight (TLB), caused by the fungus *Phytophthora colocasiae* (Chan *et al.*, 1998). Chaudhary and Rai (1988) reported 5 varieties of taro to be immune and 1 moderately resistant in Arunachal Pradesh (India). Ivancic *et al.* (1995) studied segregating populations for resistance to *P. colocasiae* in taro. Populations analyzed in this study were developed from three groups of crosses: (a) resistant x resistant, (b) resistant x susceptible, and (c) susceptible x susceptible. The most frequent segregation ratios (resistant: susceptible) were 3: 1, 9:7, 7:9 and 13:3, suggesting that the number of genes controlling resistance to *P. colocasiae* in taro may be relatively low. The appearance of resistant genotypes in populations resulting from crosses between two (partially) susceptible genotypes indicates that minor genes associated with partial resistance may be involved. Pillai *et al.* (1996) selfed and intercrossed four *P. colocasiae* leaf blight tolerant taro cultivars and their open pollinated progenies to isolate resistant recombinants from the segregating generations. The highest numbers of resistant genotypes were obtained from selfed tolerant lines, followed by open pollination of progenies of tolerant lines. A high proportion of second generation hybrids of tolerant cultivars also showed resistance. Selfed susceptible genotypes did not produce resistant lines, whereas crosses involving different susceptible lines gave a low percentage of resistant progeny. Genotypes with combined resistance, high yields and good cooking quality were identified. From a third cycle of segregating generations, 4 genotypes were indentified with a high degree of resistance under field conditions and after artificial inoculation. The isolation and characterization of leaf blight resistance linked genes were carried out and thereby two molecular markers *viz. prg 1* and *prg 8* were identified as very useful marker against *Phytophthora* blight resistant genes. The total finding implies that the transcripts of *prg 1* and *prg 8* have role in disease tolerance and the already designed primers may be used as markers in future for screening taro genotypes at early stages of development (Anon., 2017).

To date twenty nine taro cultivars from Guam, American Samoa, Yap, Pohnpei, and Thailand have been propagated *in vitro*, planted in pots and screened at the University of Guam for susceptibility to taro leaf blight. The most resistant were: Gilin, Kugfel, Oglang, Pwetepwet, Thailand, Sushi, Ol, and Pasdora (Wall and Wiecko, 1998). Hunter and Pouono (1998) screened four taro cultivars, Pwetepwet, PSB-G2, Pastora and Toantal in Western Samoa for their resistance to taro leaf blight and for their yield and eating quality. Disease severity levels were not significantly different for any of the four cultivars evaluated. Corm yields were highest for Pastora followed by PSB-G2, Pwetepwet and Toantal. Toantal and PSB-G2 rated highest for taste and dry weight (35 and 37 per cent, respectively).

In Papua New Guinea, Okpul *et al.* (1997) evaluated 35 varieties resistant to taro leaf blight (TLB) for yield components and eating quality in comparison with the locally preferred cultivar, Numkowec. The main factors affecting eating quality were pre-sence of conspicuous corm fibre and acidity. Leaf blight resistant varieties AN 65, 17, 50, 32, 46, 21, 12 and AN 20 had acceptable eating quality. Their corm yield ranged from 300 g/plot (AN 50) to 570 g/plot (AN 21). However, their corm yield was not significantly different from that of Numkowec (430 g/plot). These resistant varieties are recommended to farmers in the Lae area based on their resistance to TLB and their similarities in corm yield and eating quality to Numkowec. Subsequently, Singh and Okpul (2000) selected twelve taro clones resistant to taro leaf blight from the 32 elite plants resulting from the second cycle of recurrent selection at the Wet-Lowlands Mainland Research Centre, Lae. These clones were evaluated for seven morphological characteristics, reaction to *P. colocasiae* inoculation, yield, and eating quality in comparison with the locally-preferred cultivar Numkowec. All twelve lines were resistant to the local race of the taro leaf blight pathogen, to which Numkowec was susceptible. However, only seven lines were acceptable to a taste-testing panel of local consumers. Of these seven lines, four (C2-E1, C2-

E4, C2-E7 and C2-E10) significantly out-yielded Numkowec for corm and cormel production. These lines are recommended for regional evaluation in multi-location trials throughout Papua New Guinea.

11.0 Biotechnology

11.1.0 Tissue Culture

Tissue culture research for *in vitro* propagation of taro and tannia is perfected well and the summary of work done is presented in Tables 3 and 4.

11.2.0 Embryo Culture

Zygotic embryos of *Colocasia esculenta* were grown *in vitro* and the effect of growth regulators and glutamine on seedling development studied (Nyman, 1987). Zygotic embryos cultured on LS medium without the addition of growth regulators developed into mature plants only in the presence of endosperm tissue.

11.2.1 Biochemical and Molecular Markers

Hirai *et al.* (1989) studied the acrylamide gel electrophoresis of the tuber proteins of 85 Japanese cultivars of *C. esculenta* which revealed 5 patterns, designated E, I, D, T and A, for the major cultivars and an additional 3 patterns in the minor cultivars. Results from cluster analysis based on 40 morphological characters were consistent with those of the biochemical analyses, allowing the identification of 7 groups (Eguimo, Dodare (Dotare), Hasubaimo, Ishikawa-wase, Kurojiku, Akame and Tonoimo) for the non-fasciculated taro cultivars, and 2 groups, Yatsugashira and Shogaimo, for the fasciculated ones. Isoenzyme variation in 1417 cultivars and wild forms of taro collected in Asia and Oceania revealed 143 isoenzyme phenotypes, or zymotypes, each uniquely characterized by the presence or absence of 56 electromorphs (Lebot and Aradhya, 1991). Isshiki *et al.* (1995) studied sixty-nine wild taro clones collected in Bangladesh (38 diploids and 31 triploids) for isozyme variation. Five taro populations were subsequently derived by crossing or selfing amongst 5 of the diploid clones, and studies were conducted on the isozyme variation in the enzyme systems. Genetic variations were observed among the wild diploids and triploids for the isozyme loci.

Systematic studies regarding the genetic diversity of taro through molecular markers based on RAPD, AFLP, ISSR and SSR will help in understanding the genetic diversity of germplasm as well as identification, conservation and utilization of authentic and superior crop materials. Seventy six taro accessions those were collected from the different parts of India and maintained in the field gene bank of ICAR- All India Coordinated Research Project on Tuber Crops, BCKV, India have been characterized through RAPD, AFLP, ISSR and SSR which insights the distinction and inter-relationship among the taro accessions on evolution of the species with natural mutation, cross pollination, agro- climatic condition and human interference. Codominant markers of 27 SSR primers revealed the different levels of polymorphism having differential ability to identify unique multiband phenotypes among 76 taro accessions.

Isshiki *et al.* (1998) also used isozyme variation to classify cultivars of Japanese taro *C. esculenta* at five loci, *Gpi2* of glucose-6-phosphate isomerase, *Aat1* and *Aat3* of aspartate aminotransferase, *Skdh* of shikimate dehydrogenase and *Idh* of isocitrate dehydrogenase, were examined in 58 cultivars. Allele *a* at the *Aat3* was recognized specifically in all the examined triploid cultivars, and this allele clearly discriminated the triploid cultivars from the diploid ones.

Nguyen *et al.* (1998) studied leaf esterase isoenzyme polymorphism to determine phytogenetic relationships among 69 accessions of *C. esculenta* (including vars. *esculenta, aquatilis* and *illustris*), 4 of *C. gigantea*, 4 of *Xanthosoma sagittifolium*, 1 of *X. violaceum*, 2 of *Alocasia macrorrhiza* and 4 of *A. odora*. Banding patterns were used to generate a phylogenetic tree in which all *C. esculenta* accessions formed a single cluster, and *C. gigantea* was more closely related to *Alocasia* than to *C. esculenta*. *C. esculenta* accessions from Yunnan, China, shared bands with accessions from various areas, indicating that Yunnan may have been an important region for taro evolution. Nguyen *et al.* (1999) later studied nine enzyme systems of taro using polyacrylamide gel electrophoresis and isoelectric focusing, the genetic control of isozymes.

Molecular markers have also been used to classify species and varieties of *Colocasia* and allied genera. Matthews *et al.* (1992) surveyed the multigene family for 18S-5.8S-26S ribosomal DNA (rDNA) in 80 accessions. Restriction enzyme digests of total DNA were probed with a fragment cloned from the large intergenic spacer region of taro rDNA. Seven patterns of rDNA and two of mitochondrial DNA (mtDNA) were found. The resulting classification of taros was consistent with previous morphological and biochemical observations. Comparisons with taros from Nepal indicated that some Japanese triploids have received genetic influence from taro in Himalayan region. The actual diploid progenitors of Japanese triploids remain unknown. Comparisons with an Australian sample of *C. esculenta* var. *aquatilis* suggested the Japanese diploids, and some triploids, are at least partly derived from var. *aquatilis*, which is widespread in the wild in northern Australia and many parts of tropical Asia and the Pacific. Irwin *et al.* (1998) evaluated fortyfour taro, 2 tannia (*Xanthosoma* species) and one *C. gigantea* accessions were evaluated for genetic diversity by RAPD using 112 primers. Seventythree primers amplified DNA products which could be used to fingerprint the accessions.

Schnell *et al.* (1999) studied the genetic relatedness of eighteen cultivars of cocoyam (*Xanthosoma* spp.) and two cultivars of taro (*C. esculenta*) from the USDA/ARS germplasm collection using RAPD data. Seven random primers generated 40 RAPD loci. Of the 18 cultivars screened, 11 (61 per cent) were identical at all RAPD loci evaluated.

Ochiai *et al.* (2000) investigated the phylogenetic relationships among 51 accessions of taro (*C. esculenta*), *C. gigantea, Alocasia macrorrhiza, A. odora, Xanthosoma sagittifolium* and *Schismatoglottis* spp. were investigated using restriction fragment length polymorphisms (RFLPs) of chloroplast DNA. The phylogenetic tree using the Neighbour Joining (NJ) method revealed that *Xanthosoma* and *Schismatoglottis* genera were distantly related to *Colocasia* and *Alocasia* genera. Among *Colocasia* and *Alocasia* accessions, *C. esculenta* accessions formed a single cluster. However *C. gigantea* accessions were related to *Alocasia* species more closely than taro. Genetic linkage in the species and varieties of *Colocasia*, *Alocasia* and related genera using molecular markers has also been studied by Carneiro *et al.* (1993), Hirai *et al.* (1993) and Mathews *et al.* (1996).

11.3.0 Genetic Transformation

Few studies on the transformation of taro have been carried out. Fukino *et al.* (2000) reported the first transformation studies on taro with the objective of breeding a virus-resistant cultivar of taro. Since taro is a monocotyledon, *Agrobacterium*-mediated transformation is considered to be difficult and particle bombardment system was used. Transgenic taro plants were obtained by high-velocity particle bombardment. The plasmid pREXHGUS, carrying the hygromycin resistance gene, was used as a selection marker. Highly regenerative taro callus was obtained from the apical meristem, and maintained in liquid culture. The callus was chopped into small fragments with a forceps, then transformed and selected on LS medium containing BA, NAA and hygromycin. Since the transformed calli obtained showed vigorous growth on LS-BN medium with 20 mg/l hygromycin, this concentration was considered to be suitable for the selection of transgenic taro. The transformants were confirmed by amplification of the *gus* gene, and Southern hybridization. The expression of the foreign gene was demonstrated by the GUS activity.

12.0 References

Anonymous (1980) *Nutritive value of Indian Food*, National Institute of Nutrition, ICMR, Hyderabad.

Anonymous (2016) *Annual Report*, AICRP on Tuber Crops, Thiruvananthapuram, Kerala.

Anonymous (2018) *Annual Report*, AICRP on Tuber Crops, Thiruvananthapuram, Kerala.

Anonymous (2019) *Annual Report*, AICRP on Tuber Crops, Thiruvananthapuram, Kerala.

Anonymous (2020) *GlobeNewswire*, Dublin, Aug. 20, 2020.

Ashokan, P.K. and Nair, R.V. (1984) *J. Root Crops*, **10**: 59-63.

Asokan, M.P., O'Hair, S.K. and Litz, R.E. (1984) *HortSci.*, **19**: 885-886.

Austin, M.T., Constantinides, M. and Miyasaka, S.C. (1994) *Commun. Soil Sci. Plant Anal.*, **25**: 2159-2169.

Babu, B., Hegde, V., Makeshkumar, T. and Jeeva, M.L. (2011) *J. Plant Pathol.*, **93**: 199-203.

Bailey, L.H. (1925) *Manual of Cultivated Plants*, The MacMillan Co., London, pp. 135-136.

Banerji, I. (1937) *J. Indian Bot. Soc.*, **16**: 159-164.

Barrett, O.W. (1928) *The Tropical Crops*, The MacMillan Co., New York, pp. 375-377.

Bentham, G. and Hooker, J.D. (1883) *Genera Plantarum*, Reeve and Co., London, pp. 955-978.

Butler, E.J. and Kulkarni, G. S. (1913) *Memoirs of the Department of Agriculture in India*, **5**: 233-259.

Carneiro, M., Coutinho, M.V. and Rodrigues, C.A. (1993) *Ciencia e Cultura Sao Paulo*, **45**: 231-235.

Cedeno-Maldonado, A., Fierro, C. and Bosques, A. (1988) *J. Agril. Univ. PuertoRico*, **72**: 515-520.

Chan, E., Milne, M. and Fleming, E. (1998) *Trop. Agril.*, **75**: 93-98.

Chand, H., Pearson, M.N. and Lovell, P.H. (1998) *Plant Cell Tissue Organ Cult.*, **55**: 223-226.

Chang, R.C.O. and Goh, C.J. (1994) *Plant Sci. Limerick*, **104**: 93-100.

Chaudhary, R.G. and Rai, M. (1988) *Haryana J. Hortic. Sci.*, **17**: 278-279.

Chen, J.Y. (1996) *Zhejiang Nongye Kexue*, **6**: 285-286.

Coates, D.J., Yen, D.E. and Gaffey, P.M. (1988) *Cytologia*, **53**: 551-560.

Conner, J.E. (1908) *Bureau Plant Indian Bull.*, **142**: 38.

Coursey, D.G. (1968) *World Crops*, **20**: 25-50.

De Candolle, A. (1885) *Geographie Botanique Raisonee*, Paris, p. 827.

de la Pena, R.S. and Plucknett, D.L. (1969) *Proc. Int. Symp. Trop. Root Crops*, Trinidad, pp. 70-85.

Ebida, A.I.A. (1995) *Alexandria J. Agril. Res.*, **40**: 457-474.

Escalada, R.G. and Ratilla, B.C. (1998) *Agrofor. Syst.*, **41**: 251-266.

Ezumah, H.C. (1973) *Diss. Abstr. Int.*, B **34**: 24.

Ezumah, H.C. and Plucknett, D.L. (1981) *J. Root Crops*, **7**: 41-52.

Ezumah, H.C. and Plucknett, D.L. (1982) *J. Root Crops*, **8**: 17-26.

Fedrov (1974) *Chromosome Number of Flowering Plants*, Otto Koeltz Science Publishers, D-264, West Germany.

Fukino, N., Hanada, K., Ajisaka, H., Sakai, J., Hirochika, H., Hirai, M., Hagio, T. and Enomoto, S. (2000) *JARQ*, **34**: 59-165.

Ghosh, S.K. and Hasan, M.A. (1992) *Indian Agriculturist*, **36**: 57-59.

Goenaga, R. and Chardon, U. (1995) *J. Plant Nut.*, **18**: 1037-1048.

Gomez, L., Monge, M., Valverde, R., Arias, O. and Thorpe, T. (1989) *Turrialba*, **39**: 155-161.

Gupta, P.P. (1985) *Plant Cell Rep.*, **4**: 88-91.

Halavatau, S.M., O'Sullivan, J.N., Asher, C.J. and Blamey, F.P.C. (1998) *Trop. Agril.*, **75**: 6-12.

Hartemink, A.E. and Johnston, M. (1998) *Trop. Agril.*, **75**: 1-5.

Hartemink, A.E., Johnston, M., John, P., Julius, W. and Kerru, A. (1997) *Papua New Guinea J. Agril. For. Fish.*, **40**: 6-12.

Hayashi, H. (1973) *Bull. Agric. Res. Inst. Kanagawa Prefecture*, No. 113, pp. 1-10.

Heath, M. (1991) *Phytopathol.*, 81: 127-130.

Hirai, M., Nakamura, K., Imai, T. and Sato, T. (1993) *Japanese J. Genet.*, **68**: 229-236.

Hirai, M., Sato, T. and Takayanagi, K. (1989) *Japanese J. Breed.*, **39**: 307-317.

Hodnett, G.E. (1958) *Trop. Agric.*, Trinidad, **35**: 205-207.

Hunter, D. and Pouono, K. (1998) *J. South Pacific Agril.*, **5**: 39-43.

Igbokwe, M.C. and Ogbonnaya, J.C. (1981) Yield and Nitrogen Uptake by Cocoyam as Affected by Nitrogen Application and Spacing. In: *Tropical Root Crops, Research Strategies for the 1980s*, (Ed. E. Terry, K.A. Odura and F. Caveness). Ottawa, Ont., IDRC, p. 279.

Irwin, S.V., Kaufusi, P., Banks, K., De la Pena, R. and Cho, J.J. (1998) *Euphytica*, **99**: 183-189.

Isshiki, S., Nakada, M., Nakamura, N., Tashiro, Y. and Miyazaki, S. (1995) *J. Japanese Soc. Hort. Sci.*, **67**: 521-525.

Isshiki, S., Nakamura, N., Tashiro, Y., Miyazaki, S. and Kumazawa, S. (1998) *J. Jap. Soc. Hort. Sci.*, **67**: 521-525.

Isshiki, S., Otsuka, K., Tashiro, Y. and Miyazaki, S. (1999) *J. Japanese Soc. Hort. Sci.*, **68**: 774-779.

Ivancic, A. (1995) *Aust. J. Bot.*, **43**: 475-489.

Ivancic, A. and Lebot, V. (1999) *Pacific Sci.*, **53**: 273-285.

Ivancic, A. and Lebot, V. (2000) *The genetics and breeding of taro*, CIRAD, Montpellier, France, p. 194.

Ivancic, A., Kokoa, P., Simin, A. and Gunua, T. (1995) *J. South Pacific Agril.*, **2**: 17-21.

Jackson, G.V.H., Ball, E.A. and Arditti, J. (1977) *J. Hortic. Sci.*, **52**: 373-382.

Jacobs, B.C. and Clarke, J. (1993) *Field Crops Res.*, **31**: 317-328.

Johnston, M. and Gendua, P.A. (1998) *Trop. Agril.*, **75**: 13-17.

Joseph, K., Kunju, U.M. and Nair, K.P.M. (1981) *Agric. Res. J.*, Kerala, **19**: 27-31.

Karimata, A., Mizuta, A., Komamura, M. and Yoneyasu, A. (1994) *J. Agril. Sci., Tokyo, Nogyo-Daigaku*, **38**: 293-302.

Klee, R. (1999) *BioCycle*, **40**: 66-67.

Komamura, M., Karimata, a., Mizuta, A. and Yoneyasu, A. (1994) *J. Agril. Sci., Tokyo, Nogyo-Daigaku*, **39**: 1-9.

Kubuabola, S.K., Morrison, R.J. and Singh, U. (2000) *Commun. Soil Sci. Plant Anal.*, **1**: 2273-2282.

Kumar, R., Singh, V.N. and Jain, B.P. (1998) *South Indian Hort.*, **46**: 157-160.

Kurakubo, Y. (1940) *Arten. Bot. and Zool.*, Tokyo, **8**: 1492.

Kurschner, E. and Koch, W. (1986) *Mededelingen van de Faculteit-Landbouwwetenschappen, Rijksuniversiteit Gent.*, **51**: 439-448.

Lebot, V. (1997) *Cahiers de la Recherche Development*, **43**: 20-34.

Lebot, V. and Aradhya, K.M. (1991) *Euphytica*, **56**: 55-56.

Leon, J. (1976) *Proc. 4th Int. Symp. Trop. Root Crops*, CIAT, Colombia, pp. 19-68.

Liu, L.J.Y., Rosa Marquez, E., Licha, M. and Biascoechea, M.L. (1988) *J Agril. Univ. Puerto Rico*, **72**: 413-426.

Malvesin-Fabre, G. (1945) *Contribution a la caryologie des. Aracees, These Science (Bordeaux)*, pp. 1-275.

Mandal, S., Mandal, N., Mitra, S., Mukharjee, A. and Tarafdar, T. (2015) *App. Biol. Res.*, **17**: 70-76.

Mandi, S., Tarafadar, J and Sinha, B. (2006) *Indian J. Plant Prot.*, **34**: 236-238

Mandi, S., Tarafdar, J., Banerjee, A. and Roy, S. (2009) *J. Root Crops*, **35**: 196-205.

Massal, E. and Barrau, J. (1955) *Pacific Sustenance Crops - Taro*, S.P.C.Q. Bull., April, **5**: 33-75.

Mathews, A., Llewellyn, D.J., Wu, Y.R. and Dennis, E.S. (1996) *Plant Mol. Biol.*, **30**: 1035-1039.

Matthews, P., Matsushita, Y., Sato, T. and Hirai, M. (1992) *Japanse J. Breed.*, **42**: 825-833.

Mauny, R. (1953) *Bull. Inst. Franc. Afrique Noir.*, **16**: 684-730.

McCartan, S.A., Van Staden, J. and Finnie, J.F. (1996) *J. Southern African Soc. Hortic. Sci.*, **6**: 1-3.

Medhi, R.P. and Pathasarathy, V.A. (1998) *J. Hill Res.*, **11**: 126-127.

Medhi, R.P., Parthasarathy, V.A. and Verma, S.P. (1994) *J. Root Crops*, **20**: 57-59.

Mitra, S. (2013) *Trop. Agric. (Trinidad)*, **90**: 22-29.

Mitra, S. and Tarafdar, J. (2013) *Postharvest Management of Tropical Tuber Crops*, AICRP on Tuber Crops, Directorate of Research, BCKV, Kalyani, West Bengal, Westville Publishing House, Paschim Vihar, New Delhi.

Mitra, S. and Tarafdar, J. (2018) *Tropical Tuber Crops*, (ed. James George and S. Sunitha). Westville Publishing House, Paschim Vihar, New Delhi.

Mitra, S., Sinha, B., Pal, H. and Tarafdar, J. (2007) *Acta Hort.*, **752**: 219-224.

Miyaji, R., Shirazawa, N. and Nakhonogi, K. (1976) *Bull. Kagoshima-ken Agric. Exp. Sta.*, **4**: 49-51.

Mohan Kumar, B., Kabeerathumma, S. and Nair, P.G. (1984) *Indian Fmg.*, **33**: 35-37.

Mohan Kumar, C.R. and Sadanandan, N. (1989) *J. Root Crops*, **15**: 103-108.

Mohan Kumar, C.R. and Sadanandan, N. (1990) *J. Root Crops*, **16**: 92-97.

Mohan Kumar, C.R., Mondal, R.C. and Hrishi, N. (1976) *J. Root Crops*, **2**: 57-69.

Mohan Kumar, C.R., Saraswathy, P. and Sadanandan, N. (1990) *J. Root Crops*, **16**: 140-141.

Mohandas, P.N. and Sethumadhavan, P. (1980) *Proc. Nat. Sem. Tuber Crops Prodn. Technol.*, Tamil Nadu Agricultural University, India.

Murakami, K. and Matsubara, S. (1998) *Scientific Rep. Fact. Agril.*, Okayama Univ., **87**: 123-126.

Murakami, K., Kimura, M. and Matsubara, S. (1995) *J. Japanese Soc. Hort. Sci.*, **63**: 773-778.

Murakami, K., Nishioka, J. and Matsubara, S. (1998) *Scientific Rep. Fac. Agril.*, Okayama Univ., **87**: 127-132.

Ndoumou, D.O., Tsala, G.N., Kanmegne, G. and Balange, A.P. (1995) *C. R. Academie des Sci., Serie III, Sciences de la Vie.*, **318**: 773-778.

Nguyen Thi, Q. and Nguyen Van, U. (1987) *HortSci.*, **22**: 671-672.

Nguyen, V.X., Yoshino, H. and Tahara, M. (1998) *Scientific Rep. Fac. Agril.*, Okayama Univ., **87**: 133-139.

Nguyen, V.X., Yoshino, H. and Tahara, M. (1999) *Breeding Sci.*, **49**: 179-185.

Ni, W.Z., Zhang, Y.S. and Lin, X.Y. (1997) *Acta Agriculturae Zhejiangensis*, **9**: 143-148.

Nwufo, M.I. (1988) *J. Root Crops*, **14**: 1-4.

Nwufo, M.I. and Fajola, A.O. (1981) *J. Root Crops*, **7**: 53-59.

Nyman, L.P. and Arditti, J. (1984) *Ann. Bot.*, **54**: 459-466.

Nyman, L.P. and Irawati Arditti, J. (1987) *Aroideana*, **10**: 26-28.

Nyochembeng, L.M. and Garton, S. (1998) *Plant Cell Tissue Organ Cult.*, **53**: 127-134.

Ochiai, T., Tahara, M. and Yoshino, H. (2000) *Scientific Rep. Fac. Agril.*, Okayama Univ., **89**: 15-21.

Odawara, C. (1959) *Kyushu Agric. Res.*, **21**: 129-133.

Okpul, T., Ivancic, A. and Simin, A. (1997) *Papua New Guinea J. Agril. For. Fish.*, **40**: 13-18.

Onwueme, I.C. (1978) *The Tropical Tuber Crops*, John Wiley and Sons, New York, pp. 199-226.

Palupe, a., Taylor M.B., Powaseu, I. and Thorpe, P. (1999) *Proc. Regional Tissue Cult. Workshop, Apia, Samoa, 5-13 June, 1997*, PRAP-Report Pacific Regional Agricultural Programme, **7**: 33-34.

Pandey, G. and Dobhal, V.K. (1997) *Indian J. Genet.*, **57**: 262-265.

Pandey, G., Dobhal, V.K. and Sapra, R.L. (1996) *J. Hill Res.*, **9**: 299-302.

Pardales, J.R. Jr. (1985) *Ann. Trop. Res.*, **7**: 27-38.

Parthasarathy, V.A., Medhi, R.P. and Rao, V.S. (1983) *South Indian Hort.*, **31**: 201-205.

Pillai, M.R.C. (1975) *J. Root Crops*, **1**: 98.

Pillai, S.V., Thankappan, M. and Misra, R.S. (1996) In: *Tropical tuber crops: problems, prospects and future strategies* (Eds. G.T. Kurup, M.S. Palaniswami, V.P. Potty, G. Padmaja, S. Kabeerathumma and S.V. Pillai), pp. 164-168.

Plucknett, D.L. (1976) *Evolution of Crop Plants* (Ed. N.W. Simmonds), Longman, London, and New York, pp. 10-12.

Plucknett, D.L. and de la Pena, R.S. (1971) *World Crops*, **23**: 244-249.

Plucknett, D.L., de la Pena, R.S. and Obrero, F. (1970) *Field Crops Abst.*, **23**: 413-426.

Plucknett, D.L., Saiki, D.F. and Motooka, P.S. (1967) *Proc. Asian-Pacific Weed Interchange*, Univ. of Hawaii, Coll. Trop. Agric., East-West Centre ITI, Honolulu, pp. 90-93.

Porterfield, M. Jr. (1944) *J.N.Y. Bot. Gdn.*, **45**: 11-16.

Portieles, J.M., Ruiz, L.a., Sanchez, E.E. and Milian, J.O. (1986) *Ciencia y Tecnica en la Agricultura, Viandas Tropicales*, **9**: 1-275.

Prabhakar, M. and Nair, G.M. (1984) *J. Root Crops*, **10**: 23-27.

Purea, M. and Mataora, T. (1996) In: *Mineral nutrient disorders of root crops in the Pacific: Proceedngs of a workshop* (Eds. E.T. Craswell, C.J. Asher and J.N. O'Sullivan), Nukualofa, Kingdom of Tonga, 17-20 April, 1995, ACIAR Proceedings No. 65, pp. 130-132.

Purewall, S.S. and Dargan, K.S. (1957) *Indian J. Agric. Sci.*, **27:** 151-162.

Purseglove, J.W. (1968) *Tropical Crops-Monocotyledons*, Longman Group Ltd., London, pp. 62-68.

Raciborski (1900) *Bull. N. Y. St. Mus.*, **19**: 189.

Ramnanan, N., Ahmad, N., Griffith, S.M. and Ahmad, N. (1995) *Fertilizer Res.*, **42**: 109-115.

Rao, N.S. (1947) *Curr. Sci.*, **16**: 229.

Ravi, V. and Chowdhury, S.R. (1993) *J. Root Crops*, **19**: 1-7.

Ravi, V. and Choudhury, S.R. (1990) *J. Root Crops, ISRC Nat. Sym. Special*, pp. 129-133.

Rekha, M.R., Padmaja, G. and Pillai, P.K.T. (1997) *J. Root Crops*, **23**: 39-44.

Rodriguez Manzano, A., Rodriguez Nodals, A., Fundora Mayor, Z. and Castineiras Alfonso, L. (1999) *Revista del Jardin Botanico Nacional.*, **20**: 91-154.

Rodriguez Manzano, A., Rodriguez Nodals, A.A., Roman, M.I. and Manzano, M.J. (1998) *Revista del Jardin Botanico Nacional.*, **19**: 113-120.

Rodriguez, A., Rodriguez, A. and Roman, M.I. (1994) In: *Origin evolution and diversity of Cuban plant genetic resources* (Eds. K. Hammer, M. Esquivel and H. Knupffer), **3**: 629-636.

Ruiz, L., Milian, O., Portieles, M. and Hernandez, P. (1989) *Agrotecnia de Cuba*, **21**: 67-73.

Sabapathy, S. and Nair, H. (1992) *Plant Cell Rep.*, **11**: 290-294.

Sarkar, D, Adhikary, N.K. and Tarafdar, J. (2017) *Int. J. Curr. Microbiol. App. Sci.*, **6**: 1399-1407.

Sato, D. and Silva, J.A. (1990a) *Research Extension Series Hawaii Institute of Tropical Agriculture and Human Resources, 1990*, No. 114, pp. 78-85.

Sato, D., Silva, J. and Kuniyoshi, J. (1990b) *Research Extension Series Hawaii Institute of Tropical Agriculture and Human Resources, 1990*, No. 114, pp. 71-77.

Scheffer, J.J.C., Douglas, J.A. and Triggs, C.M. (1999) *Agron. New Zealand*, **29**: 41-46.

Schnell, R.J., Goenaga, R. and Olano, C.T. (1999) *Sci. Hortic*, **80**: 267-276.

Sen, H. and Das, P.K. (1991) *J. Root Crops*, **17:** 154-155.

Sen, H. and Roychoudhury, N. (1988) *Environ. Eco.*, **6**: 431-433.

Shanmugavelu, K.G. and Vijayakumar, M. (1980) *Proc. Nat. Sem. Tuber Crops Prodn. Technol.*, Tamil Nadu Agric. Univ., Coimbatore, p. 189.

Sharma, A.K. (1956) *Caryologia*, **9**: 93-100.

Sharma, A.K. and Das, N.K. (1954) *Agron.*, Lusiana, **16**: 23-43.

Silva, J.A., Sato, D., Leung, P.S., Santos, P.S., Santos, G. and Kuniyoshi, J. (1990) *Research Extension Series Hawaii Institute of Tropical Agriculture and Human Resources*, 1990, No. 114, pp. 60-70.

Silva, J.F. Da., Couto, F.A.A. and Tigchelaar, E. (1971) *Experientia*, **12**: 134-154.

Singh, D. and Okpul, T. (2000) *SABRAO J. Breed. Genet.*, **32**: 39-45.

Sivan, P. (1979) *Proc. 5th Symp. Inter. Soc. Trop. Root Crops*, Philippines, pp. 637-646.

Sivan, P. (1983) *Proc. 6th Symp. Inter. Soc. Trop. Root Crops*, CIP, Lima, Peru, pp. 103-107.

Spier, R.F.G. (1951) *South-western J. Anthropol.*, **7**: 69-76.

Sreekumari, M.T. and Mathew, P.M. (1991a) *Cytologia*, **56**: 215-218.

Sreekumari, M.T. and Mathew, P.M. (1991b) *Cytologia*, **56**: 399-402.

Sreekumari, M.T. and Pillai, P.K.T. (1994) *J. Root Crops*, **20**: 20-25.

Tanabe, I., Kitayama, T. and Ikeda, K. (1980) *Soil Sci. Plant Nutr.*, **26**: 343-351.

Tanimoto, T. (1990) *Japanese J. Breed.*, **40**: 233-243.

Tekle Haimanot, A. and Doku, E.V. (1995) *Trop. Agril.*, **72**: 70-72.

Thankappan, M. (1985) *J. Root Crops*, **11**: 1-8.

Thankappan, M. and Malathi, V.G. (1984) *Indian Fmg.*, **33**: 47-49.

Torres, G.S. and Vazquez, B.E. (1981) *Cento Agricola*, **8**: 3-19.

Unnikrishnan, M. and Sreekumari, M.T. (1984) *Indian Fmg.*, **33**: 21-23.

Vasudevan, K., Jos, J.S. and Padmaja, G. (1989) *Mutation Breed. Newslett.*, **34**: 5.

Vazquez, B.E. and Torres, G.S. (1981) *Centro Agricola*, **8**: 43-69.

Velez, A.G. and Lopez, J.O. (1999) *J. Agril. Univ. Puerto Rico*, **83**: 41-46.

Venkateswarlu, T. (1980) *Proc. Nat. Sem. Tuber Crops Prodn. Technol.*, Faculty of Hort., Tamil Nadu Agril. Univ., Coimbatore.

Wall, G.C. and Wiecko, A.T. (1998) *J. South Pacific Agril.*, **5**: 9-12.

Whitney, L.D., Bowers, F.A.I. and Takahashi, M. (1939) *Bull. Hawaii Agric. Exp. Sta.*, **84**: 1-86.

Widdowson, J.P. (1966) *New Zealand J. agric. Res.*, **9**: 44-58.

Yam, T.W., Hsu, G.I. and Arditti, J. (1990) *Plant Cell Rep.*, **9**: 229-232.

Yam, T.W., Ichihashi, S. and Arditti, J. (1991a) *Ann. Bot.*, **67**: 317-323.

Yam, T.W., Young, J.L.P., Fan, K.P.L. and Arditti, J. (1991b) *Plant Cell Rep.*, **9**: 459-462.

Yamasaki, T. (1952) *Bull. Nat. Inst. Agric. Sci.*, Tokyo (Ser B).

Yang, B.G. and Kong, Q.D. (1994) *Adv. Hort.*, 147-150.

Zettler, F.W. and Hartman, R.D. (1987) *Plant Dis.*, **71:** 958-963.

Zettler, F.W. and Hartman, R.D. (1986). Food and Fertilizer Technology Center Book Series, **33**: 91-100.

Zhang, G.M. and Zhang, D.X. (1990) *Euphytica*, **47**: 25-27.

Zhou, S.P., He, Y.K. and Li, S.J. (1999) *Plant Cell Tissue Organ Cult.*, **57**: 173-178.

AMORPHOPHALLUS

H. Sen, N. Roychoudhury, S. Mitra and J. Tarafdar

1.0 Introduction

Amorphophallus, commonly known as Elephant foot yam [*Amorphophallus paeoniifolius* (Dennst.) Nicolson] is an important member of tropical tuber crops. Having high nutritive and medicinal values, it is gaining popularity as a starchy vegetable. Now, it is also considered as a food security crop as well as a cash crop due to its very high yield and considerable monetary return.

The crop is grown as vegetable in tropical and subtropical regions in many countries particularly in South East Asia. It is commercially grown in India, Sri Lanka, China, Malaysia, Thailand, Indonesia, and Philippines and in tropical regions of Africa. In India, elephant-foot-yam is cultivated on large scale in Andhra Pradesh, Bihar, Gujarat, Maharashtra, West Bengal, Uttar Pradesh and Jharkhand. Konjac (*A. konjac*) is also grown as a vegetable and for preparing processed food in Korea, Japan, China and in some countries in South America and South-East Asia.

The area and production of elephant foot yam in India is reported as 26,000 ha and 6.59 lakh metric tons, respectively (NHB, 2017). In the present scenario of climate change, elephant foot yam along with some other tropical tuber crops has gained more importance due to its ability to produce considerable yield during adverse climatic conditions (Mitra and Tarafdar, 2016).

2.0 Composition and Uses

2.1.0 Composition

Elephant foot yam is a cheap source of carbohydrate and rich in certain minerals like calcium, phosphorus, potassium and iron, and some vitamins, mainly B complex (Bradbury and Holloway, 1988; Chattopadhyay *et al.*, 2010) and has medicinal and therapeutic values (Chattopadhyay and Nath, 2007). Though poor in protein, it exhibits a high degree of variation in its protein content depending on cultivar.

Edible aroids have not realized their full potential because of palatability problems associated with antinutritional factors like oxalate, trypsin inhibitors, and acridity. The acridity is believed to be due to a combination of calcium oxalate crystals (Bradbury and Holloway, 1988; Saikia and Borah, 1994) and a chemical irritant, as yet undetermined but believed to be either a diglucoside of 3,4-dihydroxy benzaldehyde (Saha and Hussain, 1983), an unidentified proteinase (Bradbury and Hammer, 1990), or a hormone or sapotoxin (Walter and Khanna, 1972). Intake of oxalate-containing foods is known to produce chronic effects, including deposition of calcium oxalate crystals in the kidneys (Connor, 1977), occurrence of renal stones (Passmore and Eastwood, 1986), and reduction in bioavailability of calcium (Kelsay, 1985). Humans excrete crystals of calcium oxalate in the urine. Most urinary oxalate is thought to be derived from ascorbic acid and glycine and only a relatively small fraction from dietary oxalate (Passmore and Eastwood, 1986). Increases in urinary oxalate, calcium, uric acid, and cystine increase risk of stones; ingestion of large amounts of oxalate or calcium may exacerbate stone formation (Passmore and Eastwood, 1986). The corms of wild types contain crystals of calcium oxalate on account of which they are highly irritant. In cultivated types, calcium oxalate content is less and therefore, suitable for consumption as vegetable. The nutritive value of elephant foot yam corm is presented in Table 1.

Table 1: Composition of Elephant Foot Yam Corm (per 100 g of edible portion)*

Particulars	*Amount*	*Particulars*	*Amount*
Moisture (g)	78.7	Phosphorus (mg)	34
Protein (g)	1.2	Iron (mg)	0.6
Fat (g)	0.1	Vitamin A	434 IU
Carbohydrate (g)	18.4	Thiamine (mg)	0.06
Energy (kcal)	79	Riboflavin (mg)	0.07
Calcium (mg)	50	Vitamin C (mg)	0

*Choudhary (2014)

2.2.0 Uses

The corm is mainly used as vegetable and a good secondary food during the period of vegetable scarcity. Besides, being used as a vegetable, the tubers can also be used for making pickles, dry and fry cubes (Singh *et al.*, 2018). Its use as chips, flour, and thickening agents is also gaining popularity. Osmo-dehydrated slices prepared from fresh corm and bread from flour of elephant foot yam corms are good source of both carbohydrate and protein (Singh and Wadhwa, 2014).

These corms are consumed by many people as a food and also widely used in many Ayurvedic preparations recommended for the cure of piles and gastrointestinal disorders. Pseudo stem and petioles are also cooked as vegetable particularly of the non-irritant type having smooth corm.

Liu *et al.* (1998) reported that traditionally, konjac was grown in China for use by the grower as konjac curd for food, but in the year 1980, area under production increased to about 30, 000 hectares, and processing on a larger scale occurred. Flour is prepared from corms of *A. konjac* which improves lipid metabolism (Tsujihara and Tani, 1998). The corms are also used as animal feed.

Fresh corms are difficult to store due to their perishable nature, and deteriorate in quality during storage. Therefore, making processed products *viz.*, dry and fry cubes will be an alternative value added product with high nutritional value. It has been reported that elephant foot yam dry cubes prepared from corms soaked in 2 per cent salt solution followed by blanching of the cultivar BCA-1(Bidhan Kusum) had the highest ascorbic acid and total phenol throughout the storage period. Highest β-carotene was observed in dry cubes prepared from corms of the cultivar IGAM-1 soaked in 2 per cent alum solution. The cultivar BCA-1 had the highest starch in dry cubes (1.5 per cent alum) and fry cubes (2 per cent salt) at 0 and 13 months after storage, respectively. The protein and organoleptic values were the highest throughout the storage in cv. BCA-1, both in dry and fry cubes prepared after soaking in 1.5 per cent alum and blanching (Singh *et al.*, 2020).

Uses of Elephant Foot Yam (EFY)

Sweet Dish (Payasam) from EFY

Mashed EFY (left), and Fritters from EFY (right)

Curry from EFY (left), and Pickles from EFY (right)

3.0 Origin, Taxonomy and Morphology

The genus *Amorphophallus* belongs to the family Araceae, and includes more than 170 species, which are mainly distributed in the tropics or subtropics of West Africa and South Asia (Wittenbach, 1996). It is indigenous to tropical parts of Old World, Australia and the Pacific Islands and about 14 species occur in India. Amongst them *Amorphophallus paeoniifolius* and *A. konjac* (*A. rivieri*) are the two important species cultivated for their edible corms. *A. oncophyllus* and *A. variabilis* are also edible species (Bailey, 1950). Six new species and two new subspecies: *A. barthlottii* (from Cote d'Ivoire and Liberia), *A. canaliculatus* (Gabon), *A. netterscheidii* (Gabon, Zaire), *A. richardsiae* (Zambia), *A. abyssinicus* subsp. *akeassii* (Cote d' Ivoire, Ghana and Nigeria) and *A. calabaricus* subsp. *mayoi* (Zaire, Uganda and Kenya) were recorded in Africa by Ittenbach and Lobin (1997). Four species, *A. henryi, A. hirtus, A.kiusianus*and, *and A. paeoniifolius* were recognized in Taiwan by Hetterscheid and Peng (1995). The first two species are native to Taiwan. They are cultivated for their underground modified stem.

New species, *A. zengianus* was described from Yunnan, China (Long and Li, 2000). It is similar to *A. krausei* in having a erect, boat-shaped spathe with a sterile portion between male and female parts, but differs in haivng a spathe without spots, an appendix shorter than the spathe, and a pleasant odour.

In China, people of 34 nationalities have indigenous knowledge of konjac. Its history is as long as 2000-3000 years. Some indigenous perceptions, traditional cultivation systems and propagation methods of konjac are reported. Local people in SW China uses nine types of konjac for food, medicine, cultural symbols, animal fodder, gum or wine-making.

Amorphophallus is either a perennial or an annual herb, generally bearing one broad, long petiolate leaf. Leaves simple; petioles smooth or warty and variously spotted; blade simple, 3 parted, the divisions pinnatifid. Spadix long; spathe funnel or bell-shaped at the base, springing from the great bulb like corm in advance of the leaves, the latter usually pedately compound, differs from arum and related genera by bota-nical characters. Flowers are unisexual with disagreeable odour which is supposed to attract carrion loving insects, which pollinate the flowers and bring about the production of seeds (Bailey, 1950). Hybridization as well as open pollination were equally good in inducing genetic variability in elephant foot yam.

Elephant Foot Yam Plant (Left), Flower (Middle) and Corm (Right)

In general, hybrid clones developed by planting corms of seedling were found to produce more yield (Thankama Pillai, 1996).

4.0 Cultivars

Though many cultivars of this crop have been developed under All India Coordinated Research Project on Tuber Crops, only a few cultivars are commercially grown in India. Among these, Gajendra, also known as Kovur in some places due to its release from Kovvur region of Andhra Pradesh, Bidhan Kusum, Sree Padma are popular in many parts of India and Bidhan Kusum is grown in the eastern region (Table 2). The chief characters of these cultivars are given below:

Gajendra

It is a high yielding cultivar with vigorous growth habit. Corms are smooth, non-acrid, uniform in shape and have flesh of good quality. It does not produce daughter corms. This cultivar is most popular, and widely cultivated in most of the elephant foot yam growing regions in India. Average corm yield is 50-60 tonnes per hectare.

Bidhan Kusum

This variety has been developed from Bidhan Chandra Krishi Viswavidyalaya, West Bengal, India under AICRP on Tuber Crops through selection. It is characterised by its vigorous growth habit and high yield without daughter corms. Corms are smooth, non-acrid with good quality creamy flesh. This variety is commercially grown mainly in West Bengal and eastern parts of India, and the average yield ranges from 50-55 tonnes per hectare.

Sree Padma

This variety has been developed from the Central Tuber Crops Research Institute (CTCRI), Thiruvananthapuram, Kerala, India through selection from indigenous germplasm of elephant foot yam collected from Wayanad district of Kerala. It performs well in Kerala and some southern parts of India. It produces non-acrid corm, matures in 8-9 months, and the yield varies from 40-45 tonnes per hectare. It is field tolerant to collar rot and fairly tolerant to mosaic disease.

Table 2: Cultivars of Elephant Foot Yam

Year of Release	*Originating Centre*	*Name of Cultivar*	*Area Suitable for Cultivation*
1991	APHU	Gajendra	Andhra Pradesh, Tamil Nadu, Kerala, West Bengal, 70 to 80 per cent area in India
1998	CTCRI TVM	Sree Padma	Kerala
2002	BCKV	BCA-1 (Bidhan Kusum)	West Bengal, Andhra Pradesh and Bihar
2014	NDUAT	NDA-9 (Narendra Asha)	Uttar Pradesh, Uttarakhand and Himachal Pradesh
2015	TNAU	Appakkudal	Tamil Nadu
2015	BCKV	BCA-3	West Bengal

Cultivars of Elephant Foot Yam

Bidhan Kusum

Gajendra

Sree Padma

BCA-3

This cultivar has been developed through selection from Bidhan Chandra Krishi Viswavidyalaya, West Bengal under AICRP on Tuber Crops in the year 2014. It produces smooth, non-acrid corm with whitish creamy flesh, and free from daughter corms. The cooking quality is very good. It performs well in West Bengal and eastern parts of India.

Sree Athira

It is the first genetically improved elephant foot yam variety (selection from the cross Am-15 x Am-45) developed from the Central Tuber Crops Research Institute (CTCRI), Thiruvananthapuram, Kerala, India. It matures in 9-10 months and the average yield is 40 t/ha. The corms are excellent in cooking quality.

In Japan, Uchida *et al.* (1998) reported that new konjak [*A. kowjac*] cv. Myougiyutaka, registered in 1997, is higher yielding than cv. Shina-shu and has similar levels of resistance to leaf blight [*Pseudomonas pseudoalcaligenes* subsp. *konjaci*] and root rots. Myougiyutaka greatly exceeds cvs. Shina-shu and Harunakuro for flour yield and flour viscosity.

5.0 Soil and Climate

5.1.0 Soil

Selection of site plays a major role in the success of amorphophallus cultivation. Since corms are highly susceptible to poor drainage conditions, these need well drained and well aerated soil. The growth of corms in heavy soil is very much restricted which is reflected on the production. Deep sandy loam soil rich in nutrient is considered suitable for high corm production (Katyal, 1977). Soils with higher amount of clay or silt should be amended by incorporating organic matter or compost. The crop prefers a rich loamy soil with a pH range of 5.5-7.0.

5.2.0 Climate

Amorphophallus is a tropical and subtropical crop. It requires a well distributed rainfall, humid and warm weather during vegetative growth. The difference between the maximum and minimum temperatures should not be wide during its growth period. It requires fairly high temperature with ample moisture during shoot emergence, while comparatively moderate temperature with humid weather promotes the development of corm. Water logging is detrimental during any phases of growth.

Rise in temperature from 13.20°C for 15 days increased yield from 5.9 to 22.2 per cent (Wang *et al.*, 1997). In an experiment on shading, Seo *et al.* (1988) recorded that total corm yields were 5.83, 6.89, 6.98 and 7.91 t/ha for full sunlight, 35 per cent shade, 55 per cent shade and 75 per cent shade, respectively.

6.0 Propagation and Planting Material

There are some cultivars of amorphophallus, particularly smooth corm types, which do not produce daughter corms for further multiplication, hence the mother corm is used as planting material. In such cases, corms of small size or bigger corms cut into pieces are used for planting. Corms of 2-3 kg each should be divided vertically in such a way that each piece contains a portion of central bud from where the future shoot will initiate after planting. In a trial on seed corm weight at planting, Sen *et al.* (1984) obtained maximum corm yield with 1 kg seed corm but considering the per cent increase in corm yield, 500 g seed corm was recommended (Table 3).

Table 3: Total Corm Yield (q/ha) of Amorphophallus as Affected by Seed Corm Weight*

Seed Corm Weight (g)	*1980*	*1981*	*1982*	*Mean*	*Increase in Yield (Per cent)*
250	612.0	665.9	323.5	533.8	-
500	817.3	845.6	524.6	729.1	36
750	987.6	865.9	605.7	819.7	53
1000	1126.3	1053.6	662.7	947.7	77
C.D. at 5 per cent	115.6	149.4	44.7	-	-

*Sen *et al.* (1984)

Smaller corms (125 or 250 g) though showed highest tuber bulking efficiency, were less profitable owing to lower total yields. Based on the cost benefit analysis, corm pieces of 500 g to 1 kg proved economical as planting material (Misra *et al.*, 1983). Corms weighing 500 g were found to be optimum planting material resulting in higher net income and gave the highest cost-benefit ratio of 1.62 (Alam and Narzary, 1997). The different morphological and yield attributes were influenced by the seed corm size and higher values were obtained when whole seed corms were used as planting materials over cut seed corms of the sizes. However, the corm production was not proportionate to the seed corm size and the percentage increase was higher in smaller seed corms as compared to larger ones (Sen and Das, 1990). They found that corm yield, number of shoots/corm, shoot length, shoot girth and canopy spread were highest when corm size was 1000 g and corms were planted whole rather than cut. Inaba (1992) observed that amorphophallus grown from seed corms with high nitrogen content had larger leaf area and root, new corm and total plant dry weight at all growth stages and had higher N uptake than when grown from seed corms with low nitrogen content.

The role of seed corm on the yield was also investigated at CTCRI, India and it was observed that seed piece weighing approximately 1000 g produced an economic yield (Anon., 1983a).

6.1.0 Seed Corm Production

Inadequate supply of planting material is a major constraint in extending the cultivation of elephant foot yam. South Indian smooth type which is of good quality and commercially acceptable, does not produce daughter corms necessary for rapid multiplication (although plenty of daughter corms arise in local irritant type). The improved type is usually propagated by cutting large corms into small pieces for

Planting Materials and Preparation of Planting Materials

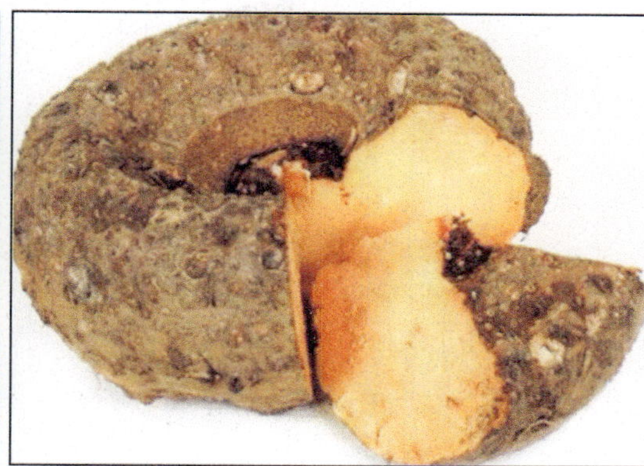

Whole Seed Corm

Cutting of Seed Corms into Small Pieces and Soaking in Cow Dung Slurry for Planting

Production of Whole Seed Corms through Mini Corm Setts

Mini Sett Technique for Whole Seed Corm Production of Elephant Foot Yam (Mitra and Tarafdar, 2013)

planting. The cut corms may fail to sprout due to dehydration at high temperature prevailing during the planting season and rotting caused by infection of fungus and bacteria. These hazards can be avoided by planting whole seed corms which not only sprout early and show better growth rate but also produce higher corm yield as compared to cut seed corms of same weight. The whole seed corm can be produced by planting mini corm setts obtained from apical portion of mother corm which sprouts readily. Sen *et al.* (1987) described the method of whole seed corm production for large scale planting. They suggested that for obtaining desirable size of seed corm, 100 g setts should be planted at a distance of 30 x 30 cm. Corm yield was highest with 250 g setts grown at a spacing of 50 x 50 cm (Das *et al.*, 1995; Nath *et al.*, 2007). Verma *et al.* (1994) observed that whole seed corm production increased to 36.3, 46.1 and 50.4 t/ha by planting 150, 300 and 450 g setts, respectively and applying 150 kg N/ha. Das *et al.* (1997) revealed that maximum corm yield (85.8 t/ha) was obtained when the largest setts (350 g) were planted at the highest density. The basal portion of the seed corm is not commonly used as planting material due to low percentage of sprouting. Sprouting percentage was more with top portion of corm than the cut corms from the basal portion. However, soaking of the basal cut corms with different chemicals significantly improved sprouting, subsequent growth and whole seed corm yield. Thiourea (200 mg/l), KNO_3 (1000 mg/l) and CCC (0.10 ml/l) showed an increase in sprouting by 36.78, 30.82 and 26.72 per cent, respectively compared to untreated bottom setts and this was at par with the cowdung slurry treatment, whereas top cut portion of corms showed 46.6 per cent increased sprouting (Dhua *et al.*, 1988). Corm yield was highest with 200 ppm GA_3 application (Das *et al.*, 1997). The fully matured seed corms are lifted from the ground when the leaves turn yellow or become dry, mostly at 8-9 months after planting. Whole seed corm thus produced from about 0.15 ha in nursery bed is sufficient for planting in 1 ha of land. The corm sett requirement per hectare is about 100 q which ultimately increases to 500 to 600 q/ha at the end of the season. The corm setts in the nursery bed should be fertilized with 10 t FYM and 80: 60: 80 kg N, P and K/ha of which N and K should be applied in two equal splits. Differential responses on light interception, yield attributes and corm production of elephant foot yam under varying plant densities and sett sizes were investigated. The result revealed that higher the plant density, greater the light interception and greater photosynthate accumulation resulted in higher seed corm production (Das *et al.*, 1995).

6.2.0 Micropropagation

The edible variety of *A. paeoniifolius* was brought into *in vitro* culture using modified MS medium. Lateral buds excised from cormels treated with antibiotic solution overnight were found to develop sterile cultures. Callus development occurred on the initial medium that contained NAA as additional ingredient. Plantlets developed on transfer of the callus to fresh medium of the same composition (Irawati *et al.*, 1986; Nyman *et al.*, 1987). Shoot tips of *A. konjac* produced protocorm-like calluses (PLC) on medium containing NAA, and the continuous division of PLC yielded plantlets in large quantities. DMV (Dasheen mosaic potyvirus) and KMV (Konjac mosaic virus) were eliminated at a high rate when shoot tips were cut off at 0.3-0.5 mm (Shishida *et al.*, 1991).

Takayama and Akita (1991) described a method of obtaining planting material on a large scale by tissue culture of corms of konjac (*A. konjac*). It comprises culturing explants in a liquid medium till the plantlet reaches an average size of 15-30 cm, sinking >90 per cent of the plantlet in the liquid medium, and then reducing the amount of the liquid medium with time so that 50-98 per cent of the plantlet is exposed to the air. A large amount of corm is thus obtained in a short time.

Explants from tubers, shoot tips, leaf stalks and floral organs of *A. konjac* were cultured on MS media supplemented with different concentrations of BA and NAA at 26 to 28°C. Green plantlets were regenerated following culture on de-differentiation medium. Cultured female flowers produced seeds through the expansion of ovaries. After a dormant period, the seeds were germinated on culture media and grew into green plantlets. Plant regeneration was directly achieved via multishoot formation. On medium containing 2 mg BA and 0.5 mg NAA/l tuber explants expanded and formed brown coloured spherical masses. Shoots developed on the surface of these masses but only a few grew into complete plantlets; most could not elongate on the original medium. When the explants were transferred to medium containing 0.5-1 mg BA and 0.2 mg NAA/l, the shoots began to elongate, formed roots and grew into complete plantlets after 5-6 weeks in culture (Anon., 1994).

Calluses derived from explants of the tubers of *A. rivieri* and *A. albus* were cultured on differentiation media containing different cytokinins. Shoot formation appeared to be closely related to the endogenous level of iPA (isopentenyladenosine). The type and concentration of cytokinins in the culture media affected cell differentiation, organogenesis and plantlet regeneration from amorphophallus callus (Huang and Li, 1993). In addition to conventional adventitious shoot differentiation from cultured tissue of *A. rivieri*, Huang and Li (1994) reported 4 other forms of regeneration: (i) root formation before shoot formation, (ii) small tuber formation, (iii) embryoid formation, and (iv) plantlet development from a dormant bud on the explant.

Huang *et al.* (1995) reported that plantlets from subculture decreased the frequency of normal diploid cells (77.27-52.97 per cent) but increased the frequency of abnormal aneuploid cells such as sub-diploid (14.7-15.99 per cent), ultra-diploid (20.20 per cent) and sub-haploid (10.00-17.65 per cent). Only one tetraploid cell was observed.

7.0 Cultivation

7.1.0 Land Preparation and Planting

The land is prepared by ploughing twice or thrice to a depth of 20-22 cm followed by making holes (50 x 50 x 30 cm) according to the seed corm weight. The pits are filled up with sufficient amount of rotted farmyard manure (FYM), thoroughly mixed with soil and allowed to settle to avoid depression after planting. The crop is also grown on flat beds or broad ridges. In that case, FYM at the rate of 30 tonnes per hectare is applied. However, the crop grown on ridges gives better

Sprouting and Irrigation of EFY

Sprouting of Seed Corm

Drip Irrigation in Elephant Foot Yam

corm yield than in flat beds. Broad ridge method is generally practised in high rainfall areas for proper drainage.

The time of planting which primarily determines the growth and yield of *Amorphophallus* depends on climate and soil and generally February-April is recommended for its planting in most parts of India (Katyal, 1977). It is also planted in March-April in certain parts of the country provided facility for irrigation during summer months is available. In South India, early planting (15 February) resulted in the highest corm yield while planting after mid-April significantly reduced the growth and yield (Philip and Nair, 1994). In the Gangetic alluvium, Verma *et al.* (1992) suggested planting in March for high yield. Time of planting plays an important role for sprouting of elephant foot yam. Corms planted from March to April, sprouted faster and were followed by better plant vigour possibly due to higher temperatures immediately after planting and other climatic factors including higher rainfall and relative humidity. As corm bulking was most rapid between June and August, the establishment of full plant vigour with high biomass before June was necessary to allow a longer period for maximum bulking and higher corm production. It is recommended that planting should be restricted between mid-March and mid-April to obtain higher corm production (Das and Sen, 1995) (Table 4). In China, Lee (1992) recorded higher tuber yields of *A. konjac* with early planting on 5 April than with the normal planting date 20 April. Lee (1992) also reported higher yield of konjac corms with polythene film mulching than in unmulched crops or crops protected by polythene tunnels.

Table 4: Corm Bulking Rate (kg/ha/day) of *Amorphophallus**

Dates of Planting	*90-120 DAP*	*120-150 DAP*	*150-180 DAP*	*180-240 DAP*
15th February	265	640	402	225
2nd March	318	622	365	206
17th March	387	578	342	195
1st April	405	565	320	148
16th April	425	560	286	98
1st May	465	488	275	-
16th May	498	435	247	-
1st June	485	412	215	-
15th June	415	362	192	-
1st July	360	325	-	-

*Das and Sen (1995)

7.1.1 Pre-planting Treatment

Pre-planting treatment by removing the central apical bud, drying the corms and then exposing to smoke for 6 h/day for 4 weeks showed 100 per cent sprouting at 16 days after planting. Heat treatment (35, 40 and 45°C) induced early sprouting, particularly in top sections (Rabi and Mohankumar, 1999). Sprouting of *A. paeoniifolius* corms was hastened by storage in diffused light and by pre-soaking in aqueous solutions of gibberellic acid, ethrel [ethephon] or thiourea. Thiourea was most effective in breaking dormancy with 92 per cent sprouting in 75 days compared to 18 per cent sprouting in the control. Darkness had an adverse effect on sprouting (Kumar *et al.*, 1998). In some areas, the corms are dipped in cowdung slurry and dried in shade before planting is done vertically in the prepared pit.

There is definite advantage of mulching the pits immediately after planting for better sprouting, conservation of moisture and reduction of the temperature around the corm. A significant increase in yield and the size of individual corm due to mulching was recorded by Mohan Kumar *et al.* (1973).

7.2.0 Spacing

Planting distance depends on the size of planting material and it varies from 50 cm to 100 cm each way. Spacing has an important role on the production of corm when the crop is to be harvested in one season. In a trial on planting distance, Sen *et al.* (1984) obtained highest yield by planting seed corm at a distance of 50 x 50 cm (Table 5). Most growth parameters (leaf area index, crop growth rate, net assimilation rate, dry matter accumulation in different plant parts and total dry matter production) were increased by higher plant density (Das *et al.*, 1997). Liu *et al.* (1998) reported the highest yield of *A. konjac* from planting 0.5 kg corms with a crop density of 90 000 plants/ha.

Table 5: Interaction Effect of Seed Corm Weight and Spacing on Corm Yield of *Amorphophallus* (kg/plant)*

Seed Corm Weight (g)	*Spacing*			
	50 x 50 cm	*75 x 75 cm*	*100 x 100 cm*	*Mean*
250	3.41	3.83	4.25	3.83
500	3.47	4.14	5.56	4.39
750	3.63	4.46	5.07	4.38
1000	4.00	5.37	6.48	5.28
Mean	3.62	4.45	5.34	
	Seed Corm Weight	*Spacing*	*Seed Corm Weight x Spacing*	
S.Em ±	0.16	0.18	0.32	
C.D. at 5 per cent	0.50	0.58		NS

*Sen *et al.* (1984)

7.3.0 Manuring and Fertilization

In general, the plant requires a high dose of nitrogen and potassium. Mohan Kumar *et al.* (1984) reported removal of 121.9 kg N, 30.5 kg P and 176.4 kg K/ha from the soil when the yield was 36 t/ha. The absorption of nutrients by amorphophallus was high at the 3rd month for N and 6th month for P, while K uptake was steady through-out the crop growth. Mandal and Saraswat (1968) found that the economic fertilizer rates were 40 kg N, 40 kg P and 80 kg K in addition to 25 tonnes of FYM per hectare in laterite soil of Kerala. Hrishi and Nair (1972) also reported that 40 kg each of N and P and 80 kg K in combination with 12 tonnes of FYM produced high yield of good quality corms. From the results of the experiments at CTCRI, Mohan Kumar *et al.* (1988) recommended 25 tonnes of FYM and 80, 60 and 100 kg/ha of N, P and K, respectively. Amorphophallus prefers high dose of organic manure as basal appli-cation in the presence of NPK fertilization, a linear response of FYM up to 25 t/ha has been observed in acid laterite soils of Kerala under rainfed condition (Ghosh *et al.*, 1985). In a similar soil condition, Nair *et al.* (1990) recommended 107 kg N, 47 kg P and 147 kg K from the experiment on crop response to fertilizers. Corm yield increased by 15.5 per cent with 30 tonnes FYM/ha and by 6.5 per cent with 150 kg N/ha (Patel and Mehra, 1984). Application of 112.5 t FYM, 450 kg urea, 1125 kg superphosphate, 2250 kg rape seed cake and 450 kg KCl/ha produced the highest yield of corms in *A. konjac*.

Mukhopadhyay and Sen (1986) suggested that nitrogen up to 150 kg/ha influenced the growth and yield of amorphophallus and potassium application beyond 50 kg/ha did not show any significant effect on corm yield; highest corm yield of 110.6 t/ha was recorded by the application of NPK at 150, 60, 50 kg/ha in a crop duration of six months (Table 6). Under partially shaded condition (coconut plantation), appli-cation of FYM and NPK @ 12.5 t/ha, 27:20:33 kg/ha respectively was found to be adequate for producing optimum corm yield (Ravindran and Kabeerathumma, 1990). Sen *et al.* (1995) observed that corm yield was highest with 150 kg N and 75 kg K/ha.

Table 6: Effect of Nitrogen, Potassium and their Interactions on Shoot Length, Mean Corm Weight, Corm Bulking Rate and Corm Yield of *Amorphophallus**

Treatments	*Shoot Length (cm)*	*Mean Corm Weight (kg)*	*Corm Bulking Rate kg/ha/day*			*Corm Yield (t/ha)*
			0-120	*120-150*	*150-180*	
N50	116.80	2.11	442	708	349	84.63
N100	120.82	2.28	493	749	313	91.13
N150	129.81	2.56	564	816	386	102.27
S.Em ±	0.68	-	-	-	-	1.91
C.D. (0.05)	2.02	-	-	-	-	5.67
K50	121.40	2.41	519	796	347	96.73
K100	123.56	2.38	505	739	380	92.68
K150	123.24	2.22	476	738	332	88.63
S.Em ±	0.64	-	-	-	-	1.91
C.D. (0.05)	NS	-	-	-	-	5.67

NS: Not significant.

*Mukhopadhyay and Sen (1986)

In *Amorphophallus*, fertilizer should be applied in split doses. Since it is normally a long duration crop and grown in the rainy season, fertilizer application in a single dose may result in excessive loss and low nutrient status may adversely affect the growth and development at the later stage. The first application at planting with full dose of P and half of N and K in the pits along with FYM helps in rapid development at the early stage. The second application of remaining dose of N and K should be done 60-70 days after planting. About 8.33 per cent and 11.11 per cent more seed corm yields of elephant foot yam were obtained when 110 kg each of N and K were applied in two and three equal splits, respectively over full basal application (Sen *et al.*, 1991). It should be followed by earthing up for better bulking of the corm and good drainage.

Boron and zinc @ 10 and 20 kg/ha significantly increased the corm yield of amor-phophallus, cultivated in tarai soil (Anon., 2002).

The application of biofertilizer like *Azotobacter* on the nitrogen use efficiency as well as growth and yield of amorphophallus proved effective. The use of azotobacter @ 2 per cent solution as corm dipping and soil application @ 9 kg/ha at 35 DAP increased the corm yield (Mukhopadhyay and Sen, 1999).

Marked variation in plant growth characters of elephant foot yam under different organic treatments was observed. Plants under organic treatment combinations (Green manure cowpea (12-13 t $ha^{-1)}$ + corm treatment with slurry containing cowdung, neem cake (1-2 kg per bucket of slurry) and *Trichoderma harzianum* (5 g per kg seed) + *Trichoderma harzianum* (2.5 kg per tonne of FYM- neem cake mixture) incorporated FYM @ 36 t ha^{-1} (3 kg per pit) and neem cake 1 t ha^{-1} (80-85 g per pit + green manure cowpea (12-13 t $ha^{-1)}$ and ash @ 3 t ha^{-1} (250 g per pit) at 2 MAP) showed superior vigor. Among three treatments under validation programme of organic farming system in elephant foot yam, the highest yield of 50.43 tha^{-1} was recorded with organic treatment combinations followed by 47.66 tha^{-1} with the conventional treatment (FYM @ 25 t ha^{-1}+ NPK @ 100:50:150 kg ha^{-1}) The B:C ratio was also recorded highest (2.44) with the organic treatment combinations (Anon, 2020).

7.4.0 Irrigation and Inter-culture

The crop is to be irrigated frequently to initiate sprouting of corms and subsequent irrigation are needed (20 days intervals) in the dry months before the onset of monsoon. Mulching reduces the number of irrigation and at least 8 to 10 irrigations normally required throughout the growth period. When the crop approaches maturity, irrigation should be light and at long intervals in the absence of optimum moisture in the soil. Care should be taken to prevent stagnation of water during the rainy season. Philip and Nair (1994) observed that irrigation in dry periods during crop growth was uneconomic though it had significant effects on growth.

Mulching with black polyethylene film, barley straw, leaf litter and hay gave total corm yields of 6.84, 9.13, 8.23 and 7.90 t/ha, respectively, compared with 5.65 t/ha in the unmulched control (Seo *et al.*, 1988).

Generally, the crop requires 2-3 weedings before the onset of monsoon. After the spread of the leaves, weeds do not grow in shade. Hand weeding can be effectively replaced by a single preplanting application of fluchloralin @ 2.0 l/ha. Planting followed by complete earthing up showed a better response in comparison to flat planting and hand weeding and the first method reduced the crop weed competition which ultimately enhanced corm production (Bhowmik *et al.*, 1988). From an experiment to find out the most effective method of weed management in elephant foot yam, Suresh Kumar *et al.* (2020) reported that *Cyperus rotundus, Cynodon dactylon* and *Commelina benghalensis* were the predominant weeds at most of the locations. Among different treatments, hand weeding thrice at 30, 60, 90 days after planting (DAP) recorded taller plants (80.85 cm) with more pseudo stem girth, canopy spread (97.07 cm), leaf area (5435.37 cm^2), corm yield (38.0 t/ha), and net returns (Rs.387253/ha), which was at par with weed control ground cover mat mulching and application of glyphosate thrice at 30, 60 and 90 DAP. Lower weed density and biomass were recorded in treatment with weed control ground cover mat mulching, which was at par with glyphosate applied at 30, 60 and 90 DAP.

7.5.0 Cropping System

In the eastern and southern parts of India where amorphophallus is cultivated in large scale, cropping systems have been standardized depending on availability of irrigation, soil and commercial importance of crops. In South India, *Amorphophallus* was grown as an intercrop, under partial shade of coconut plantation, and with cowpea without significant reduction in corm yield (Ravindran and Kabeerathumma, 1990). A tuber based cropping system with amorphophallus + cowpea (mulch and green manure)/cowpea for pod production gave the highest net returns followed by cowpea (pod)/cassava (Mohankumar and Nair, 1990). In eastern India, Sen *et al.* (1993) identified cowpea, French bean, okra and basella as suitable intercrops with row ratio of 1:1 between intercrop and main crop (amorphophallus). The intercrops significantly reduced the main crop yield to a variable extent (upto 22.4 per cent) (Table 6).

Table 7: Economics of Amorphophallus based Intercropping System*

Intercrops (Treatments)	*Yield (t/ha)*		*Cost of Cultivation (Rs./ha)*		*Gross Returns (Rs./ha)*	*Net Returns (Rs./ha)*	*Benefit Cost Ratio*
	Inter Crops	*Main Crop*	*Inter Crops*	*Main Crop*			
Cowpea (CO 2)	5.312	48.65	9,600	34,500	1,60,000	1,15,000	3.62
Bhindi (Parbhani Kranti)	10.560	45.44	13,600	34,500	1,57,000	1,10,900	3.26
Cluster bean (Pusa Navbahar)	6.281	41.54	8,000	34,500	1,38,000	95,500	3.24
Onion (CO 4)	9.660	43.63	11,500	34,500	1,58,500	1,12,500	3.50
Amaranthus	6.602	38.55	7,000	34,500	1,28,700	87,200	3.10
Control	-	47.36	-	34,500	1,35,000	1,00,500	3.61

*Anon. (2000-01)

8.0 Harvesting, Yield and Storage

The crop is harvested after 6-7 months from planting. Corm yield increased gradually with delayed harvesting but the maximum net return was obtained

Field Growing and Harvesting of Corm as Vegetable

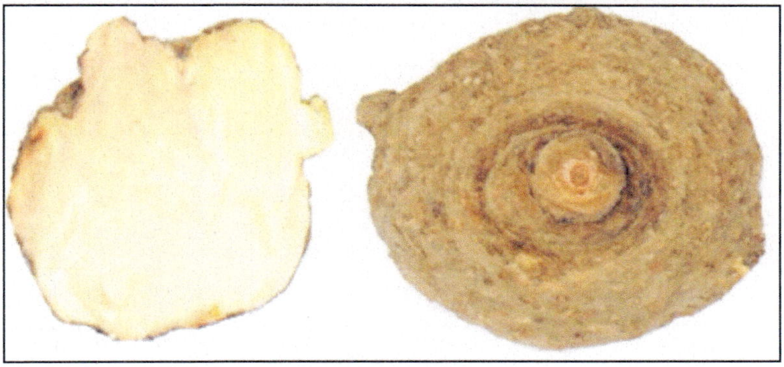

Non-acrid Corm

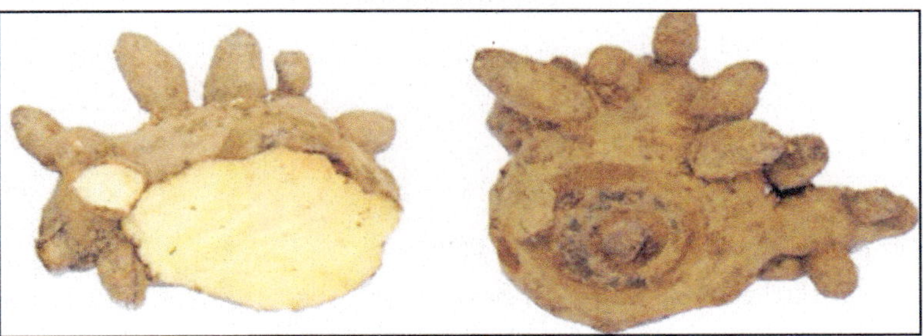

Acrid Corm

Harvesting and Storage of Seed Corm

Storage and Bulk Shipment of Seed Corm

Transportation, Packaging and Selling of EFY

when harvested 180 days after planting (Sen *et al.*, 1996). Yellowing and drooping down of the leaves are the signs of maturity of the crop. The crop can be harvested earlier before full maturity for early market due to higher price. In such case, the lower yield of corms can be compensated by higher market price. The corms can be retained in the soil even after full maturity. Though the crop can be harvested after 6 months from planting for the use as vegetable, for the purpose of planting material, matured corms are harvested at 8-9 months after planting (Mitra, 2009).

The corm yield ranges from 50 to 80 t/ha depending on the variety, weight of seed corm, soil type and nutrients applied.

8.1.0 Storage

It is very difficult to store immature corms harvested for vegetable purpose. Matured corms after proper curing can be stored for a few months in well-ventilated rooms (Mitra and Tarafdar, 2018). Locally available storage materials like dry ash, sand, rice bran, dry soil and saw dust were effective for lengthening self-life of seed corms in storage (Sen *et al.*, 1996). Storage of seed corms under dry soil or sand significantly checked the moisture loss (Das, 1992). It is better to harvest the crop at right stage, as the rate of loss of moisture from the corms for the first 4 days is about 3-4 per cent per day, the total extending to even 25 per cent or more in the first month of storage. Lower percentage of rotting with covered media as well as in corms treated with maleichydrazide was reported by Das *et al.* (1995).

8.1.1 *In vitro* Storage

A clonal propagation system for *A. konjac* [*A. rivieri*] was also established from tubers cultured *in vitro* (Zhang *et al.*, 1999). Cuttings of vigorous adventitious buds 0.5 and 1.5 cm in length were taken and cultured on an MS-NH_4 medium supplemented with BA, NAA and sucrose and with ABA or mannitol. Adventitious buds 1.5 cm long cultured on MS-NH_4 medium enriched with mannitol at 20 g/l, BA at 0.5 mg/l, NAA at 0.05-0.1 mg/l and 3 per cent sucrose gave the best results. After storage at 4°C for 180 days, 100 per cent of the buds survived. The adventitious buds stored at low temperature soon resumed growth when transferred to ambient temperatures.

9.0 Diseases and Pests

9.1.0 Diseases

Fungal Diseases

Amorphophallus is often found to suffer from collar rot disease caused by *Sclerotium rolfsii*. The symptoms appear when the plant is about 2-3 months old. The collar region (the part just above the ground) of the plant is attacked by the pathogen and water soaked lesions appear on the stem. The whole plant soon turns yellow. The stem shrinks and collapses due to rotting of the collar region. A thick white spreading mat of mycelia with globular sclerotia, the fruiting bodies of the pathogen, can be seen on the affected tissue. Heavy rains and high relative humidity, heavy soils, high organic matter and poor drainage are some of the pre-disposing factors.

The disease incites during the onset of monsoon in June- August in Easter part of India and severe water logging favours the stem rot infection. However, collar rot is the major problem in low land areas. The disease predisposes in high temperature and humidity with high rain fall. Also, close spacing aggravates the collar rot infection due to quick spread of the pathogenic propagules. The pathogen overwinters in the crop debris and follows up crop after elephant foot yam cultivation which act as a primary source of infection. Moreover, infected seed corms are the major source of infection. However, proper management of this disease should be taken from proper seed treatment followed by soil drenching of suitable plant protection chemicals and organics. As the disease occurs during onset of rain, early planting of corms escapes the disease during corm formation time. Treatment of the corms in cow dung slurry + *Trichoderma* @ 5 g/kg corm- 3 days before planting followed by drench twice with Carbendazim 25 per cent + Mancozeb 50 per cent (75 per cent WP) during earthing up is recommended in Indian condition for better management of collar rot (Anon., 2018). Application of less amount of nitrogenous chemical fertilizer also reduces the chance of infection. In severe infection, the field of elephant foot yam may be drenched/sprayed with Azoxystrobin 12.5 + Tebuconazole 12.5 SC formulated fungicide @ 800 ml/ha is very effective against collar rot pathogen (Mitra and Tarafdar, 2018) In organically grown crop, dipping of corms in 0.7 per cent Nanma (Cassava leaf extract formulation, developed in ICAR-CTCRI, Thiruvananthapuram) for 10 minutes before storing the corms followed by seed corm treatment cow dung slurry + *Trichoderma* @ 5g/kg corm 3 days before planting is to be done in the pit amended with *Trichoderma* enriched FYM @ 2.0-2.5 kg/pit. During crop growth time, twice application of *Trichoderma* + vermicompost (150-200 g) after intercultural operations is effective against this disease (Anon., 2018).

Severe corm rot in the field and in the storage is the major constraint for elephant foot yam and taro and the pathogen *Pythium helicoids* is associated with this rot. Besides this *Sclerotium rolfsii* also causes severe damage of the corms. Treatment of corms with Carbendazim 25 + Mancozeb 50 per cent WS give satisfactory control of *S. rolfsii* but the black/soft rot (*Pythium helicoids*) could be controlled by application of Metalaxyl-M + Chlorothalonil 440 SC (Folio Gold) @ 1.5ml/L of water or Metalaxyl-M + Mancozeb (4 + 64) WG (Ridomil Gold) or Cymoxanil + Mancozeb. Similarly rhizome treatment with Carbendazim 25 + Mancozeb 50 per cent WS (Sprint) and Metalaxyl-M + Mancozeb (4 + 64) WG (Ridomil Gold) is very effective against the rhizome rot of ginger and other crops. Bacterianashak (2-bromo–2-Nitro-Prppane-1,3-Diol) @ 20 g/60 L of water is very effective for controlling the bacterial infection in the corms, rhizome, bulbs *etc.*

The reproduction of two fungivours nematodes – *Aphelenchus avenae* and *Aphelenchoides* sp. isolated from fields of konjak was compared on *Rhizoctonia solani*, *Pythium ultimum*, *Fusarium oxysporum* f. sp. *raphani* and *Botrytis cinerea* (Okad, 1995).

Viral Disease

A mosaic disease has been observed on amorphophallus in most of the areas where it is cultivated. Dashen Mosaic Virus under the family of Potyviridae is very common in *Amorphophallus* which especially spread through the seed corms collected from infected plants. The characteristic symptom is mosaic mottling on the leaves. The disease causes more proliferation of lateral buds, separation of buds from the mother corms and poor growth of roots. The mottled plants produce small corms. The disease spreads through the aphids, *Myzus persicae* and *Aphis gossypii*. Since the crop is vegetatively propagated, corms from healthy plants should be used for seed purpose. Roguing of infected plants will help in minimising the secondary spread. Thankappan (1993) also reported amorphophallus mosaic virus in South India. Konjak mosaic poty-virus was studied by Shimoyama *et al.* (1992). Dasheen mosaic potyvirus (DsMV) was isolated from *A. konjac* plants showing leaf mosaic and malformation symptoms and small tubers in the region of Sao Paulo, Brazil (Chagas *et al.*, 1993). Plant free from this virus can easily be obtained through meristem-tip culture when small (0.5mm or less) meristem tips are used. Micropropagation of the growing buds on the corm (0.5cm size or less) also gives virus free plantlets of elephant foot yam (Anon., 2019).

Bacterial Diseases

Hyashi (1989) found that the residues of diseased leaflets and infested corms provide the main infection sources for bacterial leaf blight caused by *Pseudomonas pseudoalcaligenes* subsp. *konjaci.*

9.2.0 Pests

There are a number of pests attacking amorphophallus of which aphids, spider mites, thrips, caterpillars and mealy bugs are important. To control the mealy bugs use of pest-free seed corms are suggested for planting. If mealy bug infestation is noted during storage of seed corms, dipping in 0.05 per cent monocrotophos should be done to check the pest in field (Anon., 1983a). Furadon is commonly used during planting by the farmers to control white grubs, termites and weevils.

Yield losses of amorphophallus up to 29 per cent were stated to be the result of infestation by *Spodoptera litura, Aphis gossypii* and spider mites [Tetranychidae]. Control of these pests is suggested by quinalphos and dimethoate. It is also attacked by 2 other pests, *Pseudococcus citriculus* and *Rhizoecus* sp., both in the field and in storage and monocrotophos is recommended for their control (Pillai *et al.*, 1993).

Mealybugs (*Pseudococcus citriculus; Rhizoecus sp.*) nymphs and adults and their powdery meal cover the tuber surface of elephant foot yam. It leads to poor germination and also affects cooking quality. If pest infestation is severe, treatment spray of Dimethoate (0.05 per cent) is effective in checking its further spread (Mitra and Tarafdar, 2018).

10.0 Crop Improvement

The plants are characterized by protogyny and predominant cross pollination in nature. Self-fertilization is prevented mainly by strong self-incompatibility mechanisms. The crop cycle of one to three years makes the occurrence of inflorescence very rare. In the tropics, close to equator, the inflorescence will open fully and release pollen after 16-25 days, depending on the genotype, environmental

Diseases of Elephant Foot Yam

***Amorphophallus* Collar Rot**

Root Rot and Leaf Blight of Elephant Yam.
(a) Field showing symptoms in various stages of disease progression.
(b) Single plant showing severely blighted symptoms.
(c) Severely affected corm with rot symptoms

Disease and Pest of Elephant Foot Yam

Dasheen Mosaic Disease Symptom

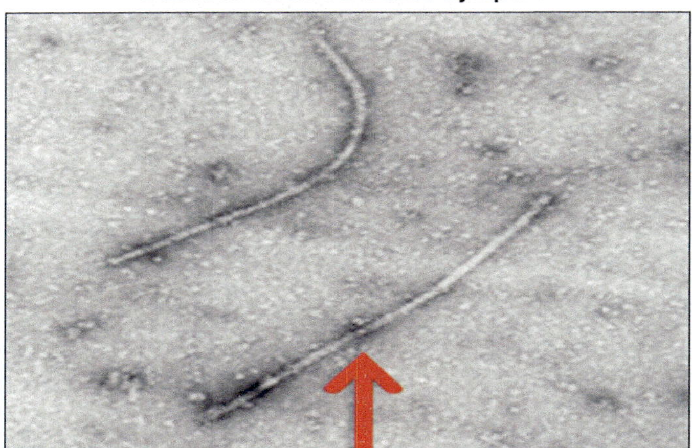

Virus Particles (Flexuous rods) of Dasheen Mosaic Virus

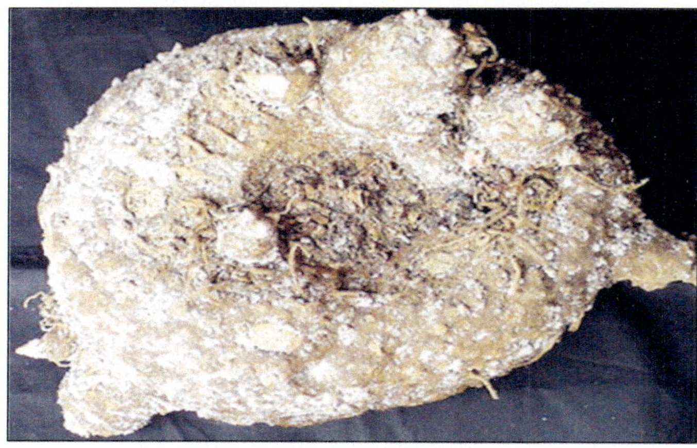

Mealy Bug Infestation on Elephant Foot Yam Tuber

condition and the size of the inflorescence. The inflorescence is produced without a leaf. They appear on the thick peduncles which may be short or long, depending on the genotypes and species. A single spadix is subtended by a large blotched spathe. Since the flowers are naked, the opening of the spathe is considered as anthesis.

Flowers are unisexual. The female part is not separated from the male by a sterile band Ovules are anatropous or semi- anatropous. The spadix terminated in a large appendix which can be elongated or short or wide. Stigmas become receptive before pollen is released. The period of stigma receptivity and shedding of pollen usually overlap within the same inflorescence. The main pollinators are insects attracted by the bad odour released by the inflorescence. The chemical responsible for the bad odour is dimethyl oligo sulphide. The release of odour is associated with the beginning of stigma receptivity, occurring in the afternoon before the spathe opens.

The odour attracts insects mainly flies from neighbouring areas, and also from other inflorescence of the same species. The latter are the most important because they bring pollen and distribute it on the stigma of female flowers which are already receptive. The odour remains intense for several days, sometimes for more than a week. It is usually so strong that it can attract flies from several hundreds of meters away. In Java, the presence of small beetles (Nitidulidae) gathered around the base of the spadix at the time of receptivity indicated that they may be the primary pollinators. Presence of extreme protogyny coupled with the delay in the opening of the spathe is responsible for the lack of seed set in *A. paeoniifolius* under natural conditions. Although there is a profuse seed set in the wild species under natural conditions. Fruits are numerous, ovoid, red, containing one or three seeds. The seeds have high germination capacity (about 75 per cent) although they have dormancy which can be removed by washing running water for six days. The number of sprouts varied from one to six per seed indicates the presence of polyembryony. Since the seed set itself has been achieved only in recent times, the genetic improvement of this crop is started.

In *Amorphophallus*, the genetic variability is very much limited due to the flowering behaviour. Seed-set was obtained at CTCRI, India by controlled pollination and the seedlings showed a good amount of variability.

Guidelines for the Conduct of Test for Distinctiveness, Uniformity and Stability on Elephant Foot Yam (*Amorphophallus paeoniffolius*) has been notified in India (DUS, PPV and FRA 2018); these test guidelines shall apply to all varieties, hybrids and parental lines of elephant foot yam for future crop improvement program.

The molecular characterization of EFY by RAPD, SSR has been done with 18 Indian varieties of EFY. Three distinct clusters have been formed in the dendrogram where Bidhan Kusum, Gajendra, (Kavuur), NDA-5 and AC-14 have the same in their origin. The similarity matrix of 18 cultivar showed that the cultivars Ranchi, BCA-5 and AC-28 are different in origin and coefficient value is very high (Anon., 2017)

Seedlings from 2 families of elephant foot yam (*A. paeoniifolius*) *viz.*, one full-sib (family 1) and one half-sib (family 2), were screened for various morphological characters (Sreekumari and Pillai, 1994). Single sprouts were produced only in very few seeds in both families. The rest of the seeds produced multiple sprouts (2-10).

The sprouts which established seedlings exhibited variation for many features. The frequency of uniform seedling types from single seed was 64.1 and 53.2 per cent in hybrid and open pollinated types, respectively. There were distinct types among seedlings from a single seed.

In a comparative study of the sexual progenies generated through hybridization (family 1) and open pollination (family 2) of elephant foot yam, Pillai (1996) reported high variability among the progenies. Yield in the seedling progenies ranged from 10 g to 1.5 kg/plant. In both families more than 50 per cent of plants yielded less than 100-500 g/plant. Both methods were equally good in inducing genetic variability. Yield of the first clonal progenies ranged from 100 g to 3.2 kg/plant.

In China, under natural conditions, 4- to 5-year-old plants of *A. konjac* [*A. rivieri*] flower in April. By crossing with *A. corrugatus*, normal hybrid seeds were harvested (Zhang *et al.*, 1998). When *A. konjac* corms with flower buds were stored at 6-9oC, the flowering period was delayed by about 4 months, thus giving an overlap with the flowering period of *A. albus*, allowing crosses between the two species to be made. Hybrid seeds of reciprocal crosses were obtained. By pre-treating the *A. konjac* corms with no flower buds with 100 x 10-6 GA3, male-sterile flowers were induced. When pollinated with pollen of *A. albus* and other species, normal seeds were obtained. Cultivation of the hybrids obtained showed that the F_1 generation was matroclinous in character. Lines with excellent agronomic characters were selected from the hybrid population.

Two high yielding strains of *Amorphophallus*, Am-6 and Am-16 were identified at CTCRI. By treatment with 0.2 per cent colchicine for 16 hours, a tetraploid plant was obtained which showed vigorous growth and broader leaves (Unnikrishnan and Sreekumari, 1984).

An experiment initiated to access the utility of SSR markers in differentiating some Indian elephant foot yam accessions to identify markers could be useful for future genetic mapping and improvement programs. Various efficiency parameters, namely, Polymorphism Information Content (PIC), Marker Index (MI), Resolving Power (RP) and Diversity Index (DI) were studied for 11 primers. And the investigation showed that among the13 primers, 11 primers were suitable for the detection of genetic diversity in *A. paeoniifolius* and finally, it was found that Amor 51 was the most efficient SSR marker (Mandal *et al.*, 2016).

11.0 References

Alam, S. and Narzary, B.D. (1997) *J. Agril. Sci. Soc. of North East India*, **10**: 82-84.

Anonymous (1983a) *Two Decades of Research*, CTCRI, Trivandrum.

Anonymous (1994) *IPGRI Newsletter for Asia, the Pacific and Oceania*, No. 15, p. 5.

Anonymous (2000-01) *Annual Report*, All India Co-ordinated Research Project on Tuber Crops, CTCRI, Trivandrum.

Anonymous (2002) *Annual Report of ad-hoc project "Standardization of elephant foot yam whole seed corm production and its commercial cultivation techniques in terai zone climate*, BCKV, Kalyani, India.

Anonymous (2017) *Annual Report,* All India Co-ordinated Research Project on Tuber Crops, CTCRI, Trivandrum, Kerala.

Anonymous (2018) *Annual Report,* All India Co-ordinated Research Project on Tuber Crops, CTCRI, Trivandrum, Kerala.

Anonymous (2019) *Annual Report,* All India Co-ordinated Research Project on Tuber Crops, CTCRI, Trivandrum, Kerala.

Anonymous (2020) *Annual Report,* All India Co-ordinated Research Project on Tuber Crops, CTCRI, Trivandrum, Kerala.

Bailey, L.H. (1950) *The Standard Cyclopedia of Horticulture,* The MacMillion Co., New York, **1**: 276-277.

Bhowmik, S.K., Sen, H. and Bhattacharya, S.P. (1988) *J. Root Crops,* **14**: 23-26.

Bradbury, J.H. and Hammer, B.C. (1990) *J. Agric. Food Chem.,* **38:** 1448–1453.

Bradbury, J.H. and Holloway, W.D. (1988) Chemistry of tropical root crops: significance for nutrition and agriculture in the Pacific. Australian Centre for International Agricultural Research, Canberra, Australia.

Chagas, C.M., Colariccio, A., Galleti, S.R. and Kitajima, E.W. (1993) *Fitopatologia Brasileira*, **18**: 551-554.

Chattopadhyay, A. and Nath, R. (2007) *Agriculture Update Hind. Agri-Hort. Soc.,* **2**: 80–82.

Chattopadhyay, A., Saha, B., Pal, S., Bhattacharya, A. and Sen, H. (2010) *Intl. J. Veg. Sci.,* **16:** 73-84.

Choudhary, B. (2014) *Vegetables,* Kalyani publishers, New Delhi. ISBN 9789327246148.

Connor, H.E. (1977) The poisonous plants in New Zealand. Government Printer, Wellington, New Zealand.

Das, P.K., Sen, H., Banerjee, N.C. and Panda, P.K. (1995) *J. Root Crops,* **21**: 90-96.

Das, P.K. (1992) *Ph.D. Thesis,* Bidhan Chandra Krishi Viswavidyalaya, West Bengal, India.

Das, P.K. and Sen, H. (1995) *Tropical Agri.,* **72**: 13-17.

Das, P.K., Sen, H., Banerjee, N.C. and Panda, P.K. (1995) *Hort. J.,* **8**: 131-139.

Das, P.K., Sen, H., Banerjee, N.C. and Panda, P.K. (1997) *Indian J. Agric. Sci.,* **67**: 197-200.

Dhua, R.S., Ghosh, S.K., Biswas, J., Mitra, S.K. and Sen, H. (1988) *J. Root Crops,* **14**: 47-49.

DUS Guide lines of Elephant Foot Yam (2018) Protection of Plant Varieties and Farmers' Rights Authority (PPV and FRA) Government of India

Ghosh, S.P., Ramanujam, T., Jos, J.S., Moorthy, S.N. and Nair, R.G. (1985) In: *Tuber Crops,* Oxford and IBH Publishing Co. Pvt. Ltd., p. 271.

Hayashi, N. (1989) *Annals Phytopath. Soc.,* Japan, **55**: 609-614.

Hetterscheid, W.L.A. and Peng, C.I. (1995) *Botanical Bull. Academia Sinica*, **36**: 101-112.

Hrishi, N. and Nair, R.G. (1972) *Indian Fmg.*, **22**: 33-37.

Huang, D.F. and Liu, P.Y. (1994) *J. Shanghai Agricultural College*, **12**: 25-30.

Huang, D.F. and Liu, W.C. (1993) *J. Southwest Agricultural University*, **15**: 522-526.

Inaba, K. (1992) *Japanese J. Crop Sci.*, **61**: 551-554.

Irawati, Arditti, J. and Nyman, L.P. (1986) *Ann. Bot.*, **57**: 11-17.

Ittenbach, S. and Lobin, W. (1997) *Willdenowia*, **27**: 147-160.

Katyal, S.L. (1977) *Vegetable growing in India*, Oxford and IBH Pub. Co. Ltd., pp. 116-117.

Kelsay, J.L. (1985) In: Nutritional bioavailability of calcium (Ed. C. Kies), American Chemical Society, Washington, D.C., pp. 105–116.

Kumar, D.A., Indira, P. and Nambisan, B. (1998) *Tropical Science*, **38**: 187-189.

Lee, H.D. (1992) *Korean J. Crop Sci.*, **37**: 117-122.

Liu, P.Y., Su, C.A. and Zhang, X.G. (1998) *Advances Horticulture*, **2**: 658-662.

Long, C.L. and Li, D.Z. (1995) *Acta Botanica Yunnanica*, pp. 89-92.

Long, C.L. and Li, H. (2000) *Novon*, **10**: 125-127.

Mandal, R.C. and Saraswat, V.N. (1968) *Indian Agric.*, **12**: 25-28.

Mandal, R., Nag, S., Tarafdar, J. and Mitra, S. (2016) *Braz. Arch. Biol. Technol.*, **59**: *e160492.*

Misra, S., Misra, S.S. and Ray, P.K. (1983) *Indian J. Agric. Sci.*, pp. 227-229.

Mitra, S. (2009) *Proceedings of International Conference on "Seed quality and food security"* Bangladesh Agricultural University, Mymensing, Bangladesh, pp. 167-171.

Mitra, S. and Tarafdar, J. (2013) In: *Postharvest Management of Tropical Tuber Crops*, AICRP on Tuber Crops, Directorate of Research, Bidhan Chandra Krishi Viswavidyalaya, Kalyani, West Bengal, India.

Mitra, S. and Tarafdar, J. (2016) *J. Agric. Engin.Food Technol.*, **3** (3): 210-204.

Mitra, S.anon

and Tarafdar, J. (2018) In: *Tuber Crops of India* (Eds. J. George, and H.P. Singh), Thiruvananthapuram, Kerala, India.

Mohan Kumar, B., Kabeerathumma, S. and Nair, P.G. (1984) *Indian Fmg.*, **33**: 35-37.

Mohan Kumar, C.R., Mandal, R.C. and Singh, K.D. (1973) *Indian J. Agron.*, **18**: 62-66.

Mohan Kumar, C.R. and Nair, P.G. (1990) *Indian J. Agron.*, **35**: 44-49.

Mukhopadhyay, S.K. and Sen, H. (1986) *J. Root Crops*, **12**: 103-106.

Mukhopadhyay, S.K. and Sen, H. (1999) *J. Root Crops*, **25**: 65-68.

Nath, R., Kundu, C.K., Majumder, A., Gunri, S., Biswas, T., Islam, S.J., Chattopadhyay, A., and Sen, H. (2007) *J. Root Crops*, **33**: 30-37.

Nair, P.G., Mohan Kumar, C.F., Saraswathy, P. (1990) *J. Root Crops*, **17**: 83-86.

NHB (National Horticulture Board) (2017) *Horticultural Statistics at a Glance*, Government of India. P. 16.

Nyman, L.P., Irawati and Arditti, J. (1987) *Aroideana*, **10**: 26-28.

Okada, H. (1995) *Japanese J. Nematology*, **25**: 56-58.

Passmore, R. and Eastwood, M.A. (1986) Human nutrition and Dietetics. 8th ed. Churchill Livingstone, Edinburgh.

Patel, B.M. and Mehta, H.M. (1984) *Indian J. Agron.*, **29**: 569-570.

Philip, J. and Nair, R.R. (1994) *Crop Res.*, **8**: 552-557.

Pillai, K.S., Palaniswami, M.S., Rajamma, P., Mohandas, C. and Jayaprakas, C.A. (1993) *Indian Hort.*, **38**: 20-23.

Pillai, P.K.T. (1996) *J. Root Crops*, **22**: 121-123.

Ravi, V. and Mohankumar, C.R. (1999) *South Indian Hort.*, **47**: 284-287.

Ravindran C.S. and Kabeerathumma, S. (1990) *J. Root Crops*, **17**: 112-115.

Saha, B.P. and Hussain, M. (1983) *Indian J. Agr. Sci.*, **53:** 833–836.

Saikia, T. and Borah, R.C. (1994) J. *Agric. Sci. Soc. Northeast India*, **7:** 90–91.

Sen, H. and Das, P.K. (1990) *Abstracts of National Symposium on Recent Advances in the Production and Utilization of Tropical Tuber Crops*, November 7-9, 1990.

Sen, H., Das Adhikary, D. and Das, P.K. (1993) *J. Root Crops*, **19**: 55-56.

Sen, H., Das, P.K. and Goswami, S.B. (1996) In: *Tropical Tuber Crops: problems, prospects and future strategies*, pp. 298-305.

Sen, H., Mukhopadhyay, S.K. and Das, P.K. (1991) *J. Potassium Res.*, **7**: 120-125.

Sen, H., Roychoudhury, N. and Bose, T.K. (1984) *J. Root Crops*, **10**: 37-39.

Sen, H., Roychoudhury, N. and Mukhopadhyay, S.K. (1987) *J. Root Crops*, **13**: 121-123.

Seo, G.S., Lee, J.Y.,Song, N.H., Sin, B.W. and Lee, H.D. (1988) *Research Reports of the Rural Development Administration*, Upland and Industrial Crops, **30**: 74-78.

Shimoyama, J., Hanada, K., Tsuda, S., Kameya-Iwaki, M. and Gunji, T. (1992) *Annals Phytopath. Soc.*, Japan, **58**: 713-718.

Shishida, Y., Shimoyama, J. and Ushiama, M. (1991) *J. Agric. Res.*, **8**: 1-10.

Singh, A. and Wadhwa, N.A. (2014) *Int. J. Pharma. Sci. Rev. Res.*, **24**: 55-60.

Singh, A.K., Chourasiya, A.K. and Mitra, S. (2018) *J. Pharmacog. Phytochem.*, **7**: 2905-2909.

Singh, A.K., Chourasiya, A.K. and Mitra, S. (2020) *Indian J. Experiment. Biol.*, **58:** 206-211.

Sreekumari, M.T. and Pillai, P.K.T. (1994) *J. Root Crops*, **20**: 60-63.

Suresh Kumar, J., Sunitha, S., Sreekumar, J., Nedunchezhiyan, M., Mamatha, K., Das, B., Sengupta, S., Kamalkumaran, P.R., Mitra, Surajit, Tarafdar, J., Damodaran,

V., Singh, R.S., Narayan, Ashish, Prasad, Rabindra, Gudadhe, P., Singh, R., Desai, K. and Srikanth, B. (2020) *Indian J. Weed Sci.*, **52**: 69–73.

Takayama, S. and Akita, M. (1991) *United States Patent*, US 5 034 327, P. 6.

Thankappan, M. (1993) *Indian Hort.*, **38**: 25-26.

Tsujihara, N. and Tani, Y. (1998) *Japanese Soc. Nutri. Food Sci.*, **51**: 157-163.

Uchida, S., Saito, H., Saito, Y., Katoh, A. and Kogure, S. (1998) *Bull. Gunma Agricultural Experiment Station*, **4**: 1-17.

Unnikrishnan, M. and Sreekumari, M.T. (1984) *Indian Fmg.*, **33**: 21-23.

Verma, P.K., Sen, H., Roychoudhury, N. and Mukhopadhyay, S.K. (1994) *The Hort. J.*, **7**: 37-44.

Verma, S.P., Sen, H., Raychoudhury, N. and Mukhopadhyay, S.K. (1992) *The Hort. J.*, **5**: 49-54.

Walter, W.G. and Khanna, P.N. (1972) *Econ. Bot.*, **26:** 364–372.

Wang, X.F., Wu, T.X. and Gong, Z.H. (1997) *China Vegetables*, **4**: 7-9.

Wittenbach, S. (1996) *Aroideana,* **9**: 7–131.

Zhang, S.L., Liu, P.Y. and Sun, Y.M. (1998) *Acta Hortic.*, **10**: 62-66.

Zhang, Y.J., Zhang, X.G. and Liu, P.Y. (1999) *J. Southwest Agricultural University*, **21**: 303-306.

COLEUS
J. Tarafdar and S. Mitra

1.0 Introduction

Coleus (*Solenostemon roundfolia* Poir. Syn. *Coleus parviflorus* Benth or *Plectraanthus roundifolia*) is an important minor tuber grown extensively as a vegetable in most of homestead gardens in Kerala and Tamil Nadu (Doraipandian, 1973). Coleus belongs to the family Labiatae, contains about 200 species which includes ornamental and medicinal plants and the pants with edible tubers. Coleus is distributed in tropics, sub-tropics of Asia, Africa and pacific Islands. Coleus (*Coleus parviflorus* Benth or *Solenostemon roundfolia*) is commonly known as Chinese Potato, produces edible tubers which can be used as a vegetable. It is a native of Tropical Africa.

Coleus amboinicus is a native of Indonesia and *C. blumei* Benth is also native of Indonesia (Purseglove, 1974). *Coleus forskohli* is perennial member of mint family, first discovered in lower elevations of India. It is now grown around the world as an ornamental plant and its tuberous root has medicinal properties.

2.0 Composition and Uses

Tubers are high in calories and essential micronutrients with good socio-economic potential to alleviate hunger (Alleman and Coertze, 1997). The tubers of this crop provide indispensable dietary and energy requirements.

The taste of the tuber is similar to that of Irish potato and trifoliate yam and in combination with legumes, rice and vegetables, it can be eaten as the main starchy staple (Schippers, 2000; Wulueng *et al.*, 2000). The crop is a vital food source during the lean periods in Africa (Archana *et al.*, 2015; Bennett-Lartey *et al.*, 2008). The tubers can be boiled, baked or fried (Schippers, 2000; Wulueng *et al.*, 2000). In comparison to other tuber crops such as cassava, yams and sweet potato, *S. rotundifolius* has a high content of protein, calcium, magnesium, fibre, and iron (Chevalier and Perrot, 1905; NRI, 1987; Gouado *et al.*, 2003; Prematilake, 2005). According to Struik *et al.* (1989), the tuber eaten within a month after harvest provides more vitamins than apples, and the biological values of its protein are higher than vegetable protein. Tubers can be used to make aromatic, alcoholic beverages (Schippers, 2000; Phungpanya *et al.*, 2013) while the leaves sometimes may also be eaten as potherbs (PROTA, 2013). In Ghana, an alcoholic drink is brewed from *S. rotundifolius* tubers (Opoku Agyeman *et al.*, 2007).

The tubers of *Solenostemon roundfolia* and *Coleus parviflorus* are used as vegetable having special flavor and taste. In India, tubers are used as potato in curry preparation. It is preferred for its particular aromatic flavor and sweetness. It can be baked like potato or made in chips.

The Coleus plants contain saponin (2-20 per cent) and alkaloids (15-25 per cent) which are primary source of its significant medicinal properties. In Africa, the Hausa potato is sometimes used in the treatment of dysentery and in the treatment of certain eye disorder.

Coleus forskohliI is a perennial member of mint and the root is used for myriad of medicinal purpose. As recorded in ancient Sanskrit texts, coleus has been used

Uses of Chinese Potato

Different Value Added Products of Chinese Potato

as a medicinal herb to treat heart and lung diseases, intestinal disorders, insomnia and convulsions. Forskolin is a chemical found in coleus herb that activates the enzyme adenylate cylase. Adenylate cylase compound initiates a myriad of critical events and processes within all the cells of the human body and are responsible for executing a number of important hormonal processes. Some studies have indicated that Forskolin in coleus may help to reduce the blood pressure and improved heart function in people with cardiomyopathy.

The leaves of *Coleus amboinicus* Lour, is used in the local folk medicine for whopping cough and anti-inflammatory treatment (Rejesus *et al.*, 1999). The juice of leaves of this species mixed with sugar is a powerful aromatic carminative. It is employed in dyspepsia, although it is said to have intoxicating properties.

The fresh fragrant roots of *Coleus vettiveroides*, K.C. Jacob are used for the decoration of temple images and also for dressing hair (Gopala Chetty, 1939). The decoction of the leaves of *C. blumei* Benth is said to be given internally for dyspepsia (Burkill, 1935).

The composition of the edible tuber of *Coleus* has been quoted as:

Water	75 per cent
Protein	1.4 per cent
Fat	0.5 per cent
Carbohydrate	21 per cent
Fibre	0.7 per cent
Ash	1.0 per cent
Calcium	17.0mg 100 g^{-1}
Iron	6.0mg 100 g^{-1}
Thiamin	0.05 mg100 g^{-1}
Riboflavin	0.02mg 100 g^{-1}
Niacin	1.0mg 100 g^{-1}
Ascorbic acid	1.0mg 100 g^{-1}

The principal amino acids in the protein are arginine, aspartic acid and glutamic acid (Kay, 1987).

3.0 Origin and Distribution

Solenostemon rotundifolius originates from tropical Africa, where it is still found in the wild in East Africa. It was widely cultivated in the savanna region from Senegal to western Sudan and in South Africa, but now-a-days there are only relics of former cultivation in Mali, Ghana, Nigeria and South Africa. As a crop it is now more important in South-East Asia, including India, Sri Lanka, Malaysia and Indonesia, where it is cultivated on a small scale (Tindall, 1983).

4.0 Cultivars

Mostly local cultivars are grown in most of the areas but two high yielding varieties of Coleus have been developed in India which are widely grown in Tamil Nadu and Kerala.

Important Cultivars of Coleus

Year of Release	*Name of Originating Centre*	*Name of Cultivar*	*Area Suitable for Cultivation*
1991	TNAU	CO 1	Tamil Nadu
1993	CTCRI	Sree Dhara	Kerala and Tamil Nadu

5.0 Soil and Climate

A fertile well-drained, sandy loam to alluvial soil with adequate organic matter is ideally suited for its cultivation. Water logging reduces tuber yield considerably. The best soil pH requirement ranges from 6.5-7.0.

Coleus grows well in hot-humid climate but the subtropical and hot temperature areas are also suitable for its cultivation. A comparatively lower temperature at night than day time favours better tuber development. Coleus requires an evenly distributed rainfall for its cultivation and cannot withstand drought conditions nor frost.

6.0 Cultivation

6.1.0 Propagation and Planting

Coleus is vegetatively propagated through suckers raised from healthy tubers of previous harvest or stem cutting or tubers. About 0.5-1.2 tonnes of tubers are planted in 0.2 ha of land to produce enough planting materials (suckers) for one hectare crop.

The healthy tubers are planted at 5 x 15 cm spacing and 4 cm depth in well prepared nursery beds of 15 cm height and 1m widths. These tubers start sprouting within 15 days, which grow to a height of 15-20 cm in about three weeks and can be readily transplanted to the field. Stem cutting of 15-20 cm length from these sprouts are also used as planting material.

6.1.1 Time of Planting

Coleus is mostly grown as a monsoon crop during August-October. It can also be planted in January and September as an irrigated crop. Under Hyderabad conditions the optimum time of planting was found to be between June and August but planting can be done from 15th June to 1st September. For UP best planting time is June to July end.

6.1.2 Method of Planting

There are three different methods, generally practiced for planting coleus

(a) *Ridge method*: Stem cuttings of 15-20 cm length with 5 leaves are planted on ridges at 15-20 cm spacing.

(b) *Coiled planting method*: Stem cuttings of 22 cm length are used and about 12 cm of the mature portion is coiled and placed in holes of about 7 cm diameter and 5 cm depth.

(c) *Horizontal planting method*: Two pieces of 30 cm long stem cuttings are placed side by side across the ridges in opposite directions. Planting pairs of cutting at 7-10 cm apart requires a lot of planting material, whereas dense planting produces smaller tubers but suppressed weeds. Among these three methods, coiled planting gives highest tuber yield.

6.1.3 Spacing and Plant Population

The planting space varies according to the methods of planting as mentioned above. The optimum spacing for coleus cultivation is 60´30 cm, which accommodates about 55,500 plants/ha.

6.2.0 Manuring and Fertilization

About 10 tonnes FYM or compost must be added at the time of field preparation. In addition to FYM, 60 kg N, 60 kg P_2O_5 and 100 kg K_2O are to be applied for one hectare of land. Half dose of nitrogen and potassium along with full dose of phosphorus should be applied at the time of planting. Another half dose of nitrogen and potassium are to be applied along with second earthing up. In Tamil Nadu, 60:60:150 kg NPK ha^{-1} is recommended.

According to Ogedegbe *et al.* (2015), application of NPK fertilizer significantly influences fresh weight and girth of *S. rotundifolius* tubers, implying that fertilizer is required to increase the yield of Frafra potato. Adequate NPK fertilizer ensures

Field Growing of Chinese Potato

high yield of potato tubers (Naz *et al.*, 2011). Lower rates of NPK application to potato amplify both fats and ash contents whereas higher rates increase protein, fibre and dry matter contents. However, *S. rotundifolius* may be produced without NPK fertilizer if soil nitrogen rating is high (Naz *et al.*, 2011). Moreover, according to Gupta (2015), the tuber yield of *S. rotundifolius* after treatment with 50:50 N and K fertilizer are statistically superior to tubers which receive absolute K or K and N in a 100:75 proportions. This could be attributed to the beneficial interaction of N and K in plant cells leading to stimulation in all of the cells and tissues.

6.3.0 Interculture

A weeding along with earthing up and top dressing of fertilizers is done at about 3 weeks after planting to cover a portion of the stem with soil and to improve the physical conditions of soil, and promote better tuberization and tuber development. The second weeding and earthing up should be done at about 6 weeks after planting.

6.4.0 Irrigation

It is essential to give irrigation immediately after planting to ensure adequate soil moisture for proper establishment of stem cuttings. During the tuber development phase proper moisture must be maintained to encourage better growth and tuber development.

7.0 Harvesting, Yield and Storage

7.1.0 Harvesting

Coleus is a short duration crop and can be harvested in about 5-6 months. With the maturity of tubes, leaves start falling from stems, which acts as an indicator of crop maturity. Harvesting should not be delayed as the mature tubers decay rapidly when left in the soil.

7.2.0 Yield

The tuber yield varies from 10-20 t ha^{-1} depending on the management practices.

7.3.0 Storage

The tubes can be stored in well-ventilated room after curing for 2-3 days under shade conditions.

8.0 Pests and Diseases

8.1.0 Pests

Several insect pests attack Coleus (Chinese potato) but no significant loss is reported due to disease.

8.1.1 Leaf Feeders

The caterpillar of *Hymenia recurvalis* (Fab.) seriously damages the leaves. They form a silken thread and web and leaves either folding the same leaf or leaves together or feed upon the leaf tissue within. This insect feed upon the green tissue

Harvesting of Chinese Potato

of the leaves and resulting of reduction of leaf area which affect photosynthesis. This insect pest is very common in Tamil Nadu and west Coast area of India. The larvae are parasitized in nature by braconid, *Apanteles delhiensis* and *Cardiochilis* sp. (Palaniswami and Tarafdar, 2009).

8.1.2 Top Shoot Folder

Phostria piasusalis Wlk is a pale brown small moth. The larva folds top shoots and leaves with silken webs and feed within the green matter of the leaves. Damage is similar to leaf folder. His insect occurs mainly during September – December in India. Naturally this insect is parasitoised by *Leptobatopsis indica*, Cameron (Palaniswami, 1999).

8.1.3 Bihar Hairy Caterpillar

Bihar Hairy Caterpillar (*Diacrisia oblique* Wlk) is observed as a pest on coleus, young larvae feed upon the tender leaves and shoots gregariously. They scrape the chlorophyll content and skeletonize the leaves. Full grown larvae feed upon the leaves and vine also. The field infestation reported was 14.5 per cent in India. The caterpillar possesses long thick hairs and fully grown ones measure up to 30-40mm in length. The larvae pupate in cocoons of silk and hairs in dried leaves and branches. It is found as a sporadic pest during September to March.

Tussok caterpillar (*Dasychira mendosa* Hubn.) lays egg on the leaves and feed upon the green tissues like Bihar Hairy Caterpillar. This insect is less serious in coleus.

8.1.4 Black Tingids

Cochlochila bullita S. is a pest of Tamil Nadu and Kerala causing substantial damage to the leaves. Adult lace wing bugs are black in colour with hyaline wings and hollow globular out growth on the dorsal side of the thorax. Adults are active and short distance fliers and life cycle periods are in the range of 2-3, 11 to 15, 6-8, 12-14 and 18-22 days respectively. Population appears maximum during August to December. During heavy infestation, the infested plants show a burnt up appearance. This insect also attacks *Ocium sanctum* (Tulsi) and *Mentha spicata* (Mint) (Palaniswami, 1994).

8.1.5 Root-knot Nematode

Meliodogyne incognita (Kofoid and White) Chitwood is serious pest of economic important. Nematode infests the roots and produce profuse root knots and tuberization is serious reduced, plants remain stunted growth and results wilting. The infected tubers are malformed and hypertrophied in size due to heavy infection which causes tuber rot before harvest (Ann., 2013). Cracking on the surface of tuber and blackening are observed. The starch content of the tubers is considerably reduced. The nematode has wide host range and adaptability (Palaniswami, 1999).

Insect pest can be controlled using a spray of Malathion (0.05 per cent). For effective control of nematode, use of nematode free seed tuber for raising nursery, keeping the field from weeds and rotating crops with non-host or resistant varieties of sweet potato (Sree

Bhadra) are recommended (Mohondas and Palaniswami, 1991). As coleus is minor tuber crops and its medicinal properties, practice of chemical control has certain limitations and hence emphasis should be given to maximum utilization or preservation of available natural enemies, use of field tolerant varieties and manipulation of cultural practices with need based application of bio-pesticides.

8.2.0 Diseases

There is no important disease infects coleus. Leaf spot disease caused by *Cercospora* spp. and mosaic symptoms has been reported.

9.0 Crop Improvement

Although many of the underutilized root and stem tuber crops flower, a greater chunk of them do not produce seeds which limits their scope for breeding and further improvement (Prematilake, 2005). Consequently, this tuber crop is propagated by vegetative methods using tubers of varying shapes and sizes. Unfortunately, knowledge on the genetic improvement potential of *Solenostemon rotundifolius* is still scanty, and not much is known on how to develop improved cultivars (Nkansah, 2004; Schipperss, 2000). This is because the crop has suffered research neglect for years (Enyiukwu *et al.,* 2014a, b). Thus, there is the need for a concerted effort at conserving, documenting and promoting the dynamic use of the crop (Olojede *et al.*, 2005). Due to the rapid increase in human population and consequent shortages of grain crops, collection, improvement and utilization of this underutilized tuber crop is of paramount importance (Vimala and Nambisan, 2005).

There is still a wide range of diversity of both wild and cultivated germplasm of this crop throughout the African continent even though it has become a rare food crop in Africa and seems uncommon in the wild (Nkansah, 2004). *In vitro* facilities exist in some institutions in Ghana for the conservation of genetic resources. Among such institutions is the Botany Department (BD) of the University of Ghana. This institution is the sole institution with facilities for cryopreservation of this crop on experimental basis in Ghana (Bennett-Lartey *et al.*, 2008). *S. rotundifolius* recorded high survival rates upon treatment with plant vitrification solution II (PVS2) and subsequently, 85 per cent survival rate on exposure to liquid nitrogen (Wang *et al.*, 2003; Turner *et al.*, 2001; Niino *et al.*, 2000). Nevertheless, according to Quain *et al.* (2012), *S. rotundifolius* explants easily become hyperhydric, and are impossible to dehydrate adequately for cryopreservation. Research is needed into the cryopreservation of genetic resources.

Outside Africa, collections are maintained at the Plant Genetic Resources Centre, Gannoruwa, Peradeniya, Sri Lanka, and at the Central Tuber Crops Research Institute, Trivandrum, India (Nkansah, 2004). Archana *et al.* (2015) reported that there are eighty-seven collected accessions of *S. rotundifolius*. He added that, significant difference existed in tuber size within the accessions and not between the accessions. Evaluation of the eighty-seven accessions showed that not much variation existed between the accessions in morphological or tuber characters and this may be the consequence of vegetative propagation and sterility of the crop.

Vasudevan and Jos (1988) observed no variation for yield contributing traits in germplasm collection of *Coleus parviflorus*. The crop does not produce seeds. They treated underground tubers with 1 kR, 2 kR, 3 kR and 4 kR gamma rays which resulted in 50 morphologically different mutants, which are maintained as mutant clones. In the M_1V_1 generation, suspected mutant sprouts, were carefully removed and grown separately. The most interesting mutant types recorded by them are (i) erect mutant with spoon shaped light green leaves, 30 cm long inflorescences against 20 cm in the control, cylindrical tubers measuring ca. 7.0 cm long and 3 cm girth against 4 cm and 2.5 cm in the control (ii) early mutants 1 and 2, one having less leaf serration, the other having light green small leaves and dwarf type (iii) fleshy leaf mutant, dark green, thick and smooth leaves. In the early mutants tuber formation is mainly restricted to the base of the plant, which makes harvest easier. The crop usually matures within 150 - 160 days, the early mutants are ready for harvest 100 days after planting. As the mutants are less spreading, the yield could be increased by closer spacing.

The response of *S. rotundifolius* to mutagens varies with concentration of mutagens as well as with the genotypes, which may be as a result of the variations in the genetic constitution of the plant. An increase in the concentration of mutagens results in delayed sprouting (Abraham and Radhakrishnan, 2009). Thus, chemical mutagens can be employed to change the size, number of tubers and plant height, both positively and negatively in *S. rotundifolius*.

10.0 Biotechnology

There are many minor tuber crops cultivated as vegetable or medicinal crop. In Chinese potato (*Solenostemon roundfolia*) a widely grown vegetable crop, standardized protocol for *in vitro* conservation and micro propagation is available. Callus culture regeneration of *S. rotundifolius* has been reported by Prematilake (2005). From his study, media protocols were developed to regenerate plants from leaf explants (direct organogenesis) of *S. rotundifolius* in order to generate variants. The results of the study showed differences in plant structure, leaf colour and tuber size among some of the regenerants. Variation occurred mainly in those plants which regenerated through indirect organogenesis. Likewise, Nkansah (2004) also reported that, experimental *in vitro* multiplication by tissue culture was successful, using stem meristems, apices and nodes.

Prematilake (2005) reported that *in vitro* plant regeneration of *S. rotundifolius* using indirect regeneration (*i.e.* via callus phase) could induce more variants *vis-a-vis* direct regeneration. Thus, tissue culture avenues could be exploited to induce somaclonal variants of this crop, so as to broaden the genetic base of the crop for future breeding exploitations. In terms of generating bigger tuber sizes *in vitro*, Prematilake (2005) reported that tubers regenerated from callus were generally larger in terms of tuber weight and tuber diameter than those from the direct methods of regeneration, as he recorded an average tuber weight range of 4.9–8.9 g from direct regeneration methods *vis-a-vis* 7.1–23.7 g obtained from callus regeneration, as well as an average tuber diameter range of 2.8–3.4 cm^2 from direct regeneration and 3.0–4.4 cm^2 from callus regeneration. This indicates the possibility of generating bigger tuber sizes of *S. rotundifolius* by tissue culture means, via callus regeneration.

According Mukherjee *et al.* (2015), the crop is highly totipotent and can easily be propagated through tuber sprouts and cuttings. In a tissue culture medium, *S. rotundifolius* regenerates either via axillary shoot proliferation, organogenesis or somatic embryogenesis. An elaborate study of *S. rotundifolius in vitro* by Mukherjee *et al.* (2015) revealed that when the plants are transplanted in the field after they have been regenerated through callusing and somatic embryogenesis, they produce bigger tubers as well as higher yield in the range of 100– 220 g/plant. Also, the average starch content and dry matter values are within the region of 16.5–20 per cent and 26.4–35.6 per cent, respectively.

Direct regeneration has been standardized from leaf lamina explants and intermodal cuttings at higher level of BAP (1-10mM). Higher level of callus is initiated with the supplement of NAA (1-10mM) and direct regeneration. Nodal explants could develop shoot and root directly in MS medium. There is need to develop protocols such as use of microarray technology, development of large number of ESTs and molecular markers.

11.0 References

Abraham, M. and Radhakrishnan, V.V. (2009) Food and Agriculture Organization of the United Nations, Rome, pp 283-285.

Alleman, J. and Coertze, A.F. (1997) Vegetable and Ornamental Plant Institute, Pretoria.

Anonymous (2013) *Annual Report,* AICRP on Tuber Crops, Thiruvananthapuram, Kerala.

Archana, D., Singh, B.K., Dutta, J. and Dutta, P.K. (2015) *Int. J. Biol. Macromol.,* **73**: 49–57.

Bennett-Lartey, S.O., Oteng-Yeboah, A.A. and Bunso, G. (2008) CSIR *Plant Genet. Resour. Res.,* **1**:14–23.

Burkill, H.M. (1935) The useful plants of west Troical Africa, 2nd Ed. Vol. 3, Royal Botanic Gardens, Kew.

Doraipandian, A. (1973) *Indian Hort.,* **18**: 14-15.

Chetty, G. (1939) *Madras Agric. J.,* **27**: 369.

Chevalier, A. and Perrot, E. (1905) Les Coleus a ' tubercules alimentaires. Les vegetaux utiles de l'Afrique tropicale francaise 1.

Enyiukwu, D.N., Awurum, A.N. and Nwaneri, J.A. (2014a) *Net. J. Agric. Sci.,* **2**(1): 30-46.

Enyiukwu, O.N., Awurum, A.N. and Nwaneri, J.A. (2014b) *Greener J. Agron. For. Hortic.,* **2**(2): 27-37.

Gouado, I., Fotso, M. and Djampou, E.J. (2003) In: *2e 'me Atelier International, Voies alimentaires d'ame ´lioration des situations nutritionnelles,* Ouagadougou, 13-28, November, 2003, pp 85-90.

Gupta, V. (2015) *Int. J. Res. Stud. Biosci.,* **35**: 74-78.

Kay, D.E. (1987) Root Crops, 2nd ed. Tropical Development and Research Institute, London, XV, p. 380.

Mohandias, C. and Palaniswami, M.S. (1996) In: *Tropical Tuber crops.* Oxford IBH, New Delhi, pp.360-369.

Mukherjee, A., Vimala, B., Nambisan, B., Chakrabarti, S.K., George, J. and Gowda, H. (2015) *Int. J. Trop. Agric.,* **33**(4): 3803-3815.

National Research Council (2006) National Academies Press, Washington, pp. 269-285.

Naz, F., Ali, A., Iqbal, Z., Akhtar, N., Asghar, S. and Ahmed, B. (2011) *Sarhad J. Agric.,* **27**(3): 353-356.

Niino, T., Hettiarachchi, A., Takahashi, J. and Samarajeewa, P.K. (1999) *Cryo. Lett.,* **21**(6): 349-356.

Nkansah, G.O. (2004) Plant Resources of Tropical Africa/Ressources ve ´ge ´tales de l'Afrique tropicale, J.K. Morton. Wageningen, Netherlands.

Ogedegbe, S.A., Safwan, I.I. and Ajala, B.A. (2015) *Int. J. Agri. Rural Development*, **18**(2): 2240-2245.

Olojede, A.O., Iluebbey, P. and Dixon, A.G.O. (2005) In: *IITA/NRCRI collaborative germplasm and data collection on root and tuber crops in Nigeria*. Annual Report National Root Crops Research Institute, Umudike, pp. 77.

Opoku-Agyeman, M.O., Bennett-Lartey, S.O., Vodouhe, R.S., Osei, C., Quarcoo, E., Boateng, S.K. and Osekere, E.A. (2007) In: *Plant genetic resources and food security in West and Central Africa*, Vodouhe R, Atta-Krah K, Achigan-Dako GE, Eyog-Matig O, Avohou H (eds). Regional conference, Ibadan, Nigeria, pp 26-30.

Palaniswami, M.S. (1994) In: *Advances in Horticulture*. Vol: 8 Tuber Crops (Eds. K.L. Chadha and G.G. Nayar), Malhotra Publishing House, New Delhi, pp. 485-511.

Palaniswami, M.S. (1999) In: *IPM System I Agriculture*. Vol.VI. (Eds. R.R. Upadhyay, K.G. Mukherjee and O.P. Duey)

Palaniswami, M.S and Shirly, R.A. (2008) In: *Tuber and Root Crops* (Horticultural series-9) (Ed. Prof. K.V. Peter), New India Publishing Agency, pp. 85-118.

Palaniswami, M.S and Tarafdar, J. (2008) In: *Tuber and Root Crops* (Horticultural series-9) (Ed. K.V. Peter), New India Publishing Agency, pp. 179-245.

Phungpanya, C., Jumpatong, K., Pripdeevech, P., Thongpoon, C. and Machan, T. (2013) In: *Pure and applied chemistry international conference*.

Prematilake, D.P. (2005) *J. Natl. Sci. Found Sri Lanka*, **33**(2): 123-131.

PROTA (Plant Resources of Tropical Africa) (2013) Dicotyledons. Vol 1 and 2 combine. London.

Quain, M.D., Acheampong, E., Egnin, M. and Berjak, P. (2012) INTECH Open Access Publisher.

Rejesus, B.M., Punzalan, *E.G.*, Vasquez, E.A. and Kraus, W. (1999) *J. Crop Sci.*, **24**(1): 81.

Schippers, R.R. (2000) National Resources Institute/ACP-EU. Technical Centre for Agricultural and Rural Cooperation, p.12.

Shela, M.N. and Rajmohon, K (2009) In: *Tuber and Root Crops* (Horticultural series-9). (Ed. K.V. Peter), New India Publishing Agency, pp. 349-379.

Struik, P.C., Geertsema, J. and Custers, C.H. (1989) *Pot. Res.*, **32**(2): 151-158.

Tindall, H.D. (1983) The Macmillan Press Limited, Basingstoke, pp. 242–245.

Turner, S.R., Touchell, D.H., Senarata, T., Bunn, E., Tan, B. and Dixon, K.W. (2001) *Cryo. Lett.*, **22**: 163-174.

Vasudevan, K. and Jos, J.S. (1988) In: Joint FAO/IAEA Division of Nuclear Techniques in Food and Agriculture, Vienna (Austria); 36 p;) p. 5.

Vimala, B. and Nambisan, B. (2005) *Tech. Bull.*, **44**: 48.

Wang, Q., Li, P., Batuman, O., Gafny, R. and Mawassi, M. (2003) *Cryo. Lett.*, **24**: 293-302.

Wulueng, W., Busson, F. and Jardin, C. (2000) United States Department of Health, Education and Welfare Nutrition division, pp. 36-306.

Index

A

B

S

T

U

V

W

Y

Z